HANDBOOK OF RELIABILITY ENGINEERING

HANDBOOK OF RELIABILITY ENGINEERING

Edited By

IGOR A. USHAKOV

Updated American Edition By

IGOR A. USHAKOV

The George Washington University
and SOTAS, Inc.

and

ROBERT A. HARRISON

Harrison and Company

A Wiley-Interscience Publication

JOHN WILEY & SONS, INC.

New York / Chichester / Brisbane / Toronto / Singapore

This text is printed on acid-free paper.

Library of Congress Cataloging in Publication Data:

Spravochnik nadezhnost' tekhnicheskikh sistem. English

Handbook of Reliability Engineering / Igor A. Ushakov, editor; this edition updated and edited by Robert A. Harrison and Igor A. Ushakov.

p. cm.

Based on the 1985 Russian edition of: Spravochnik nadezhnost' tekhnicheskikh sistem.

Includes bibliographical references and index.

ISBN 0-471-57173-3 (alk. paper)

1. Reliability (Engineering) I. Ushakov, I. A. (Igor' Alekseevich) II. Harrison, Robert A.

TA169.S6713 1994

620'.00452--dc20 93-830

Printed in the United States of America

10 9 8 7 6 5 4 3 2 1

CONTRIBUTORS

VLADIMIR V. BOLOTIN: Sections 11.3 and 11.4 (co-author)

GUENNADI N. CHERKESOV: Chapter 9

VICTOR P. CHIRKOV: Sections 11.3 and 11.4 (co-author)

ERNEST V. DZIRKAL: Chapter 13, Sections 15.7 and 15.8

VADIM A. GADASIN: Section 7.6 (co-author)

ROBERT A. HARRISON: Sections 12.1 through 12.5 (co-author), Chapter 18, Section 21.1, Appendix F

VLADIMIR V. KALASHNIKOV: Sections 6.3 and 19.1 through 19.5

MARK P. KAMINSKY: Chapter 17

GUENNADI P. KLIMOV: Chapter 10 (co-author)

BORIS A. KOZLOV: Appendix B (co-author)

EUGENE I. LITVAK: Section 7.7 (co-author)

VICTOR F. MATVEEV: Chapter 10 (co-author)

IGOR V. PAVLOV: Chapter 14, Sections 15.1 through 15.5

GRIGORI B. RUBALSKY: Sections 22.1 through 22.5, Section 22.6 (co-author)

VLADIMIR P. SAVCHUK: Chapter 16

ALEXANDER D. SOLOVYEV: Sections 6.1, 6.2, 6.4 and 6.5

OLEG I. TYOSKIN: Section 15.6

Igor A. Ushakov: Chapters 1 through 5, Sections 7.1 through 7.5, Sections 7.6 and 7.7. (co-author), Chapter 8, Sections 11.1 and 11.2, Sections 12.1–12.5 (co-author), Sections 12.6, 19.6, Chapter 20, Section 21.1 (co-author), Sections 21.2 and 21.3, Section 22.6 (co-author), Chapters 23 and 24, Appendix A, Appendix B (co-author), Appendixes C through E.

CONTENTS

FOREWORD

Success in an increasingly global business environment requires an integrated product development strategy with all business as well as engineering functions represented in the critical decision making processes. Effective product development requires the satisfaction of not only specified performance and cost requirements, but also satisfaction of attributes such as reliability, maintainability, supportability, producibility, schedule and cost. Due to the large amount of knowledge required to be truly proficient in any one of these attribute fields, the life cycle product development approach must be highly cross departmentalized, and the tools necessary to accommodate the wide variety of engineering and management tasks must be integratable and flexible enough to be used by a variety of experts with specialized concerns.

While reliability engineering is a key component to product development, today's engineer has a heightened level of skepticism about the traditional reliability methods, and a concern that the reliability theories have emphasized "placing a number" on products, rather than on improving the reliability of products. The use of traditional reliability methods in the design and operation of equipment has been very controversial, especially those associated with reliability prediction methods. In fact, many of the military handbooks which have been developed to model reliability often do not accurately predict field failures, cannot be used for comparative purposes, and present misleading trends and relations.

This book presents the techniques and practical applications associated with reliability theory and models. The purpose of this book is to get past the opinion and anecdotal stories, and instead present those theories and modeling methods which will make a difference in the products we design, manufacture and operate.

This is an interdisciplinary work that will be useful to anyone involved in analysis or optimization of equipment or systems. The basic subject is reliability itself, with many chapters on related subjects, including probability theory, statistics, and operations research. The sections on operations research cover not only optimization from the viewpoint of reliability and cost, but optimization with respect to other considerations, such as weight and volume. Design engineers will be able to use this book as a reference source on probability analysis, cost analysis, and operations research. Similarly, cost analysts and operations research analysts can use this volume as a reference on design engineering.

Simple step-by-step numerical examples are used to explain specialized techniques to readers without background in the speciality. I recommend this handbook as an encyclopedia reference that will be helpful over the years to practitioners in all these fields. I also recommend it to professors of both undergraduate and graduate courses as a supplementary text.

MICHAEL PECHT, PH.D., PE

Professor, University of Maryland

PREFACE

In this update and revision of the *Handbook of Reliability Engineering*, we have extended the scope from the traditional, narrow emphasis on engineering to the broader fields of operations research, cost analysis, and project management. This has meant, of course, writing for a broader audience, most of whom will have little or no background in some of the many technical disciplines covered. We have accordingly included background explanations of fundamentals, stressing comprehensibility rather than classical rigor. Every topic has been introduced wherever appropriate with simple, step-by-step, numerical examples. In addition readers can purchase personal computer programs from Harrison and Company to apply the techniques to their own work or study. Both individual disks and a set of disks for the entire handbook are available, as described in Appendix F.

The chapters have been organized into three groups: Chapters 1 through 11 are on probabilistic reliability, Chapters 12 through 19 are on statistical reliability, and chapters 20 through 24 are on optimization. Each topic is written to stand on its own as much as possible, so a reader can start at almost any point. The appendixes, in addition to the usual tabular material, have summary reviews of mathematical and statistical fundamentals. We have included many techniques that have not been published in the United States before and some techniques that have not been published anywhere before; in all cases the applicable sections were written by the researchers who developed the new techniques. Extensive use has been made of material from previous publications, as indicated by the references that follow the applicable chapters.

This handbook, therefore, has had some 20 coauthors including the two of us. Coaching a team of this size, particularly a team of accomplished scientists, engineers, and mathematicians, is complicated. We understand clearly that a team wins because it is a good team. If there are weaknesses here, they may be due to the coaches, not the team. As editors of this handbook, we realize that a certain number of mistakes, errors, and inaccuracies are inevitable. The two of us should bear the criticism for any that have occurred. We are confident, however, that this group effort will be very helpful to practicing engineers, reliability analysts, and students.

Igor A. Ushakov
Robert A. Harrison

Rockville, Maryland
Fairfax, Virginia
December 1993

PREFACE TO 1993 AMERICAN EDITION

The position of reliability theory in modern engineering becomes stronger every year. Every year hundreds of original research projects appear in this relatively young field of applied science. In many design departments and research laboratories, the reliability of technical systems is analyzed in searches for the best engineering solutions. A natural result is a demand for special literature to support this activity, including handbooks for the computation and empirical estimation of reliability. Many such handbooks have been published in the United States and the former USSR, as well as in other countries.

My opinion is that the place of distinction among these books belongs to the handbook edited by Professor Igor A. Ushakov and published in 1985 in Moscow. Now this completely new version is being published in the United States. I had a chance to look at it during my recent visit there.

Why is this book unique? First of all, this book was prepared by a group of prominent Russian specialists in reliability. But it is not a mere collection of papers or separate results. Professor Igor Ushakov and his American colleague, Robert Harrison, have coordinated all the best material. As an erudite and internationally recognized expert in the reliability field, Professor Ushakov has written a good part of the handbook himself. At the same time he gave others the possibility to express their own original ideas and results.

This handbook differs from its predecessors in a number of important features. First, this book has a universally complete coverage of reliability topics. Second, this book contains an incomparable volume of results of practical usefulness to reliability analysts. Third, a substantial portion of the results are new and were originated by the authors. Fourth, this book is not the product of the desire of the authors to express themselves spontaneously, but is instead the outcome of long and scrupulous selection and development of contents.

This book has come to the American reader after continuing and painstaking labor by Professor Ushakov. The first edition of the handbook, which was assembled by B. Kozlov and I. Ushakov, was published in 1966. It reflected the experience of the authors in engineering and applied mathematics in the reliability field. In the Preface of the first edition I wrote: "The authors have performed a tremendous job. They are pioneers in collecting and systematization of methods, ideas and results in reliability field." Precisely because of this uniqueness and innovation, one might have objected to the first edition of the handbook—among other matters, the material was incomplete, and there was too much use of exponential mathematical models. Despite this, the work received a very positive response. It was one of the first attempts in the area. In 1970 the book was translated and published in the United States by Holt, Rinehart and Winston.

In 1975 the authors completely rewrote the book, responding to all remarks by readers, and adding a lot of modern and completely new original results, including their own results. It was not simply a second edition; it was a new book. About 40,000 copies—a large number even for the Soviet Union—were sold in the first months (there were no second printings at that time in the USSR). This fact alone shows that the authors did their best. This edition was translated and published in both Germanys—in Berlin in 1978 and in Munich in 1979—and later in Czechoslovakia.

At last, in 1985, Professor Ushakov undertook a new attempt with a completely new idea. He managed a collective effort by 41 authors from different countries to prepare a new version of the *Reliability Handbook*. This American version of the handbook is fundamentally changed but is based on the last Russian edition.

The editors very seriously revised all the material and added new chapters. One example is the extensive coverage of reliability software dispersed through the text.

This handbook is divided into three parts and appendixes. The parts are probabilistic methods of computation, statistical inferences, and optimization problems in reliability engineering. Appendixes give to the reader a lot of useful knowledge: a brief review of probability, stochastic processes, mathematical statistics; information and reliability software; standard mathematical formulas used in reliability; and useful numerical tables.

The text is divided into 24 chapters covering the most important areas of modern reliability engineering problems. Everything I have seen shows that the editors tried to find new ways to satisfy real needs in practical engineering.

The purposes of the American version of the handbook remain the same as before: to give the reader a basic understanding of reliability, to show the main practical problems in reliability, to present the main solution techniques, to provide a collection of computation formulas, and to illustrate application of the methods for the solution of problems with simple examples.

I know that the editors had a multitude of problems in writing the handbook. How do we choose appropriate and useful materials for a virtually boundless ocean of monographs and journal papers on reliability? How do we arrange it in a systematic way? Of course there are many different ways of selecting and arranging material. Everyone has his own viewpoint on the subject based on his personal experience, his professional interests, and even his tastes and habits. Criticizing someone else's work and finding his mistakes is always simpler than doing something positive yourself. There will of course be errors and mistakes in this handbook, but I am convinced that their total weight will be almost infinitesimal in comparison to the positive features. It is not necessary to discuss this. Some of this book may be unsatisfactory because the reliability theory itself is not developed to the necessary level. An example is failure prediction: There is no appropriate theory, and a lot of ideas demand experimental checking. The same can be said about the important area of accelerated tests; a lot of very useful attempts have been made, but there are many enigmas and dark niches in the field. A lot of results for renewal systems can be found only for exponential failure distributions. But this is not the fault of the editors—it is the state of modern reliability theory.

I recommend this handbook to American readers. I hope they become friends with this book in spite of its inevitable deficiencies. If readers will send their remarks to the editors, it will help them in the future to revise material, to select it, and to improve it. In a work of this nature the improvement possibilities have no limit.

Professor Igor Ushakov informed me that this handbook is open for contributors. I find that it is a very good idea; new co-authors will expose new aspects, new ideas and new methods in such a dynamic field as reliability.

I firmly believe that the handbook will be appreciated deeply by a wide range of American readers. The coverage of reliability has not been intended solely for engineers and operations research analysts. The processes, techniques, and mathematical models also apply to biology, medicine, sociology, and insurance.

BORIS V. GNEDENKO

Chairman, Department of Probability Theory
Moscow, State University, and Consultant to SOTAS, Inc.
Washington–Moscow
December, 1993

ACKNOWLEDGMENTS

Extensive use has been made of material from the previous Russian edition (1985).

We appreciated the personal and direct contributions of Vladimir V. Bolotin, Gennadi N. Cherkesov, Victor P. Chirkov, Ernest V. Dzirkal, Vadim A. Gadasin, Vladimir V. Kalashnikov, Mark P. Kaminsky, Gennadi P. Klimov, Boris A. Kozlov, Eugene I. Litvak, Victor F. Matveev, Igor V. Pavlov, Grigori B. Rubalsky, Vladimir P. Savchuk, Alexander D. Solovyev, and Oleg I. Tyoskin. Details about the participation of all the authors are in a separate list. Appendix C is based primarily on a contribution courteously prepared for the Russian edition (1985) by Richard E. Barlow and Frank Proschan.

GLOSSARY

Reliability indexes are not given here because they are treated in subsequent sections.

Accelerated tests: Tests of an item under excessive stress or with lower tolerance limits in order to decrease the time to failure. Accelerated tests must simulate as realistically as possible the physical processes of aging, deterioration, or destruction in the actual environment.

Active redundant unit: A redundant unit that is under the same stress and thus has the same reliability characteristics as the main one. It makes no difference whether the redundant unit is a permanent or a replacement unit.

Aging: The process of monotonic degradation of the reliability properties of an item in time as measured by the parameters.

Availability: The capability of a system to be ready to perform its functions when required.

Common redundancy: Redundancy with a group of reserve units, each of which can replace any failed basic unit.

Complex system: A multiple-component item with subsystems that can suffer from partial, as well as complete, degradation. (See Remark 3.)

Component: See *unit*.

Degradation failure: A failure characterized by a gradual change of one or several system parameters. (See Remark 4.)

Dependent failure: Failure of a unit caused by previous failure of one or more other units.

Down time: The time interval during which an object is not available for its function.

Dual object: See *duplicate object*.

Duplicate object: An object consisting of one main and one redundant unit.

Durability: The ability of an object to perform required functions under given conditions of use and maintenance until its use becomes unsafe or uneconomical.

Duration of renewal: The duration of interruption of operation of a repairable system due to the detection, analysis, and elimination of failure, including the delivery of spare units if needed.

Duration of repair: The duration of the repair of a failed unit. This may be equal to the duration of renewal.

Failure: The total or partial loss of the capability of a system.

Failure criterion: An observable condition or combination of conditions regarded as conclusive evidence of failure.

Functional redundancy: Redundancy of capability to perform a function given by a means other than the main unit for the function, such as a redundant unit, a structural rearrangement, or an internal operational change.

Independent failure: Failure of a system unit that is not caused by other units or an operational change.

Informational redundancy: Capability to perform a function properly with information from a secondary source when information from the main source is inadequate or incorrect.

Inoperability: Inability of a system to meet at least one of the demands made on it.

Markov chain: A set of object states arranged in an ordered sequence with only one-step transitions between them.

Object: See *technical object*.

Operability: A state of a system in which it can fulfill all requirements for performance.

Output: The duration or amount of work completed by a system under specified conditions.

Parallel connection (*or parallel structure*): Connection of units so that all must fail before the system fails.

Partially loaded redundant unit: A redundant unit that is active but is under less stress than a main unit or an active redundant unit and hence is less apt to fail.

Performance effectiveness: The measurable ability of a system to fulfill specified functions.

Permanent redundant unit: A redundant unit that requires no special activation switch and that is connected to the basic one during the entire operating time.

Preventive maintenance: Maintenance performed on a periodic schedule or other regular schedule determined by operating or equipment conditions. Given before failure occurs in order to prevent failure.

Readiness: See *availability*.

Redundancy: System reliability improvement with secondary units or other means, such as stronger structure or additional operating time. (See *informational redundancy*, *functional redundancy*, *structural redundancy*, and *time redundancy*.)

Reliability: The measurable capability of an object to perform its intended function in the required time under specified conditions. Reliability is a complex property that involves availability, operability, repairability, and maintainability. The properties considered in a reliability analysis depend on the goal of the analysis.

Reliability index: A quantitative measure of reliability properties. See Section 1.2.

Renewal: Repair or replacement of a failed item. (See Remark 2.)

Renewable object: An object whose operation can be restored after failure with either repair or replacement. (See Remark 2.)

Repair: Restoration of an object. In many analyses, "repair" means restoration to a condition as good as new.

Repairable object: An object that can be repaired. Although not strictly correct, "repairable object" sometimes is used for any renewable object. (see Remark 2.)

Replacement redundancy: Redundancy in which an active, partially loaded, or standby redundant item replaces the main item if it fails. Replacement is carried out with special switching devices.

Series connection (or *series structure*): Connection of units so that failure of one causes a system failure.

Standby unit: An inactive unit that becomes active on failure of another component performing the same function. The unit is subject to no operating stress and hence does not fail from system operation. Spares usually are considered to be standby.

Structural redundancy: Redundancy achieved with special connections between the main units or subsystems and the secondary ones.

Sudden failure: A failure characterized by an abrupt change in values of one or more basic parameters of the system. (See Remark 4.)

System: A structure of units that functions as a whole. Designation of a collection of units as a system or other item can depend on the phase of design, production, test, or use. (See *unit*, *subsystem*, and Remark 1.)

Subsystem: A part of a system that can be divided into smaller parts. As a rule, a subsystem performs its own subfunction. (See *System* and *Unit*, as well as Remark 1.)

Test: An experiment performed to determine reliability characteristics.

Time redundancy: Redundancy achieved by supplying a system with extra time to perform a function.

Technical object: General term for a system, subsystem, or unit, depending on context.

Total failure: A failure that makes any use of a system as intended impossible.

Total redundancy: Redundancy in which the redundant equipment is equivalent to the whole main system.

Unit: Any part of a system that is indivisible for the purposes of the current stage of the reliability analysis. (See Remark 1.)

Unrenewable object: An object whose operation after failure is considered impossible, at least under the conditions of the analysis.

Up state: See *operational state*.

Yield: See *output*.

Remark 1 The concepts of system and unit are mutually dependent. Each helps to define the other. Both concepts are relative because a technical object can be considered a system in one analysis but a unit in another. Moreover division of a system into units can depend on the accuracy of the analysis, knowledge of the system, understanding of the goal, and measures of performance. Division of a system also can depend on the purpose of the analysis. For example, a maintenance analysis can require division into constructional assemblies, while a performance analysis can require division into functional structures.

Remark 2 A renewable object can be restored after failure by restoration or repair. Whether an object is renewable can change with the specific purpose or usage conditions.

In practice a renewable or repairable object is understood to be an object that can continue to perform its functions after the elimination of the defects that led to a cessation of successful operation. Renewal may or may not involve only repair of a part. In many cases the object is replaced with a new one. In these cases objects that are basically unrepairable can be considered as renewable.

If an object performs a function that cannot be interrupted (e.g., an airplane in flight), it is considered to be unrenewable. Nevertheless, an airplane can be regarded as renewable with respect to many partial failures.

Sometimes whether a technical object can be renewable depends on its utilization. For example, a computer used for routine calculations can be regarded as renewable, while the same computer used to support a continuous chemical or manufacturing process has to be considered to be unrenewable.

Remark 3 The concept of a complex system is controversial because it depends on the researcher's viewpoint. In standard reliability theory researchers usually consider a system as having only two extreme states—a state of complete operational ability and a state of complete failure—but a lot of complicated objects cannot reasonably be characterized this way. Examples are communication networks, power and transportation systems, and military equipment. These systems are comparable to living organisms that can have many states of health between perfect health and death.

A complex system usually can have partial failures that partially decrease a system's ability to perform.

Remark 4 The concept of sudden and degradation failures is relative. Any physical process such as a failure develops over a finite period of time. One can consider a sudden failure to be a process with a relatively high speed of development, like an avalanche of positive feedbacks. A failure can also develop gradually but become known suddenly.

Degradation failures may be observed with the help of instruments that measure technical parameters (and may be used for regression analysis predictions) or may be discovered indirectly by their effects. Partial failures also may be degradation failures.

HANDBOOK OF RELIABILITY ENGINEERING

PART 1

PROBABILITY

CHAPTER 1

BASIC CONCEPTS, MEASURES, AND DEFINITIONS

This chapter is divided into five sections: Probabilistic definitions of the main reliability measures are given in Section 1.1. Statistical interpretations of these measures are given in Section 1.2. Section 1.3 is devoted to indexes when partial failures can occur. Section 1.4 discusses index selection, and 1.5 covers subsystems, units, and components.

1.1 RELIABILITY INDEXES (PROBABILISTIC DEFINITIONS)

1.1.1 Unrenewable Objects

The notation for unrenewable objects is

$f_1(t)$ = density function of the distribution $F_1(t)$,
$F_1(t) = P(\xi_1, \leq t)$ = distribution function of the time to failure,
ξ_1 = random time to failure.

The probability of failure-free operation of an object in the time interval from 0 to t_0 is written

$$P(t_0) = P(0; t_0) = P(\theta_1 \geq t_0) = 1 - F_i(t_0) , \tag{1.1}$$

where $P(t_0)$ is the probability that the object operates without failure during the time interval from 0 to t_0 and θ_i is the time of the first failure.

The probability of failure-free operation of an object in the time interval from t to $t + t_0$, given failure-free operation up to time t, is

$$P(t; t + t_0) = P\left(\frac{\xi_i \geq t + t_0}{\xi_0 > t}\right)$$

$$= \frac{P(0; t + t_0)}{P(0; t)}$$

$$= \frac{P(t + t_0)}{P(t)}, \tag{1.2}$$

where $P(t; t + t_0)$ is the probability that the object operates without failure during the required time interval of length t_0 given its operability at time t, namely the conditional probability that the random operating time of the object up to failure is greater than the quantity $t + t_0$ given that the object has operated without failure during the time interval of length t.

The probability of failure of an object in the time interval from 0 to t_0 is

$$Q(t_0) = Q(0; t_0) = P(\xi_1 < t_0) = F_1(t_0), \tag{1.3}$$

where $Q(t_0)$ is the probability that the object fails before time t_0 starting at $t = 0$, namely the probability that the random operating time of an initially operational object is shorter than t_0.

It follows from (1.1) that

$$Q(t_0) = 1 - P(t_0). \tag{1.4}$$

The probability of failure of an object in the time interval from t to $t + t_0$, given failure-free operation up to time t is

$$Q(t; t + t_0) = 1 - P(t; t + t_0) = 1 - \frac{P(t + t_0)}{P(t)}, \tag{1.5}$$

where $Q(t; t + t_0)$ is the probability that the object fails during the interval of length t_0 starting at time t, given its operability at time t, namely the conditional probability that the random operating time of an object is smaller than $t + t_0$, given that the object has operated failure-free during the interval of length t.

The density function of failure time of an object operational at time 0 is written

$$f_1(t) = \frac{d}{dt} F_1(t) = \frac{d}{dt} Q(t). \tag{1.6}$$

The function $f(t)$ is the density of the probability that the random time of failure-free operation of an object is less than t; namely the probability density of failure at moment t.

It follows from (1.6) that

$$f_1(t) = -\frac{d}{dt} P(t) . \qquad (1.7)$$

The probability of failure and the probability of failure-free operation of an initially operational object during the time interval $(t, t + t_0)$ are expressed in terms of the failure density as follows:

$$Q(t; t + t_0) = \int_t^{t+t_0} f_1(x)\, dx$$

$$R(t; t + t_0) = 1 - \int_t^{t+t_0} f_1(x)\, dx .$$

The failure rate of an object at time t is

$$\lambda(t) = \frac{1}{1 - F_1(t)} \frac{d}{dt} F_1(t) = \frac{f_1(t)}{P(t)} , \qquad (1.8)$$

where $\lambda(t)$ is the instantaneous conditional density of the probability of failure of an initially operational object at time t given that the object has not failed up to time t. Using (1.8), we can obtain the expression for the probability of failure-free operation of the object:

$$P(t_0) = e^{-\int_0^{t_0} \lambda(y)\, dy} , \qquad (1.9)$$

and for the conditional probability of failure-free operation from t to $t + t_0$, given operation at time t_0,

$$P(t; t + t_0) = e^{-\int_t^{t+t_0} \lambda(y)\, dy} . \qquad (1.10)$$

The mean time to failure (MTFF) is written

$$T_1 = M(\xi_1) , \qquad (1.11)$$

$$T_1 = \int_0^{\infty} x f_1(x)\, dx , \qquad (1.12)$$

$$T_1 = \int_0^{\infty} x\, dQ(x) , \qquad (1.13)$$

$$T_1 = \int_0^{\infty} P(x)\, dx , \qquad (1.14)$$

where T_1 is the expected value (mean value) of the random operating time of the object.

Knowledge of any one of the functions $P(t)$, $Q(t)$, $f_1(t)$, or $\lambda(t)$ makes it possible to determine the remaining three listed in Table 1.1.

Table 1.1 Determination of functions $P(t)$, $Q(t)$, $f_1(t)$, and $\lambda(t)$

Given Function	$P(t)$	$Q(t)$	$f_1(t)$	$\lambda(t)$
$P(t)$	—	$1-P(t)$	$-\frac{d}{dt}P(t)$	$-\frac{1}{P(t)}\frac{d}{dt}P(t)$
$Q(t)$	$1-Q(t)$	—	$\frac{d}{dt}Q(t)$	$\frac{1}{1-Q(t)}\times\frac{d}{dt}Q(t)$
$f_1(t)$	$\int_t^\infty f_1(x)\,dx$	$\int_0^t f_1(x)\,dx$	—	$\frac{f_1(t)}{\int_t^\infty f_1(x)\,dx}$
$\lambda(t)$	$e^{-\int_0^t \lambda(x)\,dx}$	$1-e^{-\int_0^t \lambda(x)\,dx}$	$\lambda(t)e^{-\int_0^t \lambda(x)\,dx}$	—

1.1.2 Renewable Objects

A renewable object is an object that can be repaired or replaced with an operable object. The term "repairable" may be used generically to include items that are replaced rather than repaired.

The notation for renewable objects is

$f_j(t)$ = density of the distribution function $F_j(t)$,

$F_j(t)$ = distribution function of the failure time, which is a random variable,

$g(t)$ = density of the distribution function $G(t)$,

$G(t)$ = distribution function of the repair time, which is a random variable,

ξ_1 = random operating time of the initially completely operational object up to the first failure.

Remark. For systems consisting of several repairable units this random value is different from the random value for an unrepairable system despite the same notation. ξ_j is the random operating time of the object between failures $j-1$ and j. The initial state of the object is the state after renewal $j-1$ failure. η_j is the random interval of idleness after the jth failure. The process of utilization of a repairable object can be represented as a sequence of intervals of operability alternating with intervals of idleness:

$$\xi_1, \eta_1, \xi_2, \eta_2, \ldots, \xi_n, \eta_n, \ldots .$$

In general, the distributions $F_1(t)$, $F_2(t)$, $F_3(t), \ldots$ of the respective random variables ξ_1, ξ_2, $\xi_3, \ldots$ are different. The reason is the initial state of the object for each interval. For example, if a system consists of two redundant, independent, repairable units, its initial state $H_0(0)$ is generally the state in which both units are operational. Each time a unit is repaired, however, the repaired unit starts anew with the failure distribution for $t = 0$. The other unit, unless it happened to be repaired at the same time, has a failure distribution that starts at a $t > 0$. Ordinarily with independent repair of units the probability that both units will be repaired at the same moment is practically zero. A diagram of passages of such a system from state to state over a period of time is shown in Figure 1.1*a*.

Figure 1.1*b* presents a diagram for a system in which each period of operation begins with both units operational. In this case both units are completely new at the start of each period of operation. Generally speaking, the initial states may be different before each period of failure-free operation. However, most models considered below have the same initial state before every period of operation that they did before the first period. Failure time and repair time distributions are the same for every period. The basic reliability indexes are examined under these assumptions because the assumptions are widely applicable.

The probability of failure-free operation of an object in the time interval from 0 to t_0 is written

$$P(t_0) = P(0; t_0) = P(\theta_1 \geq t_0) = 1 - F_1(t_0), \tag{1.15}$$

where ξ_1 is the random operating time of the (initially operational) object up to the first failure and $F(t)$ is the distribution function of the random variable ξ_1.

The probability of failure-free operation of an object during the time

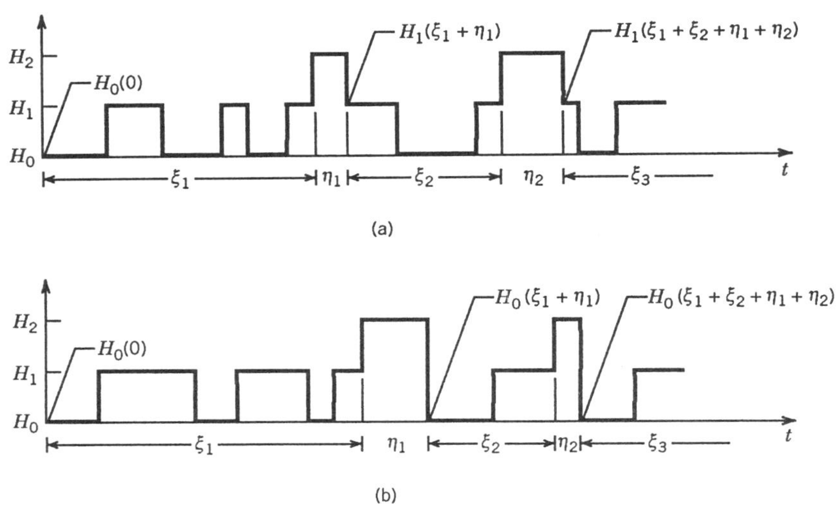

Figure 1.1 Time diagrams for a system of two redundant units: $H_i(t)$ = state of inoperability of i units at time t.

interval of length t_0, beginning at the moment of completion of the $(j-1)$st repair, is

$$P_j(t_0) = P(\theta_j \geq t_0)\,, \tag{1.16}$$

$$P_j(t_0) = 1 - F_j(t_0)\,. \tag{1.17}$$

The probability of failure of an object in the time interval from 0 to t_0 is

$$Q(t_0) = Q(0, t_0) = P(\xi_1 < t_0) = F(t_0)\,. \tag{1.18}$$

It follows from (1.15) that

$$Q(t_0) = 1 - P(t_0)\,. \tag{1.19}$$

The probability of failure of an object during an interval of duration t_0 starting at completion of the $(j-1)$st renewal is

$$Q_j(t_0) = 1 - P_j(t_0)\,, \tag{1.20}$$

$$Q_j(t_0) = P(\xi_j < t_0)\,, \tag{1.21}$$

$$Q_j(t_0) = F_j(t_0)\,. \tag{1.22}$$

The mean time to failure (MTTF), or the mean operating time of an object up to failure, T_1, is

$$T_1 = E(\xi_1)\,, \tag{1.23}$$

$$T_1 = \int_0^\infty x f(x)\, dx = \int_0^\infty x\, dF_1(x)\,, \tag{1.24}$$

$$T_1 = \int_0^\infty x\, dQ(x)\,, \tag{1.25}$$

$$T_1 = \int_0^\infty P(x)\, dx\,. \tag{1.26}$$

The mean time between failures (MTBF) is the mean time of failure-free operation of an object as a whole from the moment of completion of the $(j-1)$st renewal to the jth failure:

$$T_j = E[\xi_j]\,, \tag{1.27}$$

$$T_j = \int_0^\infty x f_j(x)\, dx\,, \tag{1.28}$$

$$T_j = \int_0^\infty x\, dQ_j(x)\,, \tag{1.29}$$

$$T_j = \int_0^\infty P_j(x)\, dx\,, \tag{1.30}$$

where T_j is the expected value (mean value) of the random length of time of failure-free operation of the object from the moment of renewal after failure $(j-1)$ to failure j.

The repair rate for an object at time t is

$$\mu(t) = \frac{1}{1-G(t)} \frac{d}{dt} G(t) = \frac{g(t)}{1-G(t)}\,, \tag{1.31}$$

where $g(t)$ is the density of the distribution $G(t)$ and $\mu(t)$ is the density of the conditional probability of repair of an object at time t, given that the object has not been repaired up to time t.

The mean repair time for an object is

$$\tau = E(\eta)\,, \tag{1.32}$$

$$\tau = \int_0^\infty xg(x)\, dx = \int_0^\infty x\, dG(x)\,, \tag{1.33}$$

$$\tau = \int_0^\infty (1 - G(x))\, dx\,, \tag{1.34}$$

where τ is the expected value (mean value) of the random repair time.

The nonstationary readiness coefficient (or availability coefficient) of an object is

$$K(t) = \sum_{i=0}^{\infty} P\left\{ \sum_{i=0}^{i} (\xi_j + \eta_j) < t < \sum_{j=0}^{i} (\xi_j + \eta_j) + \xi_{i+1} \right\}, \tag{1.35}$$

where $K(t)$ is the probability that the object is in an operational state at time t.

The stationary readiness (availability) coefficient of an object is

$$K = \lim_{t\to\infty} K(t)\,. \tag{1.36}$$

In practice K is the probability of finding the object in an operable state after enough time has passed for this probability to become almost constant. For any distributions $F(t)$ and $G(t)$ of the failure time and the repair time, having finite mean values T and τ, respectively, we can write

$$K = \frac{T}{T+\tau}\,. \tag{1.37}$$

The nonstationary idleness coefficient is

$$k(t) = 1 - K(t)\,, \tag{1.38}$$

where $k(t)$ is the probability that the object is in a state of failure (or undergoing repair) at time t.

The stationary idleness coefficient of an object is

$$k = \lim_{t\to\infty} k(t)\,, \tag{1.39}$$

where k is the probability for a stationary random process that the object is in a state of failure, and is an asymptotic value analogous to K (see 1.37). It is obvious that

$$k = 1 - K\,. \tag{1.40}$$

The nonstationary operational availability coefficient of an object in the interval from t to $t + t_0$ is

$$R(t, t+t_0) = \sum_{i=1}^{\infty} P\left[\sum_{j=1}^{i} (\xi_j + \eta_j) < t < t + t_0 \le \sum_{j=1}^{i} (\xi_j + \eta_j) + \xi_{i+1}\right], \tag{1.41}$$

where $R(t, t+t_0)$ is the probability that an object is operational at the moment t and then operates failure-free up to the moment $t + t_0$.

The stationary operational availability coefficient of an object is

$$R(t_0) = \lim_{t\to\infty} R(t, t+t_0)\,, \tag{1.42}$$

where $R(t_0)$ is the probability that an object is operational at time t (t is any moment during stationarity) and will operate failure-free during the succeeding time interval of length t_0.

1.2 RELIABILITY INDEXES (STATISTICAL DEFINITIONS)

1.2.1 Unrenewable Objects

The notation for unrenewable objects is

$d(t)$ = number of objects that fail before time t,
$d(t, t^*)$ = number of objects that fail in the interval $[t, t^*]$,
$N(O)$ = the number of objects that are operational at $t = 0$,
$N(t)$ = number of objects still operational at time t,
ξ_1 = random time up to the first failure of the object,
$\xi_1^{(i)}$ = ith observed value of a random value ξ_1,
$\xi_1^{*(k)}$ = kth order statistic formed from the random operating times up to the first failure of the object.

Estimators

Different statistical estimators are used for different purposes and different sets of data. In this subsection the estimators are given for complete sets of observations of time to failure.

The main purpose of this subsection is to convey the sense of using each index for practical purposes rather than to give methods of estimation. All estimators presented here should be considered to be illustrative examples.

An estimator of the probability of failure-free operation of an object in the time interval from 0 to t_0 is given by

$$\hat{P}(t_0) = \frac{N(t_0)}{N(0)} = 1 - \frac{d(t_0)}{N(0)}, \tag{1.43}$$

where $\hat{P}(t_0)$ is the ratio of the number of objects operating without failure up to the moment of time t_0 to the number of operational objects at the initial moment of time $t = 0$. This is the frequency with which the random duration of failure-free operation of the object is greater than required.

An estimator of the probability of failure of an object in the time interval from 0 to t_0 is

$$\hat{Q}(t_0) = \frac{d(t_0)}{N(0)}, \tag{1.44}$$

where $\hat{Q}(t_0)$ is the number of objects that have failed by time t_0 divided by the number of objects that were operational at $t = 0$. This is the frequency with which the random duration of failure-free operation of the object is shorter than required. It follows from (1.43) that

$$\hat{Q}(t_0) = 1 - \hat{P}(t_0). \tag{1.45}$$

An estimator of the probability of failure-free operation of an object in the time interval from t to $t + t_0$, given failure-free operation up to time t, is

$$\hat{P}(t; t + t_0) = \frac{N(t + t_0)}{N(t)}, \tag{1.46}$$

where $N(t)$ is the number of objects still operational at the moment of time t, and where $\hat{P}(t; t + t_0)$ is the ratio of the number of objects operating without failure up to the moment $t + t_0$ to the number of objects that were operational at the moment of time t. This is the frequency with which the length of the random period of failure-free operation of the object is greater than $t + t_0$, given that it is greater than t.

An estimator of the probability of failure of a system in the time interval from t to $t + t_0$, given failure-free operation up to time t, is

$$\hat{Q}(t; t + t_0) = \frac{d(t + t_0) - d(t)}{N(t)} = 1 - \frac{N(t + t_0)}{N(t)}, \tag{1.47}$$

where $\hat{Q}(t; t + t_0)$ is the ratio of the number of objects that have failed in the time interval $[t; t + t_0]$ to the number of objects still operational at the moment of time t_2. In other words, $Q(t; t + t_0)$ is the frequency with which the random duration of failure-free operation of the object is smaller than $t + t_0$, given that it is greater than t.

An estimator of the density at time t of the failure probability distribution of an object initially operational at time 0 is

$$\hat{f}(t) = \frac{d(t + \Delta t) - d(t)}{N(0)\,\Delta t}, \tag{1.48}$$

where $\hat{f}(t)$ is the ratio of failures per unit of time to the number of operational objects at time $t = 0$, and the unit of time Δt is as short as is possible with a requirement for a statistically significant number of failures. In other words, $\hat{f}(t)$ is the difference between the number of failures up to time $t + \Delta t$ and the number of failures up to time t divided by the product of the number of operational objects at time $t = 0$ and the length Δt of the time interval. It follows from (1.48) that

$$\hat{f}(t) = \frac{N(t + \Delta t) - N(t)}{N(0)\,\Delta t}. \tag{1.49}$$

An estimator of the hazard rate (failure rate) of an object at time t is

$$\hat{\lambda}(t) = \frac{d(t + \Delta t) - d(t)}{N(t)\,\Delta t} = \frac{N(t + \Delta t) - N(t)}{N(t)\,\Delta t}. \tag{1.50}$$

In order for $\hat{\lambda}(t)$ to provide a good approximation of $\lambda(t)$, $N(t)$ should be large and Δt should be small.

$\hat{\lambda}(t)$ is the number of failures per unit time divided by the number of operational objects at time t. In other words, it is the number of failures during Δt divided by the product of the number of objects still operational at time t and the length of the time interval Δt.

An estimator of mean time to failure is

$$\hat{T}_1 = \frac{1}{N(0)} (\xi_1^{(1)} + \xi_1^{(2)} + \cdots + \xi_1^{(N(0))}) = \frac{1}{N(0)} \sum_{i=1}^{N(0)} \xi_1^{(i)}. \tag{1.51}$$

Alternatively, where it is assumed that $0 = \xi_1^{*(0)} < \xi_1^{*(1)} < \xi_1^{*(2)} < \cdots < \xi_1^{*N(0)}$, $\hat{T}_1$ is the arithmetic mean of the random operating times of the objects up to failure:

$$\hat{T}_1 = \xi_1^{*(1)} + \frac{N(0) - 1}{N(0)} [\xi_1^{*(2)} - \xi^{*(1)}] + \cdots + \frac{1}{N(0)} [\xi^{*(N(0)} - \xi^{*(N(0)-1)}]$$

$$= \sum_{i=1}^{N(0)} \frac{N(0) - i + 1}{N(0)} [\xi^{*(i)} - \xi^{*(i-1)}]. \tag{1.52}$$

1.2.2 Renewable Objects

The notation for renewable objects is

$b(t)$ = number of objects repaired by the moment t when repair starts at $t = 0$ for all objects,

$d(t, t^*)$ = number of objects that fail during the interval $[t, t^*]$,

$d(t + t_0/t)$ = number of objects operational at the moment t that fail at least once in the interval from t to $t + t_0$,

d_t = number of objects in a state of failure at time t ($d_\infty = d_t$ for $t = \infty$),

$D(t)$ = number of objects failed at time 0 and not repaired by time t,

N = total number of objects under observation,

$N(0)$ = number of objects operational at the moment $t = 0$,

$N(t)$ = number of objects in $N(0)$ that do not fail before time t,

$N(t + t_0/t)$ = number of objects operational at the moment t that do not fail in the interval from t to $t + t_0$,

N_t = number of objects operational at time t,

ξ_j = random variable, the time between the end of renewal $j - 1$ and failure j (ξ_1 is the random time to the first failure),

$\xi_j^{(i)}$ = ith-observed value of a random value ξ_j,

$\xi_j^{*(k)}$ = kth-order statistic formed from the random values $\xi_j^{(i)}$,

η = random duration of the renewal time for an object, where the random variable does not depend on the number of repairs,

$\eta^{(i)}$ = ith observed value of the random variable η,

$\eta^{*(k)}$ = kth-order statistic formed from the random values $\eta^{(i)}$.

Estimators

An estimator of the probability of failure-free operation of an object in the time interval from 0 to t_0 is in (1.43). An estimator of the probability of failure of an object in the time interval from 0 to t_0 is in (1.44). An estimator of the probability of failure-free operation of an object during the time interval of length t_0, beginning from the moment of completion of the $(j - 1)$st repair, is

$$\hat{P}_j(t_0) = \frac{N_j(0) - d_j(t_0)}{N_j(0)}, \tag{1.53}$$

$$\hat{P}_j(t_0) = \frac{N_j(t_0)}{N_j(0)}, \tag{1.54}$$

where $\hat{P}_j(t_0)$ is the number of objects that operated without failure during period t_0 from the end of repair $j - 1$ to failure j divided by the total number of observed objects operating at the end of repair $j - 1$.

An estimator of the probability of failure of an object during the time interval of length t_0 starting from the moment of completion of the $(j - 1)$st repair is

$$\hat{Q}_j(t_0) = \frac{d_j(t_0)}{N_j(0)}, \tag{1.55}$$

$$\hat{Q}_j(t_0) = \frac{N_j(0) - N_j(t_0)}{N_j(0)}, \tag{1.56}$$

where $\hat{Q}_j(t_0)$ is the complement of $\hat{P}_j(t_0)$, and thus equals $1 - \hat{P}_j(t_0)$.

An estimator of the probability of failure-free operation of an object in the time interval from t to $t + t_0$, given failure-free operation up to time t, is

$$\hat{P}(t; t + t_0) = \frac{N(t) - d(t + t_0/t)}{N(t)}, \tag{1.57}$$

$$\hat{P}(t; t + t_0) = \frac{N(t + t_0/t)}{N(t)}, \tag{1.58}$$

where $P(t; t + t_0)$ is the number of objects that are operational at time t and operate failure-free up to time $t + t_0$ divided by the number of objects that are operational at time t.

An estimator of the probability of failure of a system in the time interval from t to $t + t_0$, given failure-free operation up to time t, is

$$\hat{Q}(t; t + t_0) = \frac{d(t + t_0/t)}{N(t)}, \tag{1.59}$$

where $Q(t; t + t_0)$ is the number of objects that are operational at time t and then fail at least once in the interval t_0 divided by the total number of objects operational at t.

An estimator of mean time to failure may be seen in (1.51) and (1.52). An estimator of the mean time to failure after the $(j - 1)$th repair of an object is

$$\hat{T}_j = \frac{1}{N}\,(\xi_j^{(1)} + \xi_j^{(2)} + \cdots + \xi_j^{(N)}) = \frac{1}{N}\sum_{i=1}^{N} \xi_j^{(i)}\,. \tag{1.60}$$

Alternatively,

$$\begin{aligned}\hat{T}_j &= \xi_j^{*(1)} + \frac{N-1}{N}\,[\xi_j^{*(2)} - \xi_j^{*(1)}] + \cdots + \frac{1}{N}\,[\xi_j^{*(N)} - \xi_j^{*N-1)}] \\ &= \sum_{i=1}^{N} \frac{N-i+1}{N}\,[\xi_j^{*(i)} - \xi_j^{*(i-1)}]\,,\end{aligned} \tag{1.61}$$

where it is assumed that $0 = \xi_j^{*(0)} < \xi_j^{*(1)} < \xi_j^{*(2)} < \cdots < \xi_j^{*N(0)}$. $\hat{T}_j$ is the arithmetic mean of the random intervals of failure-free operating times of the $N(0)$ objects from the moment of completion of the $(j - 1)$st renewal to the jth failure.

An estimator of the repair rate for an object at time t is

$$\hat{\mu}(t) = \frac{b(t + \Delta t) - b(t)}{D(t)\,\Delta t} . \tag{1.62}$$

In practice, for $\mu(t)$ to provide a good approximation of $\mu(t)$, $N_\tau(t)$ is made large and Δt is kept small. $\mu(t)$ is the number of repairs per unit time during the interval Δt divided by the number of objects in repair at time t when all these objects start being repaired at time $t = 0$.

An estimator of the mean repair time for an object is

$$\hat{\tau} = \frac{1}{N}\,(\eta^{(1)} + \eta^{(2)} + \cdots + \eta^{(N)}) = \frac{1}{N} \sum_{i=1}^{N} \eta^{(i)} . \tag{1.63}$$

Alternatively,

$$\begin{aligned} \hat{\tau} &= \eta^{(1)} + \frac{N-1}{N}\,[\eta^{(2)} - \eta^{(1)}] + \cdots + \frac{1}{N}\,[\eta^{(N)} - \eta^{(N-1)}] \\ &= \sum_{i=1}^{N} \frac{N-i+1}{N}\,[\eta^{(i)} - \eta^{(i-1)}] , \end{aligned} \tag{1.64}$$

where it is assumed that $0 = \eta^{(0)} < \eta^{(1)} < \eta^{(2)} < \cdots < \eta^{(N)}$. $\hat{\tau}$ is the arithmetic mean of the random renewal time of the N objects.

An estimator of the nonstationary readiness coefficient of an object is

$$\hat{K}(t) = \frac{N_t}{N} = \frac{N - d_t}{N} , \tag{1.65}$$

where $\hat{K}(t)$ is the ratio of the number of objects that are in an operational state at time t to the total number of objects.

An estimator of a nonstationary idleness coefficient is

$$\hat{k}(t) = \frac{n_t}{N} , \tag{1.66}$$

where $\hat{k}(t)$ is the ratio of the number of objects that are in a failed state at time t to the total number of objects.

An estimator of a stationary readiness coefficient of an object is

$$\hat{K} = \frac{N_\infty}{N} = \frac{N(0) - d_\infty}{N} , \tag{1.67}$$

or, for one object,

$$\hat{K} = \frac{\hat{T}}{\hat{T} + \hat{\tau}} . \tag{1.68}$$

When several objects are in the data base, $\hat{K}$ is the ratio of the number of objects in an operational state at a random moment of time after stationarity has been achieved to the total number of objects. With one object, $\hat{K}$ is the time the object is in an operational state divided by the total test time, given that stationarity has begun.

An estimator of a stationary idleness coefficient of an object is

$$\hat{k} = \frac{d_\infty}{N} . \tag{1.69}$$

When several objects are observed, k is the ratio of the number of objects that are in a failed state at an arbitrary moment of time during stationarity to the total number of objects. When a single object is observed, $\hat{k}$ is the ratio of the time in which the object is in a state of failure to the total test time, given that the test has started after attainment of stationarity.

An estimator of a nonstationary coefficient of availability for an object in the interval from t to $t + t_0$ is

$$\hat{R}(t, t_0) = \frac{N(t + t_0/t)}{N} , \tag{1.70}$$

where $\hat{R}(t, t_0)$ is the ratio of the number of objects that were operational at time t and operated failure-free in the time interval from t to $t + t_0$ to the total number of objects.

An estimator of a stationary coefficient of availability of an object is

$$\hat{R}(t_0) = \frac{N(t_\infty + t_0/t_\infty)}{N} , \tag{1.71}$$

where $\hat{R}(t_0)$ is the ratio of the number of objects that were operational at an arbitrary moment of time during stationarity and then operated failure-free during the succeeding time interval of length t_0 to the total number of objects.

1.3 SPECIAL INDEXES

Many modern technical systems, especially computer systems or computer-controlled systems, require special reliability indexes. The reason for these new indexes is the absence of a strict failure criterion. Often no one local event at a certain point in time can be considered to be a failure. One may have to wait until an operation is complete to determine success or failure. This section has five widely applicable indexes of this type. The five definitions are followed by sample applications.

1. The probability $A(a, t)$ that the total system up time $s(t)$ will exceed a given value a during a calendar period t is as follows:

Probabilistic definition:

$$A(a, t) = Pr\{s(t) \geq t\} .$$

Statistical definition:

$$\hat{A}(a, t) = \frac{1}{N(0)} \sum_{1 \leq i \leq N(0)} [s_i(t)]_a ,$$

where $N(0)$ is the number of operational objects at $t = 0$, $s_i(t)$ is the total operating time of the ith object during calendar time t, and

$$[s_t(t)]_a = \begin{cases} 1 & \text{if } s_i(t) \geq a \\ 0 & \text{otherwise} . \end{cases}$$

2. The probability $B^*(b, t^*)$ that the system time between failures θ will exceed a given value b at least once during a total system up time t^*, $t^* = s(t)$ is as follows:

Probabilistic definition:

$$B^*(b, t^*) = Pr\{\exists \xi_i \geq b \mid s(t) = t^*, \xi_i \in [0, t^*]\} ,$$

where $\exists \xi_i$ means "there exists an ξ_i" and where ξ_i is the duration of one successful period of operation (including the last, truncated period) of a system during a total system operating time t_*.
Statistical definition:

$$\hat{B}^*(b, t^*) = \frac{1}{N(0)} \sum_{1 \leq i \leq N(0)} [\{\xi\}_i]_b^* ,$$

where $\{\xi\}_i$ is a sequence of times between failures (TBF) of object i during a total up time $s(t) = t^*$,

$$[\{\xi\}_i]_b^* = \begin{cases} 1 & \text{if in the sequence } \{\xi\} \text{ there is at least one } \xi_i \\ & \text{(including an incomplete one) larger than } b, \\ 0 & \text{otherwise} . \end{cases}$$

3. The probability $B(b, t)$ that the system TBF ξ_i will exceed a given value b at least once during a calendar period t is as follows:

Probabilistic definition:

$$B(b, t) = Pr\{\exists \xi_i \geq b \mid \xi_i \in [0, t]\} ,$$

where ξ_i is one of the TBF in the observed sequence $\{\xi\}$ (including the last, truncated one) during calendar period t.

During this period of time there also are downtimes. $B(b, t)$ and $B^*(b, t^*)$ are related by the expression

$$B(b, t) = \int_0^t B^*(b, x)\, dF(x) ,$$

where $F(x)$ is the distribution function of a random value $s(x)$, the total system operating time, during the calendar period t. In general, determination of $F(x)$ may be difficult.

Statistical definition:

$$\hat{B}(t_0, t) = \frac{1}{N(0)} \sum_{1 \le i \le N(0)} [\{\xi, \eta\}_i]_b^* ,$$

where $\{\xi, \eta\}_i$ is the observed sequence of alternating up times and downtimes for object i during calendar period t,

$$[\{\xi, \eta\}_i]_b^* = \begin{cases} 1 & \text{if in the sequence } \{\xi, \eta\} \text{ at least one observed value of } \xi \text{ (including an incomplete one) is larger than } b , \\ 0 & \text{otherwise .} \end{cases}$$

4. The probability $C^*(c, t^*)$ that there will not be a period of system downtime η_i that will exceed a given value c during a total system downtime t^* is as follows:

Probabilistic definition:

$$C^*(c, t^*) = Pr\{\forall \eta_i < c \mid s_B(t) = t^*, \eta_i \in [0, t^*]\}$$

where $\forall \eta_i$ means "for all η_i that" and where η_i is one of the downtimes for a system (including the last, truncated one) during a system total downtime t^*.

Statistical definition:

$$\hat{C}^*(c, t^*) = \frac{1}{N(0)} \sum_{1 \le i \le N(0)} [\{\eta\}]_c^*$$

where $\{\eta\}_i$ is the observed sequence of random downtimes during a total downtime $s_B(t) = t_*$,

$$[\{\eta\}_i]_c^* = \begin{cases} 1 & \text{if in the sequence } \{\eta\}_i \text{ there is no observed value of } \eta \text{ (including an incomplete one) larger than } c , \\ 0 & \text{otherwise .} \end{cases}$$

5. The probability $C(c, t)$ that there will not be a system downtime η_i exceeding given value c during a calendar period t is as follows:

Probabilistic definition:

$$C(c, t) = Pr\{\forall \eta_i < c \mid \eta_i \in [0, t]\} ,$$

where η_i is one of the up times of a system (including the last, truncated one) during a calendar period t.

During this period of time there also are up periods. $C(c, t)$ and $C^*(c, t^*)$ are related by the expression

$$C(c, t) = \int_0^t C^*(c, t-x)\, dF(x)$$

where $F(x)$ is the distribution function of the random variable $s(x)$ is the distribution function of the random variable $s(x)$, the total system up time during a calendar period.

Statistical definition:

$$\hat{C}^*(c, t) = \frac{1}{N(0)} \sum_{1 \le i \le N(0)} [\{\xi, \eta\}_i]^*_{c^*} ,$$

where $\{\xi, \eta\}_i$ is the observed sequence of alternating up times and downtimes for the ith object during a total calendar time t,

$$[\{\xi, \eta\}_i]^*_c = \begin{cases} 1 & \text{if in } \{\xi, \eta\}_i \text{ there is no observed downtime } \xi \\ & \text{including an incomplete one) larger than } c , \\ 0 & \text{otherwise} . \end{cases}$$

The first index can be used, for example, for a computer performing a routine task with interruptions by a flow of priority tasks. In this case time to perform the loading task will be accumulated in intervals between priority tasks. The loading task of course could have its own required completion time. This index can also be useful for investigation of time sharing with computers.

The second and third indexes can be useful for a system with random errors that eliminate all previous results; after each interruption the system has to start from the beginning. The two last indexes can be used for systems with restricted resources such as conveyer lines with limited stocks and power systems with redundant, but limited, fuel resources. An example is a computer with batteries that have limited lifetimes.

1.4 SELECTION OF RELIABILITY INDEX

The choice of a reliability index depends on the nature of the technical object, the purpose, and the expected results of the operation. Reliability indexes can be divided into two main classes: operational and technological. The division is a relative one and depends on the circumstances.

An operational index is useful to characterize the acceptability of the functioning of a technical object from the user's viewpoint. Examples include the probability of successful operation, the availability coefficient, and the mean time between failures.

Technological indexes (or characteristics) are used for characterization of different subsystems or separate parts of a system. They are useful for the development of system operational indexes. They also can be needed for reliability analysis of a system as a whole. Most are functions of time rather than numerical parameters. The most clear representative of such a characteristic is a failure rate. Of course a distribution function itself is also such a characteristic.

An example can be given with a redundant group that can be characterized by an availability coefficient. In general, one needs to know the time to failure and repair time distribution for each unit in the group, as well as the service discipline for failed units. Knowledge of only the means is not sufficient.

The choice of a reliability index can be influenced not only by the designation of the technical object but by the importance of each function of the object. The following are general guidelines:

- The indexes should be as few as possible. Too many parameters can be difficult to understand or even contradictory.
- Complex indexes are generally undesirable, especially "weighted" indexes without real physical meaning.
- An index should have a clear physical definition.
- An index should be quantitatively formulable.
- An index should be estimable with analytical computations or computer simulation.
- An index should be verifiable with tests or real use.

1.5 REQUIREMENTS FOR SYSTEMS, SUBSYSTEMS, AND UNITS

For reliability requirement purposes, equipment may be divided into three groups: systems, subsystems, and units. The classification can depend upon the circumstances.

A system is an object with its own final goals and functions. It is characterized with operational reliability indexes. A subsystem is an assembly of objects within a system that performs its own functions. Subsystems can be characterized by technological and operational indexes. A unit, or component, is the smallest indivisible part of a technical object. The term "unit" sometimes is used also as a generic term for one physically separate item.

1.5.1 Reliability Requirements for a System

Sometimes a reliability requirement is based on practical experience and engineering intuition. This requires no special comment.

Reliability requirements can be based on prototypes of a new technical system. This usually requires statistical data about similar objects assembled from similar components. The determination requires both expertise and statistics.

Ideal conditions for the determination of a reliability requirement exist when

- the system outcome can be measured in the same units as the costs of system design, production, and maintenance;
- the system structure, operating characteristics, and maintenance requirements are well known in advance;
- necessary statistical data for all components are determined with satisfactory confidence.

The objective function for a system may take the form:

$$W_k(R) = E_k(R) - C_k(R) ,$$

where R is the system reliability index, $E_k(R)$ is an outcome of the kth variant of the designed system ($k = 1, \ldots, m$), and $C_k(R)$ is the cost to design, product and maintain the system with index R. The value of R depends on the variant and of the component reliability indexes, $r_i^{(k)}$. R itself in general form can be written as

$$R = R(S_k, r_i^{(k)}; 1 \le k \le m; 1 \le i \le n_k) ,$$

where m is the number of variants of the system and n_k is the number of different components in the kth variant.

For any variant the optimal level of the reliability index R usually is given for differentiable functions by

$$\frac{dE_k(R)}{dR} = \frac{dC_k(R)}{dR} .$$

One might estimate each optimum index, R_k^0, and then select the k^* with the highest value of $W_k(R_k^0)$. In practice, this ideal situation is extremely rare.

1.5.2 Reliability Requirements for a Subsystem

In this section the system reliability requirements are assumed to have been determined. Four methods are discussed.

The Method of Uniform Distribution of Reliability Requests

An index, R, which may be a probability of success or an availability coefficient, is given for a system. The system may have a series structure and consist of subsystems that are similar in complexity. The reliability requirement (given as a probability of success) for each subsystem then is

$$R_i = \sqrt[N]{R}\,, \qquad i = 1, \ldots, N\,,$$

where N is the number of subsystems. If requirements can be given as a function of the mean time,

$$T_i = nT\,,$$

where T is the system MTTF and T_i is the ith subsystem MTTF.

The Method of Distribution of Reliability Requirements in Proportion to the Number of Units

The same conditions exist as before, but in addition subsystem i consists of n_i units that are practically similar in reliability. In this case the requirement in terms of the success probability is

$$R_i = \sqrt[a_i]{R}\,,$$

where

$$a_i = \frac{n_i}{\sum_{1 \le i \le N} n_i}\,, \qquad 1 \le i \le N\,.$$

When all distributions are exponential, requirements can be formulated in terms of the MTTF:

$$T_i = \frac{T}{a_i}\,.$$

This method can be useful if different subsystems of the system are designed by different subcontractors.

The Method of Distribution of Reliability Requirements in Proportion to an Expected Total Failure Rate

If the unit failure rates are known (perhaps from previous experience or even by engineering intuition), and a hypothesis about an exponential distribution can be considered as valid, this method can be applied with

$$a_i = \frac{\sum_{1 \le j \le M} \lambda_j n_{ji}}{\sum_{1 \le i \le N} \sum_{1 \le j \le M} \lambda_j n_{ji}}\,,$$

where M is the number of types of units, and n_{ij} is the number of units of the jth type in the ith subsystem.

The Method of Optimal Reliability Requirements

This method can be applied if one knows the designed system structure S and can predict the cost curve for the reliability improvement of each subsystem. There are two types of optimization problems:

1. Find the subsystem indexes for the probability of success, R_i's, $i = 1, \ldots, n$, that maximize the system reliability index under a restriction on the total system cost C^0.

   This problem can be written formally as

   $$\max\left\{ R(S, R_i(C_i)) \,\middle|\, \sum_{1 \le i \le N} C_i \le C^0 \right\},$$

   where $R_i(C_i)$ is the reliability index of the ith subsystem as a function of the cost C_i.
2. Find the values of R_i, $i = 1, \ldots, n$, that yield a required reliability index R^0 at the lowest cost.

   This problem can be written formally as

   $$\min\left\{ \sum_{1 \le i \le n} C_i(R_i) \,\middle|\, R(S, R_i(C_i)) \ge R^0 \right\},$$

The solutions of these problems are considered in Chapter 20. The optimal distribution of reliability requirements among subsystems is a type of optimal redundancy problem.

1.5.4 Reliability Requirements for a Component

Methods for components are close to those in Section 1.5.1 from a methodological point of view.

CHAPTER 2

UNITS

In reliability theory a "unit" is any physically separable and functionally independent part of an item of equipment that does not have to be considered as a group of smaller parts for the analytical purpose at hand. Thus, depending on the analytical purpose, a unit may be a part, a module, or an entire system. There are three frequently used meanings, however: (1) a piece that cannot be disassembled, (2) an assembly of pieces that is removed and replaced as a whole, and (3) an assembly that is manufactured and used as a whole. Unless otherwise apparent, the second definition will be the applicable one in this chapter.

2.1 UNREPAIRABLE UNITS

2.1.1 All Distributions

Let us assume that the distribution function (d.f.) of a random time to failure $F(t)$ is known. All reliability indexes are expressed through this distribution and its parameters. In Table 2.1 the main reliability variables (r.v.'s) of a unit are given. There a discrete d.f. is given by

$$F(t) = \sum_{1 \le i \le k(t)} \delta_i ,$$

$$\sum_{\forall i} \delta_i = 1 , \qquad (2.1)$$

where t_i is the time of the ith change of level of the discrete process, $k(t)$ is the number of jumps up to point t, and ν_i is the magnitude of the change at time t_i.

Table 2.1 Reliability indexes for an unrenewable unit

Variable	Continuous Function	Discrete Function
$P(t_0)$	$1 - F(t_0)$	$1 - \sum_{i=1}^{k(t_0)} \nu_i = \sum_{i=k(t_0)+1}^{N} \nu_i$
$Q(t_0)$	$F(t_0)$	$\sum_{i=1}^{k(t_0)} \nu_i$
$P(t, t + t_0)$	$\dfrac{1 - F(t + t_0)}{1 - F(t)}$	$\left(1 - \sum_{i=1}^{k(t+t_0)} \nu_i\right) \Big/ \left(1 - \sum_{i=1}^{k(t)} \nu_i\right)$
$Q(t, t + t_0)$	$\dfrac{F(t + t_0) - F(t)}{1 - F(t)}$	$\left(\sum_{i=k(t)}^{k(t+t_0)} \nu_i\right) \Big/ \left(1 - \sum_{i=1}^{k(t)} \nu_i\right)$
T	$\int_0^\infty [1 - F(t)]\, dt = \int_0^\infty P(t)\, dt$	$\sum_{i=1}^{N} \nu_i t_i = \sum_{i=1}^{N} \left[\sum_{i=k+1}^{N} \nu_i (t_{i+1} - t_i) \right]$
$\lambda(t)$	$\dfrac{d \ln P(t)}{dt} = \dfrac{1}{1 - F(t)} \cdot \dfrac{dF(t)}{dt}$	$\lambda(t) = \dfrac{F(t_{i+1}) - F(t_i)}{(t_{i+1} - t_i) F(t_i)}, \; t_i \le t < t_{i+1}$

2.1.2 Exponential Distributions

In Table 2.2 the main reliability indexes are given for an exponential distribution of time to failure. All approximate expressions are given under the condition that $\lambda t_0 \ll 1$. There is a systematic error of 0.5 (λt_0). For practical purposes these approximations are valid even for λt_0 in the range from 0.05 to 0.10.

Table 2.2 Reliability indexes for unrenewable items, exponential distributions

Variable	Exact Value	Approximate Value	Systematic Error
$P(t_0)$	$e^{-\lambda t_0}$	$1 - \lambda t_0$	$\delta_+ < \dfrac{(\lambda t_0)^2}{2}$
$Q(t_0)$	$1 - e^{-\lambda t_0}$	λt_0	$\delta_- < \dfrac{(\lambda t_0)^2}{2}$
$P(t, t + t_0)$	$e^{-\lambda t_0}$	$1 - \lambda t_0$	$\delta_+ < \dfrac{(\lambda t_0)^2}{2}$
$Q(t, t + t_0)$	$1 - e^{-\lambda t_0}$	λt_0	$\delta_- < \dfrac{(\lambda t_0)^2}{2}$
T	$\dfrac{1}{\lambda}$	—	—
$\lambda(t)$	λ	—	—

2.1.3 Aging Unit

Appendix B shows various distributions with monotonic failure rates. Two important classes of "aging" distributions are distributions with increasing failure rates (IFR) and distributions with increasing failure rate averages (IFRA). Useful estimates of the limits of these distributions can be based on knowledge of only the mean and the variance. Three cases are considered here:

1. The upper and lower bounds of an IFR distribution with a known MTTF are presented in Figure 2.1. In Table 2.3 the upper bound is calculated with

$$1-\omega_t T = e^{-\omega_t t} \qquad \text{for } t > T \,. \tag{2.2}$$

The lower bound is given by

$$\underline{P}(t) = \begin{cases} e^{-t/T} & \text{for } t \le T \,, \\ 0 & \text{for } t > T \,. \end{cases} \tag{2.3}$$

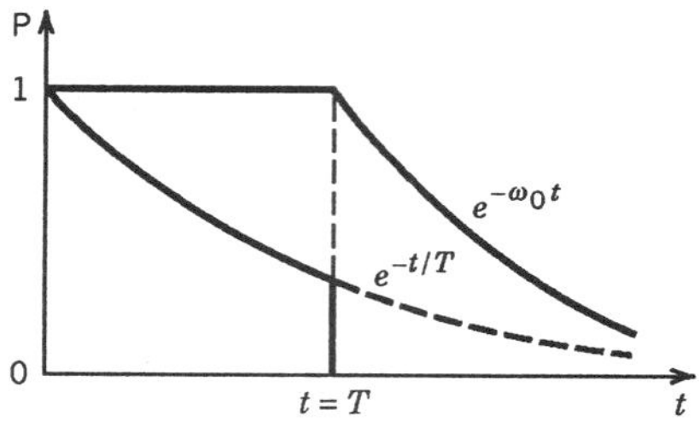

Figure 2.1 Upper and lower bounds, failure probability as a function of time, increasing failure rate distribution with known MTTF but unknown variance.

Table 2.3 Equation (2.2) values, $1.0 \le t/T \le 2.0$ (upper bound of probability of successful operation until time t, increasing failure rate, MTTF = T)

t/T	0.00	0.02	0.04	0.06	0.08
1.0	1.000	0.961	0.924	0.889	0.856
1.1	0.824	0.794	0.765	0.738	0.711
1.2	0.686	0.662	0.640	0.618	0.597
1.3	0.577	0.558	0.540	0.522	0.505
1.4	0.489	0.474	0.459	0.444	0.430
1.5	0.417	0.404	0.392	0.380	0.369
1.6	0.358	0.348	0.337	0.328	0.318
1.7	0.309	0.300	0.292	0.283	0.275
1.8	0.268	0.260	0.253	0.246	0.239
1.9	0.233	0.226	0.220	0.214	0.209
2.0	0.203	0.198	0.193	0.188	0.183

Table 2.4 Parameters of an increasing failure rate distribution of known MTTF but unknown variance

Parameter Index	Lower Bound	Upper Bound	Validity Range
$P(t)$	$e^{-t/T}$	1	$0 \leq t < T$
	0	$e^{-\omega_t t}$	$t \geq T$
	$e^{-\alpha t}$	—	$t \leq \xi_p$
	—	$e^{-\alpha t}$	$t \geq \xi_p$
ξ_p	$-\ln(1-p)T$	$-\dfrac{\ln(1-p)}{p}T$	$p \leqslant 1 - e^{-1}$
	T	$-\dfrac{\ln(1-p)}{p}T$	$p \geq 1 - e^{-1}$

Note: $P(t)$ = Probability of successful operation until time t, ξ_p = quantile.

Equations for the upper and lower bounds of the probability and quantile are given in Table 2.4.

2. When an IFR distribution has both a known MTTF and a known variance, the upper and lower bounds can be found with sophisticated methods. The numerical values of the bounds are given in Tables 2.5 and 2.6.
3. Table 2.7 gives numerical values of the lower bound of an IFRA distribution with a known MTTF.

2.1.4 Probability of Success with Random Operational Time

A probability of interest is

$$P = Pr\{\xi \geq t\} = \int_0^\infty [1 - F(t)\, dG(t) = \int_0^\infty G(t)\, dF(t)\,, \tag{2.4}$$

where $G(t)$ is the d.f. of the operational time.

The problem is discussed in detail in Chapter 11, which covers mechanical systems. There X is the strength of a unit and Y is the load. The problem is to find the probability $\Pr\{X \geq Y\}$. Mathematically this problem and that one are equivalent.

We present here only the very general result for an exponential distribution of time to failure (TTF). In this case

Table 2.5 Upper bound of probability of successful operation until time t, $P(t)$ for an increasing failure rate distribution of known MTTF and known variance (subspace of combinations of t/T and σ/T^2)

	σ^2/T^2						σ^2/T^2				
t/T	0.10	0.20	0.30	0.40	0.50	t/T	0.60	0.70	0.80	0.90	1.00
0.30	1.000	1.000	1.000	1.000	0.993	0.05	1.000	1.000	1.000	1.000	0.951
0.33	1.000	1.000	1.000	1.000	0.964	0.10	1.000	1.000	1.000	9.952	0.905
0.36	1.000	1.000	1.000	1.000	0.936	0.15	1.000	1.000	0.957	0.906	0.861
0.39	1.000	1.000	1.000	0.978	0.909	0.20	1.000	0.964	0.910	0.862	0.819
0.42	1.000	1.000	1.000	0.951	0.883	0.25	0.976	0.917	0.866	0.820	0.779
0.45	1.000	1.000	1.000	0.924	0.857	0.30	0.930	0.873	0.824	0.780	0.741
0.48	1.000	1.000	0.974	0.897	0.833	0.35	0.884	0.831	0.784	0.742	0.705
0.51	1.100	1.000	0.947	0.872	0.809	0.40	0.842	0.790	0.746	0.706	0.670
0.54	1.000	1.000	0.921	0.847	0.785	0.45	0.801	0.752	0.709	0.671	0.638
0.57	1.000	0.985	0.896	0.823	0.763	0.50	0.762	0.716	0.675	0.639	0.606
0.60	1.000	0.959	0.871	0.800	0.741	0.55	0.726	0.681	0.642	0.608	0.577
0.63	1.000	0.933	0.847	0.777	0.719	0.60	0.691	0.648	0.611	0.578	0.549
0.66	1.000	0.909	0.824	0.755	0.699	0.65	0.658	0.616	0.581	0.550	0.522
0.69	0.995	0.885	0.801	0.734	0.678	0.70	0.626	0.587	0.553	0.523	0.487
0.72	0.972	0.862	0.779	0.713	0.659	0.75	0.596	0.558	0.526	0.497	0.472
0.75	0.949	0.840	0.758	0.693	0.640	0.80	0.567	0.531	0.500	0.473	0.450
0.78	0.927	0.819	0.737	0.674	0.622	0.85	0.540	0.506	0.476	0.450	0.427
0.81	0.906	0.798	0.718	0.655	0.604	0.90	0.514	0.481	0.453	0.428	0.407
0.84	0.887	0.778	0.698	0.636	0.587	0.95	0.490	0.458	0.431	0.407	0.387
0.87	0.868	0.758	0.680	0.619	0.570	1.00	0.467	0.436	0.410	0.388	0.368
0.90	0.850	0.740	0.661	0.601	0.554	1.05	0.444	0.415	0.390	0.369	0.350
1.00	0.796	0.682	0.608	0.548	0.503	1.10	0.423	0.395	0.371	0.351	0.333
1.05	0.774	0.656	0.579	0.523	0.479	1.15	0.402	0.376	0.363	0.334	0.317
1.10	0.756	0.632	0.554	0.499	0.457	1.20	0.383	0.358	0.336	0.317	0.301
1.15	0.742	0.609	0.531	0.476	0.435	1.25	0.365	0.340	0.320	0.302	0.286
1.20	0.633	0.588	0.509	0.455	0.415						
1.25	0.514	0.570	0.488	0.435	0.396	1.30	0.348	0.324	0.304	0.287	0.272
1.30	0.411	0.554	0.469	0.416	0.377	1.35	0.331	0.308	0.289	0.273	0.259
1.35	0.326	0.519	0.451	0.397	0.360	1.40	0.316	0.293	0.275	0.260	0.247
1.40	0.258	0.444	0.434	0.380	0.343	1.45	0.300	0.279	0.262	0.247	0.235
1.50	0.163	0.321	0.405	0.348	0.313	1.50	0.286	0.266	0.249	0.235	0.223

$$P = \int_0^\infty e^{-\lambda t}\, dG(t) = \varphi_g(s)\big|_{s=\lambda},$$

where $\varphi_g(s)$ is the Laplace-Stieltjes transform (LST) of the density function $g(t) = (d/dt)G(t)$. The result can be found easily from the table of Laplace-Stieltjes transforms in Appendix D.

able 2.6 Lower bound of probability of successful operation until time t, $P(t)$, for n increasing failure rate distribution of known MTTF and known variance (subspace of ombinations of t/T and σ/T^2)

	σ^2/T^2						σ^2/T^2				
$/T$	0.10	0.20	0.30	0.40	0.50	t/T	0.60	0.70	0.80	0.90	1.00
.05	0.987	0.978	0.970	0.965	0.961	0.05	0.957	0.955	0.953	0.952	0.951
.10	0.974	0.956	0.941	0.931	0.923	0.10	0.917	0.912	0.909	0.906	0.905
.15	0.962	0.934	0.913	0.893	0.886	0.15	0.878	0.871	0.866	0.863	0.861
.20	0.949	0.913	0.886	0.866	0.851	0.20	0.840	0.832	0.826	0.821	0.819
.25	0.937	0.892	0.860	0.836	0.818	0.25	0.804	0.794	0.787	0.782	0.779
.30	0.925	0.872	0.834	0.806	0.686	0.30	0.770	0.758	0.750	0.744	0.741
.35	0.913	0.853	0.810	0.778	0.755	0.35	0.737	0.724	0.715	0.708	0.705
.40	0.901	0.834	0.786	0.751	0.725	0.40	0.706	0.692	0.682	0.674	0.670
.45	0.886	0.813	0.762	0.724	0.696	0.50	0.647	0.631	0.619	0.611	0.606
.50	0.868	0.790	0.737	0.698	0.669	0.55	0.619	0.602	0.590	0.582	0.577
.55	0.847	0.763	0.709	0.671	0.642	0.60	0.592	0.575	0.563	0.554	0.549
.60	0.819	0.731	0.677	0.639	0.613	0.65	0.564	0.548	0.536	0.527	0.522
.65	0.787	0.695	0.643	0.608	0.583	0.70	0.534	0.521	0.511	0.502	0.497
.70	0.747	0.655	0.605	0.574	0.551	0.75	0.505	0.494	0.485	0.478	0.472
.75	0.699	0.609	0.567	0.538	0.518	0.80	0.475	0.466	0.460	0.454	0.449
.80	0.640	0.562	0.524	0.502	0.486	0.85	0.446	0.440	0.435	0.431	0.427
.85	0.569	0.510	0.482	0.466	0.454	0.90	0.419	0.415	0.411	0.409	0.407
.90	0.502	0.460	0.411	0.431	0.424	0.95	0.392	0.390	0.389	0.388	0.387
.95	0.452	0.412	0.403	0.399	0.395	1.00	0.368	0.368	0.368	0.368	0.368
.00	0.368	0.368	0.368	0.368	0.368	1.05	0.345	0.347	0.348	0.349	0.350
.05	0.315	0.329	0.336	0.340	0.342	1.10	0.324	0.326	0.329	0.331	0.333
.10	0.270	0.292	0.305	0.312	0.318	1.15	0.303	0.308	0.311	0.314	0.317
.15	0.230	0.264	0.280	0.290	0.298	1.20	0.284	0.290	0.294	0.298	0.301
.20	0	0.236	0.256	0.268	0.277	1.25	0.266	0.273	0.278	0.283	0.286
.25	0	0.211	0.233	0.248	0.258	1.30	0.250	0.257	0.263	0.268	0.272
.30	0	0.188	0.213	0.229	0.241	1.35	0.234	0.242	0.249	0.254	0.259
.35	0	0	0.194	0.212	0.224	1.40	0.220	0.228	0.235	0.241	0.247
.40	0	0	0.177	0.196	0.209	1.45	0.206	0.215	0.222	0.229	0.235
.45	0	0	1.162	0.181	0.195	1.50	0.193	0.202	0.210	0.217	0.223
.50	0	0	0.146	0.167	0.182						

The result for a discrete function is

$$P = 1 - \sum_{1 \le k < \infty} (-1)^k \frac{\lambda^k M_k}{k!}, \tag{2.6}$$

where M_k is the kth initial moment of the distribution $G(t)$. In particular, $M_1 = t_0$, and $M_2 = t_0^2 + \sigma^2$.

Table 2.7 Lower bounds of probability of successful operation until time t, unit with IFRA distribution and known MTTF

t/T	Lower Bound	t/T	Lower Bound	t/T	Lower Bound	t/T	Lower Bound
0.00	1.0000	0.10	0.9043	0.20	0.8155	0.40	0.6488
0.01	0.9900	0.11	0.8952	0.22	0.7983	0.45	0.6080
0.02	0.9801	0.12	0.8861	0.24	0.7813	0.50	0.5671
0.03	0.9704	0.13	0.8771	0.26	0.7644	0.55	0.5258
0.04	0.9607	0.14	0.8681	0.28	0.7476	0.60	0.4839
0.05	0.9511	0.15	0.8592	0.30	0.7310	0.70	0.3964
0.06	0.9416	0.16	0.8504	0.32	0.7144	0.80	0.3005
0.07	0.9322	0.17	0.8416	0.34	0.6979	0.90	0.1865
0.08	0.9228	0.18	0.8329	0.36	0.6815	0.99	0.0341
0.09	0.9136	0.19	0.8242	0.38	0.6651		

2.2 REPAIRABLE UNIT

Operation of a repairable unit can be described as an alternating renewal process, a sequence of alternating random mutually independent intervals of up times and downtimes: $\xi_1, \eta_1, \xi_2, \eta_2, \ldots,$. All r.v.'s ξ belong to the same d.f., $F(t)$, and all r.v.'s η belong to the same d.f., $G(t)$. This process can be described partially with the transition diagram in Figure 2.2.

2.2.1 All Distributions

In Table 2.8 the main reliability indexes are given for arbitrary distributions of a random TTF, $F(t)$, and a random idle time $G(t)$. Notice that the expression for the operational availability

$$R(t) = KP(t_0)$$

is not general. This equation is correct only for an exponential distribution of up time. If the unit is aging and hence has an IFR or IFRA distribution, the two bounds can be written

Figure 2.2 Transition diagram, renewable unit.

Table 2.8 Reliability indexes of renewable unit, all $F(t)$ and $G(t)$ distributions, integral expressions

Index	Precise Value[a]	Approximate Value[a]	Approximate Value Systematic Error[a]
T	$\int_0^\infty xf(x)\,dx = \int_0^\infty x\,dF(x) = \int_0^\infty P(x)\,dx$	—	—
τ	$\int_0^\infty xg(x)\,dx = \int_0^\infty x\,dG(x) = \int_0^\infty [1 - G(x)]\,dx$	—	—
$P(t_0)$	$1 - F(t_0)$	—	—
$Q(t_0)$	$F(t_0)$	—	—
K	$\frac{T}{T+\tau} = \frac{1}{1+\gamma}$	$1-\gamma$	$\delta_+ < \gamma^2$
k	$\frac{\tau}{T+\tau} = \frac{\gamma}{1+\gamma}$	γ	$\delta_- < \gamma^2$
$R(t_0)$	$K\frac{1}{T}\int_{t_0}^\infty [1 - F(x)]\,dx = \frac{1}{T+\tau}\int_{t_0}^\infty P(x)\,dx$	$1 - \frac{t_0+\tau}{T}$	$\delta_+ < 2.5\left[\frac{\max(t_0,\tau)}{T}\right]^2$

[a] $\gamma = t/T$.

$$K\left(1 - \frac{t_0}{T}\right) \le R(t_0) \le Ke^{-t_0/T}, \tag{2.7}$$

where T is the MTTF, which is also the expected value of ξ, $E\{\xi\}$.

2.2.2 Exponential $F(t)$ and $G(t)$ Distributions

In Table 2.9 the main reliability indexes are given when $F(t) = 1 - e^{-\lambda t}$ and $G(t) = 1 - e^{-\mu t}$. Approximate equations for $P(t_0)$ and $Q(t_0)$ are given under the condition that $\lambda t_0 \ll 1$ and for the other indexes under the condition that $\gamma = (\lambda/\mu) < 1$. An availability coefficient $K(t)$ is written for a unit that is up at $t = 0$, with another availability coefficient $K^0(t)$ for a unit that is shown at $t = 0$. The availability coefficients are displayed graphically in Figure 2.3.

Table 2.9 Reliability indexes of renewable unit, all $F(t)$ and $G(t)$ distributions, algebraic expressions ($\gamma \approx \lambda/\mu \approx \tau/T$)

Index	Precise Value	Approximate Value	Range of Validity	Systematic Error
$P(t_0)$	$e^{-\lambda t_0}$	$1-\lambda t_0$	$\gamma t_0 \ll 1$	$\delta_+ < \frac{1}{2}(\lambda t_0)^2$
$Q(t_0)$	$1-e^{-\lambda t_0}$	λt_0	$\lambda t_0 \ll 1$	$\delta_- < \frac{1}{2}(\lambda t_0)^2$
T	$\dfrac{1}{\lambda}$	—	—	—
τ	$\dfrac{1}{\mu}$	—	—	—
K	$\dfrac{\mu}{\lambda+\mu}=\dfrac{T}{T+\tau}=\dfrac{1}{1+\gamma}$	$1-\gamma$	$\gamma \ll 1$	$\delta_+ < \gamma^2$
k	$\dfrac{\lambda}{\lambda+\mu}=\dfrac{\tau}{T+\tau}=\dfrac{\gamma}{1+\gamma}$	γ	$\gamma \ll 1$	$\delta_- < \gamma^2$
$K(t)$	$K+ke^{-(\lambda+\mu)t}$	$1-\gamma(1-e^{-\mu t})$	$\gamma \ll 1$	$\delta_+ \sim \gamma^2(1-e^{-\mu t})$
$K^0(t)$	$K(1-e^{-(\lambda+\mu)t})$	$(1-\gamma)(1-e^{-\mu t})$	$\gamma \ll 1$	$\delta_- \sim \gamma^2(1-e^{-\mu t})$

2.2.3 Partially Monitored Repairable Unit

Let us consider a unit with a part that is monitored only periodically. A failure in this part remains undiscovered until the periodical test. A failure in another part of the unit is discovered immediately. The main unit repair time equals τ, and the MTTF equals T. The probability that a failure is in the unmonitored part is α, and the probability that it is in the other part is $1-\alpha$. The period between unit tests is constant and equals t_v.

The main reliability indexes are given in Table 2.10, where $\gamma_0 = t_0/T$, $\gamma = \tau/T$ and $\gamma_v = t_v/T$. All approximate formulas are valid when all γ's are much less than one. For some cases only the upper and lower bounds are given. An arbitrary function of the type $P(t)$ is assumed to be IFR.

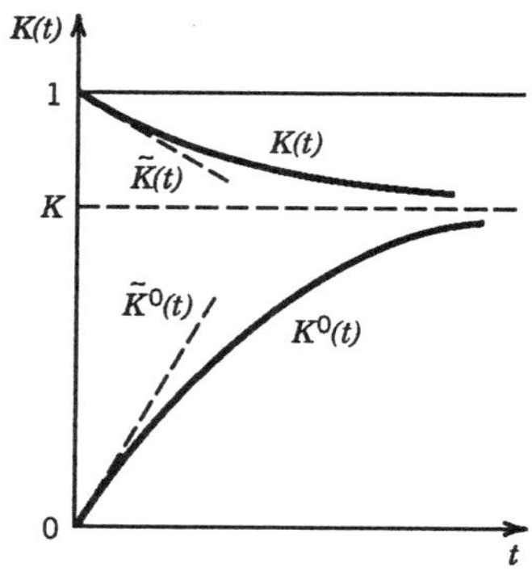

Figure 2.3 Expected availability during nonstationary phase, up initial state, and down initial state.

Table 2.10 Reliability indexes of partially monitored renewable unit ($\gamma_0 = t_0/T = \lambda t_0$; $\gamma = \tau/T = \lambda\tau$; $\gamma_\nu = t_\nu/T = 1/\nu T$; α = probability that a failure is in the unmonitored part)

Index		$P(t), t_\nu$	$P(t), \nu$	$e^{-\lambda t}, t_\nu$	$e^{-\lambda t}, \nu$
T_0		T	T	$\frac{1}{\lambda}$	$\frac{1}{\lambda}$
τ_0	Lower estimate	$0.5t_\nu\alpha + \tau$	$\frac{\alpha}{\nu} + \tau$	$0.5t_\nu\alpha + \tau$	$\frac{\alpha}{\nu} + \tau$
	Upper estimate	$\alpha t + \tau$		$t_\nu\alpha + \tau$	
$P_0(t_0)$		$P(t_0)$	$P(t_0)$	$e^{-\gamma_0}$	$e^{-\gamma_0}$
K	Lower Estimate	$1 - (\alpha\gamma_\nu + \gamma)$	$1 - (\alpha\gamma_\nu + \gamma)$	$1 - (\alpha\gamma_\nu + \gamma)$	$1 - (\alpha\gamma_\nu + \gamma)$
	Upper estimate	$1 - (0.5\alpha\gamma_\nu + \gamma)$		$1 - (0.5\alpha\gamma_\nu + \gamma)$	
$R(t_0)$	Lower estimate	$(1 - \gamma_0)[1 - (\alpha\gamma_\nu + \gamma)]$	$(1 - \gamma_0)[1 - (\alpha\gamma_\nu + \gamma)]$	$(1 - \gamma_0)[1 - (\alpha\gamma_\nu + \gamma)]$	$[1 - (\alpha\gamma_\nu + \gamma)]e^{-\gamma_0}$
	Upper estimate	$[1 - (0.5\alpha\gamma_\nu + \gamma)]e^{-\gamma_0}$	$[1 - (0.5\alpha\gamma_\nu + \gamma)]e^{-\gamma_0}$	$[1 - (0.5\alpha\gamma_\nu + \gamma)]e^{-\gamma_0}$	

2.3 REPAIRABLE UNIT WITH MAINTENANCE

2.3.1 Description of the Maintenance Rules

Let a unit have no continuous tests of operability. Unit tests and preventive replacement are performed periodically in time θ. Any maintenance procedure that cannot be postponed interrupts normal unit operation.

The following are notations for six categories used in analysis and maintenance.

1. *Operational regimes during maintenance.* A_1 is a regime during preventive maintenance in which a unit is on but does not operate. A_2 is a regime during preventive maintenance in which a unit is off, and its failure cannot appear.
2. *Types of preventive maintenance.* B_1 is regular maintenance with duration time Δ_1 and only a unit test. B_2 is regular maintenance with duration time Δ_2 and preventive replacement of the unit. B_3 is regular maintenance with a unit test and preventive replacement every mth maintenance.
2. *Types of B_3 maintenance.* C_1 is the first regular maintenance for purposes of planning preventive replacement of the unit. Counting starts either at replacement or at failure repair. C_2 is a failure repair that does not change the schedule for preventive replacement.
4. *Types of failure.* D_1 is a failure that can be found only by tests performed during regular maintenance. D_2 is a failure that can be found either by tests performed during regular maintenance or, with a random delay, by unit characteristics during operation.
5. *Failure discovery categories.* E_1 is failure discovery that occurs at the very beginning of regular maintenance and is followed immediately by repair. E_2 is failure discovery that occurs at a random moment during regular maintenance and is followed immediately by repair. E_3 is failure discovery that occurs during regular maintenance but is followed by repair only after maintenance has been completed.
6. *Test confidence.* F_1 is a test with a certainty level of 1. F_2 is a test with a certainty level <1 performed after each regular maintenance, with tests after different maintenances being independent and identical events. The probability of successful failure discovery equals $1 - \beta$.

In some cases performance of maintenance makes operation impossible. One can then abort the maintenance and try to resume operation. Let the probability of successful operation after the interruption of the maintenance equal $1 - \gamma$. For these circumstances we also define the following categories.

7. *Categories of regular maintenance.* G_1 is maintenance that is obligatory and interrupts unit operation. G_2 is maintenance that can be postponed until current operations are completed.

8. *Switching from maintenance to operation.* H_1 is a state in which switching is impossible. H_2 is a state in which switching is possible.

After maintenance or repair the unit will be considered to be completely new. After an interruption of maintenance, the unit characteristics will be the same as before maintenance.

2.3.2 Structure of the Main Reliability Indexes

The stationary availability coefficient is

$$R(t_0) = K^* P^*(t_0)\,,$$

where K^* is the stationary probability that the unit can start to operate at the required moment and $F^*(t_0)$ is the probability that from this moment the unit can operate successfully for duration t_0 until a failure or obligatory maintenance.

The probability K^* can be written approximately as

$$K^* \approx \kappa^* K\,,$$

where K is a stationary availability coefficient. In this particular case it equals

$$K = \frac{T}{T + \zeta + \chi} \tag{2.8}$$

where T = the MTTF, ζ = the mean time of the unit renewal, χ = the mean time that a hidden failure remains hidden, κ^* = a value similar to the coefficient of unit usage that is defined in Table 2.11. Expressions of terms in this table for various categories are given in Tables 2.12 through 2.14. In Table 2.12 ν^* is a random value with distribution

$$Pr\{\nu^* = \theta(0.5 + k)\} = (1 - \beta)\beta^k\,.$$

Table 2.11 χ^* **formulas**

Regime	H_1	H_2
A_1	$\dfrac{T - \tau - \zeta - \chi}{T - \zeta - \chi}$	$1 - \left(1 - \dfrac{T - \pi - \zeta - \chi}{T - \zeta - \chi}\right)(1 - \gamma)$
A_2	$\dfrac{T - \zeta - \chi}{T - \zeta - \chi + \pi}$	$1 - \left(1 - \dfrac{T - \zeta - \chi}{T - \epsilon - \chi + \pi}\right)(1 - \gamma)$

Table 2.12 Formulas for χ, Table 2.11

Regime	D_1	D_2 Exact	D_2 Approximate
F_1	0.5θ	$M\min(\delta, \nu)$	$\left(\frac{1}{\delta}+\frac{2}{\theta}\right)^{-1}$
F_2	$\theta(0.5+\beta)$	$M\min(\delta, \nu^*)$	$\left(\frac{1}{\delta}+\frac{1}{\theta(0.5+\beta)}\right)^{-1}$

Table 2.13 Expected values of renewal delay time ζ, Table 2.11

Regime[a]	E_1	E_2	E_2
B_1	τ	$0.5\Delta_1+\tau$	$\Delta_1+\tau$
B_2	τ	$0.5\Delta_2+\tau$	$\Delta_2+\tau$
B_3	τ	$0.5\Delta^*+\tau$	$\Delta^*+\tau$

[a]Regime definitions are in text.

Table 2.14 Formulas for π, Table 2.11

Regime[a]	c_1	c_2
B_1	$\frac{T}{\theta}\Delta_1$	$\frac{T}{\theta}\Delta_1$
B_2	$\frac{T}{\theta}\Delta_2$	$\frac{T}{\theta}\Delta_2$
B_3	$\frac{T}{m\theta}\Delta_2+\frac{(m-1)}{m\theta}\Delta_1$	$\left[\frac{T}{m\theta}\right]\Delta_2+\left(\left[\frac{T}{\theta}\right]-\left[\frac{T}{m\theta}\right]\right)\Delta_1$

[a]See text for regime definitions.

In all expressions $\Delta_1 \leq \Delta_2$ because regular maintenance with preventive replacement is almost always longer than maintenance that includes only a test of the unit. In addition repair after a failure is longer than preventive replacement. These circumstances were considered when Table 2.13 was compiled. The term Δ^* can be calculated approximately with

$$\Delta^* \approx \frac{1}{m}\Delta_1(m-1)+\frac{1}{m}\Delta_2 . \tag{2.9}$$

In the tables $[k]$ means an integer part of k. The difference between C_1 and C_2 may be seen to be insignificant. The expressions for calculation of $P^*(t_0)$ are in Table 2.15. The upper and lower bounds are given for a random TTF with an IFR distribution.

Table 2.15 $P^*(t_0)$ **formulas**

	G_2	
G_1	Lower Estimate	Upper Estimate
$1 - \dfrac{t_0(T+\theta)}{\theta T}$	$1 - \dfrac{t_0}{T}$	$e^{-t_0/T}$

2.4 SPECIAL RELIABILITY INDEXES

2.4.1 A Given Total Operating Time During a Fixed Calendar Period

A failure may not end unit operation but only delay its completion. An interruption of operation may be caused by either failure or maintenance. Sometimes an interruption involves only a portion of the unit operation. In a computer, for example, a failure or a random computer error may destroy only the last part of the computations. The lost time may be only the time spent recomputing the lost part. This problem and similar problems can be formulated as described in this section.

Let there be an alternating stochastic renewal process consisting of a sequence of independent values:

$$\xi_1, \eta_1, \xi_2, \eta_2, \ldots, \xi_n, \eta_n, \ldots .$$

As above, ξ's are up intervals and η's are down intervals.

During the time interval $[0, t]$ several ξ's may be observed. If the unit is up at time t, one should consider the last interval ξ as truncated from the start until t. Denote this incomplete r.v. duration with ξ'. The problem is to find the probability $A(t, t_0)$ that, during the fixed calendar time t, the total operating time

$$S(t) = \xi_1 + \xi_2 + \cdots + \xi_{k-1} + \xi_k'$$

will exceed a given required time t_0. It is assumed that there are k up intervals inside the total calendar time t:

$$A(t_0, t) = Pr\{S(t) \geq t_0\} . \tag{2.10}$$

Solution for All Distributions

Let $F(t) = Pr\{\xi \leq t\}$ and $G(t) = Pr\{\eta \leq t\}$. Then

$$A(t_0, t) = \sum_{0 \leq k < \infty} [G^{*k}(t - t_0) - G^{*(k+1)}(t - t_0)]F^{*k}(t_0) , \tag{2.11}$$

where the superscript "$*k$" means the k-time convolution:

$$Z^{*k}(x) = \int_0^x Z^{*(k-1)}(x-y)\, dZ(y) ,$$

and $F^{*0}(x) = G^{*0}(x) = 1$.

For large values of t the values of $A(t_0, t)$ are calculated with many iterations. But in this case one can use the asymptotic formula

$$A(t_0, t) \approx \Phi\left(\frac{t_0 - M_{S(t)}}{\sigma_{S(t)}}\right) , \tag{2.12}$$

where

$$M_{S(t)} \approx \frac{T}{T+\tau}\, t , \quad \sigma^2_{S(t)} \approx \frac{T^2\tau^2[(\sigma^2/T^2) + (\sigma_g^2)/\tau^2)]}{(T+\tau)^2}$$

and where T and σ^2 are the mean and the variance of the distribution $F(t)$, τ and σ_g^2 are the mean and the variance of the distribution $G(t)$, and $\phi(x)$ is a standardized normal d.f. (see Table E.3, Appendix E).

Solutions for Exponential Distributions

When $F(t) = 1 - e^{-\lambda t}$, and $G(t) = 1 - e^{-\mu t}$, Equation 2.10 takes the form

$$A(t_0, t) = e^{-\lambda t_0 - \mu(t - t_0)} \sum_{1 \le k < \infty} \frac{[\mu(t - t_0)]^k}{k!} \sum_{k \le j < \infty} \frac{(\lambda t_0)^j}{j!} . \tag{2.13}$$

Direct calculations with this formula are difficult. For large t, however, one can use the general asymptotic formula (2.12), where in this case

$$M_{S(t)} \approx \frac{\mu}{\lambda + \mu}\, t ,$$

$$\sigma^2_{S(t)} \approx \frac{2\lambda\mu}{(\lambda+\mu)^3}\, t .$$

2.4.2 A Given Up Time Interval During a Total Operational Time

Often a failure does not lead to disruption of the unit's operation because there is some redundant time permitting resumption of a task from the beginning. If the extra time is more than the time t_0 needed for the operation, failure may be avoided. Even if several failures occur during a period with a total operating time t, the final result will be success as long as any one continuous operating interval is greater than or equal to t_0.

As a simple case one can consider idle intervals that are negligibly small. An example might be a computer system with random errors. The process starts at $t = 0$. The reliability index of interest can be described as

$$B^*(t_0, t) = Pr\left\{\exists \xi_i \geq t_0 \text{ (including } \xi_k'\text{)} \,\Big|\, \sum_{1\leq i\leq k-1} \xi_i < t \leq \sum_{1\leq i\leq k} \xi_i \right\}, \tag{2.14}$$

where ξ_k' is the last truncated up period, as introduced above.

Solution for All Distributions

Consider

$$B^*(t_0, t) = [1 - F(t_0)]\delta(t) + \int_0^{t_0} B^*(t - x)\delta(t - x)\, dF(x)\,, \tag{2.15}$$

where $\delta(t)$ is the indicator function

$$\delta(t) = \begin{cases} 1 & \text{if } t \geq t_0\,, \\ 0 & \text{if } t < t_0\,. \end{cases}$$

Only numerical solutions are applicable to all distributions, but if $t \gg t_0$ and consequently $F(t_0)$ is close enough to 1 [i.e., $P(t_0) \ll 1$], one may use the approximation

$$B^*(t_0, t) \approx 1 - e^{-tP(t_0)/T^*}\,, \tag{2.16}$$

where T^* is the mean of the truncated r.v. ξ under the condition that $\xi < t_0$, and, in accordance with the special conditions on $F(t_0)$, $T^* \approx T$. A numerical table for a standard exponential distribution is available in Appendix E.

Solution for Exponential Distributions

If $F(t) = 1 - e^{-\lambda t}$, the solution is given by

$$B^*(t_0, t) = e^{-\lambda t} \sum_{0\leq k<\infty} \frac{(\lambda t)^k}{k!} \sum_{0\leq j\leq k-1} (-1)^j \binom{j}{k+1} \left(1 - \frac{(k+1)t_0}{t}\right)_+^k, \tag{2.17}$$

where the value of $(\cdot)_+$, the last term on right, is given by

$$(x)_+ = \begin{cases} x\,, & \text{if } x \geq 0\,, \\ 0\,, & \text{if } x < 0\,. \end{cases}$$

If $t \gg t_0$ and λt_0 is large enough, one can use the asymptotic approximation

$$B^*(t_0, t) \approx 1 - e^{-\lambda t e^{-\lambda t_0}}\,. \tag{2.18}$$

2.4.3 At Least One Up Interval of a Given Duration

This index can be considered as a generalization of the two previous indexes, $A(t_0, t)$ and $B(t_0, t)$. The index is the probability of at least one up interval that equals or exceeds a given duration during a fixed calendar period. In the general case this index has the form

$$B(t_0, t) = \int_0^t B^*(t_0, t)d[1 - A(y, t)] . \tag{2.19}$$

For an exact and general solution, one must use numerical methods. For practical purposes, however, one may use summation:

$$B(t_0, t) \approx \sum_{t_0 = y_k < \cdots < y_n = t} B^*\left(t_0, \frac{y_{k+1} + y_k}{2}\right)[A(y_k, t) - A(y_{k+1}, t)] . \tag{2.20}$$

If $t \gg t_0$ and $P(t_0) \ll 1$, one can use the asymptotic formula

$$B(t_0, t) \approx 1 - e^{-[t/(T^* + \tau)]P(t_0)} , \tag{2.21}$$

where T^* and τ are the means of the truncated r.v.'s ξ and the renewal time, respectively. As before, $T \approx T^*$.

2.4.4 No Idle Period of More Than a Given Duration

Sometimes an analyst wants to know the probability of no idle period of more than a given duration during a given total idle time. Let there be a sequence of idle intervals:

$$\eta_1, \eta_2, \ldots, \eta_k, \ldots .$$

To find the probability $b^*(\tau_0, t)$ of no idle interval η equal to or longer than a given interval τ_0 during a period with total idle time t, let us introduce a complimentary function:

$$\tilde{b}^*(\tau_0, t) = 1 - b^*(\tau_0, t) . \tag{2.22}$$

Finding $\tilde{b}^*(\tau_0, t)$ is the same mathematically as finding $B^*(t_0, t)$. $G(t)$ replaces $F(t)$ and τ_0 replaces t_0, yielding

$$b^*(\tau_0, t) = 1 - \tilde{b}^*(\tau_0, t) = 1 - B_G^*(\tau_0, t) . \tag{2.23}$$

where the subscript G means that $G(t)$ replaces $F(t)$.

Solution for All Distributions

The precise solution can be found only numerically. If $t \gg \tau_0$ and $G(t)$ is close to 1, however, one may use the asymptotic approximation

$$b^*(\tau_0, t) \approx 1 - e^{-\mu t e^{-\mu \tau_0}} . \tag{2.24}$$

2.4.5 Probability of No Idle Period of More Than a Given Duration During a Given Calendar Time

The process is an alternating stochastic process

$$\xi_1, \eta_1, \xi_2, \eta_2, \ldots, \xi_k, \eta_k, \ldots .$$

We are interested in the probability that there is no r.v. η equal to or larger than some given value τ_0 during time t. This index is very useful for a system with a "time of inertia," which is a maximum time that the system can be idle without an undesirable consequence. An example might be emergency equipment in a hospital. In practice, the most important case occurs when $T \gg \tau$. In this case one can use the asymptotic formula

$$b(\tau_0, t) \approx e^{-t[1-G(\tau_0)]/T+\tau} . \tag{2.25}$$

CHAPTER 3

UNRENEWABLE EQUIPMENT

3.1 SERIES SYSTEMS

In reliability analysis a series system is a system that fails if any one of the units fails. Figure 3.1 is an example of a structural diagram of parts in series. A series may or may not form a physical string or chain of units. For example, a physically parallel connection of capacitors is a series for short-circuit purposes. A short connection of any capacitor leads to system failure. In this chapter all units are independent unless there is a statement to the contrary.

3.1.1 Arbitrary Distribution of a Random TTF

The time to failure (TTF) of a series system of m units is the shortest TTF of its units:

$$\xi^{(1)} = \min_{1\le i\le m} \xi_i^{(1)} ,$$

where $\xi^{(1)}$ = the random time to failure of the system and $\xi_i^{(1)}$ = the random time to failure of the ith unit. Table 3.1 gives reliability equations for independent units for arbitrary distribution functions (d.f.'s). All approximations are given for the condition

$$\max_{1\le i\le m} F_i(t_0) \ll \frac{1}{m} .$$

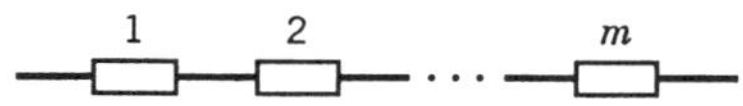

Figure 3.1 Structural diagram of series system.

3.1.2 Exponential Distribution

Table 3.2 has the main reliability indexes when each unit has an exponential d.f.

$$P_i(t) = e^{-\lambda_i t} .$$

Approximations have satisfactory accuracy when

$$\Lambda t_0 = t_0 \sum_{1 \leq i \leq m} \lambda_i \ll 1 .$$

3.1.3 Increasing Failure Rate (IFR) Distribution

Table 3.3 gives equations for the lower and upper bounds for a series system of increasing-failure-rate (IFR) units. The MTTF of each unit T is known. In the table $T_{(k)}$ denotes the kth ordered value of T_i,

Table 3.1 Unrenewable series system, arbitrary distribution $F(t)$

Index	Accurate	Approximate
$p(t_0)$	$\prod_{i=1}^{m} p_i(t_0)$	$1 - \sum_{i=1}^{m} Q_i(t_0)$
$Q(t_0)$	$1 - \prod_{i=1}^{m} p_i(t_0)$	$\sum_{i=1}^{m} Q_i(t_0)$
$P(t, t + t_0)$	$\prod_{i=1}^{m} p_i(t, t + t_0)$ $= \dfrac{\Pi_{i=1}^{m} P_i(t + t_0)}{\Pi_{i=1}^{m} P_i(t)} = \dfrac{P(t + t_0)}{P(t)}$	$1 - Q(t + t_0) + Q(t)$
$Q(t, t + t_0)$	$1 - \prod_{i=1}^{m} P_i(t, t + t_0)$ $= 1 - \dfrac{P(t + t_0)}{P(t)}$	$Q(t + t_0) - Q(t)$
T	$\int_0^{\infty} P(t)\, dt$	—

Table 3.2 Unrenewable series system, exponential distribution $F(t)$

Index	Exact	Approximate
$P(t_0)$	$\sum_{i=1}^{m} e^{-\lambda_i t_0} = e^{-t_0 \Sigma_{i=1}^{m} \lambda_i} = e^{-\Lambda t_0}$	$1 - \Lambda t_0$
$Q(t_0)$	$1 - e^{-\Lambda t_0}$	Λt_0
$P(t, t + t_0)$	$e^{-\Lambda t_0}$	$1 - \Lambda t_0$
$Q(t, t + t_0)$	$1 - e^{-\Lambda t_0}$	Λt_0
T	$\frac{1}{\Lambda}$	—

Note: $\Lambda = \Sigma_{i=1}^{m} \lambda_i$.

$$T_{(1)} = \min_{1 \le i \le m} T_i \,, \quad T_{(2)} = \min_{1 \le i \le m} \{T_i \mid T_i > T_{(1)}\} \,, \quad \ldots .$$

If the variances also were known, the more restrictive equations in Chapter 2 could be used.

3.1.4 Dependent Units in Series

In practice, units are often dependent. The most common reason for positively correlated units is environment, including temperature, humidity,

Table 3.3 Upper and lower bounds of reliability indexes of series system consisting of "aging" units

Index	Lower Bound	Upper Bound	Domain of Validity	Remarks
$P(t)$	$e^{-t} \sum_{i=1}^{m} \frac{1}{T_t}$	1	$t < \min_{1 \le i \le m} T_i$	$\omega_t^{(1)}$ depends on t and is determined from $1 - \omega_t^{(1)} \min_{1 \le i \le m} T_i = e^{-\omega_t^{(1)} t}$
	0	$e^{-k\omega_t^{(1)} t}$	$t \ge \min_{1 \le i \le m} T_i$	k = number of units with MTTF coinciding with t
		Corrected value $e^{-\omega_t^{(1)} t}$ $e^{-[\omega_t^{(1)} + \omega_t^{(2)}]t}$ $\cdots$ $e^{-\Sigma_{i=1}^{m} \omega_t^{(j)} t}$	$T_1 \le t < T_2$ $T_2 \le t < T_3$ $\cdots$ $T_m \le t$	$T_1 < T_2 < \cdots < T_m$ $\omega_t^{(j)}$ depends on t and is determined from $1 - \omega_t^{(j)} T_j = e^{-\omega_t^{(j)} t}$
T	$\frac{1}{\Sigma_{i=1}^{m} \frac{1}{T_i}}$	$\min T_i$		

radiation, vibration, loads, and power supply. Usually these external influences tend to work together and change different reliability parameters in the same direction. For systems with correlated units, the probability of successful operation and the MTTF are, correspondingly,

$$P(t_0) \geq \prod_{1 \leq i \leq m} P_i(t_0) ,$$

$$T_1 \geq \int_0^\infty \prod_{1 \leq i \leq m} P_1(t)\, dt .$$

3.2 ACTIVE REDUNDANCY

We will assume that all failures are found immediately and that switching to a redundant unit is instantaneous. Monitoring is ideal, and the switch is absolutely reliable.

3.2.1 Redundancy with One Main Unit

Figure 3.2 is a structural diagram of a unit with m active redundant units.

The probability of successful operation is

$$P(t_0) = 1 - \prod_{1 \leq i \leq m+1} q_i(t_0) ,$$

where $q_i(t_0)$ is the probability of the failure of the ith unit during time t_0. For an exponential distribution,

$$q_i(t_0) = 1 - e^{\lambda_i t_0} . \tag{3.1}$$

For small values of $\lambda_i t_0$ a simple approximation may be used:

$$\tilde{P}(t_0) \approx 1 - t_0^{m+1} \prod_{1 \leq i \leq m+1} \lambda_i . \tag{3.2}$$

For identical units, which is the usual case for active redundancy, one may write

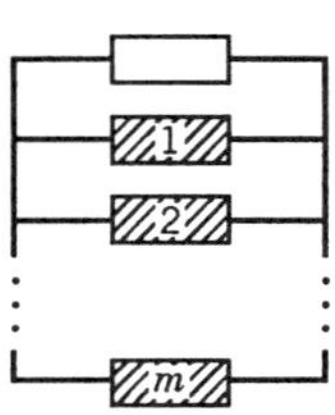

Figure 3.2 Structural diagram of a redundant group of one main and m active redundant units.

$$\tilde{P}(t_0) \approx 1 - (\lambda t_0)^{m+1} .$$

Integration of (3.1) is required to find the MTTF in the general case, that is, for any distribution whatsoever. For exponential distributions, however, one may write

$$T = \sum_{1 \le j \le m+1} \frac{1}{\lambda_j} - \sum_{(i,j) \in g_s} \frac{1}{\lambda_i + \lambda_j} + \cdots + (-1)^{s+1} \sum_{(i_1, \ldots, i_s) \in g_s} \frac{1}{\lambda_{i_1} + \cdots + \lambda_{i_s}} + \tag{3.3}$$

where

g_s = a set of exactly s different subscripts of the set $(1, \ldots, m+1)$;
$\bar{g}_s$ = the complementary set to g_s,

$$g_s \cap \bar{g}_s = \varnothing , \quad g_s \cup \bar{g}_s = (1, \ldots, m+1) ;$$

G_s = a set of all g_s (the number of all possible combinations equals a binomial $(n+m, s)$.

For identical units, when each distribution is $F(t) = 1 - e^{-\lambda t}$,

$$T = \frac{1}{\lambda} \sum_{1 \le k \le m+1} \frac{1}{k} .$$

If m is large enough, one can use the approximation

$$T \approx \frac{1}{\lambda} \left[C + \ln(m+1) + \frac{1}{2(m+1)} \right],$$

where C is Euler's constant (0.577 . . .).

In the case of a Weibull-Gnedenko distribution and identical units with

$$r_i(t_0) = r(t_0) = e^{-\lambda t_0^{\alpha}} , \quad 1 \le i \le m+1 ,$$

the MTTF is defined as

$$T = \frac{1}{\sqrt[\alpha]{\lambda}} \Gamma\left(1 + \frac{1}{\alpha}\right) \sum_{1 \le i \le m+1} (-1)^{i-1} \binom{m+1}{i} \frac{1}{\sqrt[\alpha]{i}}$$

or for large m

$$T \approx \sqrt[\alpha]{\frac{\ln(m+1)}{\lambda}} .$$

For identical aging units with a convex function $r(t)$, that is, when

$$0 < \frac{d^2 r(t)}{dt^2} < \frac{1}{r(t)} \left[\frac{dr(t)}{dt} \right]^2 ,$$

the MTTF is given by

$$\frac{0.56}{m+1} \leq r(T) \leq \frac{1}{m+1} .$$

3.2.2 Multiple Main Units

A redundant group may consist of n main units and m active redundant units, as diagramed in Figure 3.3. The formulas for $P(t_0)$ and T are in Table 3.4, which has the following notation:

$$P = \prod_{1 \leq i \leq n+m} p_i(t_0) ,$$

where

p_i = compact notation for $p_i(t_0)$;
$\tilde{p}_i = p_i$ for an exponential distribution

$$\tilde{p}_i = e^{-\lambda_i t_0} ;$$

$\tilde{q}_i = 1 - \tilde{p}_i$,
$q_{(i)}$ and $\tilde{q}_{(i)}$ = values in nonincreasing order,

$$\gamma_i = \frac{1 - p_i(t_0)}{p_i(t_0)} ;$$

$\tilde{\gamma}_i$ is given by the same expression with $\tilde{p}_i$ substituted for p_i; $\lambda_{(i)}$ are in nonincreasing order $-\lambda_{(1)} \geq \lambda_{(2)} \geq \cdots$,

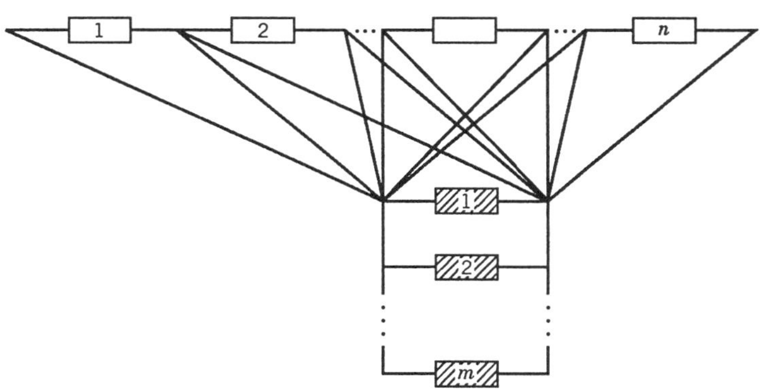

Figure 3.3 Structural diagram of a series system of n main and m redundant units.

Table 3.4 Reliability indexes of a system with "sweeping" active redundancy

Distribution	Units	Index	Accurate Expression	Approximation: Lower Estimate	Approximation: Upper Estimate
Arbitrary	Different	$P(t_0)$	$P\Big(+\sum_{1\le i\le n+m}\gamma_i+\sum_{1\le i<j\le n+m}\gamma_i\gamma_j+\cdots+\sum_{g_m\in G}\prod_{k\in g}\gamma_k\Big)$	$1-C_{n+m}^{m+1}\prod_{1\le i\le m+1}q_{(i)}$	$1-C_{n+m}^{m+1}\Big((n+m)^{-1}\sum_{1\le i\le n+m}q_i\Big)^{m+1}$
		T		$T_{(m)}$	$\sum_{0\le k\le m}\Big(\Lambda-\sum_{1\le i\le k}\lambda_{(i)}\Big)^{-1},\ \lambda_{(0)}=0$
	Identical	$P(t_0)$	$\sum_{0\le j\le m}C_{n+m}^{j}p^{n+m-j}q^{j}=1-\sum_{m+1\le j\le n+m}C_{n+m}^{j}p^{n+m-j}q^{j}$	$1-C_{n+m}^{m+1}q^{m+1}$	
		T		$\lambda^{-1}\sum_{0\le j\le m}(n+m-j)^{-1}$	
Exponential	Different	$P(t_0)$	$P\Big(1+\sum_{0\le i\le n+m}\tilde{\gamma}_i+\sum_{1\le i<j\le n+m}\tilde{\gamma}_i\tilde{\gamma}_j+\cdots+\sum_{g_m\in G_m}\prod_{k\in g_m}\tilde{\gamma}_k\Big)$	$1-C_{n+m}^{m+1}t_0^{m+1}\sum_{1\le i\le m+1}\lambda_{(i)}$	$1-C_{n+m}^{m+1}\Big((n+m)^{-1}\sum_{1\le i\le n+m}\lambda_i t_0\Big)^{m+1}$
		T		$\sum_{0\le k\le m}\Big(\Lambda-k(n+m)^{-1}\times\sum_{1\le i\le n+m}\lambda_i\Big)^{-1}$	$\sum_{0\le k\le m}\Big(\Lambda-\sum_{1\le i\le k}\lambda_{(i)}\Big)^{-1},\ \lambda_{(0)}=0$
	Identical	$P(t_0)$	$\sum_{0\le j\le m}C_{n+m}^{j}\tilde{p}^{n+m-j}\tilde{q}^{j}=1-\sum_{m+1\le j\le n+m}C_{n+m}^{j}\tilde{p}^{n+m-j}\tilde{q}^{j}$	$1-C_{n+m}^{m+1}\tilde{q}^{m+1}$	
		T	$\lambda^{-1}\sum_{0\le j\le m}(n+m-j)^{-1}$	$\ln\dfrac{n+m}{n}-\dfrac{m}{n(n+m)}$	

$$T_{(i)} = \frac{1}{\lambda_{(i)}},$$

$$\Lambda = \sum_{1 \le i \le n+m} \lambda_i .$$

When units are not identical and any m units must be operational, $P(t_0)$ is given by

$$P(t_0) = \prod_{1 \le i \le m+n} p_i + \sum_{1 \le i \le m+n} q_i \prod_{\substack{1 \le j \le n+m \\ j \ne i}} p_j$$

$$+ \sum_{1 \le i < j < +m+n} q_i q_j \prod_{\substack{1 \le k \le m+n \\ k \ne i,j}} p_k + \cdots + \sum_{g_m \in G_m} \prod_{i \in \bar{g}_m} q_i \prod_{k \in g_m} p_k , \quad (3.4)$$

where n and m are as shown in Figure 3.3. The previously introduced formulas for P and γ are used to get the results in Table 3.4.

An approximation can be made for small q_i by truncating

$$P(t_0) = 1 - \left(\sum_{g_{m+1} \in G_{m+1}} \prod_{i \in \bar{g}_{m+1}} q_i \prod_{k \in g_{m+1}} p_k + \cdots + \prod_{1 \le i \le n+m} q_i \right), \quad (3.5)$$

which is equivalent to (3.4). Equation (3.5) can be shortened to

$$P(t_0) \approx 1 - \sum_{g_{m+1} \in G_{m+1}} \prod_{i \in \bar{g}_{m+1}} q_i \prod_{k \in g_{m+1}} p_k$$

if

$$\max_{1 \le i \le m+n} q_i \ll \frac{1}{n+m} .$$

Table 3.4 in some positions has no formula for calculating T. In such a case one should integrate the corresponding $P(t)$ function directly. An accurate value is given for the exponential case, with

$$T = \frac{1}{\Lambda} + \sum_{1 \le i \le n+m} \frac{\lambda_i}{\Lambda} \frac{1}{\Lambda - \lambda_i} + \sum_{1 \le i < j \le n+m} \frac{\lambda_i \lambda_j}{\Lambda^2} \frac{1}{\Lambda - \lambda_i - \lambda_j} + \cdots$$

$$+ \sum_{g_{m+1} \in G_{m+1}} \frac{\prod_{i \in g_{m+1}} \lambda_i}{\Lambda^{m+1}} \frac{1}{\Lambda - \sum_{i \in g_{m+1}} \lambda_i} .$$

3.3 REDUNDANCY WITH SPARES

Again we will assume that all failures are found immediately and that switching time is 0. System monitoring is ideal, and the switch is absolutely reliable. Another assumption of convenience is that a redundant spare unit starts to operate immediately. This assumption is the most unrealistic one; usually a spare unit cannot operate immediately as a main one because time is required for activities such as transportation and installation. Yet another assumption is that a spare does not lose reliability with time.

If the delay time between failure and replacement cannot be neglected, one should consider making the problem tractable by treating it as a renewal system with a restricted number of repairs. Spare units usually are identical to the main units, and we will consider only this most important case.

3.3.1 Spares for One Main Unit

The probability of successful operation of a redundant group can be found with a recurrent equation for the probability of $n + 1$ failures before time t_0:

$$P(t_0) = r_{m+1}(t_0) = \int_0^{t_0} r_m(t_0 - x) f(x)\, dx ,$$

where $f(x)$ is the density of the failure distribution $F(x)$. In other words,

$$P(t_0) = 1 - F^{*(m+1)}(t_0) ,$$

where

$$F^{*k}(t) = \int_0^t F^{*(k-1)}(t - x)\, dF(x) = \int_0^t F(t - x)\, dF^{*(k-1)}(x) .$$

Here $F^{*k}(t)$ is the kth convolution of the distribution $F(t)$. When units have an increasing failure rate, one can use the results of Chapter 2, taking into account that the variance of the redundant group is $m + 1$ times as large as the variance of a single unit. The MTTF for any d.f., T_{syst} is $(m + 1)T$. For an exponential distribution we use the Poisson formula

$$P(t_0) = \sum_{0 \le k \le m} \frac{(\lambda t_0)^k}{k!} e^{-\lambda t_0} .$$

3.3.2 Multiple Main Units

A group of m main and n spare units is considered. All units are identical. The accurate formulas are very complicated and are not given here. Table 3.5 gives rough bounds. The lower bound is for the least desirable situation; all spare units are divided equally, as can occur with division of integers. The upper bound corresponds to the case when the total operating time for spare units is distributed in the best possible way, as in computer time-sharing. An IFR distribution is assumed.

Table 3.5 Reliability indexes of a system with "sweeping" spare redundancy

Distribution		Index: $P(t_0)$	Index: T
Approximation arbitrary	Lower estimate	$1-[F^{*(s+1)}(t)]^{n+m(1-s)}[F^{*s}(t)]^{sm-n}$	sT
	Upper estimate	$F^{*(m+n)}(nt_0)$	$\dfrac{n+m}{n}T$
Accurate exponential		$e^{-n\lambda t_0}\sum_{0\le k\le m+1}\dfrac{(n\lambda t_0)^k}{k!}$	$\dfrac{m+1}{n}T$

Note: s = integer part of $(m+n)/n$. $F^{*(k)}$ = kth convolution of $F(t)$.

3.3.3 Exponential Distribution

Table 3.6 gives approximate formulas for $P(t_0)$ when there is one main unit. Redundant units may be active, partially active, or inactive.

3.4 THE "DEATH PROCESS"

Sometimes a redundant system has multiple states and the time of remaining in a state has an exponential distribution with an λ that depends on the state. A diagram of possible transitions is presented in Figure 3.4. If the system has k main and n redundant units, the following formal description can be considered:

Table 3.6 Approximate formulas for $P(t_0)$ for a system with a single main unit and n redundant ones

Regime of Redundant Units	Approximation	Domain of Validity
Active	$1-(\lambda t_0)^{n+1}$	$\lambda t \ll 1$
Partially active (waiting)	$1-\dfrac{(\lambda t_0)^{n+1}}{(n+1)!}\prod_{i=0}^{n}(1+\nu i)$	$\lambda t_0 \ll 1$
Spare	$1-\dfrac{(\lambda t_0)^{n+1}e^{-\lambda t_0}}{(n+1)!\left(1-\dfrac{\lambda t_0}{n+2}\right)}$	$\lambda t_0 \ll 1$

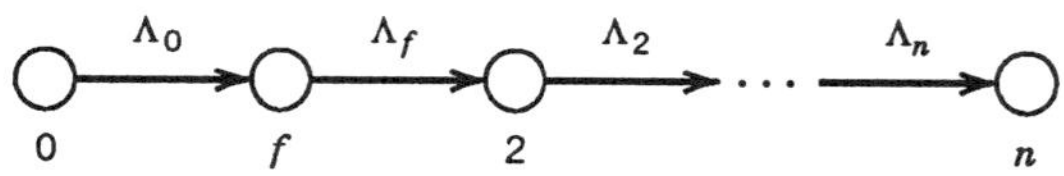

Figure 3.4 Transition graph describing the "death process".

1. If at moment t the system is in the state j (i.e., there were j failures before time t), transition to state $j+1$ will have probability $\Lambda_j \Delta t + o(\Delta t)$ of occurring in the time interval $[t, t + t\Delta]$.
2. After the failure $n+1$, the system goes into a failure state that is absorbing, and "death" occurs.

The number of states equals the number of failed units.

In accordance with the transition graph presented in Figure 3.4, one can write the following system of differential equations:

$$\frac{d}{dt} p_i(t) = \Lambda_{j-1} p_{j-1}(t) - \Lambda_j p_j(t), \qquad 0 \le j \le n-1;\ \Lambda_{-1} = \Lambda_{n+1} = 0 .$$

Here $p_i(t)$ is the probability that the system is in state H_j at the moment t. The initial condition is assumed to be

$$p_0(t^0) = 1 ,$$
$$p_j(0) = 0 , \qquad i \le j \le m+1 .$$

Thus the system initially is fully operational.

Finally, the probability of successful operation can be given in a form especially convenient for computer calculation:

$$P(t_0) = \frac{\begin{vmatrix} e^{-\Lambda_0 t} & e^{-\Lambda_1 t} & \cdots & e^{-\Lambda_n t} \\ \Lambda_0 & \Lambda_1 & \cdots & \Lambda_n \\ \Lambda_0^2 & \Lambda_1^2 & \cdots & \Lambda_n^2 \\ \vdots & \vdots & & \vdots \\ \Lambda_0^n & \Lambda_1^n & \cdots & \Lambda_n^n \end{vmatrix}}{\begin{vmatrix} 1 & 1 & \cdots & 1 \\ \Lambda_0 & \Lambda_1 & \cdots & \Lambda_n \\ \Lambda_0^2 & \Lambda_1^2 & \cdots & \Lambda_n^2 \\ \vdots & \vdots & & \vdots \\ \Lambda_0^n & \Lambda_1^n & \cdots & \Lambda_n^n \end{vmatrix}}$$

The MTTF for such a system can be found very simply as

$$T = \sum_{0 \le i \le n} \frac{1}{\Lambda_i} .$$

Table 3.7 Formulas for $P(t_0)$ for a system described with a "death process"

Index	Accurate	Approximate	Domain of Validity
$P(t_0)$	$\prod_{0\le j\le m} \Lambda_j \sum_{0\le i\le m} \frac{e^{-\Lambda_i t_0}}{\Lambda_i \sum_{0\le k\le m, k\ne i} (\Lambda_k - \Lambda_i)}$	$1 - \frac{t_0^{m+1}}{(m+1)!} \prod_{0\le i\le m} \Lambda_j$	$\max_{1\le i\le m} \Lambda_i t_0 \ll 1$
		$1 - \Phi\left(\frac{t_0 - \Sigma_{0\le i\le m} \Lambda_i^{-1}}{\sqrt{\Sigma_{0\le i\le m} \Lambda_i^{-2}}}\right)$	m is large (an approximation of the distribution of a sum of r.v.'s by a normal distribution)

If the system consists of identical units, the equation for active redundancy is

$$\Lambda_j = k\lambda + (n-j)\lambda = (N-j)\lambda , \qquad 0 \le j \le n ,$$

where λ is the failure rate of a single unit, and $N = k + n$.

For partially active redundancy,

$$\Lambda_j = k\lambda + (n-j)\nu\lambda , \qquad 0 \le j \le n ,$$

where $\nu\lambda$ is the failure rate of a single redundant unit, and ν is the ratio of the load on a redundant unit to the load on a main unit, $0 < \nu < 1$, where the load is proportional to the failure rate. For standby units,

$$\Lambda_j = k\lambda , \qquad 0 \le j \le n .$$

Table 3.7 gives exact and approximate formulas for $P(t_0)$.

CHAPTER 4

RENEWABLE SYSTEMS

4.1 SERIES SYSTEMS

4.1.1 Failure Rate, Repair Rate, and Regime

When distributions of time to failure (TTF) are not exponential, distribution of the random time to the second failure of a system will differ from the distribution of the random time to the first failure. In general, if stationarity has not been reached, the kth and $(k+1)$th failure will have different d.f.'s. Furthermore, when distributions of TTF and of repair time are unknown, analysis may be so complex that a Monte Carlo simulation may be the only practical investigative technique.

This chapter is devoted mostly to exponential d.f.'s for failure, $F_i(t)$, and repair, $G_i(t)$:

$$F(t) = 1 - e^{-\lambda t} ,$$

$$G(t) = 1 - e^{-\mu t} ,$$

where λ_i and μ_i are the parameters of the distributions of failure and repair, respectively.

Reliability indexes depend on the type of regime during system repair. There are two main regimes. In one, after a failure of any unit, all the rest are switched off. The failure rate equals 0 during repair. Only one repair facility is needed, and there is no repair queue. In the other common regime, only units in repair or in the repair waiting line are switched off.

Table 4.1 Series system consisting of n different repairable units. The ith unit failure rate is λ_i, and repair rate is μ_i. All units are switched off during the system idle time

$(\gamma_i = \lambda_i/\mu_i;\ \Lambda = \Sigma_{i=1}^{n} \lambda_i)$

Index	Exact Expression	Approximate Expression
T	$\dfrac{1}{\Lambda}$	—
τ	$\dfrac{1}{\Lambda}\sum_{i=1}^{n} \gamma_i$	—
$P(t_0)$	$e^{-\Lambda t_0}$	$1 - \Lambda t_0$
K	$\left(1 + \sum_{i=1}^{n} \gamma_t\right)^{-1}$	$1 - \sum_{i=1}^{n} \gamma_i$
$K(t)$	$K + (1-K)e^{-t(\Lambda + 1/\tau)}$	$1 - \Lambda\tau(1 - e^{-t/\tau})$
$R(t_0)$	$Ke^{-\Lambda t_0}$	$1 - \Lambda(\tau + t_0)$

4.1.2 Exponential Distribution with Independent Units

Tables 4.1 and 4.2 give reliability indexes for a series system of n units. Table 4.1 is for a system that is switched off during repair. Table 4.2 has results for the case when only failed units are switched off and the number of repair facilities is unrestricted. For restricted repair, when the number of

Table 4.2 Series system consisting of n different repairable units. The ith unit failure rate is λ_i, and repair rate is μ_i. All units are switched in during the system idle time

$(\gamma_i = \lambda_i/\mu_i;\ \Lambda = \Sigma_{i=1}^{n} \lambda_t)$

Index	Exact Expression	Approximate Expression
T	$\dfrac{1}{\Lambda}$	—
τ	$\dfrac{1}{\Lambda}\left(\prod_{i=1}^{n}(1+\gamma_i) - 1\right)$	$\dfrac{1}{\Lambda}\sum_{i=1}^{n} \gamma_i$
$P(t_0)$	$e^{-\Lambda t_0}$	$1 - \Lambda t_0$
K	$\prod_{i=1}^{n}(1+\gamma_i)^{-1}$	$1 - \sum_{i=1}^{n} \gamma_i$
$K(t)$	—	$1 - \Lambda\tau(1 - e^{-t/\tau})$
$R(t_0)$	$Ke^{-\Lambda t_0}$	$1 - \Lambda(\tau + t_0)$

Table 4.3 Series system consisting of n identical repairable units. For all units the failure rate is λ, and the repair rate is μ. All units are switched off during the system idle time
$(\gamma = \lambda/\mu)$

Index	Exact Expression	Approximate Expression
T	$\frac{1}{n\lambda}$	—
τ	$\frac{1}{\mu}$	—
$P(t_0)$	$e^{-n\lambda t_0}$	$1 - n\lambda t_0$
K	$(1 + n\gamma)^{-1}$	$1 - n\gamma$
$K(t)$	$K + (1 - K)e^{-t[n\lambda + (1/\tau)]}$	$1 - n\lambda\tau(1 - e^{-t/\tau})$
$R(t_0)$	$Ke^{-n\lambda t_0}$	$1 - n\lambda(\tau + t_0)$

Table 4.4 Series system consisting of n identical repairable units. For all units the failure rate is λ, and the repair rate is μ. All units are switched in during the system idle time. Unrestricted repair facilities
$(\gamma = \lambda/\mu)$

Index	Exact Expression	Approximate Expression
T	$\frac{1}{n\tau}$	—
τ	$\frac{1}{n\lambda}[(1+\gamma)^n - 1]$	$\frac{1}{\mu}$
$P(t_0)$	$e^{-n\lambda t_0}$	$1 - n\lambda t_0$
K	$(1 + \gamma)^{-n}$	$1 - n\gamma$
$K(t)$	—	$1 - n\lambda\tau(1 - e^{-t/\tau})$
$R(t_0)$	$Ke^{-n\lambda t_0}$	$1 - n\lambda(\tau^* + t_0)$

facilities is smaller than the possible number of failed units, the results for any fixed service discipline can be found with Laplace transforms. These are usually very bulky, and solutions are obtained mostly by numerical methods. In Tables 4.3 through 4.5 results are given for a system with identical units.

4.2 REDUNDANT SYSTEMS

4.2.1 General Markovian Model for Identical Units

Let us consider a system consisting of k main operating units and $n = n_1 + n_2 + n_3$ redundant units, where n_1 is the number of internal loaded

Table 4.5 Series system consisting of n identical repairable units. For all units the failure rate is λ, and the repair rate is μ. All units are switched in during the system idle time. Restricted repair facilities

$(\gamma = \lambda/\mu)$

Index	Exact Expression	Approximate Expression
T	$\frac{1}{n\lambda}$	—
τ	$\frac{(n-1)!}{\lambda} \sum_{i=1}^{n} \frac{\gamma_i}{(n-i)!}$	$\frac{1}{\mu}$
$P(t_0)$	$e^{-n\lambda t_0}$	$1 - n\lambda t_0$
K	$\left[n! \sum_{i=0}^{n} \frac{\gamma^j}{(n-i)!} \right]^{-1}$	$1 - n\gamma$
$R(t_0)$	$Ke^{-n\lambda t_0}$	$1 - n(\lambda t_0 + \gamma)$

units, n_2 is the number of standby internal units, and n_3 is the number of spare units. All units are identical and have a random TTF with an exponential distribution. A partially loaded unit has a parameter $\nu\lambda$, where $0 < \nu < 1$. When a main unit fails, it goes to repair. The failed operating unit is replaced with an active redundant unit, which in turn is replaced with a standby one. There are l repair facilities, $1 \leq 1 \leq n + k$. All units have an exponential repair time distribution. H_j is a state with j failed units. A transition from state H_j to state H_{j+1} in time interval $[t, t + \Delta]$ occurs with probability $\Lambda_j \Delta t + o(\Delta t)$. Transition to state H_{j-1} in the same time interval occurs with probability $M_j \Delta t + o(\Delta t)$. With probability $1 - \Lambda_j \Delta t - M_j \Delta t - o(\Delta t)$, no change occurs. All Λ_j and M_j are positive constants with a very simple physical basis. For underloaded units a coefficient of loading may be ν, $0 < \nu < 1$. Then for all possible levels of failure from 0 to $n + k$ for the system Λ_j, $0 \leq j \leq n + k = n_1 + n_2 + n_3 + k$, one can write

$$\Lambda_o = k\lambda + n_1\lambda + n_2\nu\lambda ,$$

$$\Lambda_1 = k\lambda + n_1\lambda + n_2\nu\lambda = \Lambda_0 ,$$

$$\vdots$$

$$\Lambda_{n_3} = \Lambda_0 ,$$

$$\Lambda_{n_3+1} = k\lambda + n_1\lambda + (n_2 - 1)\nu\lambda ,$$

$$\Lambda_{n_3+2} = k\lambda + n_1\lambda + (n_2 - 2)\nu\lambda ,$$

$$\vdots$$

$$\Lambda_{n_3+n_2} = k\lambda + n_1\lambda ,$$

$$\Lambda_{n_3+n_2+1} = k\lambda + (n_1 - 1)\lambda ,$$

$$\Lambda_{n_3+n_2+2} = k\lambda + (n_1 - 2)\lambda ,$$
$$\vdots$$
$$\Lambda_{n_3+n_2+n_1} = k\lambda ;$$
$$\Lambda_{n_3+n_2+n_1+1} = (k-1)\lambda ,$$
$$\Lambda_{n_1+n_2+n_3+2} = (k-2)\lambda ,$$
$$\vdots$$
$$\Lambda_{n_3+n_2+n_1+k-1} = \lambda ,$$
$$\Lambda_{n_3+n_2+n_1+k} = 0 ,$$

and for all M_j, $0 \leq j \leq n + k = n_1 + n_2 + n_3 + k$,

$$M_1 = \mu, M_2 = 2\mu, \ldots, \quad M_l = l\mu, M_{l+1} = l\mu, \ldots, \quad M_{n+k} = l\mu .$$

A system with n redundant units has $n+k+1$ states H_0, H_1, $H_2, \ldots, H_{n+k}$. States H_{n+j}, with $n+j$ failed units, $1 \leq j \leq k$, are states of a system failure. If state $n+1$ is absorbing, a corresponding equation system can be used for calculating the probabilities of successful operation, the interval availability coefficient, the MTTF, and the MTBF. If state $n+k$ is reflecting, the corresponding equation system can be used for calculating the nonstationary and stationary availability coefficients.

H_j is the state with j failed units, $0 \leq j \leq n+k$; Λ_j is the transition rate from state H_j to state H_{j+1}, $0 \leq j \leq n+k-1$; M_j is the transition rate from state H_j to state H_{j-1}, $1 \leq j \leq n+k$.

A system with an absorbing H_{n+1} state is a system that operates until a first failure. This system can be analyzed with the following system of differential equations:

$$\frac{dp_i(t)}{dt} = \Lambda_{j-1}p_{j-1}(t) - (\Lambda_j + M_j)p_j(t) + M_{j+1}p_{j+1}(t) , \qquad 0 \leq j \leq n+1 ,$$
$$\Lambda_{-1} = \Lambda_{n+1} = M_0 = M_n = \cdots = M_{n+k} = 0 ,$$

where $p_j(t)$ is the probability at the moment t that the system is in state H_j. The normalization condition is

$$\sum_{0 \leq j \leq n+1} p_j(t) = 1 .$$

A system with a reflecting H_{n+k} state can be described with the following system of differential equations:

$$\frac{dp_i(t)}{dt} = \Lambda_{j-1}p_{j-1}(t) - (\Lambda_j + M_j)p_j(t) + M_{j+1}p_{j+1}(t) , \qquad 0 \leq j \leq n+k ,$$
$$\Lambda_{-1} = \Lambda_{n+k} = M_0 = M_{n+k+1} = 0 ,$$

with the normalization equation

$$\sum_{0 \le j \le n+k} p_j(t) = 1 .$$

Table 4.6 gives the reliability indexes that result from solution of these equations. The table has the following superscripts:

1. A superscript "(0)" is used when the system starts in a state with all units operational, namely $p^0(0) = 1$.

Table 4.6 System consisting of *n* redundant units, general case, exact expressions

Index	Exact Expression
$P(t_0)$	$\sum_{i=0}^{n} p_i(0) \prod_{j=i}^{n} \Lambda_j \sum_{s=1}^{n+1} \frac{\Pi_{m=1}^{i} (x_m^{(i)} - x_s^{(n+1)})}{x_s^{(n+1)} \Pi_{\substack{r=1 \\ r \ne s}}^{n+1} (x_r^{(n+1)} - x_s^{(n+1)})} e^{-x_s^{(n+1)} t_0}$
$P^{(0)}(t_0)$	$\prod_{j=0}^{n} \Lambda_j \sum_{i=1}^{n+1} \frac{e^{-x_i^{(n+1)} t_0}}{x_i^{(n+1)} \Pi_{\substack{s=1 \\ s \ne 1}}^{n+1} (x_s^{(n+1)} - x_i^{(n+1)})}$
$P^{(n)}(t_0)$	$\sum_{i=1}^{n+1} \frac{\Lambda_n}{x_i^{(n+1)}} \frac{\Pi_{\alpha=1}^{i-1} (x_i^{(n+1)} - x_\alpha^{(n)}) \Pi_{\beta=i}^{n} (x_\beta^{(n)} - x_i^{(n+1)})}{\Pi_{\gamma=1}^{i-1} (x_i^{(n+1)} - x_\gamma^{(n+1)}) \Pi_{\delta=i+1}^{n+1} (x_\delta^{(n+1)} - x_i^{(n+1)})} e^{-x_i^{(n+1)} t_0}$
$R(t_0)$	$\frac{\Theta_n}{\Sigma_{r=0}^{n+1} \Theta_r} \sum_{i=1}^{n+1} \left(\frac{\Lambda_n}{x_i^{(n+1)}} \right)^2 \frac{\Pi_{\alpha=1}^{i-1} (x_i^{(n+1)} - x_\alpha^{(n)}) \Pi_{\beta=i}^{n} (x_\beta^{(n)} - x_i^{(n+1)})}{\Pi_{\gamma=1}^{i-1} (x_i^{(n+1)} - x_\gamma^{(n+1)}) \Pi_{\delta=i+1}^{n+1} (x_\delta^{(n+1)} - x_i^{(n+1)})} e^{-x_i^{(n+1)} t_0}$
T	$\sum_{j=0}^{n} \frac{\Sigma_{s=0}^{j} p_s(0) \Sigma_{r=0}^{j} \Theta_r}{\Lambda_j \Theta_j}$
$T^{(0)}$	$\sum_{j=0}^{n} \frac{\Sigma_{r=0}^{j} \Theta_r}{\Lambda_j \Theta_j}$
$T^{(n)}$	$\frac{\Sigma_{r=0}^{n} \Theta_r}{\Lambda_n \Theta_n}$
$T^{(R)}$	$\frac{1}{\Sigma_{s=0}^{n+1} \Theta_s} \sum_{j=0}^{n} \frac{(\Sigma_{r=0}^{j} \Theta_r)^2}{\Lambda_j \Theta_j}$
p_j	$\frac{\Theta_j}{\Sigma_{i=0}^{n+1} \Theta_i}$
K	$\left[1 + \left(\frac{\Theta_{n+1}}{\Sigma_{i=0}^{n} \Theta_i} \right) \right]^{-1}$
τ	$\frac{1}{M_{n+1}}$

Note: $\Pi_{m=i+1}^{i} (x_m^{(i)} - x_s^{(n+1)}) = 1$, $\Theta_i = \Lambda_0 \Lambda_1 \ldots \Lambda_{i-1} / M_1 M_2 \ldots M_i$, $\Theta_0 = 1$;

$$x_s^{(m)} = -x_s^{(m)}, \quad 1 \le s \le m ; \quad 0 \le m \le n+1 ,$$

where $x_s^{(m)}$ is a root of polynomial $\Delta_m(x)$ determined by the recurrent expression $\Delta_m(x) = -(\Lambda_{m-1} + M_{m-1} + x)\Delta_{m-1}(x) - \Lambda_{m-2} M_{m-1} \Delta_{m-2}(x)$; $\Delta_1(x) = -(\Lambda_0 + x)$, $\Delta_0(x) = 1$.

Table 4.7 System consisting of *n* redundant units, general case, approximate expressions

Index	Approximate Expression
$P(t_0)$	$\frac{T}{T^{(0)}} e^{-t_0/T^{(0)}}$
$P^{(0)}(t_0)$	$e^{-t_0/T^{(0)}}$
$P^{(n)}(t_0)$	$\frac{T^{(n)}}{T^{(0)}} e^{-t_0/T^{(0)}}$
$R(t_0)$	$Ke^{-t_0/T^{(0)}}$
T	$\frac{\pi_n(0)}{\Lambda_n \Theta_n}$
$T^{(0)}$	$\frac{1}{\Lambda_n \Theta_n}$
$T^{(n)}$	$\frac{1}{\Lambda_n \Theta_n}$
$T^{(R)}$	$\frac{1}{\Lambda_n \Theta_n}$
p_j	$\Theta_j(1-\Theta_1)$
K	$1-\Theta_{n+1}$

Note: $\Theta_n = \Lambda_0 \Lambda_1 \ldots \Lambda_{N-1}/M_1 M_2 \ldots M_N$, $\Theta_0 = 1$, $\pi_j(0) = \Sigma_{0 \le s \le j}\, p_s(0)$.

2. A superscript "(n)" is used when the system starts to operate just after exiting from the failure state, namely $p^n(0) = 1$.
3. A superscript "(i)" is used when the initial conditions correspond to a stationary regime of operation, namely each $p^i(t)$ is a constant that equals a corresponding stationary probability.

The approximate expressions in Table 4.7 are valid when

$$\min_{1 \le j \le n+1} M_j \gg n \max_{0 \le j \le n} \Lambda_j \, .$$

4.2.2 Active and Standby Redundancy

Let a system consist of k operating and n redundant units. Redundant units can be either active or standby. All the units are identical, and there might be either of two repair regimes, restricted or unrestricted.

Table 4.8 Active redundancy, unrestricted repair

Index	Exact Expression	Approximate Expression
$T^{(0)}$	$\frac{1}{N\lambda}\sum_{s=0}^{n}\frac{\Sigma_{i=0}^{s} C_n^i \gamma^i}{C_{N-1}^s \gamma^s}$	$\frac{1}{k\lambda C_N^k \gamma^n}$
$T^{(n)}$	$\frac{1}{k\lambda C_N^k \gamma^n}\sum_{i=0}^{n} C_N^i \gamma^i$	$\frac{1}{k\lambda C_N^k \gamma^n}$
τ	$\frac{1}{(n+1)\mu}$	—
$P^{(0)}(t_0)$	—	$e^{-t/T}$
K	$\left(1+\frac{C_N^{n+1}\gamma^{n+1}}{\Sigma_{i=0}^{n} C_N^i \gamma^i}\right)^{-1}$	$1 - C_N^{n+1}\gamma^{n+1}$
$R(t_0)$	—	$KP^{(0)}(t_0)$

C_n^i = binomial coefficient.

Reliability indexes for both cases are presented in Tables 4.8 through 4.11. In these tables the superscripts have the same meanings as those in Table 4.6. $N = k + n$; $\gamma = \lambda/\mu$; C_m^r is a binomial coefficient, and $\tilde{T}$ is an approximate expression for $T^{(0)}$ or $T^{(n)}$. Indexes $P^{(0)}(t_0)$ and $R(t_0)$ have no simple form and are referred to the general case.

Duplicate units are considered separately in Chapter 5.

Table 4.9 Active redundancy, restricted repair

Index	Exact Expression	Approximate Expression
$T^{(0)}$	$\frac{1}{\lambda}\sum_{s=0}^{n}\sum_{i=0}^{s}\frac{1}{(i+1)!C_{N-s+i}^{i+1}\gamma^i}$	$\frac{1}{\lambda}\frac{1}{\gamma^n (n+1)! C_N^{n+1}}$
$T^{(n)}$	$\frac{1}{\lambda}\sum_{i=0}^{n}\frac{1}{(i+1)!C_{k+i}^{i+1}\gamma^i}$	$\frac{1}{\lambda}\frac{1}{\gamma^n (n+1)! C_N^{n+1}}$
τ	$\frac{1}{\mu}$	—
$P^{(0)}(t_0)$	—	$e^{-t_0/\tilde{T}}$
K	$\left[1+\left(\sum_{i=1}^{n+1}\frac{1}{i!C_{N-n+i+1}^{i}\gamma^i}\right)^{-1}\right]^{-1}$	$1-(n+1)!C_N^{n+1}\gamma^{n+1}$
$R(t_0)$	—	$KP^{(0)}(t_0)$

Table 4.10 Spare redundancy, unrestricted repair

Index	Exact	Approximate
$T^{(0)}$	$\frac{1}{k\lambda}\sum_{i=0}^{n} C_{n+1}^{i+1}\frac{i!}{(k\gamma)^i}$	$\frac{1}{k\lambda}\frac{n!}{(k\gamma)^n}$
$T^{(n)}$	$\frac{1}{k\lambda}\sum_{i=0}^{n} C_{n}^{i}\frac{i!}{(k\gamma)^i}$	$\frac{1}{k\lambda}\frac{n!}{(k\gamma)^n}$
τ	$\frac{1}{(1+n)\mu}$	—
$P^{(0)}(t_0)$	—	$e^{-t_0/\tilde{T}}$
K	$\left[1+\left(\sum_{i=1}^{n+1} C_{n+1}^{i}\frac{i!}{(k\gamma)^i}\right)^{-1}\right]^{-1}$	$1-\frac{(k\gamma)^{n+1}}{(1+n)!}$
$R(t_0)$	—	$KP^{(0)}(t_0)$

4.2.3 Equations for Specific Conditions

Tables 4.12 through 4.31 give the expressions for the reliability indexes of the most common redundancy and repair conditions.

4.3 MONITORING AND SWITCHING PROBLEMS

4.3.1 System Description

In modern complex technical equipment, there are many influences that must be considered. The parameters of the switching devices, the regimes of

Table 4.11 Spare redundancy, restricted repair

Index	Exact	Approximate
$T^{(0)}$	$\frac{1}{k\lambda}\frac{1-(k\gamma)^{n+1}[1+(n+1)(1-k\gamma)]}{(k\gamma)^n(1-k\gamma)}$	$\frac{1}{k\lambda}\frac{1}{(k\gamma)^n}$
$T^{(n)}$	$\frac{1}{k\lambda}\frac{1-(k\gamma)^{n+1}}{(k\gamma)^n(1-k\gamma)}$	$\frac{1}{k\lambda}\frac{1}{(k\gamma)^n}$
τ	$\frac{1}{\mu}$	—
$P^{(0)}(t_0)$	—	$e^{-t_0/\tilde{T}}$
K	$\left[1+\left(\frac{(k\gamma)^{n+1}(1-k\gamma)}{1-(k\gamma)^{n+1}}\right)^{-1}\right]^{-1}$	$1-(k\gamma)^{n+1}$
$R(t_0)$	—	$KP^{(0)}(t_0)$

Table 4.12 System consisting of k operational units and one redundant unit; active redundancy, unrestricted repair

$x_{1,2} = \frac{\lambda}{2\gamma}\{[1 + (2k+1)\gamma] \pm \sqrt{1 + 2(2k+1)\gamma + \gamma^2}\}$, $\epsilon_{1,2} = \frac{\lambda}{2\gamma}\{[3 + (2k+1)\gamma] \pm \sqrt{1 + 2(2k-1)\gamma + \gamma^2}\}$, $\gamma_c = \frac{k(k+1)\gamma^2}{2[1 + (k+1)\gamma]}$, $\gamma = \frac{\lambda}{\mu}$.

Index	Exact	Approximate
$T^{(0)}$	$\frac{1}{k\lambda}\frac{1 + (2k+1)\gamma}{(k+1)\gamma}$	$\frac{1}{k\lambda}\frac{1}{(k+1)\gamma}$
$T^{(1)}$	$\frac{1}{k\lambda}\frac{1 + (k+1)\gamma}{(k+1)\gamma}$	$\frac{1}{k\lambda}\frac{1}{(k+1)\gamma}$
τ	$\frac{1}{2\mu}$	—
$P^{(0)}(t_0)$	$\frac{1}{x_1 - x_2}(x_1 e^{-x_2 t_0} - x_2 e^{-x_1 t_0})$	$e^{-\{[(k+1)\gamma]/[1+(2k+1)\gamma]\}k\lambda t_0}$
$K(t)$	$1 - \frac{k(k+1)\lambda^2}{\epsilon_1\epsilon_2}$	$1 - \frac{k(k+1)}{2}\gamma^2$
	$\times\left[1 - \frac{1}{\epsilon_1 - \epsilon_2}(\epsilon_1 e^{-\epsilon_2 t} - \epsilon_2 e^{-\epsilon_1 t})\right]$	$\times[1 - (2 - e^{-(\lambda/\gamma)t})e^{-(\lambda/\gamma)t}]$
K	$\frac{1}{1 + \gamma_0}$	$1 - \frac{k(k+1)}{2}\gamma^2$
$R(t_0)$	$\frac{2[x_2(k\lambda - x_2)e^{-x_1 t_0} - x_1(k\lambda - x_1)e^{-x_2 t_0}]}{\mu(x_1 - x_2)[2 + 2(k+1)\gamma + k(k+1)\gamma^2]}$	$\frac{1}{1+\gamma_c} e^{-\{[(k+1)\gamma]/[1+(2k+1)\gamma]\}k\lambda t_0}$

the redundant units, and the monitoring methods and period are examples. These factors often influence reliability more than the "main" parameters λ and μ. This section contains methods and equations for these factors. The results are presented so that they can be used in engineering practice without a computer. The contents are considered essential during the first stage of system design when an engineer should develop a "feel" for his or her decisions through the use of simple and understandable mathematical models.

After a failure of a main operational unit, let the first available redundant unit be switched into operation with probability κ in Sections 4.3.2 and 4.3.3 the probability is κ that failure of a main operational unit will be discovered instantly and that the switch will function perfectly. With a complementary probability $1 - \kappa$ the system will fail, and the failure will remain undiscovered for a duration of mean θ. If $\theta \ll 1/k\lambda$, knowledge of the distribution function of a random duration for the undiscovered failure period is unimportant.

Table 4.13 System consisting of k operational units and one redundant unit; active re- dundancy, completely restricted repair

$$x_{1,2} = \frac{\lambda}{2\gamma}\{[1+(2k+1)\gamma] \pm \sqrt{1+2(2k+1)\gamma+\gamma^2}\},\ \epsilon_{1,2} = \frac{\lambda}{2\gamma}\{[2+(2k+1)\gamma] \pm \sqrt{4k\gamma+\gamma^2}\}$$

$$\gamma_c = \frac{k(k+1)\gamma^2}{1+(k+1)\gamma},\ \gamma = \frac{\lambda}{\mu}.$$

Index	Exact	Approximate
$T^{(0)}$	$\frac{1}{k\lambda}\frac{1+(2k+1)\gamma}{(k+1)\gamma}$	$\frac{1}{k\lambda}\frac{1}{(k+1)\gamma}$
$T^{(1)}$	$\frac{1}{k\lambda}\frac{1+(k+1)\gamma}{(k+1)\gamma}$	$\frac{1}{k\lambda}\frac{1}{(k+1)\gamma}$
τ	$\frac{1}{\mu}$	—
$P^{(0)}(t_0)$	$\frac{1}{x_1-x_2}(x_1e^{-x_2t_0}-x_2e^{-x_1t_0})$	$e^{-\{[(k+1)\gamma]/[1+(2k+1)\gamma]\}k\lambda t_0}$
$K(t)$	$1-\frac{k(k+1)\lambda^2}{\epsilon_1\epsilon_2}$	$1-k(k+1)\gamma^2$
	$\times\left[1-\frac{1}{\epsilon_1-\epsilon_2}(\epsilon_1e^{-\epsilon_2t}-\epsilon_2e^{-\epsilon_1t})\right]$	$\times\left[1-\left(1+\frac{\lambda}{\gamma}t\right)e^{-(\lambda/\gamma)t}\right]$
K	$\frac{1}{1+\gamma_0}$	$1-k(k+1)\gamma^2$
$R(t_0)$	$\frac{x_2(k\lambda-x_2)e^{-x_1t_0}-x_1(k\lambda-x_1)e^{-x_2t_0}]}{\mu(x_1-x_2)[1+(k+1)\gamma+k(k+1)\gamma^2]}$	$\frac{1}{1+\gamma_c}e^{-\{[(k+1)\gamma]/[1+(2k+1)\gamma]\}k\lambda t}$

In Section 4.3.3 there is a probability κ that failure of an operational unit will be determined within some permissible time. With probability $1-\kappa$ system failure will remain undiscovered for an average time τ.

Failure of a redundant unit is discovered immediately with probability η. With probability $1-\eta$ failure discovery will have an exponential random delay with mean h. When monitoring is continuous and control is complete, the mean of the random interval between controls is h. If periodic control is incomplete with a depth $\eta_k<1$, all formulas will hold when h is multiplied by $1/\eta_k$.

All distributions of time to failure and repair times are exponential. The failure rates of an operational unit and its redundant unit are λ and λ_1, respectively, where $\lambda_1 \le \lambda$. The repair rate equals μ. If a failure is not discovered by the monitoring device, it will exist without change until a later failure is discovered. During repair no new failure can appear. When $\theta \ll 1/\mu$, κ influences the MTTF and T, but does not significantly affect K. For practical purposes one can take $\tau=0$.

Table 4.14 System consisting of *k* operational units and one redundant unit, underloaded redundancy, unrestricted repair

$x_{1,2} = \frac{\lambda}{2\gamma}\{[1+(2k+\nu)\gamma] \pm \sqrt{1+2(2k+\nu)\gamma+\nu^2\gamma}\}$, $\epsilon_{1,2} = \frac{\lambda}{2\gamma}\{[3+(2k+\nu)\gamma] \pm \sqrt{1+2(2k-\nu)\gamma+\nu^2\gamma}\}$, $\gamma_c = \frac{k(k+\nu)\gamma^2}{2[1+(k+\nu)\gamma]}$, $\gamma = \frac{\lambda}{\mu}$.

Index	Exact	Approximate
$T^{(0)}$	$\frac{1}{k\lambda}\frac{1+(2k+\nu)\gamma}{(k+\nu)\gamma}$	$\frac{1}{k\lambda}\frac{1}{(k+\nu)\gamma}$
$T^{(1)}$	$\frac{1}{k\lambda}\frac{1+(k+\nu)\gamma}{(k+\nu)\gamma}$	$\frac{1}{k\lambda}\frac{1}{(k+\nu)\gamma}$
τ	$\frac{1}{2\mu}$	—
$P^{(0)}(t_0)$	$\frac{1}{x_1-x_2}(x_1e^{-x_2t_0}-x_2e^{-x_1t_0})$	$e^{-\{[(k+\nu)\gamma]/[1+(2k+\nu)\gamma]\}k\lambda t_0}$
$K(t)$	$1-\frac{k(k+\nu)\lambda^2}{\epsilon_1\epsilon_2} \times \left[1-\frac{1}{\epsilon_1-\epsilon_2}(\epsilon_1e^{-\epsilon_2t}-\epsilon_2e^{-\epsilon_1t})\right]$	$1-\frac{k(k+\nu)}{2}\gamma^2 \times [1-(2-e^{-(\lambda/\gamma)t})e^{-(\lambda/\gamma)t}]$
K	$\frac{1}{1+\gamma_0}$	$1-\frac{k(k+\nu)}{2}\gamma^2$
$R(t_0)$	$\frac{2[x_2(k\lambda-x_2)e^{-x_1t_0}-x_1(k\lambda-x_1)e^{-x_2t_0}]}{\mu(x_1-x_2)[2+2(k+\nu)\gamma+k(k+\nu)\gamma^2]}$	$\frac{1}{1+\gamma_c}e^{-\{[(k+\nu)\gamma]/[1+(2k+\nu)\gamma]\}k\lambda t_0}$

4.3.2 System of *k* Main Units and One Redundant Unit

One assumes that it is possible to repair at once m failed units ($m = 1, 2$). The corresponding expressions are

$$T^{(0)} = T^{(1)} + \frac{\kappa}{\mu(1-\nu+A_1+A_2)},$$

$$T^{(1)} = \frac{1+A_1+A_2}{k\lambda(1-\kappa+A_1+A_2)},$$

$$K = \frac{1+A_1+A_2}{1+A_1+A_2+k(A_3+A_4+A_5)},$$

$$\tau = \frac{A_3+A_4+A_5}{\lambda(1-\kappa+A_1+A_2)},$$

where

Table 4.15 System consisting of k operational units and one redundant unit, underloaded redundancy, completely restricted repair

$x_{1,2} = \frac{\lambda}{2\gamma}\{[1+(2k+\nu)\gamma] \pm \sqrt{1+2(2k+\nu)\gamma+\nu^2\gamma}\}$, $\epsilon_{1,2} = \frac{\lambda}{2\gamma}\{[2+(2k+\nu)\gamma] \pm \sqrt{4k\gamma+\nu^2\gamma^2}\}$, $\gamma_c = \frac{k(k+\nu)\gamma^2}{1+(k+\nu)\gamma}$, $\gamma = \frac{\lambda}{\mu}$.

Index	Exact	Approximate
$T^{(0)}$	$\frac{1}{k\lambda}\frac{1+(2k+\nu)\gamma}{(k+\nu)\gamma}$	$\frac{1}{k\lambda}\frac{1}{(k+\nu)\gamma}$
$T^{(1)}$	$\frac{1}{k\lambda}\frac{1+(k+\nu)\gamma}{(k+\nu)\gamma}$	$\frac{1}{k\lambda}\frac{1}{(k+\nu)\gamma}$
τ	$\frac{1}{\mu}$	—
$P^{(0)}(t_0)$	$\frac{1}{x_1-x_2}(x_1e^{-x_2t_0}-x_2e^{-x_1t_0})$	$e^{-\{[(k+\nu)\gamma]/[1+(2k+\nu)\gamma]\}k\lambda t_0}$
$K(t)$	$1-\frac{k(k+\nu)\lambda^2}{\epsilon_1\epsilon_2}\times\left[1-\frac{1}{\epsilon_1-\epsilon_2}(\epsilon_1e^{-\epsilon_2t}-\epsilon_2e^{-\epsilon_1t})\right]$	$1-k(k+\nu)\gamma^2\times\left[1-\left(1+\frac{\lambda}{\gamma}t\right)e^{-(\lambda/\gamma)t}\right]$
K	$\frac{1}{1+\gamma_0}$	$1-k(k+\nu)\gamma^2$
$R(t_0)$	$\frac{x_2(k\lambda-x_2)e^{-x_1t_0}-x_1(k\lambda-x_1)e^{-x_2t_0}}{\mu(x_1-x_2)[1+(k+\nu)\gamma+k(k+\nu)\gamma^2]}$	$\frac{1}{1+\gamma_c}e^{\{[(k+\nu)\gamma]/[1+(2k+\nu)\gamma]\}k\lambda t_0}$

$$A_1 = \frac{k\lambda+\lambda_1}{\lambda},$$

$$A_2 = \frac{(1-\eta)\lambda_1 h}{1+h(k\lambda+\eta\lambda_1)},$$

$$A_3 = \frac{(1-\kappa)\lambda\theta}{1-\lambda_1\tau},$$

$$A_4 = \frac{[(1-\kappa)\lambda A_2+(1-\eta)\lambda_1A_3]h\theta}{h+\tau+\eta\lambda_1h\theta},$$

$$A_5 = \frac{\lambda(A_1+A_2)+\lambda_1A_3}{m\mu}.$$

Indexes $P(t)$ and R can be calculated on the basis that the d.f. of a random TTF for a repairable redundant system with ideal control and a perfect switch is close to exponential. This approximation, however, is

Table 4.16 System consisting of k operational units and one redundant unit, spare redundancy, unrestricted repair

$$x_{1,2} = \frac{\lambda}{2\gamma}\left[(1+2k\gamma) \pm \sqrt{1+4k\gamma}\right], \quad \epsilon_{1,2} = \frac{\lambda}{2\gamma}\left[(3+2k\gamma) \pm \sqrt{1+4k\gamma}\right],$$
$$\gamma_c = \frac{k^2\gamma^2}{2(1+k\gamma)}, \quad \gamma = \frac{\lambda}{\mu}.$$

Index	Exact	Approximate
$T^{(0)}$	$\frac{1}{k\lambda}\left(2+\frac{1}{k\gamma}\right)$	$\frac{1}{k\lambda}\,\frac{1}{k\gamma}$
$T^{(1)}$	$\frac{1}{k\lambda}\left(1+\frac{1}{k\gamma}\right)$	$\frac{1}{k\lambda}\,\frac{1}{k\gamma}$
τ	$\frac{1}{2\mu}$	—
$P^{(0)}(t_0)$	$\frac{1}{x_1-x_2}(x_1e^{-x_2t_0} - x_2e^{-x_1t_0})$	$e^{-\{1/[2+(1/k\gamma)]\}k\lambda t_0}$
$K(t)$	$1-\frac{k^2\lambda^2}{\epsilon_1\epsilon_2}\left[1-\frac{1}{\epsilon_1-\epsilon_2}\right.$ $\left.\times(\epsilon_1e^{-\epsilon_2t} - \epsilon_2e^{-\epsilon_1t})\right]$	$1-\frac{k^2}{2}\gamma^2$ $\times[1-(2-e^{-(\lambda/\gamma)t})e^{-(\lambda/\tau)t}]$
K	$\frac{1}{1+\gamma_0}$	$1-\frac{k^2}{2}\gamma^2$
$R(t_0)$	$\frac{2[x_2(k\lambda-x_2)e^{-x_1t_0} - x_1(k\lambda-x_1)e^{-x_2t_0}]}{\mu(x_1-x_2)[2+2k\gamma+k^2\gamma^2]}$	$\frac{1}{1+\gamma_c}e^{-\{1/[2+(1/k\gamma)]\}k\lambda t_0}$

accurate only for reliable systems. Poor control or switching equipment can make any type of redundancy practically useless.

When $k=1$ and there is an active redundant unit ($\lambda=\lambda_1$) without periodic control ($h=\infty$), one may write

$$T^{(1)} = \frac{1}{\lambda\{1-(1+\eta)[(\kappa/2)-(\lambda/\mu)]\}}.$$

4.3.3 System of k Operational and n Redundant Units

Let all units be repaired independently so that there is no repair queue. The first available redundant unit relaces a failed operational unit. Also let $\tau \ll 1/k\lambda$. The results can be found with a recurrent procedure that is convenient for both computer and hand calculations.

Two types of situations are covered. One is switch failure discovered only during the attempt to replace a failed main unit when there are redundant

Table 4.17 System consisting of *k* operational units and one redundant unit, spare redundancy, completely restricted repair

$x_{1,2} = \frac{\lambda}{2\gamma}[(1+2k\gamma) \pm \sqrt{1+4k\gamma}]$, $\epsilon_{1,2} = \frac{\lambda}{\gamma}[(1+k\gamma) \pm \sqrt{k\gamma}]$, $\gamma_c = \frac{k^2\gamma^2}{1+k\gamma}$, $\gamma = \frac{\lambda}{\mu}$.

Index	Exact	Approximate
$T^{(0)}$	$\frac{1}{k\lambda}\left(2+\frac{1}{k\gamma}\right)$	$\frac{1}{k\lambda}\frac{1}{k\gamma}$
$T^{(1)}$	$\frac{1}{k\lambda}\left(1+\frac{1}{k\gamma}\right)$	$\frac{1}{k\lambda}\frac{1}{k\gamma}$
τ	$\frac{1}{\mu}$	—
$P^{(0)}(t_0)$	$\frac{1}{x_1-x_2}(x_1e^{-x_2t_0}-x_2e^{-x_1t_0})$	$e^{-\{1/[2+(1/k\gamma)]\}k\lambda t_0}$
$K(t)$	$1-\frac{k^2\lambda^2}{\epsilon_1\epsilon_2}$	$1-k^2\gamma^2$
	$\times\left[1-\frac{1}{\epsilon_1-\epsilon_2}(\epsilon_1e^{-\epsilon_2t}-\epsilon_2e^{-\epsilon_1t})\right]$	$\times\left[1-\left(1+\frac{\lambda}{\gamma}t\right)e^{-(\lambda/\gamma)t}\right]$
K	$\frac{1}{1+\gamma_0}$	$1-k^2\gamma^2$
$R(t_0)$	$\frac{x_2(k\lambda-x_2)e^{-x_1t_0}-x_1(k\lambda-x_1)e^{-x_2t_0}}{\mu(x_1-x_2)[1+k\gamma+k^2\gamma^2]}$	$\frac{1}{1+\gamma_c}e^{-\{1/[2+(1/k\gamma)]\}k\lambda t_0}$

units available. The other is system failure when $n+1$ or more units are in a down state at the same time. The mean times to occurrence of these events are denoted as T_s and T_f, resectively. The average durations of the two idle times are denoted as τ_s and τ_f, respectively. The system reliability indexes are

$$T^{(n)} = \frac{1}{1/T_s + 1/T_f},$$

$$\tau = T^{(n)}\left(\frac{\tau_s}{T_s} + \frac{\tau_f}{T_f}\right).$$

The computational procedure consists of initial steps 1–4, recurrent calculation steps 5–9 that are repeated n times, and final steps 10–11:

Table 4.18 System with a single redundant unit, general case

$$x_{1,2} = \frac{1}{2}\left[(\Lambda_0 + \Lambda_1 + M_1) \pm \sqrt{(\Lambda_0 + \Lambda_1 + M_1)^2 - 4\Lambda_0\Lambda_1}\right];\ \epsilon_{1,2} = \frac{1}{2}\left[(\Lambda_0 + \Lambda_1 + M_1 + M_2) \pm \sqrt{(\Lambda_0 + \Lambda_1 + M_1 + M_2)^2 - 4(\Lambda_0\Lambda_1 + \Lambda_0 M_2 + M_1 M_2)}\right],\ \gamma_c = \frac{\Lambda_0\Lambda_1}{M_2(\Lambda_0 + M_1)}.$$

Index	Exact	Approximate
$T^{(0)}$	$\dfrac{\Lambda_0 + \Lambda_1 + M_1}{\Lambda_0\Lambda_1}$	$\dfrac{M_1}{\Lambda_0\Lambda_1}$
$T^{(1)}$	$\dfrac{\Lambda_0 + M_1}{\Lambda_0\Lambda_1}$	$\dfrac{M_1}{\Lambda_0\Lambda_1}$
τ	$\dfrac{1}{M_2}$	—
$P^{(0)}(t_0)$	$\dfrac{1}{x_1 - x_2} \times (x_1 e^{-x_2 t_0} - x_2 e^{-x_1 t_0})$	$e^{-\{\Lambda_0\Lambda_1/[\Lambda_0+\Lambda_1+M_1]\}t_0}$
$K(t)$	$1 - \dfrac{\Lambda_0\Lambda_1}{\epsilon_1\epsilon_2} \times \left[1 - \dfrac{1}{\epsilon_1 - \epsilon_2}(\epsilon_1 e^{-\epsilon_2 t} - \epsilon_2 e^{-\epsilon_1 t})\right]$	$1 - \dfrac{\Lambda_0\Lambda_1}{M_1 M_2} \times \left[1 - \dfrac{1}{M_2 - M_1}(M_2 e^{-M_1 t} - M_1 e^{-M_2 t})\right]$; if $M_1 = M_2 = M$, then $1 - \dfrac{\Lambda_0\Lambda_1}{M^2}[1 - (1 + Mt)e^{-Mt}]$
K	$\dfrac{1}{1 + \gamma_c}$	$1 - \dfrac{\Lambda_0\Lambda_1}{M_1 M_2}$
$R(t_0)$	$\dfrac{M_2[x_2(\Lambda_1 - x_2)e^{-x_1 t_0} - x_1(\Lambda_1 - x_1)e^{-x_2 t_0}]}{(x_1 - x_2)(\Lambda_0\Lambda_1 + \Lambda_0 M_2 + M_1 M_2)}$	$\dfrac{1}{1 + \gamma_c} e^{-\{\Lambda_0\Lambda_1/[\Lambda_0+\Lambda_1+M_1]\}t_0}$

Table 4.19 System consisting of one operational unit and n redundant units, active redundancy, unrestricted repair

$$\gamma = \frac{\lambda}{\mu}, \quad C_m^r = \frac{m!}{r!(m-r)!}, \quad \gamma_c = \frac{1}{\sum_{i=1}^{n+1} C_{n+1}^i/\gamma^i}.$$

Index	Exact	Approximate
$T^{(0)}$	$\frac{1}{(1+n)\lambda}\sum_{s=0}^{n}\frac{\sum_{i=0}^{s} C_{n+1}^i\gamma^i}{C_n^s\gamma^s}$	$\frac{1}{(1+n)\lambda\gamma^n}$
$T^{(n)}$	$\frac{1}{(1+n)\lambda\gamma^n}\sum_{i=0}^{n} C_{n+1}^i\gamma^i$	$\frac{1}{(1+n)\lambda\gamma^n}$
τ	$\frac{1}{(1+n)\mu}$	—
$P^{(0)}(t_0)$	$e^{-\Lambda t_0}$	$\exp\left\{-\frac{(1+n)\lambda t_0}{\sum_{s=0}^{n}[(\sum_{i=0}^{s} C_{n+1}^i\gamma^i)/C_n^s\gamma^s]}\right\}$
K	$\frac{1}{1+\gamma_c}$	$1-\gamma^{n+1}$
$R(t_0)$	$Ke^{-\Lambda t_0}$	$\frac{1}{1+\gamma_c} \times \exp\left\{-\frac{(1+n)\lambda t_0}{\sum_{s=0}^{n}[(\sum_{i=0}^{s} C_{n+1}^i\gamma^i)/C_n^s\gamma^s]}\right\}$

Table 4.20 System consisting of one operational unit and n redundant units, active redundancy, completely restricted repair

$$C_m^r = \frac{m!}{r!(m-r)!}, \quad \gamma = \frac{\lambda}{\mu}, \quad \gamma_c = \frac{1}{\sum_{i=1}^{n+1} 1/i\gamma^i}.$$

Index	Exact	Approximate
$T^{(0)}$	$\frac{1}{\lambda}\sum_{s=0}^{n}\sum_{i=0}^{s}\frac{1}{(i+1)!C_{n-s+i+1}^{i+1}\gamma^i}$	$\frac{1}{\lambda}\frac{1}{\gamma^n(n+1)!}$
$T^{(n)}$	$\frac{1}{\lambda}\sum_{i=0}^{n}\frac{1}{(i+1)!\gamma^2}$	$\frac{1}{\lambda}\frac{1}{n!\gamma^n(n+1)!}$
τ	$\frac{1}{\mu}$	—
$P^{(0)}(t_0)$	$e^{-\Lambda t_0}$	$\exp\left\{-\frac{\lambda t_0}{\sum_{s=0}^{n}\sum_{i=0}^{s} 1/(i+1)!C_{n-s+i+1}^{i+1}\gamma^i}\right\}$
K	$\frac{1}{1+\gamma_c}$	$1-\gamma^{n+1}(n+1)!$
$R(t_0)$	$Ke^{-\Lambda t_0}$	$\times \exp\left\{-\frac{\lambda t_0}{\sum_{s=0}^{n}\sum_{i=0}^{s} 1/(i+1)!C_{n-s+i+1}^{i+1}\gamma^i}\right\}$

Table 4.21 System consisting of one operational unit and n redundant units, underloaded redundancy, unrestricted repair

$$\gamma_c = \frac{\dfrac{\gamma^{n+1}}{(1+n)!} \Pi_{r=1}^{n+1} [1 + (n+1-r)\nu]}{\Sigma_{i=0}^{n} \dfrac{\gamma^i}{i!} \Pi_{r=1}^{i} [1 + (n+1-r)\nu]} ; \ \gamma = \frac{\lambda}{\mu} ; \ \prod_{r=1}^{0} [1 + (n+1-r)\nu] = 1.$$

Index	Exact	Approximate
$T^{(0)}$	$\dfrac{1}{\lambda} \sum_{s=0}^{n} \dfrac{s! \, \Sigma_{i=0}^{s} \gamma^i/i! \, \Pi_{r=1}^{i} [1 + (n+1-r)\nu]}{\gamma^s \, \Pi_{r=1}^{s+1} [1 + (n+1-r)\nu]}$	$\dfrac{1}{\lambda} \dfrac{n!}{\gamma^n} \dfrac{1}{\Pi_{r=1}^{n+1} [1 + (n+1-r)\nu]}$
$T^{(n)}$	$\dfrac{1}{\lambda} \dfrac{n!}{\gamma^n} \dfrac{\Sigma_{i=0}^{s} \gamma^i/i! \, \Pi_{r=1}^{i} [1 + (n+1-r)\nu]}{\Pi_{r=1}^{n+1} [1 + (n+1-r)\nu]}$	$\dfrac{1}{\lambda} \dfrac{n!}{\gamma^n} \dfrac{1}{\Pi_{r=1}^{n+1} [1 + (n+1-r)\nu]}$
τ	$\dfrac{1}{(n+1)\mu}$	—
$P^{(0)}(t_0)$	$e^{-\Lambda t_0}$	$e^{-t_0/T^{(0)}}$
K	$\dfrac{1}{1+\gamma_c}$	$1 - \dfrac{\gamma^{n+1}}{(n+1)!} \times \prod_{r=1}^{n+1} [1 + (n+1-r)\nu]$
$R(t_0)$	$Ke^{-\Lambda t_0}$	$\dfrac{1}{1+\gamma_c} e^{-t_0/T^{(0)}}$

Table 4.22 System consisting of one operational unit and n redundant units, underloaded redundancy, completely restricted repair

$$\gamma_c = \frac{\gamma^{n+1} \Pi_{r=1}^{n+1} [1 + (n+1-r)\nu]}{\Sigma_{i=0}^{n} \gamma^i \, \Pi_{r=1}^{i} [1 + (n+1-r)\nu]} ; \ \gamma = \frac{\lambda}{\mu} ; \ \prod_{r=1}^{0} [1 + (n+1-r)\nu] = 1.$$

Index	Exact	Approximate
$T^{(0)}$	$\dfrac{1}{\lambda} \sum_{s=0}^{n} \dfrac{\Sigma_{i=0}^{s} \gamma^i \, \Pi_{r=1}^{i} [1 + (n+1-r)\nu]}{\gamma^s \, \Pi_{r=1}^{s+1} [1 + (n+1-r)\nu]}$	$\dfrac{1}{\lambda} \dfrac{1}{\gamma^n \, \Pi_{r=1}^{n+1} [1 + (n+1-r)\nu]}$
$T^{(n)}$	$\dfrac{1}{\lambda} \dfrac{\Sigma_{i=0}^{n} \gamma^i \, \Pi_{r=1}^{i} [1 + (n+1-r)\nu]}{\gamma^n \, \Pi_{r=1}^{n+1} [1 + (n+1-r)\nu]}$	$\dfrac{1}{\lambda} \dfrac{1}{\gamma^n \, \Pi_{r=1}^{n+1} [1 + (n+1-r)\nu]}$
τ	$\dfrac{1}{\mu}$	—
$P^{(0)}(t_0)$	$e^{-\Lambda t_0}$	$e^{-t_0/T^{(0)}}$
K	$\dfrac{1}{1+\gamma_c}$	$1 - \gamma^{n+1} \times \prod_{r=1}^{n+1} [1 + (n+1-r)\nu]$
$R(t_0)$	$Ke^{-\Lambda t_0}$	$\dfrac{1}{1+\gamma_c} e^{-t_0/T^{(0)}}$

Table 4.23 System consisting of one operational unit and n redundant units, spare redundancy, unrestricted repair

$$\gamma_c = \frac{1}{\Sigma_{i=1}^{n+1} C_{n+1}^i (i!/\gamma^i)}, \quad \gamma = \frac{\Lambda}{\mu}, \quad C_m^r = \frac{m!}{r!(m-r)!}.$$

Index	Exact	Approximate
$T^{(0)}$	$\frac{1}{\lambda} \sum_{i=0}^{n} C_{n+1}^{i+1} \frac{i!}{\gamma^i}$	$\frac{1}{\lambda} \frac{n!}{\gamma^n}$
$T^{(n)}$	$\frac{1}{\lambda} \sum_{i=0}^{n} C_n^i \frac{i!}{\gamma^i}$	$\frac{1}{\lambda} \frac{n!}{\gamma^n}$
τ	$\frac{1}{(1+n)\mu}$	—
$P^{(0)}(t_0)$	$e^{-\Lambda t_0}$	$\exp\left\{-\frac{\lambda t_0}{\Sigma_{i=0}^{n} C_{n+1}^{i+1}(i!/\gamma^i)}\right\}$
K	$\frac{1}{1+\gamma_c}$	$1 - \frac{\gamma^{n+1}}{(n+1)!}$
$R(t_0)$	$e^{-\Lambda t_0}$	$\frac{1}{1+\gamma_c} \exp\left\{-\frac{\lambda t_0}{\Sigma_{i=0}^{n} C_{n+1}^{i+1}(i!/\gamma^i)}\right\}$

Table 4.24 System consisting of one operational unit and n redundant units, spare redundancy, completely restricted repair

$$\gamma_c = \frac{\gamma^{n+1}(1-\gamma)}{1-\gamma^{n+1}}, \quad \gamma = \frac{\lambda}{\mu} < 1.$$

Index	Exact	Approximate
$T^{(0)}$	$\frac{1}{\lambda} \frac{1-\gamma^{n+1}[1+(n+1)(1-\gamma)]}{\gamma^n(1-\gamma)^2}$	$\frac{1}{\lambda} \frac{1}{\gamma^n}$
$T^{(n)}$	$\frac{1}{\lambda} \frac{1-\gamma^{n+1}}{\gamma^n(1-\gamma)}$	$\frac{1}{\lambda} \frac{1}{\gamma^n}$
τ	$\frac{1}{\mu}$	—
$P^{(0)}(t_0)$	$e^{-\Lambda t_0}$	$\exp\left\{-\frac{(1-\gamma)^2\gamma^n\lambda t_0}{1-\gamma^{n+1}[1+(n+1)(1-\gamma)]}\right\}$
K	$\frac{1}{1+\gamma_c}$	$1-\gamma^{n+1}$
$R(t_0)$	$Ke^{-\Lambda t_0}$	$\frac{1}{1+\gamma_c} \exp\left\{-\frac{(1-\gamma)^2\gamma^n\lambda t_0}{1-\gamma^{n+1}[1+(n+1)(1-\gamma)]}\right\}$

Table 4.25 System consisting of *k* operational and *n* redundant units, active redundancy, unrestricted repair

$$N = k + n,\ C_m^r = \frac{m}{r!(m-r)!},\ \gamma = \frac{\lambda}{\mu},\ \gamma_c = \frac{C_N^{n+1}\gamma^{n+1}}{\sum_{i=0}^{n} C_N^i \gamma^i}.$$

Index	Exact	Approximate
$T^{(0)}$	$\dfrac{1}{N\lambda}\sum_{s=0}^{n}\dfrac{\sum_{i=0}^{s} C_N^i\gamma^i}{C_{N-1}^s\gamma^s}$	$\dfrac{1}{k\lambda C_N^k\gamma^n}$
$T^{(n)}$	$\dfrac{1}{k\lambda C_N^k\gamma^n}\sum_{i=0}^{n} C_N^i\gamma^i$	$\dfrac{1}{k\lambda C_N^k\gamma^n}$
τ	$\dfrac{1}{(n+1)\mu}$	—
$P^{(0)}$	$e^{-\Lambda t_0}$	$\exp\left\{-\dfrac{N\lambda t_0}{\sum_{s=0}^{n}[(\sum_{i=0}^{s} c_N^i\gamma^i)/C_{N-1}^s\gamma^s]}\right\}$
K	$\dfrac{1}{1+\gamma_c}$	$1 - C_N^{n+1}\gamma^{n+1}$
$R(t_0)$	$Ke^{-\Lambda t_0}$	$\dfrac{1}{1+\gamma_c}\exp\left\{-\dfrac{N\lambda t_0}{\sum_{s=0}^{n}[(\sum_{i=0}^{s} c_N^i\gamma^i)/C_{N-1}^s\gamma^s]}\right\}$

Table 4.26 System consisting of *k* operational and *n* redundant units, active redundancy, completely restricted repair

$$N = k + n,\ C_m^r = \frac{m!}{r!(m-r)!},\ \gamma = \frac{\lambda}{\mu},\ \gamma_c = \frac{1}{\sum_{i=1}^{n+1}(i!C_{N-n+i+1}^i\gamma^i)^{-1}}.$$

Index	Exact	Approximate
$T^{(0)}$	$\dfrac{1}{\lambda}\sum_{s=0}^{n}\sum_{i=0}^{s}\dfrac{1}{(i+1)!C_{N-s+i}^{i+1}\gamma^i}$	$\dfrac{1}{\lambda}\,\dfrac{1}{\gamma^n(n+1)!C_N^{n+1}}$
$T^{(n)}$	$\dfrac{1}{\lambda}\sum_{i=0}^{n}\dfrac{1}{(i+1)!C_{k+i}^{i+1}\gamma^i}$	$\dfrac{1}{\lambda}\,\dfrac{1}{\gamma^n(n+1)!C_N^{n+1}}$
τ	$\dfrac{1}{\mu}$	—
$P^{(0)}(t_0)$	$e^{-\Lambda t_0}$	$\exp\left\{-\dfrac{\lambda t_0}{[\sum_{s=0}^{n}\sum_{t=0}^{s}[(i+1)!C_{N-s+i}^{i+1}\gamma^i]^{-1}}\right\}$
K	$\dfrac{1}{1+\gamma_c}$	$1-(n+1)!C_N^{n+1}\gamma^{n+1}$
$R(t_0)$	$Ke^{-\Lambda t_0}$	$\dfrac{1}{1+\gamma_s}\exp\{-\lambda t_0[\sum_{s=0}^{n}\sum_{t=0}^{s}[(i+1)!C_{N-s+i}^{i+1}\gamma^i]^{-1}]^{-1}\}$

Table 4.27 System consisting of k operational and n redundant units, underloaded redundancy, unrestricted repair

$$\gamma_c = \frac{[\gamma^{n+1}(1+n)!]\,\Pi_{r=1}^{n+1}[k+(n+1-r)\nu]}{\Sigma_{i=0}^{n}(\gamma^i/i!)\,\Pi_{r=1}^{i}[k+(n+1-r)\nu]};\ \gamma = \frac{\lambda}{\mu};\ \prod_{r=1}^{0}[k+(n+1-r)\nu]=1.$$

Index	Exact	Approximate
$T^{(0)}$	$\frac{1}{\lambda}\sum_{s=0}^{n}\frac{s!\,\Sigma_{i=0}^{s}(\gamma^i/i!)\,\Pi_{r=1}^{i}[k+(n+1-r)\nu]}{\gamma^s\,\Pi_{r=1}^{s+1}[k+(n+1-r)\nu]}$	$\frac{1}{\lambda}\frac{n!}{\gamma^n}\frac{1}{\Pi_{r=1}^{n+1}[k+(n+1-r)\nu]}$
$T^{(n)}$	$\frac{1}{\lambda}\frac{n!}{\gamma^n}\frac{\Sigma_{i=0}^{n}(\gamma^i/i!)\,\Pi_{r=1}^{i}[k+(n+1-r)\nu]}{\Pi_{r=1}^{n+1}[k+(n+1-r)\nu]}$	$\frac{1}{\lambda}\frac{n!}{\gamma^n}\frac{1}{\Pi_{r=1}^{n+1}[k+(n+1-r)\nu]}$
τ	$\frac{1}{(n+1)\mu}$	—
$P^{(0)}(t_0)$	$e^{-\Lambda t_0}$	$e^{-t_0/T^{(0)}}$
K	$\frac{1}{1+\gamma_c}$	$1-\frac{\gamma^{n+1}}{(n+1)!}\times\prod_{r=1}^{n+1}[k+(n+1-r)\nu]$
$R(t_0)$	$Ke^{-\Lambda t_0}$	$\frac{1}{1+\gamma_c}e^{-t_0/T^{(0)}}$

Table 4.28 System consisting of k operational and n redundant units, underloaded redundancy, completely restricted repair

$$\gamma_c = \frac{\gamma^{n+1}\,\Pi_{r=1}^{n+1}[k+(n+1-r)\nu]}{\Sigma_{i=0}^{n}\gamma^i\,\Pi_{r=1}^{i}[k+(n+1-r)\nu]};\ \gamma = \frac{\lambda}{\mu};\ \prod_{r=1}^{0}[k+(n+1-r)\nu]=1.$$

Index	Exact	Approximate
$T^{(0)}$	$\frac{1}{\lambda}\sum_{s=0}^{n}\frac{\Sigma_{i=0}^{s}\gamma^i\,\Pi_{r=1}^{i}[k+(n+1-r)\nu]}{\gamma^s\,\Pi_{r=1}^{s+1}[k+(n+1-r)\nu]}$	$\frac{1}{\lambda}\frac{1}{\gamma^n\,\Pi_{r=1}^{n+1}[k+(n+1-r)\nu]}$
$T^{(n)}$	$\frac{1}{\lambda}\frac{\Sigma_{i=0}^{s}\gamma^i\,\Pi_{r=1}^{i}[k+(n+1-r)\nu]}{\gamma^n\,\Pi_{r=1}^{n+1}[k+(n+1-r)\nu]}$	$\frac{1}{\lambda}\frac{1}{\gamma^n\,\Pi_{r=1}^{n+1}[k+(n+1-r)\nu]}$
τ	$\frac{1}{\mu}$	—
$P^{(0)}(t_0)$	$e^{-\Lambda t_0}$	$e^{-t_0/T^{(0)}}$
K	$\frac{1}{1+\gamma_c}$	$1-\gamma^{n+1}\times\prod_{r=1}^{n+1}[k+(n+1-r)\nu]$
$R(t_0)$	$Ke^{-\Lambda t_0}$	$\frac{1}{1+\gamma_c}e^{-t_0/T^{(0)}}$

Table 4.29 System consisting of k operational and n redundant units, spare redundancy, unrestricted repair

$$\gamma_c = \frac{1}{\Sigma_{i=1}^{n+1} C_{n+1}^{i}[i!/(k\gamma)^i]}, \quad \gamma = \frac{\lambda}{\mu}, \quad C_m^r = \frac{m!}{r!(m-r)!}.$$

Index	Exact	Approximate
$T^{(0)}$	$\frac{1}{k\lambda} \sum_{i=0}^{n} C_{n+1}^{i+1} \frac{i!}{(k\gamma)^i}$	$\frac{1}{k\lambda} \frac{n!}{(k\gamma)^n}$
$T^{(n)}$	$\frac{1}{k\lambda} \sum_{i=0}^{n} C_{n}^{i} \frac{i!}{(k\gamma)^i}$	$\frac{1}{k\lambda} \frac{n!}{(k\gamma)^n}$
τ	$\frac{1}{(1+n)\mu}$	—
$P^{(0)}(t_0)$	$e^{-\Lambda t_0}$	$\exp\left\{-\frac{k\lambda t_0}{\Sigma_{i=0}^{n} C_{n+1}^{i+1}[i!/(k\gamma)^i]}\right\}$
K	$\frac{1}{1+\gamma_c}$	$1 - \frac{(k\gamma)^{n+1}}{(1+n)!}$
$R(t_0)$	$Ke^{-\Lambda t_0}$	$\frac{1}{1+\gamma_c} \exp\left\{-\frac{k\lambda t_0}{\Sigma_{i=0}^{n} C_{n+1}^{i+1}[i!/(k\gamma^i]}\right\}$

Table 4.30 System consisting of k operational unit and n redundant units, spare redundancy, completely restricted repair

$$\gamma_c = \frac{(k\gamma)^{n+1}(1-k\gamma)}{1-(k\gamma)^{n+1}}, \quad \gamma = \frac{\lambda}{\mu} < \frac{1}{k}.$$

Index	Exact	Approximate
$T^{(0)}$	$\frac{1}{k\lambda} \frac{1-(k\gamma)^{n+1}[1+(n+1)(1-k\gamma)]}{(k\gamma)^n(1-k\gamma)^2}$	$\frac{1}{k\lambda} \frac{1}{(k\gamma)^n}$
$T^{(n)}$	$\frac{1}{k\lambda} \frac{1-(k\gamma)^{n+1}}{(k\gamma)^n(1-k\gamma)}$	$\frac{1}{k\lambda} \frac{1}{(k\gamma)^n}$
τ	$\frac{1}{\mu}$	—
$P^{(0)}(t_0)$	$e^{-\Lambda t_0}$	$\exp\left\{-\frac{(k\gamma)^n(1-k\gamma)^2 k\lambda t_0}{1-(k\gamma)^{n+1}[1+(n+1)(1-k\gamma)]}\right\}$
K	$\frac{1}{1+\gamma_c}$	$1-(k\gamma)^{n+1}$
$R(t_0)$	$Ke^{-\Lambda t_0}$	$\frac{1}{1+\gamma_c} \times \exp\left\{-\frac{(k\gamma)^n(1-k\gamma)^2 k\lambda t_0}{1-(k\gamma)^{n+1}[1+(n+1)(1-k\gamma)]}\right\}$

Table 4.31 System with *n* redundant units, general case

$$\Theta_i = \frac{\Lambda_0 \Lambda_t \ldots \Lambda_{t-1}}{M_1 M_2 \ldots M_i}, \; \Theta_\theta = 1, \; \gamma_c = \frac{\Theta_{n+1}}{\Sigma_{i=0}^{n} \Theta}.$$

Index	Exact	Approximate
$T^{(0)}$	$\sum_{s=0}^{n} \frac{\Sigma_{i=0}^{s} \Theta_i}{\Lambda_s \Theta_s}$	$\frac{1}{\Lambda_n \Theta_n}$
$T^{(n)}$	$\frac{\Sigma_{i=0}^{n} \Theta_t}{\Lambda_n \Theta_n}$	$\frac{1}{\Lambda_n \Theta_n}$
τ	$\frac{1}{M_{n+1}}$	—
$P^{(0)}(t_0)$	$e^{-\Lambda t_0}$	$\exp\left\{-\frac{t_0}{\Sigma_{s=0}^{n} [(\Sigma_{i=0}^{s} \Theta_i)/\Lambda_s \Theta_s]}\right\}$
K	$\frac{1}{1+\gamma_c}$	$1 - \Theta_{n+1}$
$R(t_0)$	$Ke^{-\Lambda t_0}$	$\frac{1}{1+\gamma_c} \exp\left\{-\frac{t_0}{\Sigma_{s=0}^{n} [(\Sigma_{i=0}^{s} \Theta_i)/\Lambda_s \Theta_s]}\right\}$

1. Compute

$$\alpha = \frac{1}{\tau_c} + \eta\lambda_1,$$

$$y_1 = \tfrac{1}{2}(\mu + \lambda_1 + \alpha + \sqrt{(\mu + \lambda_1 + \alpha)^2 - 4\alpha\mu - 4\alpha\mu - 4\alpha\lambda_1 - 4(1-\eta)\lambda_1\mu}),$$

$$y_2 = \tfrac{1}{2}(\mu + \lambda_1 + \alpha - \sqrt{(\mu + \lambda_1 + \alpha)^2 - 4\alpha\mu - 4\alpha\mu - 4\alpha\lambda_1 - 4(1-\eta)\lambda_1\mu}),$$

$$\beta_1 = \frac{\mu\alpha}{y_1 y_2},$$

$$\beta_2 = \frac{\mu(\alpha + y_1)}{y_1(y_1 - y_2)},$$

$$\beta_3 = \frac{\mu(\alpha + y_2)}{y_2(y_2 - y_1)},$$

$$\delta_1 = \frac{(-\kappa)(1-\eta)\mu\lambda_1}{y_1 y_2},$$

$$\delta_2 = \frac{(-\kappa)(1-\eta)\mu\lambda_1}{y_1(y_1 - y_2)},$$

$$\delta_3 = \frac{(-\kappa)(1-\eta)\mu\lambda_1}{y_2(y_2 - y_1)},$$

In rare cases calculated values will be complex, but the complex values can be handled by writing a suitable computer program.

2. A triangular table is completed for $i, j = 0, \ldots, n, i + j \leq n$. The numbers in the table are $F_{ij}^0 = k\lambda/(k\lambda - iy_2 - jy_1)$.
3. Compile a data file

$$G_{i0},\ i = -1, \ldots, n+2;\ G_{i0} = 0;\ G_{00} = \kappa;\ G_1^0 = 1 - \kappa;$$
$$G_2^0 = \cdots G_{n+2}^0 = 0 .$$

4. Introduce the variables for the recurrent procedure: $\lambda^0 = k\lambda$; $A^0 = B^0 = 0$ (A and B are auxiliary counters); l is an auxiliary counter of the number of iterations and equals one initially.
5. Compute

$$\varphi^l = 1 - (\beta_1 F_{00}^{l-1} + \beta_2 F_{01}^{l-1} + \beta_3 F_{10}^{l-1}) ,$$
$$\chi^l = \delta_1 F_{00}^{l-1} + \delta_2 F_{01}^{l-1} + \delta_3 F_{10}^{l-1} ,$$
$$\lambda^l = \varphi^1 \lambda^{l-1} ,$$
$$\sigma = \lambda^{l-1} - \lambda^l ,$$
$$\gamma^l = \left(\frac{\varphi^l}{\chi^l}\right)\nu .$$

6. Initialize

$$A^l = A^{l-1} + \sigma(1 - G_0^{l-1}) ,$$
$$B^l = B^{l-1} + \sigma\tau \sum_{1 \leq i \leq n} iG_i^{l-1} .$$

7. Fill a data file

$$G_{i1},\ i = -1, 0, \ldots, n+2;\ G_{-1}^l = 0 ;$$
$$G_1 l = G_i^{l-1}(1 - \gamma^2) + G_{i-1}^{l-1}\gamma^l,\ i = 0, \ldots, n+2 .$$

8. Fill the triangular table

$$F_{ij}^l,\ i, j = 0, \ldots, n-1;\ i + j \leq n - 1 ;$$
$$F_{ij}^l = \frac{F_{ij}^{l-1}(1 - \beta_1) - \beta_2 F_{i,j+1}^{l-1} - \beta_3 F_{i+1,j}^{l-1}}{1 - \beta_1 F_{ij}^{l-1} - \beta_2 F_{i,j+1}^{l-1} + \beta_3 F_{i+1,j}^{l-1}} .$$

9. Either return to step 5, changing l to $l + 1$, or go to step 10.
10. Calculate the parameters of the events determined by failed switching:

$$T_s = \frac{1}{A^n} ,$$

$$\tau_s = \frac{B^n}{A^n} .$$

11. Calculate the parameters of the events determined by system failures with $n + 1$ or more units failed:

$$T_f = \frac{1}{\gamma^n} ,$$

$$\tau_f = \frac{1}{(n+1)\mu} + \tau \sum_{1 \le i \le n+1} iG_i^l .$$

4.3.4 *k* Operational and *n* Redundant Units, Unreliable Switch

A switch is considered a separate unit with its own parameters

$$\tilde{\lambda}, \quad \tilde{\mu}, \quad \tilde{\eta}, \quad \tilde{\tau}_c, \quad \tilde{\eta}_c .$$

Switching from a failed operational unit to a redundant one takes place only if a switch is operational. If a switch is failed at the needed moment, it operates after repair. In this case the system has an idle time with mean $\tau_* = 1/\tilde{\mu}$. T_0, T_c, τ_0, and τ_c should be calculated with the procedure in Section 4.3.3 modified as described in the following paragraphs. In the procedure ν is the probability of discovery of a failure of an operational unit in a given permissible time, and τ is the mean discovery delay for an operational unit failure that is not discovered within the permissible time.

The computational procedure should have the additional calculation:

$$\tilde{\gamma} = \frac{\tilde{\gamma}[k\lambda + \tilde{\alpha} + (1 - \tilde{\eta})\tilde{\mu}]}{k\lambda(\tilde{\lambda} + \tilde{\alpha} + \tilde{\mu} + k\lambda) + \tilde{\lambda}\alpha + \tilde{\alpha}\tilde{\mu} + (1 - \tilde{\eta})\tilde{\mu}\tilde{\lambda}} ,$$

where

$$\tilde{\alpha} = \frac{\tilde{\eta}_k}{\tilde{\tau}_k} .$$

If no new failure is possible in the monitored part of a switch after a failure in the unmonitored part

$$\tilde{\alpha} = \tilde{\eta}\tilde{\lambda} + \frac{\tilde{\eta}_c}{\tilde{\tau}_c} .$$

If a new failure can appear in the monitored part after a failure in the unmonitored part, one can calculate the switch MTBF with

$$T_{sw} = \frac{1}{[k\lambda - (1/T_f) - (1/T_s)]\gamma} .$$

The mean idle time for this system failure is $\tau_* = 1/\tilde{\mu}$. The system reliability indexes are

$$T^{(n)} = \frac{1}{(1/T_f) + (1/T_s) + (1/T_{sw})} ,$$

$$\tau = T^{(n)}\left[\frac{1}{T_f}(\tau_f + \tilde{\gamma}\tau_{sw}) + \frac{1}{T_s}(\tau_s + \tilde{\gamma}\tau_{sw}) + \frac{\tau_{sw}}{T_{sw}} \right] .$$

4.4 SEMI-MARKOV MODELS OF REPAIRABLE SYSTEMS

4.4.1 Stationary Indexes

One of the most common practical problems is determination of the system MTTF and the system mean time to repair when the process is semi-Markovian and stationary.

Let a system consist of n repairable units, and consequently have 2^n states in the set E. The states are divided by some failure criterion into two classes: operational, E_+, and failure, E_-. The system operation can be described as transitions from one state to another and from one subset to another. A transition from subset E_+ to subset E_- can be made only from a "boundary" subset e_+ of E_+. A transition from E_- into E_+ can be made only from a "boundary" subset of states e_- of E_-.

In a semi-Markovian process the state changes form a Markov chain. Each state has an arbitrary distribution of the duration of time in the state.

The solution requires knowledge of the values in the transition matrix

$$\mathbf{p} = \begin{bmatrix} p_{11} & p_{12} & \cdots & p_{1n} \\ p_{21} & p_{22} & \cdots & p_{2n} \\ \vdots & \vdots & \cdots & \vdots \\ p_{n1} & p_{n2} & \cdots & p_{nn} \end{bmatrix} ,$$

where p_{ij} is the probability of transition from state H_i to state H_j. One also must know, for each i and j, the d.f., $F_{ij}(t)$, of the duration of the stay in state H_i when the next state is H_j.

The system MTTF equals

$$T = \frac{\sum_{i\in E_+} P_i a_i}{\sum_{i\in e_+} P_i \sum_{j\in E_-} p_{ij}}$$

or

$$T = \frac{\sum_{i \in E_-} P_i a_i}{\sum_{i \in e_-} P_i \sum_{j \in E_+} p_{ij}} .$$

The mean idle time equals

$$\tau = \frac{\sum_{i \in E_-} P_i a_i}{\sum_{i \in e_+} P_i \sum_{j \in E_-} p_{ij}}$$

or

$$\tau = \frac{\sum_{i \in E_-} P_i a_i}{\sum_{i \in e_-} P_i \sum_{j \in E_+} p_{ij}} .$$

Here

$$a_i = \sum_{j \in E} p_{ij} T_{ij}$$

is the mean duration of the staying in state H_i.

$$P_i = \frac{D_i}{\sum_{1 \le i \le n} D_j}$$

is the stationary probability of state H_i with transition matrix P, where D_i is a minor obtained from the matrix D by eliminating the ith row and the ith column

$$\mathbf{D} = \begin{vmatrix} 1-p & -p_{12} & \cdots & -p_{1n} \\ -p_{21} & 1-p_{22} & \cdots & -p_{2n} \\ \vdots & \vdots & -p_{ij} & \vdots \\ -p_{n1} & -p_{n2} & \cdots & 1-p_{nn} \end{vmatrix} .$$

For a Markovian process, the transition from state H_i to state H_j is defined by its intensity, λ_{ij}. The system MTTF is

$$T = K\Big(\sum_{i \in e_+} P_i \sum_{j \in E_-} \lambda_{ij}\Big)^{-1}$$

or

$$T = K\Big(\sum_{i \in e_-} P_i \sum_{j \in E_+} \lambda_{ij}\Big)^{-1} .$$

The system mean repair time is given by

$$\tau = k\Big(\sum_{i\in e_+} P_i \sum_{j\in E_-} \lambda_{ij}\Big)^{-1},$$

or

$$\tau = k\Big(\sum_{i\in e_-} P_i \sum_{j\in E_+} \lambda_{ij}\Big)^{-1}$$

where K is the availability coefficient of the system, and $k = 1 - K$

$$K = \sum_{i\in E_+} \frac{P_i}{\sum_{j\in E} \lambda_{ij}} \Big(\sum_{k\in E} \frac{P_k}{\sum_{j\in E} \lambda_{kj}}\Big)^{-1},$$

$$k = 1 - K = \sum_{i\in E_-} \frac{P_i}{\sum_{j\in E} \lambda_{ij}} \Big(\sum_{k\in E} \frac{P_k}{\sum_{j\in E} \lambda_{kj}}\Big)^{-1}.$$

4.4.2 Fictitious State Analysis of Nonstationary Phases

An assumption of exponentiality of all distributions is rather unrealistic. If a system under consideration is not highly reliable, no asymptotic methods can be applied. However, a special technique with fictitious states can permit semi-Markovian processes to be analyzed with a Markov process that has a larger number of states.

The method is based on two phenomena: (1) a mixture of exponential d.f.'s results in a distribution with a decreasing failure rate (DFR), and, (2) a convolution of exponential d.f.'s results in a distribution with an increasing failure rate (IFR).

The simplest transition graphs for a two-phase approximation of nonexponential distributions of time to failure and repair time are given in Figure 4.1. The accuracy of such an exponential approximation of an arbitrary distribution should be assessed for each case individually. Even these simple cases show the necessity for a computer. Sometimes Monte Carlo simulations are better than calculations, as exemplified by the following.

EXAMPLE 4.1

Consider a unit with an exponential distribution of TTF with parameter λ_0 and an IFR repair time distribution approximated by a convolution of two exponential distributions with parameters μ_1 and μ_2:

$$g(t) = \int_0^t \mu_1\mu_2 e^{-\mu_1(t-y)} e^{-\mu_2 y}\, dy .$$

Find an expression for the nonstationary availability coefficient.

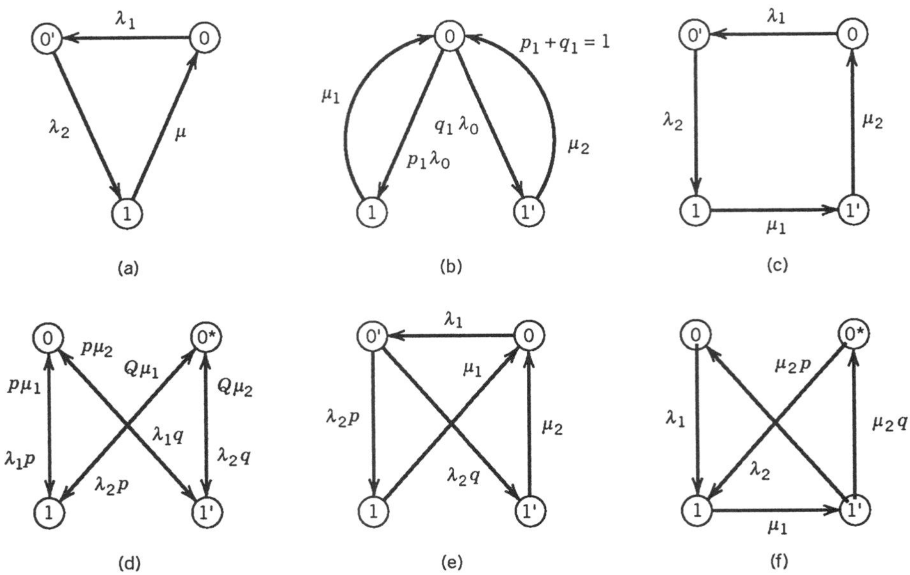

Figure 4.1 Transition graphs for renewable equipment: (*a*) $F(t)$ is IFR, $G(t)$ is exponential; (*b*) $F(t)$ is exponential, $G(t)$ is DFR; (*c*) $F(t)$ is IFR, $G(t)$ is IFR; (*d*) $F(t)$ is DFR, $G(t)$ is DFR; (*e*) $F(t)$ is IFR, $G(t)$ is DFR; (*f*) $F(t)$ is DFR, $G(t)$ is IFR.

Solution

A system of differential equations for this case is

$$\frac{dP_0(t)}{dt} = -\lambda_0 P_0(t) + \mu_2 P_2(t) ,$$

$$\frac{dP_1(t)}{dt} = \lambda_0 P_0(t) - \mu_1 P_1(t) ,$$

$$P_0(t) + P_1(t) + P_2 = 1 ,$$

$$P_0(0) = 1 .$$

The Laplace-Stieltjes Transform (LST) leads to a system of algebraic equations:

$$(\lambda_0 + s)\varphi_0(s) - \mu_2\varphi_2(s) = 1 ,$$

$$-\lambda_0\varphi_0(s) - (\mu_1 + s)\varphi_1(s) = 0 ,$$

$$s\varphi_0(s) + s\varphi_1(s) + s\varphi_2(s) = 1 ,$$

which, after simple transformations, gives

$$\varphi_0(s) = \frac{s^2 + (\mu_1 + \mu_2)s + \mu_1\mu_2}{s[s^2 + (\lambda_0 + \mu_1 + \mu_2)s + \lambda_0\mu_1 + \lambda_0\mu_2 + \mu_1\mu_2]} .$$

Denote the roots of the denominator s_1, s_2, s_3:

$$s_{1,2} = -\frac{\alpha}{2} \pm \sqrt{\frac{\alpha^2}{4} - \beta} ,$$

$$s_3 = 0 ,$$

where $\alpha = \lambda_0 + \mu_1 + \mu_2$, $\beta = \lambda_0\mu_1 + \lambda_0\mu_2 + \mu_1\mu_2$.

For any λ_0, μ_1 and μ_2 the roots s_1 and s_2 are complex conjugates. Express $\varphi_0(s)$ as

$$\varphi_0(s) = \frac{a}{s} + \frac{B}{s - s_1} + \frac{C}{s - s_2}$$

$$= \frac{(A + B + C)s^2 - (A(s_1 + s_2) + Bs_2 + Cs_1)s + AS_1s_2}{s(s - s_1)(s - s_2)} .$$

From the last expression one gets a system of equations for A, B, and C:

$$A + B + C = 1 ,$$

$$\mu_1 + \mu_2 = -A(s_1 + s_2) - Bs_2 - Cs_1 ,$$

$$As_1s_2 = \mu_1\mu_2 .$$

Finally, one gets

$$A = \frac{\mu_1\mu_2}{\lambda_0\mu_1 + \lambda_0\mu_2 + \mu_1\mu_2} ,$$

$$B = C = \frac{1}{2}\left(1 - \frac{\mu_1\mu_2}{\beta}\right) .$$

After the use of the inverse LST,

$$K(t) = P_0(t) = \frac{\mu_1\mu_2}{\beta} + \frac{1}{2}\left(1 - \frac{\mu_1\mu_2}{\beta}\right)(e^{-s_1t} + e^{-s_2t}) .$$

Taking the complexity of s_1 and s_2 into account, one gets

$$K(t) = \frac{\mu_1\mu_2}{\beta} + \left(1 - \frac{\mu_1\mu_2}{\beta}\right)\mathrm{e}^{-\alpha t/2} \cos t\sqrt{\beta - \frac{\alpha^2}{4}} .$$

4.5 HEURISTIC APPROXIMATION FOR HIGHLY RELIABLE SYSTEMS

4.5.1 Proposed Method

A method is proposed that is very close to the strict asymptotic method, but there are no strong conditions. The only restriction is that the system must be highly reliable. The probability of several failures in a short time is small. In most practical cases this means that the MTBF of each system unit is much larger than its mean idle time. The idle time might include repair, substitution of a redundant unit, or time for a failure discovery. In any case the more reliable the system, the more precise will be the result.

The arguments for the method can be stated heuristically. One considers a flow of all events that can lead to a system failure. The intensity of each such flow of events can be estimated. Then one can calculate the probability that any potential situation will develop into a system failure. In a series system each unit failure leads to a system failure with probability 1, but in a redundant structure system failure can develop only if certain failures occur while the first failure condition is existing.

For each type of system failure one can estimate the mean duration. For redundant structures therefore, one can use Renyi's Theorem about thinning of point processes and Grigelionis' Theorem about the superposition of point processes. In both cases the resulting point process is asymptotically Poisson. One should not assume that all d.f.'s are exponential.

Knowledge of the different causes of system failure and average durations of idle periods allows for the derivation of expressions for all reliability indexes of interest. The method can be explained best with several typical examples.

4.5.2 Computation of a System Failure Rate

Some system failures occur only if several unit failures are combined. No one unit failure by itself leads to a system failure. In highly reliable systems each system unit operation can be described by a sequence of brief and rare "pulses" of idle time with long up-time intervals in between. Failure of a redundant system can result from a coincidental overlap of several pulses. If original flows are formed by these alternating processes, the result can be regarded as a flow of "near-point" events. The "pulses" of idle time are rarer and shorter.

In a series structure the overall flow of near-point events can be regarded as a summation of unit flows. This section has some typical situations.

1. If a unit failure by itself leads to a type of system failure and the unit failure rate is λ, the unit rate is a component of the total system failure rate

$$\Lambda_a \approx \lambda = \frac{1}{T_a} ,$$

where T_a is the average time between unit failures. An approximation symbol is used because the expression for Λ_a does not include repair time.

2. If a type of system failure occurs if and only if two independent units are in a failure state at the same time, the corresponding component of the total system failure rate is

$$\Lambda_b \approx \frac{\tau_1 + \tau_2}{T_1 T_2} ,$$

where T_1 and τ_1 are, correspondingly, the MTBF and the mean repair time of the ith unit, $i = 1, 2$.

3. A system failure can occur if a redundant unit is a standby and is in repair when a main unit fails. If the operational one has mean repair time τ_1, the corresponding component of the total system failure rate is

$$\Lambda_c \approx \frac{\tau_1}{T_1 T_2} .$$

In this case we assume that a current main unit has a subscript 1. By the way, notice that if units have different repair times, the unit with the shorter mean repair time should stand for the current main one if it is available.

4. A case similar to (3) occurs when the units are used in turn; that is, after a repair a unit becomes a spare one. Then the corresponding component of the total system failure rate is

$$\Lambda_d \approx \frac{0.5(\tau_1 + \tau_2)}{T_1 T_2} .$$

5. A system failure can occur after failure of a main unit if the redundant unit has a hidden failure. Let the units be identical. This event can happen if the $1 - \alpha$ part of the redundant unit is continually monitored and a failure in the other part can be found only if the system as a whole has failed. In this case

$$\Lambda_e \approx \begin{cases} \dfrac{1-\alpha}{2T} & \text{if } \alpha \approx 1 , \\[2ex] \dfrac{1-\alpha}{T} & \text{if } \alpha = 0 . \end{cases}$$

6. Both units may be partly monitored. A failure in the unmonitored part of a unit can be found only if the system has failed or the monitored part of the unit has failed. Units again are identical.

$$\Lambda_f \approx \frac{(1-\alpha)(\tau + T)}{T^2}$$

7. A system may be similar to (6) but have a $1-\alpha$ part of the redundant unit monitored periodically in time Δ. Units are identical. A system failure occurs if a redundant unit is in a state of hidden failure or is under repair when an operational unit fails. For this case

$$\Lambda_g \approx \frac{(1-\alpha)[(0.5)\delta + \tau]}{T^2} \, .$$

8. An intermittent event with an average time between occurrences of T can lead to a system failure with probability ρ. The corresponding component of the total system failure rate is

$$\Lambda_h \approx \frac{\rho}{T} \, .$$

Examples of such events are the following:

A switch that has probability ρ of not performing its function.

A random duration of switching, η, from a failed main unit to a redundant one that has probability ρ of exceeding a given time, η_0,

$$\rho = Pr\{\tilde{\eta} > \eta_0\}$$

where $\tilde{\eta}$ is a residual part of η.

A standby unit (see the third example in the list) that is switched into an operational position but fails during repair of the main unit,

$$\rho = \int_0^\infty (1 - e^{-\lambda_2 t}) \, dG(t) \approx \lambda_2 \tau_1 \, .$$

If there are n rare and independent causes of system failure, the approximate resulting flow of system failures Λ_Σ is

$$\Lambda_\Sigma \approx \sum_{1 \le i \le n} \frac{1}{T_i}$$

4.5.3 Computation of System Idle Time

The first three cases in this section concern two events that coincide. The mean value of the system idle time depends on the regime of the redundant unit (active, underloaded, or standby), on the type of renewal (repair or replacement), and on the d.f.'s of the random values $\tilde{\tau}_1$ and $\tilde{\tau}_2$. Let F_1 and F_2 denote d.f.'s of r.v.'s $\tilde{\tau}_1$ and $\tilde{\tau}_2$, correspondingly, let E denote a class of exponential d.f.'s, and let D denote a class of nonrandom values.

1. Unrestricted repair:

$$\tau^{(1)} = \frac{1}{\tau_1 + \tau_2} [\tau_1 E\{\min(\tilde{\tau}_1^*, \tilde{\tau}_2)\} + \tau_2 E\{\min(\tilde{\tau}_1, \tilde{\tau}_2^*)\}] ,$$

where

$$E\{\min(x^*, y)\} = \begin{cases} \dfrac{xy}{x+y} & \text{if } (F_x, F_y) \in E , \\ x(1 - e^{-y/x}) & \text{if } F_x \in E,\ F_y \in D , \end{cases}$$

$$\frac{\frac{1}{2}xy}{\frac{1}{2}x + y} \qquad \text{if } F_x \in D,\ F_y \in E ,$$

$$\frac{1}{2}x \qquad \text{if } x > y \text{ and } (F_x, F_y) \in D ;$$

$$\frac{\Delta(2\tau - \Delta)}{2\tau} \qquad \text{if } x > y \text{ and } (F_x, F_y) \in D .$$

2. Restricted repair (one repair facility for two units) with absolute priority for the first unit:

$$\tau^{(2)} = \frac{\tau_1(\tau_1^* + \tau_2)}{\tau_1 + \tau_2} .$$

3. Restricted repair with a first-in–first-out (FIFO) serving discipline:

$$\tau^{(3)} = \begin{cases} \dfrac{\tau_1^2 + \tau_2^2}{\tau_1 + \tau_2} & \text{if } (F_{\tau_1}, F_{\tau_2}) \in E , \\ \dfrac{\tau_1\tau_1^* + \tau_2\tau_2^*}{\tau_1 + \tau_2} & \text{if } (F_{\tau_1}, F_{\tau_2}) \in D , \\ \dfrac{\tau_1\tau_1^* + \tau_2^2}{\tau_1 + \tau_2} & \text{if } F_{\tau_1} \in D,\ F_{\tau_2} \in E . \end{cases}$$

Now let us consider cases when a system failure can occur only if the second unit fails during a first-unit idle period.

4. Restricted repair:

$$\tau^{(4)} = \tau_1^* .$$

5. Unrestricted repair:

$$\tau^{(5)} = E\{\min(\tilde{\tau}_1^*, \tilde{\tau}_2)\} .$$

If a failure occurs in an unmonitored part of a unit, and monitoring has a period Δ, two cases can occur.

6. Perfect periodic monitoring (with a confidence level of 1):

$$\tau^{(6)} = \begin{cases} \frac{1}{2}\,\Delta & \text{if } F_\Delta \in D\,, \\ \Delta & \text{if } F_\Delta \in E\,. \end{cases}$$

7. Periodic monitoring with a confidence level $1-\epsilon$:

$$\tau^{(7)} = \begin{cases} \frac{1}{2}\,\Delta\,\frac{\epsilon}{1-\epsilon} & \text{if } F \in D\,, \\ \Delta\,\frac{\epsilon}{1-\epsilon} & \text{if } F \in E\,, \end{cases}$$

   where ϵ is the probability of a monitoring mistake.
8. Hidden failure of a unit is found in a random time $\tilde{\zeta}$. The system is in a failure state during this period as well as during a subsequent repair. The average system idle time is equal to:

$$\tau^{(8)} = \zeta + \tau\,.$$

9. Failure of a unit leading to immediate failure of the system. The duration of the system idle time equals the duration of the unit repair time:

$$\tau^{(9)} = \tau\,.$$

If there are several failure flows with different idle times, the average system idle time is a "weighted" value

$$\tau_\Sigma \approx \frac{\sum_{1 \le i \le n} \tau_i / T_i}{\sum_{1 \le i \le n} 1/T_i}\,.$$

EXAMPLE 4.2

Consider a duplicated system with imperfect monitoring of an operational unit. A switch can fail at any time during system operation. System failure can occur from any of four events:

An operational unit has failed (event $\tilde{O}$) and at the moment a switch is in a failure state (event $\tilde{S}$).

An operational unit has failed (event $\tilde{O}$) and, although the switch is operational at the moment, the redundant unit is in repair (event $\tilde{R}$).

An operational unit has switched off because of a false command from the monitor (event O^*), and the switch is failed although a redundant unit is up.

An operational unit has switched off because of a false command from the monitor (event O^*) and, although the switch is operational, at the moment a redundant unit is in repair (event $\tilde{R}$).

Assume that all d.f.'s of the main unit, the redundant unit, and the switch are exponential with corresponding parameters λ, μ, λ_1, μ_1, λ_5, and μ_5. The operational unit is monitored at random moments, and intervals between checks are distributed exponentially with parameter ν. The probability of a mistake during a monitoring check equals α. Failure rates and idle time durations for the different failure events are given in Table 4.32. Calculation of idle times may be illustrated with $(\tilde{O}, \tilde{S})$. The duration of the idle time depends on the state of the redundant unit when switch repair is completed. If the redundant unit is operable, it is switched into an operational position immediately. If both units are failed, the system will be out of its failure state when any failed unit has been repaired.

The reliability indexes for this system are

$$\Lambda_\Sigma \approx \lambda_1 + \lambda_2 + \lambda_3 + \lambda_4 ,$$

$$\tau_\Sigma \approx \frac{1}{\Lambda_\Sigma} (\lambda_1\tau_1 + \lambda_2\tau_2 + \lambda_3\tau_3 + \lambda_4\tau_4) ,$$

$$P(t) \approx e^{-\Lambda_\Sigma t} ,$$

$$K \approx \frac{1}{1 + \Lambda_\Sigma \tau_\Sigma} .$$

Table 4.32 Failure rates and repair times for Example 4.2

Type of Failure	Failure Rate	Mean Idle Time
$\tilde{O}\tilde{S}$	$\lambda_1 = \lambda \dfrac{\lambda_p}{\lambda_p + \mu_p}$	$\tau_1 = \dfrac{\mu_p}{(\mu_p + \mu)^2} \left(1 + \dfrac{\mu_p}{\lambda_p + \mu_p}\right)$
$\tilde{O}\tilde{R}$	$\lambda_2 = \lambda \dfrac{\lambda_p}{\lambda_p + \mu_p}$	$\tau_2 = \dfrac{1}{\mu + \mu_p}$
$O^*\tilde{S}$	$\lambda_3 = \alpha\nu \dfrac{\lambda_p}{\lambda_p + \mu_p}$	$\tau_3 = \dfrac{1}{\nu(1-\alpha)^{-1} + \mu_p}$
$O^*\tilde{R}$	$\lambda_4 = \alpha\nu \dfrac{\lambda_p}{\lambda_p + \mu_p}$	$\tau_4 = \dfrac{1}{\nu(1-\alpha)^{-1} + \mu_p}$

EXAMPLE 4.3

Consider a duplicated system an operational unit of which 0.99 is monitored continuously and 0.01 is unmonitored. Monitoring is always complete and takes a constant two-hour period. A redundant unit is in an active regime and is monitored continuously and completely. Switching from a failed unit to an operable redundant one takes a constant time, $t_n = 0.1$ hour. There is one repair facility, and it has a FIFO discipline. Each unit has an exponential TTF with parameter $\lambda = 0.01$ per hour. The repair time τ is almost constant at three hours. Find the reliability indexes.

Solution

In this case system failure occurs each time an operating unit fails because switching requires a finite time. A redundant unit permits a reduction of system idle time to 0.1 hour, but does not produce failure-free operation. The system idle time always must be included in the total idle time. One can write immediately

$$\lambda_\Sigma = \lambda = 0.01\left(\frac{1}{\text{h}}\right).$$

The system can be presented as a parallel-series structure (Figure 4.2).

We use the following notation:

M = continuously monitored part of the operating unit. Its failure rate is $(1 - \alpha)\lambda$, and its repair time is constant at τ.

R = redundant unit. Its failure rate is λ, and the constant repair time is τ.

N = unmonitored part of the operating unit. Its failure rate is $\alpha\lambda$, and the average residual time of repair is half the monitoring period, $t_k/2$.

S = hypothetical unit representing an instantly operating and absolutely reliable switch with a finite switching time. Its failure rate is λ, and the idle time is t_n.

The first step is to form a unit equivalent to units M and R, using the formulas given previously for active redundancy, completely restricted repair, and constant repair time:

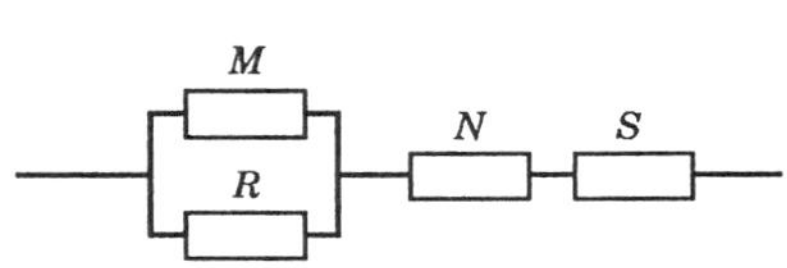

Figure 4.2 Structure for Example 4.3.

$$\lambda_1^* \approx 2(1-\alpha)\lambda^2\tau = 0.00059\left(\frac{1}{\mathrm{h}}\right),$$

$$\tau_1^* \approx \frac{\tau}{2} = 1.5h .$$

The second step is to calculate for the equivalent unit, unit S, and unit N connected in series:

$$\lambda_1^* \approx 2(1-\alpha)\lambda^2\tau + \alpha\lambda = 0.116\left(\frac{1}{\mathrm{h}}\right);$$

$$\tau^* \approx \frac{\lambda t_{sw} + 2(1-\alpha)\lambda^2\tau(\tau/2) + \alpha\lambda(t_c/2)}{\lambda + 2(1-\alpha)\lambda^2\tau + \alpha\lambda} = 0.241(\mathrm{h}) .$$

Finally, one may write for the system

$$P(t_0) \approx e^{-(0.0116)t_0},$$

$$K \approx \frac{1}{1+\lambda^*\tau^*} = 0.9971 .$$

CHAPTER 5

REPAIRABLE DUAL SYSTEMS

The most common type of redundancy is the use of one redundant component either in parallel or in standby parallel. This entire chapter therefore is devoted to repairable equipment of this type. All results in this chapter have been obtained with the heuristic method described in Section 4.5. Special cases not covered by the formulas in this chapter can be analyzed with the heuristic method.

5.1 IDENTICAL COMPONENTS IN PARALLEL

Tables 5.1 through 5.6 give formulas for calculation of reliability indexes for equipment with twin redundant components. Monitoring is assumed to be perfect. Reswitching is assumed to be failure free and instantaneous. All approximation formulas are for a mean time to failure that is orders of magnitude smaller than the mean repair time.

5.2 DEPENDENT UNITS

To increase reliability, a backup component that has the same function as the primary component may be different in structure or even in internal operation. If the primary component is mechanical or optical, the backup may be electronic. An example is equipment for coupling two space capsules.

Table 5.1 Two-unit system with active redundant unit and unrestricted repair

$$x_{1,2} = \frac{\lambda}{2\gamma}\,(1 + 3\gamma \pm \sqrt{1 + 6\gamma + \gamma^2}),\ \epsilon_{1,2} = \frac{\lambda}{2\gamma}\,(1+\gamma)(3 \pm 1),\ \gamma_c = \frac{\gamma^2}{1+2\gamma},\ \gamma = \frac{\lambda}{\mu}.$$

Index	Accurate	Approximate	Error
$T^{(0)}$	$\frac{1}{\lambda}\frac{1+3\gamma}{2\gamma}$	$\frac{1}{\lambda}\frac{1}{2\gamma}$	$\delta_+ = \frac{3}{2}\frac{1}{\lambda}$
$T^{(1)}$	$\frac{1}{\lambda}\frac{1+2\gamma}{2\gamma}$	$\frac{1}{\lambda}\frac{1}{2\gamma}$	$\delta_+ = \frac{1}{\lambda}$
τ	$\frac{1}{2\mu}$	—	—
$P^{(0)}(t_0)$	$\frac{1}{x_1 - x_2}\,(x_1 e^{-x_2 t_0} - x_2 e^{-x_1 t_0})$	$e^{-[2\gamma/(1+3\gamma)]\lambda t_0}$	$\delta \sim 2\gamma^2$
$K(t)$	$1 - \frac{2\lambda^2}{\epsilon_1\epsilon_2}$		
	$\times\left[1 - \frac{1}{\epsilon_1 - \epsilon_2}\,(\epsilon_1 e^{-\epsilon_2 t} - \epsilon_2 e^{-\epsilon_1 t})\right]$	$1 - \gamma^2[1 - (2 - e^{-\lambda t/\gamma})e^{-\lambda t/\gamma}]$	$\delta \sim 8\gamma^3$
K	$\frac{1}{1+\gamma_c}$	$1 - \gamma^2$	$\delta_+ \sim 2\gamma^3$
$R(t_0)$	$\frac{x_2(\lambda - x_2)e^{-x_1 t_0} - x_1(\lambda - x_1)e^{-x_2 t_0}}{\mu(x_1 - x_2)(1+\gamma)^2}$	$\frac{1}{1+\gamma_c}\,e^{-[2\gamma/(1+3\gamma)]\lambda t_0}$	$\delta \sim 2\gamma^2$

Table 5.2 Two-unit system with standby redundant unit and restricted repair

$$x_{1,2} = \frac{\lambda}{2\gamma}\,(1 + 3\gamma \pm \sqrt{1 + 6\gamma + \gamma^2}),\ \epsilon_{1,2} = \frac{\lambda}{2\gamma}\,[(2+3\gamma) \pm \sqrt{4\gamma + \gamma^2}],\ \gamma_C = \frac{2\gamma^2}{1+2\gamma},\ \gamma = \frac{\lambda}{\mu}.$$

Index	Accurate	Approximate	Error
$T^{(0)}$	$\frac{1}{\lambda}\frac{1+3\gamma}{2\gamma}$	$\frac{1}{\lambda}\frac{1}{2\gamma}$	$\delta_+ = \frac{3}{2}\frac{1}{\lambda}$
$T^{(1)}$	$\frac{1}{\lambda}\frac{1+2\gamma}{2\gamma}$	$\frac{1}{\lambda}\frac{1}{2\gamma}$	$\delta_+ = \frac{1}{\lambda}$
τ	$\frac{1}{\mu}$	—	—
$P^{(0)}(t_0)$	$\frac{1}{x_1 - x_2}\,(x_1 e^{-x_2 t_0} - x_2 e^{-x_1 t_0})$	$e^{-[2\gamma/(1+3\gamma)]\lambda t_0}$	$\delta \sim 2\gamma^2$
$K(t)$	$1 - \frac{2\lambda^2}{\epsilon_1\epsilon_2}$	$1 - 2\gamma^2$	
	$\times\left[1 - \frac{1}{\epsilon_1 - \epsilon_2}\,(\epsilon_1 e^{-\epsilon_2 t} - \epsilon_2 e^{-\epsilon_1 t})\right]$	$\times\left[1 - \left(1 + \frac{\lambda}{\gamma}\,t\right)e^{-\lambda t/\gamma}\right]$	$\delta \sim 8\gamma^3$
K	$\frac{1}{1+\gamma_c}$	$1 - 2\gamma^2$	$\delta_+ \sim 4\gamma^3$
$R(t_0)$	$\frac{x_2(\lambda - x_2)e^{-x_1 t_0} - x_1(\lambda - x_1)e^{-x_2 t_0}}{\mu(x_1 - x_2)(1+2\gamma+2\gamma^2)}$	$\frac{1}{1+\gamma_c}\,e^{-[2\gamma/(1+3\gamma)]\lambda t_0}$	$\delta \sim 2\gamma^2$

Table 5.3 Two-unit system with partially active redundant unit and unrestricted renewal

$$x_{1,2} = \frac{\lambda}{2\gamma}\{[1+(2+\nu)\gamma] \pm \sqrt{1+2(2+\nu)\gamma+\nu^2\gamma^2}\}, \qquad \epsilon_{1,2} = \frac{\lambda}{2\gamma}\{[3+(2+\nu)\gamma] \pm \sqrt{1+2(2-\nu)\gamma+\nu^2\gamma^2}\},$$

$$\gamma_c = (1+\nu)\gamma^2/2[1+(1+\nu)\gamma], \ \gamma = \lambda/\mu.$$

Index	Accurate	Approximate	Error
$T^{(0)}$	$\frac{1}{\lambda}\frac{1+(2+\nu)\gamma}{(1+\nu)\gamma}$	$\frac{1}{\lambda}\frac{1}{(1+\nu)\gamma}$	$\delta_+ = \frac{2+\nu}{1+\nu}\frac{1}{\lambda}$
$T^{(1)}$	$\frac{1}{\lambda}\frac{1+(1+\nu)\gamma}{(1+\nu)\gamma}$	$\frac{1}{\lambda}\frac{1}{(1+\nu)\gamma}$	$\delta_+ = \frac{1}{\lambda}$
τ	$\frac{1}{2\mu}$	—	—
$P^{(0)}(t_0)$	$\frac{1}{x_1-x_2}(x_1e^{-x_2t_0} - x_2e^{-x_1t_0})$	$\exp\left\{-\frac{(1+\nu)\gamma}{1+(2+\nu)\gamma}\lambda t_0\right\}$	$\delta \sim (1+\nu)\gamma^2$
$K(t)$	$1 - \frac{(1+\nu)\lambda^2}{\epsilon_1\epsilon_2}$	$1 - \frac{(1+\nu)}{2}\gamma^2$	
	$\times\left[1 - \frac{1}{\epsilon_1-\epsilon_2}(\epsilon_1e^{-\epsilon_2t} - \epsilon_2e^{-\epsilon_1t})\right]$	$\times[1-(2-e^{-\lambda t/\gamma})e^{-\lambda t/\gamma}]$	$\delta \sim (1+\nu)^3\gamma^3$
K	$\frac{1}{1+\gamma_c}$	$1 - \frac{1+\nu}{2}\gamma^2$	$\delta_+ \sim \frac{(1+\nu)^2}{2}\gamma^3$
$R(t_0)$	$\frac{2[x_2(\lambda-x_2)e^{-x_1t_0} - x_1(\lambda-x_1)e^{-x_2t_0}]}{\mu(x_1-x_2)[2+2(1+\nu)\gamma+(1+\nu)\gamma^2]}$	$\frac{1}{1+\gamma_c}\exp\left\{-\frac{(1+\nu)\gamma}{1+(2+\nu)\gamma}\lambda t_0\right\}$	$\delta \sim (1+\nu)\gamma^2$

Table 5.4 Two-unit system with partially loaded redundant unit and restricted renewal

$$x_{1,2} = \frac{\lambda}{2\gamma}\{[1+(2+\nu)\gamma] \pm \sqrt{1+2(2+\nu)\gamma+\nu^2\gamma^2}\};\ \epsilon_{1,2} = \frac{\lambda}{2\gamma}\{[2+(2+\nu)\gamma] \pm \sqrt{4\gamma+\nu^2\gamma^2}\};$$

$$\gamma_c = \frac{(1+\nu)\gamma^2}{1+(1+\nu)\gamma},\ \gamma = \frac{\lambda}{\mu}.$$

Index	Accurate	Approximate	Error
$T^{(0)}$	$\frac{1}{\lambda}\frac{1+(2+\nu)\gamma}{(1+\nu)\gamma}$	$\frac{1}{\lambda}\frac{1}{(1+\nu)\gamma}$	$\delta_+ = \frac{1}{\lambda}\frac{2+\nu}{1+\nu}$
$T^{(1)}$	$\frac{1}{\lambda}\frac{1+(1+\nu)\gamma}{(1+\nu)\gamma}$	$\frac{1}{\lambda}\frac{1}{(1+\nu)\gamma}$	$\delta_+ = \frac{1}{\lambda}$
τ	$\frac{1}{\mu}$	—	—
$P^{(0)}(t_0)$	$\frac{1}{x_1 - x_2}(x_1 e^{-x_2 t_0} - x_2 e^{-x_1 t_0})$	$\exp\left\{-\frac{(1+\nu)\gamma}{1+(2+\nu)\gamma}\lambda t_0\right\}$	$\delta \sim (1+\nu)\gamma^2$
$K(t)$	$1-(1+\nu)\gamma^2\left[1-\left(1+\frac{\lambda}{\gamma}t\right)e^{-\lambda t/\gamma}\right]$	$1-(1+\nu)\gamma^2 \times \left[1-\left(1+\frac{\lambda}{\gamma}t\right)e^{-\lambda t/\gamma}\right]$	$\delta \sim (1+\nu)^3\gamma^3$
K	$\frac{1}{1+\gamma_c}$	$1-(1+\nu)\gamma^2$	$\delta_+ \sim (1+\nu)^2\gamma^3$
$R(t_0)$	$\frac{x_2(\lambda - x_2)e^{-x_1 t_0} - x_1(\lambda - x_1)e^{-x_2 t_0}]}{\mu(x_1 - x_2)[1+(1+\nu)\gamma+(1+\nu)\gamma^2]}$	$\frac{1}{1+\gamma_c}\exp\left\{-\frac{(1+\nu)\gamma}{1+(2+\nu)\gamma}\lambda t_0\right\}$	$\delta \sim (1+\nu)\gamma^2$

Table 5.5 Two-unit system with unloaded redundant unit and unrestricted renewal

$$x_{1,2} = \frac{\lambda}{2\gamma}\,[(1+2\gamma) \pm \sqrt{1+4\gamma}],\ \epsilon_{1,2} = \lambda[(3+2\gamma) \pm \sqrt{1+4\gamma}]/2\gamma,\ \gamma_c = \gamma^2/2(1+\gamma),\ \gamma = \lambda/\mu.$$

Index	Accurate	Approximate	Error
$T^{(0)}$	$\frac{1}{\lambda}\left(2+\frac{1}{\gamma}\right)$	$\frac{1}{\lambda\gamma}$	$\delta_+ = \frac{2}{\lambda}$
$T^{(1)}$	$\frac{1}{\lambda}\left(1+\frac{1}{\gamma}\right)$	$\frac{1}{\lambda\gamma}$	$\delta_+ = \frac{1}{\lambda}$
τ	$\frac{1}{2\mu}$	—	—
$P^{(0)}(t_0)$	$\frac{1}{x_1-x_2}\,(x_1 e^{-x_2 t_0} - x_2 e^{-x_1 t_0})$	$\exp\left\{-\frac{1}{2+(1/\gamma)}\,\lambda t_0\right\}$	$\delta \sim \gamma^2$
$K(t)$	$1-\frac{\lambda^2}{\epsilon_1\epsilon_2}\left[1-\frac{1}{\epsilon_1-\epsilon_2}\,(\epsilon_1 e^{-\epsilon_2 t} - \epsilon_2 e^{-\epsilon_1 t})\right]$	$1-\frac{\gamma^2}{2} \times [1-(2-e^{-\lambda t/\gamma})e^{-\lambda t/\gamma}]$	$\delta \sim \gamma^3$
K	$\frac{1}{1+\gamma_0}$	$1-\frac{\gamma^2}{2}$	$\delta_+ \sim \frac{\gamma^3}{2}$
$R(t_0)$	$\frac{2[x_2(\lambda-x_2)e^{-x_1 t_0} - x_1(\lambda-x_1)e^{-x_2 t_0}]}{\mu(x_1-x_2)(2+2\gamma+\gamma^2)}$	$\frac{1}{1+\gamma_c}\exp\left\{-\frac{1}{2+(1/\gamma)}\,\lambda t_0\right\}$	$\delta \sim \gamma^2$

Table 5.6 Two-unit system with unloaded redundant unit and restricted renewal

$$x_{1,2} = \frac{\lambda}{2\gamma}\,[(1+2\gamma) \pm \sqrt{1+4\gamma}],\ \epsilon_{1,2} = \lambda[(1+\gamma) \pm \sqrt{\gamma}]/\gamma,\ \gamma_c = \gamma^2/(1+\gamma),\ \gamma = \lambda/\mu.$$

Index	Accurate	Approximate	Error
$T^{(0)}$	$\frac{1}{\lambda}\left(2+\frac{1}{\gamma}\right)$	$\frac{1}{\lambda\gamma}$	$\delta_+ = \frac{2}{\lambda}$
$T^{(1)}$	$\frac{1}{\lambda}\left(1+\frac{1}{\gamma}\right)$	$\frac{1}{\lambda\gamma}$	$\delta_+ = \frac{1}{\lambda}$
τ	$\frac{1}{\mu}$	—	—
$P^{(0)}(t_0)$	$\frac{1}{x_1-x_2}\,(x_1 e^{-x_2 t_0} - x_2 e^{-x_1 t_0})$	$\exp\left\{-\frac{1}{2+(1/\gamma)}\,\lambda t_0\right\}$	$\delta \sim \gamma^2$
$K(t)$	$1-\frac{\lambda^2}{\epsilon_1\epsilon_2}\left[1-\frac{1}{\epsilon_1-\epsilon_2}\,(\epsilon_1 e^{-\epsilon_2 t} - \epsilon_2 e^{-\epsilon_1 t})\right]$	$1-\gamma^2\left[1-\left(1+\frac{\lambda}{\gamma}\,t\right)e^{-\lambda t/\gamma}\right.$	$\delta \sim \gamma^3$
K	$\frac{1}{1+\gamma_0}$	$1-\gamma^2$	$\delta_+ \sim \gamma^3$
$R(t_0)$	$\frac{x_2(\lambda-x_2)e^{-x_1 t_0} - x_1(\lambda-x_1)e^{-x_2 t_0}]}{\mu(x_1-x_2)(1+\gamma+\gamma^2)}$	$\frac{1}{1+\gamma_c}\exp\left\{-\frac{1}{2+(1/\gamma)}\,\lambda t_0\right\}$	$\delta \sim \gamma^2$

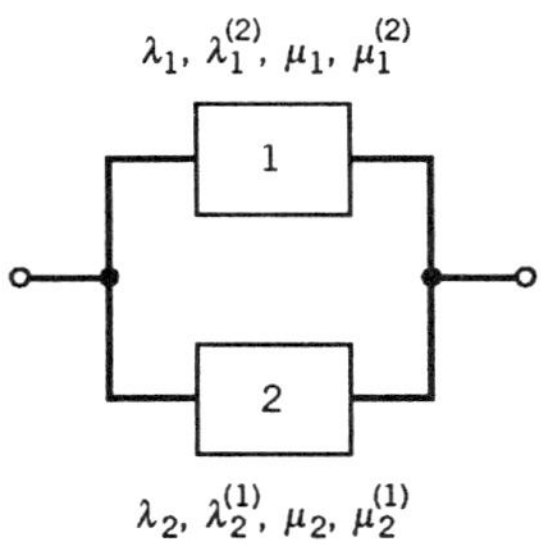

Figure 5.1 Structural diagram of two-unit system with two dependent units of different reliability.

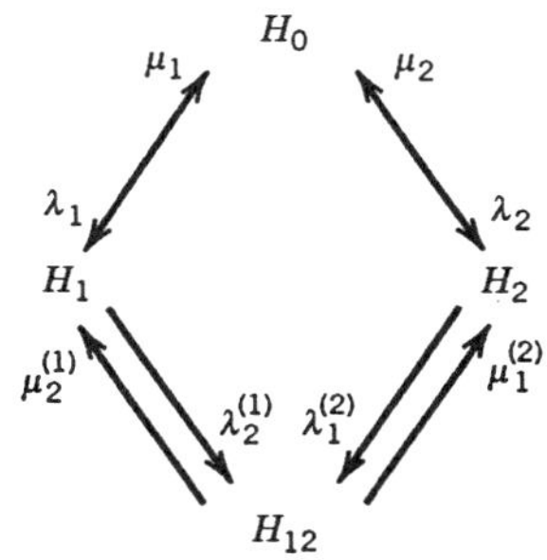

Figure 5.2 Transition graph for the system in Figure 5.1.

The reliabilities of the primary and secondary units can be different and interdependent. The situations are represented in Figure 5.1, which has the following notation for i, $j = 1, 2$, and $i \neq j$:

λ_i = failure rate of i when j has not failed,
$\lambda_i^{(j)}$ = failure rate of i when j has failed,
μ_i = renewal rate of i when j has not failed,
$\mu_i^{(j)}$ = renewal rate of i when j has failed.

Figure 5.2 depicts the transitions for the item in Figure 5.1 and has the following notations for system states: H_0 means neither unit has failed, H_1 means only unit 1 has failed, H_2 means only unit 2 has failed, and H_{12} means both units have failed.

Table 5.7 gives exact and approximate formulas for calculating reliability

Table 5.7 Two-unit system with different, dependent units and unrestricted renewal

Index	Accurate	Approximate
$T^{(0)}$	$\dfrac{1+\alpha_1\beta_1+\alpha_2\beta_2}{\alpha_1\lambda_1+\alpha_2\lambda_2}$	$\dfrac{\mu_1\mu_2}{\lambda_2\lambda_1^{(2)}\mu_1+\lambda_1\lambda_2^{(1)}\mu_2}$
$T^{(1)}$	$\dfrac{\mu_1^{(2)}\psi_1+\mu_2^{(1)}\psi_2}{(\mu_1^{(2)}+\mu_2^{(1)})(\alpha_1\lambda_1+\lambda_2\lambda_2)}$	$\dfrac{\mu_1\mu_2}{\lambda_2\lambda_1^{(2)}\mu_1+\lambda_1\lambda_2^{(1)}\mu_2}$
$P(t_0)$	—	$e^{-2\bar{\nu}\bar{\lambda}t_0(1+3\bar{\nu})}$
τ	$(\mu_1^{(2)}+\mu_2^{(1)})^{-1}$	—
K	$\left(1+\dfrac{\alpha_1\lambda_1+\alpha_2\lambda_2}{\mu_1^{(2)}\psi_1+\mu_2^{(1)}\psi_2}\right)^{-1}$	$\left[1+\dfrac{\lambda_2\lambda_1^{(2)}\mu_1+\lambda_1\lambda_2^{(1)}\mu_2}{\mu_1\mu_2(\mu_1^{(2)}+\mu_2^{(1)})}\right]^{-1}$
$R(t_0)$	—	$\left(1+\dfrac{\alpha_1\lambda_1+\alpha_2\lambda_2}{\mu_1^{(2)}\psi_1+\mu_2^{(1)}\psi_2}\right)^{-1}e^{-2\bar{\nu}\bar{\lambda}t_0/(1+3\bar{\nu})}$

Notation: $\alpha_1=\lambda_2^{(1)}/(\lambda_2^{(1)}+\mu_1)$, $\alpha_2=\lambda_1^{(2)}/(\lambda_1^{(2)}+\mu_2)$, $\beta_1=\lambda_1/\lambda^{(1)}$, $\beta_2=\lambda_2/\lambda_1^{(2)}$, $\psi_1=(1-\alpha_2)(1+\alpha_1\beta_1)+\alpha_2\beta_2[1+\alpha_1(\lambda_1/\lambda_2)]$, $\psi_2=(1-\alpha_1)(1+\alpha_2\beta_2)+\alpha_1\beta_1[1+\alpha_2(\lambda_2/\lambda_1)]$.

indexes for two-unit systems. The approximate formulas are accurate enough for practical purposes when

$$\bar{\gamma} = \frac{\bar{\lambda}}{\underline{\mu}} \ll 0.5\,,$$

where

$$\bar{\lambda} = \max(\lambda_1, \lambda_2, \lambda_1^{(2)}, \lambda_2^{(1)})\,,$$
$$\underline{\mu} = \min(\mu_1, \mu_2, \mu_1^{(2)}, \mu_2^{(1)})\,.$$

5.3 MONITORING AND RESWITCHING

Theoretical equations often give much higher reliability values than are achieved in practice. One reason for this is that mathematical models often are not adjusted for monitoring and reswitching failures, even though these factors are often critical.

5.3.1 Reswitching with a Delay

Consider a duplicate system with an absolutely reliable reswitcher but a random reswitching time, t_π. All approximate formulas are given for a mean time to failure for a unit that is much more than both the mean reswitching time and the mean time for repair of the failed unit, that is, $T \gg \tau$ and $T \gg t_\pi$. There are no restrictions on the relationship between τ and t_π. In this case any failure of an operating main unit leads to interruption of dual system operation; system failure occurs. The redundant unit simply decreases system renewal time.

The formulas for different regimes of operation and renewal are given in Tables 5.8 through 5.11. An arbitrary distribution, $P(t)$, indicates an aging

Table 5.8 Loaded redundant unit, unrestricted renewal, delayed reswitching

		All Distributions	
Index	Exponential	Lower Bound	Upper Bound
T_s	$\frac{1}{\lambda}$	$\frac{T}{2}$	$\frac{T}{2}$
τ_s	$t_\pi + \gamma\tau$	$t_\pi + 0.5\gamma\tau$	$t_\pi + \gamma\tau$
$P_s(t_0)$	$e^{-\gamma_0}$	$1 - 2\gamma_0$	$e^{-\gamma}$
K	$1 - (\gamma_\pi + \gamma^2)$	$1 - 2(\gamma + 0.5\gamma^2)$	$1 - (\gamma_\pi + \gamma^2)$
$R(t_0)$	$[1 - (\gamma_\pi + \gamma^2)]e^{-\gamma_0}$	$[1 - (2\gamma_\pi + \gamma^2)](1 - \gamma_0)^2$	$[1 - (\gamma_\pi + \gamma^2)]e^{-\gamma_0}$

Table 5.9 Loaded redundant unit, restricted renewal, delayed reswitching

Index	Exponential	All Distributions Lower Bound	All Distributions Upper Bound
T_s	$\frac{1}{\lambda}$	$0.5T$	T
τ_s	$t_\pi + 2\gamma\tau$	$t_\pi + 0.5\gamma\tau$	$t_\pi + 2\gamma\tau$
$P_s(t_0)$	$e^{-\gamma_0}$	$1 - 2\gamma_0$	$e^{-\gamma_0}$
K	$1 - (\gamma_\pi + 2\gamma^2)$	$1 - (2\gamma_\pi + \gamma^2)$	$1 - (\gamma_\pi + 2\gamma^2)$
$R(t_0)$	$[1 - (\gamma_\pi + 2\gamma^2)]e^{-\gamma_0}$	$[1 - (2\gamma_\pi + \gamma^2)](1 - \gamma_0)^2$	$[1 - (\gamma_\pi + 2\gamma^2)]e^{-\gamma_0}$

Table 5.10 Unloaded redundant unit, unrestricted renewal, delayed reswitching

Index	Exponential	All Distributions Lower Bound	All Distributions Upper Bound
T_s	$\frac{1}{\lambda}$	T	T
τ_s	$t_\pi + 0.5\gamma\tau$	$t_\pi + 0.5\gamma\tau$	$t_\pi + 0.5\gamma\tau$
$P_s(t_0)$	$e^{-\gamma_0}$	$P(t_0)$	$P(t_0)$
K	$1 - (\gamma_\pi + 0.5\gamma^2)$	$1 - (\gamma + 0.5\gamma^2)$	$1 - (\gamma_\pi + 0.5\gamma^2)$
$R(t_0)$	$[1 - (\gamma_\pi + 0.5\gamma^2)]e^{-\gamma_0}$	$[1 - (\gamma_\pi + 0.5\gamma^2)](1 - \gamma_0)^2$	$[1 - (\gamma_\pi + 0.5\gamma^2)]e^{-\gamma_0}$

Table 5.11 Unloaded redundant unit, restricted renewal, delayed reswitching

Index	Exponential	All Distributions Lower Bound	All Distributions Upper Bound
T_s	$\frac{1}{\lambda}$	T	T
τ_s	$t_\pi + \gamma\tau$	$t_\pi + 0.5\gamma\tau$	$t_\pi + \gamma\tau$
$P_s(t_0)$	$e^{-\gamma_0}$	$P(t_0)$	$P(t_0)$
K	$1 - (\gamma_\pi + \gamma^2)$	$1 - (\gamma_\pi + 0.5\gamma^2)$	$1 - (\gamma_\pi + \gamma^2)$
$R(t_0)$	$[1 - (\gamma_\pi + \gamma^2)]e^{-\gamma_0}$	$[1 - (\gamma_\pi + 0.5\gamma^2)](1 - \gamma_0)^2$	$[1 - (\gamma_\pi + \gamma^2)]e^{-\gamma_0}$

item that has either an increasing failure rate (IFR) or an increasing failure rate average (IFRA). The upper and lower bounds for the reliability indexes in these tables are given for the two limit cases of IFR distributions, exponential and degenerated (constant value) distributions. In the tables $\gamma = \tau/T$, $\gamma_0 = t_0/T$, and $\gamma_\pi = \tau_\pi/T$. The subscript "s" will denote that the index is for the system, while the "π" subscript will denote the switch.

Table 5.12 Two-unit system, reswitcher that fails in time, undelayed checking of failure state

Note: π superscript or subscript indicates switch parameter. $\gamma = \tau/T$ or $\gamma = \lambda\tau$, $\gamma^{\pi} = \tau_{\pi}/T_{\pi}$ or $\gamma^{\pi} = \lambda_{\pi\tau\pi}$, $\tilde{\gamma}^{\pi} = \tau_{\pi}/T$ or $\tilde{\gamma}^{\pi} = \lambda\tau_{\pi}$, $\gamma_0 = t_0/T$ or $\gamma_0 = \lambda t_0$, $\gamma^* = \tau^*/T$ or $\gamma^* = \lambda\tau^*$.

Index[a]		Loaded		Unloaded	
		$P(t)$, $P_{\pi}(t)$	$e^{-\lambda t}$, $e^{-\lambda_{\pi} t}$	$P(t)$, $P_{\pi}(t)$	$e^{-\lambda t}$, $e^{-\lambda_{\pi} t}$
T_s		$\dfrac{T}{2\gamma + \gamma^{\pi}}$	$\dfrac{1}{\lambda(2\gamma + \gamma^{\pi})}$	$\dfrac{T}{Q(\tau) + \gamma^{\pi}}$	$\dfrac{1}{\lambda(\tau + \gamma^{\pi})}$
	Lower bound	$\dfrac{\tau^*\gamma + 0.5\tau + 0.5\tau_{\pi}\gamma^{\pi}}{2\gamma + \gamma^{\pi}}$	$\dfrac{\tau^*\gamma + 0.5\tau_{\pi}\gamma^{\pi}}{2\gamma + \gamma^{\pi}}$	$\dfrac{\tau^* Q(\tau) + 0.5\tau_{\pi}\gamma^{\pi}}{Q(\tau) + \gamma^{\pi}}$	$\dfrac{\tau^*\gamma + 0.5\tau_{\pi}\gamma^{\pi}}{\gamma + \gamma^{\pi}}$
	Upper bound	$\dfrac{\tau^*\gamma + \tau_{\pi}\gamma^{\pi}}{2\gamma + \gamma^{\pi}}$	$\dfrac{\tau^*\gamma + \tau_{\pi}\gamma^{\pi}}{2\gamma + \gamma^{\pi}}$	$\dfrac{\tau^* Q(\tau) + \tau_{\pi}\gamma^{\pi}}{Q(\tau) + \gamma^{\pi}}$	$\dfrac{\tau^*\gamma + \tau_{\pi}\gamma^{\pi}}{\gamma + \gamma^{\pi}}$
$P_s(t_0)$	Lower bound	$1 - \gamma_0(2\gamma + \gamma^{\pi})$	$e^{-\gamma_0(2\gamma + \gamma^{\pi})}$	$e^{-\gamma_0[Q(\tau) + \gamma^{\pi}]}$	$e^{-\gamma_0(\gamma + \gamma^{\pi})}$
	Upper bound	$e^{-\gamma_0(2\gamma + \gamma^{\pi})}$			
K	Lower bound	$1 - (\gamma\gamma^* + \tilde{\gamma}^{\pi}\gamma^{\pi})$	$1 - (\gamma\gamma^* + \tilde{\gamma}^{\pi}\gamma^{\pi})$	$1 - [\gamma^* Q(\tau) + \tilde{\tau}^{\pi}\gamma^{\pi}]$	$1 - (\gamma^*\gamma + \tilde{\gamma}^{\pi}\gamma^{\pi})$
	Upper bound	$1 - (\gamma\gamma^* + 0.5\tilde{\gamma}^{\pi}\gamma^{\pi})$	$1 - (\gamma\gamma^* + 0.5\tilde{\gamma}^{\pi}\gamma^{\pi})$	$1 - [\gamma^* Q(\tau) + 0.5\tilde{\tau}^{\pi}\gamma^{\pi}]$	$1 - (\gamma^*\gamma + 0.5\tilde{\gamma}^{\pi}\gamma^{\pi})$
$R(t_0)$	Lower bound	$[1 - \gamma_0(2\gamma + \gamma^{\pi})] \times [1 - (\gamma\gamma^* + \tilde{\gamma}^{\pi}\gamma^{\pi})]$	$[1 - (\gamma\gamma^* + \tilde{\gamma}^{\pi}\gamma^{\pi})]e^{-\gamma_0(2\gamma + \gamma^{\pi})}$	$\{1 - [\gamma^* Q(\tau) + \tilde{\gamma}^{\pi}\gamma^{\pi}]\} \times e^{-\gamma_0[Q(\tau) + \gamma^{\pi}]}$	$[1 - (\gamma^*\gamma + \tilde{\gamma}^{\pi}\gamma^{\pi})] \times e^{-\gamma_0(\gamma + \gamma^{\pi})}$
	Upper bound	$[1 - (\gamma\gamma^* + 0.5\tilde{\gamma}^{\pi}\gamma^{\pi})] \times e^{-\gamma_0(2\gamma + \gamma^{\pi})}$	$[1 - (\gamma\gamma^* + 0.5\tilde{\gamma}^{\pi}\gamma^{\pi})] \times e^{-\gamma_0(2\gamma + \gamma^{\pi})}$	$\{1 - [\gamma^* Q(\tau) + 0.5\tilde{\gamma}^{\pi}\gamma^{\pi}]\} \times e^{-\gamma_0[Q(\tau) + \gamma^{\pi}]}$	$[1 - (\gamma^*\gamma + 0.5\tilde{\gamma}^{\pi}\gamma^{\pi})] \times e^{-\gamma_0(\gamma + \gamma^{\pi})}$

[a] One formula is an index; two formulas are bounds.

Table 5.13 Two-unit system, reswitcher that fails in time, monitoring at reswitching time

Note: τ_S superscript or subscript indicates a switch parameter. $\gamma = \tau/T$ or $\gamma = \lambda\tau$, $\gamma^\pi = \tau_\pi/T_\pi$ or $\gamma^\pi = \lambda_\pi\tau_\pi$, $\tilde{\gamma}^\pi = \tau_\pi/T$ or $\tilde{\gamma}^\pi = \lambda\tau_\pi$, $\gamma_0 = t_0/T$ or $\gamma_0 = \lambda t_0$, $\gamma^* = \tau^*/T$ or $\gamma^* = \lambda\tau^*$.

Index[a]		Loaded		Unloaded	
		$P(t)$, $P_\pi(t)$	$e^{-\lambda t}$, $e^{-\lambda_\pi t}$	$P(t)$, $P_\pi(t)$	$e^{-\lambda t}$, $e^{-\lambda_\pi t}$
T_s	Lower bound	$(1 - e^{-2\gamma T_\pi/T})T_\pi$	$\dfrac{1}{\lambda_\pi + 2\gamma\lambda}$	$(1 - e^{-\gamma T_\pi/T})T_\pi$	$\dfrac{1}{\lambda_\pi + \gamma\lambda}$
	Upper bound	$\dfrac{TT_\pi}{T + 2\gamma T_\pi}$		$\dfrac{TT}{T + \gamma T_\pi}$	
	Lower bound	$\min\left\{(1 - e^{-2\gamma T_\pi/T})\tau^* + \tau_\pi e^{-2\gamma T_\pi/T}, \dfrac{TT_\pi}{T + 2\gamma T_\pi}(\gamma^\pi + 2\gamma\gamma^*)\right\}$	$\dfrac{2\gamma\gamma^* + \gamma^\pi}{2\gamma\lambda + \lambda_\pi}$	$\min\left\{(1 - e^{-\gamma T_\pi/T})\tau^* + \tau_\pi e^{-2\gamma T_\pi/T}, \dfrac{TT_\pi}{T + \gamma T_\pi}(\gamma^\pi + \gamma\gamma^*)\right\}$	$\dfrac{\gamma\gamma^* + \gamma^\pi}{\gamma\lambda + \lambda_\pi}$
	Upper bound	$\max\left\{(1 - e^{-2\gamma T_\pi/T})\tau^* + \tau_\pi e^{-2\gamma T_\pi/T}, \dfrac{TT_\pi}{T + 2\gamma T_\pi}(\gamma^\pi + 2\gamma\gamma^*)\right\}$		$\max\left\{(1 - e^{-\gamma T_\pi/T})\tau^* + \tau_\pi e^{-\gamma T_\pi/T}, \dfrac{TT_\pi}{T + \gamma T_\pi}(\gamma^\pi + \gamma\gamma^*)\right\}$	
$P_s(t_0)$	Lower bound	e^{-t_0/T_π} if $t_0 < T_\pi$, 0 if $t_0 \geq T_\pi$	$e^{-t_0(\lambda_\pi + 2\gamma\lambda)}$	e^{-t_0/T_π} if $t_0 < T_\pi$, 0 if $t_0 \geq T_\pi$	$e^{-t_0(\lambda_\pi + \gamma\lambda)}$
	Upper bound	$e^{-2\gamma\gamma_0}$ if $t_0 < T_\pi$, e^{-t_0/T_π} if $t_0 \geq T_\pi$		$e^{-\gamma\gamma_0}$ if $t_0 < T_\pi$, e^{-t_0/T_π} if $t_0 \geq T_\pi$	
K	Lower bound	$1 - \max\left\{\dfrac{\tau^*}{T_\pi} + \gamma^\pi(e^{2\gamma T_\pi/T} - 1)^{-1}, \gamma^\pi + 2\gamma\gamma^*\right\}$	$1 - (\gamma^\pi + 2\gamma\gamma^*)$	$1 - \max\left\{\dfrac{\tau^*}{T_\pi} + \gamma^\pi(e^{\gamma T_\pi/T} - 1)^{-1}, \gamma^\pi + \gamma\gamma^*\right\}$	$1 - (\gamma^\pi + \gamma\gamma^*)$
	Upper bound	$K_\pi = 1 - \min\left\{\dfrac{\tau^*}{T_\pi} + \gamma^\pi(e^{2\gamma T_\pi/T} - 1)^{-1}, \gamma^\pi + 2\gamma\gamma^*\right\}$		$1 - \min\left\{\dfrac{\tau^*}{T_\pi} + \gamma^\pi(e^{\gamma T_\pi/T} - 1)^{-1}, \gamma^\pi + \gamma\gamma^*\right\}$	
$R(t_0)$	Lower bound	$KP_s(t_0)$	$[1 - (\gamma^\pi + 2\gamma\gamma^*)]e^{-t_0(\lambda_\pi + 2\gamma\lambda)}$	$KP_s(t_0)$	$[1 - (\gamma^\pi + \gamma\gamma^*)]e^{-t_0(\lambda_\pi + \lambda)}$
	Upper bound	$KP_s(t_0)$		$KP_s(t_0)$	

[a] One expression is an index; two expressions are bounds.

Table 5.14 Two-unit system, reswitcher with probability q of failure at reswitching time

		Loaded		Unloaded	
Index[a]		$P(t)$	$e^{-\lambda t}$	$P(t)$	$e^{-\lambda t}$
T_s	Lower bound	$\frac{T}{2(\gamma+q)}$	$\frac{T}{2\gamma+q}$	$\frac{T}{Q(\tau)+q}$	$\frac{T}{\gamma+q}$
	Upper bound	$\frac{T}{\gamma+q}$			
τ_s	Lower bound	$\min\left\{\frac{\gamma\tau^*+q\tau_\pi}{\gamma+q}, \frac{2\gamma\tau^*+q\tau_\pi}{2\gamma+q}\right\}$	$\frac{2\gamma\tau^*+q\tau_\pi}{2\gamma+q}$	$\frac{\tau^*Q(\tau)+\tau_\pi q}{Q(\tau)+q}$	$\frac{\gamma\tau^*+\tau_\pi q}{\gamma+q}$
	Upper bound	$\max\left\{\frac{\gamma\tau^*+q\tau_\pi}{\gamma+q}, \frac{2\gamma\tau^*+q\tau_\pi}{2\gamma+q}\right\}$			
$P_s(t_0)$	Lower bound	$e^{-2\gamma_0(\gamma+q)}$	$e^{-\gamma_0(2\gamma+q)}$	$\min\{P(t_0), e^{-\gamma_0[Q(\tau)+q]}\}$	$e^{-\gamma_0(\gamma+q)}$
	Upper bound	$e^{-\gamma_0(2\gamma+q)}$		$\max\{P(t_0), e^{-\gamma_0[Q(\tau)+q]}\}$	
K	Lower bound	$1-2(\gamma\gamma^*+q\tilde{\gamma}^\pi)$	$1-(2\gamma\gamma^*+q\tilde{\gamma}^\pi)$	$1-\left(\frac{\tau^*Q(\tau)}{T}+q\tilde{\gamma}^\pi\right)$	$1-(\gamma\gamma^*+q\tilde{\gamma}^\pi)$
	Upper bound	$1-(2\gamma\gamma^*+q\tilde{\gamma}^\pi)$			
$R(t_0)$	Lower bound	$[1-2(\pi\pi^*+q\tilde{\gamma}^\pi)]e^{-2\gamma_0(\tau+q)}$	$[1-(2\gamma\gamma^*+q\tilde{\gamma}^\pi)] \times e^{-\gamma_0(2\gamma+q)}$	$\left[1-\left(\frac{\tau^*Q(\tau)}{T}+q\tilde{\gamma}^\pi\right)\right]e^{-\gamma_0[Q(\tau)+q]}$	$[1-(\gamma\gamma^*+q\tilde{\gamma}^\pi)] \times e^{-\gamma_0(\gamma+q)}$
	Upper bound	$[1-(2\gamma\gamma^*+q\tilde{\gamma})]e^{-\gamma_0(2\gamma=q)}$			

Note: π superscript or subscript indicates switch parameter.
[a]One expression is an index; two are bounds.

5.3.2 Unreliable Reswitcher

Consider a duplicate system of identical units. A reswitcher used at the moment of operating unit failure is not absolutely reliable. It has two failure modes. There is a constant probability, q, of failure during reswitching. It also can fail with the passage of time and has a failure-free operation probability, $P_\pi(t)$.

In the latter case there are two possibilities: (1) a failure of the reswitcher can be found instantly, or (2) a failure of the reswitcher can be found at the moment of operating unit failure, but the down state leads unavoidably to interruption of continuous operation and system failure.

Tables 5.12 through 5.14 give approximations that are accurate when the mean time to failure of a unit is much longer than the mean time of repair, and the same relationship holds for the reswitcher. Thus $\gamma \ll 1$ and $\gamma_\pi \ll 1$. All distributions $P(t)$ and $P_\pi(t)$ are aging distributions.

5.3.3 Monitoring Depth

Complete and flawless monitoring of both the main unit and the redundant unit sometimes is impossible. The reliability of the duplicated system in such a case will vary with the depth of the monitoring control of both the main and the redundant units and with the regime of renewal of the failed unit.

Table 5.15 has the mean idle time of the system (τ^*) for different cases when $\gamma \ll 1$. Tables 5.16 through 5.22 give indexes for a variety of operating regimes that are shown in schematic diagrams on the tables. In the diagrams, periodic monotoring is shown with blackened or partially blackened symbols. In all of these cases reswitching is absolutely reliable, and at any moment only one unit can perform the system's operation. Failure of an unmonitored part of the main unit always leads to failure of the system and remains undiscovered until the next monitoring.

The tables have the following notation:

Table 5.15 Approximate values of τ^*

Redundant Unit	Renewal Time Distribution	Restricted Renewal	Unrestricted Renewal
Loaded	All	τ	0.5τ (lower) τ (upper)
	exponential	τ	0.5τ
Unloaded	All	0.5τ	0.5τ (lower) τ (upper)
	exponential	τ	0.5τ

Table 5.16 Unmonitored units, simultaneous periodic tests of operating state

Note: $\gamma_0 = t_0/T$, $\gamma^* = \tau^*/T$, $\gamma_\nu = t_\nu/T$ or $\gamma_\nu = 1/\nu T$.

Index		$P(t)$, t_ν	$P(t)$, ν	$e^{-\lambda t}$, t_ν	$e^{-\lambda t}$, ν
T_s	Lower bound Upper bound	$0.5T$ T	$0.5T$ T	$\frac{1}{\lambda}$	$\frac{1}{\lambda}$
τ_s	Lower bound Upper bound	$0.5t_\nu(1+2\gamma^*)$ $t_\nu(1+\gamma^*)$	$t_\nu(1+\gamma^*)$	$0.5t_\nu(1+2\gamma^*)$ $t_\nu(1+\gamma^*)$	$t_\nu(1+\gamma^*)$
$P_s(t_0)$	Lower bound Upper bound	$1-\gamma_0$ $e^{-\gamma_0}$	$1-\gamma_0$ $e^{-\gamma_0}$	$e^{-\gamma_0}$	$e^{-\gamma_0}$
K	Lower bound Upper bound	$1-\gamma_\nu(1+\gamma^*)$ $1-0.5\gamma_\nu(1+2\gamma^*)$	$1-\gamma_\nu(1+\gamma^*)$	$1-\gamma_\nu(1+\gamma^*)$ $1-0.5\gamma_\nu(1+2\gamma^*)$	$1-\gamma_\nu(1+\gamma^*)$
$R(t_0)$	Lower bound Upper bound	$(1-\gamma_0)^2[1-\gamma_\nu(1+\gamma^*)]$ $[1-0.5\gamma_\nu(1+2\gamma^*)]e^{-\gamma_0}$	$(1-\gamma_0)[1-\gamma_\nu(1+\gamma^*)]$ $[1-\gamma_\nu(1+\gamma^*)e^{-\gamma_0}$	$[1-\gamma_\nu(1+\gamma^*)]e^{-\gamma_0}$ $[1-0.5\gamma_\nu(1+2\gamma^*)]e^{-\gamma_0}$	$[1-\gamma_\nu(1+\gamma^*)]e^{-\gamma_0}$

Table 5.17 Unmonitored units, independent periodic tests of operating state

Note: $\gamma = \dfrac{\tau}{T}$ or $\gamma = \dfrac{\lambda}{\mu}$ or $\gamma = \lambda\tau$, $\gamma_0 = t_0/T$, $\gamma^* = \tau^*/T$, $\gamma_\nu = t_\nu/T$ or $\gamma_\nu = 1/\nu T$.

Index		$P(t),\ t_\nu$	$P(t),\ \nu$	$e^{-\lambda t},\ t_\nu$	$e^{-\lambda t},\ \nu$
T_s	Lower bound	$0.5T$	$0.5T$	$\dfrac{1}{\lambda}$	$\dfrac{1}{\lambda}$
	Upper bound	T	T		
τ_s	Lower bound	$0.5t_\nu(1+0.5\gamma_\nu+\gamma^*)$	$t_\nu(1+\gamma_\nu+\gamma^*)$	$0.5t_\nu(1+0.5\gamma_\nu+\gamma^*)$	$t_\nu(1+\gamma_\nu+\gamma^*)$
	Upper bound	$t_\nu(1+\gamma_\nu+\gamma^*)$		$t_\nu(1+\gamma_\nu+\gamma^*)$	
$P_s(t_0)$	Lower bound	$1-\gamma_0$	$1-\gamma_0$	$e^{-\gamma_0}$	$e^{-\gamma_0}$
	Upper bound	$e^{-\gamma_0}$	$e^{-\gamma_0}$		
K	Lower bound	$1-2\gamma_\nu(1+\gamma_\nu+\gamma^*)$	$1-2\gamma_\nu(1+\gamma_\nu+\gamma^*)$	$1-\gamma_\nu(1+\gamma_\nu+\gamma^*)$	$1-\gamma_\nu(1+\gamma_\nu+\gamma^*)$
	Upper bound	$1-0.5\gamma_\nu(1+0.5\gamma_\nu+\gamma^*)$	$1-\gamma_\nu(1+\gamma_\nu+\gamma^*)$	$1-0.5\gamma_\nu(1+0.5\gamma_\nu+\gamma^*)$	
	Lower bound	$(1-\gamma_0)^2 \times[1-2\gamma_\nu(1+\gamma_\nu+\gamma^*)]$	$(1-\gamma_0)[1-2\gamma_\nu \times(1+\gamma_\nu+\gamma^*)]$	$[1-\gamma_\nu(1+\gamma+\gamma^*)]e^{-\gamma_0}$	$[1-\gamma_\nu(1+\gamma+\gamma^*)]e^{-\gamma_0}$
	Upper bound	$[1-0.5\gamma_\nu \times(1+0.5\gamma_\nu+\gamma^*)]e^{-\gamma_0}$	$[1-\gamma_\nu \times(1+\gamma_\nu+\gamma^*)]e^{-\gamma_0}$	$[1-0.5\gamma_\nu \times(1+0.5\gamma_\nu+\gamma^*)]e^{-\gamma_0}$	

Table 5.18 Unmonitored redundant unit, periodic tests of operating state

Note: $\gamma = \tau/T$ or $\gamma = \lambda\tau$ or $\gamma = \lambda/\mu$, $\gamma^* = \tau^*/T$, $\gamma_\nu = t_\nu/T$, $\gamma_0 = t_0/T$.

Index		$P(t)$, t_ν	$P(t)$, ν	$e^{-\lambda t}$, t_ν	$e^{-\lambda t}$, ν
T_s	Lower bound	$\frac{T}{\gamma+\gamma}$	$\frac{T}{\gamma+\gamma}$	$\frac{1}{\lambda(\gamma_\nu+\gamma)}$	$\frac{1}{\lambda(\gamma_\nu+\gamma)}$
	Upper bound	$\frac{T}{0.5\gamma_\nu+\gamma}$		$\frac{T}{0.5\gamma_\nu+\gamma}$	
τ_3		T*	τ^*	τ^*	τ^*
$P_s(t_0)$	Lower bound	$\max\{P(t_0), e^{-\gamma_0(\gamma_\nu+\gamma)}\}$	$\max\{P(t_0), e^{-\gamma_0(\gamma_\nu+\gamma)}\}$	$e^{-\gamma_0(\gamma_\nu+\gamma)}$	$e^{-\gamma_0(\gamma_\nu+\lambda)}$
	Upper bound	$\max\{P(t_0), e^{-\gamma_0(0.5\gamma_\nu+\gamma)}\}$		$e^{-\gamma_0(0.5\gamma_\nu+\gamma)}$	
K	Lower bound	$1-\gamma^*(\gamma_\nu+\gamma)$	$1-\gamma^*(\gamma_\nu+\gamma)$	$1-\gamma^*(\gamma_\nu+\gamma)$	$1-\gamma^*(\gamma_\nu+\gamma)$
	Upper bound	$1-\gamma^*(0.5\gamma_\nu+\gamma)$		$1-\gamma^*(0.5\gamma_\nu+\gamma)$	
$R(t_0)$	Lower bound	$[1-\gamma^*(\gamma_\nu+\gamma)][1-\gamma_0(\gamma_\nu+\gamma)]$	$[1-\gamma^*(\gamma_\nu+\gamma)][1-\gamma_0(\gamma_\nu+\gamma)]$	$[1-\gamma^*(\gamma_\nu+\gamma)e^{-\gamma_0(\gamma_\nu+\gamma)}$	$[1-\gamma^*(\gamma_\nu+\gamma)e^{-\gamma_0(\gamma_\nu+\gamma)}$
	Upper bound	$[1-\gamma^*(0.5\gamma_\nu+\gamma)]e^{-\gamma_0(0.5\gamma_\nu+\gamma)}$	$[1-\gamma^*(\gamma_\nu+\gamma)]e^{-\gamma_0(\gamma_\nu+\gamma)}$	$[1-\gamma^*(0.5\gamma_\nu+\gamma)]e^{-\gamma_0(0.5\gamma_\nu+\gamma)}$	

Table 5.19 Unmonitored main unit, simultaneous periodic tests of operating state

Note: $\gamma = \tau/T$ or $\gamma = \lambda\tau$ or $\gamma = \lambda/\mu$, $\gamma_0 = t_0/T$, $\gamma^* = \tau^*/T$, $\gamma_\nu = t_\nu/T$ or $\gamma_\nu = 1/\nu T$.

Index		$P(t)$, t_ν	$P(t)$, ν	$e^{-\lambda t}$, t_ν	$e^{-\lambda t}$, ν
T_s	Lower bound	$0.5T$	$0.5T$	$\frac{1}{\lambda}$	$\frac{1}{\lambda}$
	Upper bound	T	T		
τ_s	Lower bound	$0.5t_\nu + \gamma\tau^*$	$t_\nu + \gamma\tau^*$	$0.5t_\nu + \gamma\tau^*$	$t_\nu + \gamma\tau^*$
	Upper bound	$t_\nu + \gamma\tau^*$		$t_\nu + \gamma\tau^*$	
$P_s(t_0)$	Lower bound	$1 - \gamma_0$	$1 - \gamma_0$	$e^{-\gamma_0}$	$e^{-\gamma_0}$
	Upper bound	$e^{-\gamma_0}$	$e^{-\gamma_0}$		
K	Lower bound	$1 - 2\gamma_\nu - 2\gamma\gamma^*$	$1 - 2\gamma\nu - 2\gamma\gamma^*$	$1 - \gamma_\nu - \gamma\gamma^*$	$1 - \gamma_\nu - \gamma\gamma^*$
	Upper bound	$1 - 0.5\gamma_\nu - \gamma\gamma^*$	$1 - \gamma_\nu - \gamma\gamma^*$	$1 - 0.5\gamma_\nu - \gamma\gamma^*)$	
$R(t_0)$	Lower bound	$(1 - 2\gamma_0)^2(1 - \gamma_\nu - 2\gamma\gamma^*)$	$(1 - \gamma_0)^2(1 - 2\gamma_\nu - 2\gamma\gamma^*)$	$(1 - \gamma_\nu - \gamma\gamma^*)e^{-\gamma_0}$	$(1 - \gamma_\nu - \gamma\gamma^*)e^{-\gamma_0}$
	Upper bound	$[1 - 0.5\gamma_\nu - \gamma\gamma^*]e^{-\gamma_0}$	$[1 - \gamma_\nu - \gamma\gamma^*)e^{-\gamma_0}$	$[1 - 0.5\gamma_\nu - \gamma\gamma^*)e^{-\gamma_0}$	

Table 5.20 Partially monitored redundant unit, periodic tests of operating state

Note: $\gamma = \tau/T$ or $\gamma = \lambda\tau$ or $\gamma = \lambda/\mu$, $\gamma^* = \tau^*/T$, $\gamma_0 = t_0/T$, $\gamma_\nu = t_\nu/T$ or $\gamma_\nu = 1/\nu T$, $A_H = 2\gamma + \alpha\gamma_\nu$, $A_B = 2\gamma + 0.5\alpha\gamma_\nu$.

Index		$P(t),\ t_\nu$	$P(t),\ \nu$	$e^{-\lambda t},\ t_\nu$	$e^{-\lambda t},\ \nu$
T_s	Lower bound	$\frac{T}{A_H}$	$\frac{T}{A_H}$	$\frac{1}{\lambda A_H}$	$\frac{1}{\lambda A_H}$
	Upper bound	$\frac{T}{A_B}$		$\frac{1}{\gamma A_B}$	
τ_s	Lower bound	$\frac{2\gamma\tau^* + 0.5\alpha t_\nu \gamma_\nu}{A_B}$	$\frac{2\gamma\tau^* + \alpha t\gamma_\nu}{A_H}$	$\tau^* + 0.5\alpha t_\nu \gamma_\nu$	$\tau^* + \alpha\gamma_\nu t_\nu$
	Upper bound	$\frac{2\gamma\tau^* + \alpha t_\nu \gamma_\nu}{A_H}$		$\tau^* + \alpha t_\nu \gamma_\nu$	
$P_s(t_0)$	Lower bound	$\max\{P(t_0), e^{-\gamma_0 A_H}\}$	$\max\{P(t_0), e^{-\gamma_0 A_H}\}$	$e^{-\gamma_0 A_H}$	$e^{-\gamma_0 A_H}$
	Upper bound	$\max\{P(t_0), e^{-\gamma_0 A_H}\}$		$e^{-\gamma_0 A_B}$	
K	Lower bound	$1 - (2\gamma\gamma^* + \alpha\gamma_\gamma^*)$	$1 - (2\gamma\gamma^* + \alpha\gamma_\gamma^*)$	$1 - (2\gamma\gamma^* + \alpha\gamma_\gamma^2)$	$1 - (2\gamma\gamma^* + \alpha\gamma_\gamma^2)$
	Upper bound	$1 - (2\gamma\gamma^* + 0.5\alpha\gamma_\nu^2)$		$1 - (2\gamma\gamma^* + 0.5\alpha\gamma_\nu^2)$	
$R(t_0)$	Lower bound	$[1 - (2\gamma\gamma^* + \alpha\gamma_\nu^2)](1 - \gamma_0 A_H)$	$[1 - (2\gamma\gamma^* + \alpha\gamma_\nu^2)](1 - \gamma_0 A_H)$	$[1 - (2\gamma\gamma^* + \alpha\gamma_\nu^2)](1 - \gamma_0 A_H)$	$[1 - (2\gamma\gamma^* + \alpha\gamma_\nu^2)]e^{-\gamma_0 A_H}$
	Upper bound	$[1 - (2\gamma\gamma^* + 0.5\alpha\gamma_\nu^2)]e^{-\gamma_0 A_s}$	$[1 - (2\gamma\gamma^* + \alpha\gamma_\nu^2)]e^{-\gamma_0 A_H}$	$[1 - (2\gamma\gamma^* + 0.5\alpha\gamma_\nu^2)]e^{-\gamma_0 A_s}$	

Table 5.21 Partially monitored redundant unit, periodic tests of operating state

Note: $\gamma = \tau/T$ or $\gamma = \lambda/\mu$, $\gamma^* = \tau^*/T$, $\gamma_0 = t_0/T$, $\gamma_\nu = t_\nu/T$ or $\gamma_\nu = 1/\nu T$, $A = \alpha + \gamma(2-\alpha)(1-\alpha)$.

Index		$P(t)$, t_ν	$P(t)$, ν	$e^{-\lambda t}$, t_ν	$e^{-\lambda t}$, ν
T_s	Accurate	$\frac{T}{A}$	$\frac{T}{A}$	$\frac{T}{\lambda A}$	$\frac{T}{\lambda A}$
τ_s	Lower bound	$\frac{1}{A}[0.5\alpha t_\nu + \tau^*\gamma(2+\alpha)(1-\alpha)]$	$\frac{1}{A}[\alpha t_\nu + \tau^*\gamma(2-\alpha)(1-\alpha)]$	$\frac{1}{A}[0.5\alpha t_\nu + \tau^*\gamma(2-\alpha)(1-\alpha)]$	$\frac{1}{A}[\alpha t_\nu + \tau^*\gamma(2-\alpha(1-\alpha)]$
	Upper bound	$\frac{1}{A}[\alpha t_\nu + \tau^*\gamma(2-\alpha(1-\alpha)]$		$\frac{1}{A}[\alpha t_\nu + \tau^*\gamma(2+\alpha(1-\alpha)]$	
$P_s(t_0)$	Accurate	$\max\{P(t_0), e^{-\gamma_0 A}\}$	$\max\{P(t_0), e^{-\gamma_0 A}\}$	$e^{-\gamma_0 A}$	$e^{-\gamma_0 A}$
K	Lower bound	$1-[\alpha\gamma_\nu + \gamma\gamma^*(2-\alpha)(1-\alpha)]$	$1-[\alpha\gamma_\nu + \gamma\gamma^*(2-\alpha)(1-\alpha)]$	$1-[\alpha\gamma_0 + \gamma\gamma^*(2-\alpha)(1-\alpha)]$	$1-[\alpha\gamma_\nu + \gamma\gamma^*(2-\alpha)(1-\alpha)]$
	Upper bound	$1-[0.5\alpha\gamma_\nu + \gamma\gamma^*(2-\alpha)(1-\alpha)]$		$1-[0.5\alpha\gamma_\nu + \gamma\gamma^*(2-\alpha)(1-\alpha)]$	
$R(t_0)$	Lower bound	$\{1-[\alpha\gamma_\nu + \gamma\gamma^*(2-\alpha)(1-\alpha)]\}(1-\gamma_0 A)$	$\{1-[\alpha\gamma_\nu + \gamma\gamma^*(2-\alpha)(1-\alpha)]\}(1-\gamma_0 A)$	$\{1-[\alpha\gamma_\nu + \gamma\gamma^*(2-\alpha)(1-\alpha)]\}e^{-\gamma_0 A}$	$\{1-[\alpha\gamma_\nu + \gamma\gamma^*(2-\alpha)(1-\alpha)]\}e^{-\gamma_0 A}$
	Upper bound	$\{1-[0.5\alpha\gamma_\nu + \gamma\gamma^*(2-\alpha)(1-\alpha)]\}e^{-\gamma_0 A}$	$\{1-[\alpha\gamma_\nu + \gamma\gamma^*(2-\alpha)(1-\alpha)]\}e^{-\gamma_0 A}$	$\{1-[0.5\alpha\gamma_\nu + \gamma\gamma^*(2-\alpha)(1-\alpha)]\}e^{-\gamma_0 A}$	

Table 5.22 Partially monitored identical units, simultaneous periodic tests of operating state

Note: $\gamma = \tau/T$ or $\gamma = \lambda/\mu$, $\gamma^* = \tau^*/T$, $\gamma_\nu = t_\nu/T$, $\gamma_0 = t_0/T$, $A_L = \alpha + (1-\alpha)(2\gamma_\nu + \gamma)$, $A_u = \alpha + (1-\alpha)(\gamma_\nu + \gamma)$.

Index		$P(t),\ t_\nu$	$P(t),\ \nu$	$e^{-\lambda t},\ t_\nu$	$e^{-\lambda t},\ \nu$
T_s	Lower bound	$\frac{T}{A_L}$	$\frac{T}{A_L}$	$\frac{1}{\lambda A_L}$	$\frac{1}{\lambda A_L}$
	Upper bound	$\frac{T}{A_u}$		$\frac{1}{\lambda A_u}$	
τ_s	Lower bound	$\frac{1}{A_u}\left\{\frac{t_\nu}{2}[\alpha + (1-\alpha) - \gamma_\nu](1-\alpha)\tau^*(\gamma_\nu + \gamma)\right\}$	$\frac{1}{A_L}\{t_\nu[\alpha + 2(1-\alpha)\gamma_\nu] + (1-\alpha)\tau^*(2\gamma_\nu + \gamma)\}$	$\frac{1}{A_u}\left\{\frac{t_\nu}{2}[\alpha + (1-\alpha)\gamma_\nu + (1-\alpha)\tau^*(\gamma_\nu + \gamma)\right\}$	$\frac{1}{A_L}\{t_\nu[\alpha + 2(1-\alpha)\gamma_\nu] + (1-\alpha)\tau^*(2\gamma_\nu + \gamma)\}$
	Upper bound	$\frac{1}{A_L}\{t_\nu[\alpha + 2(1-\alpha)\gamma_\nu] + (1-\alpha)\tau^*(2\gamma_\nu + \nu)\}$		$\frac{1}{A_L}\{t_\nu[\alpha + 2(1-\alpha)\gamma_\nu] + (1-\alpha)\tau^*(2\gamma_\nu + \gamma)\}$	
$P_s(t)$	Lower bound	$\max\{P(t_0), e^{-\gamma_0 A_L}$	$\max\{P(t_0), e^{-\gamma_0 A_L}$	$e^{-\gamma_0[\alpha + (1-\alpha)(2\gamma_\nu + \gamma)]}$	$e^{-\gamma_0[\alpha + (1-\alpha)(2\gamma_\nu + \gamma)]}$
	Upper bound	$\max\{P(t_0), e^{-\gamma_0 A_L}\}$		$e^{-\gamma_0[\alpha + (1-\alpha)(\gamma_\nu + \gamma)]}$	
K	Lower bound	$1 - \{\gamma_\nu[\alpha + 2(1-\alpha)\gamma_\nu] + (1-\alpha)\gamma^*(2\gamma_\nu + \gamma)\}$	$1 - \{\gamma_\nu[\alpha + 2(1 - \alpha)\gamma_\nu] + (1-\alpha)\gamma^* \times (2\gamma_\nu + \gamma)\}$	$1 - \{\gamma_\nu[\alpha + 2(1-\alpha)\gamma_\nu] + (1-\alpha)\gamma^*(2\gamma_\nu + \gamma)\}$	$1 - \{\gamma_\nu[\alpha + 2(1 - \alpha)\gamma_\nu] + (1-\alpha)\gamma^* \times (2\gamma_\nu + \gamma)\}$
	Upper bound	$1 - \{0.5\gamma_\nu[\alpha + (1 - \alpha)\gamma_\nu] + (1-\alpha)\gamma^* \times (\gamma_\nu + \gamma)\}$		$1 - \{0.5\gamma_\nu[\alpha + (1 - \alpha)\gamma_\nu] + (1 - \alpha)\gamma^*(\gamma_\nu + \gamma)\}$	
$R(t_0)$	Lower bound	$(1 - \{\gamma_\nu[\alpha + 2(1 - \alpha)\gamma_\nu] + (1-\alpha)\gamma^*(2\gamma_\nu + \gamma)\})(1 - \gamma_0 A_L)$	$(1 - \{\gamma_\nu[\alpha + 2(1 - \alpha)\gamma_\nu] + (1-\alpha)\gamma^* \times (2\gamma_\nu + \gamma)\})(1 - \gamma_0 A_L)$	$(1 - \{\gamma_\nu[\alpha + 2(1-\alpha)\gamma_\nu] + (1-\alpha)\gamma^*(2\gamma_\nu + \gamma)\}) \times e^{-\gamma_0[\alpha + (1-\alpha)(2\gamma_\nu + \gamma)]}$	$(1 - \{\gamma_\nu[\alpha + 2(1 - \alpha)\gamma_\nu] + (1-\alpha)\gamma^* \times (2\gamma_\nu + \gamma)\}) \times e^{-\gamma_0[\alpha + (1-\alpha)(2\gamma_\nu + \gamma)]}$
	Upper bound	$(1 - \{0.5\gamma_\nu[\alpha + (1 - \alpha)\gamma_\nu] + (1-\alpha)\gamma^* \times (\gamma_\nu + \gamma)\})e^{-\gamma_0 A_u}$	$(1 - \{\gamma_\nu[\alpha + 2(1 - \alpha)\gamma_\nu] + (1-\alpha)\gamma^* \times (2\gamma_\nu + \gamma)\})e^{-\gamma_0 A_L}$	$(1 - \{0.5\gamma_\nu[\alpha + (1 - \alpha)\gamma_\nu] + (1-\alpha)\gamma^* \times (\gamma_\nu + \gamma)\}) \times e^{-\gamma_0[\alpha + (1-\alpha)(\gamma_\nu + \gamma)]}$	

$P(t)$ = probability of failure-free operation for any distribution function,

$e^{-\lambda t}$ = same probability for an exponential distribution,

t_ν = a constant period of discrete monitoring,

ν = intensity of the discrete monitoring checks, which means that the checks constitute a Poisson process with parameter ν.

All formulas are approximations that are satisfactorily accurate when each of the coefficients γ, γ_0, γ^*, and γ_ν are $\ll 1$. They have been obtained with the heuristic method described in Section 4.5. All results are given without error estimation. Formulas for the mean time between failures, T_s, and the probability of failure-free operation, $P_s(t)$, are given for the stationary phase. The upper and lower bounds are given for aging in accordance with the indicated distributions.

In Tables 5.14–5.21 the following notation is used:

A = an arbitrary d.f. of TTF for main and redundant units, monitoring is performed in constant periods of time t_ν;

B = the same as A but monitoring is performed in a random periods of time with an exponential d.f. with a parameter ν;

C = both units have an exponential d.f. of TTF; monitoring is performed in constant periods of time t_ν;

D = the same as C but monitoring is performed in a random periods of time with an exponential d.f. with a parameter ν;

α = a portion of a unit without continuous monitoring ($\alpha = 1$ means that a unit has no monitoring at all).

Additionally, $D = 2\gamma + \alpha\gamma_\nu$, $L^* = 2\gamma + 0.5\alpha\gamma_\nu$; $N = \alpha + \gamma(2-\alpha)\ (1-\alpha)$; $M = \alpha + (1-\alpha)\ (2\gamma_\nu + \gamma)$ and $M^*\alpha + (1-\alpha)\ (\gamma_\nu + \gamma)$.

CHAPTER 6

ASYMPTOTIC METHODS FOR HIGHLY RELIABLE REPAIRABLE SYSTEMS

6.1 BASIC METHODOLOGY

Special methods can be used to facilitate and improve the analysis of highly reliable systems. The system operation is considered to be a process of periods of operational states (POS) alternating with periods of incomplete operational states (PIOS). The latter are not always states of system failure. System failure can be only a possible future outcome of an incomplete operational state. If the probability of system failure is extremely low, one may use asymptotic methods for reliability analysis of repairable systems.

At first, consider the case when a PIOS is on the average much shorter than a POS. The process can be approximated accurately with a sequence of very narrow "impulses" separated by long pauses. The analyses becomes one of a simple renewal process. If the probability of a PIOS becoming a system failure is extremely low, it can be approximated as being infinitesimal. One can consider the process to be thinned and use the Renyi theorem. The asymptotic point process is expected to be Poisson. Moreover a Poisson process will appear even if different PIOS have different thinning probabilities. The only strong requirement is that all of them tend to 0 uniformly. In addition, if the mean POS is not much larger than the mean PIOS, but there is an extremely low probability that a PIOS will result in system failure, the method can be used successfully.

Let there be two processes in the same stochastic space, a main process and an additional process. In the main process, a new PIOS can appear only after the previous one has ended. In the additional process, two PIOS can overlap. In the same time interval the number of PIOS in the additional process will not be less than the number in the main one. The main process therefore serves as the lower bound of the model. As the probability of formation of a system failure on a PIOS tends to 0, both processes converge. In addition the asymptotic approach gives good results when there are never any repairs between a first unit failure and system failure.

A system may have of a large number of unreliable redundant units but be very reliable because system failure can only result from failure of a very large number of units. Such a system can be modeled as a level crossing by stochastic processes. In such a case there also will be an asymptotic Poisson process of system failures, but the parameter of the process must be computed very differently. Before an asymptotic analysis of a highly reliable system is made, the reason for the high reliability must be determined.

6.1.1 Main Reliability Characteristics of Renewable Systems

Let a system be described by a stochastic process $\zeta(t)$ in a space E. Under some failure criterion, all of the space is divided into two disjoint subsets: E_+ is a subset of system operational states; E_- is a subset of system failure states. Let ξ_k and η_k be the random durations of the POS's and PIOS's, respectively, $k = 1, 2, \ldots$. (Note that the notation in this chapter is different from the previous ones.) At $t = 0$ the system is totally operational.

Let ξ_t be the time interval from the moment t to the first entrance into subset E_-. If the process $\zeta(t)$ is ergodic, then

$$\xi_k \xrightarrow{p} \xi\,, \quad \eta_k \xrightarrow{p} \eta\,, \quad \xi_t \xrightarrow{p} \xi^*\,,$$

where $\xrightarrow{p}$ means a convergence by probability.

In practice the most useful r.v.'s are the random time to the first failure, ξ_0, the random duration of system operability during stationarity, ξ, the random duration of a system failure state during stationarity, η, the random time before the first entrance into the subset E_-, ξ_*, the means of these four r.v.'s,

$$T_0 = E\xi_0\,, \quad T = E\xi\,, \quad \tau = E\eta\,, \quad T^* = \xi^*\,,$$

and an availability coefficient, $K = T/(T + \tau)$.

6.1.2 General Model of Redundant Renewable Systems

Let a repair shop have r facilities and an unrestricted number of waiting places. The flow of failed units in a time interval $(t, t + \Delta t)$ equals $\lambda_k\, \Delta t + o(\Delta t)$ and does not depend on the process behavior before the moment t.

Let any unit have any repair time d.f. $G(t)$. The repair discipline is FIFO and all repair times are mutually independent.

Let

$$E_+ = \{0, 1, \ldots, n\} ,$$
$$E_- = \{n+1, n+2, \ldots\} .$$

A system failure thus is a transition from state n into state $n+1$. The model may be denoted with (λ_k, G, r, n). The model includes practically all standard mathematical models of systems with operational and redundant units that have exponential d.f.'s.

For a repairable redundant system of m operational units, n_1 active redundant units, n_2 underloaded redundant units, and n_3 standby units, $n = n_1 + n_2 + n_3$, one can write

$$\lambda_k = \begin{cases} \lambda_k & \text{if } 0 \le k \le m+n , \\ \lambda(m+n_1) + \lambda'(k-m-n_1) & \text{if } m+n_1+1 \le k \le m+n_1+n_2 , \\ \lambda(m+n_1) + \lambda' n_{n_2} & \text{if } m+n-n_3+1 \le k \le m+n , \\ 0 & \text{if } k \ge m+n+1 . \end{cases}$$

where λ' is the failure rate for an undetermined unit.

After an operational unit fails, its function is taken over by an active redundant unit, whose function in turn is taken over immediately by an underloaded redundant unit. The latter is replaced with a standby unit if permitted by the model rules. After repair the redundant group is replaced in the same order: first, the active redundant unit, second, the underloaded redundant unit, and, last, the standby unit.

6.1.3 Model of a Complex Repairable System

Consider a system of n units. After failure an operational unit goes to a repair shop. After repair it returns into the system. A system state at the moment t is given by a binary vector

$$\mathbf{e}(t) = (e_1(t), e_2(t), \ldots, e_n(t)) ,$$

where

$$e_1(t) = \begin{cases} 0 & \text{if the } i\text{th unit is operable at moment } t , \\ 1 & \text{otherwise} . \end{cases}$$

At the moment t let the process $\mathbf{e}(t)$ be in state $\mathbf{e}$. The probability of failure of the ith unit during the interval $(t, t+\Delta t)$ equals $\lambda_i(\mathbf{e})\,\Delta t + o(\Delta t)$ and does not depend on previously occurred events. Let

$$\lambda(\mathbf{e}) = \sum_{1 \le i \le n} \lambda_i(\mathbf{e}) .$$

If a unit i is in a failure state, $\lambda_i(\mathbf{e}) = 0$.

Let the repair time d.f. for the ith failed unit at the jth repair facility be $G_{ij(t)}$. Any service discipline may be followed, but there are two restrictions:

1. A repair cannot be interrupted by arrivals of another failed unit.
2. If a failed unit arrives at the repair shop and there is a free facility, the repair starts immediately.

Such a model of a complex renewal system is denoted here with ($\lambda_i(\mathbf{e})$, G_v, E_+).

6.2 MARKOV MODELS OF RENEWABLE SYSTEMS

6.2.1 Markov Processes of a General Type

Let an ergodic process, $\psi(t)$, model the behavior of a complex renewable system and let it be Markovian and have a finite (or countable) number of states: $E = \{0, 1, 2, \ldots\}$. Let $\Lambda = \|\lambda_{i,j}\|$ be the transition matrix. To compute the main reliability indexes, one must find

1. Stationary probabilities

$$p_i = \lim_{t\to\infty} Pr(\psi(t) = i)\,, \qquad i = 0, 1, \ldots .$$

2. Transition time from state i into subset A,

$$\theta_i(A) = \inf\{t : \psi(t) \in A \mid \psi(0) = i \notin A\}\,.$$

3. Stationary probability that the process will enter state i when it enters subset A,

$$\pi_i(A) = \lim_{t\to\infty} Pr\{\psi(t+0) = i \in A \mid \psi(t-0) \notin A\}\,.$$

These values are found in a standard way. Stationary probabilities are determined from:

$$p_{ij} = \sum_{0\le i<\infty} p_i \lambda_{ij}\,,$$

$$\sum_{0\le i<\infty} p_i = 1\,.$$

LST's for transition times $\theta_i(A)$ for $i \in \bar{A}$ are found from

$$z\varphi_i(z, A) = \lambda_i(A) + \sum_{j\in A} \lambda_{ij}\varphi_j(z, A)\,, \tag{6.1}$$

where

$$\varphi_i(z, A) = Ee^{-z\theta_i(\bar{A})},$$

$$\lambda_i(A) = \sum_{j \in A} \lambda_{ij},$$

and $\bar{A}$ is the complimentary subset to subset A. Probabilities $\pi_i(A)$ can be found in closed form

$$\pi_i(A) = \frac{\sum_{j \in \bar{A}} p_i \lambda_{ij}}{\sum_{i \in A} \sum_{j \in \bar{A}} p_i \lambda_{ij}}.$$

With these values all main reliability variables and corresponding indexes can be determined:

$$Ee^{-z\xi_0} = Ee^{-z\theta_i(E_-)},$$

$$Ee^{-z\xi} = \sum_{i \in k_+} \pi_i(E_+) Ee^{-z\theta_i(E_-)},$$

$$Ee^{-z\eta} = \sum_{i \in k_-} \pi_i(E_-) Ee^{-z\theta_i(E_+)},$$

$$Ee^{-z\xi^*} = \sum_{i \in E_+} p_i Ee^{-z\theta_i(E_-)}.$$

Notice that if the number of system states is finite, all LST's will be rational fractions and hence have inverses that are easy to find. The mean values can be found by differentiation of the LST's at the point 0. Let $T_i(A) = E\theta_i(A)$. For $i \in A$, (6.1) yields

$$0 = 1 + \sum_{j \in \bar{A}} \lambda_{ij} T_j(A),$$

$$T_0 = T_i(E_-),$$

$$T^* = \sum_{i \in E_+} p_i T_i(E_-).$$

6.2.2 Birth-and-Death Processes

A birth-and-death process is a Markov process with a finite (or countable) number of states, $0, 1, \ldots,$ for which

$$\lambda_{i,i+1} = \lambda_i, \quad \lambda_{i,i-1} = \mu_i, \quad \mu_0 = 0, \quad \lambda_{ij} = 0 \qquad \text{for } |j - i| > 1.$$

Thus every jump is to an adjacent state. This process can be ergodic if and only if

$$\sum_{1 \le i \le \infty} \frac{\lambda_0 \lambda_1 \ldots \lambda_{k-1}}{\mu_0 \mu_1 \ldots \mu_k} < \infty,$$

$$\sum_{1 \le i < \infty} \frac{\mu_1 \mu_2 \ldots \mu_k}{\lambda_0 \lambda_1 \ldots \lambda_{k-1}} = \infty.$$

Stationary probabilities can be found in closed form:

$$p_k = \frac{\lambda_0 \lambda_1 \ldots \lambda_{k-1}}{\mu_1 \mu_2 \ldots \mu_k} p_0 .$$

Let θ_{ij} be the transition time from i to j:

$$\theta_{ij} = \text{infinum}\{t : \psi(t) = j \mid \psi(0) = i\} .$$

With a Markovian process, it follows that

$$\theta_{ij} = \theta_{i,i+1} + \cdots + \theta_{j-1,j} \qquad \text{for } j > i ,$$

$$\theta_{ij} = \theta_{i,i-1} + \cdots + \theta_{j+1,j} \qquad \text{for } j < i ,$$

and all terms in the right-hand sides are mutually independent. The distribution of a r.v. θ_{0k} can be found easily from (6.1) for $A = \{k, k+1, \ldots\}$:

$$Ee^{-z\theta_{0k}} = \frac{1}{\Delta_k(z)} ,$$

where polynomials $\Delta_k(z)$ are given by the recurrent equation:

$$\Delta_{k+1}(z) = \left(1 + \frac{\mu_k + z}{\lambda_k}\right)\Delta_k(z) - \frac{\mu_k}{\lambda_k} \Delta_{k-1}(z) ,$$

where

$$\Delta_{-1}(z) = 0 ,$$

$$\Delta_0(z) = 1 .$$

Polynomials $\Delta_k(z)$ have two following properties: (1) all roots of $\Delta_k(z)$ are negative and different, (2) roots of neighboring polynomials $\Delta_k(z)$ (e.g., $x_1, \ldots, x_k$) and $\Delta_{k+1}(z)$ (e.g., $y_1, \ldots, y_{k+1}$) alternate [i.e., between two neighboring roots of a polynomial $\Delta_{k+1}(z)$ lies a root of a polynomial of $\Delta_k(z)$: $y_1 < x_1 < y_2 < \cdots < y_k < x_k < y_{k+1}$]. These properties permit the easy calculation of all the roots of the polynomial.

For $i < j$ one may write

$$Ee^{-z\theta_{ij}} = \frac{\Delta_i(z)}{\Delta_j(z)} .$$

It is easy also to find:

$$T_{0k} = \sum_{s_0 \le s \le k-1} \frac{1}{\lambda_s b_s} \sum_{0 \le i \le s} b_i ,$$

where

$$b_0 = 1\,,$$

$$b_k = \frac{\lambda_0 \lambda_1 \ldots \lambda_{k-1}}{\mu_1 \mu_2 \ldots \mu_k}\,.$$

Then for $i < j$,

$$T_{ij} = E\theta_{ij} = T_{0j} - T_{0i}\,,$$

$$\xi_0 = \theta_{0,n+1}\,,$$

$$\xi_k = \theta_{n,n+1}\,,$$

$$\eta_k = \theta_{n+1,n}\,.$$

6.2.3 Limit Theorems for r.v.'s with a Rational LST

If a system described by a Markov process $\psi(t)$ is reliable as defined previously, ξ_0, ξ, and ξ^* have asymptotically exponential d.f. This section gives the conditions under which distributions of r.v.'s with rational Laplace-Stieltjes transformations (LST) converge to exponentiality.

Let us introduce the following classes of r.v.'s:

$$Z_0 = \left\{\theta : Ee^{-z\theta} = \prod_{1 \le = k \le n} \frac{1}{1 + a_k z}\,,\ E\theta = 1\right\},$$

$$Z_1 = \left\{\theta : Ee^{-z\theta} = \frac{1}{1 + z + a_2 z^2 + \cdots + a_n z^n}\right\},$$

$$Z_2 = \left\{\theta : Ee^{-z\theta} = \frac{1 + b_1 z + \cdots + b_m z^m}{1 + a_1 z + \cdots + a_n z^n}\,,\ a_1 - b_1 = 1\right\}.$$

Let θ be a r.v. with an exponential d.f.

Analysis of a convergence $\theta \to \theta_0$ leads to comprehension of the convergence of related d.f.'s. The following three relationships are helpful:

1. If $\theta \in Z_0$, $a_2 < 1/4$, then

$$|Pr(\theta > x) - e^{-x}| \le \frac{1 - \sqrt{1 - 4a_2}}{1 + \sqrt{1 - 4a_2}}\,,$$

$$a_2 = 1 - \left(\frac{1}{2}\right) E\theta^2\,. \tag{6.1}$$

2. If $\theta \in Z_1$, then from the condition $a_2 \to 0$, it follows that $\theta \to \theta_0$, and for all $x \ge 0$,

$$|Pr(\theta > x) - e^{-x}| \le c\sqrt{a_2}\,, \qquad c < 1\,.$$

3. If $\theta \in Z_2$ and n is fixed, then from conditions $b_1 \to 0$ and $a_2 \le 0$, it follows that $\theta \to \theta_0$.

Let us show how to use these statements.

1. The LST of the time-to-failure distribution of a birth-and-death process, $\theta_{0,n+1}$, is determined by an equality

$$Ee^{-z\theta_{0,n+1}} = \frac{1}{\Delta_{n+1}(z)} .$$

All roots of $\Delta_{n+1}(z)$ are real and negative; in other words, the normed r.v., $\theta_{0,n+1}/T_{0,n+1}$, and its asymptotic d.f. can be found from (6.1). In this case

$$a_2 = \frac{1}{T^2_{0,n+1}} \sum_{1 \le k \le n} \frac{1}{\lambda_k b_k} \sum_{1 \le s \le k} b_s T_0 s .$$

For the d.f. of a time between failures $\xi_k = \theta_{n,n+1}$, the LST has the form

$$Ee^{-z\theta_{n,n+1}} = \frac{\Delta_n(z)}{\Delta_{n+1}(z)} . \tag{6.2}$$

Let the coefficients of a polynomial be expressed by

$$\Delta_k(z) = 1 + \Delta_{k1} z + \cdots + \Delta_{kk} z^k .$$

Then, by statement 3, for the d.f. of the normed r.v.,

$$\frac{\theta_{n,n+1}}{E\theta_{n,n+1}} ,$$

to converge to an exponential distribution, the following conditions have to be fulfilled:

$$\frac{\Delta_{n1}}{\Delta_{n+1,1}} = \frac{T_{01}}{T_{0,n+1}} \to 0 ,$$

$$a_2 = \frac{\Delta_{n+1,2}}{\Delta^2_{n+1,1}} \to 0 . \tag{6.3}$$

Conditions of the convergence to exponentiality of the distribution of r.v.'s

$$\eta_k = \theta_{n+1,0} \tag{6.4}$$

can be found easily from (6.3).

For convergence of the d.f. of a normed r.v., ξ_*, to exponentially, it is necessary to fulfill (6.3) and the condition that

$$\sum_{n+1 \leq k < \infty} p_k \to 0 ,$$

where p_k are stationary probabilities.

2. Consider now a birth and death process with parameters

$$\lambda_k = \lambda , \quad \mu_k = \mu , \qquad k > 0 ,$$

$$\rho = \frac{\lambda}{\mu} .$$

This process describes the behavior of one operational unit and n standby ones. A r.v., $\theta_{0,n+1}$, is the time to the first failure of this group of units under the condition that all units are fully operational at $t = 0$. It is not too difficult to find

$$a_2 = \frac{(n-1) + (n+2)\rho + \left[\frac{(n+1)(n+2)}{2}\right]\rho^n - (n-1)(n-2)\rho^{n+1} + \left[\frac{n(n-1)}{2}\right]\rho^{n+2}}{[1-(n+1)\rho^n + n\rho^{n+1}]^2} \rho^n .$$

From this expression an asymptotically correct inequality follows: $a_2 < n\rho^n$. Substituting this inequality into the right-hand side of (6.2) yields an asymptotical estimate for the d.f. of the time until the first failure:

$$\left| Pr\left\{ \frac{\theta_{0,n+1}}{T_{0,n+1}} > x \right\} - e^{-x} \right| \leq \frac{1 - \sqrt{1 - 4n\rho^n}}{1 + \sqrt{1 - 4n\rho^n}} .$$

6.3 LIMIT THEOREMS FOR REGENERATIVE MODELS

6.3.1 Statement of the Problem

Let $\psi(t)$ be a regenerative process and let $0 = S_0 < S_1 < \cdots < S_n < \cdots$, where S_i is a regeneration moment, $\theta_n = S_n - S_{n-1}$. During the nth regeneration period an event, A_n, leading to system failure can occur at a random time $S_{n-1} + \kappa_n$, $0 \leq \kappa_n \leq \theta_n$. Let the d.f. of the r.v. κ_n and the probability $q = P(A_n)$ depend neither on n nor on the events outside of the nth period.

Let τ be the time of the first occurrence of event A_n (the first system failure), for example,

$$\tau = S_{\nu-1} + \kappa_\nu ,$$

where subscript $\nu = \inf\{n : I(A_n) = 1, n \geq 1\}$, and $I(\cdot)$ is the indicator of the event $(\cdot)$. It is evident that the r.v. ν has a geometric distribution:

$$P(\nu = k) = q(1-q)^{k-1} , \qquad k \geq 1 .$$

Introduce the following r.v.'s:

$\xi_n = \min(\theta_n, \kappa_n) = I(A_n)\kappa_n + [1 - I(A_n)]\theta_n$.
θ_n^+ = an r.v. having a d.f. $P(\theta_n \leq x \mid I(A_n) = 0) = F^+(x)$.
κ_n^+ = an r.v. having a d.f. $P(\kappa_n \leq x \mid I(A_n) = 1) = F^-(x)$.

The first system failure time τ then has the same d.f. as the random sum

$$\tau' = \theta_1^+ + \cdots + \theta_{\nu-1}^+ + \kappa_\nu^- , \tag{6.5}$$

where $\{\theta_i^+\}$ and $\{\kappa_i^-\}$ are independent sequences of independent and identically distributed (i.i.d.) r.v.'s that are independent of ν. Equation (6.5) shows the relationship between the estimates of the first occurrence time and geometric random sums. To simplify notations, lower indexes of r.v.'s sometimes will be omitted in the following discussion. For example, ξ may be used instead of ξ_n if it is not misleading.

The following are moments for $s > 0$:

$$\mu_s = E\xi^s , \quad T^s = E\theta^s , \quad T_s^+ = E(\theta^+)^s , \quad T_s = E(\kappa^-)^s .$$

The variance of a r.v. ζ is $\sigma^2(\zeta)$. Let

$$W(x) = P(\tau \leq x) = P(\tau' \leq x) . \tag{6.6}$$

This d.f. can be obtained in an explicit form only for simple processes, but it is possible to obtain approximations for it in limit cases for reliable systems when $q \to 0$. The approximations have different forms, such as the form of a limit theorem when we are interested in the limit form of $W(x)$ and the form of estimates of $W(x)$. Any valid estimates must follow limit theorems as well. Quantitative estimates of $W(x)$ are desirable.

6.3.2 Limit Theorems

To formulate the limit theorem, we first define the Khinchine convergence (notation $\xrightarrow{\text{Kh}}$) of the sequence of r.v.'s $\{\zeta_n\}$ to 0: $\zeta_n \xrightarrow{\text{Kh}} 0$ when $n \to \infty$ if

$$\frac{1}{E\{\zeta_n\}} \int_x^\infty Pr\{\zeta_n > u\}\, du \to 0 \qquad \text{for any } x > 0 .$$

Let us consider the following two conditions:

$$\frac{q\xi}{\mu_1} \xrightarrow{\text{Kh}} 0 , \tag{6.7}$$

$$\frac{q\theta}{T_1} \xrightarrow{\text{Kh}} 0 , \tag{6.8}$$

where the Khinchine convergence is uniform in that all parameters of the model (including q and the d.f.'s of different r.v.'s) can change and during these changes the r.v.'s $q\xi/\mu_1$ and $q\theta/T_1$ tend to 0. The following conditions are sufficient for (6.7) and (6.8); they give the possibility of the use of moments of r.v.'s ξ and θ for checking (6.7) and (6.8):

Theorem 6.1

i. If $q\sigma^2(\xi)/\mu_1^2 \to 0$, then (6.7) is true.
ii. If $q\sigma^2(\theta)/T_1^2 \to 0$, then (6.8) is true.

The following limit theorem shows that it is reasonable to use an exponential approximation for the d.f. $W(x)$.

Theorem 6.2

i. Under condition (6.7),

$$Pr\{q\tau/\mu_1 \le x\} = W(x\mu_1/q) \to 1 - \exp(-x) .$$

ii. Under condition (6.8),

$$Pr\{q\tau/T_1 \le x\} = W(xT_1/q) \to 1 - \exp(-x) .$$

6.3.3 Quantitative Estimates of *W*(*x*)

To apply the exponential approximation from Theorem 6.2, we need to know its accuracy. Here we display different estimates of the d.f. $W(x)$ that can be used in practice and follow limit theorems. Let us denote

$$\tau'' = \frac{\theta_1^+ + \cdots + \theta_{\nu-1}^+}{T_1^+} .$$

Then

$$\tau' = T_1^+\tau'' + \kappa_\nu^- . \tag{6.10}$$

Let $N = N(x)$, the number of renewals over the segment $[0, x]$ generated by the sequence $\{(\theta_1^+ + \cdots + \theta_k^+)/T_1^+\}_{k>0}$:

$$N = \max\left\{k : \frac{(\theta_1^+ + \cdots + \theta_k^+)}{T_1^+} \le x\right\} . \tag{6.11}$$

Let

$$H(x) = EN(x) \tag{6.12}$$

be a corresponding renewal function. Let

$$\delta_H = \sup_x (H(x) - x) . \tag{6.13}$$

It is well known that in general

$$\delta_H \le \frac{T_2^+}{(T_1^+)^2} - 1 . \tag{6.14}$$

The estimate (6.10) can be improved in some cases. If θ^+ is an aging r.v., then

$$\delta_H \le 0 . \tag{6.15}$$

From (6.9) it follows that

$$Pr\{\tau'' \le x\} = q + (1 - q)V(x) , \tag{6.16}$$

where

$$V(x) = 1 - E(1 - q)^{N(x)} . \tag{6.17}$$

Let us list several estimates of the function $V(x)$.

Theorem 6.3
The upper estimate of the d.f. $V(x)$ is

$$V(x) \le 1 - (1 - q)^{\delta_H + x} = \bar{V}(x) . \tag{6.18}$$

COROLLARY

$$\sup_x [\bar{V}(x) - (1 - e^{-qx})] \le q \max(\delta_H, 1) \le q \max\left(\frac{T_2^+}{(T_1^+)^2} - 1, 1\right) . \tag{6.19}$$

The inequality (6.19) compares the "pessimistic" estimate $\bar{V}(x)$ with the exponential approximation.

Theorem 6.4
For any $a > 0$ the following "optimistic" estimate is true:

$$V(x) \geq V_{-a}(x) = \begin{cases} \left[1 - \dfrac{e^{-px}}{1-q}\right]_+ & \text{if } x \leq \dfrac{a}{q}, \\ \left[1 - \dfrac{e^{-px}}{1-q} - \dfrac{q}{(a-q)^2}\left(\dfrac{T_2^+}{(T_1^+)^2} - 1\right)\right]_+, & \text{if } x > \dfrac{a}{q}. \end{cases} \tag{6.20}$$

where $[\cdot]_+ = \max(\cdot, 0)$,

$p = q/(1 + Aq)$,

$A = \left[\dfrac{\alpha T_2^+}{(T_1^+)^2} - 1\right]_+$,

$\alpha = \exp(a) - 1 - a)/a^2.$

To apply (6.20), one must calculate an optimal value a for fixed q. These values are the following for $b_m < T_2^+/(T_1^+)^2 \leq b_M$:

b_m	b_M	a_{opt}
1.0	1.0	0.1
1.0	1.1	0.5
1.1	1.2	0.6
1.2	1.4	0.7
1.4	1.8	0.8
1.8	2.2	0.9
2.2	2.7	1.0
2.7	3.5	1.1
3.5	4.9	1.2
4.9	8.0	1.3

COROLLARY

Let

$$\Delta_a = \sup_x \{1 - \exp(-qx) - V_{-a}(x)\}.$$

If $A \leq (1-q)^{-1}$, then

$$\Delta_a \leq \max\left\{ q(1 + Aq),\ \ q\left(\frac{T_2^+}{(T_1^+)^2} - 1\right)(a-q)^{-2} + \exp(-a)\left[\exp\left(\frac{aAq}{1 + Aq}\right)(1-q)^{-1} - 1\right]\right\}. \tag{6.21}$$

If $A > (1-q)^{-1}$, then

$$\Delta_a \le q\left\{\exp\left[-\frac{A-1-Aq}{A+qA^2-qA}\right]+\left[\left(\frac{T_2^+}{T_1^+}\right)^2-1\right](a-q)^{-2}\right\}. \quad (6.22)$$

From (6.19), (6.21), and (6.22) it follows that

$$\sup_x |V(x)-(1-\exp(-qx))| \le q\max\left(1, \frac{T_2^+}{(T_1^+)^2}\right)-1\Big) + o(q). \quad (6.23)$$

The inequality (6.23) shows the accuracy of an exponential approximation of the d.f. $V(x)$. Besides,

$$\sup_x [V(x) - \sup_a \underline{V}_a(x)] \le \frac{qT_2^+}{(T_1^+)^2} + o(q). \quad (6.24)$$

The preceding estimates can be used with (6.6) for estimating the d.f. $W(x)$.

Theorem 6.5
The upper bound for $W(x)$ is given by

$$W(x) \le q + (1-q)\bar{V}\left(\frac{x}{T_1^+}\right), \quad (6.25)$$

where the function $\bar{V}$ is defined in Theorem 6.3.

Theorem 6.6
For any $a > 0$ the estimate

$$W(x) \ge \underline{W}_a(x), \quad (6.26)$$

$$\begin{aligned}\underline{W}_a(x) = &\int_0^x \left[q + (1-q)\underline{V}_a\left(\frac{x-y}{T_1^+}\right)\right]\frac{1}{T_1^-}e^{-y/T_1^-}\,dy \\ &- q\sup_x |F^-(x) - (1-e^{-x/T_1^-})| \\ &- q(1-q)\int_0^\infty |F^-(x)-(1-e^{-x/T_1^-})|\,dx, \end{aligned} \quad (6.27)$$

and $\underline{V}_a(x)$ is defined as in Theorem 6.4.
An approximation, $\tilde{W}(x)$, for the d.f. $W(x)$ is given by

$$\begin{aligned}\tilde{W}(x) = &\frac{1}{1/T_1^- - q/T_1^+}\left[\frac{1}{T_1^-}(1-e^{-qx/T_1^+}) - \frac{1}{T_1^+}q(1-e^{-x/T_1^-}\right) \\ &+ \frac{1}{T^-}q(e^{-qx/T_1^+} - e^{-x/T_1^-})\Big]. \end{aligned} \quad (6.28)$$

Theorem 6.6
Accuracy may be estimated with:

$$\begin{aligned}\sup_x |W(x) - \tilde{W}(x)| &\le q(1-q)\max\left(1, \frac{T_2^+}{(T_1^+)^2} - 1\right) \\ &+ q \sup_x |F^-(x) - (1 - e^{-x/T_1^-})| \\ &+ q(1-q)\int_0^\infty |F^-(x) - (1 - e^{-x/T_1^-})|\, dx\,. \end{aligned} \tag{6.29}$$

One more estimate of $W(x)$ can be used when the summands θ_n^+ and κ_n^- (see 6.5) are almost exponentially distributed r.v.'s.

Theorem 6.7
Let the sum $q(\theta_1^+ + \cdots + \theta_\nu^+)/T_1^+$ (with the d.f. $V(x/q)$) have a density function bounded from above by some value $\bar{p} > 0$. Then

$$\begin{aligned}\sup_x |W(x) - (1 - e^{-qx/T_1^+})| &\le q\Big[\sup_x |F^-(T_1^+ x) - 1 + e^{-x}| \\ &+ (1-q)\int_0^\infty |F^-(T_1^+ x) - 1 + e^{-x}|\, dx\Big] \\ &+ q(1-q)\Big[2\sup_x |F^+(T_1^+ x) - 1 + e^{-x}| \\ &+ (2+\bar{p})\int_0^\infty |F^+(T_1^+ x) - 1 + e^{-x}|\, dx\Big] \\ &+ \max(1, \bar{p})\int_0^\infty \left|\int_0^x (F^+(T_1^+ y) - 1 + e^{-y})\, dy\right| dx \end{aligned} \tag{6.30}$$

One can see that the right-hand side of inequality (6.30) tends to 0 when the d.f.'s F^+ and F^- tend to exponentiality (see Section 6.3.4).

The preceding estimates were obtained under the condition that the regenerating periods have second moments. This is a highly restrictive condition, but it can be weakened. The preceding relations can be improved if we impose more restrictions on regeneration periods.

6.3.4 A Special Regenerative Process

Consider a regenerative process, $\psi(t)$, with a regenerative period consisting of two parts, θ_n^1 and θ_n^2. Thus $\theta_n = \theta_n^1 + \theta_n^2$. Suppose that θ_n^1 and θ_n^2 are independent r.v.'s; the r.v. θ_n^1 (which can be treated as an interval when all elements of the system are in working states) has an exponential d.f.

$$P(\theta_n^1 \le x) = 1 - \exp(-\lambda x)\,, \qquad \lambda > 0\,,$$

the r.v. θ_n^2 (a period when at least one element of the system is being restored) has an arbitrary d.f. with moments

$$m_s = E\{(\theta_n^2)^s\}\,, \qquad s \geq 0\,.$$

Suppose that the event A_n (system failure) can occur only during the second part, θ_n^2. Thus

$$\kappa_n = \theta_n^1 + \kappa_n^2\,,$$

where κ_n^2 is a failure time defined only for the failure subperiod of the regenerative period and starts at the beginning of the second part of θ_n. Neither A_n nor κ_n^2 depend on events outside θ_n^2.

Let

$$G^+(x) = P(\theta_n^2 \leq x \mid I(A_n) = 0)\,,$$

$$G^-(x) = P(\kappa_n^2 \leq x \mid I(A_n) = 1)\,,$$

$$m_s^+ = \int_0^\infty x^s\, dG^+(x)\,,$$

$$m_s^- = \int_0^\infty x^s\, dG^-(x)\,,$$

In particular,

$$T_1^+ = \frac{1}{\lambda} + m_1^+\,,$$

$$T_1^- = \frac{1}{\lambda} + m_1^-\,,$$

Using inequality (6.30) and taking into account that $\bar{p} \leq 1 + \lambda m_1^+$, we obtain the following accuracy estimate for an exponential approximation:

$$\sup_x \left| W(x) - \left(1 - \exp\left(-\frac{qx}{1/\lambda + m_1^+}\right)\right)\right|$$
$$\leq q\{\lambda m_1^-(2 - q + \lambda m_1^+) + (1-q)\lambda m_1^+(1/2(1-q) + 13 + 6\lambda m_1^+)$$
$$+ (1-q)\lambda^2 m_2^+\}/(1 + \lambda m_1^+)\,. \tag{6.31}$$

One can see that for reliable systems the term in the brackets $\{\ldots\}$ is very small and tends to 0 when $\lambda m_1^- \to 0$, $\lambda m_1^+ \to 0$, and $\lambda^2 m_2^+ \to 0$.

6.3.5 Estimation of Probability q

To apply the preceding formulas, one needs to calculate or estimate a value q, the probability of a failure during the current PIOS. Consider a process

$\psi(t)$ with a finite number of states $E = \{0, 1, 2, \ldots\}$. Assume that this set is partially ordered: 0 is the only minimal state, and $\theta \in E_+$. This ordering may be denoted $i < j$.

Let us call $\psi(t)$ a "Markovian to up" process, if the following conditions are fulfilled:

1. For any $i < j$,

$$\begin{aligned} &Pr\{\psi(t+h) = j \mid (\psi(t) = i) \cap (\psi(x) = f(x)), x < t\} \\ &\quad = Pr\{\psi(t+h) = j \mid \psi(t) = i\} \\ &\quad = \lambda_{ij} h + o(h) . \end{aligned} \tag{6.32}$$

2. For any $i \in E_+$ the following inequality is fulfilled

$$0 < \sum_{i<j} \lambda_{ij} = \lambda_i < \infty .$$

3. Let us call an interval free when $\psi(t) = 0$ and busy when $\psi(t) > 0$. Then for any i and j, $i > j$, a probability

$$Pr\{\psi(t+h) = j \mid (\psi(t) = i) \cap (\psi(x) = f(x)), x < t\}$$

depends only on the behavior of the process during the last busy period.

This is a special regenerating process. Let an event, A, correspond to the entrance of $\psi(t)$ into subset E_-. Let the probability of the appearance of event A during a regeneration period equal q. We call the monotonic path,

$$\pi = \{0, i_1, i_2, \ldots, i_m\} ,$$
$$0 < i_1 < i_2 < \cdots < i_m,\ i_1 \in E_+, \ldots, i_{m-1} \in E_+,\ i_m \in E_- ,$$

an increasing sequence of states from the operational state until the first entrance into the subset E_-.

Let $q(\tau)$ be the probability that an event A appears after certain monotonic path. Then $q_0 = \Sigma q(\pi)$, the sum taken over all the monotonic paths, is the probability that event A occurs via any monotonic path.

Let a process, $\psi(t)$, be ergodic. Let the stationary probabilities be denoted with p_i.

$$p_i = \lim_{t \to \infty} Pr\{\psi(t) = i\} . \tag{6.33}$$

The following statement then is fulfilled:

$$q_0 \leq q \leq \frac{1}{\lambda_0 p_0} \sum_{j \in E_-} \sum_{j \in E_+} p_i \lambda_{i,j} . \tag{6.34}$$

EXAMPLE 6.1

Consider a system consisting of two units, one operational and one standby. When the operational one has failed, it goes to a repair shop and the standby instantly replaces it. After repair the unit becomes standby. Repair makes the unit as good as new. The d.f.'s of the random TTF and the random repair time are $F(t) = Pr\{\xi \le t\}$ and $G(t) = Pr\{\eta \le t\}$, respectively. Find an approximate expression for the probability of successful system operation.

Solution

A process, $\psi(t)$, equaling the number of units that have failed by moment t is regenerating except during the first period. The regeneration moments are moments of operational unit failure and replacement with a standby unit. If

$$\frac{E\psi^2}{E(\psi)^2}\, q \to 0 \,,$$

the following asymptotic inequality is valid:

$$Pr\{\tau > t\} \approx e^{-qt/T} \,, \tag{6.35}$$

where

$$q = \int_0^\infty \bar{G}(x)\, dF(x) \,,$$

$$T = \int_0^\infty x\, dF(x) = E\xi \,.$$

EXAMPLE 6.2

Let us change the mathematical model. Both d.f.'s are exponential, but the redundant unit is in an underloaded regime, and its parameter is $\lambda' < \lambda$. Find an approximate expression for the probability of successful system operation.

Solution

Again a process, $\psi(t)$, represents the number of failed units at moment t, but the regeneration moments are those when both units have failed. The first regenerating period is exponentially distributed with parameter $\lambda + \lambda'$. If

$$(\lambda' + \lambda) T_0 = (\lambda' + \lambda) \int_0^\infty \bar{G}(t)\, dt \to 0 \,,$$

then the following asymptotical formula is correct:

$$Pr\{\xi_0 > t\} \approx e^{-(\lambda' + \lambda)qt}, \tag{6.36}$$

where

$$q = \int_0^\infty (1 - e^{-\lambda t})\, dG(t) \approx \lambda T_0 .$$

Finally, for $\lambda T_0 \ll 1$,

$$Pr\{\tau > t\} \approx e^{-(\lambda' + \lambda)\lambda T_0 t} . \tag{6.37}$$

6.4 GENERAL MODEL OF RENEWABLE SYSTEM

6.4.1 Limit Theorems

For a system of the type (λ_k, G, r, n) the following three statements are true:

1. Given

   $$\lim Pr\{\lambda_0 q\tau > x \mid \bar{\lambda} T \to 0\} = e^{-x} ,$$

   where q is the probability of system failure (transition from state n to state $n + 1$) during a single busy period,

   $$\bar{\lambda} = \max_{k \le n} \lambda_k .$$

2. If

   $$\bar{\lambda} \frac{m_{n+1}}{m_1^n} \to 0 ,$$

   where

   $$m_k = \int_0^\infty x^k\, dG(x) ,$$

   then

   $$q \approx \lambda_1 \lambda_2 \ldots \lambda_n \int_0^\infty \frac{x^{n-r}}{(n-r)!} \left[\int_x^\infty \bar{G}(t)\, dt \right]^{r-1} \frac{\bar{G}(x)}{r-1}\, dx . \tag{6.38}$$

3. If

$$\bar{\lambda} \frac{m_{n+1}}{m_1^n} \to 0 \, ,$$

then

$$\lim Pr\{\Lambda \xi_0 > x\} = e^{-x} \, , \tag{6.39}$$

where $\lambda = \Lambda_0 q$ and q is defined as in (6.38).

Two important relationships should be noted:

$$\text{If} \quad r = 1 \, , \quad \text{then} \quad \Lambda = \lambda_0 \lambda_1 \ldots \lambda_n \frac{m_n}{n!} \, .$$

$$\text{If} \quad r = n \, , \quad \text{then} \quad \Lambda = \lambda_0 \lambda_1 \ldots \lambda_n \frac{m_1^n}{n!} \, . \tag{6.40}$$

Statement (6.40) shows asymptotic invariantness: the limit distribution of τ does not depend on the d.f. $G(x)$. The most difficult variable to estimate is the probability q. Asymptotically strict estimates can be found only for boundary cases: $r = 1$ and $r = n$. For these cases the following two statements are true:

1. For the system (λ_k, G, n, n),

$$\underline{q} = \lambda_1 \lambda_2 \ldots \lambda_n \bar{\lambda} \int_0^\infty e^{-\bar{\lambda} x} \frac{\varphi^n(x)}{n} \, dx \le q \le \frac{\lambda_1 \lambda_2 \ldots \lambda_n}{n!} T^n = \bar{q} \, , \tag{6.41}$$

where

$$\varphi(x) = \int_0^\infty \bar{G}(t) \, dt \, .$$

2. For the system $(\lambda_k, G, 1, n)$,

$$q_0 \le q \le q_0 A_{n-1}(\gamma) \, ,$$

$$q_0 = \int_0^\infty p_{1n}(t) \, dG(t) \, , \tag{6.42}$$

where

$$\gamma = \bar{\lambda} \int_0^\infty e^{-\lambda x} \bar{G}(x) \, dx \, ,$$

$$\bar{\lambda} = \max_{k \le n} \lambda_k \, ,$$

$$\underline{\lambda} = \min_{k \le n} \lambda_k \, .$$

$p_{1n}(t)$ is the probability of transition of a death process with parameters λ_k from state 1 to a state higher than n during time t;

$$A_{n-1}(\gamma) = \sum_{0 \le k < \infty} (k+1)^{n-1} \gamma^k . \tag{6.43}$$

6.4.2 Estimation of Reliability Indexes

The preceding reliability indexes give asymptotic and/or two-sided boundaries of the d.f. of a random TTF τ for a model (λ_k, G, r, n). Also of interest are the behavior of ξ_k and η_k and their respective means T and τ. For these values we note the following two statements:

1. If λ_i, n, and r are fixed, and
$$\frac{m_{n+1}}{M_1^n} \to 0 ,$$
then
$$Pr\{\Lambda \xi_k > x\} \to e^{-x} ,$$
where Λ is defined in Section 6.4.1.
2. If $G(x)$, r, n, and λ_{k0} *are fixed and* $\lambda_k \supset \lambda_k^0 \epsilon$, then for $\epsilon \to 0$ uniformly by the subscript k,
$$Pr_k\{\eta > x\} \to \frac{\int_x^\infty [(t-x)^{n-r}/(n-r)!]\bar{\varphi}^{r-1}(t)[\bar{G}(t)/(r-1)!]\, dt}{\int_0^\infty [t^{n-r}/(n-r)!]\bar{\varphi}^{r-1}(t)[\bar{G}(t)/(r-1)!]\, dt} , \tag{6.44}$$
where
$$\bar{\varphi}(t) = \int_t^\infty G(x)\, dx .$$

6.4.3 Any Random Time-Between-Failures (TBF) Distribution

In the preceding section the basic unit failure rate, λ_k, is constant, and the unit failure rate therefore changes only when the number of failed units changes—namely when the individual load on the unit changes after failures of others. If the unit itself is complex enough, its repair may consist in renewal of a small failed part. The random TBF of such a unit can be considered to be exponentially distributed. (See the limit theorems of Khinchine and Grigelionis.) But if the unit is simple and merely wears out with time, an exponential model can lead to mistakes.

Consider a model with one operational unit, n standby ones, and r repair facilities. The unit failure and repair d.f.'s are $F(t) = Pr < \xi < t\}$ and

$G(t) = Pr\{\eta < t\}$, respectively. After repair a unit becomes as good as new. A system failure means failure of all $n + 1$ units.

Let

$$\alpha_k = Pr\{\eta \geq \xi_1 + \cdots + \xi_k\} ,$$

where all ξ_i's are independent and

$$\alpha = \sum_{1 \leq k < \infty} \alpha_k .$$

The following four results then hold:

1. If r and n are fixed and F and G are changing in such a way that $\alpha_n > 0$, $\alpha \rightarrow 0$, $\epsilon_0 > 0$ and $p_0 > 0$, and there are such $\epsilon_0 > 0$ and $p_0 > 0$ that

$$Pr\{\eta < (1 - \epsilon_0)\xi\} \geq p_0 ;$$

then

$$Pr\left\{\frac{\xi^0}{E\xi^0} > x\right\} \rightarrow e^{-x} , \tag{6.45}$$

where ξ^0 is the random time to failure (TTF) of the system.

2. If $r = 1$, n is fixed, and F and G are changing in such a way that $\alpha_1 \rightarrow 0$, $\alpha_1 > 0$, $\epsilon_0 > 0$ and $p_0 > 0$, and there are such $\epsilon_0 > 0$ and $p_0 > 0$ that

$$Pr\{\eta < (1 - \epsilon_0)\xi\} \geq = p_0 ;$$

then (6.46) is valid.

3. If $r = n$ are fixed and d.f.'s F and G are changing so that $\alpha_m > 0$, $\alpha \rightarrow 0$, then (6.45) is valid.
4. If $dF(0)/dt = \lambda > 0$, $G(x) = G_0(x/\epsilon)$, numbers r and n and d.f.'s $F(x)$ and $G_0(x)$ are fixed, and there exists the moment

$$m_{n+1} = \int_0^\infty x^{n+1}\, dG_0(x) , \qquad m_1 = T ;$$

then for $\epsilon \rightarrow 0$,

$$\frac{1}{E\xi^0} = \frac{\lambda^n}{T} \int_0^\infty \frac{x^{n-r}}{(n-r)!} \varphi^{r-1}(x) \frac{1 - G_0(x)}{(r-1)!}\, dx , \tag{6.46}$$

where

$$\varphi(x) = \int_x [1 - G_0(x))]\, dx .$$

6.4.4 Some Results for Nonstationary Renewable Systems

When a unit is a complex device and a repair may consist of replacement of a small component, the flow of its failures (after exclusion of repair periods from the time axis) can be considered to be Poisson. But a unit can consist of a number of components, each of which has an MTTF much larger than the MTTF of the whole redundant group. Then the unit failure process cannot be considered to be a stationary one at the interval of interest. In this case one should use Grigelionis' theorem. By this theorem the newly formed point process is a nonstationary Poisson process with a parameter that changes in time. In case where the parameter changes in time slowly, the interval between two adjacent failures of a unit can be assumed to be constant. In a model for these cases, (λ_k, G, r, n), let $\lambda_k = \lambda_k(t)$ change in time slowly. Then asymptotically the probability of failure-free operation is

$$Pr\{\xi^0 > t\} \approx \exp\left[-\int_0^t \Lambda(x)\, dx\right],$$

where

$$\Lambda(x) = \lambda_0(x) \ldots \lambda_n(x) \int_0^\infty \frac{t^{n-r}}{(n-r)!}\, \bar{\varphi}^{r-1}(t)\, \frac{1-G(t)}{(r-1)!}\, dt\,,$$

$$\bar{\varphi}(t) = \int_t^\infty [1 - G(x)]\, dx\,.$$

6.5 GENERAL MODEL OF A COMPLEX RENEWABLE SYSTEM

6.5.1 General Description

Consider a model $(\lambda_i(e), G_{ij}, E_+)$. A process $e(t)$ is a regenerating process of a special type and also is "Markovian to up."

Let state $e(i)$ be the same as state e, except that the ith unit has failed. A state $e \in E_+$ is called a *boundary* state if there exists an i such that $e(i) \in E_-$. Let Γ_+ be a subset of the boundary states. A monotonic path π is a sequence of states $0 < e_{(1)} < \cdots < e_{(s)}$, where $e_{(k)} \in E_+$, $k < s$, $e_{(s)} \in E_-$, that are occupied during monotonic development of a system failure. A number s is the length of the path. In a transition from state $e^{(k)}$ into state $e^{(k+1)}$ the unit with subscript i_k fails; that is, $e^{(k+1)} = e^{(k)}(i_k)$.

Let us call a monotonic path permissible if

$$\lambda_{i_0}(0)\lambda_{i_1}(e^{(1)}) \ldots \lambda_{i_{s-1}}(e^{(s-1)}) > 0\,.$$

Let $m = \min s$, where the minimum is taken over all monotonic permissible paths; m is a minimal cut of the system. A monotonic permissible path of length m is called a *minimal* path. Π_0 is the set of all monotonic permissible minimal paths. Let $q(\pi)$ be the probability of a failure developing on this path. Let

$$q_0 = \sum_{\pi \in \Pi_0} q(\pi)$$

be the probability of system failure on any path. We also define

$$\lambda(e) = \sum_{1 \le i \le n} \lambda_i(e) ,$$

$$\bar{\lambda} = \max_{e \in E_+} \lambda(e) ,$$

$$G(x) = \min_{i,j} G_{ij}(x) ,$$

$$T = \int_0^\infty x \, dG(x) .$$

6.5.2 Limit Theorems

Given

$$\lim_{\bar{\lambda} T \to 0} Pr\left\{ \frac{\xi_0}{E\{\xi_0\}} > x \right\} = e^{-x} .$$

To estimate $E\{\xi_0\}$, additional conditions must be given. The following is a set of conditions:

- Repair of a failed unit starts immediately if there is a free repair facility.
- Repair of a unit is not interrupted by arrival of another failed unit arrival.
- The number of units n, functions $G_{ij}(x)$, the number of serving facilities, and the service discipline are fixed; $\lambda_i(e) = \lambda_i^0 \epsilon$, where λ_i^0 are fixed and $\epsilon \to 0$.
- The d.f. G has a finite moment of the mth order.

Under the preceding conditions for $\epsilon \to 0$,

$$q \approx \sum_{\pi \in \Pi_0} \frac{\lambda_{i_0}(0)}{\lambda(0)} \lambda_{i_1}(e^{(1)}) \ldots \lambda_{i_{m-1}}(e^{(m-1)}) \mathcal{F}(\pi) , \tag{6.47}$$

where

$$\mathcal{F}(\pi) = \int \cdots \int_{\Lambda} \bar{G}_{i_0 j_0}(x_0) \bar{G}_{i_{s_1} j_{s_1}}(x_{s_1}) \ldots \bar{G}_{i_{s_{I-1}} j_{s_{I-1}}}(x_{s_{I-1}}) \, dx_1 \ldots dx_{m-1} .$$

The integral is taken in a region $x_0 > x_1 \ldots > x_{m-1}$; $i_0, i_1, \ldots, i_{m-1}$, are subscripts of sequentially failed units in the path π; j_k is the subscript of a serving facility for a unit with the subscript i_k, and $1 \leq r$. Facilities are divided into groups when some units can be served only in certain service facilities.

6.5.3 Some Particular Cases

Case 1

Let $G_{ij}(x) = G(x)$. Let any service facility serve any failed unit. Then in (6.47),

$$\mathscr{F}(\pi) = \int_0^\infty \frac{x^{m-r-1}}{(m-r-1)!} \left[\int_x^\infty \bar{G}(t)\, dt \right]^{r-1} \frac{\bar{G}(x)}{(r-1)!}\, dx = \mathscr{F}_{mr} .$$

Case 2

In addition let $\lambda_i(e) = \lambda_i$ and $e_i = 0$. Then

$$\lambda_i(e) = 0 ,$$

$$\lambda = \sum_{1 \leq i \leq n} \lambda_i ,$$

and (6.47) has the form

$$q \sim \mathscr{F}_{mI} \frac{m!}{\lambda} \sum \lambda_{k_1} \lambda_{k_2} \ldots \lambda_{k_m} , \tag{6.48}$$

where $k_1 < k_2 < \cdots < k_m$ are subscripts of failed units, that is, positions in the vector $e_{(M)} \in E_-$ with 0's. The latter sum is taken not by all paths but by all boundary states $e_{(M)}$ for which m is minimal.

Case 3

Let $G_{ij}(x) = G(x)$, and let every redundant unit be a suitable replacement for every operating unit:

$$\lambda_i(e) = \lambda_0 \Big(\sum_{1 \leq i \leq n} e_i \Big) .$$

The subset E_+ is such that any permutation of e_i's kept is inside itself. Then

$$q \sim \frac{(n-1)!}{(n-m)!} \mathscr{F}_{mr} \lambda_0(1) \lambda(2) \ldots \lambda(m-1) .$$

Case 4

Let $G_{ij}(x) = G_i(x)$, $\lambda_i(e) = \lambda_i$. Let $\lambda_i(e) = 0$ if $e_i = 0$. Repair starts instantly after failure. Then

$$\lambda q \sim \sum \lambda_{k_1} \ldots \lambda_{k_m} \tau_{k_1} \ldots \tau_{k_m} \left(\frac{1}{\tau_{k_1}} + \cdots + \frac{1}{\tau_{k_m}} \right),$$

where

$$\tau_k = \int_0^\infty t \, dG(t)$$

and the sum is taken as it is for (6.48).

6.5.4 Two-Sided Boundaries of a Probability *q*

For case 4 in Section 6.5.3 it is possible to determine the following two-sided boundaries $q \leq q'$:

$$\underline{q} = \sum_{e \in T_+} \frac{\lambda_-(\mathrm{e})}{\lambda} \left[1 - \sum_{1 \leq i \leq n} \frac{\lambda m_k}{2\tau_k} e_k \right] \prod_{1 \leq i \leq n} (\lambda_i \tau_i)^{e_i} ;$$

$$\bar{q} = \sum_{e \in \Gamma_+} \frac{\lambda_-(\mathrm{e})}{\lambda} \prod_{1 \leq i \leq n} (\lambda_i \tau_i)^{e_i} ;$$

$$m_k = \int_0^\infty x^2 \, dG_k(x) ,$$

$$\tau_k = \int_0^\infty x \, dG_k(x) ,$$

where

$$\lambda_-(e) = \sum_{k \,:\, e(k) \in E_-} \lambda_k .$$

CHAPTER 7

SYSTEMS WITH NETWORK STRUCTURES

Many real systems have structures that are far more complex than a simple series-parallel or parallel-series structure, such as a complex irreducible network structure. Among these are several different territorially dispersed systems, including communication, transportation, and energy systems. Most are monotonic; a decrease in unit reliability produces a decrease, or at best no change, in system reliability, as is the case with series-parallel and parallel-series structures. Network structures are discussed separately here because of their mathematical properties.

7.1 DEFINITION OF MONOTONIC STRUCTURE

Let us introduce x_i, which is an indicator, or boolean, variable:

$$x_i = \begin{cases} 1 & \text{if the } i\text{th unit is operational ,} \\ 0 & \text{otherwise .} \end{cases}$$

The probability of an operational state of a randomly picked unit is the mean $p_i = Ex_i$, $1 \le i \le n$, where n is the number of units in the system.

Let $\mathbf{X} = (x, \ldots, x_n)$. This n-dimensional vector defines the system state. We will also use a notation

$$\mathbf{X}_i = \{x_1, x_2, \ldots, x_{i-1}, x_{i+1}, \ldots, x_n\} , \tag{7.1}$$

which is the $(n-1)$-dimensional vector that has no ith component. In general, $\mathbf{X}_\alpha$ is a vector $\mathbf{X}$ without this absent component x_i, where $i \in \alpha$. The dimension of the vector is $n - n_\alpha$. If one introduces a system failure criterion, the "structural function of the system" can be defined for all the Boolean system state space:

$$\Phi(\mathbf{X}) = \begin{cases} 1 & \text{if the system is operational}\,, \\ 0 & \text{otherwise}\,. \end{cases}$$

The probability of an operational system state is a mean of the form

$$h = E\Phi(X)\,.$$

This value can be found using a "reliability function," which itself can be defined with unit reliability indexes:

$$h(\mathbf{p}) = h(p_1, \ldots, p_n)\,,$$

where $\mathbf{p} = (p_1, \ldots, p_n)$.

A structure is monotonic if

1. $\Phi(\mathbf{1}) = 1$, where $\mathbf{1} = (1, 1, \ldots, 1)$;
2. $\Phi(\mathbf{0}) = 0$, where $\mathbf{0} = (0, 0, \ldots, 0)$;
3. $\Phi(\mathbf{X}) \geq \Phi(\mathbf{Y})$ if (a) each $x_1 \geq y_2$, $i = 1, \ldots, n$, and (b) there is at least one $x_i > y_i$.

7.2 METHOD OF DIRECT ENUMERATION

An arbitrary system of n units can be in 2^n different states if each unit has two states, operational and failed:

H_0 is the state with all units operational.
H_i is the state with the ith unit failed and all the other units operational.
H_{ij} is the state with the ith and jth units failed and all other units operational.
$H_{1,2,\ldots,n}$ is the state when all system units have failed.

If a criterion of failure is given, the set of all system states can be divided into two subsets, an operational subset, G, and a failure subset, G'. Then for each state H_α a probability is written

$$Pr\{H_\alpha \in G\} = \sum_{H_\alpha \in G} P_\alpha\,. \tag{7.2}$$

If the system consists of a large number of independent units, the probabilities of (7.2) can be calculated using

$$P_0 = \prod_{1\le i\le n} p_i \,,$$

$$P_i = q_i \prod_{\substack{1\le k\le n \\ k\ne i}} p_k = \frac{q_i}{p_i} P_0 = \gamma_i P_0 \,,$$

$$P_{ij} = q_i q_j \prod_{\substack{1\le k\le n \\ k\ne i,j}} p_k = \gamma_i \gamma_j P_0 \,,$$

$$P_{1,2,\ldots,n} = P_0 \prod_{1=i\le n} \gamma_i = \prod_{1\le i\le n} q_i \,, \tag{7.3}$$

where p_i and q_i are the probabilities of operability and failure, respectively, for the ith unit, and $\gamma_i = q_i/p_i$,

$$p_i = p_i(t) = Pr\{\xi_i \ge t\} \,,$$

where ξ_i is a random time to failure (TTF) of the ith unit. For independent units,

$$T = \int_0^\infty P(t)\, dt \tag{7.4}$$

where

$$P(t) = Pr\{\xi \ge t\} \,,$$

and ξ is a random TTF for the system as a whole. If p_i is an availability coefficient for the ith unit, the corresponding probability for the system also will be an availability coefficient.

EXAMPLE 7.1

A bridge structure (Figure 7.1) consists of identical units, each with $p(t) = e^{-\lambda t}$ and $\lambda = 0.01$ per hour. Find the probability of failure-free operation for $t = 10$ hours and find the mean time to failure (MTTF).

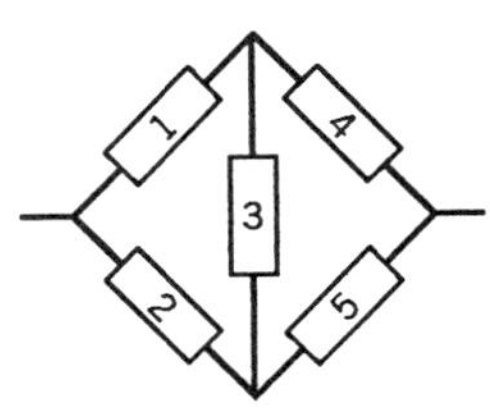

Figure 7.1 System with a bridge structure.

Solution

Construct a table of all possible states of the system (see Table 7.1) and use Figure 2.1 to assign each state to F or G. Thus

$$P(t) = p^5(t) + 5q(t)p^4(t) + 8q^2(t)p^2(t) + 2q^3(t)p^2(t)\,. \tag{7.5}$$

When the value of $P(t)$ is small, it is sometimes helpful to use the relationship $p = 1 - q$ and write

$$\begin{array}{rl} p^5 = & 1 - 5q + 10q^2 - 10q^3 + 5q^4 - q^5 \\ 5qp^4 = & \quad 5q - 20q^2 + 30q^3 - 20q^4 + 5q^5 \\ 8q^2p^3 = & \qquad 8q^2 - 24q^3 + 24q^4 - 8q^5 \\ 2q^3p^2 = & \qquad\qquad 2q^3 - 4q^4 + 2q^5 \\ \hline P = & 1 \quad -2q^2 \quad -2q^3 \quad +5q^4 - 2q^5 \end{array} \tag{7.6}$$

For $q(t = 10) = 1 - e^{-0.1} \approx 0.1$, one has $P(t = 10) = 1 - 2(0.01) - 2(0.001) + \cdots \approx 0.978$.

TABLE 7.1 Simulation results for Example 7.3

1	x_2x_1	0		$x_1x_4x_4$	—
2	x_5x_1	1		$x_5x_3x_3$	—
	x_3x_3	—	1	$x_3x_5x_1$	0
3	x_3x_1	1		$x_2x_3x_2$	—
4	x_2x_4	1		$x_3x_3x_5$	—
5	x_2x_3	1	2	$x_2x_4x_1$	0
	x_2x_2	—		$x_4x_4x_3$	—
6	x_1x_2	0	3	$x_1x_2x_3$	0
7	x_1x_3	1	4	$x_4x_2x_5$	0
8	x_4x_5	0	5	$x_2x_3x_5$	1
9	x_2x_5	1	6	$x_3x_2x_5$	1
10	x_2x_1	0	7	$x_1x_4x_5$	0
11	x_4x_1	1	8	$x_2x_3x_5$	1
	x_3x_3	—	9	$x_4x_3x_5$	0
12	x_4x_3	1	10	$x_5x_1x_3$	0
13	x_5x_2	1	11	$x_3x_4x_5$	0
14	x_3x_4	1	12	$x_5x_2x_4$	0
15	x_1x_2	0	13	$x_5x_1x_2$	0
16	x_1x_2	0	14	$x_2x_5x_1$	0
17	x_1x_3	1	15	$x_5x_1x_4$	0
18	x_1x_3	1	16	$x_5x_2x_1$	0
19	x_3x_2	1	17	$x_3x_4x_2$	0
			18	$x_4x_5x_2$	0
			19	$x_3x_1x_5$	0
			20	$x_2x_5x_1$	0

To compute the MTTF, it is more convenient to represent $P(t)$ as a function of p to compute an integral. In this case

$$P(t) = 2p^5(t) - 5p^4(t) + 2p^3(t) + 2p^3(t) . \quad (7.7)$$

After integration one has

$$T_{\text{syst}} = \frac{2}{5}\,T - \frac{5}{4}\,T + \frac{2}{3}\,T + T = \frac{49}{60}\,T \approx 82\ \text{hr} .$$

7.3 METHOD OF DECOMPOSITION WITH RESPECT TO A SPECIAL UNIT

Occasionally a decomposition rule can be used as in the algebra of logic:

$$h(p) = p_i E\Phi(X_i, x_i = 1) + q_i E\Phi(X_i, x_i = 0) . \quad (7.8)$$

Here $E\Phi(X_i, x_i = 1)$ is the probability of the operational system when unit i is absolutely reliable, and $E\Phi(X_i, x_i = 0)$ is the probability of an operational system when unit i is absolutely unreliable; X_i is as defined in (7.1).

This approach can be used to decompose the function $h(p)$ in the case of several units. For example, for two units one would perform the enumerating procedure

$$\begin{aligned} h(p) &= p_i p_j E\Phi(X_{ij}, x_i = 1, x_j = 1) + p_i q_j E\Phi(X_{ij}, x_i = 1, x_j = 0) \\ &\quad + q_i p_j E\Phi(X_{ij}, x_i = 0, x_j = 1) + q_i q_j E\Phi(X_{ij}, x_i = 0, x_j = 0) . \end{aligned}$$

Remark Unfortunately, there is seldom a good method for identifying absolutely reliable units.

EXAMPLE 7.2

Consider the same bridge structure and find the probability of failure-free operation by decomposing the structure relative to the unit x_3.

Solution

It is easy to find that for $x_3 = 1$ in Figure 7.2,

$$Pr\{\text{success} \mid x_3 = 1\} = (1 - q_1 q_2)(1 - q_3 q_4) .$$

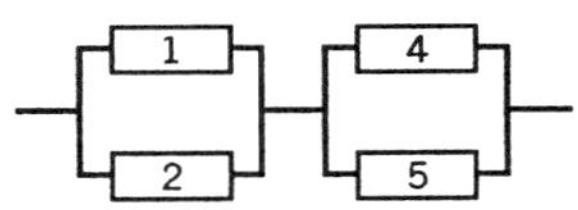

Figure 7.2 Equivalent diagram for a system with a bridge structure when the central unit is absolutely reliable.

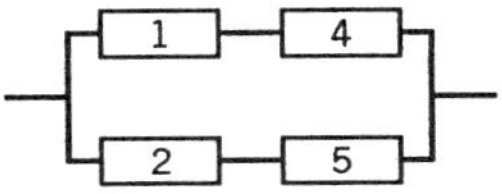

Figure 7.3 Equivalent diagram for a system with a bridge structure when the central unit is absolutely unreliable.

The structure for $x_3 = 0$ is equivalent to the one shown in Figure 7.3:

$$Pr\{\text{success} \mid x_3 = 0\} = 1 - (1 - p_1 p_4)(1 - p_2 p_5)\,.$$

Finally, we can write

$$Pr\{\text{success}\} = p_3(1 - q_1 q_2)(1 - q_3 q_4) + q_3[1 - (1 - p_1 p_4)(1 - p_2 p_5)]\,.$$

For identical units this expression becomes

$$Pr\{\text{success}\} = p(1 - q^2)^2 + q[1 - (1 - p^2)^2]\,.$$

By grouping the terms, it is easy to check that this expression is equivalent to (7.5) through (7.7).

7.4 COMBINED METHOD

None of the methods discussed thus far is effective for analyzing very complex systems with monotonic structures. Consider a communication network for which a Monte Carlo simulation is the most reasonable method of analysis. If the units are very reliable, however, the simulation takes too much computer time. Many millions of simulations may be required before obtaining a high level of confidence. Of course where the units are equally or almost equally reliable, analysis and simulation can be mixed to reduce the time.

Modelling with Fixed Failed States

One can first compute analytically the probabilities of different system states with fixed numbers of failed units. Simulation then can be used to determine conditional probabilities of system failure. The general procedure follows:

1. Analytically calculate probabilities of the $\mathrm{H}_{(k)}$ states, which have exactly k failed units

$$P_{(k)} = \binom{n}{k} q^k p^{n-k}\,. \qquad (7.9)$$

2. Use Monte Carlo simulation to estimate each conditional probability, $\hat{\Phi}_{(k)}$, that a system in state $H_{(k)}$ is operational.

3. Compute the total probability of system operation:

$$P = P_{(0)} + \sum_{1 \le k \le n} \hat{\Phi}_{(k)} P_{(k)} . \tag{7.10}$$

Remark 1 If a system is designed so that failure of any j units never leads to system failure (i.e., the minimum cut of the system has $j+1$ units),

$$P = \sum_{0 \le k \le j} P_{(k)} + \sum_{j+1 < +k < +n} \hat{\Phi}_{(k)} P_{(k)} .$$

The larger that j is, the more valuable is the simulation technique.

Remark 2 If binomial distribution is unimodal, its highest $P_{(k)}$ occurs for only one k. If q is small enough, $P_{(k)}$ begins to decrease very fast, starting from some number k. In this particular case one can find the values of k without using Monte Carlo simulation to estimate $\hat{\Phi}_{(k)}$.

Still, for small k, $\hat{\Phi}_{(k)}$ is close to 1. Minimum cuts are rare, and a summation can be used even for numbers larger than j. We consider these properties when conducting Monte Carlo experiments. We can exclude k edges from the network and check to see if the network is connected. If it is, a "success" is counted. The conditional probability, $\hat{\Phi}_{(k)}$, is estimated with

$$\hat{\Phi}_{(k)} = \frac{\nu(N_k)}{N_k} , \tag{7.11}$$

where N_k is the total number of experiments with k failed edges, and $\nu(N_k)$ is the number of "successes." An algorithm for checking networks is given in the following simple example.

EXAMPLE 7.3

Consider once more the bridge structure illustrated in Figure 7.1 and the parameters given in the previous example. For this structure, $\Phi_0 = \Phi_1 = 1$ because a single failure does not lead to system failure. Nevertheless, failures of any four units always lead to system failure: that is, $\Phi_4 = 0$.

Solution

Values Φ_2 and Φ_3 can be found using a table of random numbers. Let us designate a 0 or 1 as a failure of unit x_1, a 2 or 3 as a failure of unit x_2, a 4 or 5 as a failure of unit x_3, a 6 or 7 as a failure of unit x_4, and an 8 or 9 as a failure of the unit x_5.

We can first randomly select pairs of failed units for the $k = 2$ sample. Suppose that the first nine digits are 2 1 8 1 5 5 5 1 2. The sequence of "failed" units will be x_2, x_1, x_5, x_1, x_3, x_3, x_3, x_1, x_2. We divide the last sequence into pairs, eliminating any with two failures of the same unit. Twenty such experiments with no prohibited combinations are listed in Table 7.1. Each is given a 1 if the pair produces system failure and a 0 otherwise.

For $k = 3$ we take another column of random numbers. For each group of three we make a new set of failed units, again excluding duplicated failures. The results of the 20 experiments appear in Table 7.1. Combinations of unit failures that produce system failure can be readily discerned in Figure 7.1.

We can now estimate Φ_2 and Φ_3 from Table 7.1: $\Phi_2 = 13/20 = 0.65$ and $\Phi_3 = 3/20 = 0.15$. The $P_{(k)}$ probabilities are $P_{(0)} = (0.9)^5 = 0.5905$, $P_{(1)} = 5(0.1)(0.9)^4 = 0.3281$, $P_{(2)} = 10(0.1)^2(0.9)^3 = 0.0729$, and $P_{(3)} = 5(0.1)^3(0.9)^2 = 0.041$. Finally, $P = 0.5905 + 0.3281 + 0.0729 + 0.041 = 0.9722$. The actual value for the example is 0.978. The error results from the very small number of observations in the Monte Carlo experiment used here. A real experiment would have a larger, more adequate number, and the network would be much more complex as well, requiring the aid of a computer.

7.4.2 Simulation of Unit Failure until System Failure Occurs

System edges are excluded one by one until the system has failed. The failure criterion may or may not be loss of connectedness. When the system fails, the number of edges is stored in the computer memory. After a sufficient number of experiments, N, one uses the estimating function

$$\Phi_{(k)} = \frac{1}{N} \sum_{1 \le s \le N} \delta(k_s) , \qquad (7.12)$$

where

$$\delta(k_s) = \begin{cases} 1 & \text{if } k_s \le k , \\ 0 & \text{otherwise} . \end{cases}$$

In this technique, monotonicity is simulated directly; a new edge failure can only deteriorate the system. The final calculation is performed using (7.10).

7.4.3 Accelerated Simulation

The great majority of simulation problems are probably best handled with a direct Monte Carlo simulation. But if the system's units are highly reliable or the system itself is highly redundant, direct simulation may be ineffective, since many simulations would be required. The solution may be to do a simulation of similar but less reliable equipment and follow that with a transformation of the results. In this section we consider this alternative.

Direct Transformation of Input Data

Consider a network cut, u, that includes exactly z units (edges). The frequency of failures from this cut tends to its probability

$$P_u = \prod_{i \not\in u} p_i \prod_{j \in u} (1 - p_j) .$$

If N is a total number of system units, then

$$P_u = \Big(\prod_{1 \le i \le N} p_i \Big) \prod_{j \in u} \frac{1 - p_j}{p_j} . \tag{7.13}$$

Now, instead of a unit with probability p_j, let us use a unit with a lower reliability, p'_j, and write

$$\frac{1 - p'_j}{p'_j} = \gamma \frac{1 - p_j}{p_j} . \tag{7.14}$$

Thus

$$p'_j = \frac{p_j}{p_j + \gamma(1 - p_j)} . \tag{7.15}$$

If transformation (7.14) is made for all units of the network, the failure probability for a fixed cut will become

$$P'_u = \prod_{1 \le i \le N} \frac{p_i}{p_i + \gamma(1 - p_1)} \prod_{j \in u} \frac{\gamma(1 - p_j)}{p_j} . \tag{7.16}$$

Let us define a coefficient of simulation acceleration, δ, that characterizes the increased failure frequency for a cut:

$$\delta = \frac{P'_u}{P_u} = \sum_{1 \le I \le N} \frac{\prod_{j \in u} \gamma}{p_i + \gamma(1 - p_i)} . \tag{7.17}$$

If cut u has z units we may write

$$\delta = \gamma^z \prod_{1 \le i \le N} \frac{1}{p_i + \gamma(1 - p_i)} , \tag{7.18}$$

where the product in the right-hand side is a constant, K, which is characteristic of the system. The accelerating coefficient can be written

$$\delta = \gamma^z K . \tag{7.19}$$

The transformation thus does not change the failure frequency in direct proportion to its weight. Cuts of weight $z + w$ produce $\gamma^{z+w}/\gamma^z = \gamma^w$ times as many failures as cuts of weight z. There is no corresponding reduction, of course, in the number of simulations required for a desired statistical confidence level.

Inverse Transformation

The inverse calculation for the cut u is

$$P_u = \frac{P'_u}{\delta} = \frac{P'_u}{\gamma^z K} . \tag{7.20}$$

The relationship may be used to determine the weight required to achieve a given P_u level when other factors are constant. A cut u with weight z produces γ^{-z}/K failures for each failure produced by the unweighted cut. For m failures in the transformed system, the number for the original system can be estimated with

$$m^* = \frac{1}{K} \sum_{1 \le i < m} \frac{1}{\gamma^{-z_i}} , \tag{7.21}$$

where z_i is the number of failed units for system failure i. After m failures of the transformed system, the following estimate may be made for the original system:

$$P_m = 1 - \frac{m^*}{M} = 1 - \frac{1}{MK} \sum_{1 \le i \le m} \gamma^{-z_i} , \tag{7.22}$$

where M is a total number of experiments.

Optimal Transformation Parameters

Every accelerated simulation requires selection of the γ considered to be best. Equation (7.15) shows that $\gamma = 1$ produces no acceleration. If γ is too large, the heavily weighted cuts may not produce the types of system failures that occur in the actual system. The optimum γ is somewhere in between.

Let all units of the system be identical, and let their reliability index equal p. Let $P(z)$ be the probability of system failure with a cut weight z. The total probability of system failure is

$$Q = \sum_{1 \le z \le N} P(z) . \tag{7.23}$$

Usually for highly reliable systems the value of $P(z)$ changes very smoothly with increasing z. It is therefore possible to use the following expression to predict the cut that will be most probable for the actual system:

$$Ez = \frac{\sum_{1 \le z \le N} zP(z)}{\sum_{1 \le z \le N} P(z)} . \tag{7.24}$$

For a highly reliable system, one can neglect all cuts except the minimal cuts of weight z_0 in (7.24), unless of course the number of cuts of weight $z_0 + 1$ is large. The optimal value can be calculated in this case as

$$\gamma_0 = \frac{z_0 p}{(N - z_0)(1 - p)} . \tag{7.25}$$

There is then a new value for the unit reliability index, $p' = 1 - (z_0/N)$. In effect, for optimal acceleration of the simulation, one should calculate the unit reliability index that corresponds to the most probable cut for the original system.

Acceleration Coefficient

The acceleration of the statistical data collection is

$$\delta_m = \left[\frac{z_0}{(1-p)N}\right]^{z_0} \left(\frac{N - z_0}{pN}\right]^{N - z_0} . \tag{7.26}$$

7.5 ACCURATE CALCULATION METHODS

7.5.1 Method of Power Absorption

Let operator $\mathscr{L}\{f\}$ mean that all powers of arguments p_i in polynomial f are equal to 1. Then we write

$$\mathscr{L}\left\{1 - \prod_{1 \le j \le a} \left(1 - \prod_{i \in A_j} p_i\right)\right\} \le P \le \mathscr{L}\left\{\prod_{1 \le i \le b} \left[1 - \prod_{i \in B_k} (1 - p_i)\right]\right\} \tag{7.27}$$

Here A_j is a set of units included in the ith simple path, B_k is a set of units included in the kth cut, a is the number of all possible paths, and b is the number of all possible cuts.

Formula (7.27) is based on a Boolean presentation of the connectedness of a two-pole network. As will be shown in Section 7.7, the structural function of any two-pole network can be presented as a parallel connection of all possible simple paths or as a series connection of all possible cuts. When using (7.27), one has to remember that each unit of the system has its own probability, p_i, even if all probabilities are equal.

7.5.2 An Algorithm for Absorption of Powers

When multiplying terms R_k and R_m, we consider the relation

$$(1 - R_k) * (1 - R_m) = 1 - R_k - R_m * (1 - R_k) , \tag{7.28}$$

where $*$ means that the operator $\mathscr{L}$ is being used. The last term in (7.28) is given by

$$R_m * (1 - R_k) = R_m(1 - R_{km}) , \tag{7.29}$$

where R_{km} is the product, for all units in path k except the units that are joint with path m, of the probability of a failure-free state.

In general, when R_m is multiplied by some intermediate result, $F(R_1, \ldots, R_{m-1})$,

$$R_m * F(R_1, R_2, \ldots, R_{m-1}) = R_m F(R_{1m}, R_{2m}, \ldots, R_{m-1,m}) . \tag{7.30}$$

If all units have been excluded in a product R_{km}, then we set $R_{km} = 1$.

EXAMPLE 7.4

Figure 7.4 represents a system and has the success probabilities on the edges. Find the probability of connectedness of the system.

Solution

Assume for simplicity that all vertices of the graph are absolutely reliable. Assume also that to connect vertices A and B, we can use any path of length 3 and omit any edges. Thus we consider only paths ab, cdf, cgb, and ahf. The calculation takes the form

$$Q_{AB} = (1 - ab) * (1 - cdf) * (1 - cgb) * (1 - ahf) .$$

Multiplication is performed analogously. If an intermediate result before multiplication by the ith bracket $(1 - R_i)$ is denoted Q_{i-1}, the result, q_i, of a current multiplication can be presented as

$$Q_i = Q_{i-1} * (1 - R_i) = Q_{i-1} - Q_{i-1} * R_i = Q_{i-1,i} - Q_{i-1}R_i ,$$

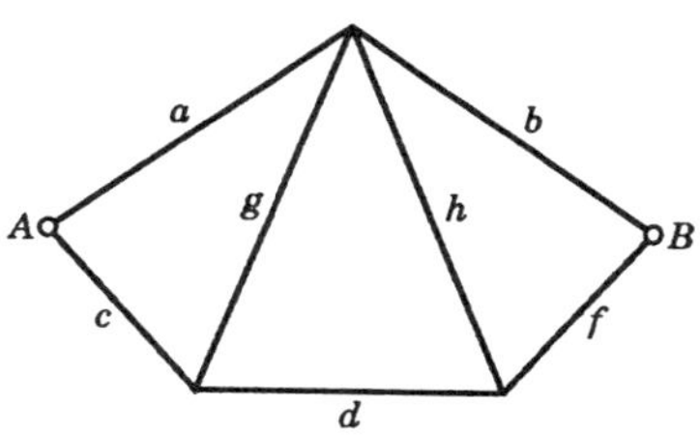

Figure 7.4 Structural diagram of the system in the example.

where $Q_{i-1,i}$ denotes Q_{i-1}, excluding any units that are joint with path i.

The equations become

$$Q_1 = 1 - ab = \overline{ab} ,$$
$$Q_2 = ab - (cdf)(\overline{ab}) = (\overline{ab})(\overline{cdf}) ,$$
$$Q_3 = (\overline{ab})(\overline{cdf}) - (cgb)(\bar{a})(\overline{df}) ,$$
$$Q_{AB} = Q_4 = (\overline{ab})(\overline{cdf}) - (cgb)(\overline{adf}) - (ah\overline{fb})(\overline{cd}) .$$

This transformation is made with basic rules of the algebra of logic:

$$1 - a = \bar{a} , \quad (a)(a) = a , \quad (a)(\bar{a}) = 0 .$$

To decrease the number of computational operations, we would use a standard procedure for obtaining different kinds of minimal forms.

7.5.3 Isotropic Structures

A system is isotropic if all of its units are identical. For a system of n independent identical units, we write

$$P = \sum_{1 \le i \le n} A_i p^{n-i}(1-p)^i , \tag{7.31}$$

where A_i is the number of system operational states when the number of failed units is i. Calculation of a value of P with (7.31) is difficult because of the necessity of finding all A_i coefficients. If we were to express (7.31) only in terms of the values of p, this would have to be done through a complex sequence of sign changes. It is more convenient to use bounds based on a transformation $\eta = q/p$:

$$\frac{1}{(1+\eta)^n} \sum_{0 \le i \le a} A_i \eta^i \le P \le 1 - \frac{1}{(1+\eta)^n} \sum_{0 \le i \le b} B_i \eta^i , \tag{7.32}$$

where

$$B_i = \binom{n}{i} - A_i .$$

In (7.32), B_i is a number of system failure states when there are i failed units, 1 is the number of units in the minimum cut, $a = n - 1$, and $b = n$. For $a = n - 1$, the left-hand side is the accurate expression for P, but for $b = n$, the right-hand side is. Because of the complexity of finding coefficients A_i and B_i for other systems, this method is restricted ordinarily to systems with symmetric structures.

Let H_j be a state when all units of a minimal path A_j, $j = 1, \ldots, r$, are operational. Let R_j be a state when all units of a minimal cut B_j, $j = 1, \ldots, s$, are failed. Then

$$P \leq \sum_{1 \leq j \leq r} P(H_j) - \sum_{1 \leq j_1 < j_2 \leq r} P(H_{j_1} H_{j_2}) + \cdots + \sum_{1 \leq j_1 < \cdots < j_m \leq r} P(H_{j_1}, \ldots, H_{i_m}), \tag{7.33}$$

$$P \geq \sum_{1 \leq j \leq r} P(H_j) - \sum_{1 \leq j_1 < j_2 \leq r} P(H_{j_1} H_{j_2}) + \cdots - \sum_{1 \leq j_1 < \cdots < j_{m+1} \leq r} P(H_{j_1}, \ldots, H_{i_{m+1}}), \tag{7.34}$$

$$P \leq 1 - \sum_{1 \leq j \leq s} P(R_j) + \sum_{1 \leq j_1 < j_2 \leq s} P(R_{j_1} R_{j_2}) - \cdots + \sum_{1 \leq j_1 < \cdots < j_{m+1} \leq s} P(R_{j_1}, \ldots, R_{i_m}), \tag{7.35}$$

$$P \geq 1 - \sum_{1 \leq j \leq s} P(R_j) + \sum_{1 \leq j_1 < j_2 \leq s} P(R_{j_1} R_{j_2}) - \cdots - \sum_{1 \leq j_1 < \cdots < j_m \leq s} P(R_{j_1}, \ldots, R_{i_m}), \tag{7.36}$$

where $P(H_j)$ and $P(R_j)$ are the probabilities of H_j and R_j, and $m = 2k + 1$. Estimations (7.38) and (7.39) are preferable for small probabilities p_i, and estimations (7.35) and (7.36) are better for large p_i.

7.6 SYSTEMS WITH RECURRENT STRUCTURES

Recurrent structures have two main properties:

1. The structures consist of subsystems that are similar in structure or at least are constructed by the same structural rules.
2. Similar units are in similar structural positions.

A network system has vertices and edges that can fail. In communication systems they are the communication nodes and channels. The nodes can be terminals or retranslation centers. We will assume that failure of a node leads to failures of all connected channels. Of course, in practice these simplified hypothetical structures are rare, but they can be useful for qualitative analysis of real systems.

7.6.1 Notation and Main Terms

A system can be diagramed as a set of nodes and channels. Channels can be oriented (noted with arrows) or unoriented. Channel numeration should be clear in the diagrams. To describe the size of a system, one introduces a system rank. The structure of the construction is strictly and formally recurrent: the structure of a rank n system is determined by the structure of the rank $(n-1)$ system.

EXAMPLE 7.5

The system in Figure 7.5 is developed recurrently. An additional system object with number n is introduced with channels that connect it with the object introduced at the previous step, $n-1$, and with objects I, II, and 1. Then the connection channel between object 1 and object $n-1$ is deleted.

Let $[\alpha, \beta]$ be a channel between objects α and β. The probability that α and β are connected is the probability that α and β are themselves operational and that there exists at least one path connecting them. If there are connections from α to β and from β to α, the connection is defined as "strong." These relationships help define the probability of interconnection of subset A of system units. The probability that the system is failure free is the probability that all system objects (nodes) are operational.

7.6.2 Isotropic Systems with Recurrent Structures

Notation

$Q\{\alpha\}$ = failure probability of object α.

$Q\{\alpha, \beta]\}$ = failure probability of channel $[\alpha, \beta]$.

$s = 1 - r$, and $\hat{s} = 1 - \hat{r}$ = actual and estimated object failure probabilities, respectively:

$$q = 1 - p\,, \quad \hat{q} = 1 - \hat{p}\,, \quad z = 1 - y\,, \quad \sigma = 1 - \hat{\pi}\,.$$

$\hat{\sigma} = 1 - \hat{\pi}$ = probability of channel failure.

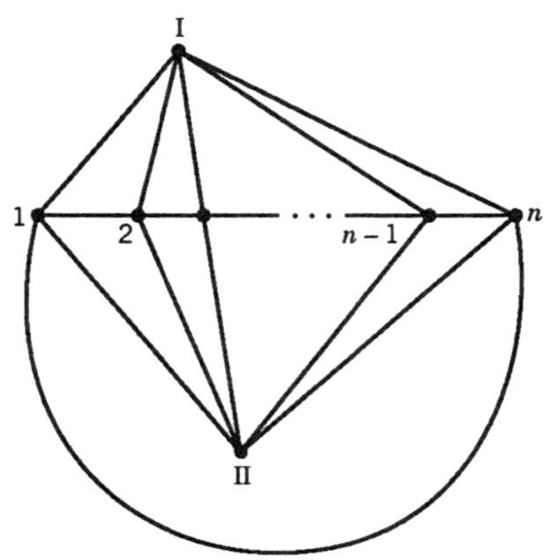

Figure 7.5 Closed recurrent radial-ring system.

$P_n\{\alpha, \beta\}$, and $P_n\{\alpha, \beta, \gamma\}$ are probabilities of existence of a path between α and β (in the first case) or between α and β, α and γ, and γ and β (in the second case) in a rank n system.

$P_n\{\alpha, m \div k\}$ = probability of a connection between object α and objects $m, m+1, \ldots, k-1, k$.

Π_n = probability of a failure-free state of a rank n system.

$P_\infty\{\alpha, \beta\}$ = asymptotic value of the probability of connection of objects α and β in a system of infinite rank.

$P_{-\infty,\infty}\{\alpha, I\}$ = asymptotic value of the probability of connection between object I and object α located in the "middle" of a system of infinite rank.

$\Phi_n^{(\alpha)}(t)$ = generating function for a number of objects with which an object α is connected in a system of the rank n:

$$\Phi_n^{(\alpha)}(t) = \sum_{0 \le i \le \infty} Pr(\xi_n = k\} t^k , \tag{7.37}$$

where $P(\xi_{n=k}\}$ is the probability that exactly k objects of a rank n system are connected with an object α, $\Phi_n^{(\alpha)}(\tau)$ is a generating function in a point τ; $M_n^{(\alpha)}$ and $D_n^{(\alpha)}$ are the mean and variance, correspondingly, of a number of objects connected with an object α in a system of rank n.

7.6.3 Isotropic System, Closed Radial-Ring Structure (Figure 7.6)

Unit Failure Probabilities

$$Q\{I\} = 0 , \quad Q\{|I, i|\} = q = 1 - p ;$$
$$Q\{|i-1, i|\} = Q\{|n, 1|\}\sigma = 1 - \pi , \qquad \{i-1, i\} \in (1, n)^- .$$

System Characteristics (see Section 7.6.2)

$$P_n\{I, m\} = \frac{1}{n} E_n^{(1)} = r - rq \frac{(1 - r\pi)^2}{(1 - r\pi q)^2} - rp \frac{2 - r\pi - r\pi q}{(1 - r\pi q)^2} (r\pi q)^n$$
$$- (n-1) r\pi p \frac{s + r\sigma q}{1 - r\pi q} (r\pi q)^{n-1} , \tag{7.38}$$

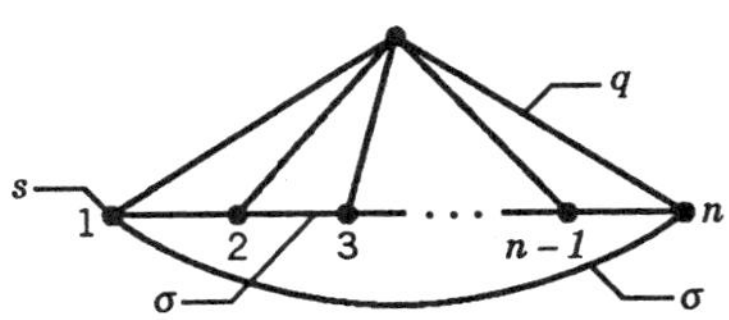

Figure 7.6 Isotropic closed recurrent radial-ring system.

$$\Pi_n = \mu_1^n + \mu_2^n - 2(r\pi q)^n , \tag{7.39}$$

where

$$\mu_{1,2} = \frac{r}{2}\,(p + 2\pi q \pm \sqrt{p^2 + 4\pi\sigma pq}) .$$

If $4\pi\sigma q \ll p$, $\mu_1 \approx r(1-\sigma^2 q)$, and $\mu_2 \approx r\pi^2 q$:

$$\Phi_n(t) = \lambda_1 + \lambda_2 + \lambda_3 - 2(rgt)^n ,$$

where λ_i's are roots of the equation $f(\lambda) = 0$,

$$f(\lambda) = \lambda^3 - (1 - rp + rpt + 2r\pi qt)\lambda^2 + r\pi qt(2 - r\pi p + r\pi t)\lambda \\ - r^2\pi^2 qt(qt + spt - sp) .$$

The asymptotic value of the variance is the same as in the previous section.

7.6.4 Isotropic System with Open Radial-Ring Structure (Figure 7.7)

Unit Failure Probabilities

$$Q\{I\} = 0 , \quad Q\{i\} = s = 1 - r , \quad Q\{|I, i|\} = Q\{|i; I|\} = q = 1 - p ;$$
$$Q\{|i-1, i|\} = Q\{|i, i-1|\} = \sigma = 1 - \pi , \qquad \{i-1, i\} \in (1, n) .$$

System Characteristics (see Section 7.6.2)

$$P_n\{I, i\} = rp\,\frac{1-(r\pi q)^n}{1 - r\pi q} . \tag{7.40}$$

$$P_n\{I, m\} = P_n\{m, I\} = P_m\{I, 1\} + (r + P_m\{I, 1\})\pi P_{n-m}\{I, 1\} . \tag{7.41}$$

$$P_n\{I, 1; 1, I\} = \frac{rp^2(1 + r\pi q)}{(1 - r\pi q)(1 - r^2\pi^2 q^2)} - \frac{2rp^2(r\pi q)^n}{(1 - r\pi q)(1 - \pi q)} \\ - \frac{rp^2(1 + \pi q)(r\pi^2 q^2)^n}{(1 - r\pi^2 q^2)(1 - \pi q)} \tag{7.42}$$

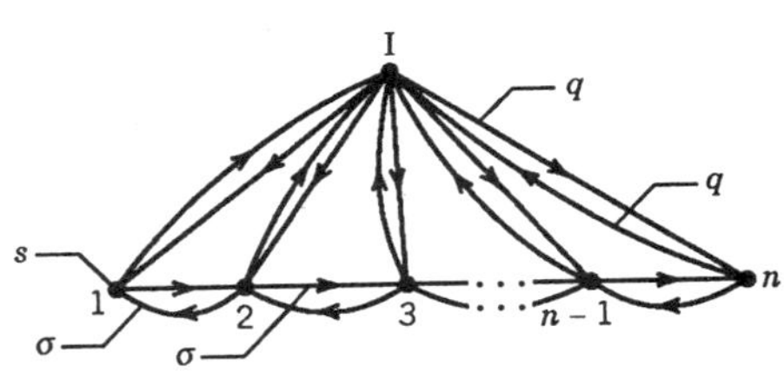

Figure 7.7 Symmetrical isotropic disclosed recurrent radial-ring system.

$$P_n\{I, m; m, I\} = rp^2 \frac{(1 - r\pi^2 q^2)(1 - r^2\pi^2 q^2)^2}{(1 - r\pi q)^4 (1 - r\pi^2 q^2)^2} + \frac{2rs\pi^2 p^2 q}{(1 - r\pi q)^2 (1 - r\pi^2 q^2)} + 0[(r\pi q)^{m-1}] + 0[(r\pi q)^{n-m}] \tag{7.43}$$

$$\Pi_n = r^n p^2 \sum_{1 \le i \le 3} \left(\left[\frac{\partial \varphi}{\partial \mu} \right]_{\mu = \mu_i} \right)^{-1} (\mu_i - \pi q)\mu_i^n , \tag{7.44}$$

where μ_i's are roots of the equation $\varphi(\mu) = 0$,

$$\varphi(\mu) = \mu^3 - [q\pi(1 + \pi q) + (1 - q\sigma)^2]\mu^2 + \pi q[\pi^2(1 + q^2 + q\pi) - p\sigma(p\sigma - 2q\pi^2)]\mu - q^3\pi^5 .$$

7.6.5 Isotropic System with a Complete Unoriented Structure

Figure 7.8 shows a structure of the fifth rank. To determine unit failure probabilities, we write

$$Q\{I\} = 0 , \quad Q\{i\} = s = 1 - r , \quad Q\{|I, i|\} = Q\{|j, i|\} = q = 1 - p ,$$
$$Q\{|j, i|\} = \sigma = 1 - \pi , \qquad \{i, j\} \in (1, n) .$$

System Parameters (see Section 7.6.2)

$$\Pi_n = r^n - \sum_{0 \le k \le n-1} \binom{n}{k} \sigma^k (rg)^{n-k} \Pi_k , \qquad \Pi_0 = 1 .$$

$$\Phi_n^{(1)}(t) = \sum_{0 \le k \le n} t^k \binom{n}{k} (s + rq\sigma)^{n-k} \Pi_k , \qquad \Phi_0 = 1 .$$

$$M_n^{(1)}(t) = n \sum_{0 \le k \le n} t^k \binom{n-1}{k-1} (s + rq\sigma)^{n-k} \Pi_k = nP_n\{I, m\} ,$$

$$\frac{\partial^m \Phi_n^{(1)}(t-1)}{\partial t^m} = n(n-1) \ldots (n-m+1) \sum_{m \le k \le n} \binom{n-m}{k-m} (s + rq\sigma^k)^{n-k} \pi_k$$
$$= n(n-1) \ldots (n-m+1) P_n\{I + m\} ,$$

where $P_n\{I + m\}$ is the probability that object I is connected with m given objects in a system of rank n:

$$D_n^{(1)} = n^2 \sum_{2 \le k \le n} t^k \binom{n-2}{k-2} (s + rq\sigma)^{n-k} \Pi_k + n \sum_{1 \le k \le n-1} \binom{n-2}{k-1} (s + rq\sigma^k)^{n-k} \Pi_k - (M_n^{(1)})^2 ,$$

$$\Pi_n \approx r^n e^{-(n+1)q\sigma^n} + 0\left(\frac{1}{n}\right) .$$

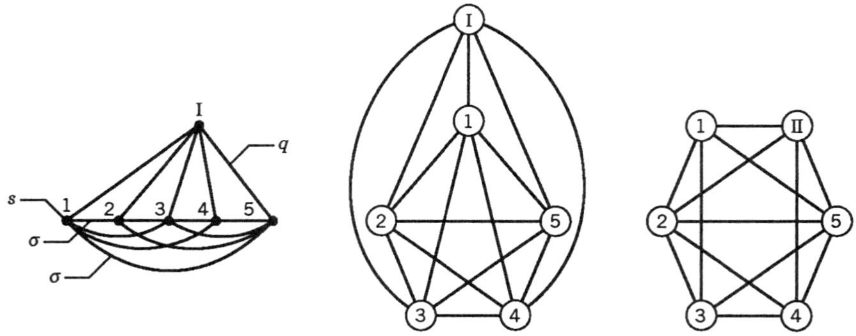

Figure 7.8 Isotropic, totally connected, nonoriented system (three variants of presentation).

7.7 UPPER AND LOWER BOUNDS OF TWO-POLE NETWORKS

7.7.1 Main Definitions and Notations

This section considers two-pole networks with absolutely reliable vertices and arbitrary edges. The symbols of logical algebra are used: $\bar{x}$ = not x, "1" is an exclusion, "1" means true.

A minimum path in a two-pole network is a set of units, X, that connects the two poles and has no subset, $X' \subset X$, that possesses this property. A minimum cut in a two-pole network is a set of units, X, that disconnects the two poles when all its units fail and that has no subset, $X' \subset X$, that possesses this property. In this section a minimum set has a limited definition with respect to a set power. A network can have many different minimum paths and cuts. In graph theory these usually are called *simple paths* and *simple cuts*.

Let the probability of an operational state of the ith unit of the system $= p_i$. Let us denote the jth minimum path as A_j, $j = 1, 2, \ldots, s$, and the kth minimum cut as B_k, $k = 1, 2, \ldots, r$.

Each minimum path A_j is defined by its structural function:

$$\alpha_j(x) = \bigcap_{i \in A_j} x_i , \tag{7.45}$$

which has the value 1 if all the path units are in an operational state and has value 0 otherwise. Thus $\alpha_j(X)$ represents a series connection of its units. Notice that the only structure with exactly one minimal path is a strictly series structure.

Analogously, any minimum cut B_k has its structural function

$$\beta_k(x) = \overline{\bigcap_{i \in B_k} \bar{x}_i} = 1 - \bigcap_{i \in B_k} (1 - x_i) , \tag{7.46}$$

which equals 0 if all the units of the cut are failed, and equals 1 if at least one unit is in an operational state. Thus a minimum cut function, $\beta_k(X)$, is a structural function of a parallel connection of its units. In general, a system can have several minimum cuts. The only system with exactly one minimum cut is a strictly parallel connection of system units.

7.7.2 Esary-Proschan Bounds

The structural function of any system can be written with its minimum paths and cuts in the following way:

$$\varphi(X) = \bigcap_{1 \le j \le = l} \alpha_j(X) = 1 - \bigcap_{i \le j \le l} [1 - \alpha_j(X)] . \tag{7.47}$$

Alternatively,

$$\varphi(X) = \bigcap_{1 \le k \le s} \beta_k(X) . \tag{7.48}$$

EXAMPLE 7.6

For the bridge in Figure 7.1 the structural functions of the paths and cuts are

$$\begin{aligned}
\alpha_1 &= x_1 x_2 , & \beta_1 &= 1 - (1 - x_1)(1 - x_4) ; \\
\alpha_2 &= x_4 x_5 , & \beta_2 &= 1 - (1 - x_2)(1 - x_5) ; \\
\alpha_3 &= x_1 x_3 x_5 , & \beta_3 &= 1 - (1 - x_1)(1 - x_3)(1 - x_5) ; \\
\alpha_4 &= x_2 x_3 x_4 , & \beta_3 &= 1 - (1 - x_2)(1 - x_3)(1 - x_4) .
\end{aligned}$$

The upper and lower bounds for the probability of two-pole network connectedness is

$$\prod_{1 \le k \le s} \left[1 - \prod_{i \in B_k} q_i\right] \le \varphi(X) \le 1 - \prod_{1 \le j \le r} \left(1 - \prod_{i \in A_j} p_i\right) . \tag{7.49}$$

If all units have exponential failure distributions, the system MTTF can be written in closed form:

$$T \le \sum_{1 \le j \le n_\alpha} T_{\alpha_j} - \sum_{1 \le j < k \le n_\alpha} \frac{1}{\Lambda_{\alpha_j} + \Lambda_{\alpha_k}} + \sum_{1 \le j < k < l \le n_\alpha} \frac{1}{\Lambda_{\alpha_j} + \Lambda_{\alpha_k} + \Lambda_{\alpha_l}} + \cdots + (-1)^{n_\alpha + 1} \frac{1}{\sum_{1 \le j \le n_\alpha} \Lambda_{\alpha_j}} , \tag{7.50}$$

where

$$T_{\alpha_j} = \frac{1}{\sum_{j \in \alpha_j} \lambda_i}$$

is the MTTF of the jth path and

$$\Lambda_{\alpha_j} = \frac{1}{T_{\alpha_j}} .$$

EXAMPLE 7.7

Let the bridge in Figure 7.1 have identical and independent units with $p_i(t) = e^{-\lambda t}$ and $\lambda = 0.01$ per hour. Find the probability of failure-free operation for $t_0 = 10$ hours and find the upper bound of the system MTTF.

Solution

The minimum paths are

1. (x_1, x_4),
2. (x_1, x_3, x_5),
3. (x_2, x_5),
4. (x_2, x_3, x_4).

The minimum cuts are

1. (x_1, x_2),
2. (x_1, x_3, x_5),
3. (x_4, x_5),
4. (x_2, x_3, x_4).

Let $p(t = 0) = e^{-0.1} = p$ and $q = 1 - p$. In this case

$$(1 - q^2)^2(1 - q^3)^2 \leq P \leq 1 - (1 - p^2)^2(1 - p^3)^2 .$$

After substitution of the approximate value of 0.1 for q,

$$0.978 \leq P \leq 0.9976 .$$

The exact solution (Example 7.1) is 0.978.

To find the upper bound of the MTTF, one finds first the failure rates of the minimum paths:

$$\Lambda_1 = 0.02\left(\frac{1}{h}\right),$$

$$\Lambda_2 = 0.03\left(\frac{1}{h}\right),$$

$$\Lambda_3 = 0.02\left(\frac{1}{h}\right),$$

$$\Lambda_4 = 0.03\left(\frac{1}{h}\right).$$

The final result is

$$T \leq 50 + 50 + 33.3 + 33.3 - 20 - 25 - 20 - 20 - 16.7 - 20$$
$$+ 14.1 + 12.5 + 14.1 + 12.5 - 10 = 97.2 \text{ hr} .$$

The exact value (see Example 7.1) is 82 hours.

7.7.3 Litvak-Ushakov Bounds

Two rules on bounds are helpful:

1. The lower bound of the probability of connectedness of a two-pole network can be obtained if one represents the system structure with a parallel connection of nonintersected minimum paths.
2. The upper bound of the probability of connectedness of a two-pole network can be obtained if one represents the system structure with a series connection of nonintersected minimum cuts.

Thus

$$\max_{1\leq v\leq V}\left\{1 - \prod_{A_j\in A^v}\left(1 - \prod_{i\in A_j} p_i\right)\right\} \leq P \leq \min_{1\leq w\leq W} \prod_{B_k\in B^w}\left(1 - \prod_{i\in B_k} q_i\right). \tag{7.51}$$

where A^v is the set of unit subscripts for the vth set of nonintersected paths, V is the number of these sets; B^w is the set of unit subscripts for the wth set of nonintersected cuts, and W is the number of these sets.

Remark To find a lower (upper) bound, one can use any set in V (W) in (7.51). The bound may or may not be the minimum (maximum), but it may suit the purpose.

EXAMPLE 7.8

For the bridge in Figure 7.1 the bounds given by (7.51) are

$$\max\{[1-(1-p_1p_2)(1-p_4p_5)],\ p_1p_3p_5,\ p_2p_3p_4\}$$
$$\leq P \leq \min\{(1-q_1q_4)(1-q_2q_5),\ 1-q_1q_3q_5,\ 1-q_2q_3q_4\}\ .$$

7.7.4 Final Results

Using both (7.49) and (7.50), we can write the resultant bounds:

$$\max\left\{\prod_{1\leq k\leq r}\left(1-\prod_{i\in B_k} q_i\right);\ \max_{1\leq v\leq V}\left[1-\prod_{A_j\in A^v}\left(1-\prod_{i\in A_j} p_j\right)\right]\right\}$$
$$\leq Pr\{\varphi(X)\}\leq \min\left\{1-\prod_{1\leq j\leq r}\left(1-\prod_{i\in A_j} p_i\right);\ \min_{1\leq w\leq W}\prod_{B_k\in B^w}\left(1-\prod_{i\in B_k} q_i\right)\right\}. \tag{7.74}$$

Remark Relationship (7.51) is easier to use than (7.49) because it does not require all minimum paths and cuts. However, one should not give an absolute preference to either one. The accuracy of each relationship depends on the network structure. If the different minimum paths have no joint units, that is, if the structure is almost strictly parallel, the best results are the upper one from (7.49) and the lower one from (7.51). If different minimum cuts have no joint units, that is, if the structure is almost strictly in series, the best results are the upper one from (7.51) and the lower one from (7.49).

We emphasize that there is no correct way to simplify the Esary-Proschan bounds to find an approximate solution. Moreover, one can easily see that some methods of "simplification" (by eliminating paths considered too long or cuts considered thick) lead to reverse Litvak-Ushakov bounds; the lower Esary-Proschan bound becomes the upper Litvak-Ushakov bound, and vice versa.

If units have exponential distributions with different parameters λ_i, the MTTF can be written in closed form:

$$\max_{1\leq v\leq V}\left\{\sum_{A_j\in A^v} T_j-\sum_{(A_j,A_k)\in A^v}\frac{1}{\Lambda_j+\Lambda_k}+\cdots+(-1)^{n_v+1}\frac{1}{\Sigma_{A_j\in A^v}\Lambda^v}\right\}$$
$$\leq T\leq \sum_{1\leq j\leq r} T_j-\sum_{1\leq j<k\leq r}\frac{1}{\Lambda_j+\Lambda_k}+\cdots+(-1)^{r+1}\frac{1}{\Sigma_{1\leq j\leq r}\Lambda_j}\ ,$$

where

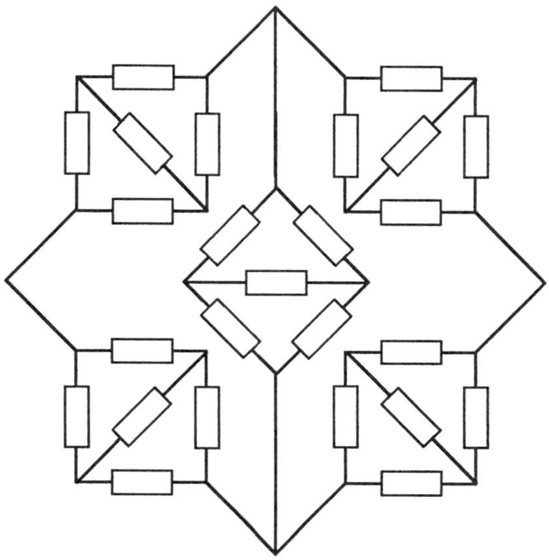

Figure 7.9 A bridge structure system with embedded bridges.

$$\Lambda_j = \sum_{i \in A_j} \lambda_i ,$$

$T_j = 1/\Lambda_j$ is the MTTF of path i, and n_0 is the number of paths in the vth set of nonintersected paths.

7.7.5 Bounds Based on a Structural Decomposition

Often it is reasonable to decompose the system structure before finding its connectedness probability bounds. Let us consider an example of such a decomposition. Compare the systems in Figures 7.1 and 7.9. Instead of using units as in the first figure, we can use subsystems with complex structures (the same bridge structures) as in Figure 7.9. The upper bound of the system structure becomes the corresponding upper bound of the subsystem. The lower bound of the system structure becomes the corresponding lower bound of the subsystem. The bounds are wider but are easier to find.

CHAPTER 8

EVALUATION OF SYSTEM EFFECTIVENESS

Most complex systems have internal redundant units and other equipment to help perform their main functions. Also, some systems can perform the same task by several different means. In this case, failure of a unit can lead to decreased capability to perform a given task, but not to complete system failure. Thus a complex system can require a measure of capability that can take partial failures into account.

The term "effectiveness" will be used generally to mean a quantitative measure of the quality of the performance of a system. For example, in a queueing system, such as a communication system, transportation, or production conveyer line, failures of serving channels can lead to increased serving time or loss of served objects. An energy system failure can result in a decrease of supply to consumers, in lower voltages for electric power systems, or in decreased quality or quantity of fuel for oil refineries.

This chapter has several measures of system output. If there are no special notes, one may assume that the more effective a system is, the higher the effectiveness index is. Otherwise, such as when effectiveness is measured by accuracy, there are special explanations. The choice of an effectiveness index is determined by the type of system, by the type of task performed, by the environment, and by other factors. In general, there are two categories of effectiveness measures, those applicable to an instant in time (or more precisely, a short operating time) and those applicable to a period of time. Corresponding systems are called *instant* and *durable*.

A durable system performs its task during some essential and extended interval of time t_0 beginning from a moment t. An interval t_0 is considered to be extended if several failures and/or renewals can occur. An instant system almost certainly will have no change of state during a task performance time t_0.

At any moment in time each unit must be in one of two states: failed or operational. The state of a system is a set of its unit states. A system transits from one state to another because of failure or renewal of a unit or because of different possible external influences. An instant system can be characterized at a moment by a coefficient of effectiveness. The coefficient numerically characterizes the system effectiveness when the system is in its current state. For a durable system the coefficient of effectiveness characterizes system effectiveness for the current sequence of state changes.

8.1 INSTANT SYSTEMS

Let $h_s(t)$ be the probability that an instant system at the moment t is in state s, and let Φ_s be the coefficient of effectiveness of this state. This effectiveness of the instant system at time t then can be defined as

$$E(t) = \sum_{\text{all } s} h_s(t)\Phi_s \, . \tag{8.1}$$

In many cases a large number of system states can be grouped into a small number of classes, each of which has only one effectiveness coefficient. Such a situation can occur, for example, when only discrete levels are chosen for the coefficient. System effectiveness then can be calculated with

$$E = \sum_{1 \leq k \leq N} \Phi_k \sum_{x \in S_k} h_s(t) \, , \tag{8.2}$$

where Φ_k is a level of effectiveness, N is the number of levels, and S_k is the set of system states with an effectiveness coefficient of Φ_k.

For a system of n mutually independent units when each unit must be either failed or operational, probabilities $h_s(t)$ can be expressed easily with the following reliability indexes.

The probability that all system units are operational is

$$h_0(t) = \prod_{1 \leq i \leq n} r_i(t) \, . \tag{8.3}$$

The probability that unit i and only unit i has failed is

$$h_s(t) = \frac{q_i(t)}{r_i(t)} \prod_{1 \leq k \leq n} r_k(t) = \gamma_i(t) h_0(t) \, . \tag{8.4}$$

The probability that units i and j and only i and j have failed is

$$h_{ij}(t) = \gamma_i \gamma_j h_0(t) \, . \tag{8.5}$$

Extension of (8.4) and (8.5) gives

$$h_s(t) = \prod_{i \in G_1^s} r_i(t) \prod_{i \in G_2^s} [1 - r_i(t)] , \tag{8.6}$$

where G_1^s and G_2^s are, correspondingly, sets of operational and failed system states for the sth system state, $r_i(t)$ is the probability that unit i is operational at the moment t, and

$$\begin{aligned} q_i(t) &= 1 - r_i(t) , \\ \gamma_i(t) &= \frac{q_i(t)}{r_i(t)} . \end{aligned} \tag{8.7}$$

For unrepairable systems, $r_i(t)$ is the probability of successful operation throughout period t. For repairable systems $r_i(t)$ is the availability coefficient for the moment t, which may be either stationary or nonstationary. We will denote the probabilities for stationary cases without t.

Expression (8.1) can be rewritten into a more open form using (8.3) through (8.5):

$$E(t) = \Phi_0 h_0(t)\Big[1 + \sum_{1 \le i \le n} \Phi_i^* \gamma_i(t) + \sum_{1 \le i \le j \le n} \Phi_{ij}^* \gamma_i(t)\gamma_j(t) + \cdots\Big] , \tag{8.8}$$

where Φ_0 is the effectiveness coefficient of the system when all units are operational, $\Phi_0 \neq 0$, and

$$\Phi_i^* = \frac{\Phi_i}{\Phi_0} , \quad \Phi_{ij}^* = \frac{\Phi_{ij}}{\Phi_0} , \ldots .$$

In (8.8), $\Phi_0 \ge \Phi_i \ge \Phi_{ij} > \Phi_{ijk\ldots}$, and $\Phi_{1,2,\ldots,n} = 0$. If

$$\max_{1 \le i \le n} q_i(t) \ll \frac{1}{n} , \tag{8.9}$$

the approximate system effectiveness, $\tilde{E}(t)$, can be calculated using

$$\tilde{E}(t) \approx \Phi_0\Big[1 - \sum_{1 \le i \le n} q_i(t)(1 - \Phi_i^*)\Big] , \tag{8.10}$$

where Φ_{i*} is a quantitative index of the relative necessity of the unit. Expression (8.10) gives a lower bound. The error of this approximation is

$$\delta = \binom{n}{2}[\max_{1 \le i \le n} q_i(t)]^2 \Phi_0 . \tag{8.11}$$

If $r_i(t)$ is the probability of failure-free operation and the distribution is exponential with parameter λ_i, one can use (8.9) and write

$$\tilde{E}(t) \approx \Phi_0 \exp\left[-t \sum_{1 \le i \le n} \lambda_i (1 - \Phi_i^*)\right]. \tag{8.12}$$

Let $\mathbf{K}_i$ be the stationary availability coefficient of unit i and all units are independent,

$$K_i = \frac{T_i}{T_i + \tau_i},$$

where T_i and τ_i are, correspondingly, mean time between failures (MBTF) and the mean time to repair (MTTR) of the ith unit. In such a case one can use (8.8) and write

$$\tilde{E}(t) \approx \Phi_0 \prod_{1 \le i \le n} \frac{T_i}{T_i + \tau_i(1 - \Phi_i^*)} \approx \Phi_0 \left[1 - \sum_{1 \le i \le n} \frac{\tau_i(1 - \Phi_i^*)}{T_i}\right]. \tag{8.13}$$

EXAMPLE 8.1

An airport radar system should cover an angle of 180°. It consists of two identical stations: one of them covers 0° to 110° and the other covers 70° to 180°. Each station has an MTBF of 95 hours and a mean time to repair of 5 hours. The availability of each station thus equals 0.95. The probability of successful automatic control of a plane by one station is 0.9. In the overlapping zone the control probability is

$$P = 1 - (1 - p)^2 = 0.99.$$

Find the probability of successful control of an arbitrary plane from an arbitrary direction at an arbitrary moment of time.

Solution

Consider the state S_0. The probability of this state is

$$h_0 = K^2 = 0.9025.$$

The overlapping zone is 70–110°. If both stations are working successfully, effectiveness coefficient is a weighted average:

$$\Phi_0 = \frac{40°}{180°} 0.99 + \frac{140°}{180°} 0.9 = 0.92.$$

Thus

$$h_0\Phi_0 = 0.83 .$$

Let us denote the stations as a and b. The probability of state S_b (with station b failed) equals

$$h_b = K(1 - K) = 0.0495 .$$

At the same time only one station is operating in a zone of 110°, so

$$\Phi_b = \frac{110}{180} 0.9 = 0.55 .$$

Finally,

$$h_b\Phi_b = 0.028 .$$

States S_a and S_b are identical, so the final result is

$$E = h_0\Phi_0 + 2h_a\Phi_a = 0.828 + 0.056 = 0.883 .$$

For highly reliable systems an approximate system effectiveness index is

$$E = \Phi_0\left[1 - \sum_{1\le i\le n} q_i(1-\Phi_i^*) + \sum_{1\le i<j\le n} q_iq_j[1-(\Phi_i^*+\Phi_j^*)+\Phi_{ij}^*] - \cdots\right] . \tag{8.14}$$

One also can write the following approximate expression

$$E = \Phi_0 \prod_{1\le i\le n} [1 - q_i(1-\Phi_i^*)] . \tag{8.15}$$

8.2 DURABLE SYSTEMS

Let us denote the probability that a durable system has its πth state change in the time interval $[t, t+t_0]$ with $dh_\pi(t, t+t_0)$. Let the conditional effectiveness coefficient for system performance during this track be Φ_π. The system effectiveness then can be written as

$$E(t, t+t_0) = \int_{G_\pi} \Phi_\pi \, dh_\pi(t, t+t_0) , \tag{8.16}$$

where the integral is taken over the space of all possible tracks of system state changes in the time interval $[t, t+t_0]$.

If a system consists of n independent units, (8.16) can be rewritten:

$$E(t, t+t_0) = \Phi_0 H_0 \left\{ 1 + \sum_{1 \le i \le n} \frac{1}{r_i} \int_t^{t+t_0} \Phi_i(x_i) f_i(x_i)\, dx_i \right.$$
$$\left. + \sum_{1 \le i < j \le n} \frac{1}{r_i r_j} \int_t^{t+t_0} f_i(x_i)\, dx_i \int_t^{t+t_0} \Phi_{ij}(x_i, x_j) f_j(x_j)\, dx_j + \cdots \right\}, \tag{8.17}$$

where Φ_0 is a conditional effectiveness coefficient for the case when no units fail during the interval $[t, t+t_0]$, $\Phi_i(x_i)$ is a coefficient for the case when the only failure is unit i at the moment x_i, and H_0 is the probability that no unit fails during the interval.

$$H_0 = \prod_{1 \le i \le n} r_i(t, t+t_0), \tag{8.18}$$

where $f_i(x_i)$ is the probability density for the failure of unit i at the moment x_i.

If

$$\max_{1 \le i \le n} q_i(t, t+t_0) = \max_{1 \le i \le n} \int_t^{t+t_0} f_i(x_i)\, dx_i \ll \frac{1}{n}, \tag{8.19}$$

an approximation may be used:

$$\tilde{E} \approx \left\{ 1 - \sum_{1 \le i \le n} \left[q_i(t, t+t_0) - \int_t^{t+t_0} \Phi_i^*(x_i) f_i(x_i)\, dx_i \right] \right\}, \tag{8.20}$$

$$\Phi_i^*(x_i) = \frac{\Phi_i(x_i)}{\Phi_0}.$$

The error of (8.20) does not exceed

$$\delta = \binom{n}{2} \max_{1 \le i \le n} [q_i(t, t+t_0)]^2 \Phi_0. \tag{8.21}$$

EXAMPLE 8.2

Consider a satellite spy system for information collecting. The system consists of two identical unrepairable receivers, a and b. System effectiveness is measured by the rate of information collection. If both receivers are operational, the system collects A units of information in a unit of time. If one receiver fails, system effectiveness decreases to level $B = 0.3A$. Failure of both receivers, of course, leads to total system failure. Assume that the probability of successful operation for each receiver has an exponential distribution:

$$r(t) = e^{-\lambda t} .$$

The system is designed to perform a task of duration $t = 0.1/\lambda$. The following conditional coefficients for the trajectories of changes of state are given:

$$\Phi_0 = At , \quad \Phi_i(x_i) = Bt + (A - B)x_i , \qquad i = a, b ,$$

$$\Phi_{ij}(x_i, x_j) + B \max(x_i, x_j) + (A - B) \min(x_i, x_j) .$$

The effectiveness coefficient thus equals the amount of information collected. Find the system's effectiveness.

Solution

For this particular case use (8.16), which gives

$$E = r^2 At + 2r \int_0^t [Bt + (A - B)x]\lambda e^{-\lambda x} \, dx$$

$$+ 2 \int_0^t \lambda e^{-\lambda x} \, dx \int_0^x [Bx + (A - B)y]\lambda e^{-\lambda y} \, dy .$$

After numerical substitutions, we have

$$E = (0.819 + 0.109 + 0.004)At = 0.932 .$$

8.3 SOME SPECIAL CASES

8.3.1 Systems with Additive Effectiveness Indexes

Some systems have simple effectiveness coefficients; system effectiveness is simply the sum of the contributions of the individual units. Examples include transportation systems and information-collecting systems.

If the ith unit contributes ϕ_i, the effectiveness of an instantaneous system at the moment t is

$$E = \sum_{1 \le i \le n} \phi_i r_i(t) , \tag{8.22}$$

where $r_i(t)$ is the probability of operability of unit i at moment t.

EXAMPLE 8.3

Consider the instantaneous system shown in Figure 8.1. The system outcome is the sum of the outcomes of all the executive units. Units in levels 0

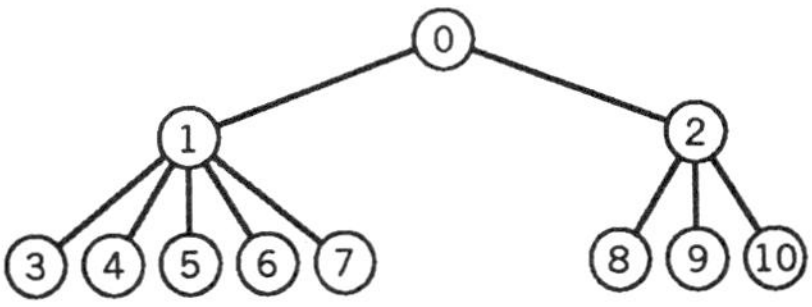

Figure 8.1 System structure for Example 8.3.

and 1 do not contribute to the system effect, but their operational or failed states influence the executive units. For example, in a power system the executive units might make the final distribution of power to customers. Failures of other units, though, might lead to the failure of a set of executive units. Availability coefficients K_i and outcomes of executive units ϕ_i are given in Table 8.1. Find the system effectiveness if all units are independent.

Solution

First, note that the ith unit brings its outcome if it is operational itself, which requires the main zero-level unit and the corresponding first-level unit to be operational. Suppose that the probability of the ith unit's outcome is $r_7 = K_0 K_1 K_7$. Using (8.22), we write

$$E = K_0 \left(K_1 \sum_{3 \le i \le 7} K_i \phi_i + K_2 \sum_{8 \le i \le 10} K_i \phi_i \right).$$

After numerical substitution,

$$E = (0.99)[(0.98)(28.1) + (0.97)(23)] \approx 49.35 .$$

The expression for durable systems is

$$E(t, t + t_0) = \sum_{1 \le i \le n} \left[r_i(t, t + t_0)\phi_{0i} + \int_t^{t+t_0} f_i(x_i)\phi_i(x_i)\, dx_i \right], \quad (8.23)$$

where $\phi_i(x_i)$ is the effectiveness coefficient when unit i fails at moment x_i, $t \le x_i \le t + t_0$ and ϕ_{0i} is the outcome of this unit operating without failures during this interval of time.

Table 8.1 Input data for Example 8.3

	0	1	2	3	4	5	6	7	8	9	10
K	0.99	0.98	0.97	0.7	0.9	0.8	0.8	0.7	0.9	0.95	0.95
φ	—	—	—	10	8	8	5	5	5	10	20

EXAMPLE 8.4

A durable system designed for information collection consists of three independent units. Capacities of these units are $\phi_1 = 100$ bit/sec, $\phi_2 = 200$ bits/sec, and $\phi_3 = 250$ bits/sec; their hourly failure rates are $\lambda_1 = 0.01$, $\lambda_2 = 0.03$, and $\lambda_3 = 0.04$. (All units have exponential failure time distributions.) Find the system effectiveness for interval [0, 50] if all units are loaded completely during this interval.

Solution

Let $\phi_{0i} = \phi_i t_0$ and $\phi_i(t_i) = \phi_i t_i$. Then

$$E(0, 50) = \sum_{1\le i\le 3} \left(e^{-\lambda_i t_0}\phi_i t_0 + \int_0^{t_0} \lambda_i \phi_i x e^{-\lambda_i x}\, dx \right)$$

$$= \sum_{1\le i\le 3} \int_0^t \phi_i e^{-\lambda_i x}\, dx = \sum_{1\le i\le 3} \frac{\phi_i}{\lambda_i} (1 - e^{-\lambda_i t}) .$$

Substitution of the numerical values of ϕ and λ in the same time units yields

$$E(0, 50) = \left[\frac{100}{0.01} (0.39) + \frac{200}{0.03} (0.78) + \frac{250}{0.04} (0.86) \right] 3600$$

$$\approx (5.2)10^7 \text{ bits} .$$

8.3.2 Multifunctional Systems

A system is called "multifunctional" if it can perform its task in different modes, which might differ in effectiveness. It is assumed that the best available mode is always chosen.

Consider a system consisting of n independent units. Let the system task be performed in m different modes. The effectiveness of the jth mode is Φ_j. Without loss of generality we will let

$$\Phi_1 > \Phi_2 > \cdots > \Phi_m .$$

Let us divide the system into m subsystems $G_1, G_2, \ldots, G_m$. Each includes the units necessary to perform the system task for its mode. Let it be possible for a unit to be involved in task performance in more than one mode. Two cases with relatively simple results are given here.

1. If each unit is only one subsystem, the probability that the system performs its task in mode j equals

$$H_j = P_j \prod_{1 \le k \le j-1} Q_k , \tag{8.24}$$

where P_j is the probability of failure-free operation of the jth subsystem, and $Q_j = 1 - P_j$. The system effectiveness is

$$E = \sum_{1 \le i \le m} \Phi_j P_j \sum_{1 \le k \le j} Q_k . \tag{8.25}$$

2. Let the best mode require all the units. Let every other mode require all the units required by the immediately less effective mode but not require at least one unit that is required by the immediately more effective mode. There are thus sets of units related by

$$G_1 \supset G_2 \supset \cdots \supset G_m .$$

Let us denote with G^*_{j-1} a set of units that belongs to subsystems through $j - 1$ but does not belong to subsystem j. We write

$$G^*_{j-1} = G_{j-1} \backslash G_j .$$

The probability of performance of the system task in mode i equals

$$H_j = P_j - P_{j-1} . \tag{8.26}$$

The system effectiveness equals

$$E = \sum_{1 \le j \le m} \Phi_j (P_j - P_{j-1}) . \tag{8.27}$$

Alternatively,

$$E = \sum_{1 \le j \le m} P_j (\Phi_{j+1} - \Phi_j) , \tag{8.28}$$

where $P_0 = 1$ and $\Phi_{m+1} = 0$.

EXAMPLE 8.5

Figure 8.2 shows a World War II system designed to perform three tasks in succession: (1) search for an enemy plane, (2) find the secret location of the plane, and (3) transmit the information. All the tasks could be performed in different modes. The data can be found in Table 8.2. For all modes the control desk had to be operational. The probabilities of operability were $r_1 = 0.80$, $r_2 = 0.99$, $r_3 = 0.80$, $r_4 = 0.95$, $r_5 = 0.90$, and $r_6 = 0.95$. Calculate the system's effectiveness.

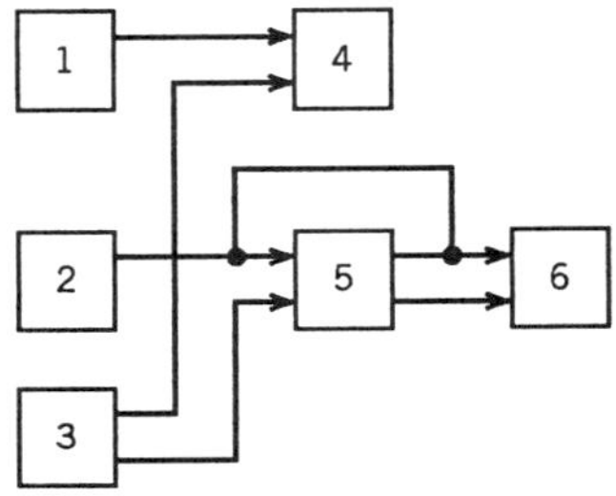

Figure 8.2 Structure of the World War II radar system in Example 8.5.

Solution

The first mode probability was

$$h_1 = r_1 r_3 r_4 r_5 r_6 \approx 0.52 \,.$$

The second mode probability was

$$h_2 = q_1 r_3 r_4 r_5 r_6 \approx 0.13 \,.$$

The third mode probability was

$$h_3 = q_3 r_1 r_4 r_5 r_6 \approx 0.13 \,.$$

The fourth mode was used in the event of failure of any of the following:

an indicator,
an indicator and a searching radar,
an indicator and a conveying radar,
an indicator and searching and conveying radars,
searching and conveying radars.

Table 8.2 Technical efficiency

Index of Method of Fulfillment	Stage: Search	Stage: Detection	Stage: Tracking	Measure of Technical Efficiency
1	Search tracer	Indicator	Tracking tracer	1.00
2	Tracking tracer	Indicator	Tracking tracer	0.60
3	Search tracer	Indicator	Optical structure and calculator	0.30
4	Optical structure	Optical structure	Optical structure and calculator	0.15
5	Optical structure	Optical structure	Optical structure	0.10

In every case the optical equipment and the computer were assumed to be operational. The probability of this mode was

$$h_4 = r_2 q_5 r_6 (q_3 + q_1 q_3 q_4) \approx 0.01 .$$

Finally, the system's effectiveness was

$$E = 0.52 + (0.13)(0.6) + (0.13)(0.3) + (0.08)(0.15) + (0.01)(0.1) \approx 0.66 .$$

The probability of failure-free operation was 0.51.

8.3.3 Systems with Symmetrical Branching Structures

A number of technical systems have executive units and a "control body." The executive unit operates successfully if it is operational itself and all its control units also are operating properly. If the system has a hierarchical structure, some control units should be controlled themselves.

Let us consider the branching systems whose effectiveness depends only on the number of successfully operating executive units. When the number of successfully operating units equals z, the system effectiveness coefficient is $\Phi(z)$. In this case one needs to know only the moments of the distribution of a random number of successfully operating executive units. The system effectiveness can be found with

$$E = \sum_{0 \le k < \infty} \frac{M^{(k)}}{k!} \left. \frac{d^k \Phi(z)}{dz^k} \right|_{z=0} , \tag{8.29}$$

where $M^{(k)}$ is the kth initial moment of the distribution of successfully operating executive units. In theory the function $\Phi(z)$ should be differentiable, but for practical purposes $\Phi(z)$ can be approximated with a polynomial of reasonably low order.

The moments can be found with a moment-generating function:

$$\varphi_n(z) = \varphi_{n-1}((r_n e^z + q_n)^{a_n}) ,$$

where a_i is called a *branching coefficient* and is the number of units at i level that depend on a control unit at the $i-1$ level. The first two initial moments are

$$M_n^{(1)} = M_{n-1}^{(1)} a_n r_n = r_0 \prod_{1 \le i \le n} a_i r_i ,$$

$$M_n^{(2)} = M_{n-1}^{(2)} (a_n r_n)^2 + M_{n-1}^{(1)} r_n q_n a_n . \tag{8.30}$$

EXAMPLE 8.6

Consider different possible variants of a symmetric branching system with executive units (Figure 8.3). As a first step, assume that $\Phi(z) = C_1 z$: the system effectiveness is a linear function of the number of successfully operating executive units. Using (8.30),

$$E = C_1 M^{(1)} = C_1 r_0 \prod_{1 \leq i \leq n} a_i r_i .$$

Introduce N_n, the total number of executive units,

$$N_n = \prod_{1 \leq i \leq n} a_i .$$

The expression for the system effectiveness then can be rewritten

$$E = C_1 N_n \prod_{1 \leq i \leq n} r_i .$$

This formula leads to an interesting observation: all variants of the system in Figure 8.3 are equivalent for a linear function $\Phi(z)$.

Now consider the case when $\Phi(z) = C_2 z^2$. Such a situation can occur, for example, when Lanchester's law is applicable. In this case

$$E = C_2 M^{(2)} .$$

For highly reliable units, when

$$q_i \ll \frac{1}{N_n} ,$$

the expression for the second initial moment can be written in the form

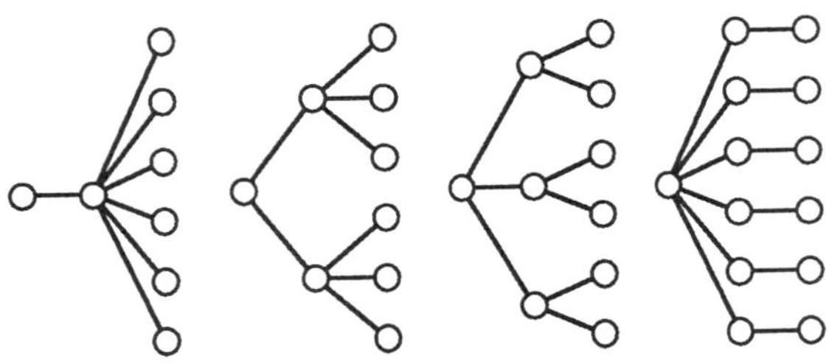

Figure 8.3 Different variants of a system with a symmetrical branching structure for Example 8.6.

$$M^{(2)} \approx 4N_n^2\left[1 - \sum_{1\le i\le n} q_i\left(1 - \frac{1}{N_i}\right)\right].$$

For simplicity let us assume that $q_0 = q_1 = q_2 = q$. Then

$$E_a \approx 2C_2N^2(1 - 0.83q) ,$$
$$E_b \approx 2C_2N^2(1 - 1.33q) ,$$
$$E_c \approx 2C_2N^2(1 - 1.5q) ,$$
$$E_d \approx 2C_2N^2(1 - 1.67q) ,$$

where subscripts of E refer to Figure 8.3. In this case the variant with the largest a_2a_1 value is preferable.

8.3.4 Systems with Redundant Functions

There are many instant systems in which a single task is performed by several executive units. If at least one unit performs its operation, the system performs the task. Antiaircraft and antimissile systems offer examples. Let the probability of success for a single executive unit be ϕ_s under the condition that the system is in state S. The system's effectiveness is of interest. Two cases are considered in this section.

First is the case when all executive units perform their operations simultaneously and independently. The probability of success for each is ϕ_s. In this case

$$E = \sum_s H_s[1 - (1 - \phi_s)^N] , \tag{8.31}$$

where H_s is the probability of state S and N is the number of executive units.

EXAMPLE 8.7

A defense system fires two shots at a time at a target. There are two modes. Mode a has a one-shot probability of success, with $\phi_a = 0.85$, and mode b has $\phi_b = 0.58$. The probability of the use of the better mode is 0.7. The shots are independent. Find the system success probability.

Solution

Write

$$E = r[1 - (1 - \phi_a)^2] + q[1 - (1 - \phi_b)^2]$$
$$= (0.7)(1 - 0.15^2) + (0.3)(1 - 0.42^2) = 0.931 .$$

In the second case executive units perform their operations at different moments, when the system is in different states. The same equation might hold for different systems, such as rings of antimissile defenses around a defended site. In this case:

$$E = 1 - \left(1 - \sum_s H_s \phi_s\right)^N . \tag{8.32}$$

EXAMPLE 8.8

As in Example 8.7 let the two shots be independent, and find the system's success probability:

$$E = 1 - [1 - (r\phi_a + q\phi_b)]^2 .$$

After numerical substitution, we have

$$E = 1 - [1 - ((0.7)(0.85) + (0.3)(0.58))]^2 = 0.947 .$$

If a symmetrical branching system has z independent redundant executive units, their effectiveness equals

$$\Phi_z = 1 - (1 - \phi)^z .$$

Then

$$E = 1 - [r_0(r_1(r_2 \ldots r_{n-1}(r_n(1-\phi) + q_n)^{a_n} + q_{n-1})^{a_{n-1}} + \cdots + q_1)^{a_1} + q_0] .$$

8.3.5 System with Multiplicative Effectiveness Coefficient

Some systems may be considered to be generalized series systems. Such systems consist of subsystems that have their own effectiveness indexes. The system's effectiveness index is the product of the separate subsystem indexes. The state of a subsystem does not influence the effectiveness of other subsystems. In this case

$$E = \prod_{1 \le i \le n} E_i , \tag{8.33}$$

where E is the effectiveness index of subsystems i.

Expression (8.33) is valid if, for any system state S and subsystem states S_i, the effectiveness coefficients are related by

$$\Phi(S) = \prod_{1 \le i \le n} \Phi_i(S_i) . \tag{8.34}$$

8.3.6 Systems with Intersecting Zones

Consider a system consisting of n units. Let the probability of an operational state for unit i be r_i. Let the i zone of action be σ_i. These zones can intersect each other to form zones of joint action σ_ω, where ω is a subset of subscripts for units acting in this joint zone. The system zone of joint action is

$$\sigma = \bigcup_{\text{all } \omega} \sigma_\omega \,. \tag{8.35}$$

Let zone σ_j be a joint zone in which n_j units act. Then (8.35) can be rewritten

$$\sigma = \sum_{1 \le j \le 2^n} \sigma_{\alpha_j} \,, \qquad \alpha_j = (i_1, i_2, \ldots, i_{n_j}) \,. \tag{8.36}$$

When units can fail, the number of effective units in each zone is random. In zone σ_α there is a random set α_*, $\alpha_* \subseteq \alpha$. In some cases the portion of the system effectiveness contributed by each zone may be proportional to its geometrical size and to the effectiveness coefficient of the zone. In such a case one can write

$$E = \sum_{1 \le j \le 2^n} \sigma_{\alpha_j} E \Phi_{\alpha_j} \,, \tag{8.37}$$

where E is the expected value.

Examples include territorial systems that spread over large areas. In most cases the number of zones is much less than 2^n. Six types of these systems are examined here.

1. *Additive coefficient inside a zone. The system S* with effectiveness E, consists of a set of subsystems $S_1, S_2, \ldots, S_N$ with effectiveness values $E_1, E_2, \ldots, E_N$. E is additive when:

$$E = \sum_{1 \le i \le N} E_i \,.$$

A necessary and sufficient condition for the validity of this equation is that, for any system state α, with subsystem states $\alpha_1, \alpha_2, \ldots, \alpha_N$,

$$\Phi_\alpha = \sum_{1 \le i \le n} \Phi_{\alpha_i} \,, \tag{8.38}$$

where Φ_i is the effectiveness coefficient of unit i.

2. *Multiplicative coefficient inside a zone*. Let $0 \le \Phi_i \le 1$ and let

$$\Phi^*_\alpha = \prod_{i \in \alpha^*} \Phi_i \,,$$

where $\Phi_{\alpha^*} = 1$ if all units in the zone have failed. This coefficient is thus a "loss coefficient." It can characterize, for example, the harmful effect of industrial pollution on environment. The system effectiveness equals

$$E = \sum_{1 \le j \le 2^n} \sigma_{\alpha_j} \prod_{i \in \alpha_j} (\Phi_i r_i + q_j) . \tag{8.39}$$

3. *Redundant action inside a zone*. In this case

$$\Phi^*_\alpha = 1 - \prod_{i \in \alpha^*} \omega_i ,$$

where

$$\omega_i = 1 - \Phi_i .$$

The system's effectiveness is

$$E = \sum_{1 \le j \le 2^n} \sigma_{\alpha_j} \Big[1 - \prod_{i \in \alpha_j} (r_i \omega_i + q_i) \Big] . \tag{8.40}$$

4. *Boolean overlapping inside a zone*. It is enough if inside a zone at least one unit operates successfully. For any $\alpha_* \subseteq \alpha$,

$$\Phi_{\alpha_*} = \Phi_\alpha . \tag{8.41}$$

5. *Maximum effectiveness coefficient for all units active in a zone*. Consider

$$\Phi_{\alpha^*} = \max_{i \in \alpha^*} \Phi_i . \tag{8.42}$$

Let the units be numbered in order of effectiveness

$$\Phi_1 > \Phi_2 > \cdots > \Phi_n .$$

Then

$$E = \sum_{\text{all } \alpha} \sigma_\alpha \sum_{i \in \alpha} r_k \Phi_k \prod_{i<k,\, i \in \alpha} q_i , \tag{8.43}$$

or

$$E = \sum_{1 \le i \le n} \Phi_i r_i \Big[\sigma_i - \sum_{\alpha_i :\, i \in \alpha_i} \sigma_{\alpha_i} \Big(1 - \prod_{j<i,\, j \in \alpha_i} q_j \Big) \Big] . \tag{8.44}$$

6. *Minimum effectiveness coefficient for all units active in a zone.* Referring to the previous procedure where

$$\Phi_1 < \Phi_2 < \cdots < \Phi_n ,$$

we use expressions (8.43) or (8.44) to calculate the system effectiveness. This coefficient also might be termed the "loss coefficient."

8.4 DECOMPOSITION OF COMPLEX SYSTEMS

An evaluation of effectiveness can result in cumbersome calculations because of the large number of states. It is therefore necessary to decompose multidimensional systems. One obvious method of decomposition is dividing the system into several independent subsystems. Another is developing a model that yields the upper and lower bounds of the index of the system effectiveness. These two methods cover a range of possibilities. Let us consider simple methods of system decomposition.

1. *A system with an additive effectiveness coefficient.* The system S with effectiveness E consists of a set of subsystems $S_1, S_2, \ldots, S_N$ with effectiveness values $E_1, E_2, \ldots, E_N$. E is additive when

$$E = \sum_{1 \le i \le N} E_i . \tag{8.45}$$

A necessary and sufficient condition for the validity of (8.45) is that for any system state α, with subsystem states $\alpha_1, \alpha_2, \ldots, \alpha_N$, the following condition must be valid:

$$\Phi_\alpha = \sum_{1 \le i \le n} \Phi_{\alpha_i} . \tag{8.46}$$

2. *A system with a multiplicative effectiveness coefficient.* A system S with effectiveness E consists of a set of subsystems $S_1, S_2, \ldots, S_N$ with values of effectiveness $E_1, E_2, \ldots, E_N$. E is multiplicative when

$$E = \prod_{1 \le i \le N} E_i . \tag{8.47}$$

A necessary and sufficient condition for 8.47 is that, for any system state α with subsystem states $\alpha_1, \alpha_2, \ldots, \alpha_N$, the following condition must be valid:

$$\Phi_\alpha = \prod_{1 \le i \le n} \Phi_{\alpha_i} . \tag{8.48}$$

3. *A system with intersecting zones.* For these systems the following inequality is valid:

$$E \leq \sum_{1 \leq i \leq N} E_i . \tag{8.49}$$

A near equality approaches the case of an additive effectiveness coefficient.

4. *A general case.* If a system effectiveness coefficient can be represented as a function of corresponding subsystem effectiveness coefficients

$$\Phi(S) = f(\Phi_1(S_1), \ldots, \Phi_N)) , \tag{8.50}$$

then

$$E = L[f(E_1, \ldots, E_N)] . \tag{8.51}$$

Different subsystems can have common units. In this case the operator L means that in an open form of (8.51), which is a sum of products, r may be substituted for any term of a form r^z if $z > 1$.

EXAMPLE 8.9

Consider the information system in Figure 8.4, which consists of a computer (1), six terminals (4–9), and two interfaces (2 and 3). The system's effectiveness is the weighted sum of the effectiveness values of both subsystems. The weighting coefficient for subsystem i equals b_i. The amounts of information collected by the subsystems are

$$\Phi_1(X) = x_1x_2(a_{11}x_4 + a_{12}x_5 + a_{13}x_6) ,$$
$$\Phi_2(X) = x_1x_3(a_{21}x_7 + a_{22}x_8 + a_{23}x_9) .$$

Here $x_i = 1$ if unit i is operable and 0 otherwise, and $X = (x_1, x_2, \ldots, x_9)$ is a system state. The total system's effectiveness equals

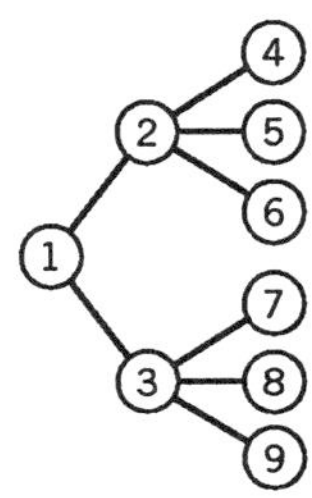

Figure 8.4 Structure of the information system in Example 8.9.

$$E = b_1[E\Phi_1(x)]^2 + b_2[E\Phi_2(X)]^3 .$$

Using (8.51) we obtain

$$\begin{aligned}
E &= L[b_1(p_1p_2(a_{11}p_4 + a_{12}p_5 + a_{13}p_6))^2 \\
&\quad + b_2(p_1p_3(a_{21}p_7 + a_{22}p_8 + a_{23}p_9))^3] \\
&= b_1p_1p_2L[(a_{11}p_4 + a_{12}p_5 + a_{13}p_6)^2] \\
&\quad + b_2p_1p_3L[(a_{21}p_7 + a_{22}p_8 + a_{23}p_9)^3] \\
&= b_1p_1p_2(a_{11}p_4 + a_{12}p_5 + a_{13}p_6 + 2a_{11}a_{12}p_4p_5 + 2a_{11}a_{13}p_4p_6 \\
&\quad + 2a_{12}a_{13}p_5p_6) \\
&\quad + b_2p_1p_3[a_{21}p_7 + a_{22}p_8 + a_{23}p_9 + 3(a_{21}^2a_{22} + a_{21}a_{22}^2)p_7p_8 \\
&\quad + 3(a_{21}^2a_{23} + a_{21}a_{23}^2)p_7p_9 + 3(a_{22}^2a_{23} + a_{22}a_{23}^2)p_8p_9 \\
&\quad + 6a_{21}a_{22}a_{23}p_7p_8p_9] .
\end{aligned}$$

8.5 EFFECTIVENESS EVALUATION METHODOLOGY

The evaluation of system effectiveness can go beyond mathematics and engineering. It can be an operations research problem that requires comprehension of equipment, environment, and purpose. In general, effectiveness evaluation of an instant system may include the following main steps:

1. Determine the system purpose.
2. Formulate possible system tasks.
3. Define the conditions of system operation.
4. Choose the most appropriate measure of effectiveness.
5. Divide the system into subsystems, taking into account structure, operating processes and repair procedures.
6. Delineate the functional structure of the system.
7. Collect reliability data on the system units.
8. Collection data on repair of units and maintenance procedures.
9. Calculate the probabilities of system states before deciding what states should be considered important for future analysis.
10. Determine the effectiveness coefficients for the system states.

In principle, the methodology for the effectiveness evaluation of durable systems is similar. The main difference is that the analysis of system states is replaced with an analysis of the process of the system state changes.

8.5.1 Effectiveness Coefficients for System States

Evaluating the effectiveness coefficients is perhaps the most difficult evaluation problem in practice. One of three methods can be used:

1. *Analytical method*. The analytical approach to determining the effectiveness coefficient is very rare. It is especially difficult in working with durable systems.
2. *Simulation method*. Both mathematical analysis and physical simulations can be very useful. This combination also enables fuller comprehension of the system.
3. *Method of physical experimentation*. Physical experimentation does not necessarily mean using developed equipment. One can use a prototype or an experimental version. The experiments should be set up to correspond to real failure conditions, with differences between experimental and real conditions corrected mathematically.

8.5.2 Types of Effectiveness Coefficients

As mentioned earlier, effectiveness coefficients can be relative, as in $0 \leq \emptyset \leq 1$. Working with relative values is convenient, since the absolute values of a system's outcome are not always known. Moreover, relative decreases in system effectiveness are generally easier to evaluate than absolute values. Still absolute effectiveness coefficients permit a comparison of different variants of a system. The system's variants operational state results can have very different components, structures, and outcomes.

Sometimes is is reasonable to consider "negative effectiveness", such as losses in communication networks and errors in computational systems. In these cases one should reformulate the related effectiveness indexes.

8.6 VULNERABILITY OF A COMPLEX SYSTEM

Vulnerability has become an issue with the increasing use of systems in hostile environments that may include such hazards as typhoons or enemy strikes. The most important aspect of a vulnerability analysis is ascertaining the time, seriousness, and location of the dangerous condition. Minimax criteria therefore may be used rather than probability-time indexes. An important variable may be the remaining value of a system's performance effectiveness index, E, after a mishap. In any case, however, the main purpose of vulnerability analysis usually is comparison of design variants of a system.

8.6.1 Formulation of the Problem

As an example, we will consider an energy system. As a measure we choose the amount of energy lost by consumers. For simplicity, we will not consider the consequences of the losses to different consumers. Using minimax criteria, we want to look at the mishap with the worst possible outcome. We need to choose the best variant of the system for this mishap. One mishap, of course, can effect different system variants differently.

A variant that is less vulnerable to one mishap may be more vulnerable to another. In general, it can be impossible to choose one best variant rigorously under more than one criteria. The final decision is almost always judgmental. An occasional exception occurs when one system, or variant, clearly dominates another for all possible mishaps.

Besides vulnerability there are other factors to consider in studying a system's effectiveness, such as safety, design cost, maintenance costs, and future modernization. Possible modernization often is an important consideration because systems evaluated for vulnerability frequently are large territorial systems. All of these factors, especially the multiple aspects of vulnerability, require final decisions based on judgmental procedures. These procedures can be applied not only to the equipment but to the external influences.

Consider a system consisting of n subsystems under K mishaps, $K \leq n$. When subsystem i is hit, the system loss equals C_i. If there are several mishaps with different subsystems, the system loss can be nonadditive in two ways:

Superadditive

$$C_{1,2,\ldots,n} \geq \sum_{1 \leq i \leq K_1} C_i \,. \tag{8.52}$$

Subadditive

$$C_{1,2,\ldots,n} \leq \sum_{1 \geq i \leq K_1} C_i \,. \tag{8.53}$$

In the first case the consequence of a set of K_i mishaps is more than the sum of the results of each mishap; in the second case, the opposite is true. The first case may occur, for example, in energy systems. The second case may arise in overlapping activity zones.

Of course the simplest case is when the system loss is strictly additive:

$$C_{1,2,\ldots n} = \sum_{1 \leq i \leq K_1} C_i \,. \tag{8.54}$$

Let us choose as a measure of vulnerability the maximum possible system loss under K mishaps. This might lead to either of two common problems in the choice of the best variant:

1. Selection of the variant with minimum loss under K mishaps.
2. Selection of the variant that requires the largest mishaps for a given level of possible system loss.

8.6.2 Some Simple Cases

Let us compare N different design variants of a system. Let the sth variant consist of n_s objects. Let the cost of loss of the ith object of the sth system variant equal C_i^*. In principle, one can find C_i^* values with methodology followed for system effectiveness coefficients for different system states.

The first problem above does not present any difficulty: for each system variant one calculates the total loss under K mishaps

$$A_s(K) = \sum_{1 \leq i \leq K} C_i^s . \tag{8.55}$$

The least vulnerable system has the smallest $A_s(K)$. In the second case above the least vulnerable system is given by

$$K_{s^*}^A = \max_{1 \leq s \leq N} K_s^A , \tag{8.56}$$

where K_s^A is the minimum number of mishaps that can make the sth variant loss exceed the value A.

To solve the problem, let us first place the variants in order:

$$C_1^s > C_2^s > \cdots > C_{n_s}^s , \qquad s = 1, \ldots, N . \tag{8.57}$$

Our objective is to consider the worst possible outcome of a mishap for each system variant. For each variant we find the value K_s^A for which

$$\sum_{1 \leq i \leq K_s^A - 1} C_i^s < A \leq \sum_{1 \leq i \leq K_s^A} C_i^s . \tag{8.58}$$

The variant with the largest K_s^A is considered the best one. For a nonadditive vulnerability index the problem ordinarily reduces to direct enumerating computation.

Remark A requirement for low vulnerability often conflicts with a requirement for high reliability. If the system is designed under cost restrictions, the problem of allocation of resources between reliability and vulnerability improvement arises. For example, to make an energy, transportation, or communication system more reliable, it may be concentrated into a smaller territory. This may permit more effective standby and internal redundancy and more effective maintenance. At the same time the concentrated system may become more vulnerable.

8.7 SAFETY

Systems sometimes can harm people or the environment. Emissions from transportation systems can endanger human health. Emissions from energy systems, especially nuclear plants and chemical plants, can devastate the environment.

A system can become unsafe or even dangerous because of internal failures, natural phenomena or hostile influences. The level of danger may depend on how well the system is protected. Some systems even may be harmful under ideal conditions. For this reason we should not just consider safety after failure or under a hostile influence; we should also consider safety under normal conditions.

8.7.1 Safety under Normal Conditions

A common hazard is release of harmful byproducts. Let $f(t)$ be the release rate as a function of time. Harmful byproducts include those collected without destruction, those destroyed naturally over time, and those checked by preventive measures. But for all of these there is yet little consensus on acceptable levels of concentration. These emission levels are often arbitrary and untestable. We consider, in turn, three typical situations involving harmful emissions.

First is the case of harmful byproduct concentration during a fixed time interval, Δ. A possible safety criterion is

$$\Psi(t) = \int_{t}^{t+\Delta t} f(t)\, dt \leq f^{0} . \tag{8.59}$$

Here the function $\Psi(t)$ is an interval average. The system is rated as safe if for all $\Delta \in [0, T_f - \Delta]$ the harmful concentration does not exceed some given level f^0, where T_f is the period of time of interest.

The function $f(t)$ is usually stochastic, so $\Psi(t, \xi)$ is a stochastic process. The safety index is thus probabilistic.

Second is the case where one might use a hazard reduction process, $\phi(t)$, to reduce the consequences of a harmful release. This could involve recycling or cleaning the air and/or water. The remaining pollution is measured using the function $f^*(t) = \max[0, f(t) - \phi(r)]$. As indicated by the notation, the protection process cannot produce a positive result; the maximum positive effect is only complete elimination of the harmful pollutants.

As in the previous case, the safety criterion can be written in the form

$$\Psi^*(t) = \int_t^{t+\Delta} f^*(t)\, dt \leq f^0\,, \qquad \Delta \in [T^* - \Delta]\,. \tag{8.60}$$

This function also depends on different random factors.

Last is the case of a harmful release from a system into the environment that decreases over time. An example is the concentration of a toxic gas such as carbon dioxide in a region where it can be absorbed or transformed by the biosphere. Usually such processes decrease exponentially and we may use:

$$\Psi(t) = \int_0^t f(x) e^{-\alpha(t-x)}\, dx \leq f^0\,, \tag{8.61}$$

where α is the intensity of concentration as a function of time.

If the investigated stochastic process $\Psi(t)$ is known, we can use a model of the high level crossing by the process. Asymptotical methods are usually suitable.

Remark Resolving a safety issue often conflicts with improving the reliability of a system. Emissions from a chemical plant are a typical example. Inevitably, choices among multiple criteria are necessary in arriving at a joint resolution of reliability, vulnerability, and safety problems.

8.7.2 Safety under Hostile Influences

Probabilistic information about external or internal dangerous influences may be known a priori or be uncertain. Among internal influences may be failure of the investigated object; external influences include power supply and transportation systems. The safety index then would almost coincide with an appropriate reliability index, and their calculation would be similar. The main difference would be the failure criteria: For a safety analysis the only events considered as failures are those that are dangerous to people or the environment. When no safety information is available, the safety analysis is conducted in the same way as a vulnerability analysis—the worst situation is determined using a minimax approach.

8.8 SECURITY

Secure systems are those protected against unsanctioned entry. Security is purely the domain of engineering design. In general, there are no special mathematical methods for evaluation of hardware security. There also are no advanced mathematical methods for software security evaluation. One exception is cryptography. Code security is evaluated by the deciphering time or by the number of computations needed. This topic lies out of the scope of this handbook.

8.9 SOFTWARE RELIABILITY

In recent years much attention has been given to software reliability. This problem is extremely important, but usually it is interpreted wrongly by software reliability specialists. They try to use various reliability indexes to describe problems that are not reliability problems.

Software reliability indexes are not probabilistic. They do not develop in time. Most "software failures" result from imperfect software design or from the use of software for the wrong tasks or in inappropriate circumstances. By nature, software reliability is very similar to the security problem. This problem also is outside the scope of current probabilistic reliability theory.

CHAPTER 9

SYSTEMS WITH TIME REDUNDANCY

9.1 TERMINOLOGY AND CLASSIFICATION

Time redundancy (TR) is the use of operational and maintenance practices to give a system reserve time to complete a task in the event of failure. A time-redundant system (TRS) typically has an operational time that is larger than the needed time to perform the system task. TR can be achieved in different ways. Sometimes a system task can be repeated. Sometimes it can be continued after elimination of a failure. Sometimes a failure of short duration does not prevent task completion. Failure of a TRS typically can be observed only after the entire designated period of time has expired; there is no one local event that can be described as a system failure.

Depending on the consequences, TRS failures can be divided into two main groups:

1. Nondepreciating failures, which lead only to delay of task performance.
2. Depreciating failures, which destroy the incomplete results of the system operation totally or partially. When destruction is partial, it is necessary to identify the useful and depreciated portions of the operating time.

9.1.1 Classification of Types of Time Redundancy

Time redundancy can be common, individual, or grouped. Common TR can be used by any system unit. Individual TR is for one unit. Grouped TR can be used by any unit in a subset of units.

The redundancy coefficient, m_t, is the ratio of the available time to the time required for task performance, t.

TR can be treated in one of three ways:

1. It can be unsupplemented, in which case τ has a duration that is determined before the task and cannot be changed afterwards.
2. It can be instantly supplemented, in which case TR is determined after each repair.
3. It can be gradually supplemented, in which case TR increases during up periods, usually linearly.

9.1.2 Classification of TR Systems

Four types of systems will be considered: series, parallel, series-parallel, and network.

There are two broad classes of series connections. In a *general series connection* any unit failure leads to interruption of system operation. In a *multi-stage series connection* there are buffer storages that give grouped TR for some system units. System failure does not occur until storage is exhausted. Parallel connections also are of two types. *Redundant connections* have extra units in parallel that do not increase system productivity but do increase reliability. *Multi-channel connection* have extra units in parallel that increase both productivity and redundant time. Network systems can possess all the previous properties.

TRS classes will be described with a four-digit code: X_1, X_2, m_3, and n, where X_1 and X_2 are, respectively, types of failure and repair time distributions. M is exponential, E is Erlang, W is Weibull-Gnedenko and G is general. Codes m and n are numbers of channels and stages, respectively. Each class has subclasses described with a five-digit code: X_1, X_2, X_3, X_4, and X_5.

The first digit describes the TR:

$$X_1 = \begin{cases} 0 & \text{for unsupplemented TR}\,, \\ 1 & \text{for instantly supplemented}\,, \\ 2 & \text{for combined supplementation}\,. \end{cases}$$

The second digit gives the type of failure:

$$X_2 = \begin{cases} 0 & \text{for nondepreciating failures}\,, \\ 1 & \text{for depreciating failures}\,, \\ 2 & \text{for combined failures}\,. \end{cases}$$

The third digit shows the other types of redundancy:

$$X_3 = \begin{cases} 0 & \text{for no other redundancy}\,, \\ 1 & \text{for structural redundancy}\,, \\ 2 & \text{for several types of redundancy}\,. \end{cases}$$

The fourth digit gives the type of monitoring:

$$X_4 = \begin{cases} 0 & \text{for continual monitoring}\,, \\ 1 & \text{for periodic monitoring}\,, \\ 2 & \text{for mixed monitoring}\,. \end{cases}$$

The fifth digit shows the type of system loading:

$$X_5 = \begin{cases} 0 & \text{for continuous loading}\,, \\ 1 & \text{for alternative or stochastic loading}\,. \end{cases}$$

If a property is irrelevant to the purpose at hand, the position is given an X.

9.1.3 Reliability Indexes

Three main random values will be used as indexes: the useful time up to a TRS failure, $T_0(\tau)$, the time of the system task performance, $T_1(t_0)$, and the useful system operation time during the interval $(0, t)$, $T_3(t)$. With these r.v.'s all reliability indexes can be introduced.

Probability of Success with No Task Disruption

$$\begin{aligned} P_0(t_0, \tau, \mathbf{w}) &= Pr\{T_0(\tau) > t_0\} \\ &= P_0^{(1)}(t_0, t_0 + \tau, \mathbf{w}) \\ &= Pr\{T_1(t_0) \le t_0 + \tau\} \\ &= P_0^{(2)}(t_0 + \tau, \tau, \mathbf{w}\} \\ &= Pr\{T_3(t_0 + \tau) \ge t_0\}\,, \end{aligned}$$

where $\mathbf{w}$ is a vector of parameters characterizing the level of redundancy.

Probability of Task Disruption

$$\begin{aligned} Q_0(t_0, \tau, \mathbf{w}) &= 1 - P_0(t_0, \tau, \mathbf{w})\,, \\ Q_0^{(1)}(t_0, t_0 + \tau, \mathbf{w}) &= 1 - P_0^{(1)}(t_0, t_0 + \tau, \mathbf{w})\,, \\ Q_0^{(2)}(t_0, t_0 + \tau, \mathbf{w}) &= 1 - P_0^{(2)}(t_0, t_0 + \tau, \mathbf{w})\,, \end{aligned}$$

Failures or Disruption Rate

$$\Lambda(t_0, \tau, \mathbf{w}) = -\frac{\partial}{\partial t_0} \ln P_0(t_0, \tau, \mathbf{w})\,.$$

Operational Availability

A_0 is the probability of success for a task that appears at a random time.

$$P(t_0, \tau, \mathbf{w}) = \sum_{I \in D} p_I P_I(t_0, \tau, \mathbf{w}) ,$$

where p_I is the probability that the task arrives when the system is in state e_I (one of the states of a Markov process), D is the set of all states, and $P_I(t_0, \tau, \mathbf{w})$ is the probability of success if the initial state is e_I.

MTTF for TRS

The mean time to failure for a time-redundant system is the time until the first violation of the redundant time restrictions:

$$\bar{T}_0(\tau, \mathbf{w}) = \int_0^{\infty} x \, dQ_0(x, \tau, \mathbf{w}) .$$

Mean Unproductively Spent Time until Task Performance

$$\bar{T}_2(t_0, \mathbf{w}) = \int_0^{\infty} y \, dP_0(y, t_0, \mathbf{w}) .$$

Mean Time of Task Completion

$$\bar{T}_1(t_0, \mathbf{w}) = t_0 + \bar{T}_2(t_0, \mathbf{w}) .$$

Availability Coefficient for a Given Time

$$K(\tau, \mathbf{w}) = P(0, \tau, \mathbf{w}) .$$

Probability of a System Starting Immediately or with an Acceptable Delay

The index showing that a task can be started either immediately or with an acceptable delay is similar to the availability index:

$$K(\tau, \mathbf{w}) = \frac{\bar{\xi}(\tau, \mathbf{w})}{\bar{\xi}(\tau, \mathbf{w}) + \bar{\eta}(\tau, \mathbf{w})} ,$$

where $\eta(\tau, \mathbf{w})$ is the first part of a long renewal cycle $(\eta > \tau)$ that equals $\eta_{N+1} - \tau$ (see Figure 9.1), $\xi(\tau, \mathbf{w})$ is all the rest of the time between neighboring intervals. In the figure the corresponding $\eta(\tau, \mathbf{w})$ is

$$\xi(\tau, \mathbf{w}) = \tau + \sum_{1 \le i \le N} \eta_i + \sum_{1 \le i \le = N+1} \xi_i ,$$

Here N is a random number of short intervals $(\eta_i \le \tau)$ of renewals between neighboring long intervals. Additional indexes will be introduced below.

9.2 ONE-CHANNEL SYSTEMS WITH NONDESTRUCTIVE FAILURES

9.2.1 General Case

We will consider a general probabilistic model for a system of the class $GG11$ with subclass $X0X0X$ (nondestructive failures and continuous monitoring). We also will consider several specific models that follow from this general model. The system consists of an arbitrary number of units connected in the general case in a network. The system is designed to perform a task of a fixed duration, t_0, and has a redundant time, τ. The system has a set of discrete states, E. Transition from state to state is semi-Markovian. This process is described with a matrix $\mathbf{P}(t) = |P_{ij}(t)|$, where $P_{ij}(t)$ is the probability of transition from state e_i to state e_j. The total space of states is divided into two subsets: E_0 and E_1. When the system is in a state in E_0, the time resource is not spent. When the system is in a state in E_1, the time resource expenditure is proportional to the time in the state. The probability of successful task performance can be determined from the following system of differential equations:

$$P_i(t_0, \tau) = 1 - F_i(t_0) + \sum_{j \in E} \int_0^{t_0} P_j(t_3 - x, \tau)\, dP_{ij}(x)\,, \qquad i \in E_0\,;$$

$$F_i(t) = \sum_{(j)} P_{ij}(t)\,,$$

$$P_i(t_0, \tau) = \sum_{i \in E} \int_0^{\tau} P_j(t_0, \tau - y)\, dP_{ij}(Y)\,, \qquad i \in E_1,\ E = E_0 \cup E_1\,. \tag{9.1}$$

The MTTF and the mean unproductive time before task performance are defined by the following equations:

$$E\{T_{0i}(\tau)\} = E\{T_i\} + \sum_{j \in E} p_{ij} E\{T_{0j}(\tau)\}\,, \qquad i \in E_0\,;$$

$$E\{T_i\} = \int_0^{\infty} x\, dF_i(x)\,,$$

$$E\{T_{0i}(t_0)\} = \sum_{j \in E} \int_0^{t_0} E\{T_{0j}(t_0 - x)\}\, dP_{ij}(x)\,, \qquad i \in E_1\,;$$

$$p_{ij} = \lim_{t \to \infty} P_{ij}(t)\,,$$

$$E\{T_{2i}(t_0)\} = \sum_{j \in E} \int_0^{t_0} E\{T_{2j}(t_0 - x)\}\, dP_{ij}(x)\,, \qquad i \in E_0\,;$$

$$E\{T_{2i}(t_0)\} = E\{T_i\} + \sum_{j \in E} p_{ij} E\{T_{2j}(t_0)\}\,, \qquad i \in E_1\,. \tag{9.2}$$

Here $E\{T_i\}$ is the mean time in state e_i.

The operational availability of the system can be found from the following integral equations, where a variable with a tilde is a residual time r.v., and π_k is a stationary probability of the embedded Markov chain.

$$\tilde{F}_{ij}(t) = \frac{1}{E\{T_{ij}\}} \int_0^t [1 - F_{ij}(x)]\, dx ,$$

$$F_{ij}(t) = \frac{P_{ij}(t)}{p_{ij}} ;$$

$$P(t_0, \tau) = \sum_{i \in E_0} P_i \left[1 - \tilde{F}_i(t) + \sum_{j \in E} \int_0^{t_0} P_j(t_0 - x, \tau)\, d\tilde{P}_{ij}(x) \right]$$

$$+ \sum_{i \in E_1} P_i \sum_{j \in E} \int_0^{\tau} P_j(t_0, \tau - y)\, d\tilde{P}_{ij}(y) ;$$

$$\tilde{F}_i(t) = \frac{1}{E\{T_i\}} \int_0^t [1 - F_i(x)]\, dx ;$$

$$\tilde{P}_{ij}(t) = \tilde{p}_{ij} \tilde{F}_{ij}(t) , \quad \tilde{p}_{ij} = \frac{p_{ij} e\{T_{ij}\}}{E\{T_i\}} , \quad E\{T_{ij}\} = \int_0^{\infty} x\, dF_{ij}(x) ;$$

$$P_i = \frac{\pi_i E\{T_i\}}{\sum_{j \in E} \pi_j E\{T_j\}} , \quad \pi = \sum_{k \in E} \pi_k p_{k_i} .$$

If the system has become stationary, the mean time until the total expenditure of time resource can be determined from:

$$E\{T_0(\tau)\} = \sum_{i \in E_0} P_i \left[E\{\tilde{T}_i\} + \sum_{j \in E} \tilde{P}_{ij} E\{T_{0j}(\tau)\} \right] + \sum_{i \in E_1} P_i \sum_{j \in E} E\{T_{0j}(\tau - x)\} ;$$

$$E\tilde{T}_i = \int_0^{\infty} x\, \tilde{d}F_i(x) ;$$

where $ET_{0i}(\tau)$ is the solution of (9.2).

We next turn to specific models.

9.2.2 A System with Unrenewable Time Resource (Subclass 00000)

Unit operation can be described with an alternating stochastic process with r.v.'s ξ_i and η_i (see Figure 9.1). All intervals ξ_i have distribution $F(t)$, and all intervals η_i have distribution $G(t)$. The system task is completed successfully if the total useful system operating time exceeds t_0 and the total downtime has been less than τ. System failure occurs when the time resource has been exhausted, and it is independent of the accumulated useful time.

For exponential distributions $F(t) = 1 - \exp(-\lambda t)$ and $G(t) = 1 - \exp(-\mu t)$ and initial system state e_0, the probability of system success is defined by

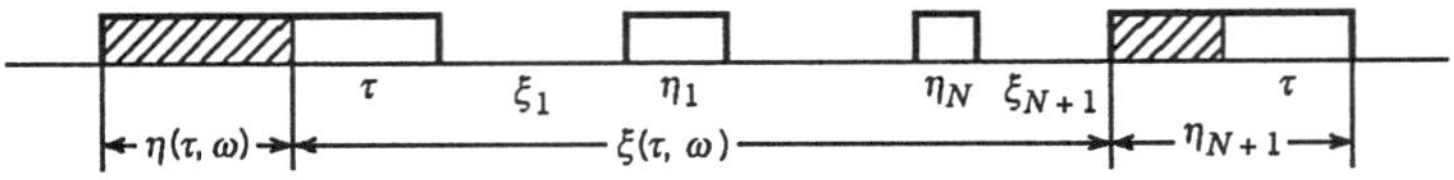

Figure 9.1 Schematic diagram of failure and renewal.

$$
\begin{aligned}
P_0(t_0, \tau) &= e^{-\lambda t_0} \sum_{0\le k<\infty} \frac{(\lambda t_0)^k}{k!} \left[1 - \sum_{0\le i\le k-1} \frac{(\mu\tau)^i}{i!} e^{-\mu\tau}\right] \\
&= 1 - e^{-\lambda t_0 - \mu\tau} \sum_{1\le k>\infty} \frac{(\lambda t_0)^k}{k!} \sum_{0\le i\le k-1} \frac{(\mu\tau)^i}{i!}, \qquad (9.3) \\
P(t_0, \tau) &= \mathrm{P}_0(\mathrm{t}_0, \tau) - \frac{\lambda}{\lambda+\mu} e^{-\lambda t_0 - \mu\tau} I_0(2\sqrt{\lambda\mu t_0 \tau}) .
\end{aligned}
$$

In the expressions above, $I_0(x)$ is a modified Bessel's Function of order 0:

$$I_0(x) = \frac{1}{(k!)^2} \sum_{0\le k<\infty} \left(\frac{x}{2}\right)^{2k} .$$

The two bounds of the probability can be found by calculating a finite number of terms in both presentations of the probability defined by (9.3).

The mean and the variance of the time to failure and the time of system task performance are

$$\bar{T}_0(\tau) = \frac{1+\mu\tau}{\lambda},$$

$$\bar{T}_1(t_0) = t_0\left(1 + \frac{\lambda}{\mu}\right);$$

$$\mathrm{Var}\{T_0(\tau)\} = \frac{1}{\lambda^2}(1 + 2\mu\tau),$$

$$\mathrm{Var}\{T_1(t_0)\} = \frac{2\lambda t_0}{\mu^2} .$$

The operational availability coefficient is

$$K(\tau) = 1 - \frac{\lambda}{\lambda+\mu} e^{-\mu\tau} .$$

For highly reliable units, when $E\{\eta\} \ll E\{\xi\}$, and for long tasks, $t_0 \gg E\{\xi\}$, the probability may be approximated with

$$P_0(t_0, \tau) \approx \Phi\left(\frac{\gamma-\rho}{\sqrt{2\rho}}\right);$$

$$P(t_0, \tau) \approx \Phi\left(\frac{\gamma-\rho+1}{\sqrt{2\gamma}}\right);$$

where

$$\rho = \lambda t_0 ,$$

$$\gamma = \mu \tau ,$$

$$\Phi(x) = \frac{1}{\sqrt{2\pi}} \int_{-\infty}^{x} e^{-y^2/2} \, dy .$$

For a fixed value $m_t = \tau / t_0$ we substitute τ for $m_t t_0$ in (9.3) and (9.4). If $m_t^* = \mu m_t / \lambda > 1$, the probability $P_0(t_0, m_t)$ first decreases with increasing t_0 but then begins to increase. The probability of success exceeds the minimum for $\rho_* = \lambda t_0^* \approx 1/m_t$ and equals

$$P_0(t_0^*, m_t) \approx e^{-1/\bar{m}_t} \left[1 + \frac{1 - e^{-1}}{\bar{m}_t} \right] .$$

Table 9.1 gives the probability value for (9.3), and Table 9.2 gives the probability values for $P_0(t_0, m_t)$.

The mean operational time during a given time interval can be found with:

$$E\{T_3(t)\} = \int_0^t \gamma(x) \, dx$$

where $\gamma(x)$ is the probability that the unit is in an up state at moment x.

For both exponential distributions:

$$\gamma(t) = \frac{\mu}{\lambda + \mu} + \frac{\lambda}{\lambda + \mu} e^{-(\lambda + \mu)t} ;$$

$$\bar{T}_3(t) = \frac{\mu t}{\lambda + \mu} + \frac{\lambda}{(\lambda + \mu)^2} (1 - e^{-(\lambda + \mu)t}) .$$

For large t the mean time in the up state is approximately $E\{T_3(t)\} \approx Kt$.

EXAMPLE 9.1

A computational task requires $t = 6$ hours on computer 1, which has productivity $C_1 = 100{,}000$ operations per second and parameters $\lambda_1 = 0.05$ per hour, and $\mu_1 = 1$ per hour. The probability of task performance, $P(t) = \exp(-\lambda_1 t) = 0.75$, is considered too low. Find a method to make reliability exceed 0.97 with structural or time redundancy.

Solution

As a structural redundancy solution let us consider the use of an additional computer of the same type in an active regime. Then

Table 9.1 **Probability of task completion for a fixed time resource**

	$\mu\tau$										
λt_0	0	1	2	3	4	5	6	7	8	9	10
0.1	0.9048	0.9632	0.9858	0.9945	0.9979	0.9992	0.9997	0.9999	—	—	—
0.2	0.8187	0.9267	0.9704	0.9881	0.9952	0.9981	0.9992	0.9997	0.9999	—	—
0.3	0.7408	0.8904	0.9539	0.9807	0.9920	0.9967	0.9986	0.9994	0.9998	0.9999	—
0.4	0.6704	0.8546	0.9364	0.9724	0.9881	0.9949	0.9978	0.9991	0.9996	0.9998	0.9999
0.5	0.6065	0.8193	0.9181	0.9633	0.9837	0.9928	0.9969	0.9986	0.9994	0.9997	0.9999
0.6	0.5488	0.7847	0.8990	0.9533	0.9787	0.9904	0.9957	0.9981	0.9992	0.9996	0.9998
0.7	0.4966	0.7508	0.8794	0.9426	0.9731	0.9875	0.9943	0.9974	0.9988	0.9995	0.9998
0.8	0.4493	0.7178	0.8591	0.9311	0.9668	0.9843	0.9926	0.9966	0.9984	0.9993	0.9997
0.9	0.4066	0.6586	0.8384	0.9189	0.9601	0.9807	0.9908	0.9956	0.9980	0.9991	0.9996
1.0	0.3679	0.6542	0.8175	0.9061	0.9528	0.9768	0.9898	0.9945	0.9974	0.9988	0.9994
2.0	0.1353	0.3943	0.6035	0.7540	0.8519	0.9139	0.9511	0.9730	0.9853	0.9921	0.9958
2.5	0.0821	0.2995	0.5041	0.6680	0.7871	0.8683	0.9208	0.9535	9.9733	0.9850	0.9917
3.0	0.0498	0.2250	0.4147	0.5833	0.7169	0.8149	0.8828	0.9278	0.9565	0.9744	0.9851
3.5	0.0330	0.1673	0.3366	0.5022	0.6443	0.7558	0.8380	0.8956	0.9344	0.9597	0.9758
4.0	0.0183	0.1234	0.2800	0.4269	0.5717	0.6930	0.7876	0.8575	0.9069	0.9406	0.9629
5.0	0.0067	0.0656	0.1686	0.2982	0.4351	0.5639	0.6755	0.7663	0.8364	0.8884	0.9256

Table 9.2 Probability of task completion for a fixed value of m_t

	m_t							
λt_0	1	2	3	4	5	6	8	10
0.1	0.9135	0.9213	0.9281	0.9349	0.9406	0.9555	0.9632	
0.2	0.8487	0.8737	0.8946	0.9121	0.9267	0.9389	0.9574	0.9704
0.3	0.7989	0.8452	0.8801	0.9078	0.9286	0.9452	0.9678	0.9808
0.4	0.7621	0.8286	0.8767	0.9114	0.9364	0.9537	0.9766	0.9881
0.5	0.7329	0.8193	0.8782	0.9181	0.9451	0.9633	0.9837	0.9928
0.6	0.7070	0.8142	0.8821	0.9248	0.9533	0.9705	0.9887	0.9957
0.8	0.6724	0.8128	0.8935	0.9395	0.9669	0.9814	0.9930	0.9984
1.0	0.6524	0.8174	0.9061	0.9528	0.9766	0.9886	0.9974	0.9994
1.5	0.6288	0.8339	0.9320	0.9691	0.0882	0.9964	0.9992	0.9999
2.0	0.6035	0.8519	0.9512	0.9853	0.9958	0.9989	0.9999	—
2.5	0.5918	0.8638	0.9647	0.9917	0.9981	0.9995	—	—
3.0	0.5833	0.8828	0.9744	0.9951	0.9992	0.9999	—	—
3.5	0.5769	0.8956	0.9814	0.9967	0.9996	—	—	—
4.0	0.5717	0.9069	0.9863	0.9984	0.9998	—	—	—
5.9	0.5669	0.9256	0.9925	0.9995	0.9999	—	—	—

$$P_0(t) \approx \exp\left(-\frac{2\lambda_i^2 t}{2\lambda_1 + \mu_1}\right) = 0.975 .$$

Another possibility is the use of computer 2 with productivity $C_2 = 200{,}000$ operations per second, which gives a t_0 of 3 hours, a τ of 3 hours, and an m_t ratio of 1. In some sense both variants are very close to each other. Using (9.3),

$$\begin{aligned} P_0(t_0, \tau) &= \exp(-0.05 \cdot 3)\{1 + 0.05 \cdot 3[1 - \exp(-3)] \\ &\quad + 0.01125[(1 - 4\exp(-3)]\} = 0.991 . \end{aligned}$$

When one uses a duplicate system the availability coefficient is

$$K = 1 - \frac{0.005}{1.105} = 0.9955 ,$$

and the probability of task performance $P(t_0, 0) = 0.97$. When one uses computer 2 for time redundancy during the same six hours, the availability coefficient is

$$K(\tau) = 1 - \frac{1}{21} e^{-3} = 0.9976 .$$

The coefficient of operational availability (i.e., the probability of task completion beginning at an arbitrary stationary moment) is

$$P(t_0, \tau) = 0.991 - \frac{1}{21}\, e^{-3.15}\left(1 + 0.45 + \frac{0.45^2}{4} + \frac{0.45^3}{36}\right) = 0.988\,.$$

If the use of computer 2 is accompanied with an increased failure rate, $\lambda_2 = 2\lambda_1$, then $P_0(t_0, \tau) = 0.981$. Notice that the probabilities $P_0(t)$ and $P_0(t, \tau)$ are equal if $\lambda_2 = 2.53\lambda_1$. If $C_2 = bC_1$, then $\rho = (0.3)/b$, $\gamma = 6(1 - 1/b)$ and the probabilities become equal for $b = 1.6$. Thus an increase in computer productivity of 60% is equivalent to the use of a second computer of type 1.

For arbitrary distributions of TTF and repair time the reliability indexes of the system can be found with asymptotic formulas:

$$P_0(t_0, \tau) \approx \Phi\left(\frac{\tau - E\{T_1\}t_0/E\{T_0\}}{\sqrt{[(E\{T_1\})^2 t_0/ET_0][\sigma_0^2/(ET_0)^2 + \sigma_2^2/(ET_1)^2]}}\right),$$

$$E\{T_0\}\bar{T}_0 = \int_0^\infty x\, dF(x)\,;$$

$$P_0(t_0, \tau) \approx \Phi\left(\frac{E\{T_0\}[1 + (\tau/E\{T_1\})] - t_0}{\sqrt{[(E\{T_0\})^2 \tau/E\{T_1\}][\sigma_0^2/(E\{T_0\})^2 + \sigma_2^2/(E\{T_1\})^2]}}\right),$$

$$E\{T_1\}\bar{T}_1 = \int_0^\infty x\, dG(x)\,;$$

$$\sigma_0^2 = \int_0^\infty x^2\, dF(x) - (E\{T_0\})^2\,;$$

$$\sigma_1^2 = \int_0^\infty x^2\, dG(x) - (E\{T_1\})^2\,.$$

The mean and variance of the useful operational time until the expenditure of the time resource, and the mean and variance of the time of the system task performance for large t_0 and τ are determined by

$$\bar{T}_0(\tau) = \left(1 + \frac{\tau}{\bar{T}_1}\right), \quad \bar{T}_1(t_0) = t_0\left(1 + \frac{\bar{T}_1}{\bar{T}_0}\right);$$

$$\operatorname{Var}\{T_0(\tau)\} = \frac{\tau\bar{T}_0^2}{\bar{T}_1}\left(\frac{\sigma_0^2}{\bar{T}_0^2} + \frac{\sigma_1^2}{\bar{T}_1^2}\right),$$

$$\operatorname{Var}\{T_1(\tau)\} = \frac{\tau\bar{T}_1^2}{\bar{T}_0}\left(\frac{\sigma_0^2}{\bar{T}_0^2} + \frac{\sigma_1^2}{\bar{T}_1^2}\right).$$

The availability coefficient for the given time is

$$K(\tau) = 1 - \frac{1}{\bar{T}_0 + \bar{T}_1}\int_0^\tau [1 - G(x)]\, dx\,.$$

Table 9.3 Reliability indexes, subclass EE11

Index	Formula
$P_0(t_0, \tau)$	$1 - I(K, \rho) + \sum_{i=1}^{\infty} I(ri, \gamma)[I(Ki, \rho) - I((i+1)K, \rho)],\ \rho = \lambda t_0,\ \gamma = \mu\tau$
$P_0^a(t_0, \tau)$	$\dfrac{\Phi(\gamma - r\rho/K)}{\sqrt{(1 + r/K)r\rho/K)}}$
	$\dfrac{\Phi(K + K\gamma/r - \rho)}{\sqrt{(1 + K/r)K\gamma/r)}}$
$\bar{T}_0(\tau)$	$\bar{T}_0 \sum_{i=0}^{\infty} I(ri, \gamma) \approx \dfrac{K}{\lambda r}\left(\gamma + \dfrac{r+1}{2}\right)$
$\bar{T}_1(t_0)$	$t_0 + \dfrac{r}{\mu} \sum_{i=1}^{\infty} I(Ki, \lambda t_0) \approx t_0\left(1 + \dfrac{r\lambda}{K\mu}\right) - \dfrac{r(K-1)}{2K\mu}$
$DT_0(\tau)$	$\dfrac{K^2}{\lambda^2}\left[1 + \dfrac{1}{K} + \sum_{i=1}^{\infty}\left(1 + \dfrac{1}{K} + 2i\right)I(ri, \gamma)\right] - \bar{T}_0^2(\tau)$
$DT_1(t_0)$	$\dfrac{r\lambda t_0[1 + (K/r)]}{K\mu^2}$
$K(\tau)$	$1 - \dfrac{\lambda}{K\mu + r\lambda}\left[r - \sum_{i=1}^{r} I(i, \gamma)\right] = K + (1 - K)\dfrac{1}{r}\sum_{i=1}^{r} I(i, \gamma),\ K = \dfrac{K\mu}{K\mu + r\lambda}$

Table 9.3 gives formulas for both gamma distributions of TTF: $F(t) = I(K, \lambda t)$ and $G(t) = I(r, \mu t)$. The corresponding formulas for the Weibull-Gnedenko TTF distribution, $F(t) = 1 - \exp\{-(t)^m\}$, and the gamma distribution of renewal time, $G(t) = I(r, \mu t)$, are in Table 9.4, where $\Gamma(x)$ is the gamma function. The coefficients a_{ij} and C_i used in these tables are given in Tables 9.5 and 9.6, respectively.

EXAMPLE 9.2

A system has a TTF distribution of the Weibull-Gnedenko type with parameters $m = 0.7$ and $\lambda = 0.0104$ per hour and an exponential distribution of repair time with parameter $\mu = 1$ per hour. The system has to perform a task that requires four days if there are no failures at all. To increase the probability of success, there is a redundant time of $\tau = 4$ hours. Find the probability of task completion, the mean useful operating time before resource exhaustion, and the mean time of the system task performance.

Table 9.4 Reliability indexes, subclass WE11

Index	Formula
$P_0(t_0, \tau)$	$e^{-\rho}\left[1 + \sum_{i=1}^{\infty} \frac{\rho^j}{i!} \sum_{j=1}^{i} a_{ji} I(ri, \gamma)\right]$
	$= 1 - \sum_{i=1}^{\infty} \frac{\rho^i}{i!} e^{-\rho} \sum_{\gamma=0}^{ir-1} e^{-\gamma} \frac{\gamma^\gamma}{\nu!} b_{[\gamma/r]+1,i}$,
	$\rho = (\lambda t_0)^m,\ \gamma = \mu\tau,\ b_{ri} = \sum_{j=r}^{i} a_{ji}$
$\bar{T}_0(\tau)$	$\bar{T}_0 \sum_{i=0}^{\infty} I(ri, \gamma) \approx \bar{T}_0\left(\frac{\gamma}{r} + \frac{r+1}{2r}\right),\ \bar{T}_0 = \frac{\Gamma(1+1/m)}{\lambda}$
$P_0^a(t_0, \tau)$	$\frac{\Phi[\gamma\Gamma(1+1/m) - r\rho]}{\sqrt{\rho r^2 A\Gamma(1+1/m)}}$,
	$A = \Gamma\left(1 + \frac{2}{m}\right) - 1 + \frac{1}{r}$,
	$\frac{\Phi[\gamma(r+1-\rho)/\Gamma(1+1/m)]}{\sqrt{\gamma A/r}}$
$\bar{T}_1(t_0)$	$t_0 + \sum_{j=1}^{\infty} \frac{r}{\mu} C_j \frac{\rho^j}{j!} e^{-\rho}$
	$\approx t_0\left(1 + \frac{r\lambda/\mu}{\Gamma(1+1/m)}\right) + \frac{r}{\mu}\left[\frac{\Gamma(1+2/m)}{2\Gamma^2(1+1/m)} - 1\right]$
$DT_1(t_0)$	$\frac{r^2}{\mu^2} \frac{2\Gamma(1+2/m)\lambda t_0}{\Gamma^3(1+1/m)} + \frac{r}{\mu^2} \sum_{i=1}^{\infty} C_i \frac{\rho^j}{i!} e^{-\rho}$
$K(\tau)$	$K + (1-K)\frac{1}{r}\sum_{i=1}^{r} I(i, \gamma),\ K = \frac{\Gamma(1+1/m)}{[\Gamma(1+1/m) + r\lambda/\mu]}$

Solution

The probability of task completion can be found with formulas from Table 9.4 for $\rho = (\lambda t)^m = (0.0104 \cdot 96)^{0.7} = 1$, $\gamma = \mu\tau = 4$. Using coefficients from Table 9.5 for calculations, we can find the upper and lower bounds:

$$
\begin{aligned}
\underline{P}_0 = {} & 0.3679(1 + 0.9817) + 0.1839(-0.3293 \cdot 0.9817 + 1.3293 \cdot 0.9084) \\
& + 0.0613(0.09358 \cdot 0.9817) - 1.1418 \cdot 0.9084 + 2.0482 \cdot 0.7619) \\
& + 0.0153(-0.9817 \cdot 0.025 + 0.9084 \cdot 0.6974 - 3.1575 \cdot 0.7619 \\
& + 3.4852 \cdot 0.5665),
\end{aligned}
$$

Table 9.5 Coefficients a_{ij} in the equation for the convolution of the Weibull-Gnedenko distribution

		m					
i	*j*	0.5	0.7	0.9	1.2	1.6	2.0
1	2	3	4	5	6	7	8
1	2	−0.57080	−0.32933	−0.10349	0.18559	0.47302	0.66667
2	2	1.57080	1.32933	1.10349	0.81441	0.52698	0.33333
1	3	0.28761	0.09358	0.00916	0.02958	0.19613	0.40000
2	3	−2.42920	−1.14178	−0.28900	0.37068	0.59912	0.53333
3	3	3.14159	2.04820	1.27984	0.59974	0.20475	0.06667
1	4	−0.13717	−0.02503	−0.00076	0.00443	0.07687	0.22857
2	4	2.70796	0.69737	0.05341	0.11826	0.47438	0.59048
3	4	−8.97300	−3.15746	−0.58723	0.46422	0.38370	0.17143
4	4	7.40220	3.48511	1.53458	0.41309	0.06504	0.00952
1	5	0.06343	0.00648	0.00006	0.00064	0.02920	0.12698
2	5	−2.64307	−0.37007	−0.00854	0.03249	0.32222	0.55873
3	5	16.8341	3.16547	0.17410	0.23060	0.48526	0.27937
4	5	−32.9937	−8.19176	−1.05219	0.46564	0.17254	0.03386
5	5	19.7392	6.38989	1.88656	0.27062	0.01777	0.00106
1	6	−0.02875	−0.00164	0.00000	0.00009	0.01088	0.06926
2	6	2.40484	0.18214	0.00126	0.00823	0.20121	0.48485
3	6	−26.2100	−2.61541	−0.04234	0.09351	0.44454	0.36825
4	6	91.4537	11.8791	0.44267	0.32117	0.27757	0.07273
5	6	−124.757	−20.8823	−1.76767	0.40767	0.06148	0.00481
6	6	58.1368	12.4381	2.36609	0.17033	0.00431	0.00010
1	7	0.01286	0.00041	0.00000	0.00001	0.00400	0.03730
2	7	−2.09699	−0.08560	−0.00018	0.00198	0.11922	0.39853
3	7	36.7466	1.93575	0.00919	0.03372	0.38200	0.42864
4	7	−202.871	−13.7020	−0.14751	0.17356	0.35050	0.12219
5	7	472.862	40.6548	0.98212	0.36232	0.12488	0.01280
6	7	−489.692	−53.2534	−2.86098	0.32472	0.01845	0.00053
7	7	186.038	25.4500	3.01735	0.01369	0.00094	0.00001
1	8	0.00568	−0.00010	−0.00001	0.00000	0.00146	0.01989
2	8	1.77770	0.03904	0.00003	0.00047	0.068180	0.31577
3	8	−48.2107	−1.33533	−0.00187	0.01126	0.30323	0.45970
4	8	394.925	13.8020	0.04279	0.08158	0.38229	0.17685
5	8	−1397.45	−61.3814	−0.42157	0.24775	0.19442	0.02605
6	8	2410.37	132.478	2.00193	0.35668	0.04539	0.00169
7	8	−1999.65	−136.953	−4.52451	0.24093	0.00484	0.00005
8	8	639.247	54.3509	3.90321	0.06135	0.00019	0.00000
1	9	0.00253	0.00001	−0.00002	−0.00001	0.00053	0.01053
2	9	−1.47712	−0.01726	0.00002	0.00011	0.03801	0.24371
3	9	60.4517	0.87721	0.00031	0.00356	0.22765	0.46530
4	9	−705.808	−12.7122	−0.01125	0.03485	0.37782	0.23139
5	9	3554.01	79.3941	0.15449	0.14416	0.25673	0.04485
6	9	−9071.39	−250.775	−1.06225	0.29601	0.08409	0.00403
7	9	12318.4	420.266	3.85514	0.31690	0.01400	0.00018

Table 9.5 (*Continued*)

		m					
i	*j*	0.5	0.7	0.9	1.2	1.6	2.0
1	2	3	4	5	6	7	8
8	9	−8491.00	−356.534	−7.04908	0.16902	0.00114	0.0000
9	9	2337.82	120.502	5.11264	0.03540	0.00004	0.0000
1	10	−0.00102	−0.00003	−0.00002	−0.00001	0.00021	0.0055
2	10	1.20876	0.00809	0.00012	0.00004	0.02079	0.1844
3	10	−73.3708	−0.55771	−0.00032	0.00105	0.16402	0.4512
4	10	1188.08	10.9936	0.00314	0.01391	0.34782	0.2814
5	10	−8173.03	−92.5091	−0.05120	0.07512	0.30275	0.0688
6	10	28933.7	400.662	0.47459	0.20607	0.13029	0.0080
7	10	−57143.8	−966.961	−2.45539	0.31054	0.03009	0.0004
8	10	63669.0	1314.41	7.13183	0.26003	0.00378	0.0000
9	10	−37438.0	−941.255	−10.8744	0.11325	0.00024	0.0000
10	10	9037.14	276.214	6.77163	0.01997	0.00001	0.0000

Table 9.6 Coefficients C_i for calculation of the MTTF and the variance of the task completion time

	m					
i	0.5	0.7	0.9	1.2	1.6	2.0
1	1	1	1	1	1	1
2	2.57080	2.32933	2.10349	1.81441	1.52698	1.33333
3	4.85398	3.95462	3.27068	2.57017	2.00862	1.66667
4	7.96857	5.83780	4.48270	3.28596	2.43691	1.96190
5	12.0010	7.94513	5.72935	3.97311	2.82745	2.22328
6	17.0035	10.2500	7.00434	4.63810	3.19049	2.45926
7	23.0006	12.7322	8.30338	5.28511	3.53218	2.67629
8	29.9994	15.3764	9.62338	5.91708	3.85653	2.87865
9	38.0000	18.1701	10.9620	6.53615	4.16636	3.06915
10	47.0005	21.1039	12.3174	7.14388	4.46369	3.24970

$$\begin{aligned}\bar{P}_0 = 1 &- 0.0067 - 0.1839(0.0183 + 1.3293 \cdot 0.0733) \\ &- 0.0613(0.0183 + 0.9064 \cdot 0.0733 + 2.0482 \cdot 0.1465) \\ &- 0.0153(0.0183 + 0.0733 \cdot 0.025 + 0.1465 \cdot 0.3276 + 0.1954 \cdot 3.4851)\,.\end{aligned}$$

The results of the calculations are given in Table 9.7.

Table 9.7 Probability of task completion

i	$\underline{P}_0$	$\bar{P}_0$	$(\underline{P}_0 + \bar{P}_0)/2$	*i*	$\underline{P}_0$	$\bar{P}_0$	$(\underline{P}_0 + \bar{P}_0)/2$
1	0.7291	0.9933	0.8612	3	0.9307	0.9484	0.9395
2	0.8931	0.9720	0.9326	4	0.9334	0.9358	0.9347

9.2.3 A System with Instantly Fulfilled Time Resources (Subclass 10000)

A system may be similar to the preceding system but have its failure time determined differently. System failure occurs when the idle time after the current failure exceeds the value τ_1, which can be, in general, an r.v. with the distribution $\Phi(t)$. Short idle periods $\eta < \tau_1$ can be either included in the useful operational time or not. In the first case the probability of successful operation is given by the equation system for the Laplace-Stieltjes transform:

$$q_0(S) = a_{01}(S)q_1(S) ,$$
$$q_1(S) = a_{10}(S)q_0(S) + a_{12}(S) ;$$
$$q_i(S) = \int_0^\infty e^{-St}\, dQ_i(t) ;$$
$$a_{ij}(S) = \int_0^\infty e^{-St}\, dP_{ij}(t) ;$$
$$P_{01}(t) = F(t) ,$$
$$P_{10} = \int_0^t [1 - \Phi(x)]\, dG(x) ,$$
$$P_{12}(t) = \int_0^t [1 - G(x)]\, d\Phi(x) . \tag{9.9}$$

The general solution of (9.9) for $E\{T_0\} \geq E\{T_1\}$ has the form

$$P_0(t) = 1 - Q_0(t) = 1 - q \sum_{0 \leq n < \infty} (1-q)^n F^{*(n+1)}(t) , \tag{9.10}$$

where

$$q = Pr\{\eta > \tau\} = \int_0^\infty [1 - G(x)]\, d\Phi(x) = p_{12} = \lim_{t\to\infty} P_{12}(t) ,$$

and $F^{*(n+1)}(t)$ is the $(n+1)$-order convolution of $F(t)$. For $F(t) = 1 - \exp(-\lambda t)$, solution (9.10) becomes

$$P_0(t) = e^{-q\lambda t}$$

For $\tau_1 = t_1 = \text{constant}$,

$$P_0(t) = \begin{cases} 1 & \text{for } t \leq t_1 , \\ \exp[-\lambda(1 - G(t_1))(t - t_1)] , & t \geq t_1 . \end{cases}$$

If the repair is fast ($E\{T_0\} \gg E\{T_1\}$) and q is small (i.e., the event $\eta > \tau_1$ is rare); the probability $P_0(t)$ can be evaluated with an asymptotic formula

$$P_0(t) \approx e^{-qt/\bar{T}_0},$$

$$\bar{T}_0 = \int_0^{\infty} x\, dF(x).$$

The MTTF can be found with (9.9) and

$$\bar{T}_0(\tau_1) = \lim_{S\to 0} \frac{1}{S}[1 - q_0(S)] = \frac{1}{q}\left\{\bar{T}_0 + \int_0^{\infty}[1 - G(t)][1 - \Phi(t)]\, dt\right\}.$$

$E\{T_0(\tau)\}$ can be approximated with the first term, which is the main one.

The availability coefficient for the given time is

$$K(\tau_1) = 1 - q\,\frac{\bar{T}_1}{\bar{T}_0 + \bar{T}_1}.$$

If the short repair intervals are not included in the useful operational time, $F(t) = 1 - \exp(-\lambda t)$ and $\tau_1 = t_1 = $ a constant, then

$$P_0(t) = e^{-\lambda t[1 - G(t_1)]}, \tag{9.12}$$

where t is the useful operating time needed to perform the task. Formula (9.12) is valid not only when $E\{T_0\} \gg E\{T_1\}$ but also for any relations between those variables. The MTTF obtained from (9.12) is

$$\bar{T}_0(t_1) = \frac{\bar{T}_0}{q} = \frac{\bar{T}_0}{1 - G(t_1)}.$$

The availability coefficient is

$$K(t_1) = \frac{\bar{T}_0 + \int_0^{t_1} x\, dG(x)}{\bar{T}_0 + \bar{T}_1}$$

EXAMPLE 9.3

A system is designed to receive, process, and transmit information in real time. A delay in information transmission of more than 20 minutes is considered to be disruption of operation.

Find reliability indexes of the system when the failure rate is $\lambda = 0.04$ per hour, the mean repair time is $\tau_1 = 1/\mu = 0.2$ hour, and the time needed for task performance is $t = 8$ hours.

Solution

The probability of successful operation is determined by (9.11):

$$P_0(t) = \exp\{-0.04(8 - 0.33)\exp(-5 \cdot 0.33)\} = 0.943 .$$

If there is no time redundancy,

$$P_0(t) = \exp(-0.04 \cdot 8) = 0.726 .$$

The gain measured in the ratio of the complementary probabilities is $G_Q = 0.274/0.057 = 4.8$. The MTTF is $E\{T_0(t_1)\} = 25\exp(5 \cdot 0.33) = 130$ hours. The gain measured in the MTTF ratio is $G_T = 130/25 = 5.2$.

The availability coefficients correspondingly are $K(t_1) = 1 - \exp(-5 \cdot 0.33)0.2/25.2 = 0.9985$ and $K(0) = 25/25.2 = 0.992$. The gain in this case measured by the complementary probabilities is $G_K = [1 - K(0)]/[1 - K(t_1)] = 0.008/0.0015 = 5.3$.

9.2.4 A System with Combined Time Redundancy (Subclass 20000)

A system has two kinds of time resources: τ and τ_1. The resource τ_1 is random with distribution $\Phi(t)$ and can be instantly replenished after repair. Resources τ is nonfulfilled, and its expenditure begins when resource τ_1 is totally exhausted. System failure occurs upon exhaustion of the unrestored time resource. All idle periods η_i that are smaller than τ_1 are included in the useful system operating time. Under these conditions the probability of successful completion of a task of duration t_0 can be found from the following equation system:

$$P_0(t_0, \tau) = 1 - F(t_0) + \int_0^{t_0} P_1(t_0 - x, \tau)\, dF(x) ,$$

$$P_1(t_0, \tau) = [1 - G(t_0)][1 - \Phi(t_0)] + \int_0^{t_0} [1 - \Phi(x)] P_0(t_0 - x, \tau)\, dG(x)$$

$$+ \int_0^{t_0} d\Phi(x) \int_x^{\tau + x} P_0(t_0 - x, \tau + x - y)\, dG(y) . \tag{9.13}$$

For $F(t) = 1 - \exp(-\lambda t)$ and $G(t) = 1 - \exp(-\mu t)$, (9.13) yields

$$P_0(t_0, \tau) = e^{-\rho}\left[1 + \sum_{1 \le K < \infty} \frac{\rho^K}{K!} I(K, \mu\tau)\right] ,$$

$$\rho = \lambda(t_0 - \bar{\tau}_1)\varphi(\mu) ,$$

$$\bar{\tau}_1 = \int_0^{\infty} x\, d\Phi(x) ,$$

$$\varphi(\mu) = \int_0^{\infty} e^{-\mu x}\, d\Phi(x) , \qquad t_0 > \bar{\tau}_1 . \tag{9.14}$$

For $\Phi(t) = 0$, formula (9.14) coincides with (9.3); for $\tau = 0$ and $\tau_1 = t_1$ it coincides with (9.11).

The mean time until the system disruption lies between the following values:

$$E\{T_0^{(1)}(\tau)\} = \frac{E\{T_0\} + E\{T_1\}}{2A^2} \int_0^\infty \int_x^\infty (u-x)^2 \, dG(u) \, d\Phi(x) + \frac{E(T_0) + E(T_1)\tau}{A} ,$$

$$E\{T_0^{(2)}(\tau)\} = (E\{T_0\} + E\{T_1\})\left(\frac{1}{q} + \frac{\tau}{A}\right),$$

where

$$A = \int_0^\infty \int_x^\infty (u-x) \, dG(u) \, d\Phi(x) ,$$

$$E\{T_0\} = \int_0^\infty x \, dF(x) ,$$

$$E\{T_1\} = \int_0^\infty [1 - G(x)][1 - \Phi(x)] \, dx .$$

For small τ the value $E\{T_0(\tau)\}$ is closer to $E\{T_0^{(2)}(\tau)\}$. For large τ this value is closer to $E\{T_0^{(1)}(\tau)\}$. For exponential distributions $G(t) = 1 - \exp(-\mu t)$ and $F(t) = 1 - \exp(-\lambda t)$, and under the condition $\tau_1 = t_1$,

$$E\{T_0(\tau, t_1)\} = \frac{1}{\mu} (1 + \mu\tau)\left[e^{\mu t_1}\left(1 + \frac{\mu}{\lambda}\right) - 1\right].$$

One may use the availability coefficient as a probability of completion of a very short task ($t \approx 0$). In addition to the previous assumptions, one should consider how the time resource is used during the repair if there is no system task on hand.

Let us consider two conditions:

(1) Time resource τ_1 is being used up even if there is no task to perform, and time resource τ is being used up only from the moment of task arrival.

(2) Time resource τ_1 is not being used up before task arrival, but for task completion it is necessary to renew the system during a time that does not exceed $\tau_1 + \tau$.

In both cases the availability coefficient can be calculated with

$$K^{(1)}(\tau) = 1 - \frac{1}{E\{T_0\} + E\{\eta\}} \int_0^\infty \Phi(x)[1 - G(x+\tau)] \, dx ,$$

$$E\{\eta\} = \int_0^\infty x \, dG(x) . \tag{9.15}$$

If the time resource τ_1 is being used up before task arrival, and if the unit must be repaired after its arrival even though τ_1 has not been exhausted, the availability coefficient is written

$$K^{(2)}(\tau) = 1 - \frac{1}{E\{T_0\} + E\{\eta\}} \int_0^\infty [1 - G(x+\tau)][\Phi(x)\,dx + x\,d\Phi(x)]\,.$$

For $G(t) = 1 - \exp(-\mu t)$, formula (9.15) has the form

$$K^{(1)}(\tau) = 1 - \frac{qe^{-\mu\tau}}{1 + \mu E\{T_0\}}\,, \qquad q = \varphi(\mu)\,.$$

If the task has duration t_0, the mean performance time lies between the two following values:

$$E\{T_1^{(1)}\} = t_0\left(1 + \frac{B}{E\{T_0\} + E\{T_1\}}\right),$$

$$\begin{aligned} E\{T_1^{(2)}(t_0)\} = E\{T_1^{(1)}\} &+ \frac{1}{E\{T_0\} + E\{T_1\}} \\ &\times \left(\left[\int_0^\infty x\,d\Phi(x)\right]\int_0^x [1 - G(u)]\,du - BE\{T_0\} - E\{\eta\}E\{\tau_1\}\right) \\ &+ \frac{B(\alpha_0 + \alpha_1 + E\{T_0\}E\{T_1\})}{2(E\{T_0\} + E\{T_1\})^2}\,, \end{aligned} \tag{9.16}$$

where

$$\alpha_0 = \int_0^\infty x^2\,dF(x)\,, \qquad B = E\{\eta\} - \int_0^\infty \mathrm{d}\phi(x)\int_0^x [1 - G(u)]\,\mathrm{d}u\,.$$

For an exponential repair time distribution,

$$\bar{T}_1 = \frac{1}{\mu}[1 - \varphi(\mu)]\,,$$

$$B = \frac{1}{\mu}\varphi(\mu)\,,$$

$$\alpha_1 = \int_0^\infty t^2 e^{-\mu t}[\mathrm{d}\Phi(t) + (1 - \Phi(t))\,d(\mu t)]$$

The mean task performance time is

$$\bar{T}_1(t_0) \approx t_0\left\{1 + \frac{\varphi(\mu)}{1 - \varphi(\mu) + \mu\bar{T}_0}\right\},$$

and for $\tau_1 = t_1$,

$$\bar{T}_1(t_0) = t_0\left\{1 + \frac{e^{-\mu t_1}}{\mu + \lambda(1 - e^{-\mu t_1})}\right\}.$$

EXAMPLE 9.4

A technological conveyer line has a failure rate $\lambda = 0.05$ per hour and mean repair time $E\{\eta\} = 10$ minutes. An interruption is acceptable if it does not exceed 15 minutes. If the idle time is longer, operation continues with the help of stock that is enough for 30 minutes.

Find the probability of successful system operation for 5 hours, and find the mean time until exhaustion of the stock.

Solution

Input data for the calculation are $\mu t_1 = 15/10 = 1.5$, $\lambda t_0 = 0.25$, $\lambda t_1 = 0.05 \cdot 0.25 = 0.0125$, $\mu\tau = 3$. The probability of successful system operation can be calculated with (9.14): $P_0(t_0, \tau) = 0.9974$. The mean time to perform for a 5-hour task is given by

$$E\{T_1(t_0) = 5\left\{\frac{1 + \exp(-1.5)}{1 - \exp(-1.5) + 120}\right\} = 5.008 \text{ hr} .$$

9.2.5 A Series System with an Unreplenished Time Resource (Subclass 00000)

A system of n units in series can continue operating after any failure if it has a time resource of value τ. All units are independent and different, the distribution of the random TTF is $R_i(t)$, and the repair time distribution is $G_i(t)$. During repair of any unit the system is switched off, so no new failure can occur during this period. Any failure is discovered immediately with monitoring equipment. The system is described with a semi-Markov model that has $n + 1$ states: e_0 is the state of the total operable system, and e_i is a state of system failure because of the ith unit. We introduce two subsets: $E_0 = e_0$ and $E_1 = \{e_i, i = 1, 2, \ldots, n\}$. Using equation system (9.1), the transition probability functions are

$$P_{0i}(t) = \int_0^t [1 - F_0(x)]\lambda_i(x)\, dx ,$$

$$F_0(t) = 1 - \prod_{1 \le i \le n} [1 - R_i(t)] ,$$

$$P_{i0}(t) = G_i(t) ,$$

$$\lambda_i(t) = \frac{R_i'(t)}{1 - R_i(t)} .$$

For exponential distributions $R_i(t) = 1 - \exp(-\lambda_i t)$ and $G_i(t) = 1 - \exp(-\mu_i t)$,

Table 9.8 Reliability indexes, series system, subclass 00000

Index	Formula
$\bar{T}_0(\tau)$	$\dfrac{(\rho\tau + \Sigma_{i=1}^{n} \rho_i/\mu_i)}{\rho^2}, \ \rho_i = \dfrac{\lambda_i}{\mu_i}, \ \rho = \sum_{i=1}^{n} \rho_i$
$\mathrm{D}T_0(\tau)$	$\dfrac{[2\tau \Sigma_{i=1}^{n} \rho_i/\mu_i + 5(\Sigma_{i=1}^{n} \rho_i/\mu_i)^2/\rho - 4 \Sigma_{i=1}^{n} \rho_i/\mu_i^2]}{\rho^3}$
$\bar{T}_1(t_0)$	$(1+\rho)t_0$
$\mathrm{D}T_1(t_0)$	$2t_0 \sum_{i=1}^{n} \dfrac{\rho_i}{\mu_i}$

$$P_0(t_0, \tau) = e^{-\lambda t_0}\left[1 + \sum_{1 \le i \le \infty} \frac{(\lambda t_0)^i}{i!} G^{*(i)}(\tau)\right],$$

$$\lambda = \sum_{1 \le i \le n} \lambda_i, \tag{9.17}$$

where $G^{*(i)}(t)$ is the ith order convolution,

$$G(t) = \sum_{1 \le i \le n} \frac{\lambda_i}{\lambda} [1 - e^{-\mu_i t}].$$

It follows that for identical $G_i(t)$ the system can be replaced with a single equivalent unit with failure rate λ. If μ_i's differ, however, numerical calculations become involved.

If the TTF is a r.v. with a Weibull-Gnedenko distribution, $R_i(t) = 1 - \exp[-(\lambda_i t)^m]$, the probability of successful performance can be found using (9.4). $I(ri, \gamma)$ must be replaced with the function $G^{*(i)}(t)$ and

$$G(t) = \sum_{1 \le i \le n} \left(\frac{\lambda_i}{\lambda}\right)^m G_i(t),$$

$$\lambda = \sum_{1 \le i \le n} \lambda_i^m. \tag{9.18}$$

Formulas for the moments of the r.v.'s $T_0(\tau)$ and $T_1(t_0)$ for both exponential distributions are given in Table 9.8. For other distributions the reliability can be calculated using (9.5) through (9.8) if the system can be reduced to a single equivalent unit.

EXAMPLE 9.5

A system of three units is characterized by the following parameters: $\lambda_1 = 0.02$ per hour, $\lambda_2 = 0.001$ per hour, $\lambda_3 = 0.0001$ per hour, $\mu_1 = 6$ per hour, $\mu_2 = 0.6$ per hour, $\mu_3 = 0.09$ per hour. A task of duration $t_0 = 16$ hours should be performed with a delay of not more than 20 minutes. Find the probability of task completion, the mean and the variance of the useful operating time before time resource exhaustion, and the mean and the variance of the task performance time.

Solution

The probability of task performance can be calculated using (9.17). To simplify the convolution calculations, let us replace $G(t)$ with $G_1(t)$ or with $(\lambda_1/\lambda)G_1(t)$, $\lambda = \lambda_1 + \lambda_2 + \lambda_3$. The first gives the upper bound, and the second gives the lower bound:

$$\begin{aligned}\overline{P_0(t_0, \tau)} = e^{-0.0211\cdot 16}\{1 &+ 0.3376(0.9479[1 - e^{-2}] \\ &+ 0.0474[1 - e^{-2}] + 0.00473[1 - e^{-0.03}]) \\ &+ 0.057(0.0479)^2[1 - 3e^{-2}] + 0.0064(0.9479)^3[1 - 5e^{-2}] \\ &+ 0.00054(0.9479)^4[1 - 6.333e^{-2}]\} \\ = 0.7135&(1.2796 + 0.9479 \cdot 0.034) \\ \underline{P_0(t_0, \tau)} = 0.7135&(1.2796 + 0.036) = 0.9387 .\end{aligned}$$

With no time resource, $P_0(t) = 0.7135$. The mean task performance time is

$$E\{T_1(t_0)\} = 16\left(1 + \frac{0.02}{6} + \frac{0.001}{0.6} + \frac{0.0001}{0.09}\right) = 16.1\ h .$$

The variance is

$$\mathrm{Var}\{T_1(t_0)\} = 32\left(1 + \frac{0.02}{36} + \frac{0.001}{0.36} + \frac{0.0001}{0.0081}\right) = 0.5017 ,$$

$$\sigma = \sqrt{\mathrm{Var}\{T_1(t_0)\}} = 0.7083 ,$$

$$r = \frac{\sigma}{ET_1(t_0)} = 0.044 .$$

The mean time until time resource exhaustion is

$$E\{T_0(\tau)\} = \frac{(0.0061}{3} + \frac{0.01568}{0.0061^2} = 474.4\ \mathrm{h} .$$

With no time resource, $E\{T_0(0)\} = 47.4$ hours. The gain is $G_T = 10$. The variance is

$$\operatorname{Var}\{T_0(\tau)\} = \left[32 \cdot 0.0157 + \frac{0.0157^2}{0.0061} - 4\left(\frac{0.02}{216} + \frac{0.001}{0.216} + \frac{0.0001}{0.00073}\right)\right]$$
$$= 593{,}000 \text{ h}^2 ,$$

$$\sigma_0 = 770 \text{ h} ,$$

$$r_0 = \frac{\sigma_0}{E\{T_0(\tau)\}} = 1.62 .$$

9.2.6 A Series System with Combined Time Resource (Subclass 20000)

The preceding system could have additional, separate, instantly replenished time resources. The resource for the ith unit is τ_{1i}. It has the distribution $\Phi_i(t)$ and can be used only by its unit. An unreplenished common time resource of τ can be spent only after an individual time resource has been exhausted. This common resource is available to all units. The probability of successful system operation can be found from the following integral equation system:

$$P_0(t_0, \tau) = 1 - F_0(t_0) + \sum_{1 \le i \le n} \int_0^{t_0} P_1(t_0 - x, \tau)\, dP_{0i}(x) ,$$

$$P_i(t_0, \tau) = [1 - G_i(t_0)][1 - \Phi_i(t_0)] + \int_0^{t_0} [1 - \Phi_i(x) P_0(t_0 - x, \tau)\, dG_i(x)$$
$$+ \int_0^{t_0} d\Phi_i(x) \int_x^{\tau + x} P_0(t_0 - x, \tau + x - y)\, dG_i(y) , \qquad i = 1, \ldots, n . \tag{9.19}$$

For exponential distributions $R_i(t) = 1 - \exp(-\lambda_i t)$ and $G_i(t) = 1 - \exp(-\mu_i t)$, the solution of (9.19) has the form

$$P_0(t_0, \tau) = e^{-\rho}\left[1 + \sum_{1 \le K < \infty} \frac{1}{K!} \rho^K G^{*(K)}(\tau)\right] ,$$
$$\rho = \Lambda t_0 - \sum_{1 \le i \le n} \Lambda_i \bar{\tau}_{1i} . \tag{9.20}$$
$$\Lambda = \sum_{1 \le i \le n} \Lambda_i ,$$
$$\Lambda_i = \lambda_i \varphi_i(\mu_i) ,$$
$$G(\tau) = \sum_{1 \le i \le n} \frac{\Lambda_i}{\Lambda} [1 - e^{-\mu_1 \tau}] ,$$
$$t_0 > \max_{(i)} \bar{\tau}_{1i} ;$$

Table 9.9 Reliability indexes, series system, subclass 20000

Index	Formula
$\bar{T}_0(\tau)$	$\left(\bar{T}_0 + \sum_{i=1}^{n} p_{0i}\bar{T}_i\right)\left(\frac{\tau}{A} + \frac{\Sigma_{i=1}^{n} \lambda_i \rho_{i,n+i}}{\lambda\mu_i^2 A^2}\right),$
	$p_{i,n+i} = \varphi_i(\mu_i),\ A = \sum_{i=1}^{n} \frac{\lambda_i p_{i,n+i}}{\lambda\mu_i},\ \varphi_i(\mu_i) = \int_0^\infty e^{-\mu_i x}\, d\Phi_i(x)$
$\bar{T}_1(r_0)$	$t_0 + \frac{1}{C} \frac{\sum_{i=1}^{n} \frac{\lambda_i}{\lambda\mu_i}\left[\varphi_i(\mu_i)\left(t_0 - \frac{1}{\lambda}\right) - \bar{C}_i\right] + \frac{1}{C^2}\sum_{i=1}^{n}\frac{\lambda_i}{\lambda\mu_i}}{}$
	$\times\, \varphi_i(\mu_i) \sum_{i=1}^{n} \frac{\lambda_i}{\lambda}\left[\alpha_{2i} + \frac{1}{\lambda}\bar{T}_i + \frac{1}{\mu_i}(\bar{T}_i - \bar{C}_i)\right],\ C = \sum_{i=1}^{n}\frac{\lambda_i}{\lambda}$
	$\times\left(\frac{1}{\lambda} + \bar{T}_i\right),\ \bar{T}_i = \frac{1}{\mu_i}[1 - \varphi_i(\mu_i)],\ \alpha_{2i} = \frac{\lambda}{2}\int_0^\infty x^2 e^{-\lambda x}\, dx,$
	$\bar{C}_i = \int_0^\infty x e^{-\mu_i x}\, d\Phi_i(x)$
$K^1(\tau)$	$1 - \frac{1}{1+\rho}\sum_{i=1}^{n} \rho_i \varphi_i(\mu_i) e^{-\mu_i \tau},\ \rho_i = \frac{\lambda_i}{\mu_i},\ \rho = \sum_{i=1}^{n} \rho_i$
$K^2(\tau)$	$k^{(1)}(\tau) - \frac{1}{\rho}\sum_{i=1}^{n} \lambda_i \bar{C}_i e^{-\mu_i \tau}$

Other reliability indexes are given in Table 9.9. Superscripts in $K^{(i)}(\tau)$ correspond to the description in Section 9.2.4. Section 9.2.4 has formulas for $n = 1$, and Section 9.2.3 has formulas for $\tau = 0$.

9.2.7 A System with an Unreplenished Time Resource and Common Structural Redundancy (Subclass 00100)

Consider a system consisting of n main and k redundant units. Suppose that all distributions of failure and repair times are exponential, there is a single repair facility, and after system failure, operation does not resume until the entire repair is completed. In this case the probability of successful task performance is

$$P_0(t_0, \tau) \approx e^{-t_0/\bar{T}_0}\left[1 + \sum \frac{1}{i!}\left(\frac{t_0}{\bar{T}_0}\right)^i I(ki, \mu\tau)\right],$$

$$\bar{T}_0 = \sum_{0 \le i \le k} \sum_{0 \le S \le i} \frac{\kappa_S}{\lambda_i \kappa_i},$$

$$\kappa_S = \prod_{1 \le i \le S} \frac{\lambda_{j-1}}{\mu_j}.$$

Here λ_i and μ_i are Markovian transition rates.

If after a failure the system begins to operate as soon as n units are ready,

$$P_0(t_0, \tau) \approx e^{-\Lambda_0 t_0}\left[1 + \sum_{1\le i<\infty} \frac{(\Lambda_k t_0)^i}{i!} I(i, \mu_{k+1}\tau) \frac{e^{\Lambda_0 t_0} - 1}{e^{\Lambda_k t_0} - 1}\right];$$

$$\Lambda_0 = \frac{1}{\bar{T}_0},$$

$$\Lambda_k = \left(\sum_{1\le i\le k} \frac{\kappa_i}{\lambda_k \kappa_k}\right)^{-1}.$$

If the redundancy is "slipping"—that is, any redundant unit can be used for any main unit—then for arbitrary distributions $F(t)$ and $G(t)$ the approximate formulas can be based on the asymptotic expressions for very reliable systems when the repair is relatively fast:

$$P_0(t_0, \tau) = \sum_{1\le i<\infty} \frac{[H_0(t_0)]^i}{i!} e^{-H_0(t_0)} I(i, \gamma),$$

$$\gamma = -\ln[1 - \tilde{G}(\tau)],$$

$$H_0(t_0) = \int_0^{t_0} h_0(x)\, dx,$$

$$h_0(t) = [h(t)]^{k+1} \int_0^\infty \left\{\int_x^\infty [1 - G(u)]\, du\right\}^{r-1} \frac{x^{k-r}}{(k-r)!} \frac{dx}{(r-1)!},$$

$$\tilde{G}(t) = 1 - \frac{\int_0^\infty \{\int_x^\infty [1 - G(u)]\, du\}^{r-1}[1 - G(x+t)]x^{k-r}\, dx}{\int_0^\infty \{\int_x^\infty [1 - G(u)]\, du\}^{r-1} x^{k-r}\, dx},$$

where $h(t)$ is an intensity function of the failure flow for a nonredundant system and r is the number of repair facilities. For common redundancy with a single active "slipping" redundant unit and $r = n + 1$,

$$h_0 \sim (n+1)\bar{\eta}^n [h(t)]^{n+1},$$

$$\tilde{G}(t) = G(t).$$

For a standby unit,

$$H_0(t_0) = \frac{q t_0}{\bar{\xi}},$$

$$q = \int_0^\infty [1 - G(x)]\, dF(x),$$

$$\bar{\xi} = \int_0^\infty x\, dF(x),$$

$$\gamma = -\ln[1 - G(\tau)].$$

EXAMPLE 9.6

A system consists of two computers with a productivity of 100,000 operations per second. One computer can perform the task in $t = 6$ hours with no failures. The redundant computer is in an active redundancy regime. Both TTF and repair time distributions are exponential, with parameters $\lambda = 0.05$ per hour and $\mu = 1$ per hour. The probability of successful task operation is

$$P(t) = e^{-2\lambda^2 t/3\lambda + \mu} = 0.974 .$$

A designer introduces a new processor that increases productivity to $C = 150{,}000$ operations per second. Find the probability of performance of the same task in the same time interval when the parameters λ and μ remain the same.

Solution

Following the input data: $k = 1$, $t_0 = 6/1.5 = 4$ hour, $\tau = 6 - 4 = 2$ hours, $\Lambda_0 = 4.35 \cdot 10^{-3}$ per hour, $\Lambda_1 = 4.55 \cdot 10^{-3}$ per hour, $\mu_2 = \mu$, $\mu\tau = 2$, and $\Lambda_1 t_0 = 0.0182$. The probability of the task performance is

$$P_0(t_0, \tau) = \exp(-0.0174)\left\{1 + \left(\frac{0.01754}{0.01836}\right)(0.0182[1 - \exp(-2)] + 1.65 \cdot 10^{-4}[1 - 3\exp(-2)])\right\} = 0.9976 .$$

The failure probability is less than 1/10 as high. For comparison, a single computer with a productivity of 150,000 operations per second performs the same task with a success probability of 0.971.

9.3 ONE-CHANNEL SYSTEMS WITH DEPRECIATING FAILURES

9.3.1 General Case

The general probabilistic model for the system of class GG11, subclass X1X0X (depreciating failures and continuous monitoring) is considered here along with some related particular models. The system has two subsets of states: up states, E_0, and down states, E_1. In a state $e \in E_0$, the system during time x performs useful work of magnitude x and does not waste a time resource. When the system passes from $e \in E_0$ to $e \in E_1$, all completed work is depreciated. At the same moment the time resource y instantly decreases to the value of $y - x$. If $x > y$, the transition immediately leads to system failure. The depreciation of the completed work is caused by incorrectable consequences of the failures. For instance, there might be destruction of input data or programs in a computer system, or there might be production of unacceptable items in a factory. In a state $e \in E_1$ useful operating time does not accumulate; thus after system transition from the E_1 into subset E_0 performance of the task starts from the very beginning.

The model is semi-Markovian, as was described in Section 9.2. $P(t) = \|P_{ij}(t)\|$ and

$$p_{ij} = \lim_{t\to\infty} P_{ij}(t) ,$$

$$F_{ij}(t) = \frac{P_{ij}(t)}{p_{ij}} .$$

The probability of successful performance of a task of duration t_0, with an unrestored time resource of volume τ under the condition that the initial state is i, can be obtained from the following system of integral equations:

$$P_i(t_0, \tau) = 1 - F_i(t_0) + \sum_{i\in E_0} \int_0^{t_0} P_j(t_0 - x, \tau)\, dP_{ij}(x)$$

$$+ \sum_{i\in E_1} \int_0^{t_0} P_j(t_0, \tau - y)\, dP_{ij}(y) , \qquad i \in E_0 ;$$

$$P_i(t_0, \tau) = \sum_{j\in E} \int_0^{\tau} P_j(t_0, \tau - x)\, dP_{ij}(x) ,$$

$$i \in E_1,\ E = E_0 \cup E_1,\ a = \min(t_0, \tau) . \tag{9.22}$$

A Laplace-Stieltjes transformation changes (9.22) into a system of algebraic equations:

$$P_i^*(t_0, \omega) = 1 - F_i(t_0) + \sum_{j\in E_1} a_{ij}(t_0, \omega) P_j^*(t_0, \omega) , \qquad i \in E_0 ,$$

$$P_i^*(t_0, \omega) = \sum_{j\in E} a_{ij}(t_0, \omega) P_j^*(t_0, \omega) , \qquad i \in E_1 , \tag{9.23}$$

where a_{ij} and P_i^* are LST defined as

$$a_{ij}(t_0, \omega) = \int_0^{t_0} e^{-\omega x}\, dP_{ij}(x) ,$$

$$P_i^*(t_0, \omega) = \omega \int_0^{\infty} e^{-\omega\tau} P_i(t_0, \tau)\, d\tau .$$

The mean expenditure of the time resource before successful completion of the task during t_0 can be found from the equation system:

$$E\{T_{2i}(t_0)\} = E\{T_i(t_0)\} + \sum_{j\in E_1} P_{ij}(t_0) E\{T_{2j}(t_0)\} , \qquad i \in E_0 ;$$

$$E\{T_i(t_0)\} = \int_0^{t_0} x\, dF_i(x) ;$$

$$E\{T_{2i}(t_0)\} = E\{T_i\} + \sum_{j\in E} p_{ij}E\{T_{2j}(t_0)\}\,, \qquad i \in E_1\,;$$

$$E\{T_i\} = \int_0^{t_0} x\, dF_i(x)\,.$$

From the general model corresponding results can easily be found for particular cases.

9.3.2 A Unit with a One-Stage Task (Subclass 01000)

A unit can be in one of two states: e_0 or e_1. Distributions of the random failure and repair times are $F(t)$ and $G(t)$, respectively. For this case:

$$P_0^*(t_0, \omega) = 1 - F(t_0) + P_1^*(t_0, \omega) \int_0^{t_0} e^{-\omega x}\, dF(x)\,,$$

$$P_1^*(t_0, \omega) = g(\omega) P_0^*(t_0, \omega)\,,$$

$$g(\omega) = \int_0^{\omega_0} e^{-\omega x}\, dG(x)\,.$$

If both distributions are exponential then

$$P_0^*(t_0, \omega) = \frac{(1+a)e^{-\lambda t_0}}{1 + ae^{-(\lambda + \omega t_0)}}\,,$$

$$a = \frac{\lambda\mu}{\omega(\omega + \lambda_\mu)}\,.$$

The formulas we use for calculations are:

$$P_0(t_0, \tau) = \sum_{0\le i\le m_t} [A_i(\tau - it_0) - A_{i+1}(\tau - it_0)]\, e^{-(i+1)\lambda t_0}\,, \qquad m_t = \frac{\tau}{t_0}\,;$$

$$A_i(t) = \binom{2i-1}{i}(pq)^i + \sum_{0\le j\le i-1} \frac{(i+j-1)!p^i q^j (\lambda t)^{i-j-1}}{i!j!(i-1-j)!}$$

$$\times\left[(-1)^{i-j}\frac{i\lambda t}{i-j} - (i+j)qe^{-(\lambda+\mu)t}\right], \qquad i \ge 1\,;$$

$$p = 1 - q = \frac{\mu}{\lambda + \mu}\,. \tag{9.24}$$

For faults of neglegible duration, $p = 1$, and $A_i(t) = (-1)^i(\lambda t)^i/i!$. The formula is

$$P_0(t_0, \tau) = \sum_{0\le i<\infty} e^{-(i+j)\lambda t_0} \frac{(-1)^i}{i!} [\lambda(\tau - it_0)]^i \left[1 + \frac{\lambda(\tau - it_0)}{i+1}\right]. \tag{9.25}$$

If we decompose $P_0^*(t_0, \omega)$ by the powers of ω, we can obtain formulas for the mean and the variance of the duration of task performance:

$$E\{T_1(t_0)\} = t_0 + \frac{E\{T_1\}F(t_0) + E\{T_0(t_0)\}}{1 - F(t_0)} ;$$

$$\mathrm{Var}\{T_1(t_0)\} = \frac{\beta_2 F(t_0) + 2E\{T_1\}E\{T_0(t_0)\} + \alpha_2(t_0)}{1 - F(t_0)} + \frac{[E\{T_1\}F(t_0) + E\{T_0(t_0)\}]^2}{[1 - F(t_0)]^2} ,$$

$$\alpha_2 = \int_0^{t_0} x^2 \, dF(x) ,$$

$$\beta_2 = \int_0^{\infty} x^2 \, dG(x) ,$$

$$E\{T_0(t_0)\} = \int_0^{t_0} x \, dF(x) ,$$

$$E\{T_1\} = \int_0^{\infty} x \, dG(x) .$$

For both exponential distributions,

$$\bar{T}_1(t_0) = (\bar{T}_0 + \bar{T}_1)[e^{\lambda t_0} - 1] ,$$

$$\mathrm{Var}\{T_1(t_0)\} = \bar{T}_1^2(t_0) + (e^{\lambda t_0} - 1)(\beta_2 + 2\bar{T}_0\bar{T}_1) + 2\bar{T}_0^2(e^{\lambda t_0} - 1 - \lambda t_0 e^{\lambda t_0}) .$$

For nonfatal self-correcting failures,

$$E\{T_1(t_0)\} = \frac{1}{\lambda} [e^{\lambda t_0} - 1] , \tag{9.26}$$

$$\mathrm{Var}\{T_1(t_0)\} = \frac{1}{\lambda^2} [(e^{\lambda t_0} - 1)^2 + 2(e^{\lambda t_0} - 1) - 2\lambda t_0 e^{\lambda t_0}] . \tag{9.27}$$

EXAMPLE 9.7

A computer with a rate of brief nonfatal failures $\lambda = 1$ per hour and depth of monitoring $\alpha = 0.9$ is designated to perform a task of duration $t_0 = 2$ hour. All brief faults found with the monitoring equipment are eliminated instantly without any consequences for task performance. The other 10% of the brief faults are found with special function tests after a delay. These brief faults destroy the previously obtained results, so the task must be restarted. Find the time resource needed to provide a probability of 0.95 of success. Find the mean and the variance of the task performance time.

Solution

The probability of task completion in the shortest possible time is $P(t_0) = \exp\{-(1-\alpha)\lambda t_0\} = e^{-0.1\cdot 2} = 0.82 < 0.95$, so a time resource is necessary. By (9.25),

$$P(t_0, \tau) = \exp(-\beta\lambda t_0)(1 + \beta\lambda\tau)\,, \qquad \beta = 1 - \alpha\,.$$

For $\tau = t_0$ the probability of interest is $P(t_0, t_0) = 0.98 > 0.95$. Thus τ can be found from

$$\tau = \frac{0.95\exp(\beta\lambda t_0 - 1)}{\beta\lambda} = 1.6\ \text{hr}\,.$$

Equations (9.26) and (9.27) yield

$$E\{T_1(t_0)\} = 10\ (e^{0.2} - 1] = 2.21\ \text{hr}\,,$$

$$\text{Var}\{T_1(t_0)\} = 100(0.221^2 + 2\cdot 0.221 - 0.4\cdot 1.221) = 0.326\ \text{hr}^2]\,,$$

$$\sigma = \sqrt{\text{Var}\{T_1(t_0)\}} = 0.57\ \text{hr}\,,$$

$$r = \frac{\sigma}{E\{T_1(t_0)\}} = 0.26\,.$$

9.3.3 A Unit with a Multi-Stage Task (Subclass 01000)

To decrease the loss after each short fault that has not been found by the monitoring equipment, we should divide the task into several stages, and develop a procedure to preserve completed stages from possible destruction. In computers this can be with regular time backup when the current complete results are stored in a secure portion of the memory. All stages usually should be equally long. For n stages of duration $\eta = t_0/n$ each, the probability of successful completion of the task is:

$$P_0(t_0, \tau, n) = e^{-\lambda t_0} \sum_{0\le i\le nm_i} e^{-i\lambda\theta}\binom{n+i-j}{i} \sum_{0\le j\le n} \binom{n}{j}(-1)^j A_{i+j}(\tau - i\theta)\,. \tag{9.28}$$

The functions $A_i(t)$ are determined by (9.24) and (9.25). The mean and the variance of the task performance time are

$$E\{T_1(t_0, n)\} = n(E\{T_0\} + E\{T_1\})(e^{\lambda\theta} - 1)\,,$$

$$E\{T_0\} = \frac{1}{\lambda}\,;$$

$$\begin{aligned}\text{Var}\{T_1(t_0, n)\} = &\frac{1}{n}(E\{T_1(t_0, n)\})^2(t_0, n) \\ &+ n(e^{\lambda\theta} - 1)(\beta_2 + 2E\{T_0\}E\{T_1\} - 2E\{T_0\}\theta) \\ &+ 2(E\{T_0\})^2(ne^{\lambda\theta} - n - n\lambda\theta) - 2n\theta E\{T_1\}e^{\lambda\theta} .\end{aligned}$$

To take into account the time, δ, needed to develop the stages, one should make $\eta = t_0/n + \delta$ and should replace m_t in (9.28) with $m_{t1} = \tau^*/t_0$, where $\tau^* = \tau - n\delta$. The optimal period, η, that gives the minimal task performance time can be defined with

$$\theta_{01} = \frac{1}{\lambda}\sqrt{2[1 - \exp(-\lambda\delta)]}$$

The optimal number of stages is

$$n_{01} = \frac{t_0}{\theta_0} = \frac{\lambda t_0}{\sqrt{2[1 - \exp(-\lambda\delta)]}} .$$

The mean time of task performance for the optimal choice of the number of stages is:

$$\bar{T}_1(t_0, n_{01}) = \frac{\lambda t_0(\bar{T}_0 + \bar{T}_1)}{\lambda\theta_{01}} e^{\lambda\theta 01 + \lambda\delta} .$$

The optimal duration of a stage and the optimal number of stages can be obtained with

$$\theta_{02} = \frac{\delta}{m_t}(\sqrt{m_t + 1} + 1) ,$$

$$n_{02} = \frac{t_0}{\delta}(\sqrt{m_t + 1} - 1) ,$$

$$m_t = \frac{\tau}{t_0} .$$

EXAMPLE 9.8

The system task in Example 9.7 is divided into 100 stages. The time required for the checking and memorizing procedure for each stage equals $\delta = 1$ second. Find the time resource and the stage duration that yield an acceptable probability of successful task completion and minimize the mean completion time.

Solution

For 100 stages $\eta = 120/100 = 1.2$ minutes. For $\tau_* = \eta$, the value $\lambda\tau_* = 0.002$, and the probability equals

$$P_0 = e^{-0.2}\left[1 + 100 \cdot 0.002 + \frac{1}{2}\binom{100}{2}0.002^2 + \frac{1}{6}\binom{100}{3}0.002^3\right] = 0.991 .$$

This value is larger than the required probability, 0.95. The necessary time resource $\tau_* < \eta$. For its calculation one uses (9.28) for $i = 0$. The calculation gives $P_0 = 0.95$ for $\tau_* = 0.0155$ hours $= 56$ seconds. Then $\tau = 56 + 101 = 157$ seconds. Thus by adding $\tau = 2.62$ minutes to the time $t_0 = 2$ hours (additional time expenses, $m_t = 2.18\%$) one increases P_0 from 0.82 to 0.95. The optimal stage duration η_{02} is 92 seconds. The optimal number of stages $n_{02} = 7200/92 = 78$. Thus $\tau_* = \tau - n_{02}\delta = 79$ seconds. The probability of success is

$$P_0(t_0, n_{02}) = e^{-0.2}\Bigg[1 + 78 \cdot 0.00217 + \frac{1}{2}\binom{78}{2}0.00217^2 + \frac{1}{6}\binom{78}{3}0.00217^3\Bigg] = 0.963 .$$

The mean task performance time for 100 stages equals $E\{T_1(t_0) = 100 \cdot 10[e^{0.00203} - 1] = 2.003$ hours. The variance of this r.v. is

$$\mathrm{Var}\{T_1(t_0)\} = \frac{2.03^2}{100} - (e^{0.002} - 1) \cdot 20 \cdot 0.02 + 200 \cdot 100[e^{0.002} - 1 - 0.002] = 4 \cdot 10^{-8}\ \mathrm{hr}^2$$

σ $2 \cdot 10^{-4}$ hr, $r = 10^{-4}$. The optimal number of stages is

$$n_{01} = \frac{7200}{\sqrt{2 \cdot 10 \cdot 3600}} = 27 .$$

The mean task completion time equals

$$E\{T_1(n_{01}, t_0)\} = \frac{2}{\sqrt{2\gamma}}\,[e^{\sqrt{2\gamma}+\gamma} - 1] = 2.015\ \mathrm{hr}\ , \qquad \gamma = \lambda\delta .$$

9.3.4 A Series System (Subclass 01000)

A system consists of N units with random TTF distributions $R_i(t)$ and repair time distributions $G_i(t)$. The equation for calculation of the probability of successful completion of a task of duration t_0 with nonrefilled time resource τ can be found using the inverse of the LST(9.23):

$$P_0(t_0, \tau) = [1 - F_0(t_0)] \sum_{0 \le n < \infty} \Phi_n(t_0, \tau) ,$$

$$1 - F_0(t_0) = \sum_{1 \le i \le N} [1 - R_i(t_0)] ,$$

where the function $\Phi_n(t_0, \tau)$ is the nth convolution by the variable τ of the function $\Phi(t_0, \tau)$. The latter function itself can be obtained from the inversion of the LST transform of the form

$$\varphi(t_0, \omega) = \sum_{1 \le i \le N} f_1(\omega) \int_0^{t_0} e^{-\omega x} \, dP_{0i}(x) ,$$

$$P_{0i}(t) = \int_0^t [1 - F_0(x)]\lambda_i(x) \, dx .$$

For the nonfatal brief faults,

$$\Phi(t_0, \tau) = F_0(\min(t_0, \tau)) .$$

For $m_t \le 1$, $\Phi(t_0, \tau) = F_0(\tau)$,

$$P_0(t_0, \tau) = [1 - F_0(t_0)][1 + H(\tau)] ,$$

where $H\tau$ is the mean number of brief nonfatal faults during the time τ. If all $G_i(t)$ are identical and the $R_i(t) = 1 - \exp(\lambda_i t)$ differ for different i, the reliability indexes must be found by the equations in Sections 9.3.2 and 9.3.3 for

$$\lambda = \sum_{1 \le i \le N} \lambda_i .$$

9.4 ONE-CHANNEL SYSTEMS WITH RANDOM ARRIVALS OF TASKS

When tasks can arrive at random moments, there exists the possibility of queue formation. Between the random arrivals of the system tasks there will be idle periods that can be used for repair. Two types of systems will be considered. In one, delay of initiation of task performance is impossible. In the other, the delay can be not more than τ_i which can have either a fixed or a random value. In the latter case the system has a time resource that can be replenished.

9.4.1 A System with a Non-replenished Time Resource

Consider two models. In model 1 the task duration is negligibly small in comparison to the system MTTF, so we will neglect it. The random TTF, the repair time, and the arrival times are independent r.v.'s with distributions $F(t)$, $G(t)$, and $A(t)$, respectively. After taking into account the time redundancy, the MTTF is determined with:

$$\bar{T}_0(z_*) = \frac{1}{q}\,[\bar{T}_0 + E\{\min(\eta, z_*0\}]\,,$$

$$\bar{T}_0 = \int_0^\infty x\,dF(x)\,, \qquad (9.29)$$

where

$$q = \int_0^\infty [1 - G(t)]\,dA^*(t)\,,$$

$$\Theta = E\{\min(\eta, z_*)\} = \int_0^\infty [1 - A^*(t)][1 - G(t)]\,dt\,,$$

$$A^*(t) = \frac{1}{\bar{z}}\int_0^t [1 - A(x)]\,dx\,,$$

$$z = \int_0^\infty [1 - A(x)]\,dx\,.$$

In (9.29), z_* is the time resource from the moment of failure until task arrival. Values of q and η for some common distributions are given in Table 9.10. For a short repair time (i.e., $E\{\eta\} \ll E\{T_0\}$, and $E\{\eta\} \ll E\{z\}$), we use the approximations

$$E\{T_0(z_*)\} \approx \frac{E\{T_0\}}{q}\,,$$

$$P(t, z_*) \approx e^{-qt/E\{T_0\}}\,, \quad t \gg E\{z\}\,. \qquad (9.30)$$

Table 9.10 Values of parameters q and η for a unit in subclass 01001

$G(t)$	$A(t)$	q	Θ
$1 - e^{-\mu t}$	$1 - e^{-\gamma t}$	$\frac{\gamma}{(\gamma + \mu)}$	$\frac{1}{(\gamma + \mu)}$
$1 - (1 + \mu t)e^{-\mu t}$	$1 - e^{-\gamma t}$	$\frac{\gamma(\gamma + 2\mu)}{(\gamma + \mu)^2}$	$\frac{(2\mu + \gamma)}{(\gamma + \mu)^2}$
$1 - e^{-\mu t}$	$1 - (1 + \gamma t)e^{-\gamma t}$	$\frac{\gamma(2\gamma + \mu)}{2(\gamma + \mu)^2}$	$\frac{(2\mu + 3\gamma)}{2(\gamma + \mu)^2}$
$1 - \sum_{i=1}^{n} v_i e^{-\mu_i t}$	$1 - e^{-\gamma t}$	$\sum_{i=1}^{n} \frac{\gamma}{\gamma + \mu_i} v_i$	$\frac{\Sigma_{i=1}^{n} v_i}{(\gamma + \mu_i)}$
$1 - e^{-\mu t}$	$1 - \sum_{i=1}^{n} v_i e^{-\gamma_i t}$	$\frac{\Sigma_{i=1}^{n} v_i/\gamma_i + \mu)}{\Sigma_{i=1}^{n} v_i/\gamma_i}$	$\frac{\Sigma_{i=1}^{n} v_i/\gamma_i(\gamma_i + \mu)}{\Sigma_{i=1}^{n} v_i/\gamma_i}$

EXAMPLE 9.9

A system consists of data processing and data transmission equipment. The input data arrive at random moments, and the information is transmitted to clients in real time. There is a limit to the delay acceptable in time available from the input data processing.

Find the mean time until the first disruption of system operation, and find the probability of successful operation throughout $t = 8$ hours. The system failure rate is $\lambda = 0.04$ per hour, and the mean repair time $E\{\eta\} = 1/\mu = 0.2$ hour. Interarrival time has an exponential distribution with the parameter $\gamma = 20$ per hour.

Solution

The system can be modeled as a single unit with the following distribution functions:

$$F(t) = 1 - e^{-\lambda t},$$
$$G(t) = 1 - e^{-\mu t},$$
$$A(t) = 1 - e^{-\gamma t}.$$

Using formulas from Table 9.10 and Equations (9.29) and (9.30), we find: $q = 20/(20 + 5) = 0.8$, $\eta = 1/(20 + 5) = 0.04$ hour, $E\{T_0\} = (25 + 0.04)/0.8 = 31.3$ hours, $P(t, z_*) \approx e^{-0.04 \cdot 0.8 \cdot 8} = 0.775$.

Model 2 differs from model 1 in the random duration of the task. The duration is a r.v., X, with distribution $B(t)$ and mean $E\{t_0\}$. For exponential distributions $F(t)$ and $A(t)$, the equation for the MTTF is

$$E\{T_0(z)\} = \frac{\lambda[1 + \lambda/\gamma(1 - g(\gamma))] = \gamma[1 - \beta(\lambda)]}{\lambda[1 - g(\gamma) + \gamma[1 - \beta(\lambda)]}, \tag{9.31}$$

where

$$g(\gamma) = \int_0^\infty e^{-\gamma t}\, dG(t),$$

$$\beta(\lambda) = \int_0^\infty e^{-\lambda t}\, dB(t).$$

For exponential $A(t) = 1 - e^{-\gamma t}$ and $B(t) = 1 - e^{-\beta t}$, and arbitrary $F(t)$ and $G(t)$, the equation has the form

$$E\{T_0(z)\} = \frac{E\{T_0\}(\gamma + \beta) + [\beta + \gamma f(\gamma + \beta)]/\gamma[1 - g(\gamma)]}{\gamma[1 - g(\gamma) f(\gamma + \beta)] + \beta[1 - g(\gamma)]}, \tag{9.32}$$

$$f(\gamma + \beta) = \int_0^\infty e^{-(\gamma+\beta)t}\, dF(t).$$

Particular cases of the formulas (9.31) and (9.32) are given in Table

Table 9.11 MTTF for a unit in subclass 01001

Special Conditions	Formula for $\bar{T}_0(z)$
$\bar{\eta} \ll \frac{1}{\lambda}$, $\bar{\eta} \ll \bar{z}$	$\frac{1}{\lambda} + \frac{1}{\gamma}[1 - \beta(\lambda)]$
$\bar{t}_0 \ll \frac{1}{\lambda}$	$\frac{1}{\gamma} + \frac{1}{\lambda}[1 - g(\gamma)]$
$\bar{\eta} \ll \bar{T}_0$, $\bar{\eta} \ll \frac{1}{\gamma}$	$\frac{\bar{T}_0(\gamma + \beta)}{\gamma[1 - f(\gamma + \beta)]}$
$\bar{T}_0(\gamma + \beta) \gg 1$	$\frac{1 - g(\gamma) + \gamma \bar{T}_0}{\gamma[1 - g(\gamma)]}$
$F(t) = 1 - e^{-\lambda t}$	$\frac{1}{\lambda\gamma} + \frac{(\gamma + \beta)(\gamma + \mu)}{\lambda\gamma(\gamma + \beta + \lambda + \mu)}$
	Approximation
$G(t) = 1 - e^{-\mu t}$	$\frac{1}{\gamma}(\lambda + \gamma + \beta)$ for $\bar{\eta} \ll \bar{T}_0$, $\bar{\eta} \ll \frac{1}{\gamma}$,
$B(t) = 1 - e^{-\beta t}$	$\frac{1}{\lambda\gamma}(\lambda + \gamma + \mu)$ for $\bar{t} \ll \frac{1}{\lambda}$

A System with a Replenished Time Resource (Subclass 02001)

The system can use both the idle time between task arrivals and the refilled time resource τ_1, which has distribution function $D(t)$. For $E\{T_0\} \gg E\{z\}$ (in practice a satisfactory ratio of these values can be 15–20), one can use (9.29) and (9.30) to find the probability of successful operation, $P(t, z_* + \tau_i)$, and the MTTF, $E\{T_0(z_* + \tau_i)$. With substitutions for η and q, (9.29) and (9.30) yield

$$\Theta = E\{\min(\eta, z_* + \tau_1)\} = \int_0^\infty [1 - S(t)][1 - G(t)]\, dt\,,$$

$$q = \int_0^\infty [1 - G(t)]\, dS(t)\,,$$

$$S(t) = \int_0^t A^*(t - x)\, dD(x)\,. \tag{9.33}$$

If τ_1 is a constant, the equations are:

$$S(t) = A^*(t - t_1)\,,$$

$$q = \int_0^\infty [1 - G(t + t_1]\, dA^*(t)\,,$$

$$\Theta = \int_0^{t_1} [1 - G(t)\, dt + \int_{t_1}^\infty [1 - G(t)][1 - A^*(t - t_1)]\, dt\,. \tag{9.34}$$

Special cases of (9.33) and (9.34) are given in Tables 9.12 and 9.13.

Table 9.12 Values of parameters q and η for a unit in subclass 02001, $P(\tau_1 < t_1) = D(t)$

Distribution	b	Θ
$G(t) = 1 - e^{-\mu t}$ $D(t) = 1 - e^{-\delta t}$	$\dfrac{\gamma\delta}{(\mu+\delta)(\mu+\gamma)}$	$\dfrac{1}{\gamma-\delta}\left(\dfrac{\gamma}{\mu+\delta} - \dfrac{\delta}{\mu+\gamma}\right)$
$G(t) = 1 - (1+\mu t)e^{-\mu t}$, $D(t) = 1 - e^{-\delta t}$	$\dfrac{\gamma\delta[3\mu^2 + \gamma\delta + 2\mu(\gamma+\delta)]}{(\mu+\delta)^2(\mu+\gamma)^2}$	$\dfrac{1}{\gamma-\delta}\left[\dfrac{\gamma(2\mu+\delta)}{(\mu+\delta)^2} - \dfrac{\delta(2\mu+\gamma)}{(\mu+\gamma)^2}\right]$
$G(t) = 1 - e^{-\mu t}$, $D(t) = 1 - (1+\delta t)e^{-\delta t}$	$\dfrac{\gamma\delta^2}{(\mu+\delta)^2(\mu+\gamma)}$	$\dfrac{(\mu+\delta)^2 + \gamma(\mu+2\delta)}{(\mu+\delta)^2(\mu+\gamma)}$
$G(t) = 1 - \sum_{i=1}^{n} v_i e^{-\mu_i t}$, $D(t) = 1 - e^{-\delta t}$	$\dfrac{\gamma\delta\, \Sigma_{i=1}^{n} v_i}{(\mu_i+\delta)(\mu_i+\gamma)}$	$\dfrac{1}{\gamma-\delta}\sum_{i=1}^{n} v_i\left(\dfrac{\gamma}{\mu_i+\delta} - \dfrac{\delta}{\mu_i+\gamma}\right)$

Table 9.13 Values of parameters q and η for a unit in subclass 02001, $\tau_1 = t_1$

Distribution	q	Θ
$G(t) = 1 - e^{-\mu t}$, $A(t) = 1 - e^{-\gamma t}$	$\dfrac{\gamma}{\gamma+\mu}\, e^{-\mu t_1}$	$\dfrac{1}{\mu}(1 - e^{-\mu t_1}) + \dfrac{1}{\gamma+\mu}\, e^{-\mu t_1}$
$G(t) = 1 - e^{-\mu t}$, $A(t) = 1 - (1+\gamma t)e^{-\gamma t}$	$\dfrac{\gamma(2\gamma+\mu)}{2(\gamma+\mu)^2}\, e^{-\mu t_1}$	$\dfrac{(2\mu+3\gamma)}{2(\gamma+\mu)^2}\, e^{-\mu t_1} + \dfrac{1}{\mu}(1 - e^{-\mu t_1})$
$G(t) = 1 - (1+\mu t)e^{-\mu t}$, $A(t) = 1 - e^{-\gamma t}$	$\dfrac{\gamma(\gamma + 2\mu + \gamma\mu 2t_1(\gamma+\mu))}{(\gamma+\mu)^2}\, e^{-\mu t_1}$	$\dfrac{\gamma + 2\mu + \mu t_1(\gamma+\mu)}{(\gamma+\mu)^2}\, e^{-\mu t_1} + \dfrac{2}{\mu}\left[1 - \left(1 + \dfrac{\mu t_1}{2}\right)e^{-\mu t_1}\right]$
$G(t) = 1 - e^{-\mu t}$, $A(t) = 1 - \sum_{i=1}^{n} v_i e^{-\gamma_i t}$	$\dfrac{\Sigma_{i=1}^{n}(v_i/\mu + \gamma_i)e^{-\mu t_1}}{\Sigma_{i=1}^{n} v_i/\gamma_i}$	$\dfrac{1}{\mu}(1 - e^{-\mu t_1}) + \sum_{i=1}^{n}\dfrac{v_i}{\gamma_i(\gamma_i+\mu)} \times \dfrac{e^{-\mu t_1}}{\Sigma_{i=1}^{n} v_i/\gamma_i}$
$G(t) = 1 - \sum_{i=1}^{n} v_i e^{-\mu_i t}$, $A(t) = 1 - e^{-\gamma t}$	$\sum_{i=1}^{n}\dfrac{\gamma v_i}{\gamma+\mu_i}\, e^{-\mu_i t_1}$	$\sum_{i=1}^{n}\left[\dfrac{v_i}{\mu_i}(1 - e^{-\mu_i t_1}) + \dfrac{v_i}{\gamma+\mu_i}\, e^{-\mu_i t_1}\right]$

EXAMPLE 9.10

In addition to the conditions of the previous example let there be a delay in task starting of $t_1 = 0.2$ hour. Find the probability of successful operation and the MTTF.

Solution

Using Table 9.13, we find that $q = [20/(20+5)]e^{-5 \cdot 0.2} = 0.294$. Then by (9.30) we find that $P(t, z_* + t_1) = e^{-0.04 \cdot 0.294 \cdot 8} = 0.91$. The MTTF equals

$$E\{T_0(z_* + t_1)\} = \frac{1}{0.294}\left[25 + 0.2(1 - e^{-5 \cdot 0.2}) + \frac{1}{20+5}\, e^{5 \cdot 0.2}\right]$$
$$= 85.5 \text{ hr} .$$

Compared with $E\{T_0\} = 25$ hours, the mean has increased 3.4 times.

9.4.4 A Series System with an Unreplenished Time Resource (Subclass 01001)

A system consists of n units. Unit i has failure rate λ_i' in the on-duty regime and λ_i in the task performance regime. The random repair time distribution is $G_1(t)$. The interarrival times have an exponential distribution with parameter γ. The task performance duration distribution is $B(t)$. The MTTF in this case equals

$$E\{T_0(z)\} = \frac{1 + \gamma[1 - \beta(\lambda)]/\lambda + \Sigma_{1 \le i \le n} \lambda_i'[1 - g_{1i}(\gamma)]/\gamma}{\gamma[1 - \beta(\gamma)] + \Sigma_{1 \le i \le n} \lambda_i'[1 - g_{1i}(\gamma)]} , \qquad (9.35)$$

$$\lambda = \sum_{1 \le i \le n} \lambda_i .$$

For $n = 1$ and $\lambda_i' = \lambda_i$, (9.31) can be obtained from (9.35). If the system does not fail in the on-duty regime, then

$$e\{T_0(z)\} = \frac{1}{\lambda} + \frac{1}{\gamma[1 - \beta(\lambda)]} .$$

For $B(t) = 1 - e^{-\beta t}$, the equation becomes

$$\bar{T}_0(z) = \frac{1}{\lambda} + \frac{1}{\gamma} + \frac{\beta}{\lambda\gamma}$$

If the task has constant duration t_0:

$$\beta(\lambda) = e^{-\lambda_0 t} ,$$

and as a result:

$$\bar{T}_0(z) = \frac{1}{\lambda} + \frac{1}{\gamma(1 - e^{-\lambda t_0})} .$$

For $t_0 \ll 1/\lambda$,

$$\bar{T}_0(z) \approx \frac{1}{\lambda}\left(1 + \frac{1}{\gamma t_0}\right) .$$

To find the availability coefficient, one also needs to know the distribution of the repair time in the state for the current task performance, $G_{2i}(t)$. The average time in the on-duty state, independently of whether the system is successfully operating, up, or down, is

$$K = C \frac{\beta(\lambda)/\gamma + [1 - \beta(\lambda)]/\lambda}{C(\beta(\lambda)/\gamma + (1 + \rho)[1 - \beta(\lambda)]/\lambda) + \beta(\lambda) \Sigma_{1 \le i \le n} \rho'_i[1 - g_{i1}(\gamma)]} ,$$

where

$$C = \gamma + \sum_{1 \le i \le n} \lambda'_i[1 - g_{1i}(\gamma)] ,$$

$$\rho = \sum_{1 \le i \le n} \lambda_i E\{\eta_{2i}\} ,$$

$$\rho' = \sum_{1 \le i \le n} \lambda'_i E\{\eta_{2i}\} ,$$

$$E\{\eta_{2i}\} = \int_0^\infty x \, dG_{2i}(x) .$$

This index is very close to, but not identical to, the availability coefficient.

The mean portion of the time when the system is in an up state (whether performing the task or not) is given by

$$K^0 = \frac{\beta(\lambda) + C[1 - \beta(\lambda)]/\lambda}{\beta(\lambda) \Sigma_{1 \le i \le n} \lambda'_i \bar{\eta}_{2i}[1 - g_{1i}(\gamma)] + C\{\beta(\lambda)/\gamma + (1 + \rho)/\lambda[1 - \beta(\lambda)]\}} .$$

The gain in reliability, G_k, because of the time redundancy, G_k, can be measured by comparing these two values:

$$G_K = \frac{K}{K^0} = 1 + \frac{(C - \gamma)\beta(\lambda)}{\gamma\beta(\lambda) + \gamma C[1 - \beta(\lambda)]/\lambda} .$$

Here $K = K^0$ and $G_K = 1$ for $\lambda'_i = 0$. For an exponential $B(t)$ with parameter β and $n = 1$,

$$K = \frac{1}{1 + [\gamma\lambda E\{\eta_2\}/(\beta + \gamma)]} .$$

If the system business ratio $\alpha = \gamma/(\beta + \gamma) \to 1$, then $K \to (1 + \rho)^{-i}$. If $\lambda'' = \lambda$ and $n = 1$, then $G_K = 2 - \alpha$.

9.4.5 A Series System with Combined Time Resources (Subclass 21001)

In the previous model let there be a replenished time resource, τ_{1i}, with distribution $D_i(t)$. This resource can be spent only if a task arrives when the system is down or if system failure occurs during task performance. The MTTF of the system equals:

$$
\begin{aligned}
E\{T_0(z_* + \tau_1)\} = {} & \Big\{\gamma\{[1-\beta(\lambda)]\Big(1 + \sum_{1\le i\le n} \frac{\lambda_i'}{\gamma+\mu_i'}\Big) \sum_{1\le i\le n} \frac{\lambda_i}{\lambda}\, d_i(\mu_i) \\
& + \beta(\lambda) \sum_{1\le i\le n} \frac{\lambda_i'}{\gamma+\mu_i'}\, d_i(\mu_i)\Big\}^{-1} \\
& \times \Big\{1 + \sum_{1\le i\le n} \frac{\lambda_i'}{\gamma+\mu_i'} \Big(1 + \frac{\gamma}{\mu_i}\,[1-d_i(\mu_i)\Big)\Big\} \\
& \times \Big\{\beta(\lambda) + [1-\beta(\lambda)] \sum_{1\le i\le n} \frac{\lambda_i}{\lambda}\, d_i(\mu_i)\Big\} \\
& + \frac{\gamma}{\lambda}\,[1-\beta(\lambda)]\Big\{1 + \sum_{1\le i\le n} \frac{\lambda_i}{\mu_i}\,[1-d_i(\mu_i)]\Big\} \\
& \times \Big\{1 + \sum_{1\le i\le n} \frac{\lambda_i'}{\gamma+\mu_i'}\,[1-d_i(\mu_i)]\Big\}\,.
\end{aligned}
$$

For $n = 1$,

$$
\begin{aligned}
E\{T_0(z_* + \tau_1)\} = {} & \frac{1}{\gamma d(\mu)[\lambda' + (\gamma+\mu')[1-\beta(\lambda)]} \\
& \times \Big\{(\gamma+\mu'+\lambda')\Big[\beta(\lambda) + \frac{\gamma}{\lambda}\,[1-\beta(\lambda)]\Big] \\
& + \frac{\gamma}{\mu}\,[1-d(\mu)][\lambda' + (\gamma+\mu')[1-\beta(\lambda)]] \\
& - \gamma\,\frac{\lambda'}{\lambda}\,[1-\beta(\lambda)]\, d(\mu)\Big\}\,.
\end{aligned} \tag{9.36}
$$

For $\lambda' = \lambda$, $\mu' = \mu$, and an exponential $B(t)$ with parameter β from (9.36):

$$
\begin{aligned}
E\{T_0(z_* + \tau_1)\} = {} & \frac{1}{1 + [\lambda/(\gamma+\mu)] - \beta K} \\
& \times \Big\{\frac{1}{\gamma}\Big[1 + \frac{\lambda}{\mu}\Big[1 - \frac{\gamma}{\gamma+\mu}\, d(\mu)\Big]\Big] \\
& + K\Big[1 + \frac{\lambda}{\mu}\,[1-d(\mu)]\Big]\Big\}\,,
\end{aligned}
$$

$$
K = \frac{1 + \lambda[1-d(\mu)]/(\gamma+\mu)}{\beta + \lambda d(\mu)}\,. \tag{9.37}
$$

If the time resource is nonrandom and equals t_1, then

$$d(\mu) = e^{-\mu t_1}$$

and

$$\bar{T}_0(z_* + \tau_1) = \frac{1}{\gamma} \frac{(K+1)[1 + \lambda(1 - e^{-\mu t_1})]\gamma/\mu + \lambda + \mu}{\gamma + \lambda + \mu - \beta K},$$

$$K = \frac{\gamma + \lambda + \mu[1 - e^{-\mu t_1}]}{\beta + \lambda e^{-\mu t_1}} \tag{9.38}$$

If the task is very short, them $1/\beta \to 0$, and (9.37) and (9.38) become the equations in Section 9.4.3.

9.5 MULTIPLE-CHANNEL SYSTEMS

Expressions for systems with more than one channel are given in Tables 9.14–9.18.

Table 9.14 Probability of task completion for an unrepairable system with a common task

m	$P_0(t_0, \tau, m)$
2	$[1 + 2\lambda \min(t_0, \tau)]e^{-2\lambda t_0}$
	$\left(1 + 2\gamma + \frac{\gamma_i^2}{2}\right)e^{-\rho_1}, \tau \le \frac{t_0}{2}, \gamma_1 = 3\lambda\tau, \rho_1 = 3\lambda t_0,$
3	$\left[1 + \rho_1 + \frac{\gamma_1^2}{2} - \frac{(2\gamma_1 - \rho_1)^2}{3!}\right]e^{-\rho_1}, \frac{t_0}{2} \le \tau \le 2t_0,$
	$\left(1 + \rho_1 + \frac{\rho_1^2}{2}\right)e^{-\rho_1}, \tau \ge 2t_0$
	$\left(1 + 3\gamma_1 + \frac{3\gamma_1^2}{2} + \frac{\gamma_1^3}{6}\right)e^{-\rho_1}, \tau \le \frac{t_0}{3}, \gamma_1 = 4\lambda\tau, \rho_1 = 4\lambda t_0,$
4	$\left[1 + \rho_1 - \frac{\rho_1^2}{4} + \frac{3\rho_1\gamma_1}{2} - \frac{3\gamma_1^2}{4} + \frac{\gamma_1^3}{6} - \frac{(3\gamma_1 - \rho_1)^3}{96}\right]e^{-\rho_1}, \frac{t_0}{3} \le \tau \le t_0,$
	$\left(1 + \rho_1 + \frac{1}{2}\rho_1^2 + \frac{1}{96}\gamma_1^3 - \frac{3}{32}\rho_1\gamma_1^2 + \frac{9}{32}\rho_1^2\gamma_1 - \frac{11}{96}\rho_1^3\right)e^{-\rho_1}, t_0 \le \tau \le 3t_0$
	$\left(1 + \rho_1 + \frac{\rho_1^2}{2} + \frac{\rho_1^3}{3!}\right)e^{-\rho_1}, \tau \ge 3t_0$

Table 9.15 Probability of task completion for an unrepairable multiple-channel system with common redundant channels ("slipping" redundancy)

m	n	Regime of Redundant Unit	$P_0(t_0, \tau, m, n)$
2	1	A	$(3-2e^{-\rho})e^{-2\rho}+6e^{-2\rho}(e^{-\rho}+\rho-1),\ t_0\le\tau,\ \rho=\lambda t_0,$ $(3-2e^{-\rho})e^{-2\rho}+6e^{-2\rho}(\gamma+e^{-\rho}-e^{-\rho+\gamma}),\ t_0\ge\tau,\ \gamma=\lambda\tau$
		S	$(1+2\rho+2\rho^2)e^{-2\rho},\ t_0\le\tau,\ \rho=\lambda t_0,\ \gamma=\lambda\tau$ $(1+2\rho+4\rho\gamma-2\gamma^2)e^{-2\rho},\ t_0\ge\tau$
3	1	A	$(28-27e^{-\rho}-24\rho+18\rho^2)e^{-3\rho},\ t_0\le\frac{\tau}{2},\ \rho=\lambda t_0,\ \gamma=\lambda\tau,$ $\left[24\rho-24\gamma+18\gamma^2-20-27e^{-\rho}+48e^{-\rho+\gamma/2}-24\left(\gamma-\frac{\rho}{2}\right)^2\right]e^{-3\rho},$ $\frac{\tau}{2}\le t_0\le 2\tau$ $(24\gamma+18\gamma^2+4-27e^{-\rho}-24e^{-\rho+2\gamma}+48e^{-\rho+\gamma/2})e^{-3\rho},\ t_0\ge 2\tau$
		S	$\left(1+3\rho+\frac{(3\rho)^2}{2!}+\frac{(3\rho)^3}{3!}\right)e^{-3\rho},\ t_0\le\frac{\tau}{2},\ \rho=\lambda t_0,\ \gamma=\lambda\tau,$ $\left[1+3\rho+\frac{(3\rho)^2}{2!}+\frac{(3\rho)^3}{3!}+\frac{3}{4}(2\rho-\gamma)^3\right]e^{-3\rho},\ \frac{\tau}{2}\le t_0\le 2\tau,$ $\left[1+3\rho+9\gamma\left(2\rho-2\gamma+\frac{3}{2}\rho\gamma-\frac{5}{4}\gamma^2\right)\right]e^{-3\rho},\ t_0\ge 2\tau$
2	2	A	$(16e^{-\rho}-3e^{-2\rho}-12+12\rho)e^{-2\rho},\ t_0\le\tau,\ \rho=\lambda t_0,\ \gamma=\lambda\tau,$ $[6-8e^{-\rho}+3e^{-2\rho}+12\gamma-24(e^{\gamma}-1)e^{-\rho}+6e^{-2\rho}(e^{2\gamma}-1)]e^{-2\rho},\ t_0\ge$
		S	$\left[1+2\rho+\frac{(2\rho)^2}{2!}+\frac{(2\rho)^3}{3!}\right]e^{-2\rho},\ t_0\le\tau,$ $\left[1+2\rho+\frac{(2\rho)^2}{2!}+4\gamma\left(\rho^2-\rho\gamma+\frac{1}{3}\gamma^2\right)\right]e^{-2\rho},\ t_0\ge\tau$

Note: A = active, S = standby.

Table 9.16 MTTF for unrepairable multi-channel system with common redundant channels

m	n	Regime of Redundant Unit	$\bar{T}_0(\tau, m, n)$
2	1	A	$\frac{4}{3\lambda} - \frac{\exp(-2\gamma)}{2\lambda}$, $\gamma = \lambda\tau$
		S	$\frac{3}{2\lambda} - \frac{\exp(-2\gamma)}{2\lambda}$
3	1	A	$\frac{5}{4\lambda} - \frac{4\exp(-3\gamma/2)}{9\lambda} - \frac{2\exp(-6\gamma)}{9\lambda}$
		S	$\frac{4}{3\lambda} - \frac{4\exp(-3\gamma/2)}{9\lambda} - \frac{3\exp(-6\gamma)}{9\lambda}$
2	2	A	$\frac{19}{12\lambda} - \frac{\exp(-2\gamma)}{2\lambda}$
		S	$\frac{2}{\lambda} - \frac{\exp(-2\gamma)}{2\lambda}$

Note: A = active, S = standby.

Table 9.17 Mean time for task completion for unrepairable multiple-channel system with common redundant channels

m	n	Regime of Redundant Unit	$\bar{T}_1(t_0, m, n)$
2	1	A	$t_0\left[1 + 6e^{-2\rho}\left(\frac{1}{\rho} - 1 + \frac{\rho}{2} - \frac{1}{\rho}e^{-\rho}\right)\right]$, $\rho = \lambda t_0$
		S	$t_0\left[1 + \frac{2}{3}\rho^2 e^{-2\rho}\right]$
3	1	A	$t_0\left\{1 + e^{-3\rho}\left[15\rho^3 + 84\left(e^{-\rho} + \rho - 1 - \frac{\rho^2}{2}\right)\right]\right\}$
		S	$t_0\left[1 + e^{-3\rho}\left(\frac{3\rho^2}{4} + \frac{11\rho^3}{128}\right)\right]$
2	2	A	$t_0\left\{1 + 6e^{-2\rho}\left[\rho + \frac{4}{\rho}(1 - \rho - e^{-\rho}) - \frac{1}{2\rho}(1 - 2\rho - e^{-2\rho})\right]\right\}$
		S	$t_0\left(1 + \frac{\rho^3 e^{-2\rho}}{3}\right)$

Note: A = active, S = standby.

Table 9.18 Probability of task completion

Index	Regime of Redundant Unit	Formula
$P_0(t_0, \tau)$	A	$\left(1 + 2\rho + \frac{(2\rho)^2}{2!} + \frac{(2\rho)^3}{2!}\right)e^{-2\rho},\ t_0 \le \tau,\ \rho = \lambda t_0,\ \gamma = \lambda\tau,$ $\left[1 + 2\rho + \frac{(2\rho)^2}{2!} + \frac{(2\rho)^3}{3!} - (1 + 2\gamma)(\rho - \gamma)^2 - \frac{4}{3}(\rho - \gamma)^3\right]e^{-2\rho},\ t_0 \ge \tau$
	S	$4(2\rho - 1)e^{-2\rho} + (5 + 4\rho)e^{-4\rho},\ t_0 \le \tau,\ \rho = \lambda t_0,\ \gamma = \lambda\tau,$ $4(1 + 2\gamma)e^{-2\rho} + (1 + 4\gamma)e^{-4\rho} + 4(1 - 2e^{2\gamma})e^{-3\rho-\gamma},\ t_0 \ge \tau$
	A	$\frac{2}{\lambda} - \frac{(3 + 2\gamma)\exp(-2\gamma)}{4\lambda}$
$\bar{T}_0(\tau)$	S	$\frac{3}{2\lambda} - \frac{2\exp(-2\gamma)}{3\lambda} + \frac{\exp(-4\gamma)}{12\lambda}$
	A	$t_0\left[\frac{1 + \rho^2(2 + 5\rho)\exp(-2\rho)}{6}\right]$
$\bar{T}_1(t_0)$	S	$t_0\left\{1 + 2e^{-2\rho}\left[2\rho - 4 + (2 + \rho)e^{-2\rho} + \frac{1}{\rho}(4 - 6e^{-\rho} + 2e^{-2\rho})\right]\right\}$
$\bar{T}(t)$	A	$\frac{2 - (2 + \lambda t)\exp(-\lambda t)}{\lambda}$
	S	$\frac{3}{2\lambda} - \frac{2\exp(-\lambda t)}{\lambda} + \frac{\exp(-2\lambda t)}{2\lambda}$

Note: A = active, S = standby.

CHAPTER 10

QUEUING SYSTEMS WITH UNRELIABLE SERVICE CHANNELS

10.1 NOTATION AND MAIN PARAMETERS

The literature on queuing theory is very broad. The main subject of this chapter is the effect of reliability on service channels in queuing systems (QS). The focus is on useful results for one-channel service, but the methods can be used for more sophisticated cases. Arrivals are Poisson throughout the chapter.

10.1.1 Definitions and Main Notation

α = arrival time (the interval between two adjacent arrivals).

β = duration of service for a single request by a single absolutely reliable service channel.

γ = duration of continuous service of a request without repair or retuning of the service channel.

δ = duration of repair of a channel that has failed during service.

ϵ = duration of a failure-free state of the service channel in an on-duty regime.

φ = duration of repair of a channel that has failed in an on-duty regime.

k = tuning time for the service channel for the first request to arrive after an idle period (time to prepare to operate).

L_{in} = number of requests arriving simultaneously in a group.

L_{out} = number of additional requests generated by servicing a single request.

FIFO = first-in-first-out service discipline (service in order of arrival).

LIFO = last-in-first-out serving discipline (service in inverse order of arrival).

h = time interval from the initiation of the service of a request to the moment of final completion (time the system is occupied by a request).

L = number of requests in a system.

ω = waiting time (the time a request waits for service to begin).

v = time of the presence of a request in a system (in all cases considered here $v = \omega + h$).

π = system's busy period (the time interval from the first arrival to the first moment of the next idle period).

Table 10.1 Notation for queues

Variable	R.V.	D.F.	LST or generating function	Mean	kth Moment
Arrival time	α	$A(t)$	$\alpha(s)$	$E\{\alpha\} = 1/a$	α_k
Time of serving	β	$B(t)$	$\beta(s)$	$E\{\beta\} = 1/b$	β_k
Time-to-failure for a channel in service	γ	$C(t)$	$\gamma(s)$	$E\{\gamma\} = 1/c$	γ_k
Repair time for a channel failed in service	δ	$D(t)$	$\delta(s)$	$E(\delta) = 1/d$	δ_k
Time-to-failure for a channel in on-duty regime (idle period)	E	$E(t)$	$E(s)$	$E\{z\} = 1/e$	ϵ_k
Repair time for a channel failed in on-duty regime (idle period)	φ	$F(t)$	$\varphi(s)$	$E\{\varphi\} = 1/f$	ψ_k
"Tuning" time	k or χ	$K(t)$	$k(s)$ or $\chi(s)$	$E\{k\}$ or $E\{\zeta\}$	k_k or χ_k
Size of a group of input demands	L_{in}	—	$\Phi(z)$	$E\{L_{in}\}$	$(E\{L_{in}\})^k$
Size of a group of additional demands	L_{out}	—	$F(z)$	$E\{L_{out}\}$	$(E\{L_{out}\})^{k*}$
Blocking time	h	—	$h(s)$	$E\{h\}$	h_k
System state	L	—	$P(z)$	$E\{L\}$	$(E\{L\})^k$
Waiting time	φ	—	$\varphi(s)$	$E\{\varphi\}$	φ_k
Time of presence of the demand in the system	v	—	$v(s)$	$E\{v\}$	v_k
System busy period	π	—	$\pi(s)$	$E\{\pi\}$	π

ρ = system's loading coefficient (the ratio of the total busy time during a period of time to the maximum service capability of the system for that period).

A request whose service has been interrupted by a channel failure may be lost, or may be completed without loss of the previous results, or may be served from the beginning after channel repair. All r.v.'s in a group are assumed to be homogeneous and independent. All r.v.'s are assumed to have means. Existence of stationary indexes is provided by the condition $\rho < 1$. Notations are given in Table 10.1.

10.1.2 Time of System Occupation with a Request

The time the channel is occupied with a request will be used for all QS's considered here. QS reliability and repair are considered in the determination of this index. The analysis requires considerations of the LST equations, $h(s) = E\{e_{sh}\}$, and the first moments, h_w, of the r.v. h for different calculations for a request whose service has been interrupted by a channel failure.

For a service that has been interrupted, the LST, $h(s)$, of the distribution function, the mean, $E\{h\}$, and the variance, $\text{Var}\{h\}$, of the channel's occupation time during a request, can be determined with the following equations. In general,

$$h(s) = \int_0^\infty e^{-su}[1 - C(u)]\,dB(u)$$

$$+ \delta(s) \int_0^\infty e^{-su}[1 - B(u)]\,dC(u)\,,$$

$$E\{h\} = -h'(0)\,,$$

$$\text{Var}\{h\} = h''(0) - (E\{h\})^2\,.$$

If $C(t) = 1 - e^{-ct}$, $c > 0$, $t \geq 0$:

$$h(s) = \frac{c\delta(s)}{s + c} + \beta(s + c)\left(1 - \frac{c\delta(s)}{s + c}\right),$$

$$Eh = \left(\frac{1}{c} + E\delta\right)[1 - \beta(c)]\,,$$

$$\text{Var}\, h = \left(\frac{1}{c^2} + \text{Var}\,\delta\right)[1 - \beta(c)] + \left(\frac{1}{c} + E\delta\right)[\beta(c)Eh + 2\beta'(c)]\,.$$

If $B(t) = 1 - e^{-bt}$, $b > 0$, $t \geq 0$:

$$h(s) = \frac{b}{s+b} + \gamma(s+b)\left[\delta(s) - \frac{b}{s+b}\right],$$

$$Eh = \frac{1}{b} + \gamma(b)\left(E\delta - \frac{1}{b}\right),$$

$$\operatorname{Var} h = \gamma(b) \operatorname{Var} \delta + \frac{1}{b^2}[1 - \gamma(b)]$$

$$+ \gamma(b)[1 - \gamma(b)]\left(E\delta - \frac{1}{b}\right)^2 - 2\gamma'(b)\left(E\delta - \frac{1}{b}\right).$$

If the interrupted service is continued after repair, then, in general:

$$h(s) = \int_0^\infty e^{-st} P(\delta(s), t)\, dB(t);$$

$$s\int_0^\infty e^{-st} P(z, t)\, dt = \frac{1 - \gamma(s)}{1 - z\gamma(s)}.$$

If $C(t) = 1 - e^{-ct}$, $c > 0$, $t \geq 0$, then:

$$h(s) = \beta(s + c - c\delta(s));$$

$$Eh = E\beta(1 + cE\delta),$$

$$\operatorname{Var} h = \operatorname{Var} \beta(1 + cE\delta)^2 + cE\beta[\operatorname{Var} \delta + (E\delta)^2].$$

If $B(t) = 1 - e^{-bt}$, $b > 0$, $t \geq 0$,

$$h(s) = \frac{b}{s+b}\, \frac{1 - \gamma(s+b)}{1 - \delta(s)\gamma(s+b)},$$

$$Eh = \frac{1}{b} + \frac{\gamma(b) E\delta}{1 - \gamma(b)},$$

$$\operatorname{Var} h = \operatorname{Var} \beta + \frac{\gamma(b) \operatorname{Var} \delta}{1 - \gamma(b)} + \frac{\gamma(b) E\delta - 2\gamma'(b)}{[1 - \gamma(b)]^2} E\delta.$$

If the service restarts after an interruption, the LST, the mean, and the variance of the r.v. are given by the following. In general:

$$h(s) = \frac{\int_0^\infty e^{-su}[1 - C(u)]\, dB(u)}{1 - \delta(s)\int_0^\infty e^{-sv}[1 - B(v)]\, dC(v)}.$$

If $C(t) = 1 - e^{-ct}$, $c > 0$, $t \geq 0$,

$$h(s) = \left[\frac{s\delta(s)}{s+c} + \left(1 - \frac{c\delta(s)}{s+c}\right)\frac{1}{\beta(s+C)}\right]^{-1},$$

$$E = \left(\frac{1}{c} + E\delta\right)\left(\frac{1}{\beta(c)} - 1\right),$$

$$\operatorname{Var} h = \left(\frac{1}{c^2} + \operatorname{Var}\delta\right)\left(\frac{1}{\beta(c)} - 1\right) + (Eh)^2$$
$$+ \left(\frac{1}{c} + E\delta\right)\left(Eh + \frac{2\beta'(c)}{[\beta(c)]^2}\right).$$

If $B(t) = 1 - e^{-bt}$, $b > 0$, $t \geq 0$, the situation is equivalent to completion of the interrupted request with loss of no previous service:

$$h(s) = \frac{b}{s+b} \frac{1 - \gamma(s+b)}{1 - \delta(s)\gamma(s+b)},$$

$$Eh = \frac{1}{b} + \frac{\gamma(b)E\delta}{1 - \gamma(b)},$$

$$\operatorname{Var} h = \operatorname{Var}\beta + \frac{\gamma(b)\operatorname{Var}\delta}{1 - \gamma(b)} + \frac{\gamma(b)E\delta - 2\gamma(b)}{[1 - \gamma(b)]^2} E\delta .$$

10.2 A STANDARD ONE-CHANNEL QUEUING SYSTEM (QS)

10.2.1 System Description

Arrivals are described with a Poisson process: $A(t) = 1 - e^{-at}$, $a > 0$, $t \geq 0$. The serving channel can fail both during service and when only on duty. The waiting line is unrestricted. The expression for the coefficient of loading has the form

$$\rho = aE\{h\} .$$

10.2.2 Number of Requests in a QS

The generating function of the stationary distribution of L has the form:

$$P(z) = P(0) \frac{1 - R(z)}{h(a - az) - z} h(a - az),$$

where

$$P(0) = \frac{(1-\rho)[1 - \epsilon(a)\varphi(a)]}{1 - \epsilon(a) + \epsilon(a)aE\varphi} .$$

The function

$$R(z) = R_0(z)$$
$$= \frac{1 - \epsilon(a)}{1 - \epsilon(a)\varphi(a)} z + \frac{\varepsilon(a)}{1 - \epsilon(a)\varphi(a)} [\varphi(a - az) - \varphi(a)]$$

is the generating function of the r.v. R, the number of requests arriving in the QS between a system idle period and the moment of service of a given request. For this case

$$ER = R'(1) = \frac{1 - \epsilon(a) + \epsilon(a) a E\varphi}{1 - \epsilon(a)\varphi(a)} ,$$

$$\text{Var } R = \frac{\epsilon(a)a^2}{1 - \epsilon(a)\varphi(a)} [\text{Var } \varphi + (E\varphi)^2] + ER(1 - ER) .$$

Then $P(0) = (i - \rho)/E\{R\}$.

The mean of L is given by

$$EL = \rho + \frac{a^2 h^2}{2(1-\rho)} + \frac{R''(1)}{2R'(1)} ,$$

or

$$E = \rho + \frac{a^2[\text{Var } h + (Eh)^2]}{2(1-\rho)} + \frac{\epsilon(a)a^2[\text{Var } \varphi + (E\varphi)^2]}{2[1 - \epsilon(a) + \epsilon(a)aE\varphi]} .$$

The variance of L is given by:

$$\text{Var } L = a^2 \text{ Var } h + EL + \frac{a^3 h^3}{3(1-\rho)} + \left[\frac{a^2[\text{Var } h + (Eh)^2]}{2(1-\rho)}\right]^2$$
$$+ \frac{\epsilon(a)a^3\varphi_3}{3[1 - \epsilon(a) + \epsilon(a)aE\varphi]} - \left[\frac{\epsilon(a)a^2[\text{Var } \varphi + (E\varphi)^2]}{2[1 - \epsilon(a) + \epsilon(a)aE\varphi]}\right]^2 ,$$
$$h_3 = -h'''(0) .$$

10.2.3 Characteristics of FIFO Service

The LST of the distribution of the time of presence of the request in the QS during stationarity, $v(s)$, is determined by the equation:

$$P(z) = v(a - az) .$$

Because $v(s) = \varphi(s)h(s)$, the LST of the waiting time for the beginning of service is given by

$$\omega(s) = \frac{1-\rho}{1 - \epsilon(a) + \epsilon(a)aE\varphi} \frac{[1 - \epsilon(a)]s + \epsilon(a)a[1 - \varphi(s)]}{s - a + ah(s)} .$$

The means and variances are related by

$$Ev = E\omega + Eh ,$$

$$\text{Var } v = \text{Var } \omega + \text{Var } h .$$

By Little's formula, $E\{L\} = aEv$, and $\text{var}\{L\} = a^2 \text{Var}\{v\} + E\{L\}$. $E\{\omega\}$ and $\text{Var}\{\omega\}$ are given by

$$E\omega = \frac{a[\text{Var}\, h + (Eh)^2]}{2(1-\rho)} + \frac{\epsilon(a)a[\text{Var}\,\varphi + (E\varphi)^2]}{2[1-\epsilon(a)+\epsilon(a)aE\varphi]},$$

$$\text{Var}\,\omega = \frac{a}{1-\rho}\left[\frac{h_3}{3} + \frac{\text{Var}\,h + (Eh)^2}{2}E\omega\right] + \frac{\epsilon(a)a}{1-\epsilon(a)+\epsilon(a)aE\varphi}\left[\frac{\varphi_3}{3} - \frac{\text{Var}\,\varphi + (E\varphi)^2}{2}E\omega\right].$$

10.2.4 Characteristics of LIFO Service

The LST of the distribution function of the service waiting time during stationarity is given by:

$$\omega^L(s) = \frac{(1-\rho)[1-\epsilon(a)]}{1-\epsilon(a)+\epsilon(a)aE\varphi} + \frac{a[1-\mu(s)]}{s+a-a\mu(s)} + \frac{(1-\rho)a\epsilon(a)}{1-\epsilon(a)+\epsilon(a)aE\varphi}\,\frac{1-\varphi[s+q-a\mu(s)]}{s+a-a\mu(s)},$$

where $\mu(s)$ is given by:

$$\mu(s) = h[s+a-a\mu(s)], \quad \text{Re}\, s > 0,\ |\mu(s)| < 1,\ \mu(0) = 1. \qquad (10.1)$$

The mean of the service waiting time coincides with that for FIFO:

$$E\omega = E\omega^L.$$

The variance of the LIFO service waiting time has the form

$$\text{Var}\,\omega^L = \frac{ah_3}{3(1-\rho)^2} + \frac{1+\rho}{1-\rho}(E\omega)^2 + \frac{a\epsilon(a)}{1-\epsilon(a)+\epsilon(a)aE\varphi}\,\frac{\varphi_3}{3(1-\rho)} - \frac{a\epsilon(a)[\text{Var}\,\varphi + (E\varphi)^2]}{1-\epsilon(a)+\epsilon(a)aE\varphi}\,\frac{E\omega}{1-\rho}.$$

Notice that the variances of the FIFO and LIFO disciplines are related by

$$\text{Var}\,\omega^L(1-\rho) = \text{Var}\,\omega + \rho(E\omega)^2.$$

The LST of the distribution of the time of the presence of the request in the QS has the form:

$$v^L(s) = \omega^L(s)h(s);$$

which is similar to the expression for the FIFO discipline. Similarly:

$$Ev^L = Ev = E\omega + h \,,$$

$$\operatorname{Var} v^L = \operatorname{Var} \omega + \operatorname{Var} h \,.$$

10.2.5 Busy Period

The LST of the distribution of the busy period has the form

$$\pi(s) = \frac{1 - \epsilon(a)}{1 - \epsilon(a)\varphi(a)} \mu(s) + \frac{a\epsilon(a)}{1 - \epsilon(a)\varphi(a)} \frac{\varphi(s + a - a\mu(s)) - \varphi(a)}{a\mu(s) - s} \mu(s) \,,$$

where $\mu(s)$ is determined by (10.1). The mean and variance of the r.v. are given by

$$E\pi = \frac{1}{a(1-\rho)} \left[\rho - \frac{\epsilon(a)}{1 - \epsilon(a)\varphi(a)} (1 - \varphi(a) - aE\varphi) \right],$$

$$\operatorname{Var} x = \frac{1}{a^2(1-\rho)} \left\{ \frac{a^2 \operatorname{Var} h + \rho^2}{P(0)} \right.$$

$$\left. + \frac{\epsilon(a)}{1 - \epsilon(a)\varphi(a)} [a^2 \operatorname{Var} \rho + (aE\varphi)^2 + 2(1 - \varphi(a) - aE\varphi)] \right\}.$$

10.3 SYSTEM WITH GROUP ARRIVALS

10.3.1 System Description

In contrast to the system in Section 10.2, the input flow is quasi-Poisson. At random moments groups of size L_{in} arrive. The time between these moments has the distribution $A(t) = 1 - e^{-at}$, $a > 0$, $t \geq 0$. The generating function of the distribution of L_{in} is given by:

$$Ez^{L_{\text{in}}} = \Phi(z) \,,$$

$$\Phi(1) = 1 \,.$$

The equation for the system loading is

$$\rho = aEhEL_{\text{in}} \,.$$

10.3.2 Number of Requests in System

The generating function of the stationary distribution of L has the form:

$$P(z) = P(0) = \frac{1 - R(z)}{h(a - a\Phi(z)) - z} h(a - a\Phi(z)) \,,$$

where

$$R(z) = R_0(\Phi(z)) = \frac{1-\epsilon(a)}{1-\epsilon(a)\varphi(a)}\,\Phi(z)$$
$$+ \frac{\epsilon(a)}{1-\epsilon(a)\varphi(a)}\,[\varphi(a - a\Phi(z)) - \varphi(a)]\,,$$
$$P(0) = \frac{1-\rho}{R'(1)} = \frac{1-\rho}{EL_{\text{in}}}\,\frac{1-\epsilon(a)\varphi(a)}{1-\epsilon(a)-\epsilon(a)aE\varphi}\,.$$

The mean and variance of L are related by

$$EL = \rho + \frac{h^2a^2(EL_{\text{in}})^2 + Eha\Phi'(1)}{2(1-\rho)} + \frac{R''(1)}{2R'(1)}\,.$$

Thus

$$EL = \rho = \frac{a^2[\text{Var}\,h + (Eh)^2]}{2(1-\rho)}\,(EL_{\text{in}})_2$$
$$+ \frac{\text{Var}\,L_{\text{in}} - EL_{\text{in}} + (EL_{\text{in}})^2}{2(1-\rho)EL_{\text{in}}} + \frac{\epsilon(a)a'^2[\text{Var}\,\varphi + (E\varphi)^2]EL_{\text{in}}}{2[1-\epsilon(a)+\epsilon(a)aE\varphi]}\,,$$
$$\text{Var}\,L = a^2\,\text{Var}\,h(EL_{\text{in}})^2 + EL + \frac{a^3h^3(EL_{\text{in}})^3 + aEh\Phi'''(1)}{3(1-\rho)}$$
$$+ \frac{a^2\,\text{Var}\,h\Phi''(1)EL_{\text{in}} + aEh\Phi''(1)}{1-\rho} + \frac{R'''(1)}{3R'(1)} - \left(\frac{R''(1)}{2R'(1)}\right)^2$$
$$+ \left[\frac{a^2[\text{Var}\,h + (Eh)^2](EL_{\text{in}})^2 + aEh\Phi''(1)}{2(1-\rho)}\right]^2.$$

10.3.3 Characteristics of FIFO Service

The LST of the stationary distribution function of the time of the presence of the request in the system, $v(s)$, is determined by:

$$P(z) = v(a - a\Phi(z))\,T(\Phi(z))\,,$$
$$T(\Phi(z)) = \frac{t(1)-t(z)}{1-z}\,,\quad |z|<1,\; T(1)=1,\; zt'(z) = \Phi(z)\,.$$

We denote the inverse of $\Phi(y)$ with $\Phi^{-1}(y)$. Then, given $v(s) = \omega(s)h(s)$, the LST of the distribution of the waiting time during stationarity is:

$$\omega(s) = \frac{1-\rho}{R'(1)}\,\frac{1-R_0(1-s/a)}{[h(s)-\Phi^{-1}(1-s/a)]T(1-s/a)}\,,$$

or

$$\omega(s) = \frac{(1-\rho)\{[1-\epsilon(a)]s + a\epsilon(a)[1-\varphi(s)]\}}{EL_{\text{in}}(s - s\epsilon(a) + s\epsilon(a)aE\varphi)[h(s)-\Phi^{-1}(1-s/a)]T(1-s/a)}\,.$$

The following relations exist between the means and the variances of the r.v.'s v, ω, h, L, and L_{in}

$$Ev = E\omega + Eh\ ,$$

$$\mathrm{Var}\ v = \mathrm{Var}\ \omega + \mathrm{Var}\ h\ ,$$

$$EL = aEvEL_{\mathrm{in}} + \tfrac{1}{2}[\Phi'(1) - 1]\ ,$$

$$\mathrm{Var}\ L = a^2\ \mathrm{Var}\ v(EL_{\mathrm{in}})^2 + \frac{\mathrm{Var}\ L_{\mathrm{in}}}{3} + \frac{(EL_{\mathrm{in}})^2 - 1}{12} + (aEv \cdot EL_{\mathrm{in}})^2 + aEv \cdot \Phi''(1)\ .$$

For $E\{\omega\}$ and $\mathrm{Var}\{\omega\}$, we have

$$aE\omega \cdot \mathrm{EL}_{\mathrm{in}} = EL - aEh \cdot EL_{\mathrm{in}} + \frac{1 - EL_{\mathrm{in}}}{2}\ ,$$

or

$$e\omega = \frac{EL}{aEL_{\mathrm{in}}} - Eh + \frac{1}{2aEL_{\mathrm{in}}} - \frac{1}{2a}\ ,$$

$$a^2\ \mathrm{Var}\ \omega(EL_{\mathrm{in}})^2 = \mathrm{Var}\ L - a^2\ \mathrm{Var}\ h(EL_{\mathrm{in}})^2 - \frac{\mathrm{Var}\ L_{\mathrm{in}}}{3} + \frac{1 - (EL_{\mathrm{in}})^2}{12} - (aEv \cdot EL_{\mathrm{in}})^2 - aEv\Phi''(1)\ ,$$

or

$$\mathrm{Var}\ \omega = \frac{\mathrm{Var}\ L}{a^2(EL_{\mathrm{in}})^2} - \mathrm{Var}\ h - \frac{\mathrm{Var}\ L_{\mathrm{in}}}{3a^2(EL_{\mathrm{in}})^2} + \frac{1}{12a^2(EL_{\mathrm{in}})^2} - \frac{1}{12a^2} - (\mathrm{Var}\ v)^2 - \frac{Ev\Phi''(1)}{a(EL_{\mathrm{in}})^2}\ .$$

10.3.4 Characteristics of LIFO Service

The LST of the distribution of the waiting time during stationarity is:

$$\omega^L(s) = \frac{(1 - p)[1 - \epsilon(a)]}{1 - \epsilon(a) + \epsilon(a)aE\varphi} + aEL_{\mathrm{in}} \frac{1 - h(s + a - a\mu(s))}{s + a - a\mu(s)}\ T_0(\mu(s)) + \frac{(1 - \rho)\epsilon(a)a}{1 - \epsilon(a) + \epsilon(a)aE\varphi}\ \frac{1 - \varphi(s + a - a\mu(s))}{s + a - a\mu(s)}\ T_0(\mu(s))\ .$$

In addition

$$\mu(s) = \Phi(h(s + a - a\mu(s)) \quad Re\, s > 0,\ |\mu(s)| \le 1,\ \mu(0) = 1,$$

and

$$T_0(z) = \frac{t(1) - t(z)}{1 - z}, \quad |z| < 1,\ T_0(1) = 1,\ zt'(z) = \Phi(z),$$

$$E\omega^L = -\frac{d}{ds}\,\omega^L(s)|_{s=0},$$

$$\mathrm{Var}\,\omega^L = \frac{d^2}{dt^2}\,\omega^L(s)|_{s+0} - (E\omega^L)^2.$$

As for FIFO service, the LST of the time of the presence of the request in the QS is related to the LST of the distribution of the waiting time by

$$v^L(s) = \omega^L(s)h(s).$$

Consequently,

$$Ev^L = E\omega^L + Eh,$$

$$\mathrm{Var}\,v^L = \mathrm{Var}\,\omega^L + \mathrm{Var}\,h.$$

It is assumed here that if a request has been interrupted by a system failure, there will be no service of the requests that arrived before the interruption.

10.3.5 Busy Period

The LST of the distribution of the busy period coincides in form with the corresponding expression in Section 10.2.5:

$$\pi(s) = \frac{1 - \epsilon(a)}{1 - \epsilon(a)\varphi(a)}\,\mu(s) + \frac{\epsilon(a)a}{1 - \epsilon(a)\varphi(a)}\,\frac{\varphi(s + a - a\mu(s)) - \varphi(a)}{a\mu(s) - s}\,\mu(s),$$

where $\mu(s)$ is determined by:

$$\mu(s) = \Phi(h(s + a - a\mu(s)), \quad Re\, s > 0,\ |\mu(s)| \le 1,\ \mu(0) = 1.$$

The mean and variance of the system busy period have the form:

$$aE\pi = \frac{1 - \epsilon(a) + \epsilon(a)aE\varphi}{(1 - \rho)[1 - \epsilon(a)\varphi(a)]} - 1,$$

$$\operatorname{Var} \pi = \frac{\operatorname{Var} h \cdot EL_{\text{in}}}{(1-\rho)^3} + \frac{\operatorname{Var} L_{\text{in}}(Eh)^2}{(1-\rho)^3}$$

$$+ \frac{\rho^3}{a^2(1-\rho)^3} + \frac{\epsilon(a)[1-\varphi(a)]}{1-\epsilon(a)\varphi(a)} \left[\left(\frac{\operatorname{Var} h \cdot EL_{\text{in}} + \operatorname{Var} L_{\text{in}}(Eh)^2}{(1-\rho)^3} \right.\right.$$

$$\left. + \frac{\rho^3}{a^2(1-\rho)^3} \right) \left(\frac{aE\varphi}{1-\varphi(a)} - 1 \right) + \frac{\operatorname{Var} \varphi}{(1-\rho)^2(1-\varphi(a))}$$

$$+ \frac{1}{a^2(1-\rho)^2} - \frac{\varphi(a)(E\varphi)^2}{(1-\rho)^2(1-\varphi(a))^2}$$

$$\left. + \frac{1-\epsilon(a)}{1-\epsilon(a)\varphi(a)} \frac{1}{(1-\rho)^2} \left(\frac{aE\varphi}{1-\varphi(a)} - 1 \right)^2 \right] .$$

10.4 SYSTEM WITH FEEDBACK

10.4.1 System Description

The difference from the standard queuing system (see Section 10.2) is that a random number, L_{out}, of new additional requests is generated after each service. The generating function of the r.v. L_{out} is given by

$$Ez^{L_{\text{out}}} = F(z) .$$

Here we assume that the additionally generated requests have to be served just immediately after the request that generated them. Except for this, there are no differences between arriving requests and those generated inside the system.

The expression for the loading coefficient has the form

$$\rho = \frac{aEh}{(1 - EL_{\text{out}})} .$$

10.4.2 Number of Requests in System

Consider the number of requests in the system at an arbitrarily chosen moment in the stationary period, $L = (L_0, L_1)$, where L_0 is the number of requests in the main (input) flow, and L_1 is the number of additional requests generated inside the system. The joint distribution of L during stationarity is given by the following generating function:

$$P(z_0, z_1) = E(z_0^{L_0} z_1^{L_1}) .$$

We also may use:

$$P(z_0, z_1) = P(0) \frac{1}{1 - \epsilon(a)\varphi(a)} \left[1 - \epsilon(a) + \epsilon(a) \frac{1 - \varphi(a - az_0)}{1 - z_0} \right]$$
$$+ \left\{ \frac{P(z_0) - P(0)}{z_0(1 - z_0)} + P(0) \frac{\epsilon(a)[\varphi(a - az_0) - \varphi(a)]}{z_0^2[1 - \epsilon(a)\varphi(a)](1 - z_0)} \right\}$$
$$\times [h(a - az_0)g(a - az_0, z_1) + z_0(1 - h(a - az_0))] ;$$

where

$$P(z) = P(0) \frac{1 - R(z)}{g(a - az) - z} g(a - az) ;$$

$$R(z) = R_0(z) = \frac{1 - \epsilon(a)}{1 - \epsilon(a)\varphi(a)} z + \frac{\epsilon(a)}{1 - \epsilon(a)\varphi(a)}$$
$$\times [\varphi(a - az) - \varphi(a)] ,$$

$$P(0) = \frac{1 - \rho}{R'(1)} = \frac{(1 - \rho)[1 - \epsilon(a)\varphi(a)]}{1 - \epsilon(a) + \epsilon(a)aE\varphi} ,$$

$$g(s, z) = z\left[\frac{1}{h(s)} - 1 \right]\left[\frac{z[g(s) - z]}{h(s)g(s)F(z) - F(0)h(s)[g(s) - z] - z^2 - 1} \right] ;$$

and the function $g(s)$ satisfies

$$g(s) = h(s)F(g(s)) , \qquad Re\, s > 0,\ |g(s)| \le 1,\ g(0) = 1 .$$

10.4.3 Waiting Time and Time of Presence in System, FIFO

For the requests in the main flow, the LST of the distribution of the waiting time is determined by

$$P(z) = \omega(a - az)g(a - az) .$$

We have:

$$\omega(s) = \frac{1 - \rho}{1 - \epsilon(a) + \epsilon(a)aE\varphi} \frac{[1 - \epsilon(a)]a + \epsilon(a)a[1 - \varphi(s)]}{s - a + ag(s)} ,$$

where, as above, $g(s)$ is determined by (10.2). The mean and the variance of the waiting time for the requests in the main flow have the form

$$E\omega = \frac{a \operatorname{Var} h}{2(1-\rho)(1-EL_{\text{out}})} + \frac{[E(h)]^2}{2(1-\rho)}\left[\frac{\operatorname{Var} L_{\text{out}}}{(1-EL_{\text{out}})^3} + \frac{1}{(1-EL_{\text{out}})^2}\right] + \frac{\epsilon(a)a[\operatorname{Var}\varphi + (E\varphi)^2]}{2(1-\epsilon(a)+\epsilon(a)aE\varphi)}$$

$$\operatorname{Var}\omega = \frac{d^2}{ds^2}(s)\Big|_{s=1} - (E\omega)^2\omega .$$

For the additional requests, the LST of the distribution of the waiting time is

$$\omega_1(s) = T(g(s)) ,$$

where $g(s)$ is determined by (10.2) and the function $T(z)$ can be found from

$$T(z) = \frac{t(1)-t(z)}{1-z}, \quad |z|<1,\ T(1)=1,\ zt'(z)=F(z) .$$

The equations for the mean and the variance of the waiting time for the additional request are:

$$E\omega_1 = \frac{Eh}{2} ;$$

$$\operatorname{Var}\omega_1 = \frac{\operatorname{Var} h}{2} + \frac{(Eh)^2}{12}\left[\frac{2\operatorname{Var} L_{\text{out}}}{(1-EL_{\text{out}})^2} + \frac{1+EL_{\text{out}}}{1-EL_{\text{out}}}\right].$$

The time of the presence of a request inside the system has a standard relationship

$$v(s) = \omega(s)h(s) ;$$

$$Ev = E\omega + Eh ;$$

$$\operatorname{Var} v = \operatorname{Var}\omega + \operatorname{Var} h .$$

10.4.4 LIFO Waiting Time

For requests in the main flow, the LST of the distribution of the waiting time is determined by:

$$\omega^L(s) = \frac{(1-\rho)[1-\epsilon(a)]}{1-\epsilon(a)+\epsilon(a)aE\varphi} + \frac{a[1-\mu(s)]}{s+a-a\mu(s)}$$
$$+ \frac{(1-\rho)\epsilon(a)a}{1-\epsilon(a)+\epsilon(a)aE\varphi} \frac{1-\varphi(s+a-a\mu(s))}{s+a-a\mu(s)} ,$$

where $\mu(s)$ is determined with

$$\mu(s) = g(s+a-a\mu(s)) , \quad Re\, s > 0,\ |\mu(s)| \le 1,\ \mu(0)=1 , \qquad (10.3)$$

and $g(s)$, as before, can be found from (10.2).

The mean and the variance of the waiting time for the requests in the main flow for the LIFO discipline have the form

$$E\omega^L = \frac{a \operatorname{Var} h}{2(1-\rho)(1-EL_{\text{out}})} \frac{1}{1-\epsilon(a)+\epsilon(a)aE\varphi}$$
$$+ \frac{a(Eh)^2}{2(1-\rho)} \frac{\operatorname{Var} L_{\text{out}} + 1 - EL_{\text{out}}}{(1-EL_{\text{out}})^3[1-\epsilon(a)+\epsilon(a)aE\varphi]}$$
$$+ \frac{\epsilon(a)a[\operatorname{Var}\varphi + (E\varphi)^2]}{2[1-\epsilon(a)+\epsilon(a)aE\varphi]} ;$$
$$\operatorname{Var}\omega^L = \left[\frac{d^2}{ds^2}\omega^L(s)\right]_{s=0} - (E\omega^L)^2 .$$

For the additional requests (because of the order of the service described above) there is no difference between the two disciplines.

10.4.5 System Busy Period

The LST of the distribution of the system busy period has the same form as in Sections 10.2.5 and 10.3.5. The corresponding function $\mu(s)$ is determined by (10.3). For the mean and the variance of the busy period, we have:

$$aE\pi = \frac{1-\epsilon(a)+\epsilon(a)aE\varphi}{(1-\rho)[1-\epsilon(a)\varphi(a)]} - 1 ;$$

$$\operatorname{Var}\pi = \frac{\operatorname{Var} g}{(1-\rho)^3} + \frac{\rho^3}{a^2(1-\rho)^3} + \frac{\epsilon(a)[1-\varphi(a)]}{[1-\epsilon(a)\varphi(a)]}$$
$$\times \left\{\left(\frac{\operatorname{Var} g}{(1-\rho)^2} + \frac{\rho^2}{a^2(1-\rho)^3}\right)\left(\frac{aE\varphi}{1-\varphi(a)} - 1\right)\right.$$

$$
+ \frac{\operatorname{Var} \varphi}{(1-\rho)^2[1-\varphi(a)]} + \frac{1}{a^2(1-\rho)^2}
$$

$$
- \frac{\varphi(a)(E\varphi)^2}{(1-\rho)^2[1-\varphi(a)]^2}
$$

$$
+ \frac{1-\epsilon(a)}{1-\epsilon(a)\varphi(a)} \frac{1}{(1-\rho)^2} \left(\frac{aE\varphi}{1-\varphi(a)} - 1 \right)^2 \Bigg\} .
$$

10.5 SYSTEM WITH A "TUNING" PERIOD

10.5.1 System Description

If a request arrives when a system is idle, it may take time for the system to go into an operational state and thus be prepared to start service. The LST of the distribution of the preparation period, which is called the "tuning" period, is given by

$$
k(s) = E\{e^{-sk}\} , \qquad Re\, s > 0 .
$$

If a failure occurs during the tuning period, after the repair the preparation continues from the point of interruption. The duration of the total preparation process, without taking into account the possibility of repair time κ, has the LST:

$$
\kappa(s) = k(s + c - c\delta(s)) .
$$

The time-to-failure distribution is $C(t) = 1 - e^{-ct}$, $c > 0$, $t \geq 0$. The LST of the distribution of the repair time, $\delta(s)$, is the same as for repairs of regular failures. By analogy with Section 10.1.2 we write:

$$
E\kappa = Ek(1 + cE\delta) ,
$$

$$
\operatorname{Var} k = \operatorname{Var} k(1 + cE\delta)^2 + cEk[\operatorname{Var} \delta + (E\delta)^2] .
$$

The expression for the loading coefficient keeps its form:

$$
\rho = aEh .
$$

10.5.2 Number of Requests in System

The generating function of the distribution of the number of requests inside the system during stationarity has the form:

$$
P(z) = P(0) \frac{1 - R(z)\kappa(a - az)}{h(a - az) - z} h(a - az) ;
$$

where

$$R(z) = R_0(z) = \frac{1 - \epsilon(a)}{1 - \epsilon(a)\varphi(a)}\, z + \frac{\epsilon(a)}{1 - \epsilon(a)\varphi(a)}\, [\varphi(a - az) - \varphi(a)] ;$$

$$P(0) = \frac{1 - \rho}{R'(1) + aE\kappa}$$

For the mean and the variance of L we have:

$$EL = \rho + \frac{a^2 h^2}{2(1 - \rho)} + \frac{R''(1) + 2R'(1)aE\kappa + a^2\kappa_2}{2[R'(1) + aE\kappa]} ,$$

or

$$EL = \rho + \frac{a^2[\text{Var}\, h + (Eh)^2]}{2(1 - \rho)} + \frac{a^2[\text{Var}\, \kappa + (E\kappa)^2]}{2[R'(1) + aE\kappa]} + \frac{R''(1) + 2R'(1)aE\kappa}{2[R'(1) + aE\kappa]} ;$$

$$\text{Var}\, L = P''(1) + P'(1) - [P'(1)]^2 .$$

10.5.3 Waiting Time and Time of Presence in the System, FIFO

As in Section 10.2.3, the LST of the distribution of the waiting time can be found from:

$$P(z) = \omega(a - az)h(a - az) .$$

The LST has the form:

$$\omega(s) = \frac{1 - \rho}{1 - \epsilon(a) + \epsilon(a)a(E\varphi + E\kappa)} \times \frac{[1 - \epsilon(a)]s + \epsilon(a)a[1 - \varphi(s)]}{s - a + ah(s)} .$$

The mean and the variance of L are:

$$aE\omega = L - aEh ;$$

$$a^2\, \text{Var}\, \omega = \text{Var}\, L - a^2\, \text{Var}\, h - EL .$$

As usual, $v(s) = \omega(s)/h(s)$.

10.5.4 Waiting Time and Time of Presence in System, LIFO

For the LST of the distribution of the waiting time for the LIFO discipline, we have:

$$\omega^L(s) = \frac{1-\epsilon(a)(1-\rho)\kappa(s+q-a\mu(S))}{1-\epsilon(a)+\epsilon(a)aE\varphi+\epsilon(a)aE\varphi}$$
$$+\frac{a[1-h(s+a-a\mu(s))]}{s+a-a\mu(s)}$$
$$+\frac{\epsilon(a)a[1-\rho][1-\varphi(s+a-a\mu(s))]\kappa(s+a-a\mu(s))}{1-\epsilon(a)+\epsilon(a)aE\varphi+\epsilon(a)aE\kappa][s+a-a\mu(s)]}$$
$$+\frac{[1-\epsilon(a)\varphi(a)]a(1-\rho)[1-\kappa(s+a-a\mu(s))]}{[1-\epsilon(a)+\epsilon(a)aE\varphi+\epsilon(a)aE\kappa][s+a-a\mu(s)]},$$

where the function $\mu(s)$ is the same as in Section 10.2.4 and is defined by

$$\mu(s) = h(s+a-a\mu(s)), \quad Re\, s > 0,\ |\mu(s)| \leq 1,\ \mu(0) = 1. \tag{10.4}$$

The waiting time means coincide for FIFO and LIFO: $E\{\omega\} = E\{\omega^L\}$. The variance of the waiting time for LIFO has the form

$$\text{Var}\, \omega^L = \left[\frac{d^2}{ds^2}\, \omega^L(s)\right]_{s=0} = (E\omega)^2,$$

where $v_L(s) = \omega_1(s)h(s)$.

10.5.5 System Busy Period

The LST of the distribution of the system busy period is determined as

$$\pi(s) = \frac{1-\epsilon(a)}{1-\epsilon(a)\varphi(a)}\, \mu(s) + \frac{\epsilon(a)a}{1-\epsilon(a)\varphi(a)}$$
$$\times \frac{\varphi(s+a-a\mu(s))-\varphi(a)}{a\mu(s)-s}\kappa(s+a-a\mu(s))\mu(s),$$

where $\omega(s)$ is function (10.4). For the mean and the variance of the system's busy period, we have:

$$aE\pi = \frac{1-\epsilon(a)aE\varphi+\epsilon(a)aE\kappa}{(1-\rho)[1-\epsilon(a)\varphi(a)]} - 1;$$

$$\text{Var}\, \pi = \left[\frac{d^2}{ds^2}\, \pi(s)\right]_{s=0} - (E\kappa)^2.$$

10.6 SYSTEM WITH GROUP ARRIVALS, TUNING PERIOD, AND FEEDBACK

10.6.1 System Description

All additions to the standard QS (see Section 10.2) introduced in Sections 10.3, 10.4, and 10.5 are considered together. A random number of requests arriving in a group, L^{in}, is given by the generating function, $\Phi(z)$. A number of additional requests generated inside the system, L^{out}, is given by the generating function, $F(z)$. The distribution of the tuning time, k, has the LST $k(s)$.

Additional requests are served just after the request that has generated them. If there is an interrupted tuning period, the total period, including possible repairs, κ, has a distribution with the LST

$$\kappa(s) = k(s + c - s\delta(s)) .$$

The distribution of the TTF under the tuning regime is $C(t) = 1 - e^{-ct}$, $c > 0$, $t \geq 0$, and the LST of the repair time is $\delta(s)$. By analogy with Section 10.1.2,

$$E\kappa = Ek(1 + cE\delta) ,$$

$$\text{Var}\,\kappa + \text{Var} \cdot \kappa(1 + cE\delta)^2 + cEk \cdot [\text{Var}\,\delta + (E\delta)^2] .$$

The expression for the coefficient of loading is

$$\rho = \frac{aEh \cdot EL_{\text{in}}}{1 - EL_{\text{out}}}$$

Consider the number of requests in the system at an arbitrarily chosen moment in the stationary period, $L = (L_0, L_1)$, where L_0 is the number of the request of the main (input) flow, and L is the number of additional requests generated inside the system. The stationary joint distribution of L is given by the following generating function:

$$\begin{aligned} P(z_0, z_1) = {} & \frac{1}{[1 - \epsilon(a)\varphi(a)][1 - \Phi(z_0)]z_0} \\ & \times \Bigg\{[(1 - \epsilon(a)\varphi(a))(P(z_0) - P(0)) \\ & + \frac{1}{z_0} P(0)\epsilon(a)[\varphi(a - a\Phi(z_0)) - \varphi(a)] \\ & \times [h(a - a\Phi(z_0))g(a - a\Phi(z_0), z_1) \\ & + z_0(1 - h(a - a\Phi(z))] \\ & + P(0)z_0[1 - \epsilon(a)\varphi(a - a - \Phi(z)]\Bigg\} , \end{aligned}$$

where

$$P(z) = P(0)\,\frac{1 - R(\Phi(z)\kappa(a - a\Phi(z))}{g(a - a\Phi(z)) - z}\,g(a - a\Phi(z))\,,$$

$$P(0) = \frac{1-\rho}{[R'(1) + aE\kappa]EL_{\text{in}}}\,,$$

$$R(z) = R_0(z) = \frac{1-\epsilon(a)}{1-\epsilon(a)\varphi(a)}\,z + \frac{\epsilon(a)}{1-\epsilon(a)\varphi(a)}\,[\varphi(a - az) - \varphi(a)]\,.$$

$R(\Phi(z))$ is the generating function of a random number of requests, R, arriving into the system at the moment of initiation of the tuning process, and:

$$ER = \frac{1 - \epsilon(a) + \epsilon(a)aE\varphi}{1-\epsilon(a)\varphi(a)}\,EL_{\text{in}}\,.$$

The function $g(s, z)$ satisfies the condition:

$$\frac{g(s,z)}{[1-h(s)]}\left\{1 - \frac{g(s)[z - h(s)F(z)]}{z[g(s) - z]} - \frac{h(s)F(0)}{z}\right\}$$
$$= \frac{g(s)[z - h(s)F(z)]}{h(s)[g(s) - z]} + F(0)\,,$$

and the function $g(s)$ is the LST of the random time that the system is blocked by a request in the main flow, g. The function $g(s)$ is determined by the equation:

$$g(s) = h(s)F(g(s))\,, \qquad Re\,s > 0,\ |g(s)| \le 1,\ g(0) = 1\,. \tag{10.5}$$

The following expressions are valid for g:

$$Eg = \frac{Eh}{1 - EL_{\text{out}}}\,,$$

$$\text{Var}\ g = \frac{\text{Var}\ h}{1 - EL_{\text{out}}} + \frac{\text{Var}\ L_{\text{out}}(Eh)^{22}}{(1 - EL_{\text{out}})^3}\,.$$

The mean and variance of the number of main and additional requests in the system during stationarity are:

$$EL_0 = \left.\frac{\partial P(z_0, 1)}{\partial z_0}\right|_{z_0=1}\,,$$

$$\text{Var}\ L_0 = \left.\frac{\partial^2 P(z_0, 1)}{\partial z_0^2}\right|_{Z_0=1} + EL_0(1 - EL_0)\,;$$

$$EL_1 = \left.\frac{\partial P(1, z_1)}{\partial z_1}\right|_{z_1-1}\,,$$

$$\text{Var}\ L_1 = \left.\frac{\partial^2 P(1, z_1)}{\partial z_1^2}\right|_{z_1=1} + EL_1(1 - EL_1)\,.$$

10.6.2 Waiting Time and Time of Presence of the System, FIFO

For the main flow requests , the LST of the distribution of the waiting time can be found with

$$P(z) = \omega(a - a\Phi(z))g(a - \Phi(z))\,T(z)\,,$$

where $g(s)$ is the function determined by the functional equation (10.5) and $T(z)$ is defined as

$$T(z) = \frac{t(1) - t(z)}{1 - z}\,, \qquad |z| < 1,\ T(1) = 1,\ zt'(z) = \Phi(z)\,.$$

The mean and the variance of the waiting time for the main flow are

$$E\{\omega\} = \frac{E\{L_0\}}{aE\{L_{\text{in}}\}} + \frac{1 - E\{L_{\text{in}}\}}{2aE\{L_{\text{in}}\}} - \frac{E\{h\}}{1 - E\{L_{\text{out}}\}}\,,$$

$$\begin{aligned} a^2\,\text{Var}\{\omega\}(E\{L_{\text{in}}\})^2 = {} & \text{Var}\{L_0\} - a^2\,\frac{\text{Var}\{h\}(E\{L_{\text{in}}\})^2}{1 - E\{L_{\text{out}}\}} \\ & - \frac{a^2\,\text{Var}\{L_{\text{out}}\}(E\{h\})^2(E\{L_{\text{in}}\})^2}{(1 - E\{L_{\text{out}}\})^3} \\ & + \frac{1 - (E\{L_{\text{in}}\})^2}{12} - \frac{\text{Var}\{L_{\text{in}}\}}{3} \\ & - \left(aE\{\omega\} + \frac{aE\{h\}}{1 - E\{L_{\text{out}}\}}\right)(\text{Var}\{L_{\text{in}}\} + (E\{L_{\text{in}}\})^2)\,. \end{aligned}$$

For the additional requests, the LST of the distribution of the waiting time has the form

$$\omega_1(s) = T_F(g(s))\,,$$

where the function $g(s)$ is determined by (10.5), and $T_F(z)$ can be found from an equation analogous to (10.6) with replacement of $\Phi(z)$ with $F(z)$.

The mean and variance of the waiting time for the additional requests are:

$$E\{\omega_1\} = \frac{E\{h\}}{1 - E\{L_{\text{out}}\}}\,\frac{1 - F'(1)}{2} = \frac{E\{h\}}{2}\,,$$

$$\text{Var}\{\omega_1\} = \frac{\text{Var}\{h\}}{2} + \frac{(E\{h\})^2}{12}\left[\frac{2\,\text{Var}\{L_{\text{out}}\}}{(1 - E\{L_{\text{out}}\})^2} + \frac{1 + E\{L_{\text{out}}\}}{1 - E\{L_{\text{out}}\}}\right].$$

As is usual for the time of the presence of the request in the QS, the requests in the main flow have $v(s) = \omega(s)h(s)$. The additional requests generated inside the system have $v(s) = \omega(s)h(s)$.

10.6.3 LIFO Waiting Time

The LST of the stationary distribution of the waiting time during stationarity for requests in the main flow is given by:

$$\omega^L(s) = \frac{(1-\epsilon(a))(1-\rho)\kappa(s+a-a\mu(s))}{1-\epsilon(a)+\epsilon(a)a(E\{\varphi\}+E\{\kappa\})}$$
$$+\frac{aE\{L_{\text{in}}\}(1-g(s+a-a\mu(s))}{s+a-a\mu(s)}$$
$$\frac{1}{1-\epsilon(a)+\epsilon(a)aE\{\varphi\}+\epsilon(a)aE\{\kappa\}](s+a-a\mu(s)]}$$
$$\times\{\epsilon(a)a(1-\rho)[1-\varphi(s+a-a\mu(s)]\kappa(s+a-a\mu(s))$$
$$+(1-\epsilon(a)\varphi(a))a(1-\rho)[1-\kappa(s+a-a\mu(s))]\}\,,$$

where $\mu(s)=\Phi(g(s+a-a\mu(s))$, $Re\, s>0$, $|\mu(s)|\leq 1$, $\mu(0)=1$, and $g(s)$, as above, is determined from (10.5).

For requests in the main input flow under the FIFO service discipline, the mean and the variance of the waiting time are:

$$E\{\omega^L\} = \frac{[1-\epsilon(a)+\epsilon(a)aE\{\varphi\}]E\{\kappa\}}{1-\epsilon(a)+\epsilon(a)a(E\{\varphi\}+E\{\kappa\})}$$
$$+\frac{\epsilon(a)a\varphi_2+[1-\epsilon(a)\varphi(a)]a\kappa_2}{1[1-\epsilon(a)+\epsilon(a)a(E\{\varphi\}+E\{\kappa\})]}$$
$$+\frac{aE\{L_{\text{in}}\}[\text{Var}\{g\}+(E\{g\})^2]}{2(1-\rho)[1-\epsilon(a)+\epsilon(a)a(E\{\varphi\}+E\{\kappa\})]}\,;$$
$$\text{Var}\{\omega^L\} = \omega_2^L-(E\{\omega^L\})^2\,;$$

$$\omega_2^L = \Bigg\{1-\epsilon(a))\left[\frac{ag_2EL_{\text{in}}E\kappa+(Eg)^2\Phi''(1)aE\kappa}{(1-\rho)^2}+\frac{\kappa^2}{1-\rho}\right]$$
$$\times\epsilon(a)a\left[\frac{\text{Var}\, gEL_{\text{in}}+(Eg)^2(\Phi''(1)+EL_{\text{in}})}{2(1-\rho)^2}(\varphi_2+aE\varphi E\kappa)\right.$$
$$+\frac{\kappa_2E\varphi_2E\kappa+\frac{1}{3}\varphi_3}{1-\rho}+(1-\epsilon(a)\varphi(a)a\left[\frac{\kappa_3}{3(1-\rho)}\right.$$
$$\left.+\frac{\text{Var}\, gEL_{\text{in}}+(Eg)^2\Phi''(1)}{2(1-\rho)^2}+\frac{(Eg)^2EL_{\text{in}}}{2(1-\rho)^2}\kappa_2\right]$$
$$+\frac{aEL_{\text{in}}}{6(1-\rho)^3}[g_3(1-\rho)2+3ag_2(\text{Var}\, gEL_{\text{in}}$$
$$+(Eg)^2(\Phi''(1)+EL_{\text{in}}))][1-\epsilon(a)$$
$$+\epsilon(a)a(E\varphi+E\kappa)]\Bigg\}[1-\epsilon(a)+\epsilon(a)a(E\varphi+E\kappa)]^{-1}\,.$$

For additional requests, because of the order of service there is no difference between the disciplines.

10.6.4 System Busy Period

The LST of the system busy period is

$$\pi(s) = \frac{1-\epsilon(a)}{1-\epsilon(a)\varphi(a)}\,\mu(s) + \frac{\epsilon(a)a}{1-\epsilon(a)\varphi(a)}$$
$$\times\,\frac{(\varphi(s+a-a\mu(s))-\varphi(a)}{a\mu(s)-s}\,\kappa(s+a-a\mu(s))\mu(s)\,,$$

where the functions $g(s)$ and $\omega(s)$ are determined by (10.5) and (10.7), respectively. The mean and variance of the system busy period are given by

$$aE\{\kappa\} = \frac{1-\epsilon(a)+\epsilon(a)aE\{\varphi\}+\epsilon(a)aE\{\kappa\}}{(1-\rho)[1-\epsilon(a)\varphi(a)]} - 1\,,$$

$$\begin{aligned}\mathrm{Var}\{\pi\} = {} & \frac{\mathrm{Var}\{g\}E\{L_{\mathrm{in}}\}}{(1-\rho)^3} + \frac{\mathrm{Var}\{L_{\mathrm{in}}\}(E\{g\})^2}{(1-\rho)^3} \\ & + \frac{\rho^3}{a^2(1-\rho)^3} + \frac{\epsilon(a)[1-\varphi(a)]}{1-\epsilon(a)\varphi(a)} \\ & \times\left\{\left[\frac{\mathrm{Var}\{g\}E\{L_{\mathrm{in}}\}+\mathrm{Var}\{L_{\mathrm{in}}\}(E\{g\})^2}{(1-\rho)^3} + \frac{\rho^2}{a^2(1-\rho)^2}\right]\right. \\ & \times\left[\frac{aE\{\varphi\}}{1-\varphi(a)} + aE\{\kappa\} - 1\right] + \frac{\mathrm{Var}\{\kappa\}}{(1-\rho)^2} \\ & + \frac{\mathrm{Var}\{\varphi\}}{(1-\rho)^2[1-\varphi(a)]} + \frac{1}{a^2(1-\rho)^2} \\ & - \frac{\varphi(a)(E\{\varphi\})^2}{(1-\rho)^2[1-\varphi(a)]^2} \\ & + \frac{1-\epsilon(a)}{1-\epsilon(a)\varphi(a)}\,\frac{1}{(1-\rho)^2} \\ & \left.\times\left[\frac{aE\{\varphi\}+aE\{\kappa\}}{1-\varphi(a)} - 1\right]^2\right\}.\end{aligned}$$

CHAPTER 11

MECHANICAL EQUIPMENT

11.1 STATIC LOAD-STRENGTH PROBLEMS

Much reliability engineering requires determination of the probability that construction or equipment will survive under extreme conditions. A construction may require determination of the probability that it will withstand a given level of external stress, such as shocks or vibration, or internal tension. Electronic equipment may require the probability that it can withstand a voltage surge or a change of input signals.

Both external loads and internal strength can be regarded as random. Loads obviously are random because they depend on many independent forces. In addition, random variation is inherent in manufacturing.

11.1.1 General Expressions

The basic problem is to find a probability R:

$$R = Pr(X \geq Y)\,, \tag{11.1}$$

where X is a random value of a system strength and Y is a load random value. Let us assume that $F_x(x) = Pr\{X \leq x\}$ and $F_x(y) = Pr\{Y \leq y\}$ are continuous and known distributions. The probability of successful operation of the system then can be calculated as

$$R = \int_{-\infty}^{\infty} [1 - F_X(y)] f_Y(y)\, dy = \int_{-\infty}^{\infty} F_Y(x) f_X(x)\, dx\,. \tag{11.2}$$

With X and Y as independent r.v.'s, it is convenient to introduce a new r.v., $W = X - Y$. Then (11.1) can be rewritten as

$$R = Pr(W \geq 0) = \int_0^{\infty} dF_W(w)\, dw\,. \tag{11.3}$$

11.1.2 Distributions of Load and Strength

We will use Z to denote any X or Y r.v. and z for any independent x and y. In the table of results these distributions appear with appropriate subscripts.

1. If F_L is a normal distribution, the density is

$$f_z(z) = \frac{1}{\sigma_z\sqrt{2\pi}} \exp\left(\frac{-(z - Ez)^2}{2\sigma_z^2}\right). \tag{11.4}$$

2. If F_z is an exponential distribution, the density is

$$f_z(z) = \lambda_z e^{-\lambda_z z}\,.$$

3. *If* F_z is a lognormal distribution, the density is

$$\frac{1}{z\sigma\sqrt{2\pi}} \exp - \frac{1}{2\sigma^2} (\ln z - E \ln Z)^2\,,$$

where

$$\sigma^2 = \mathrm{Var}(\ln Z\}\,, \qquad z > 0\,.$$

4. If F_z is a Weibull-Gnedenko distribution, the density is

$$f_z(z) = \frac{\beta}{(\theta - Z_0)^{\beta}} (Z - Z_0)^{\beta - 1} \exp - \left(\frac{z - Z_0}{\theta - z_0}\right)^{\beta}\Bigg]\,,$$

where β is the shape parameter, $\theta - Z_0$ is the scale parameter, Z_0 is the truncation parameter, and $Z \geq Z_0 \geq 0$.

5. If F_z is a gamma distribution, the density is

$$f_z(z) = \frac{\lambda_z^n z^{n-1}}{\Gamma(n)}\, e^{-\lambda_z z}\,,$$

where λ_z is a scale parameter $\lambda > 0$ and is a shape parameter $n > 0$.

Table 11.1 gives probabilities of successful operation for different distributions of load and strength.

Table 11.1 Probability equations for common combinations of strength distributions (A) and load distributions (B).

Distribution A	Distribution B	$R = P(A > B)$
Normal	Normal	$R = \Phi\left(\frac{\mu_A - \mu_B}{\sqrt{\sigma_A^2 + \sigma_B^2}}\right)$
Lognormal	Lognormal	$R = \Phi\left(\frac{\mu_A - \mu_B}{\sqrt{\sigma_A + \sigma_B^2}}\right)$, where $\sigma_x^2 = \ln\left\{1 + \frac{[\sigma(x)]^2}{x^2}\right\}$, $\mu_x = \ln \bar{x} - \frac{\sigma_x^2}{2}$
Normal	Weibull-Gnedenko	$R = \Phi\left(\frac{A_0 - \mu_B}{\sigma_B}\right) + \frac{0}{\sigma_A \sqrt{2\pi}} \int_0^\infty \exp\left[-y^\beta - \frac{1}{2}\left(\frac{\theta}{\sigma_B} y + \frac{A_0 - \mu_B}{\sigma_B}\right)^2\right] dy$
Weibull-Gnedenko	Weibull-Gnedenko	$R = \int_0^\infty e^{-y} \exp\left\{-\left[\frac{\theta_A}{\theta_B} y^{1/\beta_A} + \left(\frac{A_0 - B_0}{\theta_B}\right)\right]^{\beta_B}\right\} dy$
Normal	Exponential	$R = \Phi\left(-\frac{\mu_B}{\sigma_B}\right) + \exp\left[-\frac{1}{2}(2\mu_B\lambda_A - \lambda_A^2\sigma_B^2)\right] \times \left[1 - \Phi\left(-\frac{\mu_B - \lambda_A^2\sigma_B^2}{\sigma_B^2}\right)\right]$
Exponential	Normal	$R = 1 - \Phi\left(-\frac{\mu_A}{\sigma_A}\right) - \exp\left[-\frac{1}{2}(2\mu_A\lambda_B - \lambda_B^2\sigma_A^2)\right] \times \left[1 - \Phi\left(-\frac{\mu_A - \lambda_B\sigma_A^2}{\sigma_A}\right)\right]$
Exponential	Exponential	$R = \frac{\lambda_B}{\lambda_A + \lambda_B} = \frac{\bar{A}}{\bar{A} + \bar{B}}$

11.1.3 Numerical Method

Approximate graphical methods may be reasonable when numerical tables of distributions or histograms of empirical data are available. Direct numerical calculation may be easier than finding an analytical expression.

Let there be a set of observed values $X_1, \ldots, X_n$ and $Y_1, \ldots, Y_m$. Let there also be an ordered set $Z_1 \leq \cdots \leq Z_{n+m}$, where each Z_s is X_i or Y_j.

Calculate for each $Z_s = X_i$ a number of $Z_s = Y_j$, $r < s$. Denote this value as k_s. This value means that in k_s cases out of m possible observations of a random value Y, the load will be less than the given strength X. In other words, the conditional probability for this X that the investigated system will operate successfully is k_s/m. Thus the overall probability of success is

$$R = \frac{1}{n} \sum_{1 \leq s \leq n} k_s . \tag{11.5}$$

The same result obviously could be found by letting $Z_s = Y_j$ and by calculating for each $Z_r = Y_j$ a number of $Z_r = X_j$, $r < s$. This value may be denoted as k_s^*. This value means that in k_s^* of n observations of a random value X, the strength will be smaller than the given load Y. Thus with conditional probability k_s^*/n, the system will fail under the load Y. The complete probability of success is

$$R = 1 - \frac{1}{m} \sum_{1 \leq s \leq m} k_s^* . \tag{11.4}$$

EXAMPLE 11.1

The data are

$$X_1 = 98.1\,,\quad X_2 = 98.2\,,\quad X_3 = 99.4\,,\quad X_4 = 100.3\,,\quad X_5 = 101.2\,,$$
$$X_6 = 103.5\,,\quad X_7 = 103.9\,,\quad X_8 = 104.1, \ldots,\quad X_{16} = 110.2\,,$$
$$Y_1 = 79.1\,,\quad Y_2 = 82.4, \ldots,\quad Y_{18} = 98.0\,,\quad Y_{12} = 98.3\,,\quad Y_{20} = 98.5\,,$$
$$Y_{21} = 99.5\,.$$

Calculate the probability that the structure will not fail.

Solution

From the data we know that $k_1 = 18/21$, $k_2 = 18/21$, $k_3 = 20/21$, $k_4 = \cdots = k_{16} = 1$. Using (11.5), we have

$$R = \left(\frac{1}{16}\right)\left[\frac{18}{21} + \frac{18}{21} + \frac{20}{21} + 131\right] = \frac{329}{336} \approx 0.9792\,.$$

From the same data, we have $k_1^* = \cdots = k_{18}^* = 0$, $k_{19}^* = 2/16$, $k_{20}^* = 2/16$, $k_{21}^* = 3/16$. Therefore (11.6),

$$R = 1 - \left(\frac{1}{21}\right)\left[\frac{2}{16} + \frac{2}{16} + \frac{3}{16}\right] = 1 - \frac{7}{336} \approx 0.9792\,.$$

If tables for distributions of F_X and F_Y are available, index R can be approximated:

$$R \approx \sum_{1 \le m \le M} \left[1 - F_X\left(\left(m + \frac{1}{2}\right)\Delta\right)\right][F_Y((m+1)\Delta) - F_Y(m\Delta)]\,,$$

or

$$R \approx \sum_{1 \le m \le M} F_Y\left(\left(m + \frac{1}{2}\right)\Delta\right)[F_X((m+1)\Delta) - F_X(m\Delta)]\,,$$

where Δ is a chosen increment and M is the number of increments. A practical range for selection of the increment is 0.05 to 0.5 of the smallest standard deviation of the distributions F_X and F_Y. The smaller the increment, the greater the accuracy. For practical purposes the left bound of summation m should begin from the value

$$m = \{k : F(-k\Delta) < \epsilon\}\,,$$

where ϵ is chosen according to the required accuracy.

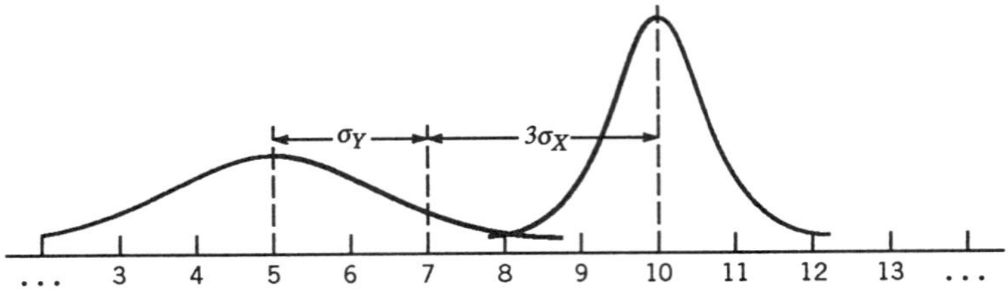

Figure 11.1 A normal load distribution overlapping a normal strength distribution.

EXAMPLE 11.2

Consider a normal strength distribution with $EX = 10$ and $\sigma_X = 1$, and a normal load distribution with $YX = 5$ and $\sigma_Y = 2$ (all values are conditional). Calculate the failure probability using the normal distribution table in Appendix E.

Solution

Figure 11.1 illustrates the solution. In the figure, point 7 equals $EY + \sigma_Y$ and also equals $EX - 3\sigma_X$. Point 9 corresponds to $EY + 2\sigma_Y$ and $EX - \sigma_X$. In Table 11.2 are compiled the corresponding numbers. We see that the probability of failure equals 0.98385. The precise calculation is

$$\Phi\left(\frac{10-5}{\sqrt{1^2+2^2}}\right) = \Phi(\sqrt{5}) = 0.9878 .$$

Large increments may result in low accuracy.

11.2 CYCLICAL FORCE-LOAD MODELS

11.2.1 Nature of Cyclical Loading

Static force-load models cover only very restricted cases involving a single cycle of loading. The cycles also must be short enough for the strength of

Table 11.2 Probability that a random load will be higher than a random strength, Example 11.1

Interval $[k, k+1]$	Value of $F_Y(k+1) - F_Y(k)$	Argument $k+1/2$	Value of $F_X(k+1/2)$	Intermediate Product
[7, 8]	0.0920	7.5	0.00621	0.00057
[8, 9]	0.0441	8.5	0.0668	0.00295
[9, 10]	0.0164	9.5	0.692	0.01114
[10, 11]	0.0049	10.5	0.308	0.00149

material to remain constant and show no deterioration or fatigue. Practical analyses usually require consideration of at least independent and identical cycles. The probability in Section 11.1 can be considered as applicable to only a fixed load during one cycle.

In real life the strength of a mechanical construction can change monotonically in time. The reasons include deterioration, fatigue, aging, and environmental influences. For electronic equipment the "strength" can fluctuate: tolerance limits can change in time depending on temperature, humidity, and other environmental influences. The load also can change in time. We will consider a simple case of cyclical loading that is almost instant loading.

In a cyclical situation, let strength and load be independent random variables with known distributions $F_X(x)$ and $F_Y(y)$, respectively. The mean strength value is $E\{X\}$, and the mean load value is $E\{Y\}$. Strength is assumed to be fixed, whether known or unknown and a load can be represented by a sequence of independent random values from cycle to cycle.

11.2.2 Fixed Level of Strength

Suppose that the strength level is known and equals x^0, and that load is random with distribution function F_Y. Values of the load at each cycle are mutually independent r.v.'s. Denote a random number of failure-free cycles as ν.

The probability of successful operation of the system during not less than K cycles equals

$$Pr\{\nu \geq K|x^0\} = p^K ,$$

where

$$p = Pr\{Y \leq x^0\} = F_Y(x^0) .$$

The mean number of cycles before a failure, $E\{\nu\}$, is $1/q$, where $q = 1 - p$. If $q \ll 1$, the approximation in exponential form is

$$Pr\{\nu \geq K|x^0\} \approx e^{-qK} .$$

Now let x^0 be unknown but constant during the period of interest. The only known information is a prior distribution, $F_X(x)$. In this case

$$Pr\{\nu \leq K\} = \int_L [F_Y(x)]^k \, dF_X(x) ,$$

where L is the domain of the distribution $F_X(x)$.

When probability $q(x) = 1 - F_Y(x)$ is very small:

$$\frac{EX - EY}{\sqrt{\sigma_X^2 + \sigma_Y^2}} \gg 1 ,$$

the right "tail" of the distribution $F_Y(x)$ is convex in the "essential" area of domain of the distribution $F_X(x)$. A simple bound can be written

$$Pr\{\nu \geq K\} \leq [F_Y(EX)]^K .$$

The mean number of cycles is

$$EK = \int_L [1 - F_Y(x)]^{-1}\, dF_X(x) .$$

The approximation for small q is

$$E\nu \approx [1 - F_Y(EX)]^{-1} .$$

11.3 DYNAMIC LOADING

11.3.1 Failure as Jumping Out of the Tolerance Area

Consider a mechanical system interacting with the environment. Let the external load $\mathbf{q}$ take values in a space $\mathbf{Q}$. The problem may be considered as a problem in statistical dynamics. The stochastic behavior of the system may be described as random walking in the space of the system performance quality, $\mathbf{V}$. The relationship between system quality and system states is usually given in the form of differential equations.

A subspace of permissible system states is defined by a chosen failure criterion, Ω, in the space $\mathbf{V}$. The system fails when the process $V(t)$ exits the closed subspace defined by Ω. One of the main reliability indexes, the probability of failure-free operation for a given time interval, can be written as

$$P(t) = Pr\{V(t) \in \Omega; \tau \in [0, t]\} . \tag{11.7}$$

This kind of definition of a reliability index leads in this case to the mathematical problems connected with random jumps of a stochastic process over a given boundary. Relationship (11.7) is for fixed parameters of external loading and internal strength. If the environment or the mechanical construction have random parameters, (11.7) should be rewritten:

$$P(t) = \int \int_G P(t \mid \mathbf{r}, \mathbf{s})p(\mathbf{r}, \mathbf{s})\, d\mathbf{r}\, d\mathbf{s} ,$$

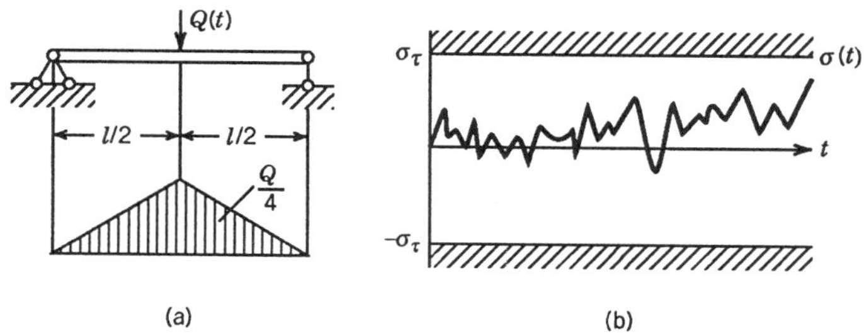

Figure 11.2 Dynamic load problem, Example 11.1.

where $P(t \mid \mathbf{r}, \mathbf{s})$ is the conditional probability of failure-free operation of the type (11.7) for some fixed pair **r** and **s**, $p(\mathbf{r}, \mathbf{s})$ is the joint distribution density of **r** and **s**, and G is the area of interest in the domain of this joint distribution.

Let us look at several examples.

EXAMPLE 11.1

Consider a console (Figure 11.2*a*) with a constant diametrical cut and a load $\mathbf{Q}(t)$ attached in the middle. The effective frequency of the random stochastic process $\mathbf{Q}(t)$ is much lower than the lowest oscillation frequency of the console, so the loading process can be considered to be quasistatic. The tension state in the console is described by the maximum bending moment $\mathbf{M}(t) = 0.25\mathbf{Q}(t)$. The system quality is described by the maximum normal tension in a dangerous cut: $\sigma(t) = \mathbf{M}(t)/W$. If normal operation consists of absence of plastic deformations, the criterion of failure-free operation has the form: $|\sigma(t)| < \sigma_\tau$, where σ_τ is the flowing limit. The space **V** is one-dimensional, and Ω is an interval, $\Omega = \{v(t): |\sigma(t)| < \sigma_\tau\}$. In this case (11.7) can be written as

$$P(t) = Pr\{-\sigma_\tau < \sigma(\tau) < \sigma_\tau; \tau \in [0, t]\}.$$

Figure 11.2*b* shows a sasmple $\sigma(t)$.

EXAMPLE 11.2

Consider a system of rods loaded with quasi-static accumulated influences $\mathbf{Q}_{1(t)}$ and $\mathbf{Q}_{2(t)}$ that form a vector of external loading $\mathbf{q}(t)$ (see Figure 11.3*a*). The system state space is described by the longitudinal force $N_1(t) = [Q_1(t) + Q_2(t)]/\sqrt{2}$ and $N_2(t) = [Q_1(t) + Q_2(t)]/\sqrt{2}$. The space of the system quality is determined by tensions $\sigma_1(t) = N_1(t)/F_1$ and

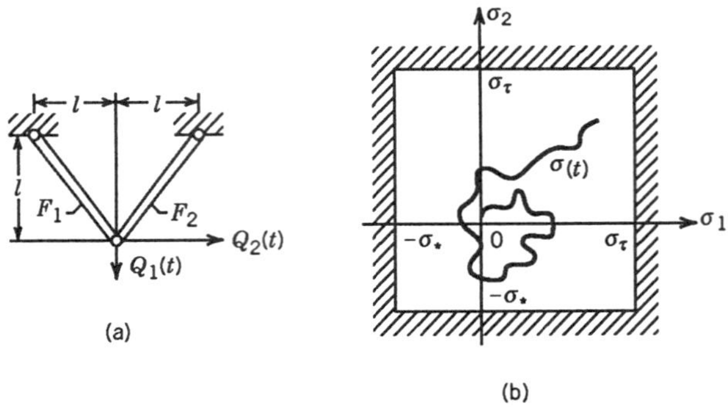

Figure 11.3 System of rods (a), and its quality space (b), Example 11.2.

$\sigma_2(t) = N_2(t)/F_1$ in the rod cuts. In the failure-free area there is no plastic deformation or loss of stability in the pressed rods:

$$\Omega = \{\mathbf{v}(t) : -\sigma_{*j} < \sigma_j(t) < \sigma_x;\ j = 1, 2\} .$$

$\sigma_{*j} F_j = \pi^2 E J_j / 21^2$ is the value of Euler's force. The quality space is two-dimensional, and the permitted area is rectangular (Figure 11.3b). The probability is

$$P(t) = Pr\{-\sigma_{*j} < \sigma_j(t) < \sigma_x;\ j + 1, 2; \tau \in [0, t]\} .$$

EXAMPLE 11.3

Consider a ferm-type system of n rods (see Figure 11.4), and a loading with random influences $Q_1(t)$, $Q_2(t)$ . . . , $Q_m(t)$. The quality space and reliability criterion are defined as in Example 11.2. The first one is n-dimensional, and the second one is an n-dimensional rectangle in this space.

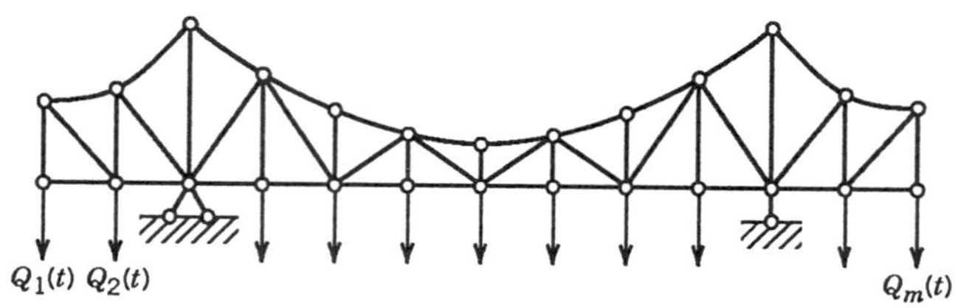

Figure 11.4 Ferm-type system of rods, Example 11.3.

EXAMPLE 11.4

Consider the problem of protecting equipment from vibration. A system has a mass M and one degree of freedom (Figure 11.5a). With the help of a viscous elastic unit with hardness c and dampening coefficient b, the system is connected to a container that oscillates with acceleration $a(t)$. Transmission $u(t)$ of the system mass satisfies:

$$M\ddot{u} + b\dot{u} + cu = -Ma .$$

For the system to operate reliably, acceleration of vibration must not exceed the limit value, a_*. The system also must not knock against the container walls. The quality vector is given by

$$\mathbf{v}(t) = \{\ddot{u}(t) + a(t), u(t)\} .$$

The tolerance area in the quality space (see Figure 11.5b) is determined using

$$\Omega = \{\mathbf{v}(t) : |v_1(t)| < a_*, -u_{**} < v_2(t) < u_*\}$$

(see Figure 11.5b). The probability of successful operation is

$$P(t) = Pr\{|\ddot{u}(\tau) + a(\tau)| < a_*, -u_{**} < u(\tau) < u_* ;\ \tau \in [0, t]\} .$$

EXAMPLE 11.5

For a two-dimensional problem in elasticity theory, the quality space is formed by the main tensions, $\sigma_1(t)$ and $\sigma_2(t)$, in some characterization

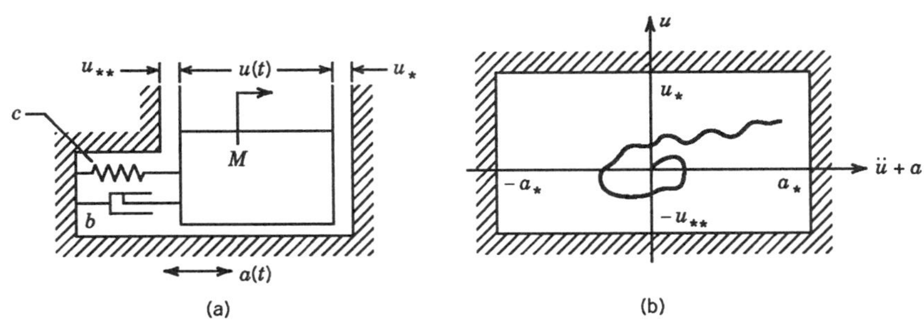

Figure 11.5 A system with one degree of freedom and its quality space, Example 11.4.

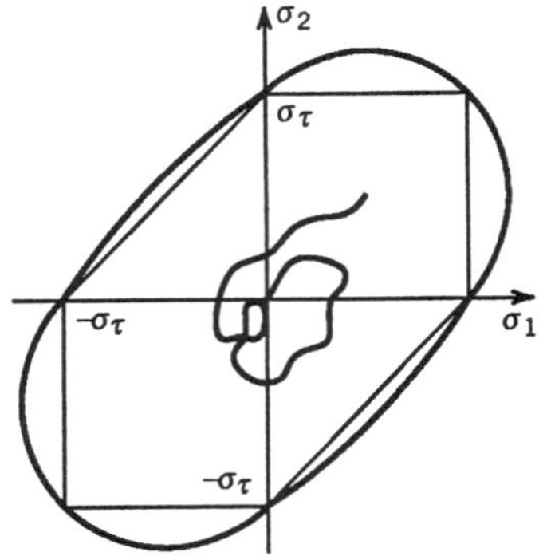

Figure 11.6 Forms of tolerance area for the two-dimensional case, Example 11.5.

point. The demand for no plastic deformation leads to tolerance area determination in the form of a Saint-Venant hexagon or Mises ellipse (see Figure 11.6). Under the Mises criterion, the probability of failure-free operation is the probability that the elliptical area will not be crossed by the two-dimensional process, $\boldsymbol{\nu}(t) = \{\boldsymbol{\nu}_1(t), \boldsymbol{\nu}_2(t)\}$:

$$P(t) = P\{\sqrt{\sigma_1^2(\tau) + \sigma_2^2(\tau) - \sigma_1(\tau)\sigma_2(\tau)} < \sigma_T; \tau \in [0, t]\} .$$

11.3.2 Markov Models

The quality vector $\mathbf{v}(t)$ represents an n-dimensional Markov process with a transition probability density $p(\mathbf{v}, t|\mathbf{v}_0, t_0)$, where $v_0 = v_0(t_0)$. This probability density, as a function of v_0 and t_0 satisfies an inverse Kolmogorov equation:

$$\frac{\partial p}{\partial t_0} = -\sum_{1 \le j \le n} \kappa_j \frac{\partial p}{\partial v_{j0}} - \frac{1}{2} \sum_{1 \le j \le n} \sum_{1 \le k \le n} \kappa_{jk} \frac{\partial^2 p}{\partial v_{j0}\, \partial v_{k0}} . \tag{11.8}$$

Intensities κ_j and κ_{jk} of the Markov process can be found with

$$\kappa_j = \lim_{\Delta t - 0} \frac{1}{\Delta t} \int (v_j - v_{j0}) p(\mathbf{v}, t|\mathbf{v}_0 t_0)\, dv ,$$

$$\kappa_{jk} = \lim_{\Delta t - 0} \frac{1}{\Delta t} \int (v_j - v_{j0})(v_k - v_{k0}) p(\mathbf{v}, t|\mathbf{v}_0, t_0)\, dv ,$$

$$\Delta t = t - t_0 .$$

For systems whose intensities do not depend on time directly, the transition probabilities depend only on differences $t - t_0$. Hence $\partial p / \partial t_0 = -\partial p / \partial t$, and (11.7) can be rewritten as

$$\frac{\partial p}{\partial t} = \sum_{i \le j \le n} \kappa_j \frac{\partial p}{\partial v_{j0}} + \frac{1}{2} \sum_{1 \le j \le n} \sum_{i \le k \le n} \kappa_{jk} \frac{\partial^2 p}{\partial v_{j0}\, \partial v_{k0}} . \tag{11.9}$$

The conditional probability of successful operation, $P(t \mid v_0, t_0)$, is determined as the probability of keeping the random vector $\mathbf{v}(t)$ in the tolerance area Ω during time $[0, t]$ under the condition that at time t_0 the system is in this area:

$$P(t \mid \mathbf{v}_0, t_0) = \int_{\mathbf{v} \in \Omega} p(\mathbf{v}, t \mid \mathbf{v}_0 t_0)\, d\mathbf{v} \, .$$

Integration of (11.9) by $\mathbf{v}$ in the area Ω gives an equation similar to the initial one:

$$\frac{\partial P}{\partial t} = \sum_{1 \le j \le n} \kappa_j \frac{\partial P}{\partial v_{j0}} + \frac{1}{2} \sum_{1 \le j \le n} \sum_{1 \le k \le n} \kappa_{jk} \frac{\partial^2 P}{\partial v_{j0}\, \partial v_{k0}} \, . \tag{11.10}$$

Equation (11.10) is solved with the following initial and boundary conditions:

$$P(0 \mid \mathbf{v}_0, t_0) = 1 \, , \qquad \text{(which is equivalent to } \mathbf{v}_0 \in \Omega) \, ,$$

$$P(t \mid \mathbf{v}_0) = 0 \qquad \text{for } \mathbf{v}_0 \in \Gamma \, , \tag{11.11}$$

where Γ is the boundary of Ω. The solution of (11.10) with conditions (11.11) gives

$$P(t) = \int_{\mathbf{v}_0 \in \Omega} P(t \mid \mathbf{v}_0, t_0) p(\mathbf{v}_0)\, d\mathbf{v}_0 \, .$$

A random time to failure $T(v_0, t_0)$, equals the time when the vector $\mathbf{v}(t)$ reaches the boundary Γ under the condition that at the time t_0 it locates inside the area Ω. This r.v. has the conditional probability density

$$p(T \mid \mathbf{v}_0, t_0) = - \left. \frac{\partial P(t \mid \mathbf{v}_0, t_0)}{\partial t} \right|_{t=T} .$$

The conditional mean of this r.v. is determined by

$$E\{T(\mathbf{v}_0, t_0)\} = - \int_0^\infty t \, \frac{\partial P(t \mid \mathbf{v}_0, t_0)}{\partial t}\, dt \, .$$

After integration by parts, this leads to

$$E\{T(\mathbf{v}_0, t_0)\} = \int_0^\infty P(t \mid \mathbf{v}_0, t_0)\, dt$$

Integrating (11.9) over t from 0 to ∞ and using the initial condition $P(0 \mid v_0, t_0) = 1$, we obtain Pontryagin's equation:

$$\sum_{1 \le j \le n} \kappa_j \frac{\partial E(T(\mathbf{v}_0, t_0\}}{\partial v_{j0}} + \frac{1}{2} \sum_{1 \le j \le n} \sum_{1 \le k \le n} \kappa_{jm} \frac{\partial^2 E\{T(\mathbf{v}_0, t_0)\}}{\partial v_{j0}\, \partial v_{m0}} = -1 , \tag{11.13}$$

which should be solved with the boundary condition

$$E\{T(\mathbf{v}_0, t_0)\} = 0 , \qquad \mathbf{v}_0 \in \Gamma . \tag{11.14}$$

Analogously for the conditional moment functions $E\{T^k(\mathbf{v}_0, t_0)\}$, we find a recurrent relationship:

$$\sum_{1 \le j \le n} \kappa_j \frac{\partial E(T^k(\mathbf{v}_0, t_0\}}{\partial v_{j0}} + \frac{1}{2} \sum_{1 \le j \le n} \sum_{1 \le k \le n} \kappa_{jm} \frac{\partial^2 E\{T^k(\mathbf{v}_0, t_0)\}}{\partial v_{j0}\, \partial v_{m0}} ,$$
$$= -kE\{T^{k-1}(\mathbf{v}_0, t_0)\} , \qquad k \ge 2 . \tag{11.15}$$

with boundary conditions

$$E\{T^k(\mathbf{v}_0, t_0)\} = 0 , \qquad \mathbf{v}_0 \in \Gamma .$$

System (11.15) can be solved approximately, for instance, by the Bubnov-Galerkin variation method. The MTTF, $E\{T\}$, and moment functions $E\{T^k\}$ can be found from

$$E\{T^k\} = \int_{\mathbf{v}_0 \in \Omega} E\{T^k(\mathbf{v}_0, t_0)\}\, p(\mathbf{v}_0)\, d\mathbf{v}_0 , \qquad k \ge 1 .$$

This method now is applied to the oscillating system of Example 11.4. The oscillation equation of the vibration-protecting system is written

$$\ddot{u} + 2\epsilon \dot{u} + \omega_0^2 u = -a(t) . \tag{11.16}$$

Let $a(t)$ be a normal stationary process with a spectral density

$$S_0(\omega) = \frac{2\sigma^2}{n} \frac{\alpha\theta^2}{(\theta^2 - \omega^2)^2 + 4\alpha^2\omega^2}$$

that corresponds to passing the stochastic process $\xi(t)$ through the linear forming filter of the second order

$$\ddot{a} + 2\alpha \dot{a} + \theta^2 a = \sqrt{s}\,\xi(t) . \tag{11.17}$$

Here $\xi(t)$ is a normal stationary "white noise" with unitary intensity $s = 4\sigma^2\alpha\theta^2$.

The stochastic differential equation (11.16) with the filter equation (11.17) describes a four-dimensional Markov process $x = \{u, \dot{u}, a, \dot{a}\}$ with intensities:

$$\kappa_1 = x_2 , \quad \kappa_2 = -2\epsilon x_2 - \omega x_2 - \omega_0^2 x_1 - x_3 ;$$

$$\kappa_3 = x_4 , \quad \kappa_4 = -2\alpha x_4 - \theta_0^2 x_3 , \quad \kappa_{44} = s .$$

All the other diffusion coefficients equal 0.

After introducing the notation

$$T_0(x) = E\{T(x_1, x_2, x_3, x_4)\}.$$

Equation (11.16) may be written

$$\frac{1}{2} s \frac{\partial^2 T_0}{\partial x_4^2} + x_2 \frac{\partial t_0}{\partial x_1} - (2\epsilon x_2 + x_3 + \omega_0^2 x_i) \frac{\partial T_0}{\partial x_2}$$
$$+ x_4 \frac{\partial T_0}{\partial x_3} - (2\alpha x_4 + \theta^2 x_3) \frac{\partial T_0}{\partial x_4} = -1 . \tag{11.18}$$

The tolerance area Ω is rectangular (see Figure 11.5*b*). On the plane of parameters $x_1 = u$ and $x_2 = \dot{u}$, this area is transformed into a parallelogram (see Figure 11.7), and in extended four-dimensional space this area becomes a cylinder $\Omega \times R^2$.

The solution of (11.18) must satisfy the boundary condition and the conditions of restriction when x_3 and $x_4 \rightarrow \pm\infty$. The approximate solution can be found in the form of the set

$$T_0(\mathbf{x}) = \sum_{1 \le k \le m} T_k \varphi_k(\mathbf{x}) ,$$

where $\varphi_k(\mathbf{x})$ is a complete system of functions inside the cylinder $\Omega \times R^2$ that satisfy all conditions placed on the function $E\{T(\mathbf{x})\}$.

The mean time to intersection of the boundary Γ is

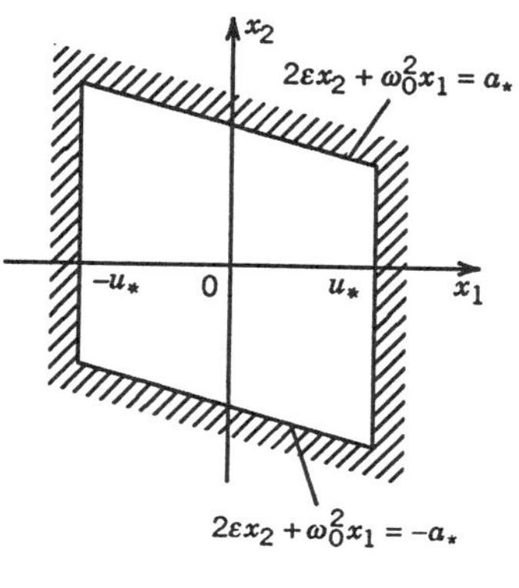

Figure 11.7 Tolerance area, Bubnov-Galerkin method.

$$E\{T\} = \iint_{\Omega} \left[\int_{-\infty}^{\infty} \int_{-\infty}^{\infty} T_0(\mathbf{x}) p(\mathbf{x}) \, dx_3 \, dx_4 \right] dx_1 \, dx_2 \, .$$

The joint probability density, $p(\mathbf{x})$, of the four-dimensional normal Markov process, $\mathbf{x}(t)$, is completely determined by its correlation matrix, $\mathbf{K}_2 = [\mathbf{x}(t)\, x^I(t)]$, with its elements found from the system of linear algebraic equations described with the help of the matrix form:

$$RK_x + K_x R^I = SS^I \, .$$

Here **R** and **S** are numerical matrices of corresponding dimensions 4×4 and 4×1, and they correspond to equations (11.16) and (11.17) in the form

$$\dot{\mathbf{x}}(t) + R\mathbf{x}(t) = S\xi(t) \, ,$$

$$\mathbf{R} \equiv \begin{bmatrix} 0 & -1 & 0 & 0 \\ \omega_0^2 & 2\epsilon & 1 & 0 \\ 0 & 0 & 0 & -1 \\ 0 & 0 & \theta^2 & 2\alpha \end{bmatrix} ,$$

$$\mathbf{R} = \begin{bmatrix} 0 & -1 & 0 & 0 \\ \omega_0^2 & 2\epsilon & 1 & 0 \\ 0 & 0 & 0 & -1 \\ 0 & 0 & \theta^2 & 2\alpha \end{bmatrix} ,$$

$$\mathbf{S} = \begin{bmatrix} 0 \\ 0 \\ 0 \\ \sqrt{s} \end{bmatrix} .$$

11.4 APPLICATION OF LEVEL-CROSSING THEORY

1.4.1 Approximation of Probability of Success

Probability (11.7) for a mechanical system whose failure consists of a crossing of the tolerance area Ω by the vector of quality $\mathbf{v}(t)$ can be expressed with the level-crossing theory. Let us denote the expected value of level-crossing on the time interval by $N(t)$. Let us assume that $P(0) = 1$; that is, the process $\mathbf{v}(t)$ is inside the area Ω at the initial moment. Let $P_k(t)$ denote the probability of k crossings from inside during the time interval $[0, t]$. The probability of successful operation is

$$P(t) = 1 - \sum_{1 \le k < \infty} P_k(t) \, .$$

The moment function of the distribution of the number of crossings is

$$N^j(t) = \sum_{1 \le k < \infty} kP_k(t) .$$

(The expected number of level-crossings corresponds to $j = 1$.)

Two-sided bounds for $P(t)$ are

$$1 - N(t) \le P(t) \le 1 - 1.5N(t) + 0.5N^2(t) .$$

More accurate bounds can be obntained with moment functions of higher orders. For very reliable systems, $P(t) \approx 1 - N(t)$. If the level-crossing forms a Poisson process, the probability of successful operation may be written

$$P(t) \approx \exp\{-N(t)\} .$$

11.4.2 Approximation of the Tolerance Area

When the tolerance area has a complex configuration, it can be simplified into a form for which calculation of characteristics of level-crossing is easier. One can find that

$$P(t) \approx P_0(t) ,$$

where $P_0(t)$ is the probability of failure-free operation for the area with boundary envelope Γ_0, which approximates the area of interest, Ω (see Figure 11.8).

To obtain two-sided bounds, one can use the method of boundary estimation. Two areas, Ω_1 and Ω_2, are chosen in such a way that $\Omega_2 \subseteq \Omega \subseteq \Omega_1$. Figure 11.8 represents a two-dimensional case with two rectangles, Γ_1 and Γ_2, that produce the bounds:

$$P_2(t) \le P(t) \le P_1(t) . \tag{11.19}$$

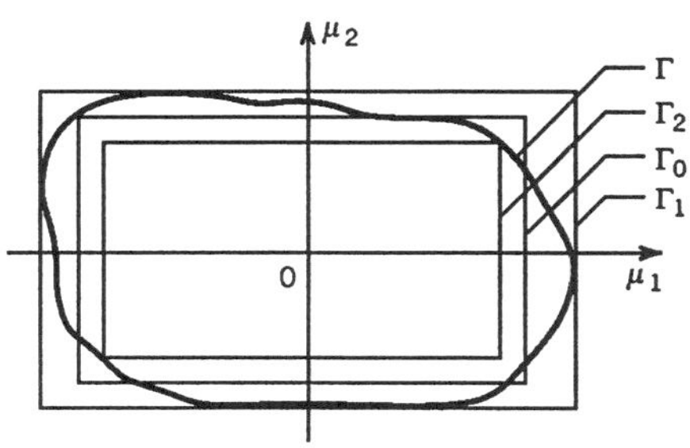

Figure 11.8 Approximation of tolerance area.

11.4.3 Level-Crossing, One-Dimensional Process

Consider a one-dimensional stochastic process $v(t)$ and a bound Γ in the form of the line $w = v$. Assume that Ω lies under this line, as in Figure 11.9. Let $v(t)$ be a continuously differentiable stochastic process with a given joint probability density, $p(v, \dot{v}, t)$, of the process and its derivative. The failure rate is defined as $\lambda(t) = dN(t)/dt$, and can be written in this case as

$$\lambda(t) = \int_0^\infty \dot{v} p(w, \dot{v}, t)\, d\dot{v} . \tag{11.20}$$

If the level $w(t)$ is a nonstochastic function of time, $\lambda(t)$ is calculated in a similar manner except that the process $v(t)$ can cross the level $w(t)$ only if $\dot{v}(t) > \dot{w}(t)$. As a result

$$\lambda(t) = \int_0^\infty (\dot{v} - \dot{w}) p(w, \dot{v}, t)\, d\dot{v} . \tag{11.21}$$

Calculation of (11.20) and (11.21) requires the density of a joint probability $p(v, \dot{v}, t)$. This probability can be found with the two-point joint probability density $p_v(v_1, v_2)$ of the stochastic process $v(t)$, where $v_1 = v(t_1)$ and $v_2 = v(t_2)$. The probability density, $p(v, \dot{v}, t)$, is determined as

$$p(v, \dot{v}, t) = \lim_{\Delta t \to 0} [p_v(v, v + \dot{v}\,\Delta t)\,\Delta t] . \tag{11.22}$$

11.4.4 Level-Crossing, Gaussian Process

The two-point joint probability density of a Gaussian stochastic process is:

$$p_v(v_1, v_2) = \frac{1}{2\pi\sigma_1\sigma_2\sqrt{1 - r^2}} \exp\left\{ -\frac{1}{2(1 - r^2)} \left[\frac{(v_1 - a_1)^2}{\sigma_1^2} + \frac{(v_2 - a_2)^2}{\sigma_2^2} - 2r\,\frac{(v_1 - a_1)(v_2 - a_2)}{\sigma_1\sigma_2} \right] \right\} ; \tag{11.23}$$

where $a(t)$ is the expected value of the process $v(t)$, $\sigma_v^2(t)$ is its variance, $a_1 = a(t_1)$, $a_2 = a(t_2)$, $\sigma_1 = \sigma_v(t_1)$, $\sigma_2 = \sigma_v(t_2)$, and $r(t_1, t_2)$ is the autocorrelation of the process $v(t)$ at moments t_1 and t_2.

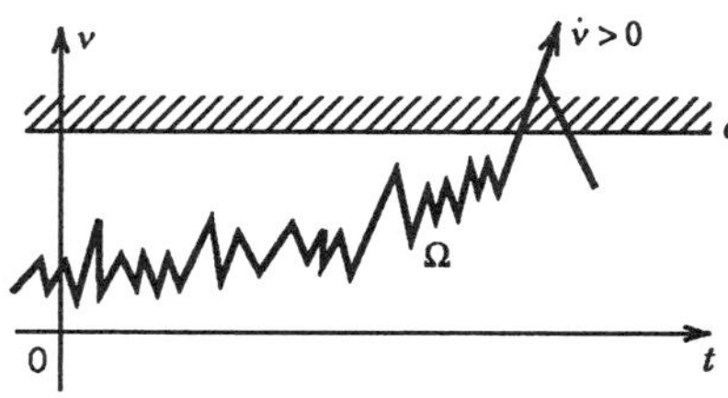

Figure 11.9 Level-crossing, one-dimensional stochastic process.

Letting $v_1 = v$, $v_2 = v + \dot{v}\,\Delta t$, $t_1 = t$, and $t_2 = t + \Delta t$, and using (11.22), we find that

$$p(v, \dot{v}, t) = \frac{1}{2\pi\sigma_v\sigma_{\dot{v}}\sqrt{1-r^2}} \exp\left\{-\frac{1}{2(1-r^2)}\left[\frac{(v-a)^2}{\sigma^{2v}} - 2\rho\,\frac{(v-a)(\dot{v}-\dot{a})}{\sigma_v\sigma_{\dot{v}}} + \frac{(\dot{v}-\dot{a})^2}{\sigma_{\dot{v}}^2}\right]\right\}.$$

Here $\dot{a}(t)$ and $\sigma_{\dot{v}}^2$ are the mean and the variance of the derivative, and $\rho(t)$ is the correlation coefficient of the process and its derivative when there is no time lag. These parameters are determined using

$$\dot{a}(t) = \lim_{\Delta t\to 0} \frac{a_2 - a_1}{\Delta t},$$

$$\sigma_{\dot{v}}^2(t) = \lim_{\Delta t\to 0} \frac{\sigma_2^2 + \sigma_1^2 - 2r\sigma_1\sigma_2}{(\Delta t)^2} = \lim_{\Delta t\to 0} \frac{1-r}{(\Delta t)^2}\, 2\sigma_v^2(t),$$

$$\rho(t) = \lim_{\Delta t\to 0} \frac{r\sigma_2 - \sigma_1}{\sigma_{\dot{v}}\,\Delta t} = \frac{\sigma}{\sigma_{\dot{v}}} \lim_{\Delta t\to 0} \frac{r-1}{\Delta t}.$$

An example is a stationary Gaussian process $v(t)$ for which $a(t) = a$ is constant, σ_v is constant, and the correlation coefficient $r(t_1, t_2) = \exp(-\alpha\,\Delta t)\cdot(\cos\beta\,\Delta t + (\alpha/\beta)\sin\beta\,\Delta t)$. The results are $\dot{a} = 0$, $\rho = 0$, and $\sigma_{\dot{v}}^2 = \sigma_v^2(\alpha_2 + \beta^2)$. Thus:

$$\lambda = \frac{1}{2\pi\sigma_v\sigma_{\dot{v}}} \exp\left[-\frac{(w-a)^2}{2\sigma_v^2}\right] \int_0^\infty \dot{v}\,\exp\left(-\frac{(\dot{v})^2}{2\sigma_{\dot{v}}^2}\right) d\dot{v}\,. \tag{11.24}$$

Finally:

$$\lambda = \frac{\omega_0}{2\pi} \exp\left[-\frac{(w-a)^2}{2\sigma_v^2}\right], \tag{11.25}$$

where $\omega_0 = \sigma_{\dot{v}}/\sigma_v$ is the effective frequency of the process.

For a nonstationary Gaussian process, we introduce a new variable of integration:

$$u = (1-\rho^2)^{-1/2}\left(\frac{\dot{v}-\dot{a}}{\sigma_{\dot{v}}} - \rho\,\frac{w-a}{\sigma_v}\right).$$

The final result is

$$\lambda(t) = \frac{\sigma_{\dot{v}}(1-\rho^2)^{1/2}}{2\pi\sigma_v} \exp\left[-\frac{(w-a)^2}{2\sigma_v^2}\right] \int_{-u_0}^\infty (u+u_0)e^{-u^2/2}\,du\,, \tag{11.26}$$

where

$$u_0 = (1-\rho^2)^{-1/2}\left(\frac{\dot{a}}{\sigma_{\dot{v}}} + \rho\,\frac{w-a}{\sigma_v}\right).$$

Using the Laplace integral,

$$\Phi(u) = \frac{1}{\sqrt{2\pi}} \int_{-u}^{\infty} e^{-x^2/2}\, dx\,,$$

the final expression is

$$\lambda(t) = \frac{\omega_0\sqrt{(1-\rho^2)}}{2\pi}\,[e^{-u_0^2/2} + \sqrt{2\pi}u_0\Phi(u_0)]\exp\left[-\frac{(w-a)^2}{2\sigma_v^2}\right].$$

Notice that all parameters of the stochastic process $v(t)$ are time functions. In the case of a stationary process, (11.27) transforms into (11.25).

When the level $w(t)$ changes in time, $\lambda(t)$ is found using (11.21). The integral is calculated as above for a nonstationary process. The final expression has the form of (11.26) if we replace the previous u_0 expression with

$$u_0 = \frac{1}{\sqrt{1-\rho^2}}\left(\frac{\dot{a}-\dot{w}}{\sigma_{\dot{v}}} - \rho\,\frac{a-w}{\sigma_v}\right).$$

11.4.5 Level-Crossing, Rayleigh Process

The Rayleigh process is given by

$$v(t) = \sqrt{x^2(t) + y^2(t)}\,,$$

where $x(t)$ and $y(t)$ are centered stochastically independent Gaussian random processes with equal variances, σ^2. The one-point density of the Rayleigh process is

$$p_v(v, t) = \frac{v}{\sigma^2}\, e^{-v^2/2\sigma^2}\,.$$

To obtain a two-point Rayleigh distribution, we introduce a four-dimensional stochastic process:

$$\{x_1 = x(t_1),\ x_2 = x(t_2),\ y_1 = y(t_1),\ y_2 = y(t_2)\}$$

with the joint probability density

$$p_{xv}(x_1, x_2, y_1, y_2) = p(x_1, x_2)p(y_1, y_2)\,,$$

where densities $p(x_1, x_2)$ and $p(y_1, y_2)$ are given by (11.22). To this we add new variables:

$$x_1 = v_1 \cos\theta_1 , \quad x_2 = v_2 \cos\theta_2 , \quad y_1 = v_1 \sin\theta_1 , \quad y_2 = v_2 \sin\theta_2 .$$

Then the density of interest can be written

$$p_v(v_1, v_2, \theta_1, \theta_2) = \frac{v_1 v_2}{4\pi^2(1-r^2)\sigma_1^2\sigma_2^2} \times \exp\left\{ -\frac{1}{2(1-r^2)} \left[\frac{v_1^2}{\sigma_1^2} - 2r \frac{v_1 v_2 \cos(\theta_1 - \theta_2)}{\sigma_1\sigma_2} + \frac{v_2^2}{\sigma_2^2} \right] \right\} .$$

The two-point density is found using

$$p_v(v_1, v_2) = \frac{v_1 v_2}{(1-r^2)\sigma_1^2\sigma_2^2} \exp\left\{ -\frac{1}{2(1-r^2)} \left(\frac{v_1^2}{\sigma_1^2} + \frac{v_2^2}{\sigma_2^2} \right) \right\} \times I_0\left(\frac{r}{1-r^2} \frac{v_1 v_2}{\sigma_1\sigma_2} \right) ,$$

where I_0 is the Bessel function of an imaginary variable.

Calculation with the help of (11.22) for a stationary Rayleigh process $v(t)$ gives

$$p(v, \dot{v}) = \frac{v}{\sqrt{2\pi\sigma^2}\sigma_{\dot{v}}} \exp\left\{ -\frac{1}{2} \left(\frac{v^2}{\sigma^2} + \frac{(\dot{v})^2}{\sigma_{\dot{v}}^2} \right) \right] .$$

The variance of the process $v(t)$ is related to the variance, σ^2 of the processes $x(t)$ and $y(t)$ by the relation $\sigma_v^2 = (4-\pi)\sigma^2/2$, and the variance of the derivative is

$$\sigma_{\dot{v}}^2 = (4-\pi)\sigma^2 \lim_{\Delta t - 0} \frac{1 - r(t, t+\Delta t)}{(\Delta t)^2} .$$

Calculation of (11.20) is not difficult. The final result is

$$\lambda = \frac{\sigma_{\dot{v}} w}{\sqrt{2\pi\sigma^2}} e^{-w^2/2\sigma^2} .$$

11.4.6 Level-Crossing, Multidimensional Process

The methodology for a multidimensional process is similar to the one-dimensional case. For example, the expected number of level-crossings in a unit of time can be found as:

$$\lambda(t) = \int_\Gamma d\Gamma \int_{\dot{v}_n > 0} p(\mathbf{v}_\Gamma, \dot{\mathbf{v}}, t) \dot{v}_n \, d\dot{\mathbf{v}} ,$$

where the first integral is taken over the surface Γ, $\mathbf{v}_c$ is the vector $\mathbf{v}(t)$ on Γ, and $\dot{v}_n$ is a component of the velocity $\dot{\mathbf{v}}$ that is normal to Γ.

Level-Crossing from a Rectangular Area

Let us consider a two-dimensional quality space and a rectangular tolerance area with sides $v_1^* v_1^{**}$ and $v_2^* v_2^{**}$ (see Figure 11.10). For a stationary two-dimensional Gaussian process, $v(t)$, the joint probability density is

$$p(\mathbf{v}, \dot{\mathbf{v}}) = p_1(v_1, v_2) \cdot p_2(v_1, \dot{v}_2) .$$

The density $p_1(v_1, v_2)$ is determined using (11.23) and parameters

$$a_1 , \quad a_2 , \quad \sigma_{v_1} , \quad \sigma_{v_2} , \quad r_{v_1 v_2} .$$

The density $p_2(\dot{v}_1, \dot{v}_a)$ has the analogous form with parameters

$$\dot{a}_1 = \dot{a}_2 = 0 , \quad \sigma_{\dot{v}_1} , \quad \sigma_{\dot{v}_2} , \quad \rho_{\dot{v}_1 \dot{v}_2} .$$

Expression (11.28) in this case takes the form

$$\lambda = \frac{1}{2\pi} [F(v_2^*, v_1^*) - F(v_2^{**}, v_1^*) + F(v_2^*, v_1^{**}) - F(v_2^{**}, v_1^{**}) \\ + F(v_1^*, v_2^*) - F(v_1^{**}, v_2^*) + F(v_1^*, v_2^{**}) - F(v_1^{**}, v_2^{**})] ,$$

where

$$F(v_j, v_k) = \frac{\sigma_{\dot{v}_k}}{\sigma_{v_k}} \exp\left[- \frac{(v_k - a_k)^2}{2\sigma_{v_k}^2} \right] \Phi\left[\frac{1}{\sqrt{1 - r_{v_1 v_2}^2}} \left(\frac{v_j - a_j}{\sigma_{v_j}} - r_{v_1 v_2} \frac{v_k - a_k}{\sigma_{v_k}} \right) \right] .$$

Let the area Ω be an N-dimensional rectangule with sides $v_k^* - v_k^{**}$, $k = 1, 2, \ldots, N$, and let $v(t)$ be the N-dimensional stochastic Gaussian process with independent components $v_k(t)$. In this case the joint density is

$$p(\mathbf{v}, \dot{\mathbf{v}}) = \prod_{1 \le k \le N} p_1(v_k) p_2(\dot{v}_k) .$$

Equation (11.28) gives

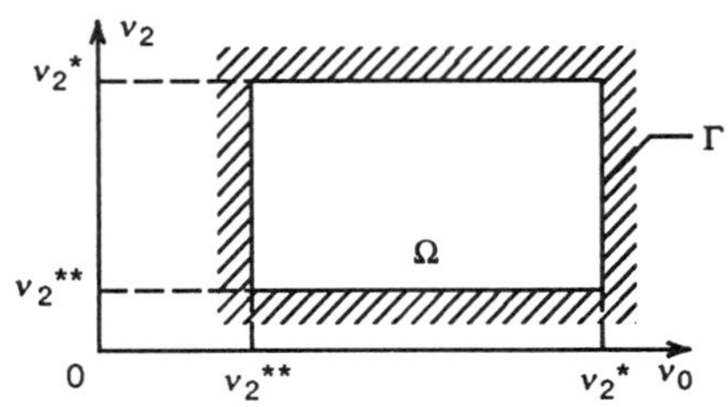

Figure 11.10 Level-crossing, two-dimensional Gaussian process.

$$\lambda = \frac{1}{2\pi} \sum_{1 \le k \le N} \frac{\sigma_{\dot{v}_k}}{\sigma_{v_k}} \left\{ \exp\left[-\frac{(v_k^* - a_k)^2}{2\sigma_{v_k}^2} \right] + \exp\left[-\frac{(v_k^{**} - a_k)^2}{2\sigma_{v_k}^2} \right] \right\}$$
$$\times \prod_{\substack{1 \le j \le N \\ j \ne k}} \left[\Phi\left(\frac{v_j^* - a_j}{\sigma_{v_j}} \right) - \Phi\left(\frac{v_j^{**} - a_j}{\sigma_{v_j}} \right) \right]. \tag{11.29}$$

If level-crossings are rare, an approximation may be used:

$$\Phi\left(\frac{v_j^* - a_j}{\sigma_{v_j}} \right) - \Phi\left(\frac{v_j^{**} - a_j}{\sigma_{v_j}} \right) \approx 1 .$$

It holds when

$$\frac{v_j^* - a_j}{\sigma_{v_j}} \gg 1 ,$$

$$\frac{a_j - v_j^{**}}{\sigma_{v_j}} \gg 1 , \qquad j = 1, 2, \ldots, N .$$

In this case (11.29) has a simpler form,

$$\lambda \approx \frac{1}{2\pi} \sum_{1 \le k \le N} \frac{\sigma_{\dot{v}_k}}{\sigma_{v_k}} \left\{ \exp\left[-\frac{(v_k^* - a_k)^2}{2\sigma_{v_k}^2} \right] + \exp\left[-\frac{(v_k^{**} - a_k)^2}{2\sigma_{v_k}^2} \right] \right\} .$$

Use of the preceding relation is called "strip approximation". It can be obtained by summation of the level-crossings of strips

$$v_k^{**} \le v_k(t) \le v_k^* , \qquad k = 1, 2, \ldots, N .$$

For each strip (11.25) is generalized for a crossing of two-sided levels:

$$\lambda_k = \frac{\sigma_{\dot{v}_k}}{2\pi\sigma_{v_k}} \left\{ \exp\left[-\frac{(v_k^* - a_k)^2}{2\sigma_{v_k}^2} \right] + \exp\left[-\frac{(v_k^{**} - a_k)^2}{2\sigma_{v_k}^2} \right] \right\} .$$

Level-Crossing from Spherical Region

Let $v(t)$ be an N-dimensional Gaussian process with independent centered components and equal variances. The variance of each component is denoted σ^2, and the variance of each derivative is denoted s^2. If the tolerance area, Ω, is an N-dimensional sphere with radius R, it is reasonable to change variables $\dot{v}_1, \dot{v}_2, \ldots, \dot{v}_N$ in (11.28) with

$$\dot{v}_n, \dot{v}_{\tau_1}, \dot{v}_{t_2}, \ldots, \dot{v}_{\tau_{N-1}} ,$$

where $\dot{v}_n$ is the component of the velocity of the vector process $v(t)$ that is normal to the sphere surface, and $\dot{v}_{\tau k}$ are tangential components of the process velocity. The result is

$$\lambda = \frac{1}{(2\pi\sigma s)^N} \int_\Gamma \exp\left[-\frac{1}{2\sigma^2} \sum_{1\le k\le N} v_{k\Gamma}^2\right] d\Gamma$$

$$\times \prod_{1\le m\le N-1} \int_{-\infty}^{\infty} \exp\left(-\frac{(\dot{v}_{\tau_m})^2}{2s^2}\right) d\dot{v}_{\tau_m} \int_0^{\infty} \exp\left(-\frac{\dot{v}_n^2}{2s^2}\right) \dot{v}_n \, d\dot{v}_n \,. \qquad (11.30)$$

Since

$$\sum_{1\le k\le N} v_{k\Gamma}^2 = R^2 ,$$

calculation of the integrals yields

$$\lambda = \frac{sR^{N-1}}{2^{(N-1/2)}\sqrt{\pi}\Gamma(N/2)\sigma^N} \exp\left(-\frac{R^2}{2\sigma^2}\right). \qquad (11.31)$$

The problem of level-crossing by a multidimensional process in a spherical area, Ω, can be reduced to a one-dimensional problem if, instead of an N-dimensional process, $v(t)$, one considers a stochastic process:

$$\varphi(t) = \sqrt{\sum_{1\le k\le N} v_k^2(t)} \,.$$

A one-dimensional process, $\varphi(t)$, that crosses the level $w = R$ corresponds to a vector process, $v(t)$, that intersects an N-dimensional sphere. The expected number of level-crossings in a unit of time can be found with (11.25).

If $v(t)$ is an N-dimensional stationary Gaussian process with independent and identical centered components, the stochastic process $\varphi(t)$ is stationary and has a χ^2 distribution. The joint probability density of the process and its derivative is

$$p(\varphi, \dot{\varphi}) = \frac{\varphi^{N-1}\eta(\varphi)}{2^{(N-1/2)}\sqrt{\pi}\Gamma(N/2)\sigma^N s} e^{(\varphi^2/2\sigma^2 - \dot{\varphi}^2/2s^2)} \,. \qquad (11.32)$$

where $\eta(\varphi)$ is the Heaviside function. After substituting (11.32), calculation of integral (11.20) results in (11.31).

For an approximate calculation of the level-crossing for a complex area, an approximation of the area can be used as well. To illustrate, we take the method of Section 11.4.2 and approximate a spherical region with a rectangular one. The bounds in Section 11.4.2 for the stationary process $v(t)$ are equivalent to $\lambda \approx \lambda_0$, $\lambda_1 \le \lambda \le \lambda_2$, where λ_0, λ_1, and λ_2 are, level-crossing parameters for areas Ω_0, Ω_1 and Ω_2. These parameters are calculated using (11.29) and (11.30). In the two-dimensional case, areas Ω_0, Ω_1, and Ω_2 are squares with sides $w_0 = R\sqrt{\pi}$, $w_2 = 2R$, and $w_2 = R\sqrt{2}$. Area Ω_0 is a square equivalent of area Ω. Calculation with (11.30) and (11.31) for $R/\sigma = 3$ gives $\lambda = 0.0133s/\sigma$, $\lambda_0 = 0.0186s/\sigma$, $\lambda_1 = 0.00707s/\sigma$, and $\lambda_2 = 0.0671s/\sigma$.

Level-Crossing from Ellipsoid

Let a surface, Γ, restrict an N-dimensional ellipsoid with axes $b_1, b_2, \ldots, b_N$. Let these axes of symmetry of the ellipsoid coincide with the coordinate axis in the space of quality. A stationary Gaussian process, $v(t)$, is determined by the vector of means, a, and correlation matrices $K_0 = E\{[\mathbf{v} \cdot \dot{\mathbf{v}}^{\mathrm{T}}]\}$ and $K = E\{(\mathbf{v} - \mathbf{a})(\mathbf{v} - \mathbf{a})^{\mathrm{T}}\}$. For this process, $p(\mathbf{v}, \dot{\mathbf{v}}) = p_1(\mathbf{v})p_2(\dot{\mathbf{v}})$, where

$$p_1(\mathbf{v}) = \frac{1}{\sqrt{(2\pi)^N \det K}} \exp\left[-\frac{1}{2}(\mathbf{v} - \mathbf{a})^T K^{-1}(\mathbf{v} - \mathbf{a})\right],$$

$$p_2(\dot{\mathbf{v}}) = \frac{1}{\sqrt{(2\pi)^N \det K_0}} \exp\left[-\frac{1}{2}\dot{\mathbf{v}}^T K_0^{-1} \mathbf{v}\right].$$

Equation (11.28) can be rewritten as

$$\lambda = \int_\Gamma p_1(\mathbf{v}_\Gamma)\left[\int_{\dot{v}_n > 0} p_2(\dot{\mathbf{v}})\dot{\mathbf{v}}_n \, d\dot{\mathbf{v}}\right] d\Gamma . \tag{11.33}$$

A normal component of the velocity $\dot{v}(t)$ on the surface Γ of the ellipse is

$$\dot{\mathbf{v}}_n(t) = \sum_{1 \le k \le N} a_{nk} \dot{\mathbf{v}}_k(t) \tag{11.34}$$

where a_{nk} are directional cosines of the unit normal to the surface Γ. The $(2N - 1)$ dimensions of the integral can be reduced if, with a normal component of the velocity $u_1 = \dot{v}_n$, one introduces mutually orthogonal tangential components of the velocity on the surface Γ:

$$u_{j+1} = \dot{v}_{\tau_j} = \sum_{1 \le k \le N} a_{\tau_j k} \dot{v}_k , \qquad j = 1, 2, \ldots, N - 1 .$$

Transition to these components can be realized with an orthogonal matrix **A** combined with the directional cosines a_{nk} and a_{rik} of the unit normal and the tangents to the surface:

$$\mathbf{u}(t) = A\dot{\mathbf{v}}(t) . \tag{11.35}$$

After the change of variables, the inner N-dimensional integral in (11.33) takes the form:

$$\int_{-\infty}^{\infty} du_2 \int_{-\infty}^{\infty} du_3 \cdots \int_{-\infty}^{\infty} du_N \int_0^{\infty} p_u(u_1, u_2, \ldots, u_N) u_1 \, du_1 .$$

This integral equals

$$\sqrt{\frac{\mathrm{Var}(u_1)}{2\pi}} ,$$

where $\mathrm{Var}(u_1)$ is the variance of the centered r.v., u_1, which is calculated as

$$\mathrm{Var}(u_1) = \sum_{i \le j \le N} \sum_{1 \le k \le N} a_{nj} a_{nk} K_0 jk .$$

Equation (11.33) can be reduced to an integral over the surface Γ:

$$\lambda = \frac{1}{\sqrt{2\pi}} \int_\Gamma p_1(\mathbf{v}_\Gamma) \sqrt{\sum_{1 \le j \le N} \sum_{1 \le k \le N} a_{nj} a_{nk} K_{0jk}}\, d\Gamma . \tag{11.36}$$

For level-crossing for an ellipsoid, (11.36) has the form:

$$\lambda = \frac{s_1 s_2}{(2\pi)^{3/2}(1-\rho^2)^{1/2}\sigma_1\sigma_2} \int_0^{2\pi} \sqrt{\frac{b_1^2 \sin^2\varphi}{s_1^2} + r\,\frac{b_1 b_2 \sin 2\varphi}{s_1 s_2} + \frac{b_2^2 \cos^2\varphi}{s_2^2}}$$
$$\times \exp\left\{-\frac{1}{2(1-\rho^2)}\left[\frac{(b_1\cos\varphi - a_1)^2}{\sigma_1^2}\right.\right.$$
$$\left.\left. -2\rho\,\frac{(b_1\cos\varphi - a_1)(b_2\sin\varphi - a_2)}{\sigma_1\sigma_2} + \frac{(b_2\sin\varphi - a_2)^2}{\sigma_2^2}\right]\right\} d\varphi .$$

Here σ_1^2 and σ_2^2 are variances of components v_1 and v_2 of the two-dimensional Gaussian process $v(t)$; s_1^2 and s_2^2 are variances of the velocities $\dot{v}_1$ and $\dot{v}_2$; ρ is a correlation coefficient between components v_1 and v_2; and r is the correlation coefficient between components $\dot{v}_1$ and $\dot{v}_2$.

When $b_1 = b_2 = R$, and $a_1 = a_2 = 0$, the expected number of level-crossings of the spherical region in one time unit by the centered Gaussian process becomes

$$\lambda = \frac{s_1 s_2 R}{(2\pi)^{3/2}(1-\rho^2)^{1/2}\sigma_1\sigma_2} \int_0^{2\pi} \sqrt{\frac{\sin^2\varphi}{s_1^2} + r\,\frac{\sin 2\varphi}{s_1 s_2} + \frac{\cos^2\varphi}{s_2^2}}$$
$$\times \exp\left\{-\frac{R^2}{2(1-\rho^2)}\left[\frac{\cos^2\varphi}{\sigma_1^2} - \rho\,\frac{\sin 2\varphi}{\sigma_1\sigma_2} + \frac{\sin^2\varphi}{\sigma_2^2}\right]\right\} d\varphi .$$

When $\rho = r = 0$, $\sigma_1 = \sigma_2 = \sigma$, $s_1 = s_2 = s$, this relationship coincides with (11.31) for $N = 2$.

14.4.7 Level-Crossing, Random Bounds

The reliability properties of a system can change during use. Consider a one-dimensional quality space with a tolerance area given by the inequality $v(t) < w(t)$, where $w(t)$ is a stochastic process with known parameters. The expected number of level-crossings in a time unit can be found from

$$\lambda(t) = \int_{-\infty}^{\infty} dw \int_{-\infty}^{\infty} d\dot{w} \int_{\dot{w}}^{\infty} (\dot{v} - \dot{w}) p_{vw}(w, \dot{v}, w, \dot{w}, t)\, d\dot{v}\,. \qquad (11.37)$$

To compute integral (11.37), one can introduce a new variable, $\dot{u} = \dot{v} - \dot{w}$. Then:

$$\lambda(t) = \int_{0}^{\infty} d\dot{u} \int_{-\infty}^{\infty} d\dot{w} \int_{-\infty}^{\infty} \dot{u} p_{vw}(w, \dot{u} + \dot{w}, w, \dot{w}, t)\, dw\,. \qquad (11.38)$$

If the stochastic process $w(t)$ is independent of the stochastic process $v(t)$, the problem of finding $\lambda(t)$ can be divided into two stages. In the first stage for fixed w and $\dot{w}$, one finds the conditional $\lambda(t \mid w, \dot{w})$ with the methods used for a nonrandom level $w(t)$. For example, in the one-dimensional case and a one-side bound, the equation takes the form of (11.21):

$$\lambda(t \mid w, \dot{w}) = \int_{\dot{w}}^{\infty} (\dot{v} - \dot{w}) p_v(w, \dot{v})\, d\dot{v}\,.$$

At the second stage,

$$\lambda(t) = \int_{-\infty}^{\infty} \int_{-\infty}^{\infty} \lambda(t \mid w, \dot{w}) p_w(w, \dot{w}, t)\, dw\, d\dot{w}\,. \qquad (11.39)$$

Equations (11.37) and (11.39) for stochastically independent processes $v(t)$ and $w(t)$ coincide. $\lambda(t \mid w, \dot{w})$ is determined, as a rule, with special functions. Equation (11.38) in many cases yields the final analytical results for $\lambda(t)$.

Nonstationary Gaussian Bound, Nonstationary Gaussian Process

Let us assume that processes $w(t)$ and $v(t)$ are stochastically independent. The direct application of (11.38), after some lengthy computations, leads to

$$\lambda(t) = \frac{\sqrt{s(1-\rho)^2}}{2\pi\sigma} \left[\exp\left(-\frac{u_1^2}{2}\right) + \sqrt{2\pi} u_1 \Phi(u_1)\right] \exp\left(-\frac{(a_v - a_w)^2}{2\sigma^2}\right), \qquad (11.40)$$

where $\Phi(u)$ is the Laplace function. The following notations are used:

$$\sigma^2 = \sigma_v^2 + \sigma_w^2\,,$$

$$s^2 = \sigma_{\dot{v}}^2 + \sigma_{\dot{w}}^2\,,$$

$$\rho = \frac{\rho_v \sigma_v \sigma_{\dot{v}} + \rho_w \sigma_w \sigma_{\dot{w}}}{s\sigma}\,,$$

$$u_1 = \frac{1}{\sqrt{1-\rho^2}} \left(\frac{a_{\dot{v}} - a_{\dot{w}}}{s} - \rho \frac{a_v - a_w}{\sigma}\right),$$

where $\sigma^2(t)$ is the variance of the difference $u(t) = v(t) - w(t)$, $s^2(t)$ is the variance of the difference of derivatives $\dot{u}(t) = \dot{v}(t) - \dot{w}(t)$, and $\rho(t)$ is the correlation coefficient between process $u(t)$ and its derivative $\dot{u}(t)$ at the same moment in time.

Equation (11.40) by its form coincides with (11.27). It leads to the possibility of using an auxiliary process $u(t) = v(t) - w(t)$ and crossings of the zero-level. The sum of two Gaussian processes generates a new Gaussian process. Hence the use of (11.20) with $w = 0$ naturally leads to (11.27), which, after substitution of (11.41) for σ_v, $\sigma_{\dot{v}}$, a, $\dot{a}$, and ρ, yields (11.40). This is the result of the preservation property of the normal distribution under linear transformations. The direct use of (11.20) in this case is restricted because of difficulties in finding the joint probability density $p(u, \dot{u}, t)$.

Rayleigh Distributed Bound, Gaussian Process

Let the bound w be random but not change in time. One can write

$$\lambda(t) = \frac{\sigma_v \sigma_{\dot{v}}}{2\pi\sigma^2} [e^{-u_2^2/2} + \sqrt{2\pi} u_2 \Phi(u_2)] e^{-a_v^2/2\sigma^2} ,$$

where $\sigma^2 = \sigma_v^2 + \sigma_1^2$, and $u_2 = a_v \sigma_1 / \sigma_v \sigma$; the parameter σ_1 is related to the Rayleigh distribution and connected with the variance σ_w^2 by

$$\sigma_w^2 = \sigma_1^2 \left(2 - \frac{\pi}{2}\right) = 0.43\sigma_1^2 .$$

Normally Distributed Bound, Rayleigh Process

If the level w is random but does not change in time, (11.38) takes the form

$$\lambda(t) = \int_0^\infty d\dot{v} \int_{-\infty}^\infty \dot{v} p_v(w, \dot{v}) p_w(w)\, dw .$$

For a Rayleigh distribution $p_v(v, \dot{v})$ with parameters σ_1 and $\sigma_{\dot{v}}$, and a normal distribution for w with parameters a_w and σ_w one can derive

$$\lambda(t) = \frac{\sigma_w \sigma_{\dot{v}}}{2\pi\sigma^2} [e^{-u_3^2/2} + \sqrt{2\pi} u_3 \Phi(u_3)] e^{-a_w^2/2\sigma^2} ,$$

where $u_3 = a_w \sigma_1 / \sigma_w \sigma$ and $\sigma^2 = \sigma_1^2 + \sigma_w^2$.

Spherical Bound with Random Radius, Multidimensional Gaussian Process

As mentioned earlier, this problem can be reduced to a one-dimensional process. Consider the N-dimensional Gaussian process with identical and independent centered components. In this case the value of the radius-vector has a χ^2 distribution with joint density (11.32). Let the radius of the sphere of the tolerance area be random and have a Rayleigh distribution with parameter σ_1. In this case

$$\lambda = \frac{\sigma_v \sigma_{\dot{v}} \sigma_1^{N-1} \Gamma[(N+1)/2]}{\sqrt{\pi}(\sigma_v^2 + \sigma_1^2)^{(N+1)/2} \Gamma(N/2)} ,$$

where σ_v and $\sigma_{\dot{v}}$ are the parameters of distribution (11.32).

Weibull-Gnedenko Distributed Bound, Gaussian Process

Let w have a Weibull-Gnedenko distribution

$$p_w(w) = \alpha c w^{\alpha-1} \eta(w) e^{-cw^{\alpha}}$$

with parameters α and c. In this case λ is expressed by an integral, which in general cannot be expressed with tabulated integrals:

$$\lambda = \frac{\alpha c \sigma_{\dot{v}} \sigma_v^{\alpha-1}}{2\pi} \int_0^{\infty} u^{\alpha-1} \exp\left[-\frac{1}{2} \left(u - \frac{a^v}{\sigma_v} \right)^2 - c\sigma_v^{a-1} u^{\alpha} \right] du .$$

PART 2

STATISTICS

CHAPTER 12

ESTIMATION OF EQUIPMENT RELIABILITY FROM TESTS

When new equipment is developed, its reliability characteristics are predicted from test data. The test data may include bench test, field test, computer simulation, or accelerated life test results. Some results may be eliminated or modified on retrospective evaluation of the test conditions. Finally, a failure time distribution is hypothesized, and the distribution parameters are estimated.

In the selection of solution algorithms, some factors usually are more important than others. Less relevant—or even irrelevant—factors may include the following:

- Whether the property to be predicted is reliability and repairability.
- Whether the data are test results, results of actual operations, or some mixture of both.
- Whether failed items are restored or replaced.

In practice it can be difficult to arrange testing that provides a designer with exactly the needed experimental data in the required form and in sufficient amount. As a rule, the problem actually consists of estimation of reliability indexes on the basis of all available statistical material. The nature of the statistical material is influenced by the testing strategy and operation mode. Combinations of factors result in a variety of testing strategies. Some of the main test conditions follow.

12.1 TEST PLANS AND NOTATION

At an initial moment of time, $t = 0$, tests of N items begin. We use letters to indicate whether failed items are replaced. The letter R means each failed item is replaced without delay with a new, identical item; this is testing with replacement. In testing under a plan of type R, the same number of items, N, always must be under a load. Observations to detect failed items are made continuously, and as a result failures are eliminated immediately. The latter W stands for plans in which items that fail during the testing period are not replaced with new ones.

The moment of termination of testing can be assigned in various ways. The common test plans follow:

T indicates a plan under which tests are made during a time interval of preassigned duration, T.

r indicates a plan under which testing is done until there are r failures.

(r, T) indicates a plan under which testing is performed either up to the moment, t_r, of occurrence of the rth failure if $t_r < T$, or up to the moment T if $t_r \geq T$.

There are thus six possible types of plans based on these rules of termination and on the replacement plans.

1. $[N\ R\ T]$—N items are placed under test, failed items are replaced, testing is performed up to a previously assigned moment of time, T.
2. $[N\ R\ r]$—N items are placed under test, failed items are replaced, testing continues until the rth failure.
3. $[N\ R\ (r, T)]$—N items are placed under test, failed items are replaced, testing continues either up to the moment, t_r, of occurrence of the rth failure if $t_r < T$, or up to a previously assigned moment of time, T, if $t_r > T$.
4. $[N\ W\ T]$—N items are placed under test, failed items are not replaced, testing continues up to a previously assigned moment of time, T.
5. $[N\ W\ r]$—N items are placed under test, failed items are not replaced, testing continues until there have been r failures.
6. $[N\ W\ (r, T)]$—N items are placed under test, failed items are not replaced, testing continues either up to the moment, t_r, of occurrence of the rth failure or up to a previously assigned moment of time, T.

We let $d(t)$ denote the number of failures that occur before the moment t. The function $d(t)$ takes on successive values $0, 1, 2, \ldots$. The function $d(t)$ has associated random moments of time t_i, $i = 1, 2, \ldots$, where t_i is the moment of the ith failure. Figure 12.1 illustrates the behavior of the function $d(t)$ for the testing plans.

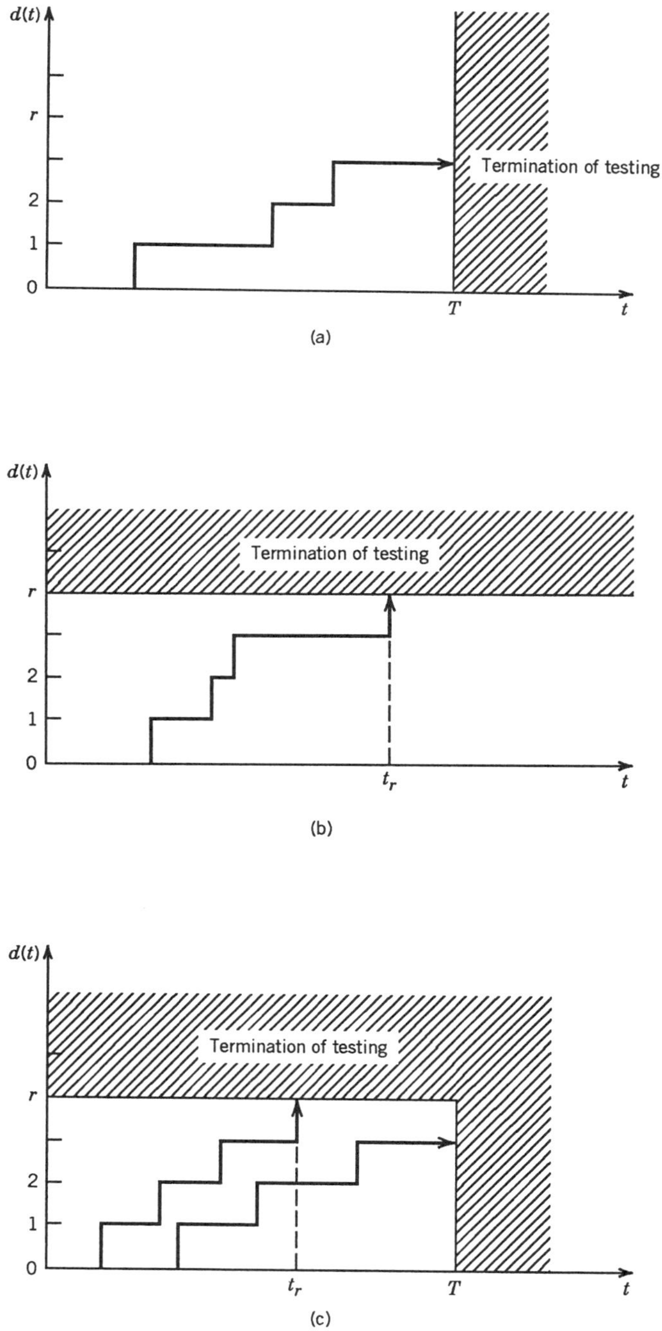

Figure 12.1 Samples of behavior of the function $d(t)$ for various testing plans: (a) plans $[N\ R\ T]$, $[N\ W\ T]$; (b) plans $[N\ R\ r]$, $[N\ W\ r]$; (c) plans $[N\ R\ (r, T)]$, $[N\ W\ (r, T)]$

Testing also can be designed to have a total running time for the tested items. $N(t)$ denotes the number of items operating failure-free up to the moment t; that is, $N(t) = N - d(t)$. $S(t)$ is the sum of the failure-free testing times for the items. Thus:

$$S(t) = \sum_{1 \le i \le d(t)} t_i + N(t)t ,$$

where t_i is the moment of failure of the ith unit, $i = 1, 2, \ldots, d(t)$.

As plans of type R always have $S(t) = Nt$, there is no separate set of plans of type R based on the total running time. For W plans, however, there are two more plans to add to the above six:

7. $[N\ W\ HS_0]$ is a plan in which N items are tested, failed items are not replaced, testing continues either until the previously assigned total running time has been reached or until all N items have failed.
8. $[N\ W\ (r, HS_0)]$ is a plan under which N items are tested, failed items are not replaced, testing is performed up to a moment, t^*, when one of two events occurs: (1) $S(t^*) = S_0$ and $d(t^*) < r$, or (2) $t^* = t_r$ but $S(t) < S_0$.

12.2 POINT ESTIMATE METHODS

12.2.1 Method of Moments

The moments of the distribution of the test results can be used to estimate the unknown parameters of the actual distribution. The number of moments equals the number of unknown parameters and the required number of equations. The method is based on the fact that, as the number of sample data increases, the values of the moments approach the actual values.

An empirical moment or order k, I_k, is given by

$$I_k = \frac{1}{n} \sum_{1 \le i \le n} t_i^k , \tag{12.1}$$

where t_i is a failure time and n is the number of failures. For example, with a two-parameter Weibull-Gnedenko distribution:

$$f(t, a, b) = \frac{a}{a^b} t^{b-1} e^{-(t/b)^b} , \tag{12.2}$$

the first and second initial moments are given by

$$m_1 = a\Gamma\left(1 + \frac{1}{b}\right) , \tag{12.3}$$

$$m_2 = a^2\Gamma\left(1 + \frac{2}{b}\right). \tag{12.4}$$

Therefore the following two equations can be used for Weibull-Gnedenko distributions with parameters $\hat{a}$ and $\hat{b}$:

$$\hat{a}\Gamma\left(1 + \frac{1}{\hat{b}}\right) = \frac{1}{n} \sum_{1 \le i \le n} t_i , \tag{12.5}$$

$$\hat{a}^2\Gamma\left(1 + \frac{2}{\hat{b}}\right) = \frac{1}{n} \sum_{1 \le i \le n} t_i^2 . \tag{12.6}$$

Estimates for parameters of common distributions obtained with the method of moments are presented in Table 12.1. In this table the first sample moment is

$$\hat{T} = \frac{1}{n} \sum_{1 \le i \le n} t_i .$$

The sample variance is

$$S^2 = \frac{1}{n^2} \left[n \sum_{1 \le i \le n} t_i^2 - \left(\sum_{1 \le i \le n} t_i \right)^2 \right].$$

The sample mean for logarithmic distributions is

$$\hat{u} = \frac{1}{n} \sum_{1 \le i \le n} \ln t_i .$$

The sample variance for logarithmic distributions is

$$\nu^2 = \frac{1}{n(n-1)} \left[n \sum_{1 \le i \le n} (\ln t_i)^2 - \left(\sum_{1 \le i \le n} \ln t_i \right)^2 \right].$$

These equations are for samples with no censored data.

Table 12.1 Estimators based on the method of moments

Type of Distribution	Parameter Estimates
Exponential	$\hat{\lambda} = \frac{1}{\hat{T}}$
Normal	$\hat{a} = \hat{T},\ \hat{\sigma} = \sqrt{S^2}$
Lognormal	$\mu = \hat{u},\ \hat{\sigma} = \sqrt{\nu^2}$

2.2.2 Method of Quantiles

Parameter estimates are obtained in the method of quantiles with the same equations used in the method of moments. Theoretical distribution quantiles are found for the empirical quantiles. The number of empirical quantiles equals the number of equations and also equals the number of parameters to be estimated. To estimate parameters $\hat{a}$ and $\hat{b}$ of a Weibull-Gnedenko distribution, two equations are used:

$$1 - e^{-(t_1/\hat{a})\hat{b}} = F_1 , \tag{12.7}$$

$$1 - e^{-(t_2/\hat{a})\hat{b}} = F_2 , \tag{12.8}$$

where t_1 and t_2 are the quantiles of the empirical distribution, and F_1 and F_2 are the values of the empirical distribution function corresponding to the quantiles t_1 and t_2.

Table 12.2 gives the equations for the most common distributions. In the table, t_1 and t_2 are quantiles of the empirical distribution, F_1 and F_2 are values of the empirical distribution function that correspond to the quantiles t_1 and t_2 and $\Phi^{-1}(F_1)$ and $\Phi^{-1}(F_2)$ are quantiles of the standard normal distribution function corresponding to levels F_1 and F_2.

The method of quantiles obviously can be used with any sample, but the estimates may have high variances. The estimates depend on the levels of the chosen quantiles.

12.2.3 Maximum Likelihood Method

This method is universal and is the most efficient. A likelihood function is developed for a fixed experimental result. The function represents the probability of obtaining the given results for any given distribution parameters. The parameters that have the highest probability are then chosen.

Table 12.2 Estimators based on the method of quantiles

Type of Distribution	Parameter Estimates
Weibull-Gnedenko	$\hat{b} = \dfrac{\ln \ln F_2 - \ln \ln F_1}{\ln t_2/t_1}$, $\hat{a} = \dfrac{t_1}{(-\ln F_1)^{1/b}}$
Exponential	$\hat{\lambda} = -\ln \dfrac{F_1}{t_1}$
Normal	$\hat{a} = \dfrac{\Phi^{-1}(F_1)t_2 - \Phi^{-1}(F_2)t_1}{\Phi^{-1}(F_1) - \Phi^{-1}(F_2)}$, $\sigma = \dfrac{t_1 - \hat{a}}{\ln \ln F_1^{-1}}$

EXAMPLE 12.1

Consider nonparametric tests when only the number of tested items and the number of failures are known. The single measure that can be estimated on the basis of such a sample is the failure probability, q, within the test period. With only this one parameter to be estimated, the maximum likelihood method may be used.

Solution

As an example, let a series of Bernoulli tests have n failures in N trials. The likelihood function is

$$L(q) = \binom{N}{n} q^n (1-q)^{N-n} , \tag{12.9}$$

which for fixed N and n is a function of only the unknown parameter q. The estimation of q is reduced to the formal procedure of searching for the coordinates of the point where (12.9) reaches its maximum. To simplify the differentiation, the logarithm of the likelihood function usually is taken:

$$\ln L = \ln \binom{n}{n} + n \ln q + (N-n) \ln(1-q) . \tag{12.10}$$

Then

$$\frac{\partial \ln L}{\partial q} = \frac{n}{q} - \frac{N-n}{1-q} . \tag{12.11}$$

Therefore the equation for the estimate of q takes the form:

$$\frac{n}{\hat{q}} - \frac{N-n}{1-\hat{q}} = 0 , \tag{12.12}$$

which yields the maximum likelihood q, $\hat{q}$:

$$\hat{q} = \frac{n}{N} . \tag{12.13}$$

The estimate given by (12.13) is unbiased and effective.

EXAMPLE 12.2

Consider a multiple-truncation sample. The tests produce n complete times to failure (TTF), t_i, and l incomplete test times, τ_i.

Solution

Let the indicated TTF have a Weibull-Gnedenko distribution:

$$F(t) = 1 - e^{-(t/a)b} \tag{12.14}$$

with density

$$f(t) = \frac{b}{a}\left(\frac{t}{a}\right)^{b-1} e^{-(t/a)^b} . \tag{12.15}$$

The logarithm of the likelihood function under the conditions of this example takes the form

$$L = \sum_{1\le i\le n} \ln f(t_i) + \sum_{1\le j\le 1} \ln[1 - F(\tau_j)] \tag{12.16}$$

or, with allowance for equations (12.15) and (12.16),

$$L = n \ln b - nb \ln a + (b-1) \sum_{1\le i\le n} \ln t_i - \frac{1}{a^b}\left(\sum_{1\le i\le n} t_i^b + \sum_{1\le j\le 1} \tau_j^b\right) . \tag{12.17}$$

The following system of equations is used to obtain the parameter estimates:

$$\frac{\partial L}{\partial a} = 0 ;$$

$$\frac{\partial L}{\partial b} = 0 . \tag{12.18}$$

Upon differentiation of expression (12.17) followed by certain algebraic transformations, we have the expressions for estimates $\hat{a}$ and $\hat{b}$:

$$\hat{a} = \sqrt[\hat{b}]{\frac{1}{n}\left(\sum_{1\le i\le n} t_i^{\hat{b}} + \sum_{1\le j\le 1} \tau_j^{\hat{b}}\right)} , \tag{12.19}$$

$$1 + \frac{b}{n} \sum_{1\le i\le n} \ln t_i = \frac{\hat{b}(\Sigma_{1\le i\le n} t_i^{\hat{b}} \ln t_i + \Sigma_{1\le j\le 1} \tau_j^{\hat{b}} \ln \tau_j)}{\Sigma_{1\le i\le n} t_i^{\hat{b}} + \Sigma_{1\le j\le 1} \tau_j^{\hat{b}}} . \tag{12.20}$$

The method of iteration is used to solve (12.20). The first approximation of b might be found from the method of moments or from a graphical method. After determination of $\hat{b}$, the value of $\hat{a}$ can be found from (12.19).

The maximum likelihood method is based on optimal or asymptotically optimal properties. However, the method often requires the solution of cumbersome equations.

In conclusion, we note that confidence in any estimate obtained with any analytical method depends essentially upon the certainty of the type of distribution function. Unless the type of distribution function is known with certainty a priori, the experimental distribution usually should be tested for agreement with the theoretical one.

12.3 POINT ESTIMATES, UNKNOWN DISTRIBUTION CLASS

In general, when the failure distribution class is unknown and must be hypothesized, the best basis for a hypothesis is knowledge of the physical processes that cause failure. Random external events, such as stress, usually produce exponential distributions. Many types of mechanical wear cause normal distributions. If the physical processes are not fully known, the known failure distributions of similar existing equipment may be a good starting point.

Under any circumstances, one can find the best fit distribution for each distribution class, and then use some statistical measure to select the best of the best fit distributions. DIANA and RELMAIN (see Appendix F) are personal computer programs that can be used for this purpose. One also can use the techniques in this section, which , in general, can be applied by an analyst even if he or she has no knowledge of the physical process, no knowledge of similar equipment, and no computer.

The statistical procedures in this section include the development of a series in nondecreasing order (variational series), the development of a histogram and a hypothesized distribution class (if the sample size is sufficient), an estimation of empirical distribution function values, a preliminary evaluation of the consistency of the experimental data with the hypothetical distribution with probability paper, an estimation of point values of the parameters after the class of distribution has been determined, and a final evaluation of the agreement of the experimental distribution with the hypothetical one.

12.3.1 Variational Series and Histograms

A variational series is a series of failure times (or other r.v.'s) in nondecreasing order. Histograms or empirical distribution functions can also be used, and they are more graphical than variational series. To develop a histogram, the range of a variational series is divided into k intervals that are not necessarily equal. For each interval one calculates

$$\bar{f}_i(x) = \frac{m_i}{n\Delta_i}, \tag{12.21}$$

where m_i, $i = 1, \ldots, k$, are the numbers within the ith interval, and Δ_i is the size of an interval.

A histogram consists of a set of rectangles with Δ_i widths and $\bar{f}_i(x)$ heights. A histogram's primarily purpose is to enable the adoption of a hypothesis describing the type of distribution involved. The information in the histogram should be presented in such a way that it can be easily extracted. The form the histogram takes depends on the number and the size of the subdivision intervals. Unfortunately, there is no ideal rule for determining the best way to subdivide. Success in developing an effective histogram comes from intuition and experience. For instance, with too few intervals, details of the distribution may not be clear, and with too many intervals, important distribution features may be obscured. In some cases the number of intervals may be approximated using

$$k \approx 1 + 3.3 \ln n \tag{12.22}$$

where n is the sample size. For samples of less than 50 observations (which is typical in reliability testing) a histogram may not be very informative, so (12.22) must be considered a tentative measure. One could try several histograms with different k. A good histogram is one with a minimum number of inversions.

Among the histograms with the same number of inversions, the best one is the histogram with the largest number of intervals. A weighted inversion significance criterion, h, for a histogram with k intervals is

$$h = \frac{1}{n} \sum_{2 \le j \le k-1} (m_i + m_{i+1})\xi_i ,$$

where n is the sample amount, m_i, m_{i+1} are the numbers of sample elements within the ith and the $(i+1)$th intervals, and

$$\xi_i = \begin{cases} 0 & \text{if intervals } i \text{ and } i+1 \text{ have no inversion ,} \\ 1 & \text{otherwise .} \end{cases}$$

Occasionally the method of *equally frequent intervals* is used. The essence of the method is division of a sample into k parts, all of which have the same number of failures. The intervals ordinarily have different widths. Values of k are given by (12.22). This method often is considered appropriate for small samples.

For censored samples, histograms are calculated using

$$\bar{f}_i(x) = \frac{m_i}{(N - l_i)\Delta_i} , \tag{12.23}$$

where m_i is the number of complete realizations within the interval Δ_i, Δ_i is the ith interval size, N is the total number of realizations, and l_i is the number of incomplete realizations below the right limit of interval i. The previous recommendations on division into intervals hold.

12.3.3 Estimation of Empirical Distribution Function Parameters

The simplest empirical distribution function (e.d.f.) estimation is

$$\hat{F}_i = \frac{i}{n}, \tag{12.24}$$

where i is the order number of a variational series term, and n is the total number of terms. Effective and almost unbiased estimates for e.d.f. values can be obtained using

$$\hat{F}_i = \frac{i - \alpha}{n - \alpha - \beta + 1}. \tag{12.25}$$

For a Weibull-Gnedenko distribution with form factor b, coefficients α and β are given by

$$\alpha = (0.52)\left(\frac{b-1}{b}\right),$$

$$\beta = 0.5 - (0.2)\left(\frac{b-1}{b}\right).$$

For an exponential distribution, $\alpha = 0$, and $\beta = 0.5$. For a normal distribution, $\alpha = \beta = 0.375$.

For a sample with a single truncation from the right-hand side, (12.26) and (12.27) are used with the total number of sample elements, N, instead of n.

For a sample with multiple truncation, the e.d.f. value for each complete realization may be calculated using

$$\hat{F}_i = \frac{r_i}{N+1}, \tag{12.26}$$

where r_i is the conditional rank of a complete realization

$$r_i = r_{i-1} + \frac{N + 1 - r_{i-1}}{M_i + 1}, \qquad r_0 = 0, \tag{12.27}$$

where M_i is the total number of realizations larger than or equal to realization i. If calculation of conditional rank seems to be unjustified, one can use

$$\hat{F}_i = \frac{i}{N - l_i + 1}, \tag{12.28}$$

where i is the order number of the complete realization within the variational series (the incomplete realizations are not numbered) and l_i is the total number of incomplete realizations below realization i.

12.4 GOODNESS-OF-FIT TESTS

A goodness-of-fit test is used to evaluate the probability that a hypothesized distribution is the correct one. The general procedure is to: (1) hypothesize the type of failure distribution (exponential, normal, etc.) the test items actually have, (2) estimate the parameters of the hypothesized distribution, (3) measure the difference between the observed failure times and the failure times predicted from the hypothesized distribution, (4) determine the probability that the hypothesized distribution would have produced the observed times under the assumption the hypothesis is correct, and (5) accept or reject (perhaps only tentatively) the hypothesis. Two of the most popular tests are the Pearson Test and the Kolmogorov-Smirnov Test. Differences are measured by the sum of the squared deviations in Pearson and by the maximum deviation in Kolmogorov-Smirnov. The empirical d.f. is given as:

$$\hat{Q}(t) = \hat{P}\{\theta \le t\} .$$

The objective is to test whether the distribution is related to any known d.f.

12.4.1 Pearson's Chi-Square Test

Pearson's chi-square test starts with calculation of a measure, U, of the divergence between the theoretical and empirical distributions. U is a r.v. with a χ^2 distribution and $l = k - 1 - S$ degrees of freedom, where S is the number of distribution function parameters estimated on the basis of the same statistic

$$\begin{aligned} U = \chi^2 &= \min \sum_{1 \le i \le k} \frac{n}{p_i} (\hat{p}_i - p_i)^2 \\ &= \min \sum_{1 \le i \le k} \frac{(m_i - np_i)^2}{np_i} , \end{aligned} \tag{12.29}$$

where n is the number of failures, k is the number of subdivisions of the variational series, m_i is the number of failures within the ith interval, $p_i = m_i/n$ is the observed frequency, and p_{0i} is the probability of failure within the ith interval for the hypothesized distribution function with previously obtained estimations of its parameters. The minimum is taken over the values of all unknown parameters of the theoretical distribution function.

For a small number of subintervals, one may use

$$\chi^2 \approx \sum_{1 \le i \le m} \frac{(\nu_i - np_i)^2}{np_i} .$$

The unknown parameters of the hypothesized probabilities are estimated. If we assume that the theoretical mean value, m_t, and the estimated mean, $\hat{m}(t)$, coincide:

$$\hat{m}_t = \sum_{1 \le i \le k} \frac{t^{(i+1)} + t^{(i)}}{2} \hat{p}_i = m_t ,$$

we assume moreover that the theoretical and empirically determined variances coincide:

$$\hat{D}_t = \sum_{1 \le i \le k} \left(\frac{t^{(i+1)} + t^{(i)}}{2} - \hat{m}_t \right)^2 \hat{p}_i = D_t .$$

Using Table E.16 (Appendix E), we can find the probability, p, that a r.v. U with a χ^2 distribution exceeds any obtained value of U and number of degrees of freedom r.

12.4.2 The Kolmogorov-Smirnov Test

To measure the deviation U between theoretical and empirical distributions, we now consider the maximum absolute value of the difference between the empirical d.f., $\hat{Q}(t)$, and the corresponding theoretical d.f.

$$U = D = \max|\hat{Q}(t) - Q(t)| .$$

For any d.f. $Q(t)$ of a continuous r.v. θ, when the number of independent observations increases infinitely, the probability of the inequality

$$D\sqrt{n} \ge y$$

approaches the limit

$$P\{D\sqrt{n} \ge y\} = P(y) = 1 - K(y) ,$$

where $K(y)$ is a function whose values are given in Table E.12 (Appendix E). Determining the goodness of fit between the theoretical and empirical distributions by the Kolmogorov-Smirnov criterion can be reduced to the following sequence of steps:

1. Construct the empirical d.f., $\hat{Q}(t)$, and the proposed theoretical d.f., $Q(t)$.

2. Find the difference, over all failures from the first to the last, between the expected and observed failure times.
3. Letting the maximum difference $= D$, determine

$$y = D\sqrt{n}\ ,$$

and from Table E.12 find the probability that the maximum deviation of $\hat{Q}(t)$ from $Q(t)$ will be not less than the observed deviation. If the probability is very small, the hypothesis is rejected as being unlikely. If it is relatively large, the hypothesis can be considered compatible with the experimental data.

12.5 CONFIDENCE INTERVALS

12.5.1 Advantages of Interval Estimations

A point estimate is simply one particular value of a random variable with no indication of the uncertainty of the estimate. Confidence limits may be shown with two error probabilities, ϵ_1 and ϵ_2, where ϵ_1 is the probability that the true value of the parameter is not less than a given lower limit, and ϵ_2 is the probability that the true value of the parameter is not more than a given upper limit.

The confidence probability, $\gamma^* = 1 - \epsilon_1 - \epsilon_2$, is taken for the measure of confidence. The probability that the real parameter lies within the confidence interval is indicated with:

$$\gamma^* = Pr\{\theta_L < \theta < \theta_U\}\ ,$$

where θ_L is the lower confidence limit for the parameter θ and θ_U is the upper confidence limit for the parameter θ.

The selected probabilities ϵ_1 and ϵ_2 usually are equal. Then $\gamma^* = 1 - 2\epsilon$, and each confidence limit is determined with the level $\epsilon = 1 - \gamma^*/2$.

If the type of distribution is known, the lower and upper confidence limits are the distribution quantiles.

12.5.2 Exponential Distribution

Table 12.3 gives expressions for point estimates and confidence intervals for a parameter, λ, for various plans. The following notation is used:

$\epsilon_1 \div \epsilon_2 = \alpha$ (generally $\epsilon_1 = \epsilon_2$).
$\Delta = \lambda NT$,

$$L_d(\Delta) = \sum_{0 \le k \le d} \frac{\Delta^k}{k!} e^{-\Delta}\ ,$$

where the quantity $\Delta_\alpha(d)$ is defined by the functional relation

Table 12.3 Confidence intervals for common testing plans

Testing Plan	Sufficient Statistic	Confidence Interval with Confidence Coefficient Not Less than $1-(\epsilon_1+\epsilon_2)$
$[N\ R\ T]$	$d(T)$ is the number of failures in time T	$\underline{\lambda} = \frac{\Delta_{1-\epsilon_2}(d-1)}{NT}$, $\bar{\lambda} = \frac{\Delta_{\epsilon_1}(d)}{NT}$, $d = d(T)$
$[N\ R\ r]$	t_r is the moment of the rth failure	$\underline{\lambda} = \frac{\Delta_{1-\epsilon_2}(r-1)}{Nt_r}$, $\bar{\lambda} = \frac{\Delta_{\epsilon_1}(r-1)}{Nt_r}$
$[N\ R\ (r, T)]$	$\begin{cases} t_r, & \text{if } t_r \le T, \\ d(T), & \text{if } t_r > T \end{cases}$	If $t_r \le T$, we use the plan $[N\ R\ r]$ for $\underline{\lambda}$ and $\bar{\lambda}$; if $t_r > T$, we use the formulas of plan $[N\ R\ T]$
$[N\ W\ T]$	$d(T)$ and $S_W(T)$	$\underline{\lambda} = \frac{\Delta'(d)}{T}$, $\bar{\lambda} = \frac{\bar{\Delta}'(d)}{T}$
$[N\ W\ r]$	$S_W(t_r)$—total operation at time t_r	$\underline{\lambda} = \frac{\Delta_{1-\epsilon_2}(r-1)}{S_W(t_r)}$, $\bar{\lambda} = \frac{\Delta_{\epsilon_1}(r-1)}{S_W(t_r)}$
$[N\ W\ (r, T)]$	$\begin{cases} d(T), S_W(T), & \text{if } t_r > T, \\ S_W(t_r), & \text{if } t_r \le T \end{cases}$	If $t_r > T$, then we use the formula for $\underline{\lambda}$, $\bar{\lambda}$ of plan $[N\ W\ T]$, but if $t_r \le T$, then $\underline{\lambda}$ and $\bar{\lambda}$ are found as solutions of the equations $\sum_{i=0}^{d} c_N^i (1-e^{-\bar{\lambda} t_0})^i e^{-(N-i)\bar{\lambda} t_0} = \epsilon_1$, $\sum_{i=d}^{N} c_N^i (1-e^{-\underline{\lambda} t_0})^i e^{-(N-i)\underline{\lambda} t_0} = \epsilon_2$; $\underline{\lambda} = \frac{\Delta_{1-\epsilon_2}(d-1)}{S_0}$, $\bar{\lambda} = \frac{\Delta_{\epsilon_1}(d)}{S_0}$, if $S_0 < S_W(t_r)$;
$[N\ W\ (r, HS_0)]$	$\begin{cases} S_W(t_r), & \text{if } t_r < T, \\ d(t_\theta), & \text{if } S_0 = S_W(t_\theta) < S_W(t_r), \end{cases}$ where t_θ is the end of testing	$\begin{cases} \bar{\lambda} = \frac{\Delta_{1-\epsilon_2}(r-1)}{S_W(t_r)}, \\ \bar{\lambda} = \frac{\Delta_{\epsilon_1}(r-1)}{S_W(t_r)}, \\ \text{if } S_0 \ge S_W(t_r) \end{cases}$

Table 12.4 Quantiles $\Delta_{1-\alpha}(d)$ of the Poisson distribution, $L_d[\Delta_{1-\alpha}(d)] = 1 - \alpha$

d \ $1-\alpha$	0.99993	0.9999	0.9993	0.999	0.993	0.99	0.95
0	0.000070	0.000100	0.000700	0.00100	0.00702	0.01005	0.05129
1	0.01188	0.01421	0.03789	0.04540	0.12326	0.14855	0.35536
2	0.07633	0.08618	0.16824	0.19053	0.38209	0.43604	0.81769
3	0.21115	0.23180	0.38894	0.42855	0.74108	0.82325	1.36632
4	0.41162	0.44446	0.68204	0.73937	1.17032	1.27911	1.97015
5	0.66825	0.71375	1.03236	1.10710	1.65152	1.78528	2.61301
6	0.97222	1.03040	1.42874	1.52034	2.17293	2.33021	3.28532
7	1.31628	1.38697	1.86297	1.97081	2.72659	2.90611	3.98082
8	1.69465	1.77758	2.32894	2.45242	3.30682	3.50746	4.69523
9	2.10271	2.19758	2.82197	2.96052	3.90942	4.13020	5.42541
10	2.53672	2.64323	3.33840	3.49148	4.53118	4.77125	6.16901
11	2.99367	3.11150	3.87531	4.04244	5.16960	5.42818	6.92421
12	3.47103	3.59988	4.43033	4.61106	5.82265	6.09907	7.68958
13	3.96672	4.10632	5.00152	5.19544	6.48871	6.78235	8.46394
14	4.47896	4.62904	5.58725	5.79398	7.16642	7.47673	9.24633
15	5.00626	5.16657	6.18615	6.40533	7.85464	8.18111	10.03596
16	5.54732	5.71762	6.79705	7.02835	8.55241	8.89457	10.83214

d \ $1-\alpha$	0.93	0.90	0.80	0.70	0.60	0.50	0.40
0	0.07257	0.10536	0.22314	0.35667	0.51082	0.69315	0.91629
1	0.43081	0.53181	0.82439	1.09735	1.37642	1.67835	2.02231
2	0.94230	1.10206	1.53504	1.91378	2.28508	2.67406	3.10538
3	1.53414	1.74477	2.29679	2.76371	3.21132	3.67206	4.17526
4	2.17670	2.43259	3.08954	3.63361	4.14774	4.67091	5.23662
5	2.85488	3.15190	3.90366	4.51714	5.09098	5.67016	6.29192
6	3.55984	3.89477	4.73366	5.41074	6.03924	6.66964	7.34265
7	4.28584	4.65612	5.57606	6.31217	6.99137	7.66925	8.38977
8	5.02895	5.43247	6.42848	7.21993	7.94661	8.66895	9.43395
9	5.78633	6.22130	7.28922	8.13293	8.90441	9.66871	10.47568
10	6.55583	7.02075	8.15702	9.05036	9.86440	10.66852	11.51533
11	7.33581	7.82934	9.03090	9.97161	10.82624	11.66836	12.55317
12	8.12496	8.64594	9.91010	10.89620	11.78972	12.66823	13.58944
13	8.92222	9.46962	10.79398	11.82373	12.75462	13.66811	14.62431
14	9.72672	10.29962	11.68206	12.75388	13.72081	14.66802	15.65793
15	10.53773	11.13530	12.57389	13.68639	14.68814	15.66793	16.69043
16	11.35465	11.97613	13.46913	14.62103	15.65651	16.66785	17.72191

d \ $1-\alpha$	0.30	0.20	0.10	0.05	0.025	0.01	0.005
0	1.20397	1.60944	2.30258	2.99573	3.68888	4.60517	5.29832
1	2.43922	2.99431	3.88972	4.74386	5.57164	6.63835	7.43013
2	3.61557	4.27903	5.32232	6.29579	7.22469	8.40595	9.27379
3	4.76223	5.51504	6.68078	7.75366	8.76727	10.04512	10.97748
4	5.89036	6.72098	7.99359	9.15352	10.24159	11.60462	12.59409
5	7.00555	7.90599	9.27467	10.51303	11.66833	13.10848	14.14976
6	8.11105	9.07538	10.53207	11.84240	13.05947	14.57062	15.65968
7	9.20895	10.23254	11.77091	13.14811	14.42268	15.99996	17.13359
8	10.30068	11.37977	12.99471	14.43465	15.76319	17.40265	18.57822
9	11.38727	12.51875	14.20599	15.70522	17.08480	18.78312	19.99842
10	12.46951	13.65073	15.40664	16.96222	18.39036	20.14468	21.39783
11	13.54798	14.77666	16.59812	18.20751	19.68204	21.48991	22.77926
12	14.62316	15.89731	17.78158	19.44257	20.96158	22.82084	24.14494
13	15.69544	17.01328	18.95796	20.66857	22.23040	24.13912	25.49669
14	16.78512	18.12509	20.12801	21.88648	23.48962	25.44609	26.83598
15	17.83246	19.23316	21.29237	23.09713	24.74022	26.74289	28.16406
16	18.89769	20.33782	22.45158	24.30118	25.98300	28.03045	29.48196

Table 12.5 Values $\Delta'_{1-\alpha}(d)$ and $\Delta'_{1-\alpha}(d)$, $1-\alpha=0.95$

c \ N	50	60	80	100	150	200
0	0.07378	0.06148	0.04611	0.03689	0.02459	0.01844
	0.00000	0.00000	0.00000	0.00000	0.00000	0.00000
1	0.11257	0.09365	0.07009	0.05600	0.03727	0.02793
	0.00051	0.00042	0.00032	0.00025	0.00017	0.00013
2	0.14750	0.12248	0.09146	0.07298	0.04849	0.03631
	0.00489	0.00407	0.00305	0.00243	0.00162	0.00121
3	0.18090	0.14994	0.11172	0.08902	0.05904	0.04417
	0.01263	0.01049	0.00783	0.00625	0.00415	0.00311
4	0.21362	0.17672	0.13136	0.10454	0.06921	0.05173
	0.02248	0.01863	0.01388	0.01106	0.00734	0.00549
5	0.24607	0.20317	0.15066	0.11972	0.07912	0.05909
	0.03384	0.02800	0.02082	0.01657	0.01097	0.00820
6	0.27853	0.22949	0.16975	0.13471	0.08886	0.06630
	0.04640	0.03832	0.02842	0.02259	0.01493	0.01115
7	0.31115	0.25582	0.18875	0.14957	0.09848	0.07341
	0.05995	0.04942	0.03657	0.02902	0.01915	0.01429
8	0.34408	0.28226	0.20772	0.16435	0.10801	0.08044
	0.07440	0.06120	0.04518	0.03580	0.02358	0.01758
9	0.37742	0.30889	0.22671	0.17910	0.11748	0.08741
	0.08966	0.07360	0.05419	0.04289	0.02819	0.02100
10	0.41126	0.33578	0.24576	0.19385	0.12691	0.09434
	0.10570	0.08657	0.06358	0.05025	0.03297	0.02453
11	0.44568	0.36297	0.26490	0.20863	0.13631	0.10123
	0.12247	0.10008	0.07331	0.05785	0.03788	0.02816
12	0.48077	0.39053	0.28417	0.22344	0.14569	0.10810
	0.13996	0.11411	0.08336	0.06568	0.04293	0.03189
13	0.51659	0.41818	0.30359	0.23831	0.15507	0.11495
	0.15818	0.12865	0.09372	0.07372	0.04809	0.03569
14	0.55322	0.44688	0.32318	0.25325	0.16445	0.12178
	0.17711	0.14368	0.10437	0.08198	0.05337	0.03957
15	0.59074	0.47577	0.34295	0.26828	0.17384	0.12861
	0.19677	0.15921	0.11531	0.09042	0.05875	0.04352
16	0.62922	0.50519	0.36294	0.28341	0.18325	0.13544
	0.21717	0.17524	0.12654	0.09906	0.06123	0.04753
17	0.66875	0.53518	0.38314	0.29864	0.19267	0.14226
	0.23833	0.19177	0.13804	0.10788	0.06980	0.05160
18	0.70942	0.56578	0.40860	0.31399	0.20213	0.14909
	0.26027	0.20881	0.14982	0.11688	0.07547	0.05574
19	0.75131	0.59704	0.42431	0.32947	0.21161	0.15592
	0.28303	0.22637	0.16188	0.12607	0.08123	0.05993

$$L_d(\Delta_\alpha(d)) = \alpha .$$

Tables 12.4 and 12.5 give the numerical computations for Table 12.3. In the tables appear both the quantiles $\Delta_{1-\alpha}(d)$ of the Poisson d.f. and the values of the quantities

$$\underline{\Delta}'_{1-\alpha}(d) \quad \text{and} \quad \bar{\Delta}'_{1-\alpha}(d)$$

for $1-\alpha = 0.95$, which are necessary for the construction of confidence intervals for an $[N\ W\ T]$ plan.

12.5.3 Weibull-Gnedenko Distribution

If the shape parameter b in the distribution

$$F(t, a, b) = 1 - e^{-(t/a)^b}$$

is known, substitution of t^* for t^b and α for a^{-b} yields an exponential distribution

$$F(t^*, \alpha) = 1 - e^{-\alpha t^*} ,$$

with scale parameter α. Confidence limits α_L and α_U for the parameter α are then determined with reference to Table 12.3 for the testing strategy being used. In the determination of the total time on test, every time, whether for a complete or incomplete realization, is raised to the power b:

$$t_\Sigma = \sum_{1\le i\le n} \tau_i^b + \sum_{1\le j\le m} t_j^b .$$

Upon determination of α_L and α_U, the confidence limits for the scale parameter, a, are determined from

$$\ln a = -\frac{\ln a}{b} .$$

If both parameters of the Weibull-Gnedenko distribution are found from experimental data, approximate confidence intervals for parameters a and b can be found by substituting a normal distribution.

12.5.4 Binomial Distribution

A binomial distribution might be hypothesized in the course of testing under a nonparametric plan in which the only available information is the number of items under test and the number of items failed within the test duration. The only measure that can be estimated is the failure probability, q, within the testing time. The maximum likelihood estimate for failure probability is given by $\hat{q} = n/N$, where n is the number of failed items and N is the total number of items.

The estimate $\hat{q}$ is considered to be a r.v. with a binomial distribution. A lower limit, q_L, and upper limit, q_U, are determined with:

$$\sum_{0 \leq i \leq n} \binom{N}{i} q_U^i (1 - q_U)^{N-i} = \epsilon , \tag{12.30}$$

$$\sum_{n \leq i \leq N} \binom{N}{i} q_L^i (1 - q_L)^{N-i} = \epsilon , \tag{12.31}$$

where 2ϵ is the significance level. Numerical solutions of these equations may be found in Tables E.14 and E.15 (see Appendix E).

12.5.5 Normal Distribution

Confidence Bounds for the Mean Value

Let there be d trials with results $\theta_1, \ldots, \theta_d$. We know the r.v. θ has a normal d.f. We can determine two-sided confidence bounds for the unknown mean, T, that correspond to a confidence coefficient $1 - \alpha$, with the expression:

$$\underline{T} = \hat{T}_0 - t_{\alpha/2}(r) \frac{s}{\sqrt{d-1}} ,$$

$$\underline{T} = \hat{T}_0 + t_{\alpha/2}(r) \frac{s}{\sqrt{d-1}} ,$$

where $\hat{T}_0$ is the arithmetic average of the θ_i variables:

$$\hat{T}_0 = \frac{1}{d} \sum_{1 \leq i \leq d} \theta_i .$$

The variable s is the square root of an unbiased estimate for the variance of θ,

$$s = \sqrt{\frac{1}{d-1} \sum_{1 \leq i \leq d} (\theta_i - \hat{T}_0)^2} ,$$

and $t_{0.5\alpha}(r)$ are the quantities determined using Table E.17, with $r = d - 1$.

Confidence Bounds for the Variance

If we have results $\theta_1, \ldots, \theta_n$, and we assume that the random variable θ has a normal distribution, we can determine the confidence bounds (with confidence coefficient $1 - \alpha$) for the unknown value of the variance by:

$$\underline{D} = \frac{s^2 d}{\chi^2_{\alpha/2}(d-1)} ,$$

$$\bar{D} = \frac{s^2 d}{\chi^2_{1-\alpha/2}(d-1)} ,$$

where s^2 is the unbiased estimate for the variance of the θ's,

$$s^2 = \frac{1}{d-1} \sum_{1 \le i \le d} (\theta_i - \hat{T}_0)^2 ,$$

and

$$\hat{T}_0 = \frac{1}{d} \sum_{1 \le i \le d} \theta_i ,$$

and χ^2 can be found in Table E.16 (Appendix E).

12.6 TRUNCATED SAMPLES

In practice, tests of individual items can be interrupted for various reasons. This leaves truncated statistical data. Such data are typical for field tests and real operations. In the latter, items can start at different times and can be excluded from operation before failure.

A general plan can be followed for tests such as these. Let an ordered complete observation (time to failure) be denoted $\boldsymbol{\xi}_i$, $1 \le i \le n$. Let an ordered time of "censored observations" of a group of $m_j, 1 \le j \le k$, items be denoted $\boldsymbol{\theta}_j$, remembering that censoring is termination of testing of an item when it has not failed.

We consider two equivalent unbiased estimates for the probability of failure-free operation, $\hat{P}(t)$. The sufficient statistics for these nonparametric tests consist of a series of the number of sequential failures before one or more censored observations, d_i, and of the number of censored observations, c_j, before the next failure. Obviously, $\Sigma\, d_i = n$, and $\Sigma c_j = k$. Censored times smaller than the time of the first observed failure are not taken into account. Thus the sequence of data $\{d_1, c_1, \ldots, d_N, c_N\}$ is sufficient to obtain an estimate $\hat{P}(t)$. Notice that c_i is the number of items that do not fail at time t.

Kaplan-Meier Estimate

The Kaplan-Meier estimate is based on the formula of conditional probability: $P(0, t) = P(0, \tau) \cdot P(\tau, t \mid 0, \tau)$. The estimate is given by

$$\hat{P}(t) = \left(1 - \frac{d_1}{n+k}\right)\left(1 - \frac{d_2}{n+k-d_1-c_1}\right) \times \left(1 - \frac{d_3}{n+k-(d_1+c_1+d_2+c_2)}\right)\cdots\left(1 - \frac{d_N}{c_N}\right) = \prod_{1\le j\le N}\left(1 - \frac{d_i}{n+k-\Sigma_{1\le i\le N-1}(d_i+c_i)}\right).$$

Pavlov-Ushakov Estimate

The Pavlov-Ushakov estimate is based on counting an equivalent number of failures for each censored observation. The estimate is based on the fact that censored observations belong to the same sample as the regular observations and consequently have the same statistical properties.

To calculate the equivalent number of failures, we use an inverse recurrent expression. Let Δ_s denote the equivalent number of failures at the first moment of the interval for the sth group of failures, d_s. Let $\tilde{\Delta}_s$ denote the equivalent number of failures at the first moment of the interval for the sth group of censored observations. Then

$$\Delta_N = d_N ,$$

$$\tilde{\Delta}_{N-1} = \Delta_N + \frac{\Delta_N}{c_N} c_{N-1} = \Delta_N\left(1 + \frac{c_{N-1}}{c_N}\right)$$

(this means that the whole group of censored trials, c_{N-1}, has the same statistical properties as the other items)

$$\Delta_{N-1} = d_{N-1} + \tilde{\Delta}_{N-1} ,$$

. . . .

$$\tilde{\Delta}_{N-2} = \Delta_{N-1}\left(1 + \frac{c_{N-2}}{c_N + d_N + c_{N-1} + d_{N-1}}\right);$$

$$\Delta_s = d_s + \tilde{\Delta}_s ,$$

$$\tilde{\Delta}_s = \Delta_s\left(1 + \frac{c_{N-2}}{\Sigma_{s+1\le i\le N}(c_i + d_i)}\right).$$

These calculations continue until Δ_1 is found. The final estimate, $\hat{P}(t)$, is given by

$$\hat{P}(t) = 1 - \frac{\Delta_1}{n+k} .$$

CHAPTER 13

ACCEPTANCE-REJECTION TESTS

13.1 INTRODUCTION

Before a new item is manufactured in production quantities, the initial production lot may be given reliability tests or trial usage under an acceptance-rejection test plan. A decision is made on whether the item does or does not conform to requirements. In this chapter the decision is based on a reliability measure, R, that increases with increasing item reliability.

A required value, R_r, is specified. The control procedure is intended to result in acceptance of items with a reliability level $R \geq R_r$ and rejection of items with a reliability level $R < R_r$.

The plot of item acceptance probability, L, against reliability, R, is presented in Figure 13.1, which shows the *operating characteristics* of the control plan. In the figure the ideal reliability acceptance-rejection level is shown with the thick line. However, realization of this ideal plan may not be practical because of the time or the number of samples required, or because of other factors. The actual operating characteristics are shown in the figure with two thin lines, one for the acceptance level, R_0, and one for the rejection level, R_1.

The values of $L(R)$ are specified for the two points. Items with reliability $R \geq R_0$ are considered to be absolutely acceptable for the customer and have to have a probability of at least $L(R_0)$ of being accepted. Items with $R \leq R_1$ are unacceptable and have to be rejected with a probability of at least $1 - L(R_1)$. Probabilities $\alpha = 1 - L(R_0)$ and $\beta = L(R_1)$ are called the *producer's risk* and the *consumer's risk*, respectively. The former is the risk of rejection of a good item; the latter is the risk of acceptance of a bad item. Thus checking the $R \geq R_r$ hypothesis against the $R < R_r$ hypothesis is

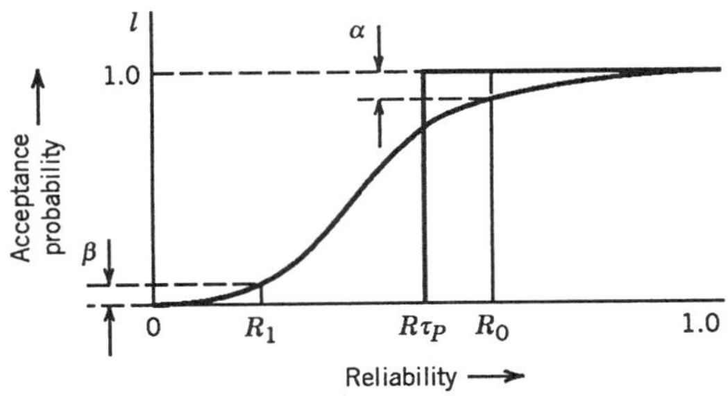

Figure 13.1 Item reliability, R, versus probability of acceptance, L.

replaced with checking the $R \geq R_0$ hypothesis against the $R \leq R_1$ hypothesis. For further simplification, complicated hypotheses $R \geq R_0$ and $R \leq R_1$ can be replaced with hypotheses $R = R_0$ and $R = R_1$. The control that provides the specified risks at these points will ensure the same or smaller risks in the regions $R \geq R_0$ and $R \leq R_1$.

The four numbers R_0, R_1, α, and β and the selected control procedure (such as single-stage, sequentional test) determine the control plan. The number of observations, V, is the only restriction that prevents arbitrary reduction of the risks and makes acceptance and rejection levels too close to each other. Therefore in control planning one has to select values of R_0, R_1, α, and β that result in use of the available capabilities and also stay within the limitations.

If the customer's risk is small compared to that of the producer, the interval $[R_1, R_0]$ is moved to the left, up toward the limiting position where $R_0 = R_r$. If, on the contrary, the producer's damage is small compared to that of the customer, then the interval is moved to the right, toward $R_1 = R_r$. The points usually are chosen to give equal risk to producer and customer.

This chapter covers just the case for which the controlled reliability measure is a one-dimensional variable, such as operating time or probability. If a number of reliability measures are specified for an item, the methods of this chapter will not provide for the item in the aggregate the risk values intended for the separate reliability measures.

Since no prior information is available for control planning, the planner has to consider the worst possible case when distinguishing between the R_0 and R_1 hypotheses. A mean indefinite result close to the midpoint between R_0 and R_1 should be envisaged. The actual result R obtained upon termination of the observation can change the confidence in the correctness of the decision. The assumed planning risks α and β simply characterize the control plan and should not be regarded as valid or adequate measures. It may be expedient to use the observed risks instead (see Section 13.2.4).

In contrast to the observed risks, risks α and β are determined prior to the tests. Both types of risks are conditional probabilities that either $R = R_0$

or $R = R_1$. Because α and β correspond to the worst case hypothesis while $\hat{\alpha}$ and $\hat{\beta}$ relate to the observed cases, $\hat{\alpha} \leq \alpha$ and $\hat{\beta} \leq \beta$.

Control of reliability measures can be performed without a preliminary plan for the number of observations when reliability control is combined with either the testing of other parameters or ordinary operation of the item. This situation is considered in Section 13.2.5.

In the course of item development, control of reliability is performed at least once during acceptance testing. As a rule the results of preliminary and other tests also are used. For full production, reliability control usually is achieved with periodic testing. For reliability control of high-reliability items produced in small batches, it is expedient to include the control procedures in the tests of changes in technology, design or components. Reliability control of complex systems with many redundant elements can be included in the acceptance tests that leave the total test duration practically unchanged.

13.2 SINGLE-STAGE CONTROL PLANS

A plan with the number of observations fixed in advance is classified as a *single-stage plan*. In this section the control plan formulas are for operating time or probability reliability measures. In addition to a general method for single-stage control, methods of control without preliminary planning also are presented.

13.2.1 Single-Stage Control with Operating Time Measures

Consider an operating time measure distributed exponentially. Control duration is limited by operating time. In control plan development, the absolute values of $R_0 = T_0$ and $R_1 = T_1$ are not important. The ratio T_0/T_1 and the risks α and β determine the plan. With the purpose of simplification of the control procedure, the acceptance rejection procedures are based on the number of failures, rather than the reliability measure level itself since the level determines the number of failures.

With control of operating time to failure, the total failure number, r, for all controlled item samples, N, is recorded along with total operating time

$$t_\Sigma = \sum_{1 \leq i \leq N} t_i ,$$

where t_1 is the operating time of the ith sample. The control plan has two numbers—the failure number for rejection, r_*, and the total operating time, t_{max}. Tests are terminated upon achievement of either value. If r items fail before time t_{max}, the item is rejected. If not, the item is accepted.

The plan is chosen in conformity with Table 13.1. Because the number of

Table 13.1 Value of T_0/T_1 for different combinations of B, α, and r, single-stage control, exponential failure rate (α = producer's risk, β = consumer's risk, r = number of failures, $T_0 = R_0$ = acceptance level, $T_1 = R_1$ = rejection level)

0.05	0.10	0.20	0.30	γ_*	t_{max}/T_0	$\bar{t}_\kappa/T_0$
			$\alpha = 0.05$			
58.82	45.45	31.25	23.25	1	0.052	0.04898
13.33	10.99	8.403	6.850	2	0.356	0.3385
7.692	6.493	5.235	4.425	3	0.817	0.803
5,682	4,878	4.032	3.484	4	1.366	1.344
4.651	4.065	3.413	2.994	5	1.970	1.944
4.032	3.546	3.030	2.681	6	2.613	2.587
3.646	3.205	2.762	2.469	7	3.285	3.252
3.350	2.958	2.570	2.315	8	3.981	3.945
3.077	2.770	2.427	2.193	9	4.695	4.652
2.898	2.618	2.309	2.102	10	5.425	5.376
2.747	2.500	2.217	2.020	11	6.169	6.120
2.631	2.398	2.137	1.957	12	6.924	6.876
2.531	2.315	2.070	1.901	13	7.689	7.635
2.445	2.242	2.012	1.855	14	8.464	8.413
2.369	2.178	1.961	1.815	15	9.246	9.190
2.096	1.961	1.779	1.658	20	13.2	13.134
1.942	1.815	1.669	1.567	25	17.3	17.230
1.835	1.721	1.597	1.515	30	21.5	21.414
			$\alpha = 0.10$			
28.57	21.74	15.38	11.36	1	0.105	0.095
8.928	7.299	5.650	4.587	2	0.532	0.503
5.714	4.831	3.891	3.278	3	1.102	1.069
4.444	3.831	3.164	2.732	4	1.745	1.694
3.769	3.289	2.762	2.421	5	2.432	2.374
3.333	2.941	2.519	2.222	6	3.152	3.082
3.039	2.703	2.331	2.083	7	3.895	3.813
2.825	2.525	2.198	1.980	8	4.656	4.568
2.659	2.392	2.096	1.897	9	5.432	5.340
2.525	2.283	2.012	1.831	10	6.221	6.121
2.415	2.193	1.945	1.776	11	7.020	6.915
2.325	2.118	1.887	1.730	12	7.829	7.711
2.247	2.057	1.838	1.692	13	8.646	8.524
2.183	2.004	1.798	1.658	14	9.469	9.336
2.127	1.953	1.760	1.628	15	10.30	10.166
1.915	1.792	1.626	1.515	20	14.52	14.346
1.792	1.672	1.538	1.445	25	18.84	18.651
1.706	1.602	1.486	1.408	30	23.23	23.020

Table 13.1 (*Continued*)

0.05	0.10	0.20	0.30	γ_*	t_{max}/T_0	$\bar{t}_\kappa/T_0$
			$\alpha = 0.20$			
13.51	10.31	7.246	5.405	1	0.223	0.197
5.747	4.717	3.636	2.958	2	0.824	0.753
4.098	3.472	2.785	2.358	3	1.535	1.430
3.378	2.907	2.404	2.074	4	2.297	2.258
2.967	2.590	2.174	1.908	5	3.089	2.925
2.695	2.375	2.024	1.795	6	3.903	3.727
2.500	2.227	1.919	1.715	7	4.733	4.529
2.358	2.110	1.835	1.653	8	5.576	5.347
2.247	2.020	1.770	1.602	9	6.428	6.190
2.155	1.949	1.718	1.562	10	7.289	7.027
2.079	1.890	1.675	1.529	11	8.157	7.880
2.016	1.838	1.636	1.501	12	9.031	8.742
1.961	1.795	1.605	1.475	13	9.910	9.603
1.916	1.757	1.577	1.453	14	10.79	10.466
1.872	1.724	1.553	1.435	15	11.68	11.341
1.718	1.608	1.460	1.360	20	16.17	15.782
1.628	1.520	1.398	1.316	25	20.72	20.264
1.565	1.468	1.362	1.292	30	25.32	24.814
			$\alpha = 0.30$			
8.403	6.451	4.525	3.378	1	0.356	0.263
4.329	3.546	2.732	2.222	2	1.097	0.935
3.289	2.785	2.237	1.890	3	1.914	1.719
2.809	2.415	1.996	1.724	4	2.763	2.509
2.519	2.202	1.852	1.621	5	3.633	3.339
2.325	2.053	1.751	1.550	6	4.517	4.187
2.188	1.945	1.678	1.499	7	5.410	5.042
2.083	1.865	1.621	1.460	8	6.312	5.908
2.000	1.798	1.577	1.426	9	7.220	6.787
1.930	1.748	1.538	1.400	10	8.133	7.661
1.876	1.703	1.508	1.377	11	9.050	8.552
1.825	1.664	1.481	1.358	12	9.971	9.442
1.786	1.631	1.460	1.342	13	10.90	10.344
1.748	1.602	1.439	1.328	14	11.82	11.229
1.715	1.577	1.420	1.314	15	12.75	12.138
1.595	1.490	1.353	1.261	20	17.44	16.742
1.524	1.422	1.309	1.230	25	22.16	21.340
1.473	1.383	1.282	1.215	30	26.90	25.985

failures is a discrete number, T_0/T_1, which corresponds to the precise solution of the problem with specified values of α and β, also has discrete values.

The mean duration (the total number of samples) of the control procedure $\bar{t}_k$ for an item with a reliability measure corresponding to the acceptance level T_0 is also presented in Table 13.1 as a reference point for different combinations of α, β, and r. For example, for $\beta = 0.05$, $\alpha = 0.10$, and $r = 30$, we see that we need a T_0/T_1 value of 1.706. Each α and r combination also has a $T_{\max}/T_0$ ratio and, for reference, a t_k/T_0 ratio, where t_k is the mean time for an item. The value of the ratio T_0/T_1 is calculated for each pair of α and β values and for 18 values of r_* in conformity with

$$\frac{T_0}{T_1} = \frac{\chi^2_{1-\beta,2r_*}}{\chi^2_{\alpha,2r_*}} ,$$

where $\chi^2_{1-\beta,2r_*}$ and $\chi^2_{\alpha,2r_*}$ are the quantiles of levels $1-\beta$ and α for a χ^2 distribution with $2r_*$ degrees of freedom.

The value $t_{\max}/T_0$ is calculated using

$$\frac{t_{\max}}{T_0} = \frac{1}{\chi^2_{\alpha,2_*}} .$$

Operating characteristics for the plans in the table can be found using

$$L(T) = \sum_{0 \le r \le r_* - 1} \frac{(t_{\max}/T)^r e^{-t_{\max}/T}}{r!} .$$

Total control duration can be varied over a wide range due to changes of the sample number, with provision only for the maximum operating time, $t_{\max}$. If the control duration, t_M, is prescribed, all samples are operated simultaneously, and the failed items are replaced or completely restored. The required sample number can be found from

$$N = \frac{t_{\max}}{t_M} .$$

If failed items are neither replaced nor restored, the total number of samples required for achievement of the same total operating time with the same control duration has to be increased to compensate for loss of operating time due to failures. The average figure is $\Delta N = t_{\max}/T$, where T represents the estimated value of the item reliability measure. In any case, ΔN does not surpass $r_* - 1$.

[N W r] and [N R r] have a duration with a predetermined number of failures, r_M. Upon achievement of this failure number, the total operating time for all samples under control is calculated, and a point estimate for time to failure is determined, $\hat{T} = t_\Sigma / r_M$. If the value of $\hat{T}$ is above the limit given by the estimation rate C, the item is accepted. Otherwise, it is rejected. There is no reason to go beyond a total operating time $t_r = Cr_M$ (where t_Σ is the sum of the t_i) because acceptance is guaranteed. When control is continued up to the specified failure number r and $t_\Sigma < Cr_M$, the item is rejected. When the operating time $t_\Sigma = Cr_M$, and $r < r_M$, the item is accepted. This corresponds to the plan $[N\ R\ r_M\ t_\Sigma]$ or $[N\ W\ r_M\ t_\Sigma]$ and is equivalent to the above.

With nonexponential distributions or more complicated test plans, such as truncated samples, general methods of single-stage control planning of the measure T should be used (see Section 13.2.3).

13.2.2 Single-Stage Control with Probability Measures

If the reliability measure represents the probability of a certain event, A, N independent trials are planned. Upon the Nth trial the item is accepted if the failure number is less than the predetermined value r_* and is rejected otherwise. The experiments can be performed with any number of items from 1 to N. If an item is used more than once, the item is completely restored before resuming the experiment. The control plan is determined by two numbers, N and r_*.

The plan is determined on the basis of the prescribed values of P_0, P_1, α, and β from Table 13.2. Values of N and r_* in the table are found from solution of the following system of equations:

$$\sum_{0 \le i \le r_* - 1} \binom{N}{i} P_0^{N-i} (1 - P_0)^i = 1 - \alpha\,, \tag{13.1}$$

$$\sum_{0 \le i \le r_* - 1} \binom{N}{i} P_1^{N-i} (1 - P_1)^i = \beta\,, \tag{13.2}$$

The number of item samples is determined by the independence requirements for all N trials.

If the measure is $P(t)$ and the distribution is exponential, N operation cycles with duration t can be planned with any number of samples from 1 to N. With this type of distribution, an item in a new cycle can be considered to be a new sample. The subsequent cycle can be started immediately upon termination of the preceding one (or upon repair in case of a failure). If the distribution is not exponential or is unknown, each cycle should be performed with a new item sample. N samples will be required.

If K is controlled, the procedure is reduced to N tests of the item availability at random moments in time. The tests of the samples should be separated in time (the control interval should exceed significantly the mean time of failure). In the control of $R(t)$ all the requirements for tests of K and $P(t)$ should be combined.

Table 13.2 Number of trials N for single-stage control

	$\alpha = \beta = 0.1$				$\alpha = \beta = 0.2$			
	$\frac{1-P_1}{1-P_0} = 2.0$	2.5	3.0	4.0	2.0	2.5	3.0	4.0
P_0	$r = 14$	8	6	4	6	4	3	2
0.999	9470	4655	3150	1745	3905	2295	1535	825
0.998	3735	2328	1575	872	1953	1148	768	413
0.997	3158	1552	1050	582	1302	765	512	275
0.996	1369	1164	788	436	977	574	384	207
0.995	1895	931	630	349	781	459	307	165
0.994	1578	776	525	291	651	382	256	138
0.993	1353	665	450	249	558	328	220	118
0.992	1884	582	394	218	489	287	192	103
0.991	1052	518	350	194	434	255	171	92
0.990	947	466	315	175	391	230	154	83
9.980	473	233	158	87	196	115	77	42
0.970	315	155	105	58	131	77	52	28
0.960	237	116	79	44	98	58	39	21
0.950	189	93	63	35	78	46	31	17
0.940	158	78	53	29	66	38	26	14
0.930	135	67	45	25	56	33	22	12
0.920	118	58	40	22	49	29	20	11
0.910	106	52	35	20	44	26	17	10
0.900	95	47	32	18	37	23	16	9

EXAMPLE 13.1

Find a reliability control plan for electrical elements under two specified levels of failure-free operation probability $P(t)$, $P_0 = 0.98$, and $P_1 = 0.96$. Let $\alpha = \beta = 0.1$, and let $t = 20$ hours.

Solution

For the given values of P_0, P_1, α and β, Table 13.2 gives $N = 473$ and $r_* = 14$. Testing ends either upon observation of the 14th failure with item rejection or upon the 473th cycle with acceptance.

13.2.3 Single-Stage Control, Scalar Measure

In all cases single-stage control has an observation amount, V, and a vector of observation results, $x_1, x_2, \ldots, x_n$. A function of the results $\hat{X}(x_1, x_2, \ldots, x_n)$ is calculated. The domain of the function has two disjoint areas, X_1 and X_2. If $\hat{X}$ is in X_1, the item is accepted. If it is in X_2, the item is rejected. There are a lot of methods for division of the sample space into two areas. One is the Neuman-Pearson criterion. For a given value of risk $\alpha = 1 - L(R_0)$, the criterion minimizes the value of the other risk $\beta = L(R_1)$. For two given risk values, the criterion requires the minimum number of observations.

The likelihood ratio $l(x_1, x_2, \ldots, x_n)$ is

$$l(x_1, x_2, \ldots, x_n) = \frac{f(x_1, x_2, \ldots, x_n \mid R_1)}{f(x_1, x_2, \ldots, x_n \mid R_0)} ,$$

where $f(x_1, x_2, \ldots, x_n \mid R_1)$ is the distribution density function for the sample $x_1, x_2, \ldots, x_n$ under the condition that the actual reliability of the item is at the rejection level, and $f(x_1, x_2, \ldots, x_n \mid R_0)$ is the same at the acceptance level. An item is accepted if

$$l(x_1, x_2, \ldots, x_n) \leq C \tag{13.3}$$

and rejected if

$$l(x_1, x_2, \ldots, x_n) > C , \tag{13.4}$$

where C is a constant.

The likelihood ratio $l(x_1, x_2, \ldots, x_n)$ also depends upon R and V. Varying the amount V and the constant C, one can ensure the given values of both risks. The values of V and C are determined from solution of the following system of equations:

$$Pr\{l\{(x_1, x_2, \ldots, x_n) > C \mid R = R_0; V\} = \alpha , \tag{13.5}$$

$$Pr\{l(x_1, x_2, \ldots, x_n) \leq C \mid R = R_1; V\} = \beta . \tag{13.6}$$

Solution of the control problem is simplified if $x_1, x_2, \ldots, x_n$ are independent values with one distribution density, $f(y, R)$. The distribution density of the sample $f(x_1, x_2, \ldots, x_n \mid R)$ then is equal to the product

$$\prod_{1 \leq i \leq n} f(x_i, R) .$$

To simplify the solution, (13.3) and (13.4) are replaced with logarithmic relationships. The rejection condition is

$$\ln l(x_1, x_2, \ldots, x_n) > \ln C = C' . \tag{13.7}$$

The acceptance condition is

$$\ln l(x_1, x_2, \ldots, x_n) \leq C' . \tag{13.8}$$

The equations for planning are

$$Pr\{\ln l(x_1, x_2, \ldots, x_n) > C' \mid R = R_0; V\} = \alpha ,$$
$$Pr\{\ln l(x_1, x_2, \ldots, x_n) \geq C' \mid R = R_1; V\} = \beta .$$

Acceptance also can be based on a point estimate of the specified reliability measure. If

$$\hat{R}(x_1, x_2, \ldots, x_n) \geq C , \tag{13.9}$$

the item is accepted. If

$$\hat{R}(x_1, x_2, \ldots, x_n) < C , \tag{13.10}$$

the item is rejected. The control plan $[V, C]$ is determined from solution of the system of equations:

$$F(C, R_1, V) = 1 - \beta ,$$
$$F(C, R_0, V) = \alpha ,$$

where $F(C, R, V) = Pr\{R < C \mid R, V\}$ is the distribution function for the sample point estimate $R(x_1, x_2, \ldots, x_n)$, with the actual item reliability, R, and observation amount, V. Operating characteristics for the chosen plan can be found from

$$L(R) = l - F(C, R, V) .$$

Criteria (13.9) and (13.10) do not possess the optimal properties of (13.3) and (13.4), but it is expedient to use them when (13.3) and (13.4) cannot be used. If an explicit analytical expression for $F(C, R, V)$ is difficult to obtain, one can use a normal approximation or a computer simulation.

If there are convenient formulas or methods for determining confidence limits $\bar{R}_{\Delta 1}(\hat{R}, V)$ and $\underline{R}_{\Delta 2}(\hat{R}, V)$, control with a point estimate of the specified reliability measure can be formulated from these limits. Acceptance condition (13.9) is transformed into

$$\underline{R}_{1-\beta}(\hat{R}, V) \geq R_1 ,$$
$$\bar{R}_{1-\alpha}(\hat{R}, V) > R_0 . \tag{13.11}$$

Rejection condition (13.10) is transformed into

$$\underline{R}_{1-\beta}(\hat{R}, V) < R_1 ,$$
$$\bar{R}_{1-\alpha}(\hat{R}, V) \leq R_0 . \tag{13.12}$$

Application of the confidence limits does not change the required number of observations. The estimation rate is also preserved as compared with ordinary single-stage control. However, it takes the form of conditions (13.11) and (13.12). Control planning is reduced to the solution of a system of two equations:

$$\bar{R}_{1-\alpha}(\hat{R}, V) = R_0 , \tag{13.13}$$

$$\underline{R}_{1-\beta}(\hat{R}, V) = R_1 . \tag{13.14}$$

The values R^* and V^* that satisfy (13.13) and (13.14) coincide with the estimation rate C and the required amount V used in single-stage control (13.9) and (13.10). This is valid for at least the case where confidence limits $\underline{R}_{\Delta 2}(\hat{R}, V)$ and $\bar{R}_{\Delta 2}(\hat{R}, V)$ are monotonic with regard to $\hat{R}$ and V. In that case $\underline{R}_{\Delta 2}$ and $\bar{R}_{\Delta 2}$ come together with increasing V, and increase with increasing $\hat{R}$. The value of R^* can be used as the value of the estimation rate C in the ordinary form (13.9) and (13.10).

From the above it follows that if formulas and tables of ordinary single-stage control are available, they can be used to determine the required control with the confidence limits without solving (13.13) and (13.14). Similarly, if these formulas are absent but a confidence limit determination method is available, (13.13) and (13.14) enable the ordinary control plan, including the estimation rate, to be determined.

13.2.4 Determination of Observed Risks upon Control

The values of risks α and β do not depend on the observation results. It is expedient to use also the observed risks $\hat{\alpha}$ and $\hat{\beta}$

$$\hat{\alpha} = Pr\{\hat{R} \leq \tilde{R} \mid R_0\}$$
$$\hat{\beta} = Pr\{\hat{R} \geq \tilde{R} \mid R_1\} ,$$

where $\tilde{R}$ is the $\hat{R}$ given by the observation results. Recall that $\alpha = Pr\{\hat{R} \leq C \mid R_0\}$, and that $\beta = Pr\{R \geq C \mid R_1\}$. In the case of acceptance, $\tilde{R}_T \geq C$. In the case of rejection, $\tilde{R} \leq C$. Thus $\hat{\alpha} \leq \alpha$, $\hat{\beta} \leq \beta$.

Utilization of observed risks does not alter the decisions on acceptance or rejection. However, it can change the estimate of the sufficient number of observations, and it can change the confidence limits of the estimate. It also permits the accepted items to be separated into groups with regard to quality.

The observed producer's risk, $\hat{\alpha}$, is determined from

$$\underline{R}_{1-\hat{\alpha}}(\tilde{R}, V) = R_0 , \tag{13.15}$$

Through selection of the confidence probability $\hat{\gamma} = 1 - \hat{\alpha}$, the upper limit of the one-sided confidence interval for the controlled measure is combined with the acceptance level R_0.

The observed customer's risk, $\hat{\beta}$, is determined from

$$\underline{R}_{1-\hat{\beta}}(\tilde{R}, V) = R_1 , \tag{13.16}$$

EXAMPLE 13.2

Let two levels of the operating time to failure be specified: T_0 and $T_1 = T_0/2$. Let $\alpha = \beta = 0.1$. Restorable items with exponential failure distributions are tested.

Solution

The single-stage control plan is found from Table 13.1: $t_{\Sigma} = 9.47\,T_0$; $r_* = 14$. Let 13 failures be observed in the first batch and 9 failures be observed in the second batch. Both batches are accepted, but in the second case the probability of an error undoubtedly is less.

13.2.5 Control without Preliminary Planning

Sometimes the reliability control can be combined in part or even completely with tests of other measures. The number of observations might be unknown in advance. Under these conditions a conventional procedure can be used. Planned risks have to be chosen so that the calculated value of V is equal to the actual one. An estimation rate, C, is calculated for making a decision on acceptance or rejection. The values of the risks α and β chosen in such a manner also can be used in estimating the upper limit for the observed risks. However, when the observations are over, we can use $\hat{\alpha}$ and $\hat{\beta}$ directly. They are determined in accordance with Section 13.2.4.

Another technique is control with confidence limits. When observations are terminated for whatever reason, the confidence interval $[\bar{R}_{\gamma_1}, \underline{R}_{\gamma_2}]$ is determined on the basis of the accumulated statistics. Values of γ_1 and γ_2 are chosen so that one of the following conditions holds:

$$\underline{R}_{\gamma_2} = R_1 , \quad \bar{R}_{\gamma_1} > R_0 ; \tag{13.17}$$

$$\underline{R}_{\gamma_2} < R_1 , \quad \bar{R}_{\gamma_1} = R_0 . \tag{13.18}$$

When changing the values of γ_1 and γ_2, a preset relation between the two has to be observed. When γ_1 and γ_2 increase, the confidence interval becomes narrower. If the values decrease, the interval expands from both sides. Condition (13.17) may hold values of $\hat{\gamma}_1$ and $\hat{\gamma}_2$. The left limit of the confidence interval coincides with the left limit of the specified interval, $[R_1, R_0]$, and the former may be broader than the latter (see Figure 13.2). The acceptance decision is made, and the observed consumer's risk $\hat{\beta} = 1 - \hat{\gamma}_2$ is found as in (13.16). If condition (13.18) holds, the interval right limits coincide (see Figure 13.2), and rejection occurs. The observed producer's risk, $\hat{\alpha} = 1 - \tilde{\gamma}_1$, is determined as in (13.15). If the confidence limit is shifted to the right from the specified one, the item is accepted. If the interval is shifted to the left, the item is rejected. It is evident that correlated variation of γ_1 and γ_2 results in the fulfillment of just one of the two conditions (13.17) and (13.18).

Strict coincidence of the limits is not necessary. Only a shift is required. Complete covering of one interval by the other has to be excluded. The following inequalities are valid:

$$\hat{\beta} \le 1 - \hat{\gamma}_2 \qquad \text{for } \underline{R}_{\hat{\gamma}_2} \ge R_1,\ \bar{R}_{\hat{\gamma}_1} > R_0\,,$$

$$\hat{\alpha} \le 1 - \tilde{\gamma}_1 \qquad \text{for } \underline{R}_{\hat{\gamma}_2} < R_1,\ \bar{R}_{\hat{\gamma}_1} \le R_0\,.$$

EXAMPLE 13.3

One sample of a restorable article with an exponential failure distribution was tested. Two levels of time to failure were established: T_0 and $T_1 = T_0/2$. Error probabilities should be equal: $\gamma_1 = \gamma_2 = \gamma$. The operating time, t_Σ, is $4T_0$. The failure number, r, is 2. Conformity or nonconcofmity with the specified requirements has to be determined, and the observed risk has to be indicated.

Solution

Using the formulas from Chapter 12 to find confidence limits, we rewrite (13.17) in the form

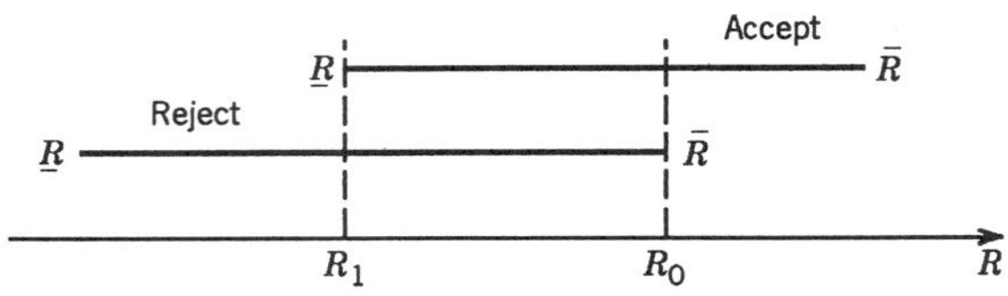

Figure 13.2 Acceptance and rejection intervals.

$$\underline{T}_{\hat{\gamma}} = \frac{T_0}{2}, \qquad \bar{T}_{\hat{\gamma}} > T_0,$$

or

$$\chi^2_{1-\hat{\gamma}}(2r+2) = 4\frac{t_\Sigma}{T_0},$$

$$\chi^2_{\hat{\gamma}}(2r) < 2\frac{t_\Sigma}{T_0}.$$

We also may write

$$\underline{T}_{\hat{\gamma}} < \frac{T_0}{2}, \qquad \bar{T}_{\hat{\gamma}} = T_0,$$

$$\chi^2_{1-\hat{\gamma}}(2r+2) > 4\frac{t_\Sigma}{T_0}$$

$$\chi^2_{\hat{\gamma}}(2r) = 2\frac{t_\Sigma}{T_0}.$$

Using a χ^2 distribution table, we see that condition (13.17) is fulfilled. $\hat{\gamma} = 0.985$, $\chi^2_{0.015}(6) = 16$, $\chi^2_{0.985}(4) = 0.36 < 8, \underline{T}_{0.985} = T_0/2$, and $\bar{T}_{0.985} = 22T_0$. The item is accepted. The observed consumer's risk = 0.015.

Consider variation of the decisions and risks under the example conditions with different numbers of failures r:

If $r = 0$, $\underline{T} = T_0/2$ for $\hat{\gamma} > 0.999$ and $\bar{T}_{\hat{\gamma}} = \infty$. The item is accepted with $\hat{\beta} < 0.001$.
If $r = 1$, $\underline{T}_{\hat{\gamma}} = T_0/2$ for $\hat{\gamma} \approx 0.995$, $\hat{\beta} \approx 0.005$, $\bar{T}_{0.995} \approx 80T_0$.
If $r = 2$ (as we have already seen), $\hat{\beta} = 0.015$, $\bar{T}_{0.985} = 22T_0$.
If $r = 3$, $\underline{T}_{\hat{\gamma}} = T_0/2$ for $\hat{\gamma} = 0.95$, $\hat{\beta} \approx 0.05$, $\bar{T}_{0.95} = 4.9T_0$.
If $r = 4$, $\underline{T}_{\hat{\gamma}} = T_0/2$ for $\hat{\gamma} = 0.90$, $\hat{\beta} = 0.10$, $\bar{T}_{0.90} = 2.3T_0$.
If $r = 5$, $\underline{T}_{\hat{\gamma}} = T_0/2$ for $\hat{\gamma} = 0.80$, $\hat{\beta} = 0.20$, $\bar{T}_{0.80} = 1.3T_0$.
if $r = 6$, rejection occurs; $\bar{T}_{\hat{\gamma}} = T_0$ for $\hat{\gamma} = 0.8$, $\hat{\alpha} = 0.20$, $T_{0.80} = 0.44T_0 < T_1$.
If $r = 7$, $\bar{T}_{\hat{\gamma}} = T_0$ for $\hat{\gamma} = 0.87$, $\hat{\alpha} = 0.13$, $\bar{T}_{0.87} = 0.36T_0$.
If $r = 8$, $\bar{T}_{\hat{\gamma}} = T_0$ for $\hat{\gamma} = 0.95$, $\alpha = 0.05$, $\bar{T}_{0.95} = 0.28T_0$.

Note that the maximum value for risks $\hat{\alpha}$ and $\hat{\beta}$ is 0.2, which occurs at $r = 5$ and $r = 6$. These are the results for which it is most difficult to distinguish between the T_0 and $T_0/2$ hypotheses. If the control value of $4T_0$ is planned in advance, and the ordinary plan of a single-stage control has to be chosen on the basis of this value (and the given ratio $T_0/T = 2$); according to Table 13.1 the plan should be just the same: $\alpha = \beta = 0.2$, $t_\Sigma = 4T_0$, and $r_* = 6$.

13.3 SEQUENTIAL CONTROL PLANS

In practice sequential control is often more effective than single-stage control. This method is especially good when probability of acceptance (or rejection) is large. But testing duration under sequential control has a random value. The required number of observations can be more than is sufficient for single-stage control. Sequential tests may be truncated, and this results in increased risks α and β. The more extreme the truncation, the greater the increased risk.

13.3.1 Operating Time Measure, Exponential Distribution

The planning (see Figure 13.3) consists of development of the acceptance and rejection areas in the grid with operating time, t_Σ / T_0, as one coordinate and the failure number, r, as the other coordinate.

The limits do not depend on the absolute values of T_0 and T_1 but on their ratio. With the limits prepared in advance, a stepwise plot of the number of failures and the total operating time can be developed. If r_{tr} is reached and $t_r < t_{tr}$, the item is rejected. If t_{tr} is reached and $r < r_{tr}$, the item is accepted. As indicated in Figure 13.3, $r = at/T_0 + r_0$ is the rejection area limit and $r = a(t/T_0 - t_0/T_0)$ is the acceptance area limit.

Values of the constants a, r_0, and t_0/T_0 are given in Table 13.3. The constants are calculated using

$$a = \frac{T_0/T_1 - 1}{\ln(t_0/T_1)},$$

$$r_0 = \frac{\ln[(1-\beta)/\alpha]}{\ln(T_0/T_1)},$$

$$\frac{t_0}{T_0} = -\frac{\ln[(\beta/(1-\alpha)]}{(T_0/T_1) - 1}.$$

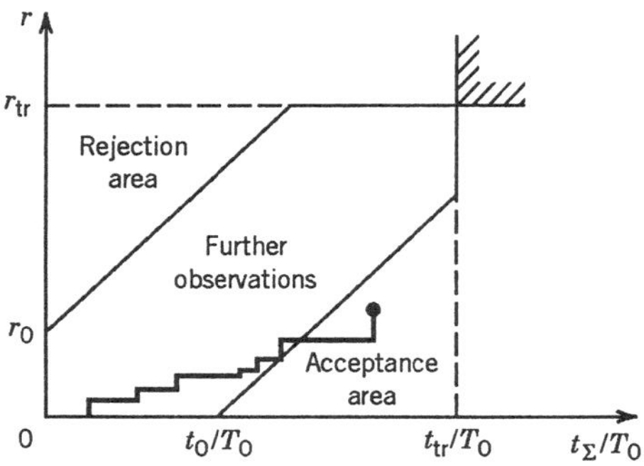

Figure 13.3 Acceptance, rejection, and further testing areas for sequential control.

Table 13.3 Values of test plan characteristics, operating time reliability measure, exponential distribution, sequential control

T_0/T_1	Test Plan a	r_0	t_0/T_0	$\bar{t}/T_0$
		$\alpha = \beta = 0.05$		
58.8	14.2	0.72	0.049	0.049
13.3	4.76	1.14	0.239	0.272
7.69	3.28	1.44	0.440	0.570
5.68	2.69	1.70	0.629	0.90
5.00	2.49	1.83	0.736	1.11
4.65	2.38	1.92	0.806	1.25
4.03	2.17	2.11	0.971	1.62
4.00	2.16	2.12	0.981	1.64
3.64	2.04	2.28	1.12	1.97
3.50	2.00	2.35	1.18	2.12
3.30	1.93	2.47	1.28	2.40
3.08	1.85	2.62	1.42	2.78
3.00	1.82	2.68	1.47	2.94
2.90	1.78	2.77	1.55	3.18
2.75	1.73	2.91	1.69	3.60
2.63	1.69	3.04	1.81	3.99
2.53	1.65	3.17	1.92	4.40
2.50	1.64	3.21	1.96	4.54
2.45	1.62	3.29	2.04	4.81
2.37	1.59	3.41	2.15	5.23
2.10	1.48	3.98	2.69	7.44
2.00	1.44	4.25	2.94	8.64
1.94	1.42	4.44	3.13	9.52
1.84	1.38	4.85	3.53	11.6
1.50	1.23	7.26	5.89	28.0
		$\alpha = \beta = 0.1$		
21.7	6.74	0.713	0.106	0.0995
7.30	3.17	1.11	0.349	0.408
5.00	2.49	1.37	0.549	0.735
4.83	2.43	1.40	0.574	0.780
4.00	2.16	1.59	0.732	1.09
3.83	2.11	1.64	0.776	1.18
3.50	2.00	1.75	0.879	1.41
3.29	1.92	1.85	0.960	1.60
3.00	1.82	2.00	1.10	1.95
2.94	1.80	2.04	1.13	2.04
2.70	1.71	2.21	1.29	2.48
2.53	1.65	2.38	1.44	2.94

Table 13.3 (*Continued*)

	Test Plan			
T_0/T_1	a	r_0	t_0/T_0	$\bar{t}/T_0$
		$\alpha = \beta = 0.1$		
2.50	1.64	2.40	1.46	3.01
2.39	1.60	2.52	1.58	3.38
2.28	1.55	2.66	1.71	3.84
2.19	1.52	2.81	1.84	4.31
2.12	1.49	2.93	1.97	4.78
2.06	1.47	3.05	2.08	5.24
2.00	1.44	3.16	2.19	5.69
1.95	1.42	3.28	2.31	6.20
1.79	1.36	3.77	2.77	8.42
1.67	1.31	4.27	3.27	11.1
1.60	1.28	4.66	3.65	13.4
1.50	1.23	5.42	4.39	18.6
		$\alpha = \beta = 0.2$		
7.25	3.15	0.70	0.222	0.195
5.00	2.49	0.712	0.347	0.348
4.00	2.16	1.00	0.462	0.515
3.64	2.04	1.07	0.526	0.618
3.50	2.00	1.11	0.555	0.667
3.00	1.82	1.26	0.693	0.923
2.79	1.74	1.35	0.777	1.09
2.50	1.64	1.51	0.924	1.42
2.40	1.60	1.58	0.987	1.58
2.17	1.51	1.79	1.18	2.09
2.02	1.45	1.97	1.35	2.61
2.00	1.44	2.00	1.39	2.71
1.92	1.43	2.13	1.51	3.11
1.84	1.38	2.28	1.66	3.65
1.77	1.35	2.43	1.80	4.18
1.72	1.33	2.56	1.93	4.70
1.68	1.31	2.69	2.05	5.23
1.64	1.29	2.81	2.18	5.79
1.61	1.28	2.93	2.29	6.31
1.58	1.27	3.04	2.40	6.85
1.55	1.26	3.15	2.51	7.86
1.50	1.23	3.42	2.77	8.80
1.46	1.22	3.66	3.47	10.2
1.40	1.19	4.13	3.48	13.2
1.36	1.17	4.48	3.83	15.7

Table 13.3 (*Continued*)

T_0/T_1	Test Plan a	r_0	t_0/T_0	$\bar{t}/T_0$
		$\alpha = \beta = 0.3$		
5.00	2.49	0.526	0.212	0.142
4.00	2.16	0.611	0.282	0.210
3.50	2.00	0.676	0.339	0.272
3.38	1.95	0.696	0.356	0.292
3.00	1.82	0.771	0.424	0.376
2.50	1.64	0.925	0.565	0.581
2.22	1.53	1.06	0.693	0.800
2.00	1.44	1.22	0.847	1.10
1.89	1.40	1.33	0.952	1.34
1.72	1.33	1.56	1.170	1.89
1.62	1.29	1.75	1.364	2.46
1.55	1.26	1.93	1.54	3.03
1.50	1.23	2.09	1.69	3.58
1.46	1.22	2.24	1.84	4.16
1.43	1.20	2.39	1.99	4.76
1.40	1.19	2.52	2.12	5.33
1.38	1.18	2.65	2.25	5.94
1.36	1.17	2.76	2.37	6.52
1.34	1.16	2.88	2.48	7.08
1.33	1.16	2.99	2.58	7.65
1.31	1.15	3.10	2.70	8.28
1.26	1.13	3.65	3.25	11.6
1.23	1.11	4.09	3.68	14.7
1.22	1.10	4.35	3.94	16.7

If $\alpha = \beta$, $r_0 = at_0/T_0$, and the equations of the lines are $r = at/T_0 + r_0$ and $at/T_0 - r_0$. Note that the slope of each line (the coefficient a) depends only on the T_0/T_1 ratio and not on the risks. The lower estimate of the mean control duration, $\bar{t}/T_0$, for a restorable item sample with $T = T_0$ is also given in Table 13.3. This value can be calculated from

$$\frac{\bar{t}}{T_0} = \frac{(1-\alpha)\ln[(1-\alpha/\beta)] - \alpha\ln[(1-\beta)/\alpha]}{(T_0/T_1) - 1 - \ln(T_0/T_1)} .$$

The chosen truncation level for sequential control should allow for both organizational and technical capabilities, on the one hand, and for increasing risks due to truncation on the other. The risk increments $\Delta\bar{\alpha}$ and $\Delta\bar{\beta}$ become smaller if the truncation is performed later (with regard to operating time and failure number). The upper estimates for $\Delta\bar{\alpha}$ and $\Delta\bar{\beta}$ are given by

$$\Delta\bar{\alpha} \le Pr\left\{ \frac{2[r_{tr} \ln(T_0/T_1) - \ln[(1-\beta)/\alpha]]}{(T_0/T_1) - 1} \le \chi^2(2r_{tr}) < \frac{2r_{tr} \ln(T_0/T_1)}{(T_0/t_1) - 1} \right\},$$

$$\Delta\bar{\beta} \le Pr\left\{ \frac{2[r_{tr}(T_0/T_1) \ln(T_0/T_1)}{(T_0/T_1) - 1} \le \chi^2(2r_{tr}) < \frac{2(T_0/T_1)[r_{tr} \ln(T_0/T_1) - \ln(\beta/1-\alpha)]}{(T_0/t_1) - 1} \right\}, \quad (13.19)$$

where $\chi^2(2t_{tr})$ is the χ^2 distribution with $2r_{tr}$ degrees of freedom.

Values of the increments $\Delta\alpha$ and $\Delta\beta$ are essentially overestimated and have a considerable spread for different control plan parameters. The values of α and β (in percentage of the rated values of α and β) are $30 - 110$ with $r_{tr} = r_*$, $10 - 50$ with $r_{tr} = 2r_*$, and $1 - 10$ with $r_{tr} = 3r$, where r_* is determined according to 13.2.1.

As the values are overestimated, truncation for most cases is recommended when $r_{tr} = (1.5 - 2.5)r_*$. A decision to accept, however, should not be made immediately upon crossing the acceptance zone limit. Observations should be continued at least through the next failure. Sometimes there is a return from the acceptance zone back into the further observations zone.

Table 13.4 Values of r_0'

	r_0' for $\alpha = \beta$ and equal to			
T_0/T_1	0.05	0.1	0.2	0.3
8.0	1.01	0.80	—	—
7.5	1.08	0.82	—	—
7.0	1.16	0.85	0.32	—
6.5	1.24	0.88	0.38	—
6.0	1.32	0.91	0.44	—
5.5	1.41	0.94	0.50	—
5.0	1.52	0.99	0.57	—
4.5	1.64	1.09	0.64	—
4.0	1.79	1.25	0.73	—
3.5	2.01	1.44	0.82	—
3.0	2.35	1.67	0.93	0.432
2.5	2.88	2.06	1.16	0.641
2.0	3.92	3.06	1.67	0.899
1.9	4.25	3.09	1.82	0.961
1.8	4.68	3.41	2.02	1.07
1.7	5.21	3.81	2.28	1.26
1.6	5.93	4.34	2.62	1.48
1.5	6.93	5.08	3.08	1.76

Sequential control under continuous time can be accomplished with a procedure to be followed when the line $r(t_\Sigma)$ crosses the acceptance zone limit. Since the method results in more effective use of the data on operating time of the controlled item samples, the mean control duration can be decreased somewhat. The acceptance zone remains unchanged, but the limit of the rejection zone moves down:

$$r = \frac{at}{T_0} + r_0' .$$

where values of r_0' can be found in Table 13.4.

13.3.2 Time-to-Failure Measures, Weibull-Gnedenko Distribution

The distribution of time to failure is

$$P\{t < y\} = F(y) = 1 - e^{-y^S/T} , \qquad (13.20)$$

where T and S are the distribution parameters. The value of the form factor, S, is assumed to be known. The procedure is similar to the one in Section 13.3.1. However, instead of total operating time, the important statistic is

$$\sum_{1 \le i \le r} \left(\frac{t_i}{T_0}\right)^S ,$$

where t_i is the ith operating time.

If items are restored, the restoration has to be complete back to the initial state. Acceptance and rejection zones are developed in the coordinates:

$$\sum_{1 \le i \le r} \left(\frac{t_i}{T_0}\right)^S , \quad r .$$

The acceptance area limit is given by

$$r = a \sum_{1 \le i \le r} \left(\frac{t_i}{T_0}\right)^S + r_0 .$$

The rejection area limit is given by

$$r = a\left(\sum_{1 \le i \le r} \left(\frac{t_i}{T_0}\right)^S - \frac{t_0}{T_0}\right) .$$

Coefficients a, t_0, and r_0 are

$$a = \frac{(T_0/T_1)^s - 1}{S \ln(t_0/T_1)} \,,$$

$$r_0 = \frac{\ln[(1-\beta)/\alpha]}{S \ln(T_0/T_1)} \,,$$

$$\frac{t_0}{T_0} = -\frac{\ln[(\beta/(1-\alpha)]}{(T_0/T_1)^s - 1} \,. \tag{13.21}$$

These coefficients can be determined from Table 13.3 with substitution of the figure $(T_0/T_1)_s$ for the value of ratio T_0/T_1.

The mean failure number sufficient for control termination (with $T = T_0$) is

$$\bar{r} \approx \frac{(1-\alpha)\ln[\beta/(1-\alpha)] + \alpha \ln[(1-\beta)/\alpha]}{1 + S\ln(T_0/T_1) - (t_0/T_1)^s} \,. \tag{13.22}$$

13.3.3 Time-to-Failure Measure, Truncated Normal Distribution

The distribution of time to failure is

$$Pr\{t < y\} = F(y) = 1 - \frac{\Phi[(T-y)/\sigma]}{\Phi(T/\sigma)} \,, \qquad y > 0 \,, \tag{13.23}$$

where $\Phi(z)$ is the standard normal distribution, T is the MTTF, and σ is a standard deviation. The value of σ is assumed to be known. A decision can be made at the moment of a failure. Control can be planned either with restoration of failed items or without restoration. When a failure occurs, the total operating time for all N samples, t_Σ, is recorded along with the total failure number r.

The limits of the acceptance and rejection areas are given by

$$\frac{\sigma^2}{T_1 - T_0} \ln\frac{1-\beta}{\alpha} + \frac{N(T_0+T_1)}{2} r + \frac{r\sigma^2}{T_0 - T_1} \times \ln\frac{\Phi(T_1\sqrt{N}/\sigma)}{\Phi(T_0\sqrt{N}/\sigma)} = t_\Sigma \,, \tag{13.24}$$

$$\frac{\sigma^2}{T_1 - T_0} \ln\frac{\beta}{1-\alpha} + \frac{N(T_0+T_1)}{2} r + \frac{r\sigma^2}{T_0 - T_1} \times \ln\frac{\Phi(T_1\sqrt{N}/\sigma)}{\Phi(T_0\sqrt{N}/\sigma)} = t_\Sigma \,, \tag{13.25}$$

The mean control duration with $T = T_0$ is

$$\bar{t} = \bar{r}\left[\frac{N(T_0 + T_1)}{2} + \frac{\sigma^2}{T_0 - T_1} \ln \frac{\Phi(T_0\sqrt{N}/\sigma)}{\Phi(T_1\sqrt{N}/\sigma)}\right] + \frac{\sigma^2}{T_0 - T_1} \ln \frac{\alpha}{1-\beta}, \tag{13.26}$$

where

$$\bar{r} = \frac{(1-\alpha)\{\ln[(1-\alpha)/\beta] - \ln[\alpha/(1-\beta)]\} + \ln[\alpha/(1-\beta)]}{N(T_0 - T_1)^2/2\sigma^2 - \ln\left[\dfrac{\Phi(T_1\sqrt{N}/\sigma)}{\Phi(T_0\sqrt{N}/\sigma)}\right]}.$$

13.3.4 Sequential Control, Probability Measures

If a reliability measure represents the probability of a certain event, A, control of the measure $Pr\{A\}$ can be planned in the form of a series of trials. After each trial a decision is made to accept, reject, or test further. The decision is based on the total number of experiments, N, and the observed failure number, r. A visual aid might be a plot similar to Figure 13.3 but with N and r coordinates (Figure 13.4). Control is terminated immediately when the stepwise function $r(N)$ crosses into either the acceptance area or the rejection area.

A distribution function of the r.v. that determines the reliability measures is not relevant because the estimate has a binomial distribution regardless of the r.v. distribution. Control can be realized with either a single sample or a number of samples as long as experiments are independent.

The equation of the rejection area limit is $r = aN + r_0$. The acceptance area limit is $r = a(N - N_0)$; r_0 and N_0 are the points where the lines cross the coordinate axes. Constants a, r_0, and N_0 are calculated from

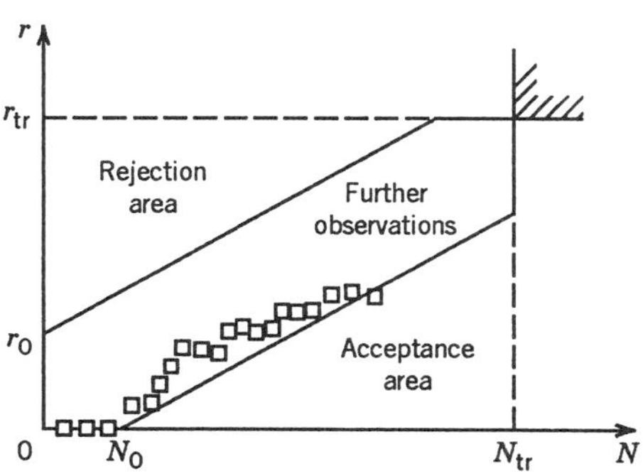

Figure 13.4 Acceptance, rejection, and further testing areas for sequential control and probability type of reliability measure.

$$a = \frac{\ln(P_0/P_1)}{\ln[(1-P_1)/(1-P_0)] + \ln(P_0/P_1)} ,$$

$$r_0 = \frac{\ln(1-\beta)/\alpha]}{\ln[(1-P_1)/(1-P_0)] + \ln(P_0/P_1)}$$

$$N_0 = \frac{\ln[(1-\beta)/\alpha]}{\ln(P_0/P_1)} .$$

If $\alpha = \beta$, $r_0 = aN_0$, and the equations of the lines assume the form $r = aN \pm r_0$.

The mean number of trials, N, with $P = P_0$, is given by

$$\bar{N} \approx \frac{(1-\alpha)\ln[(1-\alpha)/\beta] - \ln[(1-\beta)/\alpha]}{P_0 \ln(P_0/P_1) - (1-P_0)\ln[(1-P_1)/(1-P_0)]} .$$

Note that the slope of the lines depends exclusively on the values of P_0 and P_1 and is independent of the risks. Values of a, r_0, N_0, and $\bar{N}$ are presented in Table 13.5 for values of P_0, P_1, α and β that are most frequently encountered in practice. Operating characteristics may be found using

$$L(1-P) \approx \frac{1 - [(1-\beta)/\alpha]^{h(P)}}{[\beta/(1-\alpha)]^{h(P)} - [(1-\beta)/\alpha]^{h(P)}} ,$$

where the function $h(P)$ is determined by

$$\frac{[(1-P_1)/(1-P_0)]^h - 1}{[(1-P_1)/(1-P_0)]^h - (P_1/P_0)^h} = P .$$

Truncation of tests can be planned with the method in Section 13.2.2 using an r_{tr} that is $2r$ or $3r_*$.

13.3.5 Sequential Control in One-Dimensional Case

Consider a test that consists of independent observations in stages $1, 2, \ldots, m$. The result, x_m, is recorded for each observation. After the mth stage there exists the sample $x_1, x_2, \ldots, x_m$. The distribution density of the variable x_m—that is, the function $f(x_m) = Pr\{y = x_m \mid R\} = f(y_m, R)$—is assumed to be known as a function of the controlled reliability measure R. At each stage of control the following relationships are tested:

$$l = \frac{f(x_1, R_1) f(x_2, R_1) \cdots f(x_m, R_1)}{f(x_1, R_0) f(x_2, R_0) \cdots f(x_m, R_0)} \leq B , \qquad (13.27)$$

$$l \geq A , \qquad (13.28)$$

$$A > l > B , \qquad (13.29)$$

Table 13.5 Sequential control tests with probability type of reliability measure

Input Data		Test Plan			Average Volume
P_0	P_1	a	r_0	N_0	$\bar{N}$
		$\alpha = \beta = 0.05$			
0.999	0.998	0.00144	4.24	2940	8620
	0.997	0.00182	2.68	1470	2930
	0.996	0.00216	2.12	980	1640
	0.995	0.00249	1.82	730	1110
0.998	0.996	0.00289	4.24	1470	4300
	0.994	0.00364	2.67	733	1460
	0.992	0.00432	2.12	488	816
	0.990	0.00495	1.82	366	550
0.997	0.994	0.00433	4.23	977	2870
	0.991	0.00545	2.66	488	973
	0.988	0.00647	2.11	325	543
	0.985	0.00746	1.82	243	366
0.996	0.992	0.00577	4.22	732	2140
	0.988	0.00729	2.66	370	728
	0.984	0.00865	2.10	240	406
	0.980	0.00996	1.81	180	273
0.995	0.990	0.00722	4.22	584	1710
	0.985	0.00906	2.66	290	581
	0.980	0.0108	2.10	190	324
	0.975	0.01244	1.81	150	218
0.994	0.988	0.00866	4.21	490	1430
	0.982	0.0109	2.65	240	483
	0.976	0.0130	2.10	160	269
	0.970	0.0150	1.80	120	181
0.993	0.986	0.0101	4.21	420	1220
	0.979	0.0128	2.64	210	413
	0.972	0.0152	2.09	140	230
	0.965	0.0175	1.80	100	155
0.992	0.984	0.0115	4.20	370	1060
	0.976	0.0146	2.64	180	361
	0.968	0.0173	2.09	120	201
	0.960	0.0200	1.79	89.8	135
0.991	0.982	0.0129	4.19	320	945
	0.973	0.0164	2.64	160	320
	0.964	0.0196	2.08	110	178
	0.955	0.0225	1.79	79.6	119
0.990	0.980	0.0144	4.19	290	850
	0.970	0.0182	2.63	140	287
	0.960	0.0217	2.08	100	160
	0.950	0.0250	1.78	71.4	107

Table 13.5 (*Continued*)

Input Data		Test Plan			Average Volume
P_0	P_1	a	r_0	N_0	$\bar{N}$
		$\alpha = \beta = 0.05$			
0.980	0.960	0.0289	4.12	143	418
	0.940	0.0365	2.58	70.7	140
	0.920	0.0436	2.03	46.6	78
	0.900	0.0502	1.74	34.6	52
0.970	0.940	0.0434	4.06	93.7	274
	0.910	0.0550	2.53	46.1	91
	0.880	0.0656	1.98	30.2	50.1
	0.850	0.0758	1.69	22.3	33.2
0.960	0.920	0.0578	4.00	69.2	202
	0.880	0.0734	2.48	33.8	66.9
	0.840	0.0878	1.94	22.1	36.4
	0.800	0.1027	1.64	16.1	23.9
0.950	0.900	0.0723	3.94	54.5	159
	0.850	0.0919	2.43	20.4	52.2
	0.800	0.1100	1.89	17.1	28.2
	0.750	0.1280	1.60	12.5	18.4
0.940	0.880	0.0869	3.88	44.6	130
	0.820	0.111	2.38	21.6	42.4
	0.760	0.133	1.84	13.9	22.7
	0.700	0.155	1.55	10.0	14.7
0930	0.860	0.101	3.82	37.6	109
	0.790	0.129	2.33	18.0	35.4
	0.720	0.156	1.79	11.5	18.8
	0.650	0.182	1.53	8.22	12
0.920	0.840	0.116	3.76	32.4	93.8
	0.760	0.148	2.28	15.4	30.2
	0.680	0.179	1.74	9.74	15.8
	0.600	0.210	1.45	6.89	10.0
0.910	0.820	0.131	3.69	28.3	81.8
	0.730	0.167	2.23	13.4	26.1
	0.640	0.202	1.69	8.37	13.5
	0.550	0.238	1.39	5.85	8.5
0.900	0.800	0.145	3.63	25.0	72
	0.700	0. 6	2.18	11.7	23
	0.600	0.226	1.64	7.26	11.7
	0.500	0.268	1.34	5.01	7.20
0.890	0.780	0.160	3.57	22.3	64.4
	0.670	0.205	2.13	10.4	20.0
	0.560	0.250	1.59	6.36	10.2
	0.450	0.298	1.28	4.32	6.16

Table 13.5 (*Continued*)

Input Data		Test Plan			Average Volume
P_0	P_1	a	r_0	N_0	$\bar{N}$
		$\alpha = \beta = 0.05$			
0.880	0.760	0.175	3.51	20.1	57.8
	0.640	0.225	2.08	9.25	17.9
	0.520	0.275	1.54	5.60	8.93
	0.400	0.329	1.23	3.73	5.20
0.870	0.740	0.189	3.44	18.2	52.3
	0.610	0.244	2.03	8.29	16.0
	0.480	0.300	1.48	4.95	7.86
	0.350	0.361	1.17	3.23	4.55
0.860	0.720	0.204	3.40	16.6	47.5
	0.580	0.264	1.97	7.48	14.3
	0.440	0.326	1.43	4.39	6.93
	0.300	0.396	1.13	2.80	3.89
0.850	0.700	0.219	3.32	15.2	43.4
	0.550	0.284	1.92	6.76	12.9
	0.400	0.352	1.38	3.91	6.12
	0.250	0.432	1.04	2.41	3.32
		$\alpha = \beta = 0.1$			
0.999	0.998	0.00144	3.16	2190	5730
	0.997	0.00182	2.00	1100	1950
	0.996	0.00216	1.58	730	1090
	0.995	0.00249	1.36	550	734
0.998	0.996	0.00289	3.16	1110	2860
	0.994	0.00364	2.00	550	972
	0.992	0.00432	1.58	360	542
	0.990	0.00495	1.36	273	366
0.997	0.994	0.00433	3.16	730	1900
	0.991	0.00545	1.99	364	646
	0.988	0.00647	1.57	242	360
	0.985	0.00746	1.36	180	243
0.996	0.992	0.00577	3.15	546	1420
	0.988	0.00729	1.98	270	484
	0.984	0.00865	1.57	180	270
	0.980	0.00996	1.35	136	182
0.995	0.990	0.00722	3.15	440	1140
	0.985	0.00906	1.98	220	386
	0.980	0.108	1.57	145	215
	0.975	0.0124	1.35	108	145
0.994	0.988	0.00866	3.14	363	947
	0.982	0.0109	1.98	180	321
	0.976	0.0130	1.56	120	179
	0.970	0.0150	1.34	89.9	120

Table 13.5 (*Continued*)

Input Data		Test Plan			Average Volume
P_0	P_1	a	r_0	N_0	$\bar{N}$
		$\alpha = \beta = 0.1$			
0.993	0.986	0.0101	3.14	310	810
	0.979	0.0128	1.97	155	275
	0.972	0.0152	1.56	103	153
	0.965	0.0175	1.34	76.8	103
0.992	0.984	0.0115	3.13	270	708
	0.976	0.0146	1.97	135	240
	0.968	0.0174	1.56	89.7	133
	0.960	0.0200	1.34	67.2	89.6
0.991	0.982	0.0129	3.13	240	628
	0.973	0.0164	1.97	120	213
	0.964	0.0195	1.55	79.8	118
	0.955	0.0225	1.33	59.7	79.3
0.990	0.980	0.0144	3.12	216	564
	0.970	0.0182	1.96	108	191
	0.960	0.0217	1.55	71.6	106
	0.950	0.0250	1.33	53.3	71.2
0.980	0.960	0.0289	3.08	107	277
	0.940	0.0365	1.93	52.7	93.3
	0.920	0.0436	1.52	34.8	51.5
	0.900	0.0502	1.30	25.8	34.3
0.970	0.940	0.0434	3.03	69.9	182
	0.910	0.0550	1.89	34.4	60.7
	0.880	0.0656	1.48	22.6	33.3
	0.850	0.0758	1.26	16.6	22.1
0.960	0.920	0.0578	2.99	51.6	134
	0.880	0.0734	1.85	25.2	44.5
	0.840	0.0878	1.45	16.5	24.2
	0.800	0.1027	1.23	12.1	15.9
0.950	0.900	0.0723	2.94	40.7	105
	0.850	0.0919	1.82	19.7	34.7
	0.800	0.110	1.41	12.8	18.7
	0.750	0.128	1.19	9.33	12.2
0.940	0.880	0.0869	2.89	33.3	86.2
	0.820	0.111	1.78	16.1	28.1
	0.760	0.133	1.37	10.4	15.1
	0.700	0.155	1.15	7.46	9.75
0.930	0.860	0.101	2.85	28.1	72.6
	0.790	0.129	1.74	13.4	23.5
	0.720	0.156	1.34	8.58	12.5
	0.650	0.182	1.14	6.13	7.98

Table 13.5 (*Continued*)

Input Data		Test Plan			Average Volume
P_0	P_1	a	r_0	N_0	$\bar{N}$
		$\alpha = \beta = 0.1$			
0.920	0.840	0.116	2.80	24.2	62.2
	0.760	0.148	1.70	11.5	20.0
	0.680	0.179	1.30	7.24	10.5
	0.600	0.210	1.08	5.15	6.65
0.910	0.820	0.131	2.76	21.1	54.3
	0.730	0.167	1.67	9.97	17.3
	0.640	0.202	1.26	6.25	9.02
	0.550	0.238	1.04	4.37	5.64
0.900	0.800	0.145	2.71	18.6	47.8
	0.70	0.186	1.63	8.74	15.3
	0.600	0.226	1.23	5.42	7.76
	0.500	0.268	1.00	3.74	4.78
0.890	0.780	0.160	2.66	16.7	42.7
	0.670	0.205	1.59	7.74	13.0
	0.560	0.250	1.19	4.74	6.77
	0.450	0.298	0.96	3.22	4.09
0.880	0.760	0.175	2.62	15.0	38.3
	0.640	0.225	1.55	6.90	11.9
	0.520	0.275	1.15	4.18	5.92
	0.400	0.329	0.92	2.79	3.51
0.870	0.740	0.189	2.57	13.6	34.7
	0.610	0.244	1.51	6.19	10.6
	0.480	0.300	1.11	3.69	5.21
	0.350	0.361	0.87	2.41	3.02
0.860	0.720	0.204	2.52	12.4	31.5
	0.580	0.264	1.47	5.58	9.49
	0.440	0.326	1.07	3.28	4.60
	0.300	0.396	0.82	2.09	2.58
0.850	0.700	0.219	2.48	11.3	29.8
	0.550	0.284	1.43	5.04	8.56
	0.400	0.352	1.03	2.92	4.06
	0.250	0.432	0.78	1.80	2.20
		$\alpha = \beta = 0.2$			
0.999	0.998	0.00144	2.00	1380	2710
	0.997	0.00182	1.26	692	920
	0.996	0.00216	1.00	460	515
	0.995	0.00249	0.86	346	347
0.998	0.996	0.00289	1.99	691	1350
	0.994	0.00364	1.26	345	459
	0.992	0.00432	1.00	230	256
	0.990	0.00495	0.86	172	173

Table 13.5 (*Continued*)

Input Data		Test Plan			Average Volume
P_0	P_1	a	r_0	N_0	$\bar{N}$
		$\alpha = \beta = 0.2$			
0.997	0.994	0.00433	1.99	460	899
	0.991	0.00545	1.26	230	306
	0.988	0.00647	0.99	153	170
	0.985	0.00746	0.85	114	115
0.996	0.992	0.00577	1.99	344	673
	0.988	0.00729	1.25	172	229
	0.984	0.00865	0.99	114	127
	0.980	0.00996	0.85	85.7	85.8
0.995	0.990	0.00722	1.99	275	538
	0.985	0.00906	1.25	137	183
	0.980	0.0108	0.99	91.3	102
	0.975	0.0124	0.85	68.3	68.4
0.994	0.988	0.00866	1.98	230	447
	0.982	0.0109	1.25	114	152
	0.076	0.0130	0.99	75.8	84.5
	0.970	0.0150	0.85	56.5	56.8
0.993	0.986	0.0101	1.98	196	383
	0.979	0.0128	1.25	97.5	130
	0.972	0.0152	0.98	65.0	72.2
	0.965	0.0175	0.85	48.5	48.6
0.992	0.984	0.0115	1.98	171	334
	0.976	0.0146	1.24	85.2	113
	0.968	0.0174	0.98	56.5	63.0
	0.960	0.0199	0.84	42.4	42.3
0.991	0.982	0.0129	1.97	152	297
	0.973	0.0164	1.24	75.8	100
	0.964	0.0195	0.98	50.4	55.8
	0.955	0.0225	0.84	37.7	37.5
0.990	0.980	0.0144	1.97	137	267
	0.970	0.0182	1.23	67.8	90.2
	0.960	0.0217	0.98	45.2	50.1
	0.950	0.0250	0.84	33.4	33.6
0.980	0.960	0.0289	1.94	66.9	131
	0.940	0.0365	1.22	33.4	44.1
	0.920	0.0436	0.96	22.1	24.3
	0.900	0.0502	0.82	16.5	16.2
0.970	0.940	0.0434	1.91	44.3	86.0
	0.910	0.0550	1.19	21.7	28.7
	0.880	0.0656	0.93	14.1	15.7
	0.850	0.0758	0.80	10.4	10.4

Table 13.5 (*Continued*)

Input Data		Test Plan			Average Volume
P_0	P_1	a	r_0	N_0	$\bar{N}$
		$\alpha = \beta = 0.2$			
0.960	0.920	0.0578	1.88	32.6	63.4
	0.880	0.0734	1.17	15.9	21.0
	0.840	0.0878	0.91	10.4	11.4
	0.800	0.102	0.77	7.58	7.52
0.950	0.900	0.0723	1.85	25.7	49.8
	0.850	0.0919	1.15	12.4	16.4
	0.800	0.110	0.89	8.05	8.86
	0.750	0.128	0.75	5.39	5.77
0.940	0.880	0.0869	1.83	21.0	40.8
	0.820	0.111	1.12	10.2	13.3
	0.760	0.133	0.87	6.54	7.13
	0.700	0.155	0.73	4.71	4.61
0.930	0.860	0.100	1.80	17.7	34.3
	0.790	0.130	1.10	8.47	11.1
	0.720	0.156	0.84	5.41	5.90
	0.660	0.178	0.72	3.87	3.99
0.920	0.840	0.116	1.77	15.3	34.3
	0.760	0.148	1.07	7.25	8.20
	0.680	0.179	0.82	4.57	4.50
	0.600	0.210	0.68	3.25	2.91
0.910	0.820	0.131	1.74	13.3	25.7
	0.730	0.167	1.05	6.31	8.18
	0.640	0.202	0.80	3.94	4.26
	0.550	0.238	0.66	2.75	2.66
0.900	0.800	0.145	1.71	11.8	22.7
	0.700	0.186	1.03	5.31	7.15
	0.600	0.226	0.77	3.42	3.68
	0.500	0.268	0.63	2.36	2.26
0.890	0.780	0.160	1.68	10.5	20.2
	0.670	0.205	1.00	4.00	6.31
	0.560	0.250	0.75	2.99	3.20
	0.450	0.298	0.61	2.03	1.94
0.880	0.760	0.174	1.65	9.46	18.2
	0.640	0.225	0.98	4.36	5.61
	0.520	0.275	0.72	2.64	2.81
	0.400	0.329	0.58	1.76	1.66
0.870	0.740	0.189	1.62	8.57	16.4
	0.610	0.244	0.95	3.90	5.01
	0.480	0.300	0.70	2.33	2.47
	0.350	0.361	0.55	1.52	1.43

Table 13.5 (*Continued*)

Input Data		Test Plan			Average Volume
P_0	P_1	a	r_0	N_0	$\bar{N}$
		$\alpha = \beta = 0.2$			
0.860	0.720	0.204	1.59	7.82	14.9
	0.580	0.264	0.93	3.52	4.50
	0.440	0.326	0.67	2.07	2.18
	0.300	0.396	0.52	1.32	1.22
0.850	0.700	0.219	1.56	7.16	13.6
	0.550	0.284	0.90	3.18	4.05
	0.400	0.352	0.65	1.84	1.92
	0.250	0.432	0.49	1.13	1.04
		$\alpha = \beta = 0.3$			
0.999	0.998	0.00144	1.22	846	1103
	0.997	0.00182	0.77	423	375
	0.996	0.00216	0.61	282	209
	0.995	0.00249	0.52	211	141
0.998	0.996	0.00289	1.22	422	550
	0.994	0.00364	0.77	211	187
	0.992	0.00432	0.61	141	104
	0.990	0.00495	0.52	105	70.4
0.997	0.994	0.00433	1.22	281	366
	0.991	0.00545	0.72	140	124
	0.988	0.00647	0.61	93.4	69.4
	0.985	0.00746	0.52	70.0	46.8
0.996	0.992	0.00577	1.22	211	274
	0.988	0.00729	0.76	105	93.2
	0.984	0.00865	0.61	69.9	51.9
	0.980	0.00996	0.52	52.4	35.0
0.995	0.990	0.00722	1.21	168	219
	0.985	0.00906	0.76	83.7	74.3
	0.980	0.0108	0.60	55.8	41.4
	0.975	0.0125	0.52	41.7	27.9
0.994	0.988	0.00866	1.21	140	182
	0.982	0.0109	0.76	69.6	61.8
	0.976	0.0130	0.60	46.3	34.4
	0.970	0.0150	0.52	34.5	23.1
0.993	0.986	0.0101	1.21	120	156
	0.979	0.0128	0.76	59.6	52.9
	0.972	0.0152	0.60	39.7	29.4
	0.965	0.0175	0.52	29.6	19.8
0.992	0.984	0.0115	1.21	105	136
	0.976	0.0146	0.76	52.1	46.2
	0.968	0.0174	0.60	34.5	25.7
	0.960	0.0199	0.52	25.9	17.2

Table 13.5 (*Continued*)

Input Data		Test Plan			Average Volume
P_0	P_1	a	r_0	N_0	$\bar{N}$
		$\alpha = \beta = 0.3$			
0.991	0.982	0.0120	1.21	92.9	121
	0.973	0.0164	0.76	46.3	40.9
	0.964	0.0195	0.60	30.8	22.7
	0.995	0.0225	0.51	23.0	15.3
0.990	0.980	0.0144	1.20	83.5	109
	0.920	0.0182	0.76	41.4	36.8
	0.460	0.0217	0.60	27.6	20.4
	0.950	0.0250	0.51	20.4	13.7
0.980	0.960	0.0289	1.19	40.9	53.4
	0.940	0.0365	0.74	20.4	18.0
	0.920	0.0436	0.58	13.5	9.91
	0.900	0.0502	0.50	10.1	6.61
0.970	0.940	0.0434	1.17	27.0	35.0
	0.910	0.550	0.73	13.2	11.7
	0.880	0.0656	0.57	8.63	6.41
	0.850	0.0758	0.49	6.33	4.25
0.960	0.920	0.0578	1.15	19.9	25.8
	0.880	0.0734	0.71	9.73	8.56
	0.840	0.0878	0.56	6.36	4.66
	0.800	0.1027	0.47	4.63	3.06
0.950	0.900	0.0723	1.13	15.7	20.3
	0.850	0.0919	0.70	7.60	6.68
	0.810	0.110	0.54	4.93	3.61
	0.750	0.128	0.46	3.58	2.35
0.940	0.880	0.0869	1.12	12.8	16.6
	0.820	0.111	0.69	6.20	5.43
	0.760	0.133	0.53	3.99	2.91
	0.700	0.155	0.44	2.87	1.88
0.930	0.860	0.101	1.10	10.8	14.0
	0.790	0.129	0.67	5.19	4.53
	0.720	0.156	0.52	3.31	2.40
	0.660	0.178	0.44	2.47	1.63
0.920	0.840	0.116	1.08	9.31	14.0
	0.760	0.148	0.66	4.34	3.34
	0.680	0.179	0.50	2.80	1.83
	0.600	0.210	0.42	1.98	1.19
0.910	0.820	0.131	1.06	9.10	10.5
	0.730	0.167	0.64	4.05	3.33
	0.640	0.202	0.49	2.49	1.73
	0.550	0.238	0.40	1.68	1.08

Table 13.5 (*Continued*)

Input Data		Test Plan			Average Volume
P_0	P_1	a	r_0	N_0	$\bar{N}$
			$\alpha = \beta = 0.3$		
0.900	0.800	0.145	1.04	7.19	9.24
	0.700	0.186	0.63	3.37	2.91
	0.600	0.226	0.47	2.09	1.50
	0.500	0.268	0.39	1.44	0.921
0.890	0.780	0.160	1.03	6.42	8.23
	0.670	0.205	0.61	2.99	2.57
	0.560	0.250	0.46	1.83	1.30
	0.450	0.298	0.37	1.24	0.788
0.880	0.760	0.175	1.01	5.78	7.39
	0.640	0.225	0.60	2.66	2.28
	0.520	0.275	0.44	1.61	1.14
	0.400	0.329	0.35	1.07	0.677
0.870	0.740	0.189	0.99	5.24	6.69
	0.610	0.244	0.58	2.39	2.04
	0.480	0.300	0.43	1.42	1.01
	0.350	0.361	0.34	0.929	0.581
0.860	0.720	0.204	0.97	4.78	6.08
	0.580	0.264	0.57	2.15	1.83
	0.440	0.326	0.41	1.26	0.887
	0.300	0.396	0.32	0.806	0.498
0.850	0.700	0.219	0.95	4.37	5.55
	0.550	0.284	0.55	1.95	1.65
	0.400	0.352	0.40	1.13	0.783
	0.250	0.432	0.30	0.694	0.424

where A and B are constants and l is the likelihood ratio. Fulfillment of (13.27) results in item acceptance, (13.28) results in rejection, and (13.29) results in continuation.

As was shown by Wald, two inequalities hold:

$$B \geq \frac{\beta}{1-\alpha},$$

$$A \leq \frac{1-\beta}{\alpha}. \tag{13.30}$$

Control planning is reduced to development of acceptance and rejection areas in coordinates that are most convenient. In all cases the acceptance area limit as $l = B$ and the rejection area limit is $l = A$.

CHAPTER 14

SYSTEM CONFIDENCE LIMITS BASED ON UNIT TESTS

In many cases a system reliability measure has to be estimated on the basis of test results for separate system units. A variety of these are common situations. The system tested may not have the composition of the complete system. To reduce the cost of pilot samples of large systems, there may be a reduced composition, a smaller number of channels, or fewer redundancy devices. A similar problem arises in testing of items at different stages as they are developed. Last, the item may be designed in a number of versions differing in structure or composition; reliability may be required for all the versions, but testing can be performed for only one.

System reliability estimation may have to be based on tests of constituent parts. Tests of parts or subsystems may have to be performed in different places or at different times. Analysis may be required to predict the system MTTF from statistics for individual units.

A highly reliable redundant system may be tested as a system with few or even no failures. However, tests of individual units—whether basic or redundant—usually are preferred because they permit more accurate estimates of system reliability.

14.1 SYSTEM RELIABILITY MEASURES

System reliability measures thus often must be based on reliability measures of system components. For example, a series system failure-free operation probability, $P(t)$, can be a function of m measures, $p_1, \ldots, p_m$. Exact values of these measures may be unknown, but data on system unit reliability testing may be available. There might be a $[N\ W\ T]$ test plan, so that test duration and system failure-free operation time, T, coincide. If just failure numbers are recorded for the units, the tests are binomial. If d_i is the failure number for N_i units tested within the time interval T, the point estimate of failure-free operation probability for the ith unit is given by

$$\hat{p}_i = \frac{N_i - d_i}{N_i} . \tag{14.1}$$

Estimate (14.1) is both unbiased and effective. To find a point estimate for a system reliability measure, $P(T) = P(T, p_1, \ldots, P_m)$, we substitute $\hat{p}_i$ for each unknown value p_1. For a series system,

$$\hat{P}(T) = \prod_{1 \le i \le m} \frac{N_i - d_i}{N_i} . \tag{14.2}$$

Estimate (14.2) is an unbiased estimate of $P(T)$.

Often, particularly with highly reliable systems, interval estimates are more valuable than point estimates. An interval estimate has the form

$$Pr\{\underline{R} \le R \le \bar{R}\} \ge \gamma , \qquad 0 < \gamma < 1 , \tag{14.3}$$

where $\underline{R}$ is the lower confidence limit and $\bar{R}$ is the upper confidence limit.

14.2 SERIES SYSTEMS

14.2.1 Binomial Model

A series system composed of m independent units has a probability of failure-free operation for time t given by

$$R = \prod_{1 \le i \le m} (1 - q_i)^{n_i} , \tag{14.4}$$

where q_i is the failure probability within time t for unit i, $i = 1, \ldots, m$. The values of these q_i are unknown. Suppose that N_i units of type i are tested for time t and d_i failures are observed, $d_i \le N_i$; in other words, a plan $[N_i \ W \ t]$ is applied. The test results for different types of units are assumed to be independent.

The confidence limits would be developed from the test result vector $\mathbf{d} = (d_1, \ldots, d_m)$, with the specified confidence coefficient γ for the system reliability index (14.4). Note that for highly reliable systems

$$\sum_{1 \le i \le m} n_i q_i \ll 1 .$$

Equation (14.4) then has an approximation:

$$R \approx \exp\left(- \sum_{1 \le I \le m} n_i q_i \right) .$$

14.2.2 Exponential Model

An exponential model similar to the model in 14.2.1 often is useful. Each system unit i has an exponential distribution of failure-free operation time,

$$q_i(t) = 1 - e^{-\lambda_i t} ,$$

where λ_i is the unit failure rate parameter, $1 \le i \le m$. The system reliability index (14.4) takes the form

$$R = \exp\left(-t \sum_{1 \le I \le m} n_i \lambda_i\right) . \tag{14.5}$$

Test of ith type units was performed according to a $[N_i \; R \; T_i]$ plan. N_i units of type i were tested. Units that failed were restored at the moment of failure. Test duration was T_i; d_i failures were observed, $i = 1, \ldots, m$. Confidence limits have to be developed for system reliability index (14.4) on the basis of unit testing results $\mathbf{d} = (d_1, \ldots, d_m)$.

14.2.3 Failure-Free Testing

Sometimes $d_1 = d_2 = \cdots = d_m = 0$. Then the lower γ-level confidence limit for (14.4) for a binomial model is given by

$$\underline{R}_\gamma = (1 - \gamma)^{n_1/N_1} , \tag{14.6}$$

where

$$\frac{N_1}{n_1} = \min_{1 \le i \le m} \frac{N_i}{n_i}$$

is the smallest value for all the units (which is indicated by the subscript 1 here and below).

For an exponential model, the lower γ-level confidence limit for system reliability index (14.6) is determined with

$$\underline{R}_\gamma = (1 - \gamma)^{n_1/S_1} , \tag{14.7}$$

where

$$\frac{S_1}{n_1} = \min_{1 \le i \le m} \frac{S_i}{n_i} ,$$

and $S_i = N_i T_i$ is the total test time for units of type i.

14.2.4 Lloyd-Lipov Method

Usually $d_i \geq 0$, $i = 1, \ldots, m$. For series systems, the Lloyd-Lipov heuristic method is the most frequently used method when at least one failure occurs. This method yields the system statistic from the results for separate units.

To illustrate the method, let $m = 2$ and $n_1 = n_2 = 1$. Let the test magnitude satisfy the inequality $N_1 \leq N_2$. Test results are given by two sets, $\{\mathbf{x}_{1j}, j = 1, \ldots, N_1\}$ and $\{\mathbf{x}_{2j}, j = 1, \ldots, N_2\}$, where $\mathbf{x}_{ij} = 0, 1$ represents the failure indicator in tests of the jth unit of the ith type. For each test result from the first set (with smallest N_1) we establish a randomly chosen correspondence with a test result from the second set. Thus N_1 pairs are formed $(\mathbf{x}_{1j}, \mathbf{x}_{2k})$, $j = 1, \ldots, N_1$. Each of the pairs can be considered as a test result for a sample of the system. The expected number of pairs with at least one failed unit (i.e., the number of "failed" systems) is $N_1(1 - \hat{R})$, where

$$\hat{R} = \prod_{1 \leq i \leq 2} \left(1 - \frac{d_i}{N_i}\right).$$

This value, $D_1 = N_1(1 - \hat{R})$, is considered "equivalent" to a number of failures for the system for N_i tests, where

$$N_1 = \min_{i=1,2} N_i .$$

Confidence limits for the system reliability are then calculated as for a single unit tested N_1 times and failed D_1 times. When $m \geq 2$, $n_1 = n_2 = \cdots = n_m = 1$, the lower γ-level confidence limit, $\underline{R}_\gamma$, for the system reliability index is given by:

$$\underline{R}_\gamma = \underline{P}_\gamma(N_1, D_1) . \tag{14.8}$$

In (14.8), $\underline{P}_\gamma(N, d)$ is the lower γ-level Clopper-Pearson confidence limit for a failure-free operation probability of a single unit found on the basis of N trials with d failures, and N_1 is the smallest N_i. The unbiased point estimate for system reliability index R, $D_1 = N_1(1 - \hat{R})$, is

$$\hat{R} = \prod_{1 \leq i \leq m} \left(1 - \frac{d_i}{N_i}\right).$$

If D_1 is a fraction, any standard approximation, such as a piecewise-linear approximation, may be used. The upper confidence limit for R is calculated similarly.

A similar heuristic procedure is used for an exponential model to obtain the lower γ-level confidence limit for the system reliability index R when $n_1 = n_2 = \cdots = n_m = 1$:

$$\underline{R}_\gamma = e^{-\bar{\lambda}_\gamma(S_1, D_1)t} , \tag{14.9}$$

where

$$D_1 = \sum_{1 \le i \le m} \frac{S_1}{S_i} d_i \tag{14.10}$$

represents the "equivalent" failure number for the system relative to the first unit and $S_i = N_i T_i$ is the smallest test number for all i units.

The standard upper γ-level confidence limit for the parameter λ of a separate single unit on the basis of test results (S, d) is

$$S_1 = \min_{1 \le i \le m} S_i , \qquad \bar{\lambda}_\gamma(S, d) = \frac{\chi^2_\gamma(2d+2)}{2S} ,$$

where $\chi^2_\gamma(n)$ is the quantile at level γ for a χ^2 distribution with n degrees of freedom.

Formulas (14.8) and (14.9) can be rewritten

$$\underline{R}_\gamma = \min_{1 \le i \le n} \underline{P}_\gamma[N_i, N_i(1 - \hat{R})] , \tag{14.11}$$

$$\underline{R}_\gamma = \min_{1 \le i \le n} e^{-\bar{\lambda}_\gamma(S_i, S_i \hat{f})t} , \tag{14.12}$$

where

$$\hat{f} = \sum_{1 \le i \le m} \frac{d_i}{S_i}$$

is an unbiased estimate of

$$f = \sum_{1 \le i \le m} \lambda_i .$$

Expressions (14.11) and (14.12) indicate an interpretation of the Lloyd-Lipov method. For each unit, the lower confidence interval is found with an assumption that an "equivalent" failure number, D_i, was observed in the course of testing. The failure number is calculated in such a way that the point estimate of the unit reliability should coincide with the point reliability estimate for the system as a whole, that is, from the conditions $D_i/N_i = 1 - \hat{R}$ and $D_i/S_i = \hat{f}$ for binomial and exponential models, respectively. In both, the minimum unit limit is taken as the lower confidence interval limit for the system as a whole.

With this interpretation the method can be used also for series systems with $n_i \ge 1$, $i = 1, \ldots, m$ and for series-parallel systems (see Section 13.4.4). For series systems the lower confidence limit, $\underline{R}_\gamma$, in binomial and exponential models is given by:

$$\underline{R}_\gamma = \min_{1 \le i \le n} [\underline{P}_\gamma(N_i, D_i)]^{n_i} , \tag{14.13}$$

where D_i is found from

$$\left(1 - \frac{D_i}{N_i}\right)^{n_i} = \hat{R} , \quad \hat{R} = \prod_{1 \le i \le m} \left(1 - \frac{d_i}{N_i}\right)^{n_i} ,$$

and

$$\underline{R}_\gamma = \min_{1 \le i \le n} e^{-n_i \bar{\lambda}_\gamma(S_i, D_i) t} , \tag{14.14}$$

where

$$D_i = \sum_{1 \le i \le m} \frac{S_i n_j}{n_i S_j} d_j .$$

The Lloyd-Lipov method is very useful, but it has drawbacks. One drawback is insensitivity to unit test increases over the minimum value of N_1. As an example, there could be two situations for a system with $m = 2$ units, $n_1 = 1$, $n_2 = 1$. In the first situation, test results could be $N_1 = 10$, $d_1 = 0$; $N_2 = 20$, $d_2 = 2$. In the second situation, $N_1 = 10$, $d_1 = 0$; $N_2 = 100$, $d_2 = 10$. In both cases point reliability estimates for the system are the same:

$$\hat{R} = \left(1 - \frac{d_1}{N_1}\right)\left(1 - \frac{d_2}{N_2}\right) = 0.9 .$$

The confidence limits also are the same. However, it is clear that the second case has a better statistic because it involves more testing of the second unit.

14.2.5 Normal Approximation

The Lloyd-Lipov method can be improved. If there are enough tests and failures (when D_1 is approximately 20 or more), a normal approximation can be used to develop the system confidence limit. The approximation is based on the limit theorem for a sum of independent random variables. This section considers normal approximations for the exponential model. For a binomial model, an analyst can use the same formulas with $t = 1$, $S_i = N_i$, $i = 1, \ldots, m$. For the normal approximation the equation for the γ-level confidence limit corresponds to the Lloyd-Lipov equation for the reliability index of a series system:

$$\underline{R}_\gamma = \min_{1 \le i \le n} e^{-n_1 \bar{\lambda}_\gamma(S_1, D_1) t} , \tag{14.15}$$

where the equivalent number of failures is

$$D_1 = \sum_{1 \le i \le m} \frac{S_1 n_j}{n_1 S_j} d_j . \tag{14.16}$$

$$\frac{S_1}{n_1} = \min_{1 \le i \le m} \frac{S_j}{n_j} \quad \text{and} \quad \bar{\lambda}_\gamma(S, d)$$

give the standard upper γ-level confidence limit for the parameter λ of one unit. The normal approximation of λ is given by

$$\lambda - u_\gamma \sqrt{\frac{\lambda}{S}} = \frac{d}{S} , \tag{14.17}$$

where u_γ is the quantile at level γ for the standard normal distribution. The distribution and Laplace function are given by:

$$\Phi(u_\gamma) = \gamma ,$$

$$\Phi(u) = \frac{1}{\sqrt{2\pi}} \int_0^u e^{-z^2/2} \, dz .$$

Relationship (14.16) usually yields an optimistic (high) lower limit for system reliability as compared to the limit based on the exact distribution. An improvement can be made by ordering the units so that

$$\frac{S_1}{n_1} \le \frac{S_2}{n_2} \le \cdots \le \frac{S_m}{n_m}$$

with

$$\alpha_j = \frac{S_j n_1}{n_j S_1} \ge 1 , \qquad 1 \le j \le m .$$

The lower γ-level confidence limit for system reliability is found from

$$\underline{R}'_\gamma = e^{-n_1 \bar{\lambda}_\gamma (S_1, D'_1) t} , \tag{14.18}$$

where $\gamma > 0.5$.

$$D'_1 = \sum_{1 < j \le m} b_j \frac{S_1 n_j}{n_1 S_j} d_j ,$$

where $b_1 = 1$, and coefficients $b_2, \ldots b_m$ are found from:

$$b_j = \frac{\alpha_j \sqrt{A}}{\delta} \left(1 - \sqrt{1 - \frac{2\delta}{\alpha_j \sqrt{A}} \left(1 - \frac{\delta}{2\sqrt{A}} \right)} \right) , \qquad 1 \le j \le m ,$$

$$A - \delta\sqrt{A} = \sum_{1 \le j \le m} b_j n_j \frac{d_j}{S_j} , \tag{14.19}$$

where

$$\delta = \frac{u_\gamma}{\sqrt{S_1/n_1}} .$$

A from (14.19) and (14.18) are related by

$$\underline{R}'_\gamma = e^{-\lambda t} . \tag{14.20}$$

Equation (14.20) is more convenient for calculation than (14.19). In conformity with (14.19) $b_i = 1$ and $b_j \leq 1$, $j = 2, \ldots, m$. Thus the lower γ-level confidence limit given by (14.18) is under the limit given by whatever test results $\mathbf{d} = (d_1, d_2, \ldots, d_m)$ prove to be better than the results given by (14.15): $\underline{R}'_\gamma \geq \underline{R}_\gamma$.

The system of equations (14.19) has a unique solution. The numerical solution, and hence the calculation of the lower confidence limit, $\underline{R}'_\gamma$, is fairly simple. To illustrate the advantages of the limit $\underline{R}'_\gamma$ as compared to the Lloyd-Lipov limit, $\underline{R}_\gamma$, we consider the following numerical examples.

EXAMPLE 14.1

Let the number of unit types be $m = 2$. Let the number of units of the first type be $n_1 = 1$, and let the number of units of the second type be $n_2 = 1$. The tests for the first type, S_1, are 100. For the second type, $S_2 = 200$. A lower confidence limit has to be developed for the system failure-free operation probability for the time period $t = 1$. Let the confidence coefficient, γ, be 0.9. Table 14.1 gives the values of limits $\underline{R}_\gamma$ and $\underline{R}'_\gamma$ for various sets of test results (d_1, d_2) for units of the system.

EXAMPLE 14.2

Let $m = 2$, $n_1 = 1$, $n_2 = 1$, and $S_1 = 100$, $S_2 = 400$, $\gamma = 0.9$, and $t = 1$. Calculation results are in Table 14.2.

Table 14.1 Failures and confidence limits, Example 14.1

(d_1, d_2)	(1, 8)	(0, 14)	(2, 14)	(0, 10)	(1, 12)	(1, 16)	(3, 12)
$\underline{R}_\gamma$	0.915	0.893	0.871	0.915	0.893	0.871	0.871
$\underline{R}'_\gamma$	0.922	0.903	0.879	0.914	0.901	0.880	0.878

Table 14.2 Failures and confidence limits, Example 14.2

(d_1, d_2)	(2, 12)	(1, 24)	(0, 20)	(1, 16)	(2, 28)	(0, 28)	(3, 8)
$\underline{R}_\gamma$	0.915	0.893	0.915	0.915	0.871	0.893	0.915
$\underline{R}'_\gamma$	0.922	0.904	0.927	0.925	0.882	0.906	0.920

EXAMPLE 14.3

Let $m = 8$, and let $n_1 = n_2 = \cdots = n_8 = 1$. Let S_i and d_i for different types of units be as given in Table 14.3, and let $\gamma = 0.95$, $t = 1$. The lower γ confidence limits for system reliability are $\underline{R}_\gamma = 0.852$, $\underline{R}'_\gamma = 0.870$.

EXAMPLE 14.4

Let $m = 8$, $n_1 = n_2 = \cdots = n_s = 1$. Test and failures numbers for the units are given in Table 14.4. Here $\underline{R}_\gamma = 0.852$ and $\underline{R}'_\gamma = 0.868$.

14.2.6 Upper Confidence Limit, Normal Approximation

We now turn to the upper γ-level confidence limits, $\bar{R}_\gamma$ and $\bar{R}'_\gamma$, which are analogous to the lower limits, $\underline{R}_\gamma$ and $\underline{R}'_\gamma$, for the series system reliability indexes considered previously:

$$\bar{R}_\gamma = e^{-n_1 \lambda_\gamma(S_1, D_1)t}, \tag{14.21}$$

where $\underline{\lambda}_\gamma(S, d)$ is the lower γ-level confidence limit under a normal approximation for the parameter of a single unit calculated from

$$\lambda + u_\gamma \sqrt{\frac{\lambda}{S}} = \frac{d}{S}.$$

The value D_1 is again the "equivalent" number of failures found in conformity with (14.16).

The improved limit is given by the expression:

$$\bar{R}'_\gamma = e^{-n_1 \underline{\lambda}_\gamma(S_1, \tilde{D}_1)t}, \tag{14.22}$$

where $\gamma > 0.5$.

Table 14.3 Tests and failures, Example 14.3

i	1	2	3	4	5	6	7	8
S_i	100	200	300	400	500	600	700	800
d_i	1	1	2	3	4	5	6	32

Table 14.4 Tests and failures, Example 14.4

i	1	2	3	4	5	6	7	8
S_i	100	200	300	400	500	600	700	800
d_i	2	1	2	3	4	5	6	24

$$\tilde{D}_1 = \sum_{1 \le j \le m} b_j \frac{S_1 n_j}{n_1 S_j} d_j ,$$

where $b_1 = 1$, and coefficients $b_2, \ldots, n_m$ are found from the system of equations for $b_2, \ldots, b_m$, and A:

$$b_j = \frac{\alpha_j \sqrt{A}}{\delta} \left(1 + \sqrt{1 - \frac{2\delta}{\alpha_j \sqrt{A}} \left(1 + \frac{\delta}{2\sqrt{A}} - 1\right)}\right), \qquad 1 \le j \le m ,$$

$$A + \delta\sqrt{A} = \sum_{1 \le j \le m} b_j n_j \frac{d_j}{S_j} . \tag{14.23}$$

Consider the case where $\bar{R}^*_\gamma = \exp(-At)$, where A is given by (14.23). The coefficients satisfy the relations $b_1 = 1$, $b_j \ge 1$, $j = 2, \ldots, m$. Thus $\tilde{D}_1 > = D_1$, and limit (14.22) based on all test results $\mathbf{d} = (d_1, d_2, \ldots, d_m)$ is better than limit (14.21), $\bar{R}_\gamma \ge \bar{R}'_\gamma$.

14.3 CONFIDENCE LIMITS FOR SERIES-PARALLEL SYSTEMS

Let there be a series of m subsystems. The ith subsystem has n_i identical redundant units in parallel. The system failure-free operation probability for time t is

$$R = \prod_{1 \le i \le m} (1 - q_i^{n_i}) , \tag{14.24}$$

where $q_i = q_i(t)$ is the failure probability for a unit within the ith subsystem for the time t. We consider two basic testing models.

Binomial Test Model

N_i units of the ith type are tested for time t according to the plan $[N_i \ W \ t]$; as a result *di* failures are observed, $i = 1, \ldots, m$.

Exponential Test Model

The failure-free operation time for each unit of type i follows the exponential distribution $q_i(t) = 1 - \exp(-\lambda_i t)$. Units of type i are tested in conformity with the $[N_i \ R \ T_i]$ plan. As a result, d_i failures are observed, $i = 1, \ldots, m$. Testing time for the ith type unit is $S_i = N_i T_i$. The confidence limits for the system reliability index (14.24) are developed from the test results $\mathbf{d} = (d_1, d_2, \ldots, d_m)$.

14.3.1 Failure-Free Tests

Let $d_1 = d_2 = \cdots = d_m = 0$. Then the lower γ-level confidence limit, $\underline{R}_\gamma$, for (14.24) is calculated using

$$\underline{R}_\gamma = \min_{1 \le i \le m} [1 - (1 - (1 - \gamma)^{1/N_i})]^{n_i} . \tag{14.25}$$

For the exponential model the equation has the form

$$\underline{R}_\gamma = \min_{1 \le i \le m} [1 - (1 - (1 - \gamma)^{1/S_i})]^{n_i} . \tag{14.26}$$

The units within each of the subsystems are not necessarily identical; they can have different reliabilities. In this case the failure-free operation probability for the system is given by

$$R = \prod_{1 \le i \le m} \left(1 - \prod_{1 \le j \le n_i} q_{ij}\right) , \tag{14.27}$$

where $q_{ij} = q_{ij}(t)$ is the failure probability for time t for the jth unit of the ith subsystem.

When no failure has been observed, $d_{ij} = 0$, $j = 1, \ldots, n_i$; $i = 1, \ldots, m$, the lower γ-level confidence limit for the system reliability index (14.27) is given by

$$\underline{R}_\gamma = \min_{1 \le i \le m} \left(1 - \prod_{1 \le j \le n_i} \frac{v_i}{v_i + N_{ij}}\right) , \tag{14.28}$$

where the value of v_i is found from

$$\sum_{1 \le j \le n_i} N_{ij} \ln\left(1 + \frac{v_i}{N_{ij}}\right) = \ln \frac{1}{1 - \gamma} . \tag{14.29}$$

N_{ij} is the number of tests for the jth unit from the ith subsystem, $i = 1, \ldots, m$. For the exponential model, equations (14.28) and (14.29) remain valid upon substitution of N_{ij} for (S_{ij}/t).

When $m = 1$, (14.28) and (14.29) permit the expression of the lower confidence limit of the parallel system reliability index.

$$R = 1 - \prod_{1 \le j \le n} q_j$$

can be expressed in the form

$$R_\gamma = 1 - \prod_{1 \le j \le n} \frac{v}{v + N_j} , \tag{14.30}$$

where the value of v is found from

$$\sum_{1 \le j \le m} N_j \ln\left(1 + \frac{v}{N_j}\right) = \ln \frac{1}{1 - \gamma} , \tag{14.31}$$

where N_j = the number of tests for the jth unit.

Let a parallel system with n units be tested N times and no failure be observed. In this case $N_1 = N_2 = \cdots = N_n = N$, and (14.30) and (14.31) yield

$$\underline{R}_\gamma = 1 - [1 - (1-\gamma)^{1/nN}]^n . \tag{14.32}$$

Expressions for $\underline{R}_\gamma$ can not be improved for failure-free tests. Relationship (14.28) may be clearly interpreted. The best lower γ-level confidence limit for a series-parallel system reliability index in the case of failure-free testing can be found by calculation of the best γ-level confidence limits in the form of (14.30) for each separate parallel subsystem and then taking the smallest.

EXAMPLE 14.5

A system has three parallel subsystems in series. The numbers of units in the subsystems are $n_1 = 2$, $n_2 = 3$, and $n_3 = 3$. The system as a whole was tested $N = 6$ times, and no unit failure was observed. The lower γ-level confidence limits with the confidence coefficient $\gamma = 0.9$ for the three subsystems calculated with (14.32) are 0.970, 0.998, and 0.998. Applying (14.28), we find that the best lower γ-level confidence limit for system reliability index is 0.970.

EXAMPLE 14.6

In the preceding example all the units within each of the parallel subsystems are identical. The best lower γ-level confidence limit for the system reliability index can be calculated with (14.25), where $N_i = Nn_i$, since, in the course of N trials with n_i identical units with q_i, the total test number for the parameter q_i is Nn_i. The limit is the same as in the above example: $\underline{R}_\gamma = 0.970$. Thus, additional information on the identical redundant units within each subsystem does not improve the confidence limit with failure-free tests.

14.3.2 Bewler's Method

The following method for development of confidence limits was proposed for the reliability index $R = 1 - q_1 q_2$ of a parallel system of two units and a binomial testing model. Let d_1, d_2 be the observed failure numbers for the first and second units, respectively. Assume that a correspondence is established between any possible set (d_1, d_2) and that a number $K(d_1, d_2)$ is monotonically increasing (not decreasing) with respect to d_1 and d_2. For a given test result, (d_1^*, d_2^*), the lower γ-level confidence limit for the system reliability index is given by

$$\underline{R}_\gamma = \min(1 - q_1 q_2) .$$

The minimum is taken over the set of points $q = (q_1, q_2)$ that meets the restriction

$$\sum_{(d_1, d_2)} \prod_{1 \le i \le 2} \binom{N_i}{d_i} q_i^{d_i}(1 - q_i)^{N_i - d_i} \ge 1 - \gamma .$$

The summation is performed for the pairs (d_1, d_2) that satisfy the condition $K(d_1, d_2) \le K(d_1^*, d_2^*)$. In a similar way the upper confidence limit can be developed. Note that in the capacity of function $K(d_1, d_2)$ any monotonically increasing function of d_1, d_2 (not necessarily an integer one) can be used.

This approach also can be used in other more general cases. Let $R = R(q)$ be the function representing dependence of the system reliability measure upon the vector of unit parameters $\mathbf{q} = (q_1, \ldots, q_m)$. Let $\mathbf{d} = (d_1, \ldots, d_m)$ be the vector of test failures for the units; $K(\mathbf{d})$ is a function that increases monotonically with regard to each of the d_i. Then the lower γ-level confidence limit for R with a given vector of test results, $\mathbf{d}^* = (d_1^*, \ldots, d_m^*)$ is calculated from

$$\underline{R}_\gamma = \min R(q) , \tag{14.33}$$

where the minimum is taken with a restriction for q:

$$\sum_{K(d) \le K(d^*)} \prod_{1 \le i \le m} \binom{N_i}{d_i} q^{d_i}(1 - q_i)^{N_i - d_i} \ge 1 - \gamma . \tag{14.34}$$

This approach can be used similarly with exponential and other testing models.

Thus the confidence limit $\underline{R}_\gamma$ is defined by an ordering $K(\mathbf{d})$ of the test results $\mathbf{d}$. With a given ordering $K(\mathbf{d})$, expressions (14.33) and (14.34) are the best ones for the confidence limit. The basic question is the optimal ordering for $K(\mathbf{d})$. Calculation of the minimum (14.33) proves to be a rather complex problem with an increasing number of system units m and increasing failure numbers. The calculation is easy with small m and small failure numbers $d_1, d_2, \ldots, d_m$.

14.3.3 Method of Confidence Sets

Consider as an example an exponential testing model. Let $\boldsymbol{\lambda} = (\lambda_1, \ldots, \lambda_m)$ be the vector of the unit reliability parameters and $R(\lambda)$ be the function representing dependence of the system reliability index (14.24) upon the unit reliability parameters:

$$R(\boldsymbol{\lambda}) = \prod_{1 \le i \le m} [1 - (1 - e^{-\lambda_i t})^{n_i}] . \tag{14.35}$$

In the method of confidence sets, each test result, $\mathbf{d} = (d_1, \ldots, d_m)$, corresponds to a set, H, within the parameter space, $\boldsymbol{\lambda}$, for which

$$Pr\{\lambda \in H_d\} \geq \gamma \tag{14.36}$$

The totality of sets H_d satisfying inequality (14.36) is the totality of γ-level confidence sets. The lower γ-level confidence limit for the system reliability measure is calculated for a given vector of test results, $\mathbf{d}$, as the minimum value of the function $R(\boldsymbol{\lambda})$ over the confidence set:

$$\underline{R}_\gamma = \min_{\lambda \in H_d} R(\boldsymbol{\lambda}) . \tag{14.37}$$

The upper confidence limit is calculated in the same way as the maximum value of the function $R(\boldsymbol{\lambda})$ over the set $\mathbf{H}_d$.

Some of the most simple methods used in development of the confidence sets and the corresponding confidence limits for series-parallel systems follow.

Method of the "Rectangle"

Calculate the standard upper confidence limit for each parameter λ_i:

$$\bar{\lambda}_i = \frac{\chi^2_\gamma(2d_i + 2)}{2S_i} .$$

Then an m-dimensional "rectangle" of the form

$$H_d = \{\lambda : 0 \leq \lambda_i \leq \bar{\lambda}_i, 1 \leq i \leq m\} \tag{14.38}$$

comprises the confidence set with a confidence factor at least equal to γ^m. The general expression (14.37) gives the trivial confidence limit:

$$\underline{R}_\gamma = R(\bar{\lambda}_1, \bar{\lambda}_2, \ldots, \bar{\lambda}_m) = \prod_{1 \leq i \leq m} [1 - (1 - e^{-\bar{\lambda}_i t})^{n_i}] . \tag{14.39}$$

The confidence coefficient for limit (14.39) is at least γ^m. The essential drawback of this method is the abrupt decrease of the confidence factor γ^m with increasing m. This method in most cases underestimates system reliability.

Method of "Planes"

The number of failures, d_i, is a Poisson r.v. with the parameter $\Lambda_i = \lambda_i S_i$. Therefore the sum

$$\sum_{1 \leq i \leq m} d_i$$

is also a Poisson r.v. with the parameter

$$\sum_{1 \le i \le m} \Lambda_i \, .$$

Then the set in the form

$$H'_d = \left\{ \lambda : \sum_{1 \le i \le m} S_i \lambda_i \le \frac{1}{2} \chi^2_\gamma \left(2 \sum_{1 \le i \le m} d_i + 2 \right) \right\} \qquad (14.40)$$

is the γ-level confidence set. The limit of this set represents a "plane" in a space of m dimensions. The corresponding γ-level confidence limit for a series-parallel system index assumes the form

$$\underline{R}_\gamma = \min_{\lambda \in H'_d} R(\boldsymbol{\lambda}) \, . \qquad (14.41)$$

Expression (14.41) can be calculated easily since its minimum is realized in one of m "corner" points of its set:

$$\underline{R}_\gamma = \min_{1 \le i \le m} \left\{ 1 - \left[1 - \exp\left(\frac{\chi^2_\gamma (2 \sum_{1 \le j \le m} d_j + 2)}{2S_i} \right) \right]^{n_i} \right\} . \qquad (14.42)$$

Relationship (14.42) may be explained. First the lower γ-level confidence limit is calculated for each ith parallel subsystem under the assumption that the observed failure number for the ith type unit was equal to

$$\sum_{1 \le i \le m} d_i \, .$$

Then the smallest of these limits for separate subsystems is taken as a confidence limit for the reliability index of the system. This method works well enough when the system reliability index is almost equal to that of one of the subsystems and the other $m - 1$ subsystems are very unreliable. If, however, the other subsystems are of significant reliability, utilization of this method results in an underestimation system reliability.

Method of the "Truncated Rectangle"

The method of truncated rectangles is a combination of the two preceding methods. Let $\gamma = \gamma_0$ for (14.38), and let $\gamma = \gamma_1$ for (14.40). Make a new confidence set by crossing sets (14.38) and (14.40):

$$H''_d = H_d \cap H'_d = \left\{ \boldsymbol{\lambda} : 0 \le \lambda_i \le \frac{\chi^2_{\gamma_0}(2d_i + 2)}{S_i} , \qquad 1 \le i \le m , \right.$$

$$\left. \sum_{1 \le i \le m} S_i \lambda_i \le \frac{1}{2} \chi^2_{\gamma_1} \left(2 \sum_{1 \le j \le m} d_j + 2 \right) \right\} . \qquad (14.43)$$

The confidence coefficient for the sets in (14.43) satisfies the inequality

$\gamma \geq \gamma_0^m + \gamma_1 - 1$. Each set represents a "rectangle" that is truncated by a "plane" in m-dimensional space. For series-parallel systems, the reliability function minimum (14.35) within the set (14.43) is in one of the "corner" points and can be calculated easily with a computer. In some cases the confidence limit proves to be better than those given by the previous two methods.

14.3.4 Heuristic Lloyd-Lipov Method

The Lloyd-Lipov method considered earlier for series systems (the method of "equivalent" failures) can also be used for series-parallel systems. One must calculate the point estimate for system reliability, $\hat{R} = R(\hat{\lambda}_1, \hat{\lambda}_2, \ldots, \hat{\lambda}_m)$, where $\hat{\lambda}_i = d_i/S_i$ is the point estimate for parameter λ_i, $i = 1, \ldots, m$. One introduces for the ith parallel subsystem an "equivalent" failure number D_i from the condition $R(0, \ldots, 0_{i-1}, D_i/S_i, 0_{i+1}, \ldots, 0) = \hat{R}$; in other words, D_i is the failure number for type i units that corresponds to the point reliability estimate of the ith subsystem coinciding with the reliability estimate of the system as a whole. Then one calculates the lower confidence limit for the system reliability with

$$\underline{R} = \min_{1 \leq i \leq m} R\left(0_1, \ldots, 0_{i-1}, \frac{\chi_\gamma^2(2D_i + 2)}{2S_i}, 0_{i+1}, \ldots, 0_m\right).$$

Thus, at first we calculate the lower γ-level confidence limit for the reliability of each separate parallel ith subsystem under the assumption that D_i failures have been observed for type i units in the course of testing. Then the smallest of these limits for separate subsystems is taken as the lower confidence limit for the system. This method can be used similarly with binomial and other test models.

Use of this procedure for series-parallel systems is not, in general, substantiated with the exception of the normal distribution for the large number of failures. This method usually is useful just as an approximate one.

14.3.5 Optimal Algorithm

An algorithm for confidence limits is developed in two stages. A set of auxiliary hypothetical simple structures composed of units from the initial system under estimation is introduced in the first stage. Confidence limits are developed for each simple structure. The obtained set of limits for the set of simple structures generates a basic group of "trivial" confidence limit calculation algorithms for the initial complicated structure. Then operations are introduced on the set of basic algorithms that enable new algorithms to be developed. Then within this expanded class of algorithms the optimal one is found.

An example is algorithm development for an exponential test model and a series-parallel system. Let the lower γ-level confidence limit be required for the series-parallel system reliability measure (14.35). Function (14.35) can be represented by $R(\boldsymbol{\lambda}) = \exp[-f(\boldsymbol{\lambda})]$, where $f(\boldsymbol{\lambda})$ is a convex function with regard to $\boldsymbol{\lambda} = (\lambda_1, \ldots, \lambda_m)$:

$$f(\lambda) = -\sum_{1 \le i \le m} \ln[1 - (1 - e^{-\lambda_i t})^{n_i}] .$$

Thus the problem is reduced to development of an upper confidence limit for $f(\boldsymbol{\lambda})$.

Consider all possible series structures composed of units from the initial series-parallel system. Denote by $S^{(\mu)}$, where $\boldsymbol{\mu} = (\mu_1, \mu_2, \ldots, \mu_m)$, a series structure composed of μ_i units of the ith type. Then the failure-free operation probability for the series structure $S^{(\mu)}$ is given by

$$r^{(\mu)} = \exp\left(-t \sum_{1 \le i \le m} \mu_i \lambda_i\right) .$$

For each series structure $S^{(\mu)}$ develop the confidence limit using any method. For example, by the Lloyd-Lipov method (see 14.15),

$$\underline{r}_{\gamma}^{(\mu)}(\mathbf{d}) = \min_{1 \le i \le m} e^{-\mu_i \bar{\lambda}_{\gamma}(S_i, D_i)t} , \tag{14.44}$$

where $\mathbf{d} = (d_1, \ldots, d_m)$ is the vector of failures for different unit types, and

$$D_i = \sum_{1 \le j \le m} \frac{S_i \mu_j}{\mu_i S_j} d_j .$$

For given fixed μ the developed limit (14.45) gives the total set of all γ-level confidence sets within the (λ) parameter space:

$$H_d^{(\mu)} = \left\{\boldsymbol{\lambda} : \sum_{1 \le i \le m} \mu_i \lambda_i \le g_{\mu}(\mathbf{d})\right\} , \tag{14.45}$$

where

$$g_{\mu}(\mathbf{d}) = -\frac{1}{t} \ln \underline{r}_{\gamma}^{(\mu)}(\mathbf{d}) .$$

The γ-level confidence limit for the initial system reliability, $R(\lambda)$, can be developed in the form:

$$\underline{R}_{\gamma}^{(\mu)} = \min_{\lambda \in H_d^{(\mu)}} R(\lambda) , \tag{14.46}$$

where the minimum is taken with a given fixed $\boldsymbol{\mu}$ and a given fixed structure, $S^{(\mu)}$. Thus there is a (14.47) algorithm for each auxiliary series structure $S^{(\mu)}$.

For example, when μ_i is proportional to S_i, $i = 1, \ldots, m$, (14.47) coincides with the method of a "plane."

Algorithms of the (14.47) type are called "trivial." The totality of all such algorithms (with different $\boldsymbol{\mu}$) is called the basic set of trivial algorithms. The problem actually consists in determination of the optimal way of complete utilization of the information that is generated by the totality of the structures $S^{(\mu)}$. The problem can be reduced to the development of the algorithm that is optimal and is based on the set of "trivial" algorithms (14.47).

An optimal algorithm gives the following lower γ-level confidence limit for the initial system reliability index:

$$\underline{R}_\gamma = \exp[-\max_{\mu} \min_{\lambda \in \Gamma_\mu} f(\lambda)] , \qquad (14.47)$$

where Γ_μ is the limiting hyperplane for (14.45):

$$\Gamma_\mu = \left\{ \boldsymbol{\lambda} : \sum_{1 \le i \le m} \mu_i \lambda_i = g_\mu(\mathbf{d}) \right\} .$$

In particular, the confidence limit given by (14.47) for all test results d is better than any limit $\underline{R}_\gamma^{(\mu)}$ found on the basis of whatever "trivial" algorithm in the form (14.47): $\underline{R}_\gamma \ge R_\gamma^{(\mu)}$ for all $\boldsymbol{\mu} = (\mu_1, \ldots, \mu_m)$.

In the following examples, the confidence limit given by (14.47) is compared with those given by the other methods.

EXAMPLE 14.7

The number of different subsystems, $m = 10$. The numbers of redundant units, n_i, within the subsystems, test numbers, S_i, and failure numbers, d_i, for different unit types are presented in Table 14.5. Values of the lower confidence limits with $\gamma = 0.9$ calculated by different methods are presented in Table 14.6.

Table 14.5 Tests and failures, Example 14.7

i	1	2	3	4	5	6	7	8	9	10
n_i	4	4	4	2	5	5	3	3	3	2
S_i	100	100	150	500	500	400	200	400	500	500
d_i	2	1	1	3	4	8	3	5	1	1

Table 14.6 Comparison of lower confidence limits estimated with different methods, Example 14.7

Method of Lloyd-Lipov	Method of Planes	Method of the Rectangle	Method of the Truncated Rectangle	Optimal Algorithm (14.48)
0.9643	0.9671	0.9662	0.9990	0.9998

Table 14.7 Comparison of lower confidence limits estimated with different methods, Example 14.8

Method of Lloyd-Lipov	Method of Planes	Method of the Rectangle	Method of the Truncated Rectangle	Optimal Algorithm (14.47)
0.972	0.955	0.907	0.984	0.997

EXAMPLE 14.8

The number of subsystems, $m = 20$. The numbers of redundant units within the subsystems are $n_1 = n_2 = \cdots = n_{20} = 2$. The unit testing numbers are $S_1 = S_2 = \cdots = S_{20} = 100$. The failure numbers are $d_1 = d_2 = \cdots = d_5 = 0$; $d_6 = d_7 = \cdots = d_{20} = 1$. Values of the lower confidence limits with $\gamma = 0.9$ calculated by different methods appear in Table 14.7.

14.4 CONFIDENCE LIMITS FOR COMPLEX STRUCTURES

Consider a monotonic structure and exponential testing model. Let L_i be the set of unit indexes for the ith minimum cuts of the system, and let A_i be the event that consists of failure of all the units within the ith minimum cut L_i. The probability, Q, of system failure can be presented as:

$$Q = P\Big(\bigcup_{1 \le i \le M} A_i\Big), \tag{14.48}$$

where M is the number of minimum cuts within the system. With highly reliable units one may approximate:

$$q_j(t) = 1 - e^{-\lambda_j t} \approx \lambda_j t \ll 1 .$$

From (14.48) one can find the approximate relationship:

$$Q \approx \sum_{1 \le i \le M} P(A_i) = \sum_{1 \le i \le M} \prod_{j \in L_i} q_j \approx \sum_{1 \le i \le M} \prod_{j \in L_i} (\lambda_j t) . \tag{14.49}$$

In a series-parallel structure each unit is in just one cut. In complex structures a unit can be in several cuts. Therefore each λ_i can be in several products in (14.49). Also, different units can be known to be identical and have the same parameters. Function (14.49) takes the following more general form (upon a corresponding transformation of unit subscripts):

$$Q = Q(\boldsymbol{\lambda}) \approx \sum_{1 \le i \le M} \prod_{j \in L_i} (\lambda_i t)^{n_{ij}} , \tag{14.50}$$

where $\boldsymbol{\lambda} = (\lambda_1, \ldots, \lambda_m)$, n_{ij} = the number of identical units with parameter λ_j in the cut L_i, $i = 1, \ldots, M$; $j = 1, \ldots, m$; and m = the number of different types of units within the system. The validity of (14.50) is limited within the probability of simultaneous failure of two or more cuts. The right-hand side of (14.50) gives the upper estimate for the system failure probability, Q. Therefore development of the lower confidence limit for system reliability is reduced to that for the upper γ-level confidence limit for the probability of system failure (14.50). Different solution methods follow.

14.4.1 Method of "Rectangle"

Calculate for each of the separate parameters λ_i its upper γ-level confidence limit. Applying the confidence set in the form of (14.40) and (14.41), we find the following upper confidence limit for Q:

$$\bar{Q} = Q(\bar{\lambda}_1, \ldots, \bar{\lambda}_m) = \sum_{1 \le i \le m} \prod_{j \in L_i} (\bar{\lambda}_i t)^{n_{ij}} . \tag{14.51}$$

The confidence coefficient of the limit (14.51) is at least equal to γ^m, where γ is the confidence coefficient for the parameters of separate units. If there are enough system units m, the limit given by (14.51) is an underestimate.

14.4.2 Modified Method of "Rectangle" (Method of "Substitution")

The basic disadvantage of the rectangle method is a rapid decrease in the confidence coefficient, γ^m, with an increasing number of units, m. For $\gamma \ge 1 - \exp(-3/2) \approx 0.777$, and for functions of a general type (14.51), the upper confidence limit (14.51) is characterized by a confidence coefficient not less than γ. It can be used with preservation of the initial confidence coefficient, γ, for units of the system as a whole. With increasing m it yields a considerable gain over the "trivial" method.

14.4.3 Statistical Simulation (Fiducial) Method

One method is based on a fiducial approach proposed by Fisher. The test result distribution function for units of the jth type is

$$F_j(\lambda_j, d_j) = 1 - e^{-\lambda_j S_j} \sum_{0 \le 1 \le d_j} \frac{(\lambda_j S_j)^k}{k!} . \tag{14.52}$$

With given fixed values of (S_j, d_j), (14.52) has all the formal properties of a distribution function for λ_j. The distribution of the parameter λ_j is called a "fiducial distribution." The system reliability measure, $Q = Q(\lambda)$, considered as a function of fiducial random variables $\boldsymbol{\lambda} = (\lambda_1, \ldots, \lambda_m)$, then also becomes a fiducial random variable with a distribution function $\Phi(Q)$ that is generated by distributions of separate parameters of the (14.52) type. The upper γ-level fiducial limit, $\tilde{Q}$, for the system reliability index, $Q = Q(\lambda)$, is determined from the condition $\Phi(\tilde{Q}_\gamma) = \gamma$. In many cases $\tilde{Q}_\gamma$ can be found rather easily with a computer statistical simulation.

In the one-dimensional case, the γ-fiducial limit for one parameter, λ_j, coincides with the standard γ-confidence limit.

$$\bar{\lambda}_j = \frac{\chi^2_\gamma(2d_j + 2)}{2S_j}$$

However, in the course of estimation performed for the function $Q = Q(\boldsymbol{\lambda})$ of the multidimensional parameter $\boldsymbol{\lambda} = (\lambda_1, \lambda_2, \ldots, \lambda_m)$, the corresponding γ-level limit $\tilde{Q}_\gamma$ in general can differ [depending upon a type of the function $Q(\boldsymbol{\lambda})$] from the γ-level confidence limit for Q. In other words, the inequality $P_\lambda\{\tilde{Q}_\gamma \geq Q(\boldsymbol{\lambda})\} \geq \gamma$ (where $\boldsymbol{\lambda}$ represents a fixed vector parameter, and $\tilde{Q}_\gamma$ is a random value) can be wrong for some values of λ.

It can be shown that for a function of the general type (14.50) the lower γ-level fiducial limit $\tilde{Q}_\gamma$ found in conformity with the above method represents also the upper γ-confidence limit for Q:

$$\min_\lambda P_\lambda\{\tilde{Q}_\gamma \geq Q(\boldsymbol{\lambda})\}\gamma .$$

This means that, for reliability measures in the form of (14.50), a fiducial approach can be used for the upper estimate that corresponds to the approach for the lower estimate. This method yields more accurate confidence limits than the methods in Sections 14.4.1 and 14.4.2. However, the method requires more cumbersome computations. The limit in the form of (14.51) requires computation of $Q(\boldsymbol{\lambda})$ just at a single point $\bar{\boldsymbol{\lambda}} = (\bar{\lambda}_1, \ldots, \bar{\lambda}_m)$. At the same time the fiducial method requires multiple computations of $Q(\boldsymbol{\lambda})$ at different points $\boldsymbol{\lambda} = (\boldsymbol{\lambda}_1, \ldots, \lambda_m)$ formed in the course of statistical simulation.

14.4.4 Method of Majority Functions

A method similar to the method in Section 14.3.5 can be used. The estimated system reliability measure $Q = Q(\boldsymbol{\lambda})$ is represented as a maximum or a minimum over the (infinite in the general case) set of more simple majority functions whose confidence limits can be developed easily with other techniques. Then the confidence limit can be calculated for the estimated measure of system reliability Q. This method requires a rather complicated computer program (see Appendix F), but it is the most effective method for finding confidence limits.

14.5 CONFIDENCE LIMITS FOR AVAILABILITY COEFFICIENT

Denote by ξ and η the random time to failure and the random repair time, respectively, for an item for which the availability coefficient has to be estimated on the basis of test results:

$$K = \frac{T}{T + \tau} . \tag{14.53}$$

Here $T = \mathbf{M}\{\xi\}$ = the MTTF; and $\tau = \mathbf{M}\{\eta\}$ = the mean repair time. Both r.v.'s, ξ and η, are assumed to have an exponential distribution with parameters $\lambda = 1/T$ and $\mu = 1/\tau$, respectively.

The number of items under test n is known, as are the number of failed items, r, and their operating times, $\xi_1, \ldots, \xi_r$. The availability coefficient, K, is estimated after the rth failure. The number of items that failed and are subsequently repaired, m, is known, as are their repair times, $\eta_1, \ldots, \eta_d$, $d \le m$.

The lower ($\underline{K}$) and upper ($\bar{K}$) confidence limits have to be developed with corresponding confidence probabilities γ_1 and γ_2:

$$P(\underline{K} \le K) = \gamma_1 ,$$
$$P(\bar{K} \ge K) = \gamma_2 .$$

Usually $\gamma_1 = \gamma_2 = \gamma_0$. If the single-side limits $\underline{K}$ and $\bar{K}$ with levels γ_1 and γ_2, respectively, are found, the confidence interval $(\underline{K}, \bar{K})$ for the index K with level $\gamma = \gamma_1 + \gamma_2 - 1$ is also known. With $\gamma_1 = \gamma_2 = \gamma_0$, we have $\gamma = 2\gamma_0 - 1$. Hence, with the specified level γ, the calculation of the two-sided interval requires determination of the single-side limits $\underline{K}$ and $\bar{K}$ with level $\gamma_0 = (1 + \gamma)/2$.

Relation (14.53) can be rewritten:

$$K = \frac{1}{1 + \rho} , \tag{14.54}$$

where $\rho = \tau/T$. This leads to determination of the single-side confidence limits $\underline{\rho}$ (the lower one) and $\bar{\rho}$ (the upper one) with levels γ_1 and γ_2, respectively. Then the limits $\underline{K}$ with level γ_1 and $\bar{K}$ with level γ_2 will assume the forms:

$$\underline{K} = \frac{1}{1 + \bar{\rho}} , \tag{14.55}$$

$$\bar{K} = \frac{1}{1 + \underline{\rho}} , \tag{14.56}$$

To determine the limits ρ and $\bar{\rho}$ with levels γ_2 and γ_2, respectively, we have to calculate in advance the total operating time for n items according to:

$$S = \sum_{1 \le i \le r} \xi_i + (n - r)\xi_r , \tag{14.57}$$

provided that no repair of failed items took place. With repair and return to operation, we use:

$$S = n\xi_r . \tag{14.58}$$

Then the total restoration time is calculated:

$$V = \sum_{1 \le i \le d} \eta_i + (m - d)\eta_d ,$$

where d is the number of failed items from the total number, m, that have already been restored at the time at hand. Frequently $d = m$. With known values S, V, r, and d, the values of $\bar{\rho}$ and $\underline{\rho}$ are determined from:

$$\bar{\rho} = \frac{V}{S}\frac{r}{d} f_{\gamma_1}(2r, 2d) , \tag{14.59}$$

$$\underline{\rho} = \frac{V}{S}\frac{r}{d} f_{1-\gamma_1}(2r, 2d) , \tag{14.60}$$

where $f_{\gamma 1}(2r, 2d)$ and $f_{1-\gamma 2}(2r, 2d)$ are the quantiles of Fisher's F distribution with $2r$ and $2d$ degrees of freedom and levels γ_1 and $1 - \gamma_2$.

EXAMPLE 14.9

The times in hours are

$n = 5$	$m = 5$
$\xi_1 = 845$	$\eta_1 = 5$
$\xi_2 = 1035$	$\eta_2 = 8$
$\xi_3 = 1045$	$\eta_3 = 15$
$\xi_4 = 1160$	$\eta_4 = 60$
$\xi_5 = 1195$ $(r = n = 5)$	$\eta_5 = 65$ $(d = m = 5)$.

From (14.57) and (14.58) we find $S = 5280$ and $V = 153$. From the F distribution tables in Appendix E, $f_{0.9}(10, 10) = 2.32$. Using (14.59) and $\gamma_1 = 0.9$, we have

$$\bar{\rho}_{0.9} = \frac{30.5}{1056} 2.32 = 0.067 .$$

From (14.54) we find that

$$K = \frac{1}{1 + 0.067} = 0.937 . \tag{14.61}$$

EXAMPLE 14.10

Under the condition of the preceding example let $n = 25$ and $r = 5$, with the same values of ξ_i. The repair number, $m = 5$, and η^i are also the same. The two-sided confidence interval with level $\gamma = 0.8$ has to be determined for K.

From (14.57) and (14.58) we find that $S = 29{,}180$ and $V = 153$. We find first the single-side limits $\underline{K}$ and $\bar{K}$ at the level $\gamma_0 = 0.9$. The values from the tables are: $f_{\gamma 0}(2r, 2m) = f_{0.9}(10, 10) = 2.32$; $\bar{f}_{1-\gamma 0}(2r, 2m) = f_{0.1}(10, 10) = 0.43$. Note that $f_{1-\gamma}(a, b) = 1/f_\gamma(b, a)$; that is, $f_{0.1}(10, 10) = 1/f_{0.9}(10, 10) = 1/2.32 = 0.43$.

Therefore from (14.59) and (14.60), we find that

$$\bar{\rho}_{0.9} = \frac{154}{29{,}180}\,\frac{5}{5}\,2.32 = 0.012\,,$$

$$\underline{\rho}_{0.9} = \frac{153}{29{,}180}\,\frac{5}{5}\,0.43 = 0.002\,.$$

Hence $\underline{K} = 1/(1 + 0.12) = 0.988$ and $\bar{K} = 0.998$. The interval (0.988, 0.998) is the two-sided confidence interval for K at the level $\gamma = 0.8$.

14.6 CONFIDENCE LIMITS FOR SYSTEMS WITH REPAIRABLE UNITS

Fiducial probability may be used to find the confidence limits for the reliability index of a system with repairable units when only information about unit tests is available. Assume that a system includes units of m different types and that both the time to failure and the repair time are exponential. Denote by λ_i and μ_i the rates of failure and repair, respectively, of type i units. Confidence limits have to be developed for a system reliability index, R, on the basis of unit test results. Test results for the parameter λ_i are S_i, the total operating time, and d_i, the number of failures of units of type i. Test results for the parameter μ_i are similar: V_i is the sum of all restoration intervals, and k_i is the number of repairs of units of type i. Usually $d_i = k_i$. Test results for different units are assumed to be independent.

Let $R = R(\lambda_1, \ldots, \lambda_m;\ \mu_1, \ldots, \mu_m)$ represent the system reliability index dependence upon λ and μ. The procedure for obtaining confidence limits from fiducial probabilities is essentially the same as that of the preceding section. The only significant difference consists in simulation for both types of random parameters, λ_i and μ_i.

We let $z_i = \ln \lambda_i$, $y_i = \ln \mu_i$, $\mathbf{z} = (z_1, \ldots, z_m)$, $\mathbf{y} = (y_1, \ldots, y_m)$, and $\tilde{R}(\mathbf{z}, \mathbf{y}) = R(e^{z_1}, \ldots, e^{z_m}, e^{y_1}, \ldots, e^{y_m})$. The lower (upper) γ-level confidence limit for the index R can be developed on the basis of fiducial probabilities method provided that the function $\bar{R}(\mathbf{z}, \mathbf{y})$ is concave (convex) with regard to

$(\mathbf{z}, \mathbf{y})$. In this section we consider some of the widely used models of repairable systems composed of identical units.

14.6.1 Active Redundancy with Unlimited Repair

A system has $n+1$ identical active redundant units in parallel. The failure rate of each unit is λ. Any unit repair begins at the moment of a failure. The repair rate for each unit is μ. The availability coefficient for each unit is $\tilde{K} = \mu/(\lambda + \mu)$. The availability coefficient for the system is

$$K = 1 - (1 - \tilde{K})^{n+1} = 1 - \left(\frac{\lambda}{\lambda + \mu}\right)^{n+1} . \tag{14.62}$$

Estimation of K reduces to estimation of the value $\rho = \lambda/\mu$. The lower and upper confidence limits for ρ, and hence for the availability coefficient (14.63), can be developed using the fiducial probabilities method. The limits coincide with those obtained with the F-distribution for the ratio $(\lambda S)/(\mu V)$.

The MTTF of the system is

$$T = \frac{1}{(n+1)\mu}\left(\left(1 + \frac{\mu}{\lambda}\right)^{n+1} - 1\right) . \tag{14.63}$$

Function (14.63) is concave, so it can be used to find the upper γ-level confidence limit for T with the fiducial probabilities method. To determine the lower limit, the approximation

$$T = \frac{1}{n+1} \frac{\mu^n}{\lambda_{n+1}} \tag{14.64}$$

gives a good result when $\lambda \ll \mu$. Expression (14.64) also gives the lower estimate for T. The function $\ln(\mu^n/\lambda^{n+1}) = ny - (n+1)z$ is a linear function of z and y. Therefore the lower γ-level confidence limit for (14.64) can be found with the method of fiducial probabilities. For highly reliable units this lower limit for T is very effective.

Another approximate expression for highly reliable systems can be used for the operational availability index:

$$R(t_0) \approx Ke^{-t_0/T} \approx 1 - \left(\frac{\lambda}{\lambda + \mu}\right)^{n+1} - \frac{t_0}{T} \approx 1 - \left(\frac{\lambda}{\mu}\right)^{n+1} - \frac{n\lambda^{n+1}}{\mu^n} t_0 . \tag{14.65}$$

Relationship (14.65) is a convex function of z and y. Therefore the lower γ-level confidence limit can also be developed with the fiducial probabilities method.

$$T \approx \frac{1}{\Sigma_{1 \le i \le m} 1/T_i}, \tag{14.73}$$

where T_i is the MTTF of subsystem i. The lower estimation of T is reduced to the upper one of the value:

$$\frac{1}{T} \approx \sum_{1 \le i \le m} \frac{1}{T_i} \approx \sum_{1 \le i \le m} a_i' \left(\frac{\lambda_i}{\mu_i} \right)^{n_i} \lambda_i , \tag{14.74}$$

where

$$\alpha_i' = \frac{C_{n_i,i} \alpha_{n_i,i}}{C_{0,i}} .$$

Function (14.74) is convex with respect to z and y. The upper γ-level confidence limit can be developed with the fiducial probabilities method.

The system operating availability coefficient is given by

$$R(t_0) \approx Ke^{-t^0/T} \approx 1 - \sum_{1 \le i \le m} a_i \left(\frac{\lambda_i}{\mu_i} \right)^{n_i+1} - t_0 \sum_{1 \le i \le m} a_i' \left(\frac{\lambda_i}{\mu_i} \right)^{n_i} \lambda_i . \tag{14.75}$$

Function (14.75) is concave with respect to z and y. The lower γ-level confidence limit can be developed with application of the fiducial probabilities method.

Historically the methods for estimation of confidence limits for system reliability indexes have been suitable principally for systems without repair. For systems with repair, the substitution method is applicable when repair rates are known. Application of a normal approximation is accurate for large samples. For very reliable systems and small samples, the fiducial probabilities method is regarded as the most practical approach.

14.7 SEQUENTIAL CONFIDENCE LIMITS AND SYSTEM RELIABILITY CONTROL

14.7.1 Sequential Control for Systems with Redundancy

Information on system unit failures may be supplied as a sequence of data files. Let there by m parallel units. The failure-free operation probability for the time t is given by

$$R = R(q) = 1 - \prod_{1 \le i \le m} q_i , \tag{14.76}$$

where $q_i = q_i(t)$ is the probability of failure of unit i within time t, and $q = (q_1, q_2, \ldots, q_m)$.

System testing is a sequence of independent cycles of duration t each.

Within each cycle, failures of units and of the whole system are observed; the vector $\mathbf{X} = (x_1, \ldots, x_m, Y)$ is recorded, where x_i is the failure indicator for unit i and Y is the failure indicator for the system:

$$Y = \prod_{1 \le i \le m} x_i ,$$

where a failure indicator is 1 in case of failure and 0 otherwise. Thus a sequence of independent random vector variables

$$X_1, X_2, \ldots, X_n, \ldots , \tag{14.77}$$

with similar distributions has to be observed. $\mathbf{X}_n = (x_{1n}, \ldots, x_{mn}, Y_n)$, where x_{in} is the failure indicator for unit i of cycle n and Y_n is the system failure indicator for the nth cycle. With specified risks α and β (see Chapter 13), one of two decisions will be sought with regard to the unknown system reliability index, (14.76):

$$R \ge R_1 ,$$

or

$$R \le R_2 , \tag{14.78}$$

where R_1 and R_2, $R_1 > R_2$, are the specified critical levels of acceptance and rejection, respectively, as discussed in Chapter 13.

This problem can be solved exclusively on the basis of information about the system failures:

$$Y_1, Y_2, \ldots, Y_n, \ldots . \tag{14.79}$$

In this case the solution can be developed under a Wald sequential control plan (see Chapter 13). However, a sequential control plan with complete information on system and unit failures is preferable. Let $d_{in} = x_{i1} + \cdots + x_{in}$ be the vector of failures for unit i at each cycle (step) of testing;

$$\hat{q}_i(n) = \frac{d_{in} + K_i q_i^*}{n + K_i}$$

gives the estimate for q_i on the nth step, where K_i and q_i^* are plan parameters, q_i^* is a prior estimate of q_i, $K_i > 0$, $0 < q_i^* < 1$, $i = 1, \ldots, m$.

$$A_n = \sum_{1 \le i \le m} \sum_{1 \le r \le n} \{x_{ir} \ln \hat{q}_i(r-1) + (1 - x_{ir}) \ln[1 - \hat{q}_i(r-1)]\} ,$$

$$L_n(q) = \sum_{1 \le i \le m} [d_{in} \ln q_i + (n - d_{in} \ln(1 - q_i)] ,$$

$$F_n(R) = \max\{L_n(q) \mid R(q) \ge R\} , \tag{14.80}$$

$$F_n'(R) = \max\{L_n(q) \mid R(q) \leq R\} . \tag{14.81}$$

A sequential rule of decision making with the specified risks α and β is developed. After each test cycle, a decision to stop or continue is based on the inequalities

$$A_n - F_n(R_1) < \ln \frac{1}{\alpha} , \tag{14.82}$$

$$A_n - F_n'(R_2) < \ln \frac{1}{\beta} , \tag{14.83}$$

where the subscript n is the test cycle number.

Testing goes on until a stage, N^*, has one of the two inequalities violated. The item i then is accepted or rejected depending on which of the two levels is crossed. Each stage requires calculation of the maximums under (14.80) and (14.81). This control plan is asymptotically optimal with regard to the mean testing extent, $E\{N^*\}$, when $\alpha, \beta \rightarrow 0$.

14.7.2 Gain from Data on Units

There is a gain when unit failures are included. When $R(q) = R_1$, the gain coefficient is as follows: $k = E\{N_w\}/E\{N^*\}$, where $E\{N_w\}$ stands for the mean testing extent under the Wald plan in Section 14.7.1. The gain coefficient is approximated by

$$K \approx \frac{R_1/(1 - R_1)}{\Sigma_{1 \leq i \leq m} (1 - q_i)/q_i} \geq 1 .$$

The maximum value of k is realized when units are equal in reliability: $q_1 = q_2 = \cdots = q_m$. In this case

$$K \approx \frac{1}{m(1 - R_1)^{(m-1)/m}} .$$

The gain becomes larger when system reliability, R_1, increases. A numerical example follows.

EXAMPLE 14.10

Let the system have $m = 2$ units. The critical levels are $R_1 = 0.99$ and $R_2 = 0.95$. The risks are $\alpha = \beta = 0.05$. The mean test extent for the Wald plan with $R = R_1$ is $E\{N_w\} = 107$ cycles of system testing. With information on unit failures, the expected value of $E\{N^*\} = 32$. The value of $E\{N^*\}$ is found with the Monte Carlo simulation using units with equal reliability $q_1 = q_2 = \cdots = qm$ and $R(q) = R_1$.

The decision-making rule (14.78) can also be developed with the single-stage control plan when the testing extent N is fixed. The decision-making rule also can be developed either directly on the basis of system failures (14.79) or with information on unit failures. The required number of cycles for Example 14.10 is presented in Table 14.8 for all four possible test plans. There is no gain with a single-stage control, but there is a considerable gain with a sequential plan that takes into account the data on units.

14.7.3 Sequential Confidence Limits for System

The sequence $R_\gamma(n) = R_\gamma(n, X_1, \ldots, X_n)$, $n = 1, 2, \ldots$, is the sequence for the lower γ confidence limit for the system reliability index $R = R(q)$ if, for any q and for any ("independent of the future") random moment N, the following inequality holds:

$$Pr\{R_\gamma(N) \leq R(q)\} \geq \gamma .$$

In an analogous way the sequential upper confidence limit and the sequential confidence interval are determined. The sequential γ-level confidence limit for the system reliability index, R, is calculated on the nth testing step with

$$\underline{R}_\gamma(n) = \max_{1 \leq k \leq n} R_*(k) . \tag{14.84}$$

The value of $R^*(N)$ for the nth stage is given by

$$F_n'(R) = A_n - \ln \frac{1}{1-\gamma}$$

We calculate the sequential upper γ-level confidence limit for R analogously:

$$\bar{R}_\gamma(n) = \min_{1 \leq i \leq n} R^*(k) \tag{14.85}$$

where $R^*(n)$ is the solution of

Table 14.8 System failures versus element failures as a decision basis, Example 14.10

	Single-Stage Control	Sequential Control
Decision making on the basis of system failures	173	107
Decision making on the basis of data on element failures	173	32

$$F_n(R) = A_n - \ln \frac{1}{1-\gamma} .$$

The interval $[\underline{R}_{1-\beta}(n), \bar{R}_{1-\alpha}(n)]$ represents the sequential $(1-\alpha-\beta)$-confidence interval for $R = R(q)$. The lower bound, $\underline{R}_{1-\beta}(n)$, is monotonically nondecreasing, while the upper one, $\bar{R}_{1-\alpha}(n)$, is monotonically nonincreasing with n. Therefore the sequential confidence limit monotonically narrows with time. The sequential confidence limits (14.84) and (14.85) are asymptotically optimal with regard to the mean time to achieve a given accuracy as $\gamma \rightarrow 1$.

The sequential rule of decision making (the control plan in Section 14.8.1) can be specified in terms of sequential confidence limits (14.84) and (14.85). The area of testing continuation (14.82) and (14.83) is given by

$$\underline{R}_{1-\beta}(n) < R_2 , \tag{14.86}$$

$$\bar{R}_{1-\alpha}(n) < R_1 . \tag{14.87}$$

The control continues to the $(n+1)$th step if at the nth step the confidence interval $[\underline{R}_{1-\beta}(n), \bar{R}_{1-\alpha}(n)]$ includes both critical levels R_1 and R_2. The testing is terminated upon the first violation of the two presented inequalities. If (14.86) is violated, the decision $R \geq R_1$ is made. If (14.87) is violated, the decision $R \leq R_2$ is made.

Control plan specification in the form of inequalities (14.86) and (14.87) is preferable, because with the decision-making rule it gives additional information on the system reliability index, R, in the form of sequential confidence limits.

If no unit failures have been observed in the initial n steps, the sequential lower γ-level confidence limit for the system reliability index can be found explicitly:

$$\underline{R}_\gamma(n) = 1 - (1 - e^{-a/mn})^m , \tag{14.88}$$

where $a = \ln[1/(1-\gamma)]$. Expression (14.88) demonstrates the gain of using unit failure data. This gain increases with the system redundancy, m.

Similar sequential plans of control and sequential confidence limits can be developed for other types of structures and test plans, both with and without repair.

CHAPTER 15

ACCEPTANCE-REJECTION TEST PLANS FOR COMPLEX SYSTEMS

The techniques of Chapters 13 and 14 are extended in this chapter to determination of the most efficient tests and field test plans for two-level control of reliability of complex systems.

15.1 REDUNDANT UNREPAIRABLE SYSTEMS

An unrepairable system consists of n identical parallel connected redundant units (active redundancy). The system fails if all of its units fail. When unit failures are independent, the system's failure probability within time t is

$$R = q^n , \tag{15.1}$$

where q represents failure probability within t for a single unit.

Let testing of units be performed under a $[N \; W \; t]$ plan. As a result, d failures are recorded under a binomial testing model. N_c identical systems are tested. The total unit tests, N, equal $N_c n$. One of two hypotheses has to be accepted: $H_0 : R \leq R_0$ or $H_1 : R \geq R_1$, where $R_0 < R_1$ are the specified levels of acceptance and rejection for the reliability index R.

Testing under these hypotheses is equivalent to tests with the reliability index q for a single unit: $H_0 : q < q_0$ and $H_1 : q \geq q_1$, where the levels are

$$q_0 = {}^{n}\sqrt{R_0} ,$$

$$q_1 = {}^{n}\sqrt{R_1} .$$

It is a standard problem in testing hypotheses on the binomial distribution parameter within Bernoulli trials. If the observed failures $d \le C$, the hypothesis H_0 is accepted (acceptance). If $d > C$, the hypothesis H_1 is accepted (rejection). The test extent (number of tests) and constant C are chosen for given risks α and β discussed in Chapter 13. Let

$$F(q, d) = \sum_{0 \le k \le d} \binom{N}{k} q^k (1-q)^{N-k} .$$

The operating characteristic has the form $L(q) = F(q, C)$. The risk equations can be written

$$\begin{aligned} 1 - F(q, C) &\le \alpha \qquad \text{for } q \le q_0 , \\ F(q, C) &\le \beta \qquad \text{for } q \ge q_1 . \end{aligned} \tag{15.2}$$

The function $f(q, C)$ decreases monotonically with respect to q. Therefore inequalities (15.2) are equivalent to the following two inequalities:

$$\begin{aligned} 1 - F(q_0, C) &\le \alpha , \\ F(q_1, C) &\le \beta . \end{aligned} \tag{15.3}$$

The test extent and the constant C are chosen to satisfy (15.3). For small values of N, standard binomial distribution tables are used. For values of N larger than 20–25, it is convenient to use a normal approximation. The mean value and the variance of the random value d are equal, respectively, to Nq and $Nq(1-q)$. The distribution function for d is approximated with:

$$F(q, d) \approx \Phi\left(\frac{d - Nq}{\sqrt{Nq(1-q)}} \right) , \tag{15.4}$$

where $\Phi(\;)$ is a standard normal distribution function.

When employing the normal approximation, the minimal necessary testing extent, N^*, that provides for the required risks, α and β, with the specified levels q_0 and q_1 can be written in explicit form. Let U_α be the $(1-\alpha)$ quantile for the normal d.f. $\Phi(U_\alpha) = 1 - \alpha$. Inequalities (15.3), with allowance for normal approximation (15.4), can be written

$$\begin{aligned} Nq_0 + U_\alpha \sqrt{Nq_0(1-q_0)} &\le C , \\ Nq_1 - U_\beta \sqrt{Nq_1(1-q_1)} &\ge C . \end{aligned} \tag{15.5}$$

The specified values of risks α and β can be ensured with testing extent N provided that both inequalities (15.5) are fulfilled for a given value of C. The following inequality has to be satisfied:

$$Nq_0 + U_\alpha\sqrt{Nq_0(1-q_0)} \le Nq_1 - U_\beta\sqrt{Nq_1(1-q_1)}\,. \tag{15.6}$$

Then any value in the interval

$$Nq_0 + U_\alpha\sqrt{Nq_0(1-q_0)} \le C \le Nq_1 - U_\beta\sqrt{Nq_1(1-q_1)} \tag{15.7}$$

can be taken in the capacity of the constant C specifying the decision rule. If inequality (15.7) is strict, the values of the risks are less than α and β. The minimal testing extent N^* (with α and β given) is determined with:

$$N^*q_0 + U_\alpha\sqrt{N^*q_0(1-q_0)} = N^*q_1 - U_\beta\sqrt{N^*q_1(1-q_1)}\,. \tag{15.8}$$

It follows that

$$N^* = \frac{[U_\alpha\sqrt{q_0(1-q_0)} + U_\beta\sqrt{q_1(1-q_1)}]^2}{(q_1-q_0)^2}\,. \tag{15.9}$$

The constant C is determined unambiguously; it coincides with the left- and right-hand sides of expression (15.8).

15.2 GAIN FROM INFORMATION ON SYSTEM UNITS

In conformity with Section 15.1, consider the case when N_c identical samples of the system are tested within time t, or a single sample is tested N_c times with complete restoration of all failed units before each test. The decision-making rule can be developed with or without tests of separate units. When just failures of the system as a whole are registered, (15.7) may be applied. The necessary number of system tests is determined with

$$N_c^* = \frac{[U_\alpha\sqrt{R_0(1-R_0)} + U_b\sqrt{R_1(1-R_1)}]^2}{(R_1-R_0)^2}\,. \tag{15.10}$$

With unit tests, the number of equivalent "system" tests needed is N^*/n. The coefficient $k = N_c^*/(N/n)$ shows the increase in testing extent in the second method. Let $\delta = (R_1 - R_0)/R_0$. For small δ from expressions (15.7) and (15.10), with due allowance for the relationships between $R_0 = q_0^n$ and $R_1 = q_1^n$, we can approximate for R_0 and R_1 much less than 1:

$$k \approx \frac{1 + a + a^2 + \cdots + a^{n-1}}{n}\left(1 - \frac{n-1}{n}\,\frac{U_\alpha}{U_\alpha + U_\beta}\,\delta\right), \tag{15.11}$$

where $a = \bar{q}_0^{\,-1} = (1/R_0)^{1/n}$. Coefficient (15.11) can serve as a measure of testing extent gain with information on system units. The higher the system reliability requirements (i.e., the smaller are the levels R_0 and R_1), and the higher the level n, the larger is the gain. For the first multiplier in (15.11) the following estimates hold:

$$\frac{1-R_0}{R_0 \ln(1/R_0)} \sqrt[n]{R_0} \le \frac{1+a+a^2+\cdots+a^{n-1}}{n} \le \frac{1-R_0}{R_0 \ln(1/R_0)}$$

for a fixed value of R_0, the maximal possible gain (with increasing n and $R_1 \to R_0$) is determined by the coefficient $(1-R_0)/[R_0 \ln(1/R_0)]$.

15.3 SERIES-PARALLEL SYSTEM WITHOUT REPAIR

A system has a series of m groups of active redundant units. The ith redundant group consists of n_1 identical units. Time to failure for each unit within redundant group i follows an exponential d.f.: $F_1(t) = 1 - \exp(-\lambda_1 t)$, with an unknown parameter λ_1. Failures of different units are independent. The probability of system failure within time t_0 is

$$R = 1 - \prod_{1 \le i \le m} (1 - q_i^{n_i}), \tag{15.12}$$

where $q_i = 1 - \exp(-\lambda_i t_0)$ is the unit failure probability within time t_0 for redundant group i. With high reliability ($\lambda_i t_0 \ll 1$), an approximation may be used:

$$R \approx \sum_{1 \le i \le m} Q_i, \tag{15.13}$$

where

$$Q_i = q_{q_i}^{n_i} \approx (\lambda_i t_0)^{n_i},$$

is the failure probability for the ith redundant group. For standby redundant units, the failure probability for the ith redundant group, Q_i, can be found as

$$Q_i \approx \frac{(\lambda_i t_0)^{n_i}}{n_i!}.$$

Suppose that testing of type i units is performed according to the $[N_i \ R \ T_i]$ plan and that d_i failures are observed. Random variable d_i has a Poisson distribution with parameter $\Lambda_i = \lambda_i S_i$, where $S_i = N_i T_i$ is the total operating time for the ith type unit in the testing. Estimation of system reliability measure (15.13) is reduced to estimation of a power function of unknown Poisson parameters $\mathbf{\Lambda} = (\Lambda_1, \ldots, \Lambda_m)$:

$$R = R(\mathbf{\Lambda}) = \sum_{1 \le i \le m} a_i \Lambda_i^{n_i}. \tag{15.14}$$

The a_1 coefficients for the active and standby regimes are calculated from

$$a_i = \left(\frac{t_0}{S_i}\right)^{n_i},$$

$$a_i = \frac{(t_0/S_i)^{n_i}}{n_i!},$$

respectively.

We also may consider a binomial testing model of type i units which are tested according to the $[N\ W\ t_0]$ plan. Let d_i failures be observed. If there are enough tested units, N_i, the failure number, d_i, approximately follows a Poisson distribution with parameter $\Lambda_i = N_i q_i$. The problem is again reduced to estimation of the index in the form of (15.14), where

$$a_1 = \frac{1}{N_i^{n_i}} .$$

An unbiased estimate of index (15.14) is

$$\hat{R} = \sum_{1 \le i \le m} a_i d_i^{n_i} , \tag{15.15}$$

where $d^{(n)}$ denotes $d(d-1)(d-2)\cdots(d-n+1)$.

The decision-making rule is developed on the basis of the unbiased estimate, (15.15). If $\hat{R} \le C$, the H_0 hypothesis (acceptance) is adopted. If $\hat{R} > C$, the H_1 hypothesis (rejection) is adopted. The testing extent and value of the constant C are chosen so as to ensure the specified risk values, α and β.

Denote by $F(\mathbf{\Lambda}, y) = P_\Lambda(\hat{R} \le y)$ the distribution function for estimate $\hat{R}$ with the actual value of the vector parameter equal to $\mathbf{\Lambda}$. The operating characteristic assumes the form $L(\mathbf{\Lambda}) = F(\mathbf{\Lambda}, C)$. The basic inequalities are:

$$\begin{aligned} 1 - F(\mathbf{\Lambda}, C) \le \alpha \qquad & \text{for } R(\mathbf{\Lambda}) \le R_0 ; \\ F(\mathbf{\Lambda}, C) \le \beta \qquad & \text{for } R(\mathbf{\Lambda}) \le R_1 . \end{aligned} \tag{15.16}$$

We introduce the functions:

$$\begin{aligned} F_-(R, C) &= \min F(\mathbf{\Lambda}, C) \\ F_+(R, C) &= \max F(\mathbf{\Lambda}, C) , \end{aligned} \tag{15.17}$$

where both minimum and maximum are taken over all possible combinations of the $\mathbf{\Lambda}$ parameters that correspond to the fixed value of the index R. For such a set, $\mathbf{\Lambda}$ meets the condition $R(\mathbf{\Lambda}) = R$. Functions (15.17) decrease monotonically with respect to R. Therefore (15.16) is equivalent to

$$1 - F_{-}(R_0, C) \leq \alpha ,$$
$$F_{+}(R_1, C) \leq \beta . \tag{15.18}$$

Development of the decision-making rule is reduced to determining the testing extent and constant C that fulfill inequalities (15.18). For this purpose, extremums in (15.17) have to be found. Exact solution of the problem proves to be rather complicated. Therefore we make use of normal approximation for distribution of the estimate $\hat{R}$ in light of the fact that the random variable $\hat{R}$ is formed as a sum of m random independent variables with finite moments. The exact distribution function is approximated with

$$F(\mathbf{\Lambda}, y) \approx \Phi\left(\frac{y - E\{\mathbf{\Lambda}\}}{\sqrt{\mathrm{Var}\{\mathbf{\Lambda}\}}}\right), \tag{15.19}$$

where $E\{\mathbf{\Lambda}\}$ and $\mathrm{Var}\{\mathbf{\Lambda}\}$ are the mean and the variance of estimate R for a given $\mathbf{\Lambda}$ and $\Phi(\ )$ is a standard normal distribution. The accuracy of (15.19) improves with increasing test extent and with an increasing number of redundant groups, m.

By virtue of the lack of bias in the estimate, its mean coincides with the measure $E\{\mathbf{\Lambda}\} = R(\mathbf{\Lambda})$ for all values of $\mathbf{\Lambda}$. Calculation of the variance of the estimate is more complicated. It can be shown that

$$\mathrm{Var}\{\mathbf{\Lambda}\} = \sum_{1 \leq i \leq m} a_i^2 \, \mathrm{Var}\{\Lambda_i\} , \tag{15.20}$$

where $\mathrm{Var}\{\Lambda_i\}$ is the variance of the random variable $d_i^{n_1}$. This variance in turn is determined from

$$\mathrm{Var}\{\Lambda_i\} = n_i! \Lambda_i^{N_i} \sum_{0 \leq k \leq n_i - 1} \binom{n_i}{k} \frac{\Lambda_i^k}{k!} . \tag{15.21}$$

Within the set of parameters where the extreme values in (15.17) are taken, the function $E\{\mathbf{\Lambda}\}$ is constant due to the absence of bias in the estimate. Therefore these extremums can be found from:

$$\mathrm{Var}\{R\} = \max D(\mathbf{\Lambda}) , \tag{15.22}$$

where the maximum is taken over the same set of parameters, $\mathbf{\Lambda}$, that satisfies $R(\mathbf{\Lambda}) = R$. This value represents essentially the maximum variance possible with the fixed mean of the estimate equal to R.

Let us introduce the variables

$$y_i = \Lambda_i^{n_i} .$$

With these variables we reduce the maximum (15.22) to

$$\max \sum_{1\le i\le m} a_i^2 \operatorname{Var}\{\sqrt[n_i]{y_i}\}\,, \tag{15.23}$$

with linear constraints

$$\sum_{1\le i\le m} a_i y_i = R\,, \qquad y_i \ge 0,\ i=1,\ldots,m\,. \tag{15.24}$$

The function in (15.23) is convex with respect to y. The maximum in (15.23) is realized in one of the limiting points of the domain (15.24) having the form $(0,\ldots,0,R/a_1,0,\ldots,0)$ with all but one coordinate equal to zero. Therefore:

$$\operatorname{Var}_+\{R\} = \max_i a_i^2 \operatorname{Var}\left\{\sqrt[n_i]{\frac{R}{a_i}}\right\}. \tag{15.25}$$

Consider active redundancy. $V_i = S_i/t_0$ is the testing extent for type i units. In a binomial testing model we set $V_i = N_i$. The maximum variance (15.25) assumes the form

$$\operatorname{Var}_+\{R\} = R \max_i \sum_{0\le k\le n_i-1} \left(\frac{n_i!\binom{n_i}{k}}{k!}\right) \frac{R^{k/n_i}}{V_i^{n_i-k}}\,. \tag{15.26}$$

This value decreases monotonically with respect to the testing extents V_i. For equal redundancies for different types of units, $n_1 = n_2 = \cdots = n_m = n$, (15.26) is simplified:

$$\operatorname{Var}_+\{R\} = R \sum_{0\le k\le n-1} \left(\frac{n!\binom{n_i}{k}}{k!}\right) \frac{R^{k/n}}{V_m^{n-k}}\,, \tag{15.27}$$

where $V_m = \min V_i$ is the minimum testing extent for the system units.

Inequalities (15.18) within the framework of normal approximation are equivalent to

$$\begin{aligned} R_0 + u_\alpha\sqrt{\operatorname{Var}_+\{R_0\}} &\le C\,,\\ R_1 - u_\beta\sqrt{\operatorname{Var}_+\{R_1\}} &\ge C\,. \end{aligned} \tag{15.28}$$

The testing extent is specified by the vector $\mathbf{V} = (V_1,\ldots,V_m)$ of the testing extents for the system units. Risks α and β can be assured for the given value of V provided inequalities (15.28) are fulfilled for the derived value of C. If, for the given V, we have the inequality

$$R_0 + u_\alpha\sqrt{\operatorname{Var}_+\{R_0\}} \le R_1 - u_\beta\sqrt{\operatorname{Var}_+\{R_1\}} \ge C\,, \tag{15.29}$$

then any value from the interval

$$R_0 + u_\alpha\sqrt{\mathrm{Var}_+\{R_0\}} \le C \le R_1 - u_\beta\sqrt{\mathrm{Var}_+\{R_1\}} \ge C \qquad (15.30)$$

can be taken as the constant C for the decision rule. If the inequality is strict, the risks are less than α and β. The minimal required components of the testing extent vector $\mathbf{V}$ (for a given α and β) are determined from:

$$R_0 + u_\alpha\sqrt{\mathrm{Var}_+\{R_0\}} = R_1 - u_\beta\sqrt{\mathrm{Var}_+\{R_1\}} \ge C\,. \qquad (15.31)$$

The constant C is determined unambiguously. Because $\mathbf{V}$ is multidimensional, equation (15.31) actually determines a set of "minimal" vectors $\mathbf{V}^*$ rather than a single vector.

Consider system tests. Assume that N_c identical samples of a system are tested within time period T and that all failed units are replaced with new ones at the failure moments. The case testing extents for the units are proportional to the number of units in the redundant groups: $V_i = n_i V_c$, $i = 1, \ldots, m$, where $V_c = N_c T$ is the testing extent (number of tests) for the system. A similar situation occurs with a binomial model when N_c samples of the system are tested within time t_0. Here also $V_i = n_i V_c$, $i = 1, \ldots, m$, where $V_i = N_1$, $V_0 = N_c$. The minimal required testing extent for the system V_c is determined unambiguously from equation (15.31), where $\mathrm{Var}_\pm\{R\}$ is given by (15.26) with $n_1 V_c$ substituted for V_i.

15.4 AVAILABILITY CONTROL FOR REPAIRABLE SYSTEMS

Let a system have n identical active redundant units. Denote the MTTF of a unit by T. Restoration of a failed unit begins immediately regardless of the state of other units. Denote the mean renewal (repair or replacement) time of a unit by τ. Operation of a unit is a sequence of mutually independent periods of failure-free operation and renewal $\xi_1, \eta_1; \xi_2, \eta_2; \ldots; \xi_n, \eta_n; \ldots$, where $\xi_1, \xi_2, \ldots, \xi_n, \ldots$, have the same d.f. with the mean $E\{\xi_n\} = T$ and where $\eta_1, \eta_2 \ldots \eta_n \ldots$, have the same d.f. with the mean $E\{\eta_n\} = \tau$. Failures and renewals of different units are independent. The availability index of a unit is

$$\tilde{K} = \frac{T}{T + \tau}\,.$$

The availability index for the system is

$$K = 1 - (1 - \tilde{K})^n = 1 - \left(\frac{\tau}{T + \tau}\right)^n . \qquad (15.32)$$

The actual values of τ, T, and K are unknown. One of two hypotheses has to be adopted on the basis of test results. $H_0: K > K_0$ and $H_1: K \le K_1$, where K_0 and K_1 are the specified levels of acceptance and rejection, $K_0 > K_1$. The availability coefficient is determined unambiguously by the ratio of the mean values, $\rho = \tau / T$:

$$K = 1 - \left(\frac{\rho}{1+\rho} \right)^n . \tag{15.33}$$

Expression (15.33) decreases monotonically with q. Therefore the availability coefficient testing is equivalent to tests of hypotheses about ρ:

$$\begin{aligned} H_0 &: \rho \leq \rho_0 , \\ H_1 &: \rho \geq \rho_1 , \end{aligned} \tag{15.34}$$

where ρ_0 and ρ_1 are calculated with the K_0 and K_1 given by (15.33).

Suppose that N unit operating times, $\xi_1, \xi_2, \ldots \xi_N$, were observed along with l values of unit repair times, $\eta_1, \eta_2, \ldots \eta_l$. All results were independent of each other. Both the operating time and the repair time are assumed to be have "aging" distributions. One of the two hypotheses, H_0 and H_1, with specified risks α and β has to be adopted.

The decision rule is developed on the basis of

$$\hat{\rho} = \frac{\hat{\tau}}{\hat{T}} , \tag{15.35}$$

where

$$\hat{\tau} = \frac{1}{l} (\eta_1 + \cdots + \eta_l) ,$$

$$\hat{T} = \frac{1}{N} (\xi_1 + \cdots + \xi_N) .$$

The inequality $\hat{\rho} \leq C$ is equivalent to $\hat{\tau} - \widehat{CT} \leq 0$. The operating characteristic coincides with the probability $Pr\{\hat{\tau} - \widehat{CT} \leq 0\}$. The inequalities have the form

$$\begin{aligned} 1 - Pr\{\hat{\tau} - \widehat{CT} \leq 0\} \leq \alpha \qquad & \text{for } \rho \leq \rho_0 , \\ Pr\{\hat{\tau} - \widehat{CT} \leq 0\} \leq \beta \qquad & \text{for } \rho \leq \rho_1 . \end{aligned} \tag{15.36}$$

The random variable $\hat{\tau} - \widehat{CT}$ is formed as a sum of $N + l$ independent random variables. If N and l are large enough, the distribution can be approximated with a normal one that has a mean and variance given by

$$\tau - CT ,$$

$$\frac{\sigma_\tau^2}{l} + \frac{C^2 \sigma_T^2}{N} ,$$

where σ_x^2 and σ_χ^2 are the variances of renewal time τ and the time to failure T, respectively. Within the framework of the normal approximation, and with both α and β less than 0.5, inequalities (15.36) can be rewritten:

$$(\tau - CT) + u_\alpha \sqrt{\frac{\sigma_\tau^2}{l} + C^2 \frac{\sigma_T^2}{N}} \le 0 \qquad \text{for } \rho \le \rho_0 \,,$$
$$(\tau - CT) - u_\alpha \sqrt{\frac{\sigma_\tau^2}{l} + C^2 \frac{\sigma_T^2}{N}} \ge 0 \qquad \text{for } \rho \le \rho_1 \,. \tag{15.37}$$

After a simple transformation of each,

$$\rho - C + u_\alpha \sqrt{\frac{1}{l} \frac{\sigma_\tau^2}{\tau^2} \rho^2 + \frac{C^2}{N} \frac{\sigma_T^2}{T^2}} \le 0 \qquad \textit{for } \rho \le \rho_0 \,,$$
$$\rho - C - u_\alpha \sqrt{\frac{1}{l} \frac{\sigma_\tau^2}{\tau^2} \rho^2 + \frac{C^2}{N} \frac{\sigma_T^2}{T^2}} \ge 0 \qquad \textit{for } \rho \le \rho_1 \,. \tag{15.38}$$

The limit value of the variation coefficient, which is the ratio of the root mean square to the mean value for aging distributions, is 1, (the value for an exponential distribution.) Therefore inequalities (15.38) assume the following form (with $l \ge u_\alpha^2$):

$$\rho_0 - C + u_\alpha \sqrt{\frac{1}{l} \rho_0^2 + \frac{C^2}{N}} \le 0 \,,$$
$$\rho_1 - C - u_\beta \sqrt{\frac{1}{l} \rho_0^2 + \frac{C^2}{N}} \ge 0 \,. \tag{15.39}$$

With proper transformations, (15.39) can be rewritten

$$\frac{\rho_0}{(1 - u_\alpha^2/N)} \left(1 + u_\alpha \sqrt{\frac{1}{N} + \frac{1}{l} - \frac{u_\alpha^2}{Nl}}\right) \le C \,,$$
$$\frac{\rho_1}{(1 - u_\beta^2/N)} \left(1 + u_\beta \sqrt{\frac{1}{N} + \frac{1}{l} - \frac{u_\beta^2}{Nl}}\right) \ge C \,. \tag{15.40}$$

Values of risks α and β can be assured for the given values of N and l provided that both inequalities in (15.40) hold for C when $\rho_0 < C < \rho_1$. For this purpose the following inequality has to be fulfilled:

$$\frac{\rho_0}{(1 - u_\alpha^2/N)} \left(1 + u_\alpha \sqrt{\frac{1}{N} + \frac{1}{l} - \frac{u_\alpha^2}{Nl}}\right)$$
$$\le \frac{\rho_1}{(1 - u_\beta^2/N)} \left(1 + u_{\beta\alpha} \sqrt{\frac{1}{N} + \frac{1}{l} - \frac{u_\beta^2}{Nl}}\right) \tag{15.41}$$

The minimal required values of N and l have to satisfy

$$\frac{\rho_0}{(1-u_\alpha^2/N)}\left(1+u_\alpha\sqrt{\frac{1}{N}+\frac{1}{l}-\frac{u_\alpha^2}{Nl}}\right)$$
$$=\frac{\rho_1}{(1-u_\beta^2/N)}\left(1+u_\beta\sqrt{\frac{1}{N}+\frac{1}{l}-\frac{u_\beta^2}{Nl}}\right). \tag{15.42}$$

If N and l obey (15.42), the constant C for the decision-making rule with risks α and β coincides with both sides of equation (15.42).

In most tests, $N = 1$. Thus equation (15.42) can be simplified to

$$\frac{\rho_0}{(1-u_\alpha^2/N)}\left(1+\frac{u_\alpha}{\sqrt{N}}\sqrt{2-\frac{u_\alpha^2}{N}}\right)$$
$$=\frac{\rho_1}{(1-u_\beta^2 N)}\left(1+\frac{u_\alpha}{\sqrt{N}}\sqrt{2-\frac{u_\beta^2}{N}}\right). \tag{15.43}$$

Solution of (15.43) with respect to N gives the minimal required testing extent, N^* (with specified ρ_0, ρ_1, α, and β). When $(\rho_1 - \rho_0)\rightarrow 0$, the following approximation holds:

$$N^* = \frac{2\rho_0^2(u_\alpha + u_\beta)^2}{(\rho_1 - \rho_0)^2}. \tag{15.44}$$

When $\alpha = \beta$, we can simplify the foregoing:

$$N^* = \frac{u_\alpha^2}{1-\sqrt{1-[(\rho_1-\rho_0)/(\rho_1+\rho_0)]^2}} \approx \frac{2u_\alpha^2(\rho_1+\rho_0)^2}{(\rho_1-\rho_0)^2}. \tag{15.45}$$

With this procedure we obtain confidence limits for ρ and K. The upper $(1-\beta)$-confidence limit for ρ is

$$\bar{\rho}_{1-\beta} = \hat{\rho}\,\frac{1-(u_\beta^2/N)}{1-u_\beta\sqrt{(1/N)+(1/l)-(u_\beta^2/Nl)}}. \tag{15.46}$$

The lower $(1-\alpha)$-confidence limit for ρ is

$$\underline{\rho}_{1-\alpha} = \hat{\rho}\,\frac{1-(u_\alpha^2/N)}{1-u_\alpha\sqrt{(1/N)+(1/l)+(u_\alpha^2/Nl)}} \tag{15.47}$$

The interval $(\underline{\rho}_{1-\alpha}, \bar{\rho}_{1-\beta})$ is the confidence interval for ρ, with a confidence coefficient $\gamma = 1-\alpha-\beta$. As N and l increase, the following approximations hold:

$$\bar{\rho}_{1-\beta} \approx \hat{\rho}\left(1 + u_\beta \sqrt{\frac{N+1}{Nl}}\right),$$

$$\underline{\rho}_{1-\alpha} \approx \hat{\rho}\left(1 - u_\alpha \sqrt{\frac{N+1}{Nl}}\right).$$

Lower and upper confidence limits for the availability coefficient, K, are calculated through substitution of upper and lower limits for ρ in (15.33);

$$\underline{K}_{1-\beta} = 1 - \left(\frac{\bar{\rho}_{1-\beta}}{1+\bar{\rho}_{1-\beta}}\right)^n,$$

$$\bar{K}_{1-\alpha} = 1 - \left(\frac{\underline{\rho}_{1-\alpha}}{1+\underline{\rho}_{1-\alpha}}\right)^n.$$

15.5 GAIN DUE TO INFORMATION ON SYSTEM UNITS

We will consider whole system testing in conformity with Section 15.4. Simultaneous independent flows of failures and restorations for each of the n system units are observed. Each flow is a sequence of independent intervals of unit time to failure and unit repair. The system fails if all n units fail.

The system availability factor, K, can be determined directly with the observed system failures and repairs. In the preceding section the observed intervals of operation and repair were assumed to be independent, random aging variables. In the resulting system flow, the system operation and repair intervals might be interdependent. Consider the case for which the repair time for each unit follows an exponential distribution. Intervals of system operation and repair then are independent. The system repair time follows an exponential (and hence aging) distribution with mean $\tau_c = \tau/n$. In the case of high reliability one can assume that the system time to failure also follows an exponential distribution. Then one can use the results of the preceding section for making decisions on the basis of the observations of the system as a whole.

A system operation cycle consists of an interval of failure-free operation followed by an interval of restoration for the system. Suppose that N_c cycles of system operation are observed. The observed results are N_c intervals of system failure-free operation and N_c intervals of system repair. Let T_c be the MTTF, τ_c be the mean system repair time, and $R = \tau_c/T_c$. The problem is reduced to standard acceptance-rejection hypotheses for the test results: H_0 is $R \le R_0$ and H_1 is $R \ge R_1$. Here levels R_0 and R_1 are related to the initial levels of acceptance and rejection, $K_0 > K_1$, by:

$$K_0 = \frac{1}{1 + R_0}\,,$$

$$K_1 = \frac{1}{1 + R_1}\,. \tag{15.48}$$

Let $\alpha = \beta$. The required number of observations of the system operation cycles, N_c, is analogous to (15.45):

$$N_c^* = \frac{u_\alpha^2}{1 - \sqrt{1 - [(R_1 - R_0)/(R_1 + R_0)]^2}}\,. \tag{15.49}$$

The mean duration of system testing is

$$a_2 = N_c^*(T_c + \tau_c)\,.$$

Consider another decision-making procedure based on the information obtained in system tests of separate units. System testing will continue for N failure-free operation intervals and N unit repair intervals. Under this test plan, the results in the preceding section, which are based on an assumption of interval independence, are only approximate. The required failure number determined from (15.45) is

$$N_c = \frac{u_\alpha^2}{1 - \sqrt{1 - [(\rho_1 - \rho_0)/(\rho_1 + \rho_0)]^2}}\,, \tag{15.50}$$

where the levels ρ_0 and ρ_1 are related to K_0 and K_1 by

$$K_0 = 1 - \left(\frac{\rho_0}{1 + \rho_0}\right)^n,$$

$$K_0 = 1 - \left(\frac{\rho_1}{1 + \rho_1}\right)^n. \tag{15.51}$$

The mean testing duration is

$$a_1 = \frac{N^*(T + \tau)}{n}\,.$$

Applying the equation for system operating time:

$$T_c = \frac{\tau}{n}\left[\left(1 + \frac{1}{\rho}\right)^n - 1\right],$$

we obtain the approximate asymptotic expression for $(K_0 - K_1) \to 0$:

$$\frac{a_2}{a_1} \approx \left(\frac{K_2}{1 - \sqrt[n]{(1 - K_0)}} \right) \frac{1}{n^2 \sqrt[n]{(1-K)^{n-1}}} \approx \frac{1}{n^2 \sqrt[n]{(1-K)^{n-1}}} \,. \tag{15.52}$$

In (15.52), K_0 stands for the acceptance level and K is the actual value of the availability coefficient. Approximation (15.52) characterizes the testing extent gain that is due to use of the information on units. The gain increases with the value of K.

15.6 AVAILABILITY TESTING BASED ON THE *F* DISTRIBUTION

Control plans for tests of the availability coefficient, K, are presented in this section for the case where both time to failure and repair time are exponential.

A sample consists of N items ($N \geq i$). Each item is tested until the total number of operation-repair cycles equals a certain value. The item is accepted if the point estimate is not less than the control value, C. Otherwise the item is rejected. The estimate $\hat{K}$ is calculated from

$$\hat{K} = \frac{\hat{T}}{\hat{T} + \hat{\tau}} \,,$$

where

$$\hat{T} = \frac{1}{r} \sum_{1 \leq i \leq r} \xi_i \,,$$

$$\hat{\tau} = \frac{1}{r} \sum_{1 \leq i \leq r} \eta_i \,,$$

and ξ_1 and η_1 stand for the ith intervals of failure-free operation and repair, respectively. The statistic is summed for all the samples.

The control plan (the number of cycles, r), and the estimation rate C are determined on the basis of risks α and β and specified acceptance and rejection levels K_0 and K_1. The number r must fulfill the relation

$$\frac{\varphi_\beta(2r, 2r)}{\varphi_{1-\alpha}(2r, 2r)} = \frac{(-K_0)K_1}{(1 - K_1)K_0} \,, \tag{15.53}$$

where $\varphi_{1-\alpha}(2r, 2r)$ and $\varphi_\beta(2r, 2r)$ stand for quantiles of a F-distribution. When r is found, the value of C is calculated:

$$C = \frac{\hat{K}}{(1 - K_0)\varphi_{1-\alpha}(2r, 2r) + K_0} \,. \tag{15.54}$$

As an example, let $\alpha = 0.05$, $\beta = 0.05$, $K_0 = 0.97$, and $K_1 = 0.88$. The value of $(1 - K_1)/(1 - K_0)$ is 4.0. Assume the estimate of K, $\hat{K}$, is 0.981 after

Table 15.1 Availability Testing Based on F Distribution

	$(1-K_1)/(1-K_0)$									
	2.0		2.5		3.0		4.0		5.0	
K_0	r	C	r	C	r	C	r	C	r	C
					Plans for $\alpha = \beta = 0.1$					
0.9	20	0.857	11	0.840	8	0.818	—	—	—	—
0.95	25	0.928	14	0.920	9	0.913	6	0.986	4	0.886
0.96	25	0.943	14	0.935	10	0.929	6	0.918	5	0.903
0.97	25	0.958	15	0.952	10	0.948	6	0.940	5	0.929
0.98	25	0.972	15	0.968	11	0.965	7	0.959	5	0.954
0.99	27	0.986	16	0.984	11	0.983	7	0.980	5	0.978
0.995	27	0.9930	16	0.9920	11	0.9913	7	0.9900	5	0.989
0.998	28	0.9972	16	0.9968	11	0.9965	7	0.9963	5	0.9957
0.999	28	0.9986	16	0.9984	11	0.9983	7	0.9980	5	0.9978
0.9955	28	0.99929	16	0.99921	11	0.99914	7	0.9990	5	0.9989
0.9999	28	0.99986	16	0.99984	11	0.99983	7	0.99980	5	0.99978
					Plans for $\alpha = \beta = 0.05$					
0.9	33	0.857	18	0.839	12	0.822	—	—	—	—
0.95	39	0.929	22	0.920	14	0.914	9	0.898	7	0.882
0.96	40	0.943	23	0.936	16	0.930	9	0.920	7	0.908
0.97	41	0.958	24	0.952	16	0.948	10	0.949	8	0.930
0.98	42	0.972	25	0.968	17	0.965	11	0.959	8	0.955
0.99	44	0.986	25	0.984	16	0.982	11	0.980	8	0.978
0.995	44	0.993	26	0.9921	18	0.9913	11	0.9901	9	0.9885
0.998	45	0.9972	26	0.9968	18	0.9965	12	0.9959	9	0.9954
0.999	45	0.9986	26	0.9984	18	0.9983	12	0.9980	9	0.9977
0.9995	45	0.99929	26	0.99921	18	0.99914	12	0.99900	9	0.99890
0.9999	45	0.99986	26	0.99984	18	0.99983	12	0.99980	9	0.99977

$r = 10$ operation repair cycles. Using $F_{0.05}(20, 20) = 2.12$, we find that (15.54) gives a C value of 0.9491. This is very slightly higher than the critical value of C for the conditions in Table 15.1, 0.949. Further testing therefore is not mandatory. Control plans for the most common test condition are given in Table 15.1.

15.7 CALCULATION OF CONTROL PLAN

When there is no basis for calculation of a distribution function of the estimate, $F(\hat{R})$, a tentative calculation based on normal approximation is used. The variance, $\mathrm{Var}\{\hat{R}\}$, is calculated with linearization. In addition to normal distribution errors the approach may introduce errors due to bias of the estimate $\hat{R} = \hat{R}(\eta_1)$ as well as linearization errors. However, the calculation may work where more rigorous methods fail.

In estimating the reliability index under control, the normal distribution is assumed to have a mean equal to the actual value of the index. In fact, these two values are not equal due to the estimate bias.

$$F(y, \hat{R}) = F_0\left[\frac{y - R}{\alpha(\hat{R})}\right].$$

Relationship (15.28) then takes the form

$$R_1 + u_{1-\beta}\sigma_1(\hat{R}) = C\,, \quad \sigma_1(\hat{R}) = \sqrt{\mathrm{Var}_1\{\hat{R}\}}\,, \tag{15.55}$$

$$R_0 + u_\alpha\sigma_0(\hat{R}) = C\,, \quad \sigma_0(\hat{R}) = \sqrt{\mathrm{Var}_0\{\hat{R}\}}\,, \tag{15.56}$$

where $\mathrm{Var}_0\{\hat{R}\}$ and $\mathrm{Var}_1\{\hat{R}\}$ are the variances of the distribution function $F_0(y, R)$ at the points where $R = R_0$ and $R = R_1$, respectively.

The variances are calculated from

$$\mathrm{Var}\{\hat{R}\} = \sum_i \left(\frac{\partial R}{\partial \theta_i}\right)^2 \mathrm{Var}\{\theta_i\}\,. \tag{15.57}$$

The values of η_i corresponding to R_0 are substituted in the $\mathrm{Var}_0\{R\}$ expression in (15.57), while the values of η_i corresponding to R_1 are substituted in the $\mathrm{Var}_1\{\hat{R}\}$ expression. The required observation extent is included in $\mathrm{Var}\{\hat{\eta}_1\}$.

Since different combinations of η_i can correspond to the values of R_0 and R_1, one has to select the combinations for both equations that give the maximum values of variances $\mathrm{Var}_0\{\hat{R}\}$ and $\mathrm{Var}_1\{\hat{R}\}$.

EXAMPLE 15.1

A single sample of a restorable item is placed under tests in which $K_0 = 0.98$, $K_1 = 0.96$, and $\alpha = \beta = 0.1$. The tests will be performed until there are d failures. Values of C and d have to be found. Operation interval and repair time have an exponential distribution.

Solution

The availability, K, is estimated from

$$\hat{K} = \frac{\beta}{\hat{\lambda} + \beta}\,,$$

where $\hat{\lambda}$ is the estimated mean time-before-failure and $\hat{\mu}$ is the estimated mean repair time.

The variance $\mathrm{Var}_1\{\hat{K}\}$ is calculated in conformity with (15.57). With the chosen testing plan $[1Rd]$, the expression for $\mathrm{Var}_1\{\hat{K}\}$ can be presented in a form independent of λ and μ:

$$\mathrm{Var}\{\hat{K}\} = \frac{2K^4(1-K)^2}{d} . \tag{15.58}$$

Then the system of equations (15.55) and (15.56) takes the form

$$K_1 + u_{1-\beta}K_1^2(1-K_1)\sqrt{\frac{2}{d}} = C ,$$

$$K_0 = u_\alpha K_0^2(1-K_0)\sqrt{\frac{2}{d}} = C .$$

We substitute numerical values:

$$0.96 + 1.28 \cdot 0.96^2(1-0.96)\sqrt{\frac{2}{d}} = C ,$$

$$0.98 - 1.28 \cdot 0.98^2(1-0.98)\sqrt{\frac{2}{d}} = C .$$

Solving the system, we find that $C = 0.973$ and $d = 26$.

If, instead of the normal approximation and (15.58), an exact distribution is used for K (see Section 15.6), the answer becomes $C = 0.972$ and $d = 25$.

EXAMPLE 15.2

A single sample of a repairable item is placed under test. The item involves two identical devices. One is used as a loaded reserve with complete control and ideal switching. Failure interval and restoration time distributions are exponential. Two levels of operating time to failure are specified: $T_0 =$ 1300 h; $T_1 = 650$ h; $\alpha = \beta = 0.1$. Testing is performed within the specified time period t_*. Values of C and t_* have to be determined.

Solution

T is estimated from

$$\hat{T} = \frac{\hat{\mu} + 3\hat{\lambda}}{2\hat{\lambda}^2} . \tag{15.59}$$

Let us determine $\mathrm{Var}_1\{\hat{T}\}$. The item is tested in conformity with plan $[1Rt_*]$, so the values of λ and μ are estimated according to plans $[2Rt_*]$ and $[2Rd]$, respectively. For these plans $\mathrm{Var}_1\{\lambda\} = \lambda/2t_*$ and $\mathrm{Var}_1\{\hat{u}\} = \mu^2/(d-2)$. Then from (15.57),

$$\mathrm{Var}\{\hat{T}\} = \frac{(3\lambda + 2\mu)^2}{8\lambda^5 t_*} + \frac{\mu^2}{4\lambda^4(d-2)} .$$

Now the values of λ and μ that yield the maximum variance have to be found. For any value of $\hat{T}$, the estimates of λ and $\hat{\mu}$ are related by (15.59) and decrease or increase simultaneously. If, for instance, one represents μ in terms of λ and T with (15.59), an approximation is accurate when $\hat{\lambda} \ll \hat{\mu}$:

$$\mathrm{Var}\{\hat{T}\} \approx \frac{2T^2}{\lambda t_*} + \frac{2T^2}{d-2} .$$

Hence it is evident that the maximum variance is obtained with the minimum λ and μ.

As the failure number, d, initially is unknown, an approximation is used:

$$\frac{\mu^2}{d-2} \approx \frac{\mu^2}{2\lambda t_*} .$$

Then

$$\mathrm{Var}\{\hat{T}\} = \frac{(3\lambda + 2\mu)^2 + \mu^2}{8\lambda^5 t_*} .$$

It is obvious that λ and μ have certain limits, and hence it is sufficient to find just one of them. For example, technical considerations or experience with similar units may lead to a conclusion that the mean repair time should be less than 2 hours. We let $\mu \geq 0.5$ per hour. Then we use (15.59) to find $1.44 \cdot 10^{-2}$ per hour and $2.08 \cdot 10^{-2}$ per hour as values of λ that correspond to $T_0 = 1300$ h and $T_1 = 650$ h. The system of equations is written

$$1300 - 1.28\sqrt{\frac{(2.70) \cdot 10^8}{t_*}} = C ,$$

$$650 + 1.28\sqrt{\frac{(0.44) \cdot 10^8}{t_*}} = C .$$

Solving this system leads to $C = 840$ h and $t_* = 2070$ h.

The testing time is overestimated because the minimum possible values of λ and μ were taken. But this drawback is justified by a more economical test method. Indeed, testing of such an item with the same accuracy and confidence would require $9.47T_0 = 12{,}300$ hours.

An important point is that the gain is conserved even in the limiting case when, due to lack of information on λ and μ, the trivial estimate $\mu = 0$ is taken as the lower limit for μ. Equations (15.55) and (15.56) result in this case $C = 820$ h, $t_* = 5200$ h. However, considerable increase of the error due to normal approximation is possible.

EXAMPLE 15.3

A single sample of an unrepairable system is tested. The system consists of two units, and both have active redundant units with an ideal switch. Distributions of time to failure are exponential. A unit that fails in the tests can be replaced. Failure-free operation probabilities are specified:

$$P_0(t) = 0.98\,, \quad P_1(t) = 0.96\,, \quad t = 2\,\text{h}\,, \quad \alpha = \beta = 0.1\,.$$

Tests are performed for time t. The values of C and t_* have to be determined.

Solution

A calculation and experimentation method are used. From all observed failures of basic and redundant units of both types, the failure rates $\hat{\lambda}_1$ and $\hat{\lambda}_2$, are determined. Then the estimate is made:

$$\hat{P}(t) = (2e^{-\tilde{\lambda}_1 t} - e^{-2\hat{\lambda}_1 t})(2e^{-\hat{\lambda}_2 t} = e^{-2\hat{\lambda}_2 t})\,. \tag{15.60}$$

As both λ_1 and λ_2 were estimated from statistics for the two devices and the estimates were found for time t_*, the variance $\text{Var}\{\hat{\lambda}\}$ can be calculated as $\lambda/2t_*$. With allowance for the value, the equation for $\text{Var}\{\hat{P}\}$ assumes the form:

$$\text{Var}\{\hat{P}\} = (2e^{-\lambda_2 t} - e^{-2\lambda_2 t})^2(e^{-\lambda_1 t} - e^{-2\lambda_1 t})^2\,\frac{2\lambda_1 t^2}{t_*}$$
$$+ (2e^{-\lambda_1 t} - \text{e}^{-2\lambda_1 t})^2(e^{-\lambda_2 t} - e^{-2\lambda_2 t})^2\,\frac{2\lambda_2 t^2}{t_*}\,. \tag{15.61}$$

With $P = 0.98$, the variables λ_1 and λ_2 can vary from one extreme set of values, $\lambda_1 = 0$ and $\lambda_2 = 0.075$ per hour when all failures are in the second pair of systems, to the other with $\lambda_1 = 0.075$ per hour and $\lambda_2 = 0$ (all failures within the first pair). Evidently the function $\text{Var}\{\hat{P}\}$ is symmetrical with regard to the middle point where $\lambda_1 = \lambda_2 = 0.053$ per hour. The function decreases monotonically on either side of the point. Direct calculation demonstrates that $\text{Var}\{\hat{P}\}$ is maximum at the two extremes, (0, 0.075 and 0.075, 0), where it is $0.0862/t_*$. With $P = 0.96$, the maximal variance similarly is $0.0862/t_*$ for $\lambda_1 = 0.112$ per hour and $\lambda_2 = 0$, or vice versa.

Therefore the system (15.55) and (15.56) assumes the form

$$0.98 - 1.28\,\frac{0.00862}{t_*} - C = 0\,,$$

$$0.96 + 1.28\,\frac{0.0231}{t_*} - C = 0\,.$$

The solution of the system is $C = 0.972$ and $t_* = 245$ h. This approach also results in a considerably shorter time than testing the whole system. Indeed testing such a system with the same initial data requires about 470 cycles with duration t—at least 940 hours. The estimation rate still remains unchanged; up to 13 cycles with failures are allowed for acceptance corresponding to

$$C = \frac{470 - 13}{470} = 0.972 .$$

15.8 RELIABILITY INDEX CONTROL WITH CONFIDENCE LIMITS

Let R represent a reliability index with an unknown value. In general, R is likely to depend upon a set of unknown parameters: $R = R(\eta)$, $\eta = (\eta_1, \ldots, \eta_m)$. Consider a control problem for the index R with two specified levels $R_0 < R_1$ of acceptance and rejection.

The problem can be solved with confidence limits for R. Let $\xi = \xi(x)$ represent the statistic (test result) used for the development of confidence limits.

An unbiased point estimate of the index $\xi = \hat{R}(x)$ might be an initial statistic or any other measure. Let $\underline{R}$ and $\bar{R}$ represent the lower and upper confidence limits for R with confidence coefficients γ_1 and γ_2, respectively. Denote the dependence of the confidence limits upon the testing result, ξ, and the confidence coefficients with

$$\begin{aligned} \underline{R} &= \underline{R}(\xi, \gamma_1) , \\ \bar{R} &= \bar{R}(\xi, \gamma_2) . \end{aligned} \tag{15.62}$$

Let us fix the values of risks α and β and set $\gamma_1 = 1 - \alpha$ and $\gamma_2 = 1 - \beta$. The inequalities for a decision rule with confidence limits take the form

$$\begin{aligned} \underline{R}(C, 1 - \alpha) &\geq R_0 , \\ \bar{R}(C, 1 - \beta) &\leq R_1 . \end{aligned} \tag{15.63}$$

A decision-making rule may be based on (15.63). The actual risks are less than values α and β.

If, with the selected values of α and β, the system of inequalities (15.63) has no solution, development of a decision-making rule with these risks is not feasible (at least on the basis of the statistic ξ). Either the risks have to be increased or there must be more observations.

Development of a decision rule is reduced to solution of inequality system (15.53). We have a method of obtaining this solution that may prove to be convenient. With a fixed value of $\alpha = 1 - \gamma_1$, we start increasing the value of ξ until the lower limit R_* passes for the first time the level R_0. We then let:

$$C = \min\{\xi : R_*(\xi, 1-\alpha) \geq R_0\} . \tag{15.64}$$

With this value of C, we calculate the upper limit, $\bar{R} = \bar{R}(C, 1-\beta)$. If $\bar{R} \leq R_1$, β is fully developed. If $\bar{R} > R_1$ we increase β until the value of $\bar{R}$ is less than R_1. With a fixed α, we find the minimum β that can be assured in the experiment.

If, for the determined C, inequalities (15.62) prove to be strict, the risks of the developed criterion will be smaller than α and β. Exact values of the risks are then found. With fixed $\xi = C$, α and β are decreased while the confidence interval $(R, \bar{R})$ is still within the interval (R_0, R_1). The lower limit, $\underline{R}$, decreases, and the upper one, $\bar{R}$, increases. The values of α and β when inequalities (15.63) are transformed into equations represent the exact risks for the developed criterion.

EXAMPLE 15.4

Let $R = \lambda$ be an unknown failure rate parameter for a unit with a time-to-failure distribution in the form $F(t) = 1 - \exp(-\lambda t)$. Acceptance and rejection levels are $\lambda_0 = 0.1$ and $\lambda_1 = 0.25$, respectively. Specified risks are α–0.1 and $\beta = 0.1$. Tests are performed with renewal of failed units until the total operating time S reaches a preset value of 16—(under the $[N\ R\ T]$ plan, $NT = S$. The observed number of failures, d, is the testing result. Confidence limits have to be determined, and a decision has to be made on the type of hypothesis.

Solution

Lower and upper confidence limits with confidence coefficients γ_1 and γ_2 for λ are found from

$$\begin{aligned} \underline{\lambda}(d, \gamma_1) &= \frac{\chi^2_{1-\gamma_1}(2d+2)}{2S} \\ \bar{\lambda}(d, \gamma_2) &= \frac{\chi^2_{\gamma_2}(2d+2)}{2S} . \end{aligned} \tag{15.65}$$

Inequalities (15.63) take the form

$$\begin{aligned} \underline{\lambda}(C, 1-\alpha) &= \frac{\chi^2_{\alpha}(2d+2)}{2S} \geq \lambda_0 , \\ \bar{\lambda}(C, 1-\beta) &= \frac{\chi^2_{1-\beta}(2d+2)}{2S} \leq \lambda_1 . \end{aligned} \tag{15.66}$$

With $\alpha = 0.1$, we use the first inequality to find C as the minimum value of observation result d that ensures the location of the lower limit within the interval (λ_0, λ_1):

$$C = \min\{d : \underline{\lambda}(d, 1-\alpha) \geq \lambda_0\} = 3\,. \tag{15.67}$$

The corresponding value of the lower limit, $\underline{\lambda}(C, 1-\alpha)$, is $1.74/16 = 0.109$. With $\beta = 0.1$, the upper limit, $\bar{\lambda}(C, 1-\beta)$, is 0.417. This value surpasses the second level λ_1, and hence the risks $\alpha = \beta = 0.1$ cannot be assured. We need to increase the value of β to locate the upper limit, $\bar{\lambda}(C, 1-\beta)$, within the interval (λ_0, λ_1). We find that with $\alpha = 0.1$, the minimal value of β is 0.43. The H_0 decision rule with $\alpha < 0.1$ and $\beta = 0.43$ is made if $d \leq 3$. If $d > 3$, then the H_1 decision is made.

Let S be increased to 53. Repeating the above procedure with $\alpha = 0.1$ and $\beta = 0.1$, we find that when $C = 8$, the lower limit, $\underline{\lambda}\ (C, 1-\alpha) = 0.103 > \lambda_0$. The upper limit, $\bar{\lambda}(C, 1-\beta)$ is $0.245 < \lambda_1$; hence the risks $\alpha = \beta = 0.1$ can be assured. The H_0 decision is made when $d \leq 8$. If $d > 8$, the H_1 decision is made.

EXAMPLE 15.5

Consider a series system. The system failure-free operation probability for a specified time, t, is

$$P = \prod_{1 \leq i \leq m} p_1\,,$$

where $p_1 = \exp(-\lambda_1 t)$ and λ_1 is the failure rate parameter for unit i. Unit i is tested for a preset total operating time S_1. Failure numbers, d_i, $i = 1, \ldots, m$, are observed. Confidence limits have to be found and a decision on the hypothesis has to be made.

Solution

Estimation of P reduces to estimation of the weighted sum of unknown Poisson parameters on the basis of the vector of Poisson r.v. $\mathbf{d} = (d_1, \ldots, d_m)$. To find confidence limits for R, we take the failure numbers for the units $\boldsymbol{\xi} = d_1 + \cdots + d_m$. We order the unit indexes with regard to decreasing testing extent: $S_1 \geq \cdots \geq S_m$. The lower and upper limits for R with confidence coefficients γ_1 and γ_2 are determined from

$$\underline{R}(\xi, \gamma_1) = \frac{\chi^2_{1-\gamma_1}(\xi)}{S_1}\,,$$
$$\bar{R}(\xi, \gamma_2) = \frac{\chi^2_{\gamma_2}(\xi)}{S_m} \tag{15.68}$$

where S_1 and S_m are the minimum and maximum testing extents for the system units, respectively. System (15.63) takes the form

$$\frac{\chi^2_\alpha(C)}{S_1} \geq R_0$$
$$\frac{\chi^2_{1-\beta}(C)}{S_m} \leq R_1 \,. \tag{15.69}$$

Let the number of units in the system, $m = 3$. Testing extents are $S_1 = 95$, $S_2 = 80$, and $S_3 = 60$. Specified levels are $R_0 = 0.100$ and $R_1 = 0.300$. Risks of the first and second kind are $\alpha = 0.2$ and $\beta = 0.1$. With $\alpha = 0.2$, from the first inequality in (15.69) we find that

$$C = \min\left\{\xi : \frac{\chi^2_\alpha(2\xi + 2)}{S_1} \geq R_0\right\} = 12 \,. \tag{15.70}$$

The corresponding value of the lower limit is $R(C, 1-\alpha) = 0.104 > R_0$. The upper limit, $R(C, 1-\beta)$, is 0.296, which is within the interval (R_0, R_1). The criterion is developed with risks not worse than $\alpha = 0.2$ and $\beta = 0.1$. The H_0 decision is made if $d_1 + \cdots + d_m < 12$. Otherwise the H_1 decision is made.

The lower and upper confidence limits, $R(\xi, \gamma_1)$ and $R(\xi, \gamma_2)$, can in general be found from

$$F_*(R, \xi) = \gamma_1 \,,$$
$$F^*(\bar{R}, \xi) = 1 - \gamma_2 \,, \tag{15.71}$$

where

$$F_*(R, \xi) = \min_{R(0)=R} F(\theta, \xi) \,, \tag{15.72}$$

$$F^*(R, \xi) = \max_{R(0)=R} F(\theta, \xi) \,. \tag{15.73}$$

$F(\eta, \xi)$ is the distribution for statistic ξ with a given η. Functions (15.72) and (15.73) are assumed to decrease monotonically with respect to R. The assumption corresponds to increase of the limits $\underline{R}$ and $\bar{R}$ with regard to ξ. The basic system of inequalities (15.63) can be rewritten into the following equivalent form:

$$F_*(R_0, C) = 1 - \alpha \,,$$
$$F^*(R_1, C) \leq \beta \,. \tag{15.74}$$

The lower and upper confidence limits, $\underline{R}(\xi, \gamma_1)$ and $\bar{R}(\xi, \gamma_2)$ can be found from

$$\begin{aligned} \bar{K}(\bar{R}, \gamma_1) &= \xi , \\ \underline{K}(\bar{R}, \gamma_2) &= \xi . \end{aligned} \tag{15.75}$$

In equation set (15.75),

$$\bar{K}(R, \gamma) = \max_{R(\theta)=\mathrm{R}} k_\gamma(\theta) , \tag{15.76}$$

$$\underline{K}(R, \gamma) = \min_{E(\theta)=R} k_{1-\gamma}(\theta) , \tag{15.77}$$

where $k_\gamma(\eta)$ is the quantile at level γ for the distribution function $F(\eta, \xi)$ of statistic ξ with a given η. Functions (15.76) and (15.77) are assumed to increase monotonically with respect to R. The basic system of inequalities (15.63) can be written in equivalent form:

$$\begin{aligned} \bar{K}(R_0, 1-\alpha) &\le C , \\ \underline{K}(R_1, 1-\beta) &\ge C . \end{aligned} \tag{15.78}$$

This form of the basic inequalities is convenient for finding the testing extent for specified values of R_0, R_1, α, and β. We may denote the dependencies of (15.76) and (15.77) on the testing extent, n, with

$$\begin{aligned} \bar{K} &= \bar{K}(R_0, \gamma, n) , \\ \underline{K} &= \underline{K}(R_1, \gamma, n) . \end{aligned}$$

Upon eliminating C, inequalities (15.78) take the form

$$\bar{K}(R_0, 1-\alpha, n) \le \underline{K}(R_1, 1-\beta, n) . \tag{15.79}$$

In a typical case, the left-hand side of (15.79) decreases monotonically with respect to n, while the right-hand side increases. Therefore the minimum required test extent for the specified risks α and β is found as the minimum number n that fulfills inequality (15.79). This inequality is convenient in particular for cases for which statistic ξ represents the unbiased point estimate for R and a normal approximation can be used for the distribution of ξ. The values $\underline{K}$ and $\bar{K}$ might be presented in the following form:

$$\bar{K}(R, \gamma, n) = R + u_\gamma \sqrt{\mathrm{Var}_+\{R\}} , \tag{15.80}$$

$$\underline{K}(R, \gamma, n) = R - u_\gamma \sqrt{\mathrm{Var}_+\{R\}} , \tag{15.81}$$

where

$$\mathrm{Var}_+\{R\} = \max_{R(\theta)=R} \mathrm{Var}\{\theta\} . \tag{15.82}$$

$\text{Var}\{\eta\} = E_\eta(\hat{R} - R)^2$ is the variance of the unbiased estimate $\xi = \hat{R}$ with a given η, and u_γ is the γ quantile of a standard normal distribution. With independent trials, the values of $\text{Var}\{\eta\}$ and $\text{Var}_{\pm}\{R\}$ are inversely proportional to the characteristics of the testing extent, n. The problem is reduced to a search for a $\text{Var}_{\pm}\{R\}$ that is the maximum variance of the unbiased estimate for P with a given value of the index R.

CHAPTER 16

BAYESIAN RELIABILITY ESTIMATION

16.1 INTRODUCTION

The Bayesian method takes its name from the use of Bayes' theorem and the associated approach developed by Thomas Bayes in his eighteenth-century work. The method differs from classical methods in both practical and philosophical aspects. The practical distinction lies in the fact that Bayesian methods permit the combination of operating data with any other relevant information available for reliability studies. Some possible sources of supplemental information are engineering design and test data, operating data in different environments, engineering judgment, and personal experience with similar equipment. The philosophical distinctions lie in the interpretation of probability mainly as a subjective "degree of belief" rather than a relative frequency, as is the case in classical statistics.

Information that is available when the analysis first begins is called *prior information*. The analyst's prior belief about the value of a reliability parameter in a Bayesian analysis is embedded in a "prior distribution." The problem is to obtain the estimate of the reliability parameter by combining the prior information (in the form of the prior distribution) with experimental information (in the form of a sample). Such an opportunity is given by Bayes' theorem, which transforms the prior distribution into the "posterior distribution", which is the basis of a second estimate of the reliability parameters.

As is shown in this chapter, one of the main attractive aspects of a Bayesian approach is its mathematical and methodological strictness. Besides that, a Bayesian approach has an important practical advantage: it saves money and time required for experiments because it utilizes prior information.

16.2 GENERAL BAYESIAN METHOD

The general Bayesian method of reliability estimation is depicted in Figure 16.1. Four variables must be determined by an analyst.

1. The conditional probability distribution of a random variable, the time-to-failure, ξ, given by the probability density function (p.d.f.) $f(t; \boldsymbol{\theta})$, where $\boldsymbol{\theta} = (\theta_1, \ldots, \theta_k)$, denotes a vector of parameters.
2. The type of life testing that produces the ordered sample data $\mathbf{z}$. There may be censored data; that is, there may be items that had not failed when the testing stopped.
3. The joint prior probability distribution of $\boldsymbol{\theta}$—a prior p.d.f. $h(\theta)$ that is characteristic of the distribution of the ξ in (1).
4. A loss function, $L(\hat{R}, R)$, that reflects the discrepancy between the real value of a reliability index, $R(\theta)$, and the estimate $\hat{R}$. R is a function of a random parameter $\boldsymbol{\theta}$ and in consequence has to be random as well. A Bayesian reliability estimation has three stages given below.

Stage 1. Determining the likelihood function. The likelihood function generally is a joint p.d.f. of sample data considered as a function of $\boldsymbol{\theta}$. The precise form of the likelihood function depends on the type of life testing. For example, if $\mathbf{z}$ is an uncensored sample, the function can be written

$$l(\boldsymbol{\theta} \mid \mathbf{z}) = \prod_{1 \le i \le n} f(t_1; \boldsymbol{\theta}) , \qquad (16.1)$$

where t_1 denotes the observed time to failure.

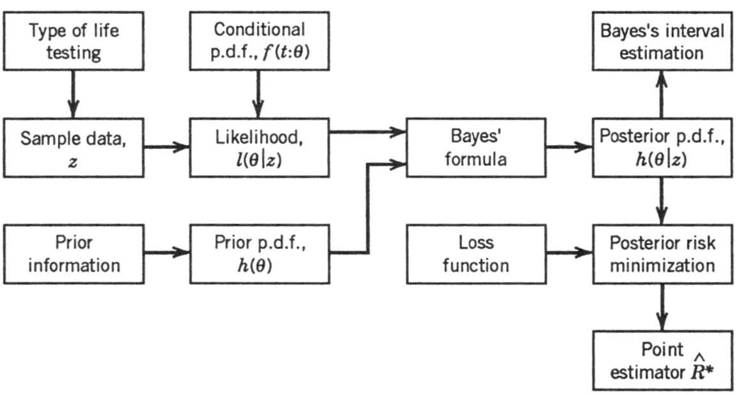

Figure 6.1 Bayesian reliability estimation.

Stage 2. Assigning the posterior distribution. Bayes' hypothesis is used in this stage. The hypothesis states that the posterior model is related to the prior model and the sampling data by

$$h(\boldsymbol{\theta} \mid \mathbf{z}) = \alpha h(\boldsymbol{\theta}) l(\boldsymbol{\theta} \mid \mathbf{z}) . \tag{16.2}$$

This fundamental relationship says that the posterior distribution is proportional to the product of the prior distribution and the likelihood function. The constant of proportionality α ensures that the posterior p.d.f. integrates (for continuous data) or sums (for discrete data) to one:

$$\beta = \int_{\theta} h(\boldsymbol{\theta}) l(\boldsymbol{\theta} \mid \mathbf{z})\, d\boldsymbol{\theta} . \tag{16.3}$$

Stage 3. Bayesian inference. Obtaining the Bayesian estimators. Two types of Bayesian estimates commonly are made: the point estimate and the interval estimate. For determining the point estimate, the posterior risk is recorded for a given loss function $L(\hat{R}, R)$ as

$$G(\hat{R}) = \int_{\theta} L(\hat{R}, R(\boldsymbol{\theta})) h(\boldsymbol{\theta} \mid \mathbf{z})\, d\boldsymbol{\theta} . \tag{16.4}$$

Of all possible estimates $R \in [0, 1]$, a single $\hat{R}^*$ is selected so that

$$\hat{R}^* = \arg \min_{\hat{R} \in [0,\, 1]} G(\hat{R}) . \tag{16.5}$$

If the squared-error loss function $L(\hat{R}, R) = (\hat{R} - R)^2$ is selected, a point estimator is the posterior mean of $R(\boldsymbol{\theta})$:

$$\hat{R}^* = \int_{\theta} R(\boldsymbol{\theta}) h(\boldsymbol{\theta} \mid \mathbf{z})\, d\boldsymbol{\theta} . \tag{16.6}$$

For Bayesian interval estimation, the following relationship is used:

$$Pr\{\underline{R} \leq R(\boldsymbol{\theta}) \leq \bar{R}\} = \gamma , \tag{16.7}$$

where $\underline{R}$ and $\bar{R}$ denote the lower and the upper limits of the interval, respectively. Relationship (16.7) may be expressed with

$$\int_{\underline{R} \leq R(\theta) \leq \bar{R}} h(\boldsymbol{\theta} \mid \mathbf{z})\, d\boldsymbol{\theta} = \gamma . \tag{16.8}$$

In addition the minimum of the length $|\bar{R} - \underline{R}|$ might be required. We will denote this interval as $[\bar{R} - \underline{R}]_{\gamma}$.

EXAMPLE 16.1

A random variable time to failure has an exponential distribution

$$f(t) = f(t; \lambda) = \lambda e^{-\lambda t}, \qquad t \geq 0,\ \lambda \geq 0 . \tag{16.9}$$

We select a gamma distribution of parameter λ as the prior distribution:

$$h(\lambda) = \frac{1}{\Gamma(\delta)} \rho^{\delta} \lambda^{\delta-1} e^{-\rho\lambda}, \qquad \lambda > 0,\ \delta \geq 0,\ \rho \geq 0 . \tag{16.10}$$

Hyperparameters δ and ρ are assumed to be known. We choose the squared-error loss function $L(R, \hat{R}) = (R - \hat{R})^2$.

During a life test, d unit failures are observed with ordered times $t_1^*, \ldots, t_d^*$. Another $k = n - d$ units are tested for times $t_1, \ldots, t_k$ without failing. We have to estimate the failure rate, λ. The estimation procedure follows the three steps mentioned previously.

Step 1. $\mathbf{z} = (z_1, \ldots, z_n)$ denotes the sample information $\mathbf{z} = (\mathbf{t}_{-*}, \bar{\mathbf{t}})$, where $\mathbf{t}^* = (t_1^*, \ldots, t_d^*)$ and $\bar{\mathbf{t}} = (t_1, \ldots, t_k)$. For the censored sample $\mathbf{z}$, we have the likelihood function

$$l(\boldsymbol{\theta} | \mathbf{z}) = \prod_{1 \leq i \leq d} f(t_i^*; \boldsymbol{\theta}) \prod_{1 \leq j \leq k} [1 - F(t_j; \boldsymbol{\theta})] \tag{16.11}$$

where $f(t; \theta)$ is the p.d.f. of the failure time ξ given parameter θ, and $F(t; \boldsymbol{\theta})$ denotes the conditional distribution function (c.d.f.) of ξ given θ. Substituting (16.9) into (16.11) yields the final equation of the likelihood function:

$$l(\lambda | \mathbf{z}) = \prod_{1 \leq i \leq d} \lambda e^{-\lambda t_i^*} \prod_{1 \leq j \leq k} e^{-\lambda t_j} = \lambda^d e^{-\lambda \kappa}, \tag{16.12}$$

$$\kappa = \sum_{1 \leq i \leq d} t_i^* + \sum_{1 \leq j \leq k} t_j .$$

Step 2. Substituting (16.10) and (16.12) into (16.2) gives the equation for the posterior p.d.f. of λ:

$$h(\lambda | \mathbf{z}) = \frac{(\rho + \kappa)^{\delta + d}}{\Gamma(\delta + d)} \lambda^{\delta + d + 1} e^{-\lambda(\rho + \kappa)} . \tag{16.13}$$

This p.d.f. is also a gamma distribution.

Step 3. After using the squared-error loss function, the point estimator can be obtained with (16.6). It is the expected value of the r.v. that has p.d.f. (16.13). The point estimator is obtained with

$$\hat{\lambda}^* = \int_0^{\infty} \lambda h(\lambda \mid \mathbf{z})\, d\lambda = \frac{\delta + d}{\rho + \kappa}\,. \tag{16.14}$$

Let us now obtain the $(100 - \gamma)$ upper limit of the interval for the estimate of λ, $\bar{\lambda}_\gamma$, that satisfies:

$$Pr\{\lambda \leq \hat{\lambda}_\gamma\} = \gamma\,.$$

With a gamma posterior p.d.f., it can be proved that this is a chi-squared r.v. with $2(\delta + d)$ degrees of freedom. Thus

$$Pr\{2\lambda(\rho + \kappa) \leq \chi^2_{\gamma, 2(\delta + d)}\} = \gamma\,,$$

where $\chi^2_{\gamma, 2(\delta + d)}$ is the $(100 - \gamma)$ percentage point of a χ^2 distribution with $2(\delta + d)$ degrees of freedom. Finally, the equation for the interval estimator can be given:

$$\bar{\lambda}_\gamma = \frac{\chi^2_{\gamma, 2(\delta + d)}}{2(\rho + \kappa)}\,.$$

Four aspects of Bayesian estimation are discussed in this chapter:

Bayesian parametric estimation has a given parametric family for the distribution of the time to failure.

Bayesian nonparametric estimation has an assumption that the probability distribution is nonparametric.

Complete prior definiteness means that there is a prior p.d.f. with a complete equation.

Partial prior definiteness means that there is an unknown prior p.d.f. with functional constraints.

16.3 PARAMETRIC BAYESIAN ESTIMATORS

The parametric estimating procedures follow the general procedure described above. This section considers different parametric families and prior distributions. A squared-error loss function is selected for each instance. The Bayesian point estimator of the reliability index is the posterior expected value of the r.v. in question. The accuracy of the estimator is indicated with a posterior variance or standard deviation.

Two types of prior distributions are used for reliability estimating, a uniform distribution on an interval and a conjugate prior. The latter has an analytical form of p.d.f. that coincides with the kernel of the likelihood function. The calculations sometimes require numerical integration or use of transcendental equations.

16.3.1 Binomial Model

A binomial model describes a number of failures, x, in n independent tests. The number of tests must be known in advance. The outcome (success or failure) is statistically independent for each test. The survival probability p (the estimated reliability parameter) is constant across the test. The number of failures has a binomial distribution:

$$f(x; n, d) = \frac{n!}{(n-x)!x!} p^{n-x}(1-p)^x . \tag{16.15}$$

The equations for the survival probability estimators—namely the posterior point estimator $\hat{p}^*$, variance $\mathrm{Var}\{p\}$, and $(100-\gamma)$ percentage lower probability limit—are given in Table 16.1.

Table 16.1 Bayesian estimators of survival probability for binomial model

Prior p.d.f.	Estimators
$h(p) = \dfrac{p^{a-1}(1-p)^{b-1}}{B(a, b)}$, $0 \le p \le 1,\ a \ge 0,\ b \ge 0$	$\hat{p}^* = \dfrac{a+n-d}{a+b+n}$, $\mathrm{Var}(p) = \dfrac{(a+n-d)(b+d)}{(a+b+n)(a+b+n+1)}$, $\underline{p}_\gamma = \left[1 + \dfrac{b+d}{a+n-d} F_{1-\gamma;\, 2(b+d),\, 2(a+n-d)}\right]^{-1}$
$h(p) = \dfrac{1}{B(a, b)(p_2 - p_1)} \times \left(\dfrac{p-p_1}{p_2-p_1}\right)^{a-1}\left(1 - \dfrac{p-p_1}{p_2-p_1}\right)^{b-1}$, $0 \le p_1 \le p \le p_2 \le 1,\ a \ge 0,\ b \ge 0$	$\hat{p}^* = p_1 + (p_2 - p_1)\dfrac{a+n-d}{a+b+n}$, $\mathrm{Var}(p) = (p_2-p_1)^2 \dfrac{(a+n-d)(b+d)}{(a+b+n)^2(a+b+n+1)}$ $\underline{p}_\gamma = p_1 + (p_2 - p_1) \times \left[1 + \dfrac{b+d}{a+n-d} \underline{F}_{1-\gamma;\, 2(b+d),\, 2(a+n-d)}\right]^{-1}$

Note:
$F_{\alpha;\nu_1,\nu_2}$ = the 100 − α percentage point of an F distribution with ν_1 and ν_2 degrees of freedom
$B(a, b)$ = a beta function
a = beta distribution parameter
b = beta distribution parameter
d = number of failures
n = number of trials
p = survival probability
p_1 = lower limit of p estimate interval
p_2 = upper limit of p estimate interval

EXAMPLE 16.2

Ten identical units are tested, and only one unit fails. The analyst hypothesizes that the survival probability lies in the interval [0.75, 1.0] and has a generalized beta distribution with $a = 8$ and $b = 1$. By substituting $n = 10$, $d = 1$, $p_1 = 0.75$, $p_2 = 1.0$, $a = 8$, and $b = 1$ into the equations in the second part of Table 16.1, we find that

$$\hat{p}^* = 0.75 + 0.25\,\frac{8 + 10 - 1}{8 + 1 + 10} = 0.9373\,,$$

$$\sigma_p = \frac{0.25}{19}\sqrt{\frac{17 \cdot 2}{20}} = 1.716 \cdot 10^{-2}\,.$$

With tables of Fisher's distribution, we obtain the value of the 10 percentage point with $2(b + d) = 4$ and $2(a + n - d) = 34$ degrees of freedom, namely $F_{0.1:4.34} = 2.12$ (the probability γ is $1.0 - 0.1 = 0.9$), and then we calculate

$$\underline{p}_{0.9} = 0.75 + \frac{0.25}{1 + (2/17) \cdot 2.12} = 0.9501\,.$$

16.3.2 Poisson Model

A Poisson model describes the number of failures, κ, in T time units, where T is known. Failures occur independently and at a constant rate, λ, for all items. The number of failures follows a Poisson distribution given by

$$f(x; T, \lambda) = \frac{(\lambda T)^x}{x!}\,e^{-\lambda T}\,, \qquad \lambda \geq 0\,. \tag{16.16}$$

Table 16.2 presents the equations for Bayesian estimators when the data consist of only the number of failures, d, that occur during T.

EXAMPLE 16.3

There are 29 systems, and each of them has 289 identical units in operation for 24 hours per day, 365 days per year. In a certain year there were a total of 32 unit failures. We need to estimate the failure rate, λ, when the prior distribution postulated for λ is uniform in the interval $[0, 10^{-5}]$ failures per hour (f/h).

First, we calculate the total time:

$$t = 8 \cdot 289 \cdot 8760 = 20{,}253,\,120 \text{ h}\,.$$

The Bayesian point estimate, from the upper part of Table 16.2, is

Table 16.2 Bayesian estimators of failure rate for Poisson model

Prior p.d.f.	Estimators
$h(\lambda) = \begin{cases} \dfrac{1}{\lambda_1}, & \text{if } 0 \le \lambda \le \lambda_1 \\ 0, & \text{elsewhere} \end{cases}$	$\hat{\lambda}^* = \dfrac{\Gamma(d+2, \lambda_1 T)}{T\Gamma(d+1, \lambda_1 T)}$, $\text{Var}(\lambda) = \dfrac{1}{T^2\Gamma(d+1, \lambda_1 T)} \times \left[\Gamma(d+3, \lambda_1 T) - \dfrac{\Gamma^2(d+2, \lambda_1 T)}{\Gamma(d+1, \lambda_1 T)}\right]$, $\Gamma(d+1, \underline{\lambda}_\gamma T) = \dfrac{1-\gamma}{2}\Gamma(d+1, \lambda_1 T)$, $\Gamma(d+1, \bar{\lambda}_\gamma T) = \dfrac{1+\gamma}{2}\Gamma(d+1, \lambda_1 T)$
$h(\lambda; \alpha, \beta) = \dfrac{\beta^\alpha \lambda^{\alpha-1} e^{-\beta\lambda}}{\Gamma(\alpha)}$, $\lambda, \alpha, \beta \ge 0$	$\hat{\lambda}^* = \dfrac{d+\alpha}{T+\beta}$, $\text{Var}(\lambda) = \dfrac{d+\alpha}{(T+\beta)^2}$, $\underline{\lambda}_\gamma = \dfrac{\chi^2(1-\gamma)/2; 2(d+\alpha)}{2(T+\beta)}$, $\bar{\lambda}_\gamma = \dfrac{\chi^2(1+\gamma)/2; 2(d+\alpha)}{2(T+\beta)}$

Note:
$\Gamma(a, z)$ = the incomplete gamma function
$\chi^2_{\alpha,\nu}$ = the $100 - \alpha$ percentage point of a chi-square distribution with ν degrees of freedom
λ_1 = upper limit of λ estimate interval

$$\hat{\lambda}^* = \frac{\Gamma(34)}{T\Gamma(33)} = \frac{8.6833 \cdot 10^{36}}{(20{,}253{,}120)(2.6313 \cdot 10^{35})}$$
$$= 16.3 \cdot 10^{-7} \text{ failures per hour}.$$

The posterior variance is $8.0451 \cdot 10^{-14}$. By solving two transcendental equations in the upper part of Table 16.2, the probability limits of the interval estimate are computed to be

$$\underline{\lambda}_{0.9} = 1.19 \cdot 10^{-6} \; f/\text{h},$$
$$\bar{\lambda}_{0.9} = 2.12 \cdot 10^{-6} \; f/\text{h}.$$

EXAMPLE 16.4

In Example 16.3 let the prior distribution be gamma with $\alpha = 2.45$ and $\beta = 183{,}200$. The prior mean and variance become $1.23 \cdot 10^{-5}$ and $7.30 \cdot 10^{-11}$, respectively. The Bayesian point estimate of λ, in accordance with the lower part of Table 16.2, is

$$\hat{\lambda}^* = \frac{32 + 2.45}{20{,}253{,}120 + 183{,}200} = 1.69 \cdot 10^{-6}\ f/\mathrm{h}\,,$$

with associated posterior variance $\mathrm{Var}\{\lambda\} = 8.25 \cdot 10^{-14}$.

Let us now find the upper one-sided 90 percent, Bayesian probability interval estimate (i.e., $\hat{\lambda}_{0.9}$). This estimate for λ is also given by the equation in the lower part of Table 16.2, in which $(1 + \gamma)/2$ is replaced by γ. Now $\chi^2_{0.9,68.9} = 84.307$, from which we have

$$\bar{\lambda}_{0.9} = \frac{84.307}{2 \cdot 20{,}436{,}320} = 2.06 \cdot 10^{-6}\ f/\mathrm{h}\,.$$

16.3.3 Exponential Model

The exponential model, which is widely used in reliability as a failure time distribution, has a p.d.f. given by

$$f(t; \lambda) = \lambda \exp(-\lambda t)\,, \qquad t \geq 0,\ \lambda \geq 0\,. \tag{16.17}$$

The reliability function is

$$R(t) = R(t; \lambda) = \exp(-\lambda t)\,. \tag{16.18}$$

The estimating procedure described here is based on the censored sample $z = (t_{-*}, \bar{t})$ shown in Example 16.1. During life testing, d failures occur for units that have been on test for ordered times $t_{-*} = (t_1^*, \ldots, t_d^*)$ and $k = n - d$ units are tested for times $\bar{t}^+ = (t_1, \ldots, t_k)$ without failing. The sufficient statistic consists of two numbers, the total time on test, $\kappa = t_1^* + \cdots + t_d^* + t_1 + \cdots + t_k$ and the number of failures, d. The expressions for the estimators of the failure rate, λ, and the survival probability, $R(t)$, are in Table 16.3.

If the prior mean value λ_0 and standard deviation σ_{16} are known, the parameters of the prior gamma distribution may be found with expressions in Table 16.3:

$$\epsilon = \frac{\lambda_0}{\sigma^2_{\lambda_0}}\,,$$

$$s = \left(\frac{\lambda_0}{\sigma_{\lambda_0}}\right)^2 - 1\,. \tag{16.19}$$

Table 16.3 Bayesian estimators of failure rate λ and survival probability $R(t)$ for exponential model

Prior p.d.f.	Estimators
$h(\lambda) = h(\lambda; s, \epsilon) = \dfrac{1}{\Gamma(s+1)} \lambda^s e^{-\lambda\epsilon}$, $s \geq 0,\ \lambda,\ \epsilon > 0$	$\hat{\lambda}^* = \dfrac{d+s+1}{\kappa+\epsilon}$, $\mathrm{Var}(\lambda) = \dfrac{d+s+1}{(\kappa+\epsilon)^2}$, $\underline{\lambda}_\gamma = \dfrac{\chi^2_{1-\gamma;2(d+s+1)}}{2(\kappa+\epsilon)}$ $\hat{R}^*(t) = \left(\dfrac{u}{u+1}\right)^{\upsilon+1}$, $u = \dfrac{\epsilon+\kappa}{t}$, $\upsilon = s + d$, $\mathrm{Var}(R(t)) = \left(\dfrac{u}{u+2}\right)^{\upsilon+1} - \left(\dfrac{u}{u+1}\right)^{2(\upsilon+1)}$, $\underline{R}_\gamma(t) = \exp\left[-\dfrac{\chi^2_{1-\gamma,2(d+s+1)}}{2u}\right]$
$h(r) = \begin{cases} 0, & r < p_1, \\ \dfrac{1}{p_2 - p_1}, & p_1 \leq r \leq p_2, \\ 0, & 0,\ r > p_2, \end{cases}$ $0 \leq p_1 < p_2 \leq 1$	$\hat{R}^*(t) = \dfrac{I_E(p_2, \omega+1, d) + I_E(p_1, \omega+1, d)}{I_E(p_2, \omega, d) + I_E(p_1, \omega, d)}$ $\mathrm{Var}(R(t)) = \dfrac{I_E(p_2, \omega+2, d) + I_E(p_1, \omega+2, d)}{I_E(p_2, \omega, d) + I_E(p_1, \omega, d)} - \hat{R}^{*2}(t)$, $I_E(R_\gamma(t), \omega, d) = (1-\gamma) I_E(p_2, \omega, d) + \gamma I_E(p_1, \omega, d)$, $I_E(x, a, n) = x^{a+1} \sum_{k=0}^{n} \dfrac{1}{(n-k)!} \dfrac{\lvert \ln^{n-k} x \rvert}{(a+1)^{k+1}}$, $\omega = \dfrac{\kappa}{t}$

EXAMPLE 16.5

During life testing one unit fails at 48 hours, another at 64 hours, and the other six units are tested without failing until hour 80. These is certain prior knowledge that $\lambda_0 = 0.4 \cdot 10^{-4}$ f/h and $\sigma_{\lambda_0} = 0.1 \cdot 10^{-4}$ f/h. We need to estimate the failure rate and the survival probability for $t = 70$ h with probability that $\gamma = 0.9$.

First, we calculate the total time on test:

$$\kappa = 48 \cdot 64 + 6 \cdot 80 = 592 \text{ h} .$$

Using equation (16.5), we estimate the values of the parameters of the prior gamma distribution, $\epsilon = 0.4 \cdot 10^6$ h, and $s = 15$. By the equations in the upper part of Table 16.3, the point estimate and standard deviation are

$$\hat{\lambda}^* = \frac{2 + 15 + 1}{0.4 \cdot 10^6 + 592} = 0.449 \cdot 10^{-4} \ f/\text{h} .$$

$$\sigma_\lambda = \frac{\sqrt{2 + 15 + 1}}{0.4 \cdot 10^6 + 592} = 0.106 \cdot 10^{-4} \ f/\text{h} .$$

Using a chi-square distribution table we can find $\chi^2_{0.1,36} = 47.212$, which gives

$$\bar{\lambda}_{0.9} = \frac{47.212}{2(0.4 \cdot 10^6 + 592)} = 0.589 \cdot 10^{-4} \ f/\text{h} .$$

Let us now estimate survival probability $R(70)$. In accordance with the first part of Table 16.3, we can calculate $u = 5772.7$ and $v = 17$. Then we compute

$$\hat{R}^*(70) = \left(\frac{5722.7}{5722.7 + 1}\right)^{17+1} = 0.9969 ,$$

$$\text{Var}\{R(70)\} = \left(\frac{5722.7}{5722.7 + 2}\right)^{17+1} - 0.9969^2 = 1.78 \cdot 10^{-6} ,$$

$$R_{0.9}(70) = e^{-47.212/2 \cdot 5722.7} = 0.9959 .$$

EXAMPLE 16.6

Eight units are tested for 80.4, 92.5, 108.3, 78.4, 96.5, 112.0, 108.4, and 89.3 hours. The seventh unit fails after 108.4 hours of normal functioning. All the other units finish without failures. According to prior judgment, the survival probability for $t = 70$ h lies in the interval [0.8, 1.01. We need to estimate $R(70)$. The estimates may be made with the lower part of Table 16.3. First, we calculate ω:

$$\omega = 80.4 + 92.5 + 108.3 + 78.4 + 96.5 + 112.0 + 108.4 + 89.3/70 = 10.94 \text{ h} .$$

Then we calculate I_E:

$$I_E(0.8, 10.94, 1) = 0.8^{11.94}\left(-\frac{\ln 0.8}{11.94} + \frac{1}{11.94^2}\right) = 1.79 \cdot 10^{-3} .$$

The other values of this function are calculated similarly:

$$I_E(1, 10.94, 1) = 7.014 \cdot 10^{-3} ,$$
$$I_E(0.8, 11.94, 1) = 1.294 \cdot 10^{-3} ,$$
$$I_E(1, 11.94, 1) = 5.972 \cdot 10^{-3} ,$$
$$I_E(0.8, 12.94, 1) = 9.428 \cdot 10^{-3} ,$$
$$I_E(1, 12.94, 1) = 5.146 \cdot 10^{-3} .$$

The posterior point estimate and standard deviation are computed to be

$$\hat{R}^*(70) = 0.8955 ,$$
$$\sigma_{\hat{R}(70)} = 0.517 \cdot 10^{-1} .$$

By solving the transcendental equation of Table 16.3 in the interval [0.8, 1.0] for $\gamma = 0.9$, we can obtain $R_{0.9}(70) = 0.8627$.

16.3.4 Linear Increasing Failure Rate (LIFR) Model

The LIFR model has an initial failure rate, λ_0, that increases linearly with time:

$$\lambda(t) = 2\alpha t + \lambda_0 , \tag{16.20}$$

where α and λ_0 denote the parameters of the model. The failure rate function (16.20) corresponds to the p.d.f.,

$$f(t) = f(t; \alpha, \lambda_0) = (2\alpha\tilde{t} + \tilde{\lambda}_0) \exp[-(\alpha t^2 + \lambda_0)] , \tag{16.21}$$

and the c.d.f.,

$$F(t) = F(t; \alpha, \lambda_0) = 1 - \exp[-(\alpha t^2 + \lambda_0)] . \tag{16.22}$$

The survival probability for time t is given by

$$R = R(t) = \exp[-(\alpha t^2 + \lambda_0)] .$$

Relationships (16.20) through (16.22) correspond to relationships for the exponential model except when $\alpha = 0$. However, these parameters as a whole have no clear engineering meaning. A clearer interpretation occurs when we use the following parameters: $v = \lambda(t)/\lambda(0)$, a parameter of degradation of a failure rate during the time t, and $r = R(t)$, an estimated survival probability. The relationship between (α, λ_0) and (v, r) may be expressed with

$$\alpha = -\frac{\ln r}{t^2}\,\frac{v-1}{v+1}\,,$$

$$\lambda_0 = -\frac{2\ln r}{t(v+1)}\,.$$

From these equations $\alpha = 0$ if the degradation parameter is one. This results in transformation of the LIFR model to the exponential one.

Table 16.4 Bayesian estimators of survival probability for LIFR model

Prior p.d.f.	Estimators
$h(r, v) = h_1(r)h_2(v)$	$\hat{R}^*(t) = \frac{1}{B}\int_{v_1}^{v_2} a(v)I(v, p_1, p_2, 1, d)\,dv$
$h_1(r) = \begin{cases} 0, & r < p_1 \\ \frac{1}{p_2 - p_1}, & p_1 \le r \le p_2 \\ 0, & r > p_2 \end{cases}$	$\mathrm{Var}(R(t)) = \frac{1}{B}\int_{v_1}^{v_2} a(v)I(v, p_1, p_2, 2, d)\,dv - \hat{R}^{*2}(t)$
$h_2(v) = \begin{cases} 0, & v < v_1 \\ \frac{1}{v_2 - v_1}, & v_1 \le v \le v_2 \\ 0, & v > v_2 \end{cases}$	$\int_{v_1}^{v_2} a(v)I(v, \underline{R}_\gamma(t), p_2, 0, d)\,dv = \gamma B$ $B = \int_{v_1}^{v_2} a(v)I(v, p_1, p_2, 0, d)\,dv$
$h(r) = \begin{cases} 0, & r < p_1 \\ \frac{1}{p_2 - p_1}, & p_1 \le r \le p_2 \\ 0, & r > p_2 \end{cases}$	$\hat{R}^*(t) = \frac{I(v, p_1, p_2, 1, d)}{I(v, p_1, p_2, 0, d)}$ $\mathrm{Var}(R(t))\ \frac{I(v, p_1, p_2, 2, d)}{I(v, p_1, p_2, 0, d)} - \hat{R}^{*2}(t)$ $I(v, \underline{R}_\gamma(t), p_2, 0, d) = \gamma I(v, p_1, p_2, 0, d)$

Note: Parameter v is assumed to be known.

$$\omega = \frac{1}{t}\left(\sum_1^d t_i^* + \sum_1^k t_j\right)$$

$$\epsilon = \frac{1}{t^2}\left(\sum_1^d t_i^{*2} + \sum_1^k t_j^2\right)$$

$$b(x) = \frac{1}{x+1}\left[(x-1)\epsilon + 2\omega\right]$$

$$a(x) = \frac{1}{(x+1)^d}\prod_1^d\left[(x-1)\frac{t_i^*}{t} + 1\right]$$

$I(x, p_1, p_2, k, d) = I_E(p_2, b(x) + k, d) - I_E(p_1, b(x) + k, d)$

Function $I_E(x, a, n)$ is given in table 6.3.

In Table 16.4 the equations for the reliability estimators are presented for a uniform prior distribution and sample data $z = (T^{-*}, \bar{t})$ described in Section 16.3. Software for numerical integration with transcendental equations may be needed to make this approach practical (see Appendix F).

EXAMPLE 16.7

During life testing, one failure occurs at $t = 130$ hours. The other units are tested for 240, 310, 280, 160, 240, 340, 250, and 300 hours without failing. We need to estimate survival probability $R(100)$ when $R(100)$ and the degradation parameter v have uniform prior distributions in the intervals $[0.9, 1.0]$ and $[1.0, 2.0]$, respectively. The posterior point estimate, standard deviation, and lower one-sided probability limit are calculated to be

$$\hat{R}^*(100) = 0.9576 ,$$

$$\sigma_{R(100)} = 0.0242 ,$$

$$R_{0.9}(100) = 0.9216 .$$

If the degradation parameter v is assumed to be equal to 2.0, these estimates may be found, in accordance with the second part of Table 16.4, to be

$$\hat{R}^*(100) = 0.9729 ,$$

$$\sigma_{R(100)} = 0.0181 ,$$

$$R_{0.9}(100) = 0.9476 .$$

16.3.5 Weibull-Gnedenko Model

The Weibull-Gnedenko distribution is widely used for failure times. It is given by

$$F(t; \lambda, \alpha) = \lambda \alpha t^{\alpha - 1} \exp(-\lambda t^{\alpha}) , \qquad t \geq 0, \lambda > 0, \alpha > 0 . \qquad (16.23)$$

In practice it is more difficult to assign a prior distribution to the parameter λ than to the parameter $r = R(t)$, the survival probability for time t. We therefore will use parameters (r, α) instead of (λ, α). There are many different types of Bayesian estimation for this case, but for simplicity Table 16.5 gives only uniform prior distributions for parameters r and α. The sample data are the same as in Section 16.3—the censored sample $z = (t^{-*}, \bar{t})$.

Table 16.5 Bayesian estimators of survival probability for Weibull model

Prior p.d.f.	*Estimators*
$h(r, \alpha) = h_1(r)h_2(\alpha)$ $h_1(r) = \begin{cases} 0, \ r < p_1 \\ \dfrac{1}{p_2 - p_1}, \ p_1 \le r \le p_2 \\ 0, \ r > p_2 \end{cases}$ $h_2(\alpha) = \begin{cases} 0, \ \alpha < \alpha_1 \\ \dfrac{1}{\alpha_2 - \alpha_1}, \ \alpha_1 \le \alpha \le \alpha_2 \\ 0, \ \alpha > \alpha_2 \end{cases}$	$\hat{R}^*(t) = \dfrac{1}{B} \displaystyle\int_{\alpha_1}^{\alpha_2} x^d \mu^{x-1} I_W(x, p_1, p_2, 1, d)\, dx$ $\text{Var}(R(t)) =$ $\dfrac{1}{B} \displaystyle\int_{\alpha_2}^{\alpha_2} x^d \mu^{x-1} I_W(x, p_1, p_2, 2, d)\, dx - \hat{R}^{*2}(t)$ $\displaystyle\int_{\alpha_1}^{\alpha_2} x^d \mu^{x-1} I_W(x, \underline{R}_\gamma(t), p_2, 0, d)\, dx = \gamma B$ $B = \displaystyle\int_{\alpha_1}^{\alpha_2} x^d \mu^{x-1} I_W(x, p_1, p_2, 0, d)\, dx$
$h(r) = \begin{cases} 0, \ r < p_1 \\ \dfrac{1}{p_2 - p_1}, \ p_1 \le r \le p_2 \\ 0, \ r > p_2 \end{cases}$	$\hat{R}^*(t) = \dfrac{I_W(\alpha, p_1, p_2, 1, d)}{I_W(\alpha, p_1, p_2, 0, d)}$ $\text{Var}(R(t)) = \dfrac{I_W(\alpha, p_1, p_2, 2, d)}{I_W(\alpha, p_1, p_2, 0, d)} - \hat{R}^{*2}(t)$ $I_W(\alpha, \underline{R}_\gamma(t), p_2, 0, d) = \gamma I_W(\alpha, p_1, p_2, 0, d)$

$$\mu = \prod_1^d \frac{t_i^*}{t}$$

$$\epsilon(\alpha) = \frac{1}{t^\alpha}\left(\sum_1^d t_i^{*\alpha} + \sum_1^k t_j^\alpha\right)$$

$$I_W(\alpha, p_1, p_2, k, d) = I_E(p_2, \epsilon(\alpha) + k, d) - I_E(p_1, \epsilon(\alpha) + k, d)$$

Function $I_E(x, a, n)$ is given in Table 16.3.

EXAMPLE 16.8

Twelve units did not fail. The times were 190, 350, 240, 310, 280, 160, 240, 340, 420, 180, 250, and 300 hours. One unit failed at 270 hours. Based on preliminary analysis, a Weibull-Gnedenko distribution was postulated. The analyst put α in the interval [1.0, 1.8]. The problem was to estimate survival probability $R(100)$ when this probability is believed to belong to the interval [0.90, 1.001]. Using special software and the first part of Table 16.5 the variables for $R(100)$ were computed to be

$$\hat{R}^*(100) = 0.9666\ ,$$
$$\sigma_{R(100)} = 0.0282\ ,$$
$$R_{0.9}(100) = 0.9291\ .$$

The scale parameter was then postulated to be 1.8. The lower part of Table 16.5 was then used and the variables were calculated to be

$$\hat{R}^*(100) = 0.9766\ ,$$
$$\sigma_{R(100)} = 0.0163\ ,$$
$$R_{0.9}(100) = 0.9564\ .$$

16.4 NONPARAMETRIC BAYESIAN METHODS

A nonparametric statistical procedure is one in which a model is not required to be specified. Thus nonparametric Bayesian estimators cover more situations than parametric ones. There are a number of theoretical Bayesian nonparametric statistical methods. Most of them are so complicated that it is difficult to use them in reliability practice. In this section the simplest methods are considered. Two classes of nonparametric estimations are common. One class has all time-to-failure probability distributions without any restrictions. The estimators are called "general." The second class is for increasing failure rate (IFR) distributions. Only approximate solutions can be found for this class.

16.4.1 General Nonparametric Bayesian Estimators

The more common Bayesian nonparametric method uses Dirichlet finite prior increments. Let $F(t)$, $t \in [0, \omega)$ be any cumulative distribution function over a lifetime. Prior information is available for the interval $[0, \omega)$, and quantity $\alpha([a, b])$ is given for each interval $[a, b]$ from $[0, \omega)$. Let $F_0(t)$ be the prior lifetime d.f. It can be represented by a parametric model. Then for any subinterval $[0, t]$ from $[0, \omega)$, the metric α is determined from

$$F_0(t) = \frac{\alpha([0, t])}{\alpha([0, \infty))} = \frac{\alpha([0, t])}{\beta}\ , \tag{16.24}$$

where $\beta = \alpha([a, b])$.

The value of β is selected on the basis of the analyst's judgment of the relative significance of the prior data and the equivalent number of trials. If the analyst believes the available prior information is of the same significance as the test data, he or she should let β equal the number of trials, n. If the prior information is as significant as half the test data, the analyst should let $\beta = n/2$. A prior metric thus is given completely by a prior judgment about the distribution of a lifetime $F_0(t)$, and the quantity of the measure of significance of the prior information. When $F_0(t)$ is given, a prior metric can be determined from (16.24):

$$\beta F_0(t) = \alpha([0, t]) . \tag{16.25}$$

16.4.2 Point Posterior Estimate with Complete Sample Data

Let the times to failure for n tests be given by the vector $\mathbf{t}^* = (t_1^*, \ldots, t_d^*)$. A posterior nonparametric estimate of $F(\mathbf{t})$ for a given $\mathbf{t}$ may be made as a weighted average of the prior form $F_0(t)$ and the empirical d.f. $F_n(\mathbf{t}|\bar{\mathbf{t}}^*)$:

$$\hat{F}^*(t|\mathbf{t}^*) = p_n F_0(\mathbf{t}) + (1 - p_n) F_n(\mathbf{t}|\mathbf{t}^*) , \tag{16.26}$$

where $p_n = \beta/(\beta + n)$.

A survival probability, $R(t)$, can be estimated with (16.26) as follows:

$$\hat{R}^*(t) = p_n R_0(t) + (1 - p_n)\left[1 - \frac{d(t)}{n}\right] , \tag{16.27}$$

where $R_0(t)$ is the prior estimate of the survival probability and d denotes the number of failures in the interval $[0, t]$.

EXAMPLE 16.9

Six life tests resulted in a vector of failure times in hours of $\hat{\mathbf{t}}^* = (124, 118, 148, 106, 92, 134)$. The prior estimate of the survival probability at 110 hours was 0.95. The prior distribution was exponential:

$$F_0(\mathbf{t}) = 1 - \exp(-\lambda_0 t) .$$

The prior information and the test data have the same significance. We need to estimate the survival probability for 100 hours. First, we estimate the value of the parameter λ_0 using the fact that $p_0(110) = 0.95$:

$$\lambda_0 = \frac{-\ln P_0(110)}{110} = 4.66 \cdot 10^{-4} \ f/\mathrm{h} .$$

Because the prior information and the test data are considered to have the same significance, $\beta = n = 6$, the parameter p_n is 0.5. The number of failures before time 100 is one; that is, $d(100) = 1$. By (16.27) the required point estimate is

$$\hat{P}^*(100) = 0.5e^{-4.66 \cdot 10^{-4} \cdot 100} + 0.5\left(1 - \frac{1}{6}\right) = 0.8939 .$$

16.4.3 Posterior Estimate with Unfailed Test Units

Suppose that a life test has d failures at ordered times $t_1^*, \ldots, t_d^*$, and $k = n - d$ units tested for times $t_1, \ldots, t_k$ without failures. You can transform the presentation of the subsample. Let $t_{(1)}, \ldots, t_{(m)}$ denote the m censoring times among the censored observations $t_1, \ldots, t_k$. Let also l_1 denote the number of observations censored at t_1. Let $N(u)$ denote the number of observations (censored or not) smaller than or equal to u. Let $N^+(u)$ be the number greater than u. Using the metric $\alpha(\cdot)$ given by (16.25) as the parameter of the prior Dirichlet process and a squared-error loss function, the Bayesian point estimate of survival probability $R(t)$ becomes

$$\hat{R}^*(t) = \frac{\beta R_0(t) + N^+(t)}{\beta + n} \prod_{1 \le j \le i} \frac{\beta R_0(t_{(j)}) + N(t_{(j)})}{\beta R_0(t_{(j)}) + N(t_{(j)}) - l_j} . \tag{16.28}$$

In the interval $t_{(1)} < t < t_{(1+i)}$, $0 \le i \le m$, with $t_{(0)} = 0$, $t_{(m+1)} = \omega$, $R_0(t) = 1 - F(t)$.

The posterior variance is

$$\operatorname{Var}\{R(t)\} = \hat{R}^*(t)\Bigg[\frac{\beta R_0(t) + N^+(t) + 1}{\beta + n + 1} \times \prod_{1 \le j \le i} \frac{\beta R_0(t_{(j)}) + N(t_{(j)}) + 1}{\beta R_0(t_{(j)}) + N(t_{(j)}) - l_j + 1} - \hat{R}^*(t)\Bigg] .$$

If $\beta = 0$ the prior information is ignored, and (16.28) is reduced to the Kaplan–Meier estimator.

EXAMPLE 16.10

During a life test the failure times in hours are $t_1^* = 36$, $t_2^* = 42$, and $t_3^* = 46$. The censoring times are $t_1 = 20$, $t_2 = 34$, $t_3 = 38$, and $t_4 = 20$. The prior model is exponential, $F_0(t) = 1 - \exp(-\lambda t)$, and the survival probability for 30 hours is known to equal 0.95. We want to obtain the point posterior estimate for $R(t)$ for a given value of β.

After reordering the subsample of censored data, we have $t_{(1)} = 20$, $t_{(2)} = 34$, and $t_{(3)} = 38$. In addition $l_1 = 2$, and $l_2 = l_3 = 1$. We can use (16.28) directly. The calculation is given in Table 16.6.

Let β equal 7, which means that the prior information and the test data are equal in significance. With $R_0(30) = 0.95$, $\lambda = -(\ln 0.95)/30 = 1.71 \cdot 10^{-3}$ f/h. From the second line of Table 16.5 we calculate

$$\hat{R}^*(30) = \frac{e^{-1.71 \cdot 10^{-3} \cdot 30} + 5}{7 + 7} \, \frac{7e^{-1.71 \cdot 10^{-3} \cdot 20} + 7}{7e^{-1.71 \cdot 10^{-3} \cdot 20} + 5} = 0.9736 .$$

The corresponding posterior variance is found to be $1.82 \cdot 10^{-3}$.

Table 16.6 Bayesian estimation of survival probability for Example 16.10

Interval	Point Estimate
$[0, 20)$	$\dfrac{\beta e^{-\lambda t} + 7}{\beta + 7}$
$[20, 34)$	$\dfrac{\beta e^{-\lambda t} + 5}{\beta + 7} \times \dfrac{\beta e^{-20\lambda} + 7}{\beta e^{-20\lambda} + 5}$
$[34, 36)$	$\dfrac{\beta e^{-\lambda t} + 4}{\beta + 7} \times \dfrac{\beta e^{-20\lambda} + 7}{\beta e^{-20\lambda} + 5} \times \dfrac{\beta e^{-34\lambda} + 5}{\beta e^{34\lambda} + 4}$
$[36, 38)$	$\dfrac{\beta e^{-\lambda t} + 3}{\beta + 7} \times \dfrac{\beta e^{-20\lambda} + 7}{\beta e^{-20\lambda} + 5} \times \dfrac{\beta e^{-34\lambda} + 5}{\beta e^{-34\lambda} + 4}$
$[38, 42)$	$\dfrac{\beta e^{-\lambda t} + 2}{\beta + 7} \times \dfrac{\beta e^{-20\lambda} + 7}{\beta e^{-20\lambda} + 5} \times \dfrac{\beta e^{-34\lambda} + 5}{\beta e^{-34\lambda} + 4} \times \dfrac{\beta e^{-38\lambda} + 3}{\beta e^{-38\lambda} + 2}$
$[42, 46)$	$\dfrac{\beta e^{-\lambda t} + 1}{\beta + 7} \times \dfrac{\beta e^{-20\lambda} + 7}{\beta e^{-20\lambda} + 5} \times \dfrac{\beta e^{-34\lambda} + 5}{\beta e^{-34\lambda} + 4} \times \dfrac{\beta e^{-38\lambda} + 3}{\beta e^{-38\lambda} + 2}$
$[46, \infty)$	$\dfrac{\beta e^{-\lambda t}}{\beta + 7} \times \dfrac{\beta e^{-20\lambda} + 7}{\beta e^{-20\lambda} + 5} \times \dfrac{\beta e^{-34\lambda} + 5}{\beta e^{-34\lambda} + 4} \times \dfrac{\beta e^{-38\lambda} + 3}{\beta e^{-38\lambda} + 2}$

16.4.4 Approximate Estimation for IFR Distribution

Increasing (more accurately nondecreasing) failure rate distributions have a function

$$\eta(t) = \frac{\Lambda(t)}{t} = \frac{1}{t}\int_0^t \lambda(x)\, dx$$

that is nondecreasing in the interval $(0, \infty)$. $\lambda(t)$ is the failure rate, and

$$\Lambda(x) = \int_0^x \lambda(t)\, dt = -\ln[1 - F(x)]\,.$$

The function $\Lambda(t)$ can be approximated with a two-parameter function:

$$\tilde{\Lambda}(x) = \delta(t - x)\lambda_0 x + \delta(x - 1)[\lambda_0 x - (\lambda_1 - \lambda_0)t]\,,$$

where t is the time for which the survival probability, $R(t)$, is estimated and the δ function is

$$\delta(y) = \begin{cases} 0 & \text{if } x < 0\,, \\ 1 & \text{otherwise}\,. \end{cases}$$

The parametric distribution

$$\tilde{F}(t; \lambda_0, \lambda_1) = 1 - e^{-\tilde{\Lambda}(t; \lambda_0, \lambda_1)}$$

is used for $F(t)$, which is unknown. The regions for parameters λ_0 and λ_1 are

$$0 \leq \lambda_0 \leq \omega\,, \quad \lambda_0 \leq \lambda_1 < \omega\,.$$

We consider the use of prior information when an interval $[R_1, R_2]$ is given such that $R(t) \in [R_1, R_2]$. We also assume that $R(t)$ has a uniform prior distribution in the interval $[R_1, R_2]$. The prior distribution of λ_0 and λ_1 is obtained under these assumptions.

The survival probability, $R(t)$, is estimated for the censored sample $z = (\mathbf{t}^*, \bar{\mathbf{t}})$ mentioned above. The procedure of obtaining estimators of the survival probability follows in general the method described in Section 16.2 for the parametric family given by $\tilde{F}(t)$. The sufficient statistic is determined with the total number of failures, d, the number, d_1, of failures that occurred after t, the total relative time on test, ω:

$$\omega = \frac{1}{t}\left[\sum_{1\leq i\leq d} t_i^* + \sum_{1\leq j\leq k} t_j\right],$$

and the relative time on test after t, ω_1:

$$\omega_1 = \sum_{1\leq i\leq d}\left(\frac{t_i^*}{t} - 1\right)\delta\left(\frac{t_i^*}{t} - 1\right) + \sum_{1\leq j\leq k}\left(\frac{t_j}{t} - 1\right)\delta\left(\frac{t_j}{t} - 1\right).$$

The final Bayesian estimators are

$$\hat{R}^*(t) = \frac{I_c(1)}{I_c(0)}\,,$$

$$\mathrm{Var}\{R(t)\} = \frac{I_c(2)}{I_c(0)} - [\hat{R}^*(t)]^2\,, \tag{16.29}$$

where the function $I_c(m)$ is given by

$$I_c(m) = \sum_{0\leq i\leq d} \frac{d_1^{(i)}}{(\omega_1 + 1)^{i+1}} \sum_{0\leq j\leq d-i} \frac{(d-i)^{(j)}}{(\omega + m + 1)^{j+1}}$$
$$\times [R_2^{\omega+m+1}|\ln R_2|^{d-i-j} - R_1^{\omega+m+1}|\ln R_1|^{d-i-j}]\,,$$

$$n^{(k)} = \frac{n!}{(n-k)!}\,. \tag{16.30}$$

The $(100 - \gamma)$ percentage confidential level lower limit can be found from the following transcendental equation:

$$\sum_{0\leq i\leq d} \frac{d_1^{(i)}}{(\omega_1 + 1)^{i+1}} \sum_{0\leq j\leq d-i} \frac{(d-i)^{(j)}}{(\omega + m + 1)^{j+1}}$$
$$\times [R_2^{\omega+m+1}|\ln R_2|^{d-i-j} - \bar{R}_\gamma(t)^{\omega+m+1}|\ln \bar{R}_\gamma|^{d-i-j}] - \gamma I_c(0) = 0\,. \tag{16.31}$$

EXAMPLE 16.11

The life testing data in hours are $\bar{t}^* = (150, 160, 180)$ and $\bar{t} = (120, 140, 90, 160, 95, 115, 105)$. On the basis of prior information, we assume that $r(100)$ is in the interval $[0.8, 1.0]$. We wished to estimate $R(100)$ using a uniform prior distribution in this interval.

First, we find the values of the variables that determine the sufficient statistic. They are calculated to be $d = 3$, $d_1 = 3$, $\omega = 13.15$, and $\omega_1 = 3.30$. Further calculations with (16.29)–(16.31) are made with special software (see Appendix F). The estimates are computed to be

$$\hat{R}^*(100) = 0.9307\ ,$$

$$\sigma_{R(100)} = 0.0523\ ,$$

$$\underline{R}_{0.9} = 0.8495\ .$$

If all the failures occur before 100 hours, the same ω and ω_1 yield worse estimates. This is shown by the results for $d = 3$ and $d_1 = 0$:

$$\hat{R}^*(100) = 0.8579\ ,$$

$$\sigma_{R(100)} = 0.0596\ ,$$

$$R_{0.9} = 0.8099\ .$$

16.5 BAYESIAN ESTIMATION WITH PARTIAL OR CONSTRAINED PRIOR INFORMATION

Lack of prior information may prevent assignment of a prior probability density function for a parameter or a reliability index. The only information about the prior distribution may be certain functional constraints. An appropriate estimating procedure cannot be called strictly Bayesian because the full information about the prior p.d.f. is unknown. Nevertheless, the problem is solved with a general Bayesian methodology. The method requires some prior information but not as much as a prior p.d.f. Under certain circumstances, the estimating procedure in this section may be much more useful than the procedure used for full prior information. The essence of the method is to use a conjugate prior distribution with unknown parameters estimated so that they give the maximum value possible to an appropriate criterion when the constraints are satisfied.

16.5.1 Types of Constraints

In practice there are a number of different kinds of restrictions on reliability estimation. This section covers two types. The first type is very simple and very widespread. Only a prior mean of survival probability, such as R_0 or some other reliability index, has to be specified. We can write the following Bayesian equation:

$$\int_0^1 rh(r)\, dp = R_0 , \tag{16.32}$$

where $h(r)$ is a prior p.d.f. This constraint is called the "mean value constraint." The second type has a given prior probability interval $[R_1, R_2]_u$. Also given is the probability $\mu = Pr\{R_1 \leq R \leq R_2\}$. This probability may be a subjective guarantee that the estimated reliability index lies in $[R_1, R_2]$. With this kind of partial prior information, we can write

$$\int_{R_1}^{R_2} h(r)\, dr = \mu . \tag{16.33}$$

This constraint is called a "prior interval constraint."

16.5.2 Criteria for Choice of a Prior Distribution

An infinite number of different prior distributions satisfy constraints (16.32) or (16.33). We need to make additional assumptions to select the prior distribution for a particular case. Two assumptions are made:

1. The prior distribution is a conjugate one—the prior and the posterior distributions are members of the same family.
2. The prior distribution should be the most pessimistic of the two criteria in this section.

Maximum Entropy Criterion

One criterion is maximum entropy of a prior p.d.f. The entropy, $H(h)$, of a prior p.d.f., h, for a reliability index such as survival probability R is defined as

$$H(h) = -\int_0^1 h(r) \ln h(r)\, dr .$$

The general problem of a prior p.d.f. satisfying some restriction can be written as follows:

$$H(h) \rightarrow \max_{h \in H^*} , \tag{16.34}$$

$$S(h(r)) = 0 , \tag{16.35}$$

where H^* is a class of acceptable p.d.f.'s $S(h(r))$ denotes the function that is determined by the certain form of partial prior information and thus can be obtained from (16.32) or (16.33). The problem in (16.34) and (16.35) is that of conditional optimization. According to assumption 1, the class H^* of the prior p.d.f. is parametric because it is the class of conjugate priors.

EXAMPLE 16.12

Consider a binomial model. Let n independent trials be carried out and d failures occur. On the basis of prior judgment, the prior mean value R_0 of the survival probability has been given. We wish to estimate the survival probability.

A beta distribution is a conjugate prior for binomial sampling and is given by

$$h(p) = \beta(p; a, b) = \frac{p^{a-1}(1-p)^{b-1}}{B(a, b)}$$

where

$$B(a, b) = \int_0^1 x^{a-1}(1-x)^{b-1}\, dx$$

and a and b are distribution parameters.

We want to find the a and b that maximize the entropy

$$\begin{aligned} H(\beta(p)) &= H_B(a, b) \\ &= -\int_0^1 \beta(p) \ln \beta(p)\, dp \\ &= \ln B(a, b) - \frac{a-1}{B(a, b)} \int_0^1 x^{a-1}(1-x)^{b-1} \ln x\, dx \\ &\quad - \frac{b-1}{B(a, b)} \int_0^1 x^{a-1}(1-x)^{b-1} \ln(1-x)\, dx\,. \end{aligned}$$

Let us now deal with the restriction. Substituting $\beta(a, b)$ into (16.32) and then integrating yields the equation for the unknown parameters a and b:

$$\frac{a}{a+b} = R_0\,.$$

We can finally present the problem of finding the a_* and b_* that maximize the entropy of the prior subject to the restriction on the mean of the prior probability:

$$H_B(a_*, b_*) = \max_{\substack{a \geq 0 \\ b \geq 0}} H_B(a, b) ,$$

$$aR_0 - b(1 - R_0) = 0 .$$

Maximum Posterior Risk Criterion

The second criterion is based on maximizing a Bayesian posterior risk function,

$$G(h(r), \hat{R} \mid data) = \int_0^1 L(\hat{R}, r) h(r \mid data)\, dr ,$$

where $h(r \mid data)$ is a posterior p.d.f. of a reliability index, R, and $\hat{R}$ is the estimate of R. The criterion is called the "maximum posterior risk criterion." The general problem is similar to (16.34) and (16.35) and can be written as follows:

$$G(h(r), \hat{R} \mid data) \rightarrow \max_{h \in H^*} , \tag{16.36}$$

$$S(h(r)) = 0 . \tag{16.37}$$

EXAMPLE 16.13

We solve the previous example under maximum posterior risk criterion. We let the loss function, $L(\hat{R}, R)$, be given. We can assign the function of Bayesian risk that is a posterior mean of $L(\hat{R}, R)$, namely

$$G(\hat{R}; a, b \mid n, d) = \int_0^1 L(\hat{R}, r) h(r; a, b, n, d))\, dr ,$$

where $h(r; a, b \mid n, d)$ denotes a posterior p.d.f. of the survival probability R, which is binomial parameter p. We replace r with p.

Under Bayesian methodology (see Section 16.2), we can write

$$h(p; a, b, n, d) = \beta(p; a + n - d, b + d) .$$

The Bayesian point estimator $\hat{p}^*$ minimizes the posterior risk. Then we use the criterion of maximum posterior risk for selecting the prior distribution. We have a joint problem of both the point estimation and obtaining the unknown values of parameters a and b, which may be denoted a_* and b_*. This problem can be presented as

$$G(\hat{p}^*; a_*, b_*, n, d) = \min_{0 \leq p \leq 1} \max_{\substack{a \geq 0 \\ b \geq 0}} G(\hat{p}; a, b, n, d) .$$

In general this minimax problem can be difficult because of nonlinearity of the functions. Significant simplification can be achieved if a squared-error loss function is used for estimating. It is easy to prove that in this case the problem coincides with that of maximizing the posterior variance, which may be denoted the function U of parameters a and b:

$$U(a, b) = \frac{(a+n-d)(b+d)}{(a+b+n)^2(a+b+n+1)} .$$

The problem of maximizing $U(a, b)$ has to be solved under the restriction on the partial prior information. For a special prior mean of p, restriction is given by (16.32) and the joint problem can be written

$$U(a_*, b_*) = \max U(a, b) , \quad a \geq 0, \; b \geq 0 ,$$
$$aR_0 - b(1 - R_0) = 0 .$$

16.5.3 Binomial Model

A beta distribution, $\beta(a, b)$, is the conjugate one for a binomial model. To find the unknown prior, we must find the values of the two parameters a and b, which will be denoted a_* and b_*. The test data are represented by the number of independent trials, n, and the number of failures, d. We consider the estimation of a survival probability that is the binomial parameter p for this model.

The point posterior estimate, $\hat{p}^*$, the posterior variance, $\operatorname{Var}(p)$, and the probability posterior one-sided limit, $\underline{p}_\gamma$, are given by (16.38) through (16.40):

$$\hat{p}^* = \frac{a_* + n - d}{a_* + b_* + n} , \tag{16.38}$$

$$\operatorname{Var}(p) = \frac{(a_* + n - d)(b_* + d)}{(a_* + b_* + n)(a_* + b_* + n + 1)} , \tag{16.39}$$

$$\underline{p}_\gamma = \left[1 + \frac{b_* + d}{a_* + n - d} F_{1-\gamma; 2(b_* + d), 2(a_* + n - d)} \right]^{-1} , \tag{16.40}$$

where $F_{g,\nu,\upsilon,\mu}$ is the $100 - \alpha$ percentage point of the F distribution with ν and μ degrees of freedom. The values of a_* and b_*, the unknown parameters of the beta prior distribution, are found by the procedures that follow.

Mean Value Constraint and Maximum Entropy Criterion

The values of a_* and b_* are obtained by solving (16.41) and (16.42).

$$\left\{ \ln B(a, b) - \frac{1}{B(a, b)} [(a-1)B'_{10}(a, b) + (b-1)B'_{01}(a, b)] \right\} \to \max_{\substack{a \geq 0 \\ b \geq 0}} , \tag{16.41}$$

$$aR_0 - b(1 - R_0) = 0\,, \tag{16.42}$$

where R_0 is the given value of the mean of the prior survival probability, $B(a, b)$ is a beta function, and

$$B'_{10}(a, b) = \frac{\partial B(a, b)}{\partial a}\,,$$

$$B'_{01}(a, b) = \frac{\partial B(a, b)}{\partial b}\,,$$

are partial derivatives of the beta function.

Mean Value Constraint and Maximum Posterior Risk Criterion

In this case the problem of obtaining a_* and b_* is solved in closed form. If the condition

$$\left(\frac{1 - R_0}{R_0} - g\right)\left(d - \frac{n}{2}\right) \geq 0 \tag{16.43}$$

is satisfied, where

$$g = \frac{d}{n - d}\,\frac{(d + 1)(n - 2d) + 2(n - d)}{(n - d - 1)(n - 2d) - 2d^2}\,,$$

then $a_* = b_* = 0$. The prior distribution is uniform on [0, 1]. Otherwise, the cubic equation

$$u^3 + Au^2 + Bu + C = 0\,, \tag{16.44}$$

where

$$A = -\,\frac{2s(2R_0 - 1)}{R_0(1 - R_0)}\,, \quad B = -\,\frac{s(3s + 2R_0 - 1)}{R_0(1 - R_0)}\,,$$

$$C = -\,\frac{2s^2}{R_0(1 - R_0)}\,, \quad s = n(1 - R_0) - d\,,$$

should be solved. For a root $u_* \geq n$, the unknown value of the parameters are calculated with

$$a_* = (u_* - n)p_0 b_* = (u_* - n)(1 - p_0)\,. \tag{16.45}$$

Probability Interval Constraint and Maximum Entropy Criterion

The values of a_* and b_* are found by solving (16.46) and (16.47):

$$\left\{\ln B(a, b) - \frac{1}{B(a, b)}\,[(a - 1)B'_{10}(a, b) + (b - 1)B'_{01}(a, b)]\right\} \to \max_{\substack{a \geq 0 \\ b \geq 0}}\,, \tag{16.46}$$

$$\int_{R_1}^{R_2} x^{\alpha-1}(1-x)^{b-1}\,dx - \mu B(a, b) = 0\,, \tag{16.47}$$

where R_1 and R_2 are the known values of the limits of the prior probability interval for the survival probability and μ is the given value of the probability that the survival probability lies in $[R_1, R_2]$.

Probability Interval Constraint and Maximum Posterior Risk Criterion The values are found with (16.48) and (16.49):

$$\frac{(a+n-d)(b+d)}{(a+b+n)^2(a+b+n+1)} \to \max_{\substack{a\geq 0\\ b\geq 0}}\,, \tag{16.48}$$

$$\int_{R_1}^{R_2} x^{a-1}(1-x)^{b-1}\,dx - \mu B(a, b) = 0\,, \tag{16.49}$$

where R_1, R_2, and μ were denoted in the previous case.

EXAMPLE 16.14

Ten identical units are tested independently, and no failures occur. The prior mean value of the survival probability is 0.9. Find the point estimate and its variance.

First, let us use a maximum posterior risk criterion. Condition (16.43) is not satisfied with $n = 10$ and $d = 0$. Therefore we have to solve the cubic equation (16.44). For the initial data the coefficients of the equation are calculated to be $A = -19.59$, $B = -15.92$, and $C = -4.082$. Substituting $z = x - A/3$ into (16.44) yields

$$x^2 + px + r = 0\,,$$

where $p = -143.86$ and $r = -665.08$. Solving the equation by the standard method gives the only root u_* that is greater than $n = 10$, namely $x_* = 13.852$ and then $u_* = 20.38$. Substituting this value into (16.45) yields the values of unknown parameters, which are $a_* = 10.1720$ and $b_* = 0.2076$. The estimates are found with (16.38) and (16.34). They are

$$\beta^* = \frac{10.172 + 10}{10.172 + 0.2076 + 10} = 0.9898\,,$$

$$\text{Var}\{p\} = \frac{(10.172 + 10)\cdot 0.2076}{(10.172 + 0.2076 + 10)^2(10.172 + 0.2076 + 10 + 1)} = 4.71\,.$$

If we wish to use a maximum entropy criterion, we have to solve (16.41) and (16.42). The estimates are computed to be

$$\hat{p}^* = 0.9619 ,$$
$$\mathrm{Var}\{p\} = 0.0475^2 .$$

Special software was used (see Appendix F).

EXAMPLE 16.15

Solve the previous example using the prior probability interval $[0.96, 0.99]_{0.9}$ for estimating the survival probability. Estimate it under both maximum entropy criterion and a maximum posterior risk criterion.

In the first case we solve (16.46) and (16.47) for $R_1 = 0.96$, $R_2 = 0.99$, and $\mu = 0.9$. Values of a_* and b_* are 254.03 and 5.8808, respectively. Use of (16.38) and (16.39) yields

$$\hat{p} = 0.9782 ,$$
$$\mathrm{Var}\{p\} = 0.0887^2 .$$

With a maximum posterior risk criterion, we solve (16.48) and (16.49). The results are

$$\hat{p}^* = 0.9770 ,$$
$$\mathrm{Var}\{p\} = 0.09137^2 .$$

Special software again was used (see Appendix F).

16.5.4 Exponential Model

An exponential model is assigned p.d.f. $f(t) = f(t; \lambda)$ given by (16.17) with parameter λ. The gamma distribution with the p.d.f.

$$h(\lambda; s, \epsilon) + \frac{\epsilon^{s+1}}{\Gamma(s+1)} \lambda^s e^{-\lambda\epsilon}$$

is a conjugate prior of the parameter λ. We estimate here both a failure rate, λ, and a survival probability, $R(t) = R(t; \lambda) = e^{-\lambda t}$, using censored sample data $\mathbf{z} = (\mathbf{t}^*, \bar{t})$, where $\bar{\mathbf{t}}^* = (t_1^*, \ldots, t_d^*)$ is a subsample of times for unfailed items. The sufficient statistic is the total time on test

$$\kappa = \sum_{1 \le i \le d} t_i^* + \sum_{1 \le j \le k} t_j$$

and the number of failures, d.

When the values of the parameters of a conjugate prior, s_* and ϵ_*, are known, the estimators of the failure rate (the point estimator $\hat{\lambda}^*$, the variance, $\mathrm{Var}\{\lambda\}$, and the upper probability limit, $\bar{\lambda}_\nu$) may be found with

$$\hat{\lambda}^* = \frac{d + s_* + 1}{\kappa + \epsilon_*}, \tag{16.50}$$

$$\mathrm{Var}\{\lambda\} = \frac{d + s_* + 1}{(\kappa + \epsilon_*)^2}, \tag{16.51}$$

$$\overline{\lambda_\gamma} = \frac{\chi^2_{1-\gamma;2(d+s_*+1)}}{2(\kappa + \epsilon_*)}, \tag{16.52}$$

where $\chi^2_{\alpha;\nu}$ is the $(100 - \alpha)$ percentage point of a χ^2 distribution with ν degrees of freedom.

The estimators for the survival probability, $R(t)$, are given by

$$\hat{R}^*(t) = \left(\frac{u_*}{u_* + 1}\right)^{\nu_* + 1}, \tag{16.53}$$

$$\mathrm{Var}\{R(t)\} = \left(\frac{u_*}{u_* + 2}\right)^{\nu_* + 1} - \left(\frac{u_*}{u_* + 1}\right)^{2(\nu_* + 1)}, \tag{16.54}$$

$$\underline{R}_\gamma(t) = \exp\left[-\frac{\chi_{1-\gamma;2(d+s_*+1)^2}}{2u_*}\right]. \tag{16.55}$$

where $u_* = (\epsilon_* + \kappa)/t$ and $\nu_* = s_* + d$. The procedure for finding s_* and ϵ_* follows.

Mean Value Constraint and Maximum Entropy Criterion

The values of s_* and ϵ_* are found by solving

$$H_E(s, \epsilon) = (s + 1) - \ln \frac{\epsilon}{\Gamma(s + 1)} - \frac{s\Gamma'(s + 1)}{\Gamma(s + 1)} \rightarrow \max_{\substack{s \ge 0 \\ \epsilon \ge 0}}, \tag{16.56}$$

$$\frac{s + 1}{\epsilon} = \lambda_0, \tag{16.57}$$

when the mean value of the failure rate, λ_0, is given, and by solving

$$H_E(s, \epsilon) \rightarrow \max_{\substack{s \ge 0 \\ \epsilon \ge 0}}, \tag{16.58}$$

$$\left(\frac{\epsilon}{\epsilon + t_0}\right)^{s+1} = R_0, \tag{16.59}$$

when the prior mean value R_0 (the mean value of the survival probability for time t_0), is given. Here $\Gamma(x)$ is a gamma function of x and $\Gamma'(x)$ is its first derivative.

Mean Value Constraint and Maximum Posterior Risk Criterion

The problem of finding s_* and ϵ_* can be presented in closed form. If

$$\lambda_0 \leq \frac{2d+1}{\kappa}, \tag{16.60}$$

$s_* = 0$ and $\epsilon_* = 1/\lambda$. Otherwise,

$$s_* = \lambda_0 - 2d - 1, \quad \epsilon_* = \kappa - \frac{2d}{\lambda_0}. \tag{16.61}$$

These expressions are used when a prior mean value, λ_0, of the failure rate is given. If we have the value of R_0, that is, the mean of the prior survival probability for a specified time t_0, we still can use (16.60) and (16.61). We just have to notice that $\lambda_0 = -\ln R_0/t_0$.

Probability Interval Constraint and Maximum Entropy Criterion

The probability interval $[R_1, R_2]_\mu$ for a survival probability $R(t_0)$ or $\{\lambda', \lambda'\}_\mu$ for a failure rate may be given. These intervals are equivalent because

$$\lambda' = -\frac{\ln R_2}{t_0}, \quad \lambda' = -\frac{\ln R_1}{t_0}.$$

The equations for finding s_* and ϵ_* are

$$H_E(s, \epsilon) = (s+1) - \ln \frac{\epsilon}{\Gamma(s+1)} - \frac{s\Gamma'(s+1)}{\Gamma(s+1)} \rightarrow \max, s \geq 0, \epsilon \geq 0, \tag{16.62}$$

$$\int_{\lambda'\epsilon}^{\lambda''\epsilon} x^s e^{-x}\, dx = \mu\Gamma(s+1). \tag{16.63}$$

Probability Interval Constraint and Maximum Posterior Risk Criterion

The values of s_* and ϵ_* are found with

$$\frac{d+s+1}{(\kappa+\epsilon)^2} \rightarrow \max, \quad s \geq 0, \epsilon \geq 0, \tag{16.64}$$

$$\int_{\lambda'/\epsilon}^{\lambda''\epsilon} x^s e^{-x}\, dx = \mu\Gamma(s+1). \tag{16.65}$$

EXAMPLE 16.15

The results of the life test of the unit in hours were (120, 128, 144, 130, 152, 104, 92). No failures occurred. Prior knowledge gave an estimate of 0.94 for the survival probability for 80 hours. We want to estimate the survival probability for 100 hours.

Solution

Use of $R(t_0) = \exp(-\lambda_0 t_0)$ yields

$$\lambda_0 = -\frac{\ln R_0}{t_0} = -\frac{\ln 0.94}{80} = 7.73 \cdot 10^{-4}\ f/\text{h}\,.$$

We decide to use a maximum posterior criterion. The total of the test times is $\kappa = 870$ hours. These quantities satisfy condition (16.60). Hence $s_* = 0$ and $\epsilon_* = 1/(7.73 \cdot 10^{-4}) = 1292.92$ h. Using (16.53) and (16.54), we find that $\hat{R}(100) = 0.9558$ and $\text{Var}\{R(100)\} = 1.788 \cdot 10^{-3}$. If we use a maximum entropy criterion, (16.56) and (16.57) yield $\hat{R}(100) = 0.9550$,

$$\text{Var}\{R(100)\} = 0.042997^{-2}\,.$$

EXAMPLE 16.16

As in the previous example, we use the prior probability interval $[0.96, 0.99]_{0.9}$ for $R(80)$. We want to estimate $R(100)$ under both a maximum entropy criterion and a maximum posterior risk criterion. For the former we use (16.62) and (16.63). The evaluated quantities of s_* and ϵ_* are 5.6753 and 20,964, respectively. Then by (16.53) and (16.54) we obtain $\hat{R}(100) = 0.9700$ and $\text{Var}\{R(100)\} = 0.01143^{-2}$. For the latter criterion, we use (16.64) and (16.65) to yield $\hat{R}(100) = 0.9684$ and $\text{Var}\{R(100)\} = 0.01143^{-2}$. Special software was used (see Appendix F).

CHAPTER 17

ACCELERATED LIFE TESTS

17.1 BASIC CONCEPTS

A "stress factor" may be defined as a stress or environmental variable such as voltage, cycling rate, load, pressure, temperature, or humidity. A "stress condition" may be defined as a vector of stress factors that influences the time to failure of an item. Operating specifications of an item usually give the region of normal stress conditions in a space of stress factors.

Let $F_1(t; z_1)$ and $F_2(t; z_2)$ be the failure time distribution functions of an item at stress conditions z_1 and z_2, respectively. Stress condition z_2 is called more severe than z_1 if $F_1(t; z_1) < F_2(t; z_2)$ for all positive values of t. The inequality means that increased severity of a stress condition reduces the time to failure. In technical specifications, reliability parameters ordinarily are estimated for the most severe conditions in the region of normal stress conditions.

A stress condition that is more severe than any in the region of normal stress conditions is called an "accelerated stress condition." To predict the reliability of an item under normal stress conditions, life tests are conducted at accelerated stress conditions. Such accelerated life testing often is performed under automodality, which usually is defined as the conservation of mechanisms that induce failures of a device under both accelerated and normal (design) stress conditions. This concept is discussed later in this chapter.

17.2 PROBABILISTIC ASSUMPTIONS

17.2.1 The Distribution Shape Assumption

The most fundamental assumption is that a change of stress condition changes only the scale, not the shape, of the failure time distribution function. This sometimes is called an accelerated life model.

Let z be a vector of stress factors that is constant in time, and let $F(t; z)$ be the failure time distribution function for this vector. Without loss of generality, we can define the basic normal stress condition as $z = 0$. If we denote a failure time distribution function under normal stress conditions by $F_0(\cdot)$, the basic assumption then is given by the following relationship:

$$F(t; z) = F_0[t \cdot \psi(z)] , \tag{17.1}$$

where $\psi(z)$ is a function relating the time to failure with a vector of stress factors, z, and always corresponds to a decrease of time-to-failure. For $z = 0$, $\psi(z)$ is assumed equal to one.

The relationships for a quantile of time to failure $t^p(z)$, and hazard rate $\lambda(z)$, can be obtained from (17.1):

$$t^p(z) = \frac{t_0^p}{\psi(z)} ,$$

$$\lambda(t; z) = \lambda_0[t \cdot \psi(z)] , \tag{17.2}$$

where t_0^p and λ_0 are the p quantile and hazard rate, correspondingly, for normal stress conditions at which $z = 0$.

17.2.2 Sedyakin's Principle

The foregoing assumption is connected closely with another assumption, Sedyakin's physical principle of reliability. This principle is widely used in accelerated tests with time-dependent stress conditions, and in time-step-stress tests. Let $F_1(t; z_1)$ and $F_2(t; z_2)$ be failure time distribution functions at stress conditions z_1 and z_2, respectively. Let z_1 be less severe than z_2. Let t_1 and t_2 be times at which $F_1(t_1; z_1) = F_2(t_2; z_2)$. Clearly, for all positive t_1 and t_2, there exists a function g such that $t_1 = g(t_2)$. Then

$$F_2(t_2; z_2) = F_1(g(t_2), z_1) .$$

Since $F_1(t; z_1) < F_2(t; z_2)$, $g(t)$ increases with time. We can write that $g(0) = 0$, and that

$$\lim_{t \to \infty} g(t) = \infty .$$

The function $g(t)$ is called the "acceleration" or the "time transformation" function. The first assumption (that a change of stress condition changes the scale of the failure time distribution function, but does not change its shape) is related to the acceleration function by

$$g(t) = \psi(z) \cdot t\,.$$

In words, the time transformation function is linear with time.

Suppose that $F_1(t; z_1)$ and $F_2(t; z_2)$ are specified. We are interested in the time-to-failure distribution function, $F_{12}(t/t_1)$, the distribution function of a device tested under stress condition z_1 until time t_1, when z_1 is changed to stress condition z_2. We write

$$F_{12}(t/t_1) = \begin{cases} F_1(t; z_1) & \text{for } t \leq t_1\,, \\ F_x(t; z_1, z_2) & \text{for } t > t_2\,, \end{cases} \tag{17.3}$$

where $F_x(t; z_1, z_2)$ needs to be specified.

The physical principle of reliability postulates the following statements:

1. Testing at stress condition z_1 does not change the reliability function at stress condition z_2; that is, $F_x(t; z_1, z_2)$ is a segment of $F_2(t; z_2)$.
2. For all $t > t_1$, $F_x(t; z_1, z_2) = F_2(t + t^* - t_1; z_2)$, where t^* is such that

$$F_1(t_1; z_1) = F_2(t^*; z_2)\,. \tag{17.4}$$

With these two statements, which constitute a mathematical expression of Sedyakin's principle, (17.3) can be written as

$$F_{12}(t/t_1) = \begin{cases} F_1(t; z) & \text{for } t \leq t_1\,, \\ F_2(t + t^* - t_1; z_2) & \text{for } t > t_1\,. \end{cases} \tag{17.5}$$

Note that under the basic assumption that the acceleration function is linear with time, Sedyakin's principle will be fulfilled automatically.

17.2.3 Automodality

Automodality is defined in the Physical Encyclopedia (Sovetskaja Encyclopedia, Vol. 1, Moscow, 1988, in Russian) as special symmetry in a physical system. This symmetry leads to the possibility of compensating for independent variable scale changes with similar transformations of other dynamic variables. The automodality results in an effective reduction of the number of independent variables. It can easily be shown that the linearity of the time transformation function not only leads to automatic fulfillment of Sedyakin's principal, but also results in automodality of failure time distribution functions.

The concept of automodality can be illustrated by using a reliability degradation model with Increasing Failure Rate Average (IFRA) distribution functions. The assumption of an IFRA failure time distribution function is widespread in accelerated life test problems. Automodality of failure time distribution functions will be fulfilled at all stress conditions if the conditions differ from each other only in the intensities of the Poisson shock processes.

17.2.4 Statistical Testing

It is possible to test the hypothesis that the time transformation function is linear with time by using time-to-failure data. The appropriate statistical procedures are called "criteria of linearity of time transformation function".

Consider one of these criteria, referring to standard test plans such as the [NWr] or [NWT] plans. The criterion is based on the use of quantile-quantile plots, which are curves in which the coordinates of every point are time-to-failure quantiles for investigated stress conditions. If, in addition to assuming 17.1, one supposes that the failure time distribution function increases monotonically with time, the quantiles will be determined simply and the quantile-quantile plot will be a straight line going through the origin.

17.3 RELIABILITY MODELS

Reliability models are defined as the relationships between the parameters of the failure time distribution and the stress factors. The models are derived on the basis of the physical considerations of the processes resulting in device failures. Some of the most popular reliability models are discussed briefly here. The following notation is used:

Y = parameter of the failure time distribution.

X = applied stress factor, such as voltage, pressure, or electrical field strength.

T = absolute temperature.

A, B, C, and D = unknown parameters to be estimated. The parameter B is usually the activation energy of an aging process.

17.3.1 Power Rule Model

The power rule model is a relationship between the parameter of the time-to-failure distribution and a mechanical or electrical stress factor level, X:

$$Y = \frac{A}{X^c}, \qquad c > 0 .$$

17.3.2 Arrhenius Reaction Rate Model

The most popular model used to examine the effect of temperature on reliability is the Arrhenius reaction rate model. It has a wide range of applications. Originally it was intended as a chemical reaction rate model:

$$Y = A \exp\left(\frac{B}{T}\right).$$

17.3.3 Combined Power Rule and Arrhenius Reaction Rate Model

The main applications of the combined model are in accelerated life testing of capacitor and polymer materials. In the physics of polymers it is known as Bruhanova's and Bartenev's model:

$$Y = AX^{-c} \exp\left(\frac{B}{T}\right).$$

17.3.4 Jurkov's Model

Jurkov's model is based on the kinetic theory of mechanical and electrical strengths of solids. The model is widely used for reliability problems involving the long-term mechanical and electrical strengths of items and materials:

$$Y = A \exp\left(\frac{B - CX}{T}\right).$$

17.3.5 Generalized Eyring Model

Like the Arrhenius model, the generalized Eyring model was proposed originally as a chemical reaction rate model. In accelerated life testing it is used as the relationship between the parameter of the failure time distribution function and thermal, electrical, or mechanical stress:

$$Y = AT \exp\left(\frac{B}{T}\right) \exp\left(CX + \frac{DX}{T}\right).$$

17.4 STATISTICAL METHODS OF RELIABILITY PREDICTION

The reliability model is not the only tool needed to predict reliability under normal conditions on the basis of accelerated life tests. An assumption must be made about the failure time distribution function for the device. One might assume merely a class of distribution functions, such as continuous, increasing failure rate, or increasing failure rate average. Alternatively, one might assume a specific distribution function, such as normal, exponential, or Weibull-Gnedenko. Each type of assumption has its statistical methods.

Nonparametric methods are used for classes of d.f.'s, and parametric methods are used with specific d.f.'s. In this section Weibull regression is used as an example of a parametric technique, and regression of failure time quantiles is used as a nonparametric technique. These procedures were chosen for their simplicity and their wide applicability.

17.4.1 Weibull Regression

Let the failure times be identically distributed according to a two-parameter Weibull-Gnedenko law. The failure time, t, has the distribution function

$$F(T \le t) = \begin{cases} 1 - \exp\left[-\left(\frac{t}{\delta}\right)^{\beta}\right], & t \ge 0, \\ 0, & t < 0, \end{cases}$$

where δ is the scale parameter and β is the shape parameter, $\delta > 0$, $\beta > 0$.

Suppose that the shape parameter β does not depend on stress factors, but the scale parameter is a loglinear function of p stress factors z_j, $j = 1, 2, \ldots, p$. We then write

$$\ln \delta = \alpha_0 + \alpha_1 z_1 + \cdots + \alpha_p z_p , \tag{17.6}$$

where $\alpha_0, \alpha_1, \ldots, \alpha_p$ are unknown parameters. All the models in Section 17.3 can be transformed easily into (17.6).

The above assumption is a parametric analogue of (17.1). It is easy to check that the transformed random value $y = \ln t$ has an extreme value distribution function on any real axis:

$$F(Y < y) = 1 - \exp\left[-\exp\left(\frac{y - \mu}{b}\right)\right], \tag{17.7}$$

where $\mu = \ln \delta$ is the location parameter and $b = 1/\delta$ is the scale parameter.

Let there be K different test stress conditions ($K > p + 2$). Each stress condition is defined by the set of stress factors $z_{1i}, z_{2i}, \ldots, z_{pi}$; $i = 1, \ldots, K$. Let that the size of sample i be N_i. Let the number of observed failure times be R_i, $t_{i(1)} < t_{i(2)} < \cdots < t_{i(R_i)}$. Let the remaining $N_i - R_i$ failure times be type II censored times.

The problem is to obtain the best linear unbiased estimates of the parameters of the failure time distribution function at normal stress conditions. Let $y_{i(r)} = \ln t_{i(r)}$, $r = 1, 2, \ldots, R_i$.

The expectation and variance of the r'th order statistic (time to r'th failure at the i'th stress condition) from distribution function (17.7) (as from the distribution function of the form $F[(x - \mu)/b]$) are given by

$$E(y_{i(r)}) = \mu_i + \gamma_{kr} b; \tag{17.8}$$

$$V(y_{i(r)}) = \sigma^2 v_{irr} ,$$

where γ_{kr} and v_{irr} are the expectation and variance of the r'th order statistic of the standardized form of distribution (17.7) with location parameter μ equal to zero and scale parameter b equal to one. In other words γ_{kr} and v_{irr} depend on the sample size $\mathbf{N}_i$ and the number of failures observed (R_i) only, and can be tabulated. Analogously the covariance of the order statistics $y_{i(r)}$ and $y_{i(r')}$ is given by

$$\mathrm{Var}(y_{i(r)}) = \delta^2 v_{i(r)} , \tag{17.9}$$

where $v_{i(r)}$ is the variance of the rth order statistic of the standardized form. Thus γ_{kr} and $v_{i(r)}$ depend only on the sample size, N_1, and on the observed failures, R_1, so they need to be computed only once. Analogously, the covariance of the order statistics $y_{i(r)}$ and $y_{i'(r')}$ is given by

$$\mathrm{Cov}(\mathbf{y}_{i(r)}, \mathbf{y}_{i(r')}) = \sigma^2 v_{iir} , \tag{17.10}$$

where $v_{i(r)}$ are known constants. Because the sample test results at different stress conditions are statistically independent, for $i \neq i'$ and any r and r':

$$\mathrm{Var}(y_{i(r)}, y_{i'(r')}) = 0 . \tag{17.11}$$

The means, variances, and covariances of order statistics from the standardized distribution often are tabulated for the extreme value distribution and for a number of other commonly used parametric distributions. Let the column vector of the logarithms of failure times at all the stress conditions be

$$\mathbf{y} = (y_{1(1)}, y_{1(2)}, \ldots, y_{1(R_1)}, \ldots, y_{k(1)}, y_{k(2)}, \ldots, y_{k(R_k)})^T , \tag{17.12}$$

where the superscript T denotes transposition of a row vector to a column vector. Let b and parameter vector (17.6) be combined in:

$$\boldsymbol{\alpha} = (\alpha_0 \alpha_1 \ldots \alpha_p \beta) . \tag{17.13}$$

Let $\mathbf{X}$ denote the regression matrix for model (17.8).

$$\mathbf{X} = \begin{bmatrix} 1 & z_{11} & z_{12} & \cdots & z_1 p & \gamma_{11} \\ \cdots & \cdots & \cdots & \cdots & \cdots & \cdots \\ 1 & z_{11} & z_{12} & \cdots & z_{1_p} & \gamma_{1R_1} \\ \cdots & \cdots & \cdots & \cdots & \cdots & \cdots \\ 1 & z_{k1} & z_{k2} & \cdots & z_{kp} & \gamma_{k1} \\ \cdots & \cdots & \cdots & \cdots & \cdots & \cdots \\ 1 & z_{k1} & z_{k2} & \cdots & z_{kp} & \gamma_{kR_k} \end{bmatrix} . \tag{17.14}$$

The regression for the (17.8) order statistics can be written as

$$\mathbf{y} = \mathbf{X}\alpha + \boldsymbol{\epsilon} ,$$

where $\boldsymbol{\epsilon} = \mathbf{y} - \mathbf{X}\alpha = (\epsilon_{11}, \ldots, \epsilon_{1R_1}, \ldots, \epsilon_{k1}, \ldots, \epsilon_{kR_k})^T$ denotes the error of y. The mean of the error vector is the zero vector. The covariance matrix of the order statistics is

$$E(\epsilon\epsilon') = \sigma^2 V ,$$

which is given by equations (17.9) through (17.11).

Because the samples at different stress conditions are independent, matrix $\mathbf{V}$ has a block diagonal form

$$\mathbf{V} = \text{diag}(V_1, \ldots, V_i, \ldots, V_k) ,$$

where the matrix $\mathbf{V}_i$ is the R_i by R_i covariance matrix of the standardized order statistics in sample i. According to (17.10) and (17.11), row r of $\mathbf{V}_i$ is

$$(v_{ir_1}, v_{ir_2}, \ldots, v_{irR_i}) .$$

Thus the regression model for censored data can be reduced to a general linear model with correlated observations. All standard least squares methods can be used. In particular, according to the Gauss-Markov theorem, the best linear unbiased estimator, $\hat{\alpha}$, for the vector of parameters, $\boldsymbol{\alpha}$, is

$$\hat{\alpha} = (\mathbf{X}'\mathbf{V}^{-1}\mathbf{X})^{-1}\mathbf{X}'\mathbf{V}^{-1}\mathbf{y} .$$

As the vector $\boldsymbol{\alpha}$ includes the vector of the parameters of the model, (17.6), and the parameter $b = 1/\beta$, the estimate of the scale parameter at normal stress can be found.

17.4.2 Regression of Failure Time Quantiles

Under the distribution shape assumption, relationship (17.2) for quantiles of failure time distribution is true, it is also assumed that the reliability model $\eta(\mathbf{z}, \mathbf{A})$ for a quantile t^p is a function of the stress factors $\mathbf{z}$ with an unknown vector of parameters, $\mathbf{A}$:

$$t^p(\mathbf{z}, \mathbf{A}) = \eta(\mathbf{z}, \mathbf{A}) . \qquad (17.15)$$

The objective is to obtain the estimate of the vector $\mathbf{A}$ and to predict the quantile t^p at normal stress on the basis of accelerated life tests at different stress conditions $z_1, \ldots, z_w$ where k is larger than the dimension of the vector $\mathbf{A}$.

Besides the standard assumptions above, it is assumed that:

1. The failure time distribution functions at both increased and normal stress conditions are increasing failure time average (IFRA\) distribution functions having continuous probability density, $f(t; z)$.
2. The test results are type II censored samples and the number of uncensored failure times, r_i $(i = 1, \ldots, k)$ and the sample size, N_1, are large enough to estimate the quantiles t^p as $\hat{t}^p$ (sample quantile).
3. The sample sizes N_i are large enough so that the asymptotic normal distribution estimates for t^p can be used.

Under these three assumptions, model (17.15) can be written

$$t^{\wedge p}(\mathbf{z}, \mathbf{A}) = \eta(\mathbf{z}, \mathbf{A})\boldsymbol{\epsilon} , \tag{17.16}$$

where ϵ is a normally distributed r.v. The multiplicative model (17.16) can be transformed logarithmically into a model with additive normally distributed error. This gives the linear regression:

$$\ln t^{\wedge p}(\mathbf{z}, \mathbf{A}) = \ln \eta(\mathbf{z}, \mathbf{A}) + \boldsymbol{\epsilon}_1 , \tag{17.17}$$

where ϵ_1 is an unknown constant. This transformation into a model with additive normally distributed error is based on the properties of quantiles of IFRA distributions and an approximation. The approximation is good when x has a normal d.f. with mean 0 and variance σ^2, with $\sigma \ll 1$. The r.v. $y = \ln(1 + x)$ then has approximately the same distribution as x.

An additional restriction for transformation of model (17.16) into (17.17) is a restriction imposed on the sample of size N_i:

$$\sqrt{\frac{p}{N_i(1-p)\ln^2(1-p)}} \ll 1 .$$

Thus the problem of quantile prediction is reduced to estimation of a standard normal regression model followed by the point and confidence prediction by means of this model. Moreover dealing with the model (17.17) one can use the regression experiment design technique for the accelerated life test planning.

CHAPTER 18

RELIABILITY GROWTH

18.1 DUANE MODEL

As newly designed equipment progresses from the prototype items to the final production stems, improvements are made that increase reliability. In 1964, J. T. Duane published the results of a regression analysis of new aerospace items that indicated that the cumulative failure rate of a new item is inversely proportional to the cumulative operating hours of the item raised to an exponent of less than one. Thus

$$\lambda_t = kt^{-m}, \tag{18.1}$$

where

λ_t = cumulative failure rate at time t (the total failures in the history of the items divided by the total operating hours, t, in the history of the item),

k = a presumed inherent rate of the first item,

m = exponent.

Duane originally estimated the value of the exponent m at 0.5, but later investigations showed the value might vary with the type of item anywhere from 0 to 0.6 or more. In general, only new technology components were found to have values as high as 0.5. The overall value for an assembly of components tended to be a weighted average of exponents that were lower for less novel components. The value of an exponent also was determined to vary with other factors, such as complexity, whether an item was electronic or nonelectronic, and the priority given to the reliability improvement program.

Equation (18.1) may be changed into an expression for the instantaneous failure rate that may be more useful for some purposes:

$$f_t = k[t^{(1-m)} - (t-1)^{(1-m)}]\,, \tag{18.2}$$

where f_t is the failure rate at time t. Another convenient conversion of (18.1) results from taking the logarithm of both sides of the equation:

$$\begin{aligned} \ln(\lambda_t) &= \ln(kt^{-m}) \\ &= \ln(k) - m \cdot \ln(t)\,. \end{aligned} \tag{18.3}$$

When plotted on log-log paper, (18.3) produces a straight line that may serve as a visual aid. Figure 18.1 is an example showing the first three years of development of the F15-A aircraft. The m value of 0.127 rather than 0.5 is typical. Although the F15-A was, of course, the first F15 aircraft, it was composed largely of mature components developed for earlier aircraft.

The basic method for estimation of the parameters of (18.1) is least-squares regression, which may be either linear or logarithmic. The basic logarithmic equations are

$$m = -\frac{n \times \Sigma_1^n (\ln t_i \times \ln c_j) - (\Sigma_1^n \ln t_i\, \Sigma_1^n \ln c_j)}{n \times \Sigma_1^n (\ln t_i)^2 - (\Sigma_1^n \ln t_i)^2}\,,$$

$$\ln k = \frac{\Sigma_1^n \ln c_j - m(\Sigma_1^n \ln t_i)}{n}\,,$$

where

n = total number of failures,

t_i = time of failure i,

c_i = cumulative mean time between failures at t_i.

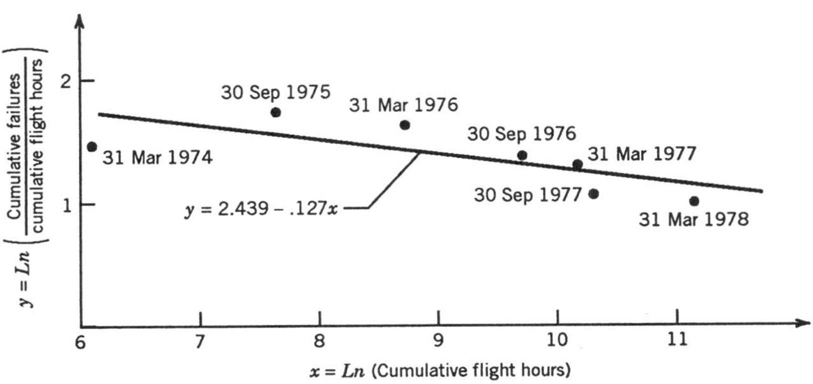

Figure 18.1 Reliability improvement of the F-15A, an example of Duane model growth.

18.2 ARMY MATERIEL MODEL

A widely used form of the Duane model was developed by L. H. Crow for the Army Materiel System Analysis Activity (AMSAA) and published in MIL HDBK 189 in 1981. The AMSA model is intended for an item developed in a succession of models, with each model having a number of components redesigned on the basis of failures of items of the previous design. The unknown Poisson failure rate for a new model is assumed to remain constant throughout the existence of the model, but each model has its own rate. A Weibull intensity function is used:

$$f(t) = \lambda\beta t^{\beta-1}, \qquad \lambda > 0,\ \beta > 0, \tag{18.4}$$

where

$f(t)$ = failure rate after t operating hours,
λ = scale parameter,
β = shape parameter.

When $\beta < 1$, reliability is increasing. When $\beta > 1$, reliability is decreasing. If $\beta = 1$, the Weibull function is the exponential function and the failure rate does not change.

The maximum likelihood method can be used to estimate both λ and β. For grouped data, β may be estimated with

$$\sum_{i=1}^{K} N_i\left[\frac{t_i^{\beta}\ln t_i - t_{i-1}^{\beta}\ln t_{i-1}}{t_i^{\beta} - t_{i-1}^{\beta}} - \ln t_K\right] = 0, \tag{18.5}$$

where

K = number of models tested,
t_i = total operating hours at end of test interval i,
N_i = number of failures during test interval i.

Equation (18.5) can be solved iteratively. Lambda can be estimated with

$$\hat{\lambda} = \frac{\Sigma_1^K N_i}{t_K^{\beta}}. \tag{18.6}$$

A chi-square goodness-of-fit test can be used to measure the closeness of an AMSAA model to the data. The expected number of failures during interval i, e_i, is given by

$$e_i = \lambda(t_i^{\beta} - t_{i-1}^{\beta}).$$

If an interval has less than five expected failures, it should be combined with an adjacent interval. However, there should be at least three intervals because the test statistic is distributed approximately as a chi-square random variable with $K - 2$ degrees of freedom. The statistic is

$$\chi^2 = \sum_{i=1}^{K} \frac{(N_i - e_i)^2}{e_i} ,$$

where the K and N_i values will be those of (18.5) after adjustment, if any, for combined intervals.

18.3 OTHER MONOTONIC GROWTH MODELS

Weibull models are probably the most widely used models of the monotonic growth class, which are based on expectations of two general trends. One is that each new design will be better than its immediate predecessor. The other is that each improvement will be less than its predecessor. This section summarizes some of the major models in this class. More complete coverage may be found in Jewell (1984), Gray (1986), Dhillon (1983), and MIL HDBK 189.

Weibull models, which include the Duane and exponential models, belong to the gamma family of models (Mann et al. 1974). Gray (1986) gives a breakdown and evaluation of reliability growth gamma models, which have a density function, $g(t)$, of the general form

$$g(t) = \frac{\beta^\alpha t^{\alpha-1}}{\Gamma(\alpha)} \exp(-\beta t) ,$$

where

t = time,
α = shape parameter,
β = scale parameter.

The gamma function, $\Gamma(\alpha)$, is

$$\Gamma(\alpha) = \int_{x=0}^{\infty} x^{\alpha-1} e^{-\alpha} \, dx ,$$

which yields the recursion formula

$$\Gamma(\alpha) = (\alpha - 1)\Gamma(\alpha - 1) ,$$

and hence, when α is an integer,

$$\Gamma(\alpha) = (\alpha - 1)!\ .$$

The distribution function, $G(t)$, gives the failure rate as a function of time:

$$G(t) = \int_{x=0}^{t} g(x)\, dx\ .$$

When $\alpha < 1$, the failure rate decreases with time. For $\alpha = 1$, the rate is constant, and $G(t) = 1 - \exp(-\beta t)$, the exponential distribution. When α is any positive integer, $G(t)$ may be expressed in closed form:

$$G(T) = 1 - \sum_{i=0}^{\alpha-1} \frac{(\beta t)^i}{i!} e^{-\beta t}\ .$$

The logarithmic model is a gamma model whose reliability growth characteristics fall between those of the Weibull and exponential models:

$$g(t) = At^{-1}e^{-t}\ ,$$

where A is the characteristic parameter.

Selection of a reliability growth model may be preceded by a logical analysis of the likely classes of results of the test-analyze-fix program, and the percentage of attempted reliability improvements that are apt to fall into each class. If every fix worked immediately and perfectly, a Weibull model with $0 < \beta < 1$ might fit best. An exponential model would fit if no attempted fix had any effect, good or bad. If a fix could produce new faults or actuate latent faults, as sometimes happens, for example, with software, a random walk or time series model (Section 18.4) could be applicable. Other modeling might be appropriate for fixes that are partially effective or have delayed results.

Limit models are based on an assumption that the item has an inherent reliability limit, the R for infinite reliability improvement time, that is due to uncorrectable causes. A limit model requires an estimate of R_{inf} that is made in advance. The general form is

$$R_j = R_{\text{inf}} - \alpha f(j)\ , \qquad j = 1, 2, 3, \ldots,$$

where

R_j = reliability of model j,

α = growth modification parameter,

$f(j)$ = a function of j with a positive first derivative with respect to j and a negative second derivative with respect to j.

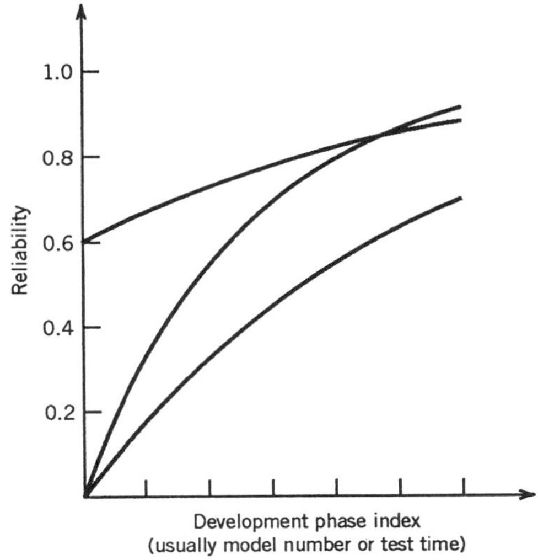

Figure 18.2 Exponential growth curves.

The Lloyd-Lipow model, which is in this class, has reliability given by

$$R_j = R_{\text{inf}} - \frac{\alpha}{j} .$$

In some cases the test stage j may be replaced with j^{β}, where β is fitted to a growth curve that is not hyperbolic.

The Gompertz model may be useful when the reliability growth is nonlinear:

$$R = R_{\infty} A^{\beta^{t}} , \qquad 0 < A < 1,\ 0 < \beta < 1 .$$

Exponential growth models (Figure 18.2) may be applied when the original reliability estimate is believed to be fairly accurate and no uncorrectable defects are anticipated:

$$R = 1 - \alpha e^{-\beta t} ,$$

where $\alpha = 1 - R_0$. If R_{inf} is believed to be less than one, the analyst may use

$$R = R_{\text{inf}} - (R_{\text{inf}} - R_0) e^{-\beta t} .$$

18.4 TIME SERIES MODELS

The monotonic growth models are based on an optimistic expectation of continual improvement that does not always hold true. Random variation can cause backsliding. Furthermore a fix of one problem sometimes can create a more serious problem. This often happens with computer software. In addition, occasionally a type of failure does not ever occur unless another type of failure has been eliminated.

A time series model may offer a solution. The long-range trend is upward, but random variation is superimposed. Some investigators have used a first-order, autoregressive Box-Jenkins model:

$$X_t = X_{t-1}^{\Theta_t} \gamma_t , \qquad t = 1, 2, \ldots , \tag{18.7}$$

where

X_t = time to failure of modification t (with X_0 = time to failure of prototype),
θ_t = exponent,
γ_t = multiplicative error term.

When $\theta_t > 1$, reliability is growing. When $\theta_t < 1$, reliability is decreasing. X_t is assumed to be lognormal for all t. γ_t also is assumed to be lognormal with known parameters 0 and α^2. Taking the natural logarithm of both sides of (18.7) yields:

$$\ln X_t = \Theta_t \ln X_{t-1} + \ln \gamma_t , \tag{18.8}$$

where both $\ln X_t$ and $\ln \gamma_t$ are normally distributed.

While reliability growth can be modeled with a Box-Jenkins model of another form, empirical support can be claimed for (18.7). A power rule model, rather than an additive model, is chosen on the basis of historical evidence. Only a first-order autoregressive term is included because each modification is a modification of only the most recent design. A θ that varies with time permits both occasional major fixes and an overall trend of expected improvement ratios that approach one as the design matures.

18.5 SYSTEM RELIABILITY GROWTH

18.5.1 Prediction of System Reliability

A system, C, of n subsystems in series has a reliability, $\rho(\mathbf{r})$, given by

$$\rho(\underline{r}) = \prod_{i=1}^{n} r_i , \qquad 0 \le r_i \le 1 , \tag{18.9}$$

where r_i = reliability of subsystem i. The kth noncentral moment of the distribution function of the random variable, $\rho(\underline{r})$, μ_C^k, is given by

$$\mu_C^k = E(\rho^k(\mathbf{r})) = \prod_{i=1}^{n} \mu_i^k ,$$

where

E = expected value,
μ_i^k = kth moment of the random variable r_i.

A system, C, of n subsystems in parallel has reliability

$$\rho(\mathbf{r}) = 1 - \prod_{i=1}^{n} (1 - r_i) \tag{18.10}$$

and kth moment

$$u_C^k = E(\rho^k(\mathbf{r})) = \sum_{j=0}^{k} \left[\binom{k}{j} (-1)^j \prod_{i=1}^{n} \left\{ \sum_{m=0}^{j} \left(\binom{j}{m} (-1)^m \mu_i^m \right) \right\} \right], \tag{18.11}$$

where two terms in parentheses indicate a binomial coefficient.

For more complex systems, reliability and μ_C^k may be found by combining the previous basic series and parallel equations. This method can be used to estimate system reliability when subsystem reliabilities are known but subsystems have not yet been combined, as can occur, for example, when subsystems are developed in different locations. The method also can be used to predict system reliability growth after subsystem reliability growth predictions have been made.

18.5.2 Confidence Limits of System Failure Rate Estimates

After the first two moments of a system failure rate, L, have been estimated, the confidence limits for L may be estimated when L is assumed to have a gamma distribution. This assumption often may be justified because the L probability distribution will be a convex combination of gamma distributions if every subsystem failure rate distribution is either uniform or gamma. The conditional distribution function of the failure rate, $g(L)$, may be predicted with:

$$g(L; \alpha, \beta) = \frac{\beta e^{-\beta L} (\beta L)^{\alpha - 1}}{\Gamma(\alpha)}, \quad L, \alpha, \beta > 0 \tag{18.12}$$

where α and β are the characteristic parameters. The parameters α and β may be estimated from the estimated first two moments of L by solving two equations simultaneously:

$$\bar{L} = \frac{\alpha}{\beta} = \mu_L^1 , \tag{18.13}$$

$$\sigma_L^2 = \frac{\alpha}{\beta^2} = \mu_L^2 - (\mu_L^1)^2 . \qquad (18.14)$$

A symmetric $(1-\gamma)100$ percent confidence interval for L is expressed by

$$Pr(L_{\text{low}} \le L \le L_{\text{high}}) = 1 - \gamma,$$

where

$$\int_0^{L_{\text{low}}} g(l)\, dL = \frac{\gamma}{2}$$

and

$$\int_{L_{\text{high}}}^{\infty} g(L)\, dL = \frac{\gamma}{2} ,$$

with

$$L_{\text{low}} = \frac{\chi^2_{1-r/2}(2s+2\alpha)}{2(t+\beta)} , \qquad L_{\text{high}} = \frac{\chi^2_{r/2}(2s+\alpha)}{2(t+\beta)} ,$$

where the numerator is the chi-square value for the indicated probability level and $2s + 2\alpha$ degrees of freedom, t is the test time, s is the number of failures, and $t = s = 0$ for a system that has not yet been tested.

18.6 ADJUSTMENTS OF BASIC PREDICTION EQUATION

18.6.1 Inherent Unreliability

Some of the failure modes of a new item may remain unaffected by the fix program. There may be failures that are inherently uncorrectable, such as those due to acts of God, acts of war, or innate characteristics of materials. Some components may be stock items off the shelf. When this is the case, the total inherent failure rate should be estimated. Then the reliability growth and development time prediction equations should be applied to the other failure modes (Crow 1986).

There also may be partially correctable failure modes. An item may have some combination of new technology, present technology in new configurations, present technology in a new environment, and present technology under present conditions. This situation may be handled with either more than one growth equation or a composite growth equation. In some cases, a composite equation may be best at the start of a fix program when failure records are sparse, but multiple equations may be better after the data base improves.

18.6.2 Aging, Cycling, and Inactivity Failures

In addition to being a function of operating hours, component failures may depend on age, the number of on–off cycles, or the idle time since last operation:

$$F_t = F_0 + F_a + F_c + F_i\,, \tag{18.15}$$

where

F_t = total failures,
F_0 = failures due to operation,
F_a = failures due to age,
F_c = failures due to cycles,
F_i = failures due to idle time.

If failures are measured only as a function of operating hours, tests of a model may yield incorrect predictions of the failures in actual use unless the operating hours per month and operating hours per cycle are the same. A discussion of the problem may be found in "True Reliability Growth Measurement" by E. Demko, in *Proceedings of the 1986 IEEE Annual Reliability and Maintainability Symposium* (© 1986 IEEE). In most cases, all types of failures will occur during operation. For example, a bearing may experience lubricant migration during the prolonged period of idleness, but not fail until it has been operating for hours.

Thus, a prototype may require determination of as many as four failure rates rather than a simple failure rate per operating hour. A new model also will require a new failure rate estimate for each rate believed to be affected significantly. Multiple regression analysis, ordinarily preceded by a logical analysis of failure modes, will be necessitated.

In the simplest case, all terms in the regression will be linear. For example, failure of a mechanical part in a motor may be a function of miles, cold engine starts, and age in years. All rates may be independent, linear, and additive. Equation (18.15) will become

$$n = C_1 \sum_{i-1}^{n} m_i + C_2 \sum_{i-1}^{n} s_i + C_3 \sum_{i-1}^{n} y_i\,, \tag{18.16}$$

where

n = number of failures,
C_j = unknown coefficient ($j = 1, 2, 3$),
m_i = previous miles on part i,
s_i = previous cold starts for part i,
y_i = age of part i in years.

Dividing through this equation by the summed mileage yields

$$\lambda_t = \lambda_m + \lambda_s \frac{s_i}{m_i} + \lambda_y \frac{y_i}{m_i} \tag{18.17}$$

where

λ_t = total failures per mile,
$\lambda_m = C_1$ = failures per mile due to mileage,
$\lambda_s = C_2$ = failures per cold start,
s_i/m_i = mean cold starts per mile,
$\lambda_y = C_3$ = failures per year,
y_i/m_i = mean years per mile.

The linear regression form of (18.17) is

$$\underline{Y} = \underline{X\beta} + \underline{\epsilon}\,, \tag{18.18}$$

where

$\mathbf{Y} = n \times 1$ vector of inverses of mileage on each component at failure,
$\mathbf{X} = n \times 3$ matrix of ratios,

$$\begin{vmatrix} 1 & \dfrac{s_i}{m_i} & \dfrac{y_i}{m_i} \\ 1 & \dfrac{s_2}{m_2} & \dfrac{y_2}{m_2} \\ \vdots & \vdots & \vdots \\ 1 & \dfrac{s_n}{m_n} & \dfrac{y_n}{m_n} \end{vmatrix},$$

$\boldsymbol{\beta} = 3 \times 1$ vector of coefficients,
$\boldsymbol{\epsilon} = n \times 1$ vector of disturbances.

The solution is

$$\boldsymbol{\beta} = (\mathbf{X}'\mathbf{X})^{-1}\mathbf{X}'\mathbf{Y}\,, \tag{18.19}$$

where

$\mathbf{X}'$ = transpose of $\mathbf{X}$,
$(\mathbf{X}'\mathbf{X})^{-1}$ = inverse of $\mathbf{X}'\mathbf{X}$.

18.6.3 Relmain-X Complex Failure Analysis Model

Analysis is more difficult when failure relationships are nonlinear or unknown. Age failures, for example, may have a distribution that is probably normal, a distribution that is probably Weibull, or an unknown distribution. An analyst frequently may not even know whether age failures are significant enough to be included in the analysis.

Analysis of such a situation can include a combination of failure groupings, analysis of variance, and solution of nonlinear equations with iterative approximation. The first step may be isolation of two or more groups of failures that have approximately the same values for one or more variables of interest. Analysis of variance determines which failure causes appear to be significant enough to be included. The final step may be a cumulative failure distribution equation of the form

$$F(t) = 1 - \prod_{i=1}^{n} (1 - G_i(\alpha_i, \beta_i, i)) , \tag{18.20}$$

where

$F(T)$ = probability of failure by time t,
n = number of significant failure causes,
G_i = probability of failure from cause i by time t,
α_i, β_i = characteristic parameters of function G_i.

Relmain-X is a personal computer program for reliability and maintainability analysis of equipment in research, development, test, or evaluation. The program includes (18.20) and may be used for any number of causes, each of which may have a function that is linear, nonlinear, or unknown. Information is available from Harrison and Company, Post Office Box 2732, Merrifield, Virginia 22116, (703) 280-2202.

18.6.4 The Bull Model, an Example of Model Refinement

As new equipment is developed and used in the field, its reliability growth model can be made increasingly accurate. The model can be made applicable to the entire life cycle of the equipment rather than the development phase alone. Forecasts can cover factors other than equipment design, such as manufacturing variables, maintenance, and customer familiarity. The Bull model for the DPS 7000 mainframe computer system is an example of a mature and verified model. Details of the model may be seen in "Dynamic Reliability Prediction: How to Adjust Modeling and Reliability Growth?" by Gerard Collas, in *Proceedings of the 1991 IEEE Annual Reliability and Maintainability Symposium*. (© 1991).

The model for a digital bipolar integrated circuit is

$$\lambda_b = \left(\frac{94}{t}\right) G^{5/t} , \tag{18.21}$$

where λ_b is the failure rate at a junction temperature T_j of 60°C, G is the number of gates, and t is the manufacturing year (with 1965 as zero). The operational failure rate is

$$\lambda = \lambda_b \pi_0 \pi_e \pi_T ,$$

where π_0 is a quality factor, π_E is an environmental factor, and π_T is a temperature factor.

The temperature factor is a function of the junction temperature

$$T_j = T_a + (\theta_{jc} + \theta_{ac}) P_{av} ,$$

where T_a is the ambient temperature, θ_{jc} is the thermal resistance (junction case), θ_{ac} is the thermal resistance (ambient case), and P_{av} is the average power dissipation based upon the maximum power dissipation. θ_{jc} is a function of the type of package, the material, and the number of pins; θ_{ac} is a function of the same factors plus airflow. The temperature factor equation is

$$\prod_T = 3.64 \cdot 10^4 e^{-3,5000/T_j} + 2.52 \cdot 10^{13} e^{-11,600/T_j} .$$

These thermal models are based on thermodynamic formulas that are validated with laboratory measurements during design, qualification, and production. For example, the models for thermal resistances of a plastic dual in-line package are:

$$\theta_{jc} = 3.3 + \left(\frac{506}{N_p + 3}\right) ,$$

$$\theta_{ac} = k\left\{6.6 + \left(\frac{1012}{N_p + 3}\right)\right\} ,$$

where N_p = number of pins, $K = (0.55V + 1.15)/(V + 0.7)$, and V is the airflow in meters per second.

A reliability growth factor is implemented at the module, electronic board, and system levels. The growth factor is based on improvements in product quality, manufacturing processes, maintenance skills of repair personnel, and user training. The growth factor also is a function of the number of years between the year of the first shipment to a customer and the manufacturing year.

The initial predictions are modified using two types of field data. One type is for components. Operational data are accumulated for electronic boards and modules returned for repair. However, failures are validated and analyzed, and failure statistics are compiled before the failure data are correlated with the prediction. The other type of data, which are at the electronic board, module, and system levels, form a continuous field reliability evaluation performed throughout the product life cycle. During the early life period, all the systems, peripherals, controllers, and central system main boards are followed. During the rest of the life period, a sample of units is tracked.

Predicted failures are checked against failures that occur during the first six months of field use. Comparisons of predicted and actual failure rates are made for each type of component, for each component family, and for each technology. Prediction equation parameters are adjusted if the actual failure rate is outside the 80 percent confidence limits.

After the initial six-month period and adjustment of equations, comparisons of predicted and actual failures continue. However, each predicted failure rate is adjusted by learning curve factors whose values generally take the form

$$L = \alpha e^{-\beta t},$$

where α and β are characteristic parameters, and t is usually either the age of the equipment or the age of the equipment design. For example, each manufacturing process change brings a multiplication of the failure rate λ_b predicted by (18.21) in accordance with an equation determined over the years by regression analysis:

$$L = \alpha e^{-\beta(1-t_p)},$$

where t_p is the process change date. Another factor is applied for the number of years an item has been manufactured; this factor adjusts for reduction of error by factory personnel, particularly assembly errors. For yet another factor, t is the number of years since the first shipment to a customer; this factor reflects improvements in repair methods and tools, better diagnosis of malfunctions, user training, and user experience with the equipment.

The final result of the Bull reliability program has been predicted aggregate reliability growth that does not closely fit a Duane, Weibull, or exponential model. Predicted reliability growth is a decreasing hyperbolic function

$$\pi_s = \frac{f(\tau)}{g(\tau)}, \qquad (18.22)$$

where the terms of both the function $f(\tau)$ and the function $g(\tau)$ are linear, and τ is the number of years between the production year and the first customer shipment year. The overall experience has been that an approximately steady state is reached about one year after first customer shipment, so model improvement efforts have been concentrated on yielding accurate predictions for this point in the equipment life cycle.

18.7 ESTIMATION OF GROWTH MODEL PARAMETERS

To date, there has been little clear agreement on the best techniques for estimating the parameters of a growth model. This section is intended to present the principal techniques that have been used rather than select techniques for the reader. Comparisons and evaluations may be found in Gray (1986), Jewell (1984), Fard (1983), and Woods (1990).

18.7.1 Selection of Dependent and Independent Variables

The fundamental dependent variable used to date has been the mean time to failure (MTTF) for each model, k, where $k = 1, 2, 3, \ldots$. Unless each model has approximately the same number of failures, however, this basic variable can bias the model. Bias can be eliminated at the expense of computation if each failure of each model is included as a separate data point. This also aids the estimation of the model variance. Alternatively, each MTTF can be multiplied by an integer for the relative number of failures. In any case the failure time may be inverted into a failure rate for computational convenience.

A dependent variable other than some form of failure time may be chosen if it is more suited to the purpose of the reliability improvement program. Either availability or probability of survival for a period of time may be more appropriate for emergency equipment. A failure cost function may be better for many industrial purposes.

In a classic growth model, the independent variable usually is either the equipment model number or the cumulative operating time. Alternatives that might improve forecasts, however, include the number of engineering changes in a model, the number of engineering changes with each change weighted for its type, the calendar time, or the money spent on the test-analyze-and-fix program.

18.7.2 Classical, Bayesian, and Nonparametric Methods

Maximum likelihood estimation is the classical method for estimating the parameters of a growth equation. The parameters selected are those that minimize the sum of the squared errors when the forecasting equation is applied to the previous values. A typical objective is:

$$\text{minimize} \quad f(i; \alpha, \beta) = \sum_{i=1}^{n} (m_i - p_i)^2 , \qquad (18.23)$$

where n is the number of models, α and β are the parameters of the forecasting equation, m_i is the observed mean time to failure of model i, and p_i is the MTTF of model i given by the equation.

When the prediction expression is a sum of linear terms or can be converted into a sum of linear terms, it ordinarily is solved with a relationship of the form of (18.18). For example, for a Duane model such as (18.3), equation (18.23) might take the form

$$\text{minimize} \quad f(i; k, m) = \sum_{i=1}^{n} (\ln \lambda_i - \ln k + m \ln t_i)^2 , \qquad (18.24)$$

where t_i = operating hours before model i was designed. Expression (18.24) is minimized when its derivatives with respect to $\ln(k)$ and m equal zero:

$$\frac{\partial f}{\partial \ln(k)} = -2 \sum_{i=1}^{n} (\ln(\lambda_i) - \ln(k) + m \ln(t_i) = 0 ,$$

$$\frac{\partial f}{\partial m} = -2 \sum_{i=1}^{n} \ln(t_i)(\ln(\lambda_i - \ln(k) + m \ln(t_i)) = 0 .$$

Hence

$$m = -\frac{\Sigma \ln(t_i) \ln(\lambda_i) - (\Sigma \ln(t_i))(\Sigma \ln(\lambda_i))/n}{\Sigma (\ln(t_i))^2 - (\Sigma \ln(t_i))^2/n} ,$$

$$\ln(k) = \ln(\lambda_i) + m \ln(t_i) .$$

Comparable closed-form solutions for other models may be found in the books in the bibliography. When a closed-form solution is not possible, a solution may be approximated with an iterative search. A personal computer program covering all the major reliability growth models is available from Harrison and Company, Post Office Box 2732, Merrifield, Virginia 22116, (703) 280-2202.

The methodology, advantages, and disadvantages of Bayesian analysis are discussed in a very large number of the references for this chapter, including Mann et al. (1974). A detailed development of a growth model may be found in Mazzuchi and Soyer (1991). The parameters are not considered to be unknown fixed values as in classical analysis. A parameter, β, is regarded as a random variable that has an assumed probability density function, $g(\beta)$. This assumed prior density is whatever seems both computationally convenient and reasonable in light of previous experience.

For illustration purposes, there may be one parameter, θ, whose value is unknown. The prior estimate of θ, θ_0, is the expected value of the assumed prior distribution:

$$\theta_0 = \int_{-\infty}^{\infty} \theta g(\theta)\, d\theta . \qquad (18.25)$$

After there have been n observations of a random variable, X, that is a function of θ, an expression is derived for the joint probability density of the n values as a function of θ, $f(x_1, \ldots, x_n/\theta)$. The joint marginal probability density, which is the probability of the occurrence of the n values summed over all possible values of θ, is

$$f(x_1, \ldots, x_n) = \int_{-\infty}^{\infty} f(x_1, \ldots, x_n, \theta)\, d\theta\,.$$

Finally, the posterior probability density function, $g(\theta/x_1, \ldots, x_n)$, gives the estimated probability of any given value of θ given the n observed values

$$g\left(\frac{\theta}{x_1, \ldots, x_n}\right) = \frac{f(x_1, \ldots, x_n, \theta)}{f(x_1, \ldots, x_n)}\,. \tag{18.26}$$

The expected value yielded by (18.26) is the posterior estimate of θ, θ_1,

$$\theta_1 = \int_{-\infty}^{\infty} \theta g\left(\frac{\theta}{x_1, \ldots, x_n}\right) d\theta\,. \tag{18.27}$$

When the Bayesian analysis consists of a succession of estimates, the posterior distribution of one estimate becomes the prior distribution for the following estimate.

An example of an application may be found in Kaplan et al. (1990). A product was given a succession of tests $(1, 2, \ldots, j)$ with each unit either passing or failing each test. The Bayesian equation was

$$P\left(\frac{Q_i}{E_i}\right) = \frac{P_0(Q_i)P(E_i/Q_i)}{\int_0^1 P_0(Q_i)P(E_i/Q_i)\, dQ_i}\,, \tag{18.28}$$

where

E_i = test i failed units, f_i, divided by total test i units, N_i,
Q_i = value E_i would approach if N_i approached infinity,
$P(Q_i/E_i)$ = posterior probability distribution hypothesized for Q_i after learning E_i,
$P_0(Q_i)$ = prior probability distribution hypothesized for Q_i before learning E_i,
$P(E_i/Q_i)$ = conditional probability that test i would have resulted in a failure ratio of E_i if Q_i had been the true value for E_i as N_i approached infinity.

By Bernoulli's theory, the last value is given by

$$P\left(\frac{E_i}{Q_i}\right) = Q_i^{F_i}(1 - Q_i)^{N_i - F_i}\,. \tag{18.29}$$

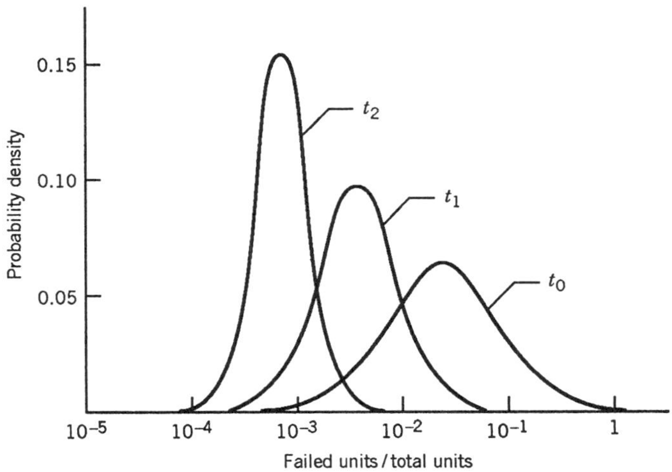

Figure 18.3 Succession of lognormal Bayesian distributions, Kaplan study (© 1990, IEEE).

A lognormal prior distribution was assumed, which is termed a "conjugal" prior distribution because it produces a posterior distribution of the same family. Figure 18.3 shows the initial prior distribution and the posterior distribution for the first two tests. With each model change, the mean value of Q_i is lower, and the confidence region is narrower.

The method of estimating Q_i in a procedure such as this can be adjusted for expected reliability growth by applying (18.28) twice for each test, once with E_i and once with an adjusted value of E_i, Z_i. Let

$$Z_i = \frac{F_i - C_i}{N_i - C_i} \,. \tag{18.30}$$

where C_i is the estimated number of failed units that would not have failed test i if all the design fixes resulting from test i had worked perfectly. Z_i yields an unrealistically optimistic estimate because fixes are seldom 100 percent effective, some fixes can even cause new problems, and $N_i - C_i$ would not be a random sample in any case. However, engineering judgment and past experience can be used to find a realistic estimate between the pessimistic E_i distribution and the optimistic Z_i distribution.

Nonparametric methods have been investigated far less extensively to date than classical and Bayesian methods, and there consequently is little empirical evidence either for or against them. However, an analyst might want to investigate them if there is reason to believe there might be hidden bias in any assumed probability distribution. Methodological background may be found in Barlow and Scheur (1971) and subsequent papers. Gray (1986) gives a very limited assessment of estimation accuracy in comparison to traditional techniques.

CHAPTER 19

MONTE CARLO SIMULATIONS

19.1 PURPOSE AND NATURE

Simulations, which usually are done on a computer, are used to predict the behavior of equipment whose operation processes are difficult or impossible to represent with analytical relationships that are known to be valid. Simulations also may be used if the mathematical relationships are known but would require more time than the simulations. In general, there are three steps: (1) development of a formal model, (2) creation, selection, or adaptation of the software, and (3) simulation itself, sometimes coupled with computation.

A reliability simulation model ordinarily is a discrete model with a governing sequence of discrete events, such as failures, repairs, and switchings. Development of a formal model requires creation of an algorithm that transforms a set of initial data into a sequence of discrete events. The events may have characteristics that are computed. Statistical inferences are made about the equipment or process after the runs are completed.

19.2 GENERAL DESCRIPTION OF A DISCRETE SIMULATION MODEL

A discrete model has as its core a set of interacting units, $A_1, A_2, \ldots,$ which usually are components of complex equipment such as an aircraft, computer, or communications network. Each unit, A_j, is defined with a collection of attributes, a_j. One attribute is the state, s_j, which completely defines the dynamics of the system. All other attributes, if any, are represented by auxiliary variables, π_j. Thus, $a_j = (s_j, \pi_j)$.

Events e_j are defined for the unit A_j. Each event is defined by the state s_j and its attributes a_j. The event e_j may follow a change of attributes of units. A set of units determining the event depends on the nature of the event.

The model is developed as a sequence of events. One of the attributes, a_j, is the real variable τ_j, which is the residual time until the occurrence of the event e_j given that there is no intervening event. Notice that τ_j can be considered also to be a component of the state s_j. At the initial time $t = 0$, $s_j = s_j^0$ and $\tau = \tau_j^0$ for each unit A_j. The first event occurs after a period of time $\theta_1 = \min \tau_j$ over all j. If j_1 = argument of $\min \tau_j$, the event occurs in unit A_j. (The possible nonuniqueness of j_1 is neglected here.) Event e_{j_1} occurring at time $t_1 = \theta_1$ changes attributes of one or more units. A new state and new residual time τ_j^1 must be found for each affected unit. For each unaffected unit the residual time is given by $\tau_j^1 = \tau_j^0 - \theta_1$. A new time, the *interoccurrence time*, $\theta_2 = \min \tau_j^1 = \tau_{j_2}^1$. The second event, e_{j_2}, occurs at time $t_1 + \theta_2$. Again, the attributes and hence the residual times of one or more units may be changed depending on the nature of the event e_{j_2}. For an unaffected unit, the new residual time is $\tau_j^2 = \tau_j^1 - \theta_2$.

These variables and relationships are in almost all simulation programs. A governing sequence $(t_1, e_{j_1}), (t_2, e_{j_2}), \ldots,$ and corresponding values $(s_j^1, \pi_j^1, \tau_j^1), (s_j^2, \pi_j^2, \tau_j^2) \ldots$ are calculated. The sequence and its values are the *trajectory* of the simulation.

19.3 DETAILED EXAMPLE OF SIMULATION

In this section we consider a simple example that illustrates some details of Monte Carlo simulation. Suppose that a system consists of m main and n standby redundant units. There are l repairing stations, $l < m + n$. The time to failure of each unit is a r.v. with d.f. $F(x)$ and the repair time is a r.v. with d.f. $G(x)$. Sample system models follow.

Model 1

The system has units $A_1, \ldots, A_{m+n}$. Each A_i is characterized by the attribute $a_i = (s_i, \pi_i)$. The state s_i can take integer values $\ldots, -2, -1, 0, 1, 2, \ldots$. If $s_i < 0$, unit i is a spare. When $s = -k$, the unit will be placed into service. Suppose that the order of placement into service corresponds to the order of completion of repair, i.e., FIFO serving discipline is considered. If $s_i = 0$, the unit is in operation. If $s_i = 1$, the unit is in repair. If $s_i = 1 + k$, $k \geq 1$, the unit is the kth one in the queue awaiting repair. The number of failures of unit i that have occurred before any given time in the simulation is π_i. The residual time τ_i is defined only for states 0 and 1. For 0, τ_i is the residual lifetime. For 1, it is the residual repairing time. For other states, τ_i may be set equal to infinity to indicate arbitrarily the absence of a recorded time. Event e_i can have only two values: $e_i = 0$ denotes a failure, and $e_i = 1$ denotes a repair.

Let the model be described by the vector $[(s_1^k, \pi_1^k, \tau_1^k), (s_2^k, \pi_2^k, \tau_2^k), \ldots, (s_{m+n}^k, \pi_{m+n}^k, \tau_{m+n}^k)]$ at the time t_k. The nearest occurrence time then is $t_{k+1} = t_k + \theta$, where θ is the shortest residual time for all $m + n$ units at time t_{k+1}:

$$\theta = \min_{1 \le i \le m+n} \tau_i^k .$$

Let $r = \arg\min \tau_i^k$. Event e_r occurs at time t_{k+1}.

Case 1 For the first case, let $e_r = 0$, so $s_r^k = 0$. Then for all standby units A_α (defined by the condition $s_\alpha^k < 0$),

$$(s_\alpha^{k+1}, \pi_\alpha^{k+1}, \tau_\alpha^{k+1}) = \begin{cases} (s_\alpha^k + 1, \pi_\alpha^k, \infty) & \text{if } s_\alpha^k < -1 , \\ (0, \pi_\alpha^k, \xi_k) & \text{if } s_\alpha^k = -1 . \end{cases}$$

Thus one of the standby units (if any) becomes operating and others advance in the queue. Here $\{\xi_k\}$ are independent identically distributed (i.i.d.) r.v.'s with a common d.f., F. ξ_k is the time to failure of the unit A_α.

Let δ^k be the number of repairing stations occupied at time t_k (and hence at time t_{k+1} because no repair is finished until t_{k+1}). Thus δ^k is the number of A_j for which $s_j^k = 1$. If $\delta < l$ (i.e., there is at least one empty repairing place), then

$$(s_r^{k+1}, \pi_r^{k+1}, \tau_r^{k+1}) = (1, \pi_r^k + 1, \eta_k) ,$$

where $\{\eta_k\}$ are i.i.d. r.v.'s with a common d.f., $G(t)$, and η_k are the repair times for the unit A_r. When a repair is completed, the auxiliary variable π_r increases by one.

If $\delta^k = l$,

$$(s_r^{k+1}, \pi_r^{k+1}, \tau_r^{k+1}) = (\max_{1 \le i \le m+n} s_i^k + 1, \pi_r^k + 1, \infty) .$$

The unit joins the repair queue in position with the number

$$\max_{1 \le i \le m+n} s_i^k .$$

For units A_β ($\beta \ne r$) that are failed and units at the main places (where $s_\beta^k \ge 0$),

$$(s_\beta^{k+1}, \pi_\beta^{k+1}, \tau_\beta^{k+1}) = (s_\beta^k, \pi_\beta^k, \tau_\beta^k - \theta) .$$

Here $\infty - \theta$ is considered to be ∞ for the units A_β from the queue to be repaired (where $s_\beta^k > 1$).

Case 2 For the second case, let $e_r = 1$. The unit A_τ has just left a repairing place: $s_r^k = 1$. The first unit in the queue, if any, occupies this repairing place. Other units advance one position. The units under repair stay at their repairing places, but all residual repairing times decrease by θ. Thus

$$(s_\alpha^{k+1}, \pi_\alpha^{k+1}, \tau_\alpha^{k+1}) = \begin{cases} (s_\alpha^k - 1, \pi_\alpha^{k+1}, \infty) & \text{if } s_\alpha^k > 2\,, \\ (1, \pi_\alpha^{k+1}, \eta_k) & \text{if } s_\alpha^k = 2\,, \\ (s_\alpha^k, \pi_\alpha^{k+1}, \tau_\alpha^k - \theta) & \text{if } s_\alpha^k = 1\,. \end{cases}$$

Let ψ^k be the number of operable units at time t_k. For these units, $s_j^k \leq 0$. If $\psi^k < m$, the desired number of main units, the repaired unit A_i becomes a main unit:

$$(s_r^{k+1}, \pi_r^{k+1}, \tau_r^{k+1}) = (0, \pi_r^{k+1}, \xi_k)\,.$$

If $\psi^k = m$, A_i becomes the last unit in the queue awaiting work:

$$(s_r^{k+1}, \pi_r^{k+1}, \tau_r^{k+1}) = (\max_{1 \leq i \leq m+n} s_i^k - 1, \pi_r^k, \infty)\,.$$

All main and standby units stay in place, but all main unit residual times decrease by θ:

$$(s_\beta^{k+1}, \pi_\beta^{k+1}, \tau_\beta^{k+1}) = (s_\beta^k, \pi_\beta^{k+1}, \tau_\beta^{k+1}) \qquad \text{for } s_\beta^k \leq 0\,.$$

Model 2

In this case we do not fix a number of failures of each unit, and we introduce auxiliary variables. Let

A_1 = set of main units,
A_2 = set of standby units,
A_3 = set of units in the queue awaiting repair,
A_4 = set of units in repair.

A_1 is characterized by a vector state $s_1 = (\kappa; s_{1i}, \ldots, s_{1\kappa})$, where κ is the number of main units at the current time, $0 \leq \kappa \leq m$, and s_{1i} is the residual lifetime of main unit i. Then

$$\tau_1 = \min_{1 \leq i \leq m+n} s_{1i}\,.$$

Note that if $\kappa = 0$, then we put $\tau_1 = \infty$.

A_2 is characterized by an integer state s_2 taking an integer value $0 \leq s_2 \leq n$, where s_2 is the current number of standby units. As a standby unit is regarded as being under no stress, and thus having an infinite expected time to failure, $\tau_2 = \infty$.

A_3 is characterized by an integer state s_3, the current number of units awaiting repair, $0 \le s_3 \le m + n$.

The state of A_4 is a vector:

$$\mathbf{s}_4 = (\nu, s_{41}, s_{42}, \ldots, s_{4\nu}),$$

where ν is the number of occupied repair stations, $0 \le \nu \le l$, and s_{4i} is the residual repair time for unit i. If $v > 0$, $\tau_4 = \infty$; otherwise,

$$\tau_4 = \min_{1 \le i \le \nu} s_{4i}.$$

This model has only two events. The failure of a unit is e_1 and the repair of a unit is e_4. The model at the time of event k is characterized by a vector:

$$\begin{aligned}&[(s_1^k, \tau_1^k), (s_2^k, \tau_2^k = \infty), (s_3^k, \tau_3^k = \infty), (s_4^k, \tau_4^k)] \\ &\quad = [(\kappa^k, s_{1,1}^k, \ldots, s_{1}^k\kappa^k, \tau_1^k), (s_2^k, \infty), (s_3^k, \infty), (\nu^k, s_{4,1}^k, \ldots, s_{4,\nu^k}^k, \tau_4^k)].\end{aligned}$$

Time $t_{k+1} = t_k + \theta = t_k + \min(\tau_1^k, \tau_4^k)$.

First, if $\tau_1^k < \tau_4^k$, then e_1 occurs at time t_{k+1}. Let $\tau_1^k = \tau_{1r}^k$. Then

$$(\kappa^{k+1}, s_{1,1}^{k+1}, \ldots, s_{1,\kappa^{k+1}}^{k+1}) = \begin{cases} Y_1 & \text{if } s_2^k > 0, \\ Y_2 & \text{if } s_2^k = 0, \end{cases}$$

where

$$Y_1 = (\kappa^k, s_{1,1}^k - \theta, \ldots, s_{1,r-1}^k - \theta, \xi_k, s_{1,r+1}^k - \theta, \ldots, s_{1,\kappa^k}^k - \theta),$$

$$Y_2 = (\kappa^k - 1, s_{1,1}^k - \theta, \ldots, s_{1,r-1}^k - \theta, s_{1,r+1}^k - \theta, \ldots, s_{1,\kappa^k} - \theta),$$

$$s_2^{k+1} = \max(0, s_2^k - 1);$$

$$s_3^{k+1} = \begin{cases} s_3^k + 1 & \text{if } \nu^k = l, \\ 0 & \text{if } \nu^k < l; \end{cases}$$

$$(\nu^{k+1}, s_{41}^{k+1}, \ldots, s_{4\nu^{k+1}}^{k+1}) = \begin{cases} (\nu^k, s_{41}^k - \theta, \ldots, s_{4\nu^k}^k - \theta) & \text{if } \nu^k = l, \\ (\nu^{k+1}, s_{41}^k - \theta, \ldots, s_{4\nu^k}^k - \theta, \eta_k) & \text{if } \nu^k < l. \end{cases}$$

Second, if $\tau_1^k > \tau_4^k$ the event at time t_{k+1} is a repair, e_4. Let $\tau_4^k = s_{42}^k$. Then

$$(\kappa^{k+1}, s_{11}^{k+1}, \ldots, s_{1\kappa^{k+1}}^{k+1}) = \begin{cases} (\kappa^k, s_{1,1}^k - \theta, \ldots, s_{1\kappa^k}^k - \theta) & \text{if } \kappa^k = m , \\ (\kappa^k + 1, s_{11}^k - \theta, \ldots, s_{1\kappa^k}^k - \theta, \xi_k) & \text{if } \kappa^k < m ; \end{cases}$$

$$s_2^{k+1} = \begin{cases} s_2^k + 1 & \text{if } \kappa^k = m , \\ 0 & \text{if } \kappa^k < m ; \end{cases}$$

$$s_3^{k+1} = \max(0, s_3^k - 1) ;$$

$$(\nu^{k+1}, s_{41}^{k+1}, \ldots, s_{1\nu^{k+1}}^{k+1}) = \begin{cases} Z_1 & \text{if } s_3^k > 0 , \\ Z_2 & \text{if } s_3^k = 0 , \end{cases}$$

$$Z_1 = (\nu^k, s_{41}^k - \theta, \ldots, s_{4,r-1}^k - \theta, \eta_k, s_{4,r+1}^k - \theta, \ldots, s_{4\nu^k}^k - \theta) ,$$

$$Z_2 = (\nu^k - 1, s_{41}^k - \theta, \ldots, s_{4,r-1}^k - \theta, s_{4,r+1}^k - \theta, \ldots, s_{4\nu^k}^k - \theta) .$$

Model 3

Model 2 can be simplified when $F(t) = 1 - \exp(-\lambda t)$. Consider the unit A_1. Let s_1 = the current number of main units. The residual time τ_1 at every occurrence time may be considered to be a r.v. that is independent of the past and has $F(t) = 1 - \exp(-\lambda t)$. All transition formulas are preserved, but all transforms connected with $s_{11}, s_{12}, \ldots$, can be deleted.

Model Assessments

The three previous models illustrate different levels of compromise between the desire for maximum validity and the desire for minimum complexity in a model. Each level of compromise might be appropriate for certain purposes.

The first model is the most complex and hence takes the longest to program and to run. It also reveals the most characteristics of the operation under study. However, no insight is given into causes of failures, types of failures, or repair station efficiency for individual units. These variables are not modeled.

The second model is a simplification of the first model that can be derived through formal algebraic equivalences. The model reveals characteristics of the four sets of units but no characteristics of individual units.

The third model is the simplest, but it can be used only with Poisson failures. In this model simplicity is increased through limitation of the process analyzed rather than through algebraic combination.

19.4 SELECTION OF COMMERCIALLY AVAILABLE SOFTWARE

A simulation model can be written in a programming language such as FORTRAN, Pascal, or C to represent any desired operational characteristics. In general, however, such a model requires a lot of time and has restricted flexibility.

Some modern simulation software, such as SIMPLEX and GEST, has the programming for a general class of mathematical models. The user has to contribute only structural parameters, numerical values, and, in some cases, transitional subroutines. A rule of thumb for the selection of software of this type is that the user programming time increases with the scope and versatility of the model.

A simulation language can be used by a person who does not know the mathematical relationships in the model. A conceptual model description is converted into algorithms automatically. Unfortunately, users without mathematical comprehension will not be abe to analyze the results statistically. They also will be unable to expand the scope of the model to new applications. There are many simulation languages, but only a few are widespread. Some of the most popular are SIMULA, GASP, and SIMSCRIPT.

Whatever software is used, simulations convert input data into output data. The three models in Section 19.3 contain both common classes of input data: factual and hypothetical. Factual inputs are those such as the number of units and repair stations. The hypothetical inputs in the three models are the distributions for ξ and η. The output data usually are large arrays that can be analyzed with the statistical inference techniques used for random samples of real life data.

One advantage of simulations over real life is that the accuracy and reliability of the estimates can be improved with special governing sequences. Methods of variance reduction that minimize the simulation time needed for a desired standard error of estimate are very important. We discuss some of them next.

19.4.1 Method of Complementary Variables

Let the output of the model be the mean of a r.v., Y, that monotonically depends on input r.v.'s $\theta_1, \theta_2, \ldots$, governed by the model dynamics. The variable $Y = Y(\theta_1, \theta_2, \ldots)$ is monotonic with respect to each of its arguments. Let $F_i(t)$ be the d.f. of a r.v., θ_i. Form a new sequence $\{\eta_i'\} = (\theta_1', \theta_2', \theta_2', \ldots)$ with r.v. $\theta_i' = F_i^{-1}(1 - F_i(\theta_i))$. Thus the distribution function of θ_i' also would be $F_i(t)$. It is convenient to generate both θ_i and θ_i' with the help of a sequence ω_i of i.i.d. r.v.'s that are distributed uniformly over $(0, 1)$:

$$\theta_i = F_i^{-1}(\omega_i)\,,$$

$$\theta_i' = F_i^{-1}(1 - \omega_i)\,.$$

It follows that r.v.'s $Y = Y(\theta_1, \theta_2, \ldots)$ and $Y' = Y(\theta_1', \theta_2', \ldots)$ have a nonpositive covariance: $\mathrm{Cov}\{Y, Y'\} \le 0$. At the same time

$$E\{Y\} = E\{Y'\} = \mu ,$$
$$\text{Var}\{Y\} = \text{Var}\{Y'\} = \sigma^2 .$$

Hence the value $(Y + Y')/2$ is a nonbiased estimate of μ with a variance σ_1^2 given by

$$\sigma_1^2 = \sigma^2 + \frac{\text{Cov}\{Y, Y'\}}{2}\Big) \leq \frac{\sigma^2}{2} .$$

Thus it is possible to reduce the variance of estimates given that there is a monotonic dependence of Y on the θ_i variables. Sometimes such monotonicity takes place in reliability models.

19.4.2 Method of Dependent Trials

The previous method is a variant of a general method of dependent trials. The efficiency of this method is implied by the dependence between Y and Y'. Another application of dependent trials occurs in a sensitivity analysis when coefficients of the relationships between an output variable and its input variables are estimated. Dependent trials also can be helpful in random searches for extreme values and in other situations.

An output variable, Y, with the mean value μ depends on an input variable, X. Y', μ', and X' are similarly related. $Y' - Y$ then can be used to estimate $\mu' - \mu$. As $\text{Var}\{Y' - Y\} = \text{Var}\{Y'\} + \text{Var}\{Y\} - 2\,\text{Cov}\{Y'Y\}$, $\text{Var}\{Y' - Y\}$ can be reduced with any technique for generating Y' and Y so that $\text{Cov}\{Y', Y\} > 0$. It can be possible to reduce $\text{Var}\{(Y' - Y\}$ relative to $\text{Var}\{Y'\} + \text{Var}\{Y\}$ by several orders.

An example might be an investigation of the sensitivity of Y to X when Y is positively and monotonically related to X and there is special interest in extremely high values of Y. One plan for the investigation might be 100 simulations with random X inputs. Certainty would be increased, however, with 50 simulations using random X inputs and 50 more using each original X plus a fixed amount.

19.4.3 Regenerative Method

The regenerative method is one of the most popular methods of simulation used with statistical inference. It can be used when a regenerative stochastic process is used for a reliability model of equipment. The simulation trajectory can be divided into identical independently distributed "cycles."

The regeneration cycles can be analyzed statistically as if they were samples of i.i.d. random variables. Statistical analysis will yield the stationary characteristics of the process, such as the expected availability, expected cycle duration, and the number of cycle simulations required for any desired accuracy of the estimates of the stationary characteristics.

19.5 ESTIMATES OF RARE EVENTS

Conventional simulations for the analysis of rare events, such as power blackouts or nuclear reactor failures, can be impractical because of the numbers required. If q is the probability of a rare event, the number of simulations, N, needed to estimate q with relative accuracy ϵ is $(e^2 q)^{-1}$. Thus, if $e = 0.1$ and $q = 10^{-5}$, then $N \approx 10^7$. As a consequence, methods of forced estimation are used. These methods have model dynamics based on knowledge of the physical processes.

19.5.1 Weighted Trials Method

The most versatile method, which has many variants, is the weighted trials method. A vector $\boldsymbol{\theta} = (\theta_1, \theta_2, \ldots, \theta_n)$ is a sequence of n random variables governed by model dynamics. The probability density function of $\boldsymbol{\theta}$, which may be either discrete or continuous, is $w(\underline{x})$. Possession of a sample $\boldsymbol{\theta}$ shows whether or not a rare event has occurred.

Let the relationship $\theta \in B$ be equivalent to the occurrence of the event

$$q = Pr\{\theta \in B\} = \int_B w(x)\, dx = \int 1_B(x) w(x)\, dx\ ,$$

where the dummy variable $1_B(x)$ is given by

$$1_B(x) = \begin{cases} 0 & \text{if } x \notin B\ , \\ 1 & \text{if } x \in B\ . \end{cases}$$

We take k samples of $\boldsymbol{\theta}$, $x_1, x_2, \ldots, x_k$, and calculate an unbiased estimate $\hat{q}$ of the probability of a rare event:

$$\hat{q} = \frac{1}{k} \sum_{1 \le j \le k} 1_B(x_j)\ .$$

Then the variance is

$$\mathrm{Var}\{\hat{q}\} = \frac{1-q}{kq}\ ,$$

which approaches infinity as q approaches zero. This means that it may be impossible to estimate a very small q with satisfactory accuracy and a practical number of samples.

The problem can be solved by selecting a random vector, $\boldsymbol{\xi}$, that has the same dimension as $\boldsymbol{\theta}$ and a density function $v(x)$. When $v(x) \neq 0$ for $x \in B$, and $g(x)$ is given by

$$g(x) = 1_B(x) \frac{w(x)}{v(x)} ,$$

$$q = \int 1_B(x) \frac{w(x)}{v(x)} v(x)\, dx \equiv \int g(x)v(x)\, dx = E\{g(\xi)\} .$$

Thus

$$\hat{q}_1 = \frac{1}{k} \sum_{1 \le j \le k} g(y_j) ,$$

also is an unbiased estimator of q where y_j is a sample of ξ. If the integral of $v(x)d(x)$ is not small, then the event $\{\xi \in B\}$ will not be rare, and the variance will be low. Although new problems of calculating $w(x)$ and $v(x)$ to obtain $g(x)$ may arise, usually the new problems are not formidable. The optimal choice is

$$v(x) = \begin{cases} \dfrac{w(x)}{q} & \text{when } x \in B , \\ 0 & \text{otherwise} . \end{cases}$$

To use this density, we need to know q.

Operation of much equipment can be described mathematically with regenerative stochastic processes. Let $\{Y(t)\}$ be such a process. It means that there exist such random times $0 = S_0 < s_1 < \cdots$, and $\theta = S_{i+1} - S_i$, $i \ge 0$, that the cycles $\{Y(S_i + t)\}$, $t \le \theta_i$ are i.i.d. r.v.'s. Many important reliability indexes depend on the probability, q, that a failure occurs during a regeneration cycle. Examples of the estimation of q follow.

EXAMPLE 19.1

Let us return to the example in Section 19.3. Let there be only one main unit and only one repairing place, that is, $l = m = 1$. Let $Y(t)$ be the number of components at time t. A regeneration point S_i is such that $Y(S_i - 0) = 0$ and $Y(S_i) = 1$ (S_i is a breakdown time of a main unit when other units have not failed). Let τ be the first breakdown time:

$$\tau = \inf\{t\colon Y(t) = n + 1\} .$$

In other words, the time when all units fail for the first time,

$$q = Pr\{ \sup_{S_{i-1} \le y \le S_i} Y(t) = n + 1\} ,$$

is the probability of system failure during a regenerative period.

EXAMPLE 19.2

Let the failure time distribution be $F(t) = 1 - \exp(-\lambda t)$ and the repair time distribution $G(t)$ be arbitrary. Suppose that the probability, ϵ, of failure of a main unit during repair of another unit:

$$\epsilon = \int_0^\infty (1 - e^{-\lambda t})\, dG(t)\,,$$

is small.

Let q_n = probability that n units will fail during the repair time for one unit. It can be shown that $q/q_n \to 1$ as $\epsilon \to 0$. Therefore q may be used as an approximation for q_n for small ϵ:

$$q_n = \sum_{n \le k < \infty} \int_0^\infty e^{-\lambda t} \frac{(\lambda t)^k}{k!}\, dG(t) \approx \frac{\lambda^n}{n!} \int_0^\infty t^n\, dG(t)\,.$$

Thus q_n can be estimated by estimating the nth moment of a repairing time.

A simple numerical illustration is given by four independent units having a common λ of 0.01 per hour and a common repair time approximated as being exactly two hours. The probability of these units failing within two hours, q_n, can be estimated with

$$q_3 \approx \frac{(0.01)^3}{3!} \int_0^\infty t^3\, dG(t) = \frac{(0.01)^3}{6}\, 2^3 = 1.33 \times 10^{-6}\,.$$

EXAMPLE 19.3

Let $F(x)$ be a general d.f. Apply a special form of weighted trials method for estimating q.

Consider possible trajectories of the random process $Y(t)$ that are piecewise constant. Consider a regenerating cycle starting with a jump $0 \to 1$ and continuing until the next similar jump. Further consider only trajectories that start with a jump $0 \to 1$ and finish at the state $n+1$ without returning to state 0. Every trajectory consists of upward jumps $(k \to k+1)$ and downward ones $(k \to k-1)$. Divide all such trajectories into disjoint classes C_w, $k \ge 0$, where k is the number of backward jumps so that every trajectory of this class has $N + 1 + 2k$ jumps. We will characterize each trajectory by a vector $\boldsymbol{v}_1(p)$ that consists of $N + 1 + 2k$ components (see Figure 19.1).

Let q_i be the probability that a trajectory belongs to class C_k. Evidently

$$q = \sum_k \sum_i q_i(k)\,.$$

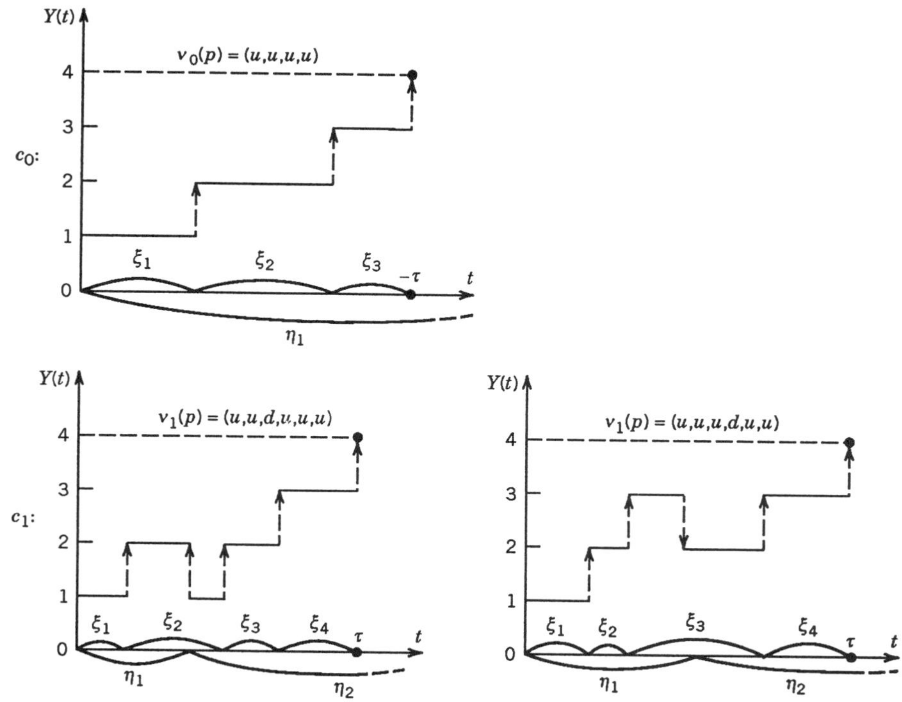

Figure 19.1 Sample failure and repair sequences.

The estimation of q_i may be illustrated with an example. Let us take a trajectory from class C_1 with a vector $\boldsymbol{\nu}_1(p) = (u, u, d, u, u)$. The procedure is given in Table 19.1.

The procedure in Table 19.1 is fast and accurate. Less than one second is required for 1000 samples on an IBM-PC/XT. Running time does not depend on the $q_i(k)$ values, which may be as small as needed. As the number of $q_i(k)$ values is infinite, they all cannot be estimated. Hence the most probable values are estimated, and an upper limit is estimated for the sum of the remaining probabilities.

The algorithm can be simplified if necessary. For example, for small t, $F(t)$ may be approximated with dt^{β}, $\beta > 0$. In step 2, the distribution function for ξ_1' would become $(t/\eta_1')^{\beta}$. In step 5, the d.f. of ξ_3' would be $[t/(\eta_1' + \eta_2' - \xi_1' - \xi_2')]^{\beta}$, and in step 6, the d.f. for ξ_4' would be $[t/(\eta_1' + \eta_2' - \xi_1' - \xi_2' - \xi_3')]^{\beta}$.

The algorithm is a combination of analytical calculations, such as those for q, and simulation, such as the selection of random failure and repair times. This type of combination requires understanding the peculiarities of the model. The general technique may be applied to many reliability models, but each requires analysis and possible special modification. I. N. Kovalenko and his colleagues have suggested many different forced methods.

Table 19.1 Example of simulation algorithm: Generation of one set of occurrence times when sequence is failure, failure, repair, failure, failure, failure

1. Generate a r.v. η_1' having a d.f. $G(x)$	
2. Generate a r.v. ξ_1' having a d.f.	$P(\xi_1 \le x \mid \xi_1 \le \eta_1')$ $= F(x)/F(\eta_1'),\ x \le \eta_1'$
Put	$Q_1 = F(\eta_1')$
3. Generate a r.v. ξ_2' having a d.f.	$P(\xi_2 \le x \mid \xi_2 > \eta_1' - \xi_1')$ $= [F(x) - F(\eta_1' - \xi_1')]/(1 - F(\eta_1' - \xi_1')]$, $x > \eta_1' - \xi_1'$
Put	$Q_2 = Q_1[1 - F(\eta_1' - \xi_1')]$
4. Generate a r.v. η_2' having a d.f.	$P(\eta_2 \le x \mid \eta_2 > \xi_1' + \xi_2' - \eta_1')$ $= [G(x) - G(\xi_1' + \xi_2' - \eta_1')]/[1 - G(\xi_1' + \xi_2' - \eta_1')]$, $x > \xi_1' + \xi_2' - \eta_1'$
Put	$Q_3 = Q_2[1 - G(\xi_1' + \xi_2' - \eta_1')]$
5. Generate a r.v. ξ_3' having a d.f.	$P(\xi_3 \le x \mid \xi_3 < \eta_1' + \eta_2' - \xi_1' - \xi_2')$ $= F(x)/F(\eta_1' + \eta_2' - \xi_1' - \xi_2')$, $x \le \eta_1' + \eta_2' - \xi_1' - \xi_2'$
Put	$Q_4 = Q_3 F(\eta_1' + \eta_2' - \xi_1' - \xi_2')$
6. Generate a r.v. ξ_4' having a d.f.	$P(\xi_3 \le x \mid \xi_3 < \eta_1' + \eta_2' - \xi_1' - \xi_2' - \xi_3')$ $= F(x)/F(\eta_1' + \eta_2' - \xi_1' - \xi_2' - \xi_3')$, $x \le \eta_1' + \eta_2' - \xi_1' - \xi_2' - \xi_3'$
Put	$Q = Q_4 F(\eta_1' + \eta_2' - \xi_1' - \xi_2' - \xi_3')$

Note: A quantity Q is in fact a sample value of $q_i(k)$. If we repeat this algorithm as many times as necessary to obtain N successive values $Q(j)$, we can use the estimate

$$q_i(k) \approx \frac{\sum_{j=1}^{N} Q(j)}{N}.$$

19.6 SAMPLE FAILURE AND REPAIR SIMULATIONS

In reliability analysis one of the most important uses of Monte Carlo simulation is predicting the availability of complex, repairable equipment before stationarity is reached. The simulations are used to select equipment and to evaluate matters of policy, such as repair priority rules and levels of spare parts.

The analyst must decide on the events to be simulated, the software to be used, the number of iterations, and the type of smoothing, if any, of the results. The simulations reported here were done with the Diana program for IBM and IBM-compatible personal computers, which is discussed in the appendix on software. Diana is available through Harrison and Company, Post Office Box 2732, Merrifield, Virginia 22116, (703) 280-2202.

19.6.1 An Example with One Repairable Item

Suppose that a repairable unit failed exponentially with a mean of one unit of time. The repair time had a truncated normal distribution with a mean of

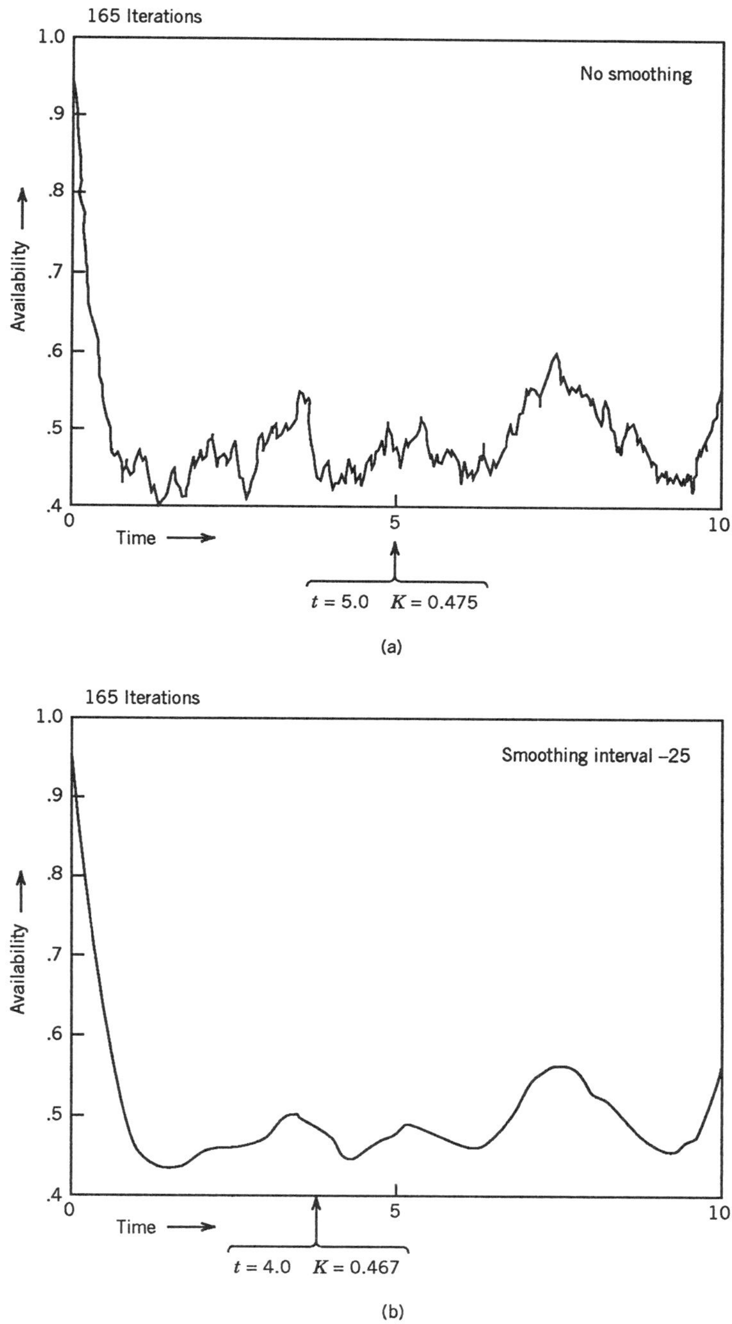

Figure 19.2 Probability of availability for hypothetical item for first ten units of time after 165 iterations: Unsmoothed (*above*); smoothed into mean of 25 observations (*below*).

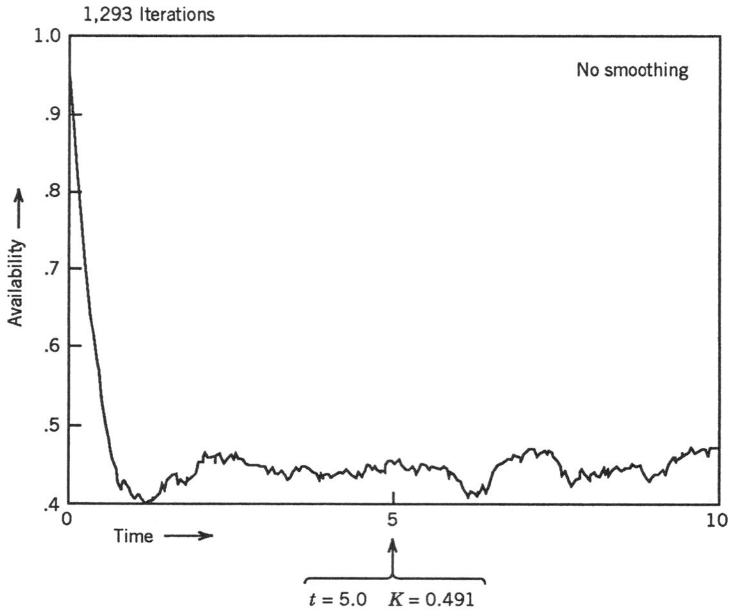

(a)

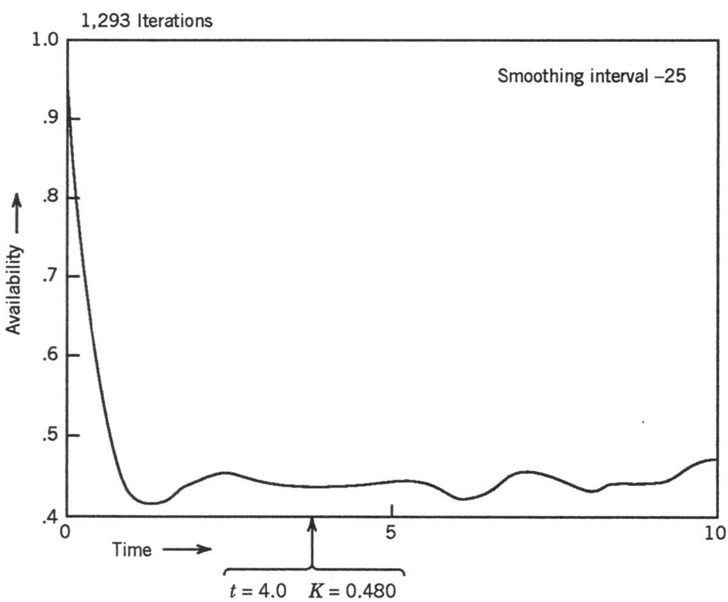

(b)

Figure 19.3 Same as Figure 19.2, after 1283 iterations.

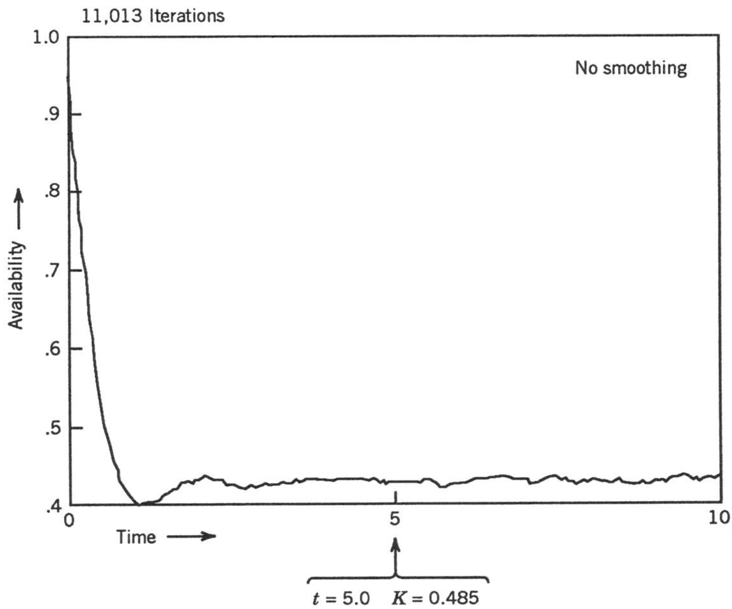

(a)

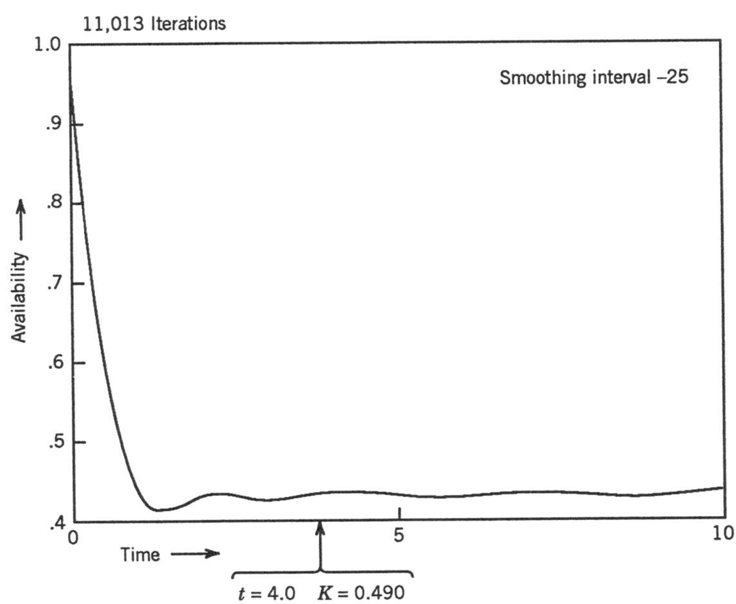

(b)

Figure 19.4 Same as Figure 19.2, after 11,013 iterations.

1 and a standard deviation of 0.5. In each iteration, the alternating failures and repairs were simulated for ten units of time. For each moment in the ten-unit period, the availability was defined as the percentage of iterations in which the unit is up.

The simulations were typical of a test. A test of simple and predictable hypothetical equipment may be seen before the real, unpredictable equipment is simulated. The test reveals whether the program accurately predicts parameters whose true values can be found easily by calculation. In addition, characteristics can be found that are difficult to predict with calculation, such as the fluctuation in expected availability in this example between the initial decline and the final approach to an asymptote of 0.5. The test also may indicate how many iterations will be needed and what smoothing, if any, will be desirable.

Results for 165 iterations are shown in Figure 19.2. An initial run of a small number of iterations indicates both the number of iterations required to find mean values and the approximate variances these mean values will have. No smoothing is above, and maximum smoothing, which yields a 25-point moving average in this case, is below. The y axis gives availability and indicates the 39 percent low point and 97 percent high point. The x axis gives the 10-unit time scale, and has displays of midpoint values selected as reference points.

Results for 1283 and 11,013 iterations are given in Figures 19.3 and 19.4, respectively. As may be seen, some 10,000 iterations were enough, but smoothing brought out no characteristics of interest.

19.6.2 An Example of an Assembly

Figure 19.5 shows an assembly that is complex enough to be difficult to predict analytically but easy to predict with Diana software (see Appendix F). Subsystem 1 has two identical units in parallel. Each unit has a constant repair time of 0.2 and a Weibull-Gnedenko distribution of time to failure with a shape parameter of 2 and a form parameter of 2. Subsystem 2 has three units in series, each having a constant repair time of 0.02. The first unit has an exponential distribution of time to failure with a mean of 5. Each of the other two units has a normal distribution of time to failure with a mean of 2 and a standard deviation of 0.1. Each of the three units has a constant repair time of 0.02.

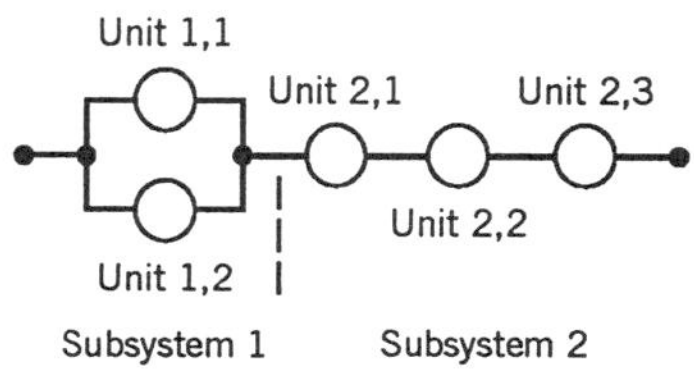

Figure 19.5 Simple hypothetical assembly that is difficult to predict analytically but easy to predict with weighted trials simulation.

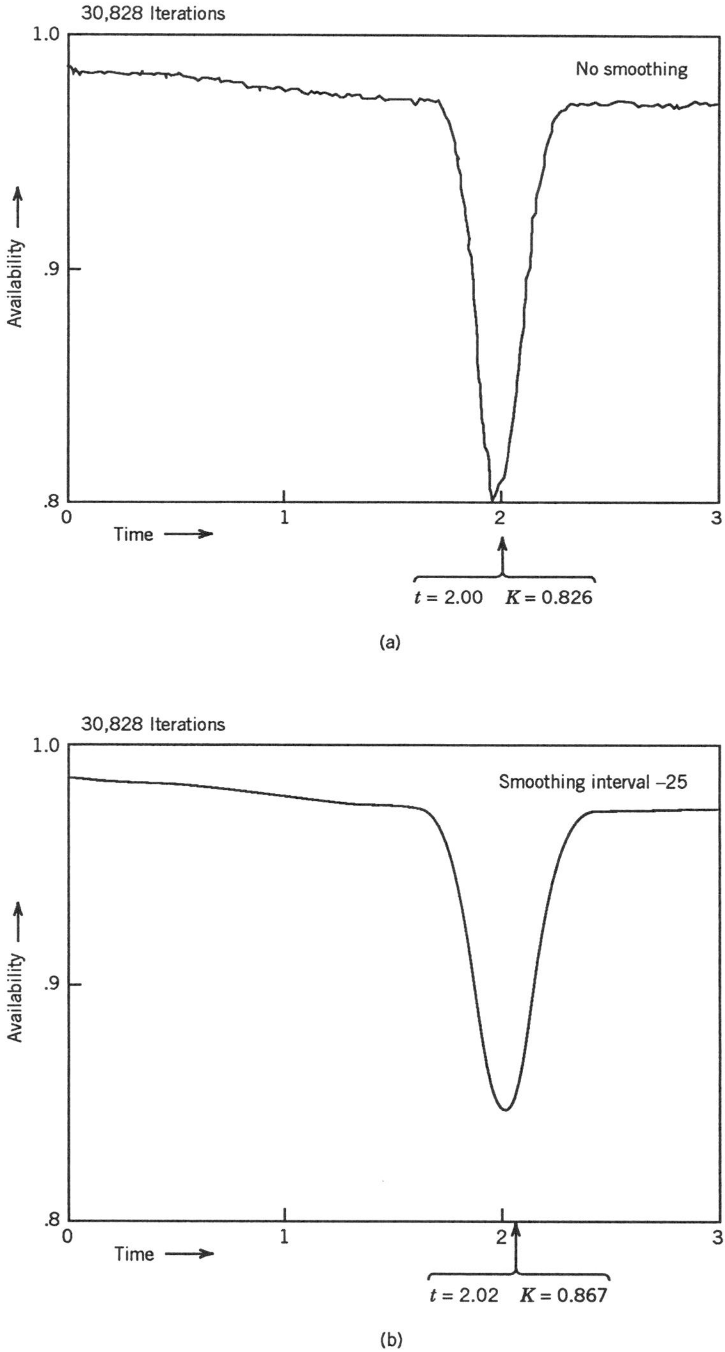

Figure 19.6 Simulation of first three time units in Figure 19.5: Assembly after 30,828 iterations.

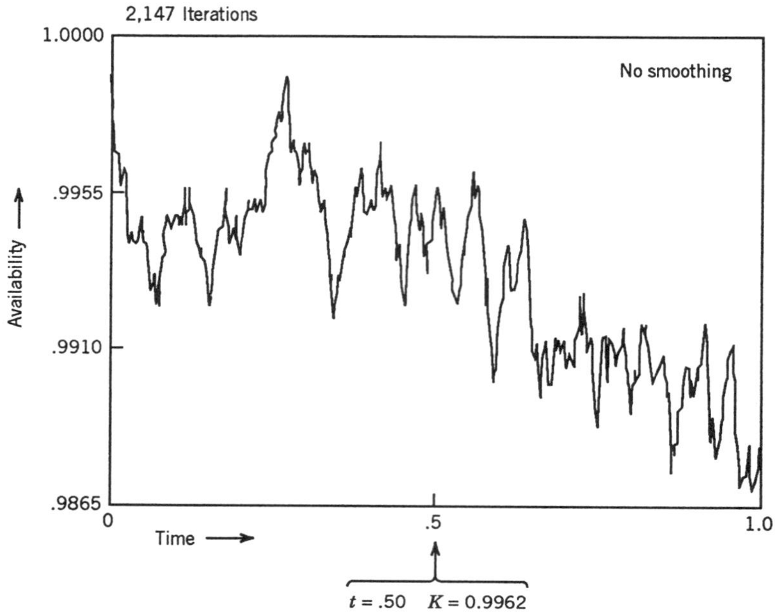

(a)

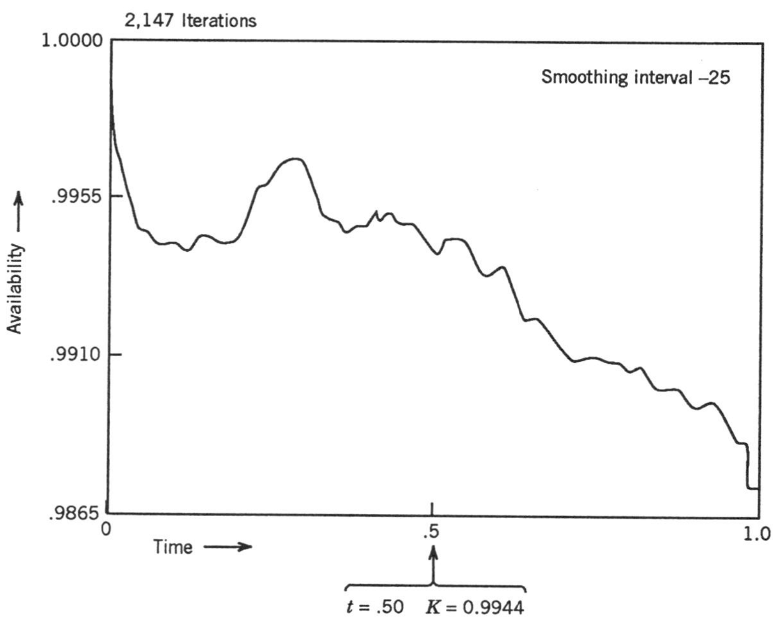

(b)

Figure 19.7 First time unit, after 2147 iterations.

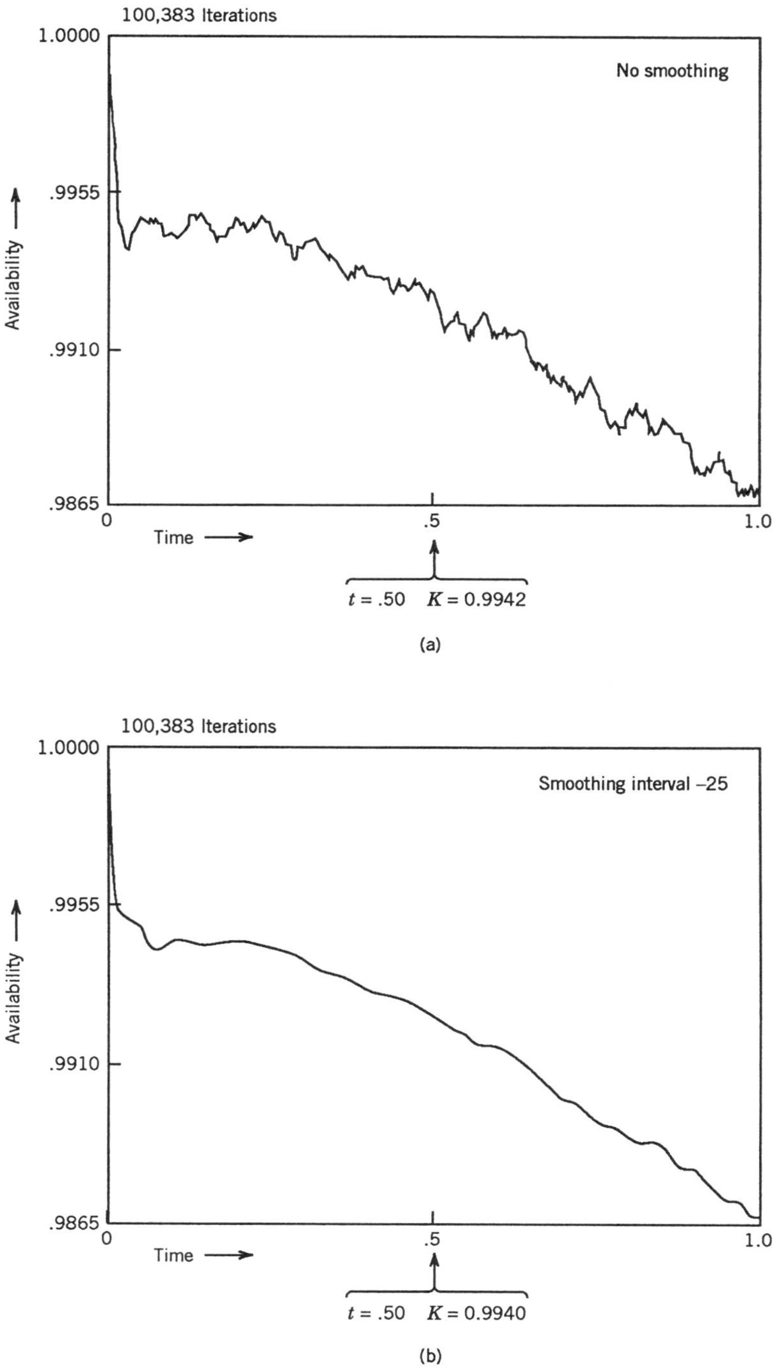

Figure 19.8 First time unit, after 100,383 iterations.

Two repair stations are available. The exponential unit has top repair priority and the other two series units are next. A first-in-first-out service discipline is followed for equal units. If repair is interrupted, it resumes from the interruption point for subsystem 2 but must start from the beginning for subsystem 1.

Figure 19.6 gives the results for 30,828 iterations of three time units each. The run was made after an initial run of 148 iterations indicated the order of magnitude of the number of iterations that probably would be required for satisfactory precision. Variation of expected availability with time is shown by the y axis reference points and the x-axis displays of the time and value of the minimum expected availability.

As may be seen, the simulation indicates a freak coincidence of the configuration, a short period of severe weakness near $t = 2$. This would call for a redesign followed by new simulations if the problem were a real one. Figure 19.6 also shows that there are enough iterations and that smoothing in this case tends to obscure the critical feature rather than make it apparent.

Sometimes there is special interest in the earliest phase of a nonstationary process. Figures 19.7 and 19.8, for 2147 and 100,383 iterations, respectively, show the characteristics of the expected availability fluctuations during the first unit of time.

PART 3

OPTIMIZATION

CHAPTER 20

OPTIMAL REDUNDANCY

Redundancy is used to increase the reliability of technical systems. The term is applied to system structural improvement, duplicate internal parts, and spare parts. In using redundancy, one should consider not only reliability factors but the total expenditure of each resource. The crucial resource is cost for most technical systems. Weight is important for aircraft and missiles, and size and volume may be considered for ships. In most cases there is a single factor, usually monetary cost. For convenience we will call the resource the "cost," no matter what its actual nature.

In this chapter we will consider mainly series systems of independent subsystems or units. In this context we will define a subsystem to be a part of the system that can use identical units for reliability improvement. The subsystem may or may not be separate in construction or function. For instance, in an inventory problem, a subsystem can consist of all units of the same type even if they are located in different places inside the system.

Optimization with resource restrictions is considered in Section 20.2.1.

20.1 NOTATION

a = the parameter of a Poisson distribution.

c_i = cost of the ith type of unit (or the ith subsystem).

c_{ij} = expenditure of the jth type of resource for a unit of the ith type.

C_0 = a given acceptable level of a single resource expenditure for the whole system.

C_{j0} = a given acceptable level of a resource expenditure of the jth type for the whole system.

$C(r_1, \ldots, r_n)$ = total system cost when its subsystems have reliability indexes $r_1, \ldots, r_n$.

$C(x_1, \ldots, x_n)$ = total system cost when its subsystems have $x_1, \ldots, x_n$ redundant units.

$C_j(x_1, \ldots, x_n)$ = total system expenditure of the resource of the jth type when its subsystems have $x_1, \ldots, x_n$ redundant units.

$C_i(x_i)$ = ith subsystem cost when it has x_i redundant units.

$C_{ij}(x_i)$ = ith subsystem cost for resource j when it has x_i redundant units.

$E(r_1, \ldots, r_n)$ = system performance effectiveness index when its subsystems have reliability indexes $r_1, \ldots, r_n$.

E_0 = a given value of the system performance effectiveness index.

$L_0 = -\log R_0$.

$L(x_1, \ldots, x_n) = -\log R(x_1, \ldots, x_n)$.

$L_i(x_i) = -\log R_i(x_i)$.

m_i = initial number of units in the ith subsystem.

n = number of subsystems (different types of units).

$P(x_i)$ = probability of failure-free operation of the ith subsystem when it has x_i redundant units.

$p_k(a)$ = probability of exactly k events for a Poisson distribution with the parameter a.

$P_k(a)$ = probability of not less than k events for a Poisson distribution with the parameter a.

$Q_0 = 1 - R_0$.

$Q(x_1, \ldots, x_n) = 1 - R(x_1, \ldots, x_n)$.

$Q(x_i) = 1 - R(x_i)$.

r_i = value of the reliability index of the ith subsystem.

R = system reliability index of the type $P(t_0)$, $R(t_0)$, or K.

R_0 = objective function value of the reliability index R.

$R(x_1, \ldots, x_n)$ = system reliability index of type R when the numbers of redundant units in its subsystems are $x_1, \ldots, x_n$.

$R_i(x_i)$ = ith subsystem reliability index when the subsystem has x_i redundant units.

$T(x_1, \ldots, x_n)$ = system MTTF when the numbers of redundant units in its subsystems are $x_1, \ldots, x_n$.

T_0 = objective function value of the system MTTF.

x_i = number of redundant units of the ith type.

$X = (x_1, \ldots, x_n)$.

λ_i = failure rate of a unit of the ith type.

20.2 FORMULATION OF OPTIMAL REDUNDANCY PROBLEMS

20.2.1 One Constraint, Reliability Index of Type *R*

The system reliability index can be expressed with the subsystem indexes in a multiplicative form

$$R(x_1, \ldots, x_n) = \prod_{1 \le i \le n} R_i(x_i) , \tag{20.1}$$

or in additive form

$$L(x_1, \ldots, x_n) = \sum_{1 \le i \le n} L_i(x_i) . \tag{20.2}$$

If the system is very reliable:

$$Q(x_1, \ldots, x_n) \ll 1 ,$$

one may use the approximation

$$Q(x_1, \ldots, x_n) \approx \sum_{1 \le i \le n} Q_i(x_i) . \tag{20.3}$$

In an optimal redundancy problem, the total system cost usually is additive:

$$C(x_1, \ldots, x_n) = \sum_{1 \le i \le n} C_i(x_i) \tag{20.4}$$

Also, each subsystem cost usually is linear:

$$C_i(x_i) = c_i x_i . \tag{20.5}$$

There are two main optimization tasks.

1. One type, *minimum cost*, in determination of the least expensive suite of redundant units that achieves an objective system reliability index, R. This "direct" type may be written

$$\min_x \{C(x_1, \ldots, x_n) \mid R(x_1, \ldots, x_n) \ge R_0\} , \tag{20.6}$$

where the inequality may be replaced by $Q(x_1, \ldots, x_n) < Q_0$ or $L(x_1, \ldots, x_n) < L_0$.
2. The other type, maximum effectiveness, is determination of the suite of redundant units that produces the highest system reliability index that can be achieved for a total system cost that is not larger than a given constraint cost, C_0. The "inverse" type may be written

$$\max_x \{R(x_1, \ldots, x_n) \mid C(x_1, \ldots, x_n) \le C_0\}, \qquad (20.7)$$

where $R(x_1, \ldots, x_n)$ might be replaced with $Q(x_1, \ldots, x_n)$ or $L(x_1, \ldots, x_n)$ when the operator *max* is changed to *min*.

For concreteness we will define the index R as the probability of failure-free operation of an unrepairable system, even though it will sometimes have some other meaning. This topic is considered in Section 20.3.

20.2.2 One Constraint, Reliability Index of Type *T*

For the one-constraint case the reliability index can be written

$$T(x_1, \ldots, x_n) = \int_0^\infty P(t \mid x_1, \ldots, x_n)\, dt$$

$$= \int_0^\infty \prod_{1 \le i \le n} P_i(t \mid x_i)\, dt, \qquad (20.8)$$

where $P(t \mid \cdot)$ is the notation of a conditional probability. There are two optimization tasks similar to (20.6) and (20.7):

1. Direct optimization,

$$\min_x \{C(x_1, \ldots, x_n) \mid T(x_1, \ldots, x_n) \ge T_0\}. \qquad (20.9)$$

2. Inverse optimization,

$$\max_x \{T(x_1, \ldots, x_n) \mid C(x_1, \ldots, x_n) \le C_0\}. \qquad (20.10)$$

These tasks are considered in Section 20.8.2.

20.2.3 One Constraint, System Performance Effectiveness Index

The index of the system performance effectiveness can be written in the form:

$$E = \sum_S \Phi_S H_S, \qquad (20.11)$$

where S is the system state, characterized by the collection of its unit states, $S = (s_1, \ldots, s_n)$; Φ_S is the effectiveness of the system when it is in a state S (i.e. the system outcome in the state S), and H_S is the probability that the system is in the state S. For independent system units

$$H_S = \prod_{1 \le i \le n} r_i^{s_i}(1 - r_i)^{1 - s_i}, \qquad (20.12)$$

where s_i is the state of the ith subsystem:

$$s_i = \begin{cases} 1 & \text{if the } i\text{th subsystem is in an up state,} \\ 0, & \text{Otherwise .} \end{cases}$$

Consider a system that can be assembled from different variants of its subsystems. Each of these variants is characterized by its own parameters of reliability and cost. Thus the problem is to find the optimal set of variants. The minimum cost objective may be written

$$\min_{j_1,\ldots,j_n} \left\{ \sum_{1 \le i \le n} c_i(j_i) \,|\, E(r_1(j_1), \ldots, r_n(j_n); \Phi_s) \ge E_0 \right\}, \qquad (20.13)$$

where $c_i(j_i)$ is the cost of the i_ith variant of the ith subsystem, $r_i(j_i)$ is the reliability index of the j_ith variant of the ith subsystem, and $E(r_1(j_1), \ldots, r_n(j_n))$ is the system performance effectiveness index for chosen subsystem variants $j_1, \ldots, j_n$ with their reliability index. The performance effectiveness index in (20.13) is chosen so that higher values mean better performance.

The maximum effectiveness objective for this case may be written

$$\max_{j_1,\ldots,j_n} \left\{ E(r_1(j_i), \ldots, r_n(j_n); \Phi_s) \,|\, \sum_{1 \le i \le n} c_i(j_i) \le C_0 \right\}. \qquad (20.14)$$

These problems are considered in Section 20.8.1.

20.2.4 Several Constraints, Reliability Index of Type *R*

Consider restrictions on several types of resources, such as cost, weight, and volume. All of these restrictions must be satisfied simultaneously. The conditional optimization problem can be written in the form

$$\max_x \{R(x_1, \ldots, x_n) \,|\, C_j(x_1, \ldots, x_n) \le C_{j0};\ j = 1, \ldots, M\}. \qquad (20.15)$$

This problem is considered in Section 20.9.

20.2.5 Multifunctional System Redundancy under One Constraint

A multifunctional system may have an objective function with a reliability index goal for each of the functions. Multiple goals are especially likely if different functions are of different importance. K different functions may not necessarily have K completely independent functional subsystems. Such subsystems may have common units. Let $R_{(j)}(x_1, \ldots, x_n)$ denote the reliability index of the jth subsystem, $j = 1, \ldots, K$. The cost minimization problem can be written in the form

$$\min_x \left\{ C(x_1, \ldots, x_n) \,|\, \prod_{i \in G_j} R_i(x_i) \ge R_0^j;\ j = 1, \ldots, K \right\}, \qquad (20.16)$$

where G_j is the subset of units whose failure-free operation is necessary to perform the jth system function. This problem is considered in Section 20.8.3.

20.3 RELIABILITY INDEX OF TYPE *R*, ONE CONSTRAINT (APPROXIMATE SOLUTION)

20.3.1 Upper and Lower Bounds, Number of Units

Before finding the problem solution, it may be important to find in advance the upper, U_i, and lower, u_i, bounds of the number, x_i, of each subsystem redundant unit in the optimal solution: $u_i \le x_i \le U_i$ for $i = 1, \ldots, n$.

For (20.6), u_i can be found directly from the given restrictions:

$$u_i = \min\{x_i \mid R(x_i) \ge R_0\} .$$

This bound follows from the clear fact that each subsystem must be more reliable than R_0. The upper bound can be found from

$$U_i = \min\left\{x:\ 1 - \left[1 - \prod_{1\le i\le n} R_i(t_0)\right]^x \ge R_0\right\}$$

for all $i = 1, \ldots, n$. This bound follows from the fact that redundancy on the system level is always less effective than redundancy on the unit level.

For (20.7) only a trivial lower bound can be written: $u_i = 0$. The upper bound for a subsystem must satisfy

$$U_i \le \frac{C_0}{C_i} . \tag{20.17}$$

20.3.2 Approximate Solution

This method may be useful when a very reliable system is considered. R_0 is close to 1, so Q_0 is small: $1 - R_0 = Q_0 \ll 1$. The core of the method is a quick and easy technique for finding the general level of cost effectiveness of a redundant subsystem at the desired reliability. To investigate problem (20.6) for a given Q_0, one first finds

$$z = \frac{Q_0}{\sum_{1\le i\le n} C_i} .$$

Then one calculates

$$y_i = c_i z = \frac{c_i Q_0}{\sum_{1\le i\le n} c_i} .$$

Finally, each x_i is approximated roughly with

$$Q_i(x_i) = y_i\ . \tag{20.18}$$

The numerical results normally are not integers, and one chooses one of the two closest integers: $x_i^* = [x_i]$ or $x_i^{**} = [x_i] + 1$, where $[x]$ is the integer part of x. It is important to note that ordinarily even the total set of 2^n possible combinations of all integer x_i^* and x_i^{**}, $i = 1, 2, \ldots, n$, does not yield the optimal solution.

EXAMPLE 20.1

A system consists of 5 independent units with the following availability coefficients and costs:

$K_1 = 0.97,\ c_1 = 1,$
$K_2 = 0.97,\ c_2 = 5,$
$K_3 = 0.90,\ c_3 = 1,$
$K_4 = 0.90,\ c_4 = 5,$
$K_5 = 0.995,\ c_5 = 0.1.$

We use only active redundant units, and all units can be repaired independently. We wish to raise the availability coefficient of the system to at least $K = 0.999$ at minimum cost.

Solution

The high reliability method is warranted since $(1 - K) \ll 1$. We first find

$$z = \frac{1 - K}{\Sigma_{1 \le i \le n} c_i} = \frac{0.001}{1 + 5 + 1 + 5 + 0.1} = 8.264 \cdot 10^{-6}\ .$$

Then

$$\begin{aligned} y_1 &= 8.26 \cdot 10^{-5}\ , \\ y_2 &= 4.13 \cdot 10^{-4}\ , \\ y_3 &= 8.26 \cdot 10^{-5}\ , \\ y_4 &= 4.13 \cdot 10^{-4}\ , \\ y_5 &= 8.26 \cdot 10^{-6}\ . \end{aligned}$$

The results can be found from

$$x_i = \frac{\log y_i}{\log(1 - K_i)} - 1\ .$$

With base 10 logarithm, for example, we find that

$$x_1 = \frac{-4.08}{-1.52} - 1 = 1.68\,,$$

$$x_2 = \frac{-3.38}{-1.52} - 1 = 1.22\,,$$

$$x_3 = \frac{-4.08}{-1.0} - 1 = 3.08\,,$$

$$x_4 = \frac{-3.38}{-1.0} - 1 = 2.38\,,$$

$$x_5 = \frac{-5.08}{-2.30} - 1 = 1.21\,.$$

From the computed values it follows that

$$1 \le x_1 \le 2\,,$$

$$1 \le x_2 \le 2\,,$$

$$3 \le x_3 \le 4\,,$$

$$2 \le x_4 \le 3\,,$$

$$1 \le x_5 \le 2\,.$$

We will not consider all possible 2^n combinations. We take the upper value for each unit: $x_1 = 2$, $x_2 = 2$, $x_3 = 4$, $x_4 = 3$, and $x_5 = 2$. These quantities yield the system unavailability coefficient calculation:

$$\begin{aligned} 1 - K &\approx \sum_{1 \le i \le 5} (1 - K_i)^{x_i + 1} \\ &= 0.03^3 + 0.03^3 + 0.1^5 + 0.1^4 + 0.005^3 \\ &= 5.04 \cdot 10^{-4}\,, \end{aligned}$$

or

$$K = 1 - 5.04 \cdot 10^{-4} \approx 0.9995\,.$$

The cost of the redundant units is 31.2. With the lowest values—$x_1 = 1$, $x_2 = 1$, $x_3 = 3$, $x_4 = 2$, $x_5 = 1$—the system reliability will be $K \approx 0.9982$ and the cost will be 19.1.

One often can find a value of K closer to the required value by picking a suite of units between $x^*{}_1$ and x_1^{**}. Such a suite, of course, may not be optimal. In this particular case, if $x_1 = 2$ and all other values are minimal ($x_2 = 1$, $x_3 = 3$, $x_4 = 2$, $x_5 = 1$), the system reliability would be $K \approx 0.998965$, which practically equals 0.999, and the cost would increase only one unit.

The following is an iterative procedure based on this method for solution of (20.7) for C_0.

First Iteration

1. Find $x_{(0)}$ from (20.17) or from

$$x = \frac{C_0}{\sum_{1 \leq i \leq n} c_i} .$$

Both values of $x_{(0)}$ obtained in this way are very rough.

2. Compute

$$Q^{(1)} = \sum_{1 \leq i \leq n} Q_i(x) .$$

3. Compute

$$z^{(1)} = \frac{Q_0}{\sum_{1 \leq i \leq n} c_i} .$$

Second Iteration

1. Using the previous value of $z^{(1)}$, compute $x_1^{(1)}, \ldots, x_n^{(1)}$ with (20.18).
2. Control the cost after the first step:

$$C^{(1)} = \sum_{1 \leq i \leq n} x_i^{(1)} c_i .$$

3. If $C^{(1)} > C_0$, choose a new value $z^{(2)} > z^{(1)}$. If $C^{(1)} < C_0$, choose a new value $z^{(2)} < z^{(1)}$.

Repeat the procedure from the second step until the solution is found.

Remark To find z^k, one can use a simple linear interpolation

$$z^{(k+1)} = z_* + (z^* - z_*) \frac{C_0 - C_*}{C^* - C_*} ,$$

where w_* and w^* are, respectively, the smallest and largest values of w^k and w^{k-1}, where w is a general notation for z and C.

20.4 STEEPEST DESCENT METHOD: TYPE *R* INDEX, ONE CONSTRAINT

The preceding approximate solutions are very crude and often yield only a start toward a more precise algorithm. If the precise unit statistical parameters are not known, or if the mathematical model is a very crude approximation to the real system, there may be little accuracy achieved with exact computational methods. The imprecision of a method may be only a secondary source of error. A practical computation method in such a case may be the steepest descent algorithm. It may be accurate enough and demand less computer memory and fewer computations than other methods. This method does not always lead to the optimal solution, but may produce a solution that is very good for practical purposes.

For good solutions, this method demands goal function properties. Usually the properties are held in reliability engineering. The cost, $C(X)$, is usually directly proportional to the number of identical units. The reliability function, $R(X)$, is usually a logarithmically convex additive function of its components $R_1(x)$. Thus

$$\Delta(x) = R(x+1) - R(x) > 0 ,$$
$$\Delta^2(x) = \Delta(x+1) - \Delta(x) < 0 \qquad \text{for all } x .$$

20.4.1 Description of the Algorithm

The algorithm consists of the following steps.

1. For the ith subsystem, a fixed interval of duration t_0, and for the alternative numbers of redundant units x_i, we compute the probabilities of failure-free operation, $P_i(x_i)$. For convenience, we can put the data into a table.

Remark Here we consider the probability of failure-free operation during a given period of time, but we could instead consider an availability or operational availability coefficient.

2. We find the logarithm $P_i(x_i)$ values. The logarithm can be taken to any base.
3. On the basis of $P_i(x_i)$ and the known unit cost, c_i, we find values $g_i(x_i)$:

$$g_i(x_i) = \frac{\log P_i(x_i) - \log P_i(x_i - 1)}{c_i}$$

for all i and the proper number of units, x_i.

Remark For values $P_i(x_i)$ close to unity, the quantities $g_i(x_i)$ can be approximated:

$$g_i(x_i) \approx \frac{Q_i(x_i - 1) - Q_i(x_i)}{c_i} = \frac{P_i(x_i) - P_i(x_i - 1)}{c_i} .$$

4. All values $g_i(x_i)$ are numbered in decreasing order.
5. We now consider a multiple-step procedure. First, choose

$$g^{(1)} = \{g_j^{(1)} : \max_{1 \le i \le n} g_i^{(1)}\} .$$

$$\text{Compute } \log P^{(1)} = \log P^{(0)} - \log P_j(0) + \log P_j((1) ,$$

where

$$\log P^{(0)} = \sum_{1 \le i \le n} \log P_i(0)$$

is the initial value of the logarithm of the probability of system failure-free operation. Compute

$$C^{(1)} = C_0 + c_j ,$$

where C_0 is the initial cost of the system.

Second, choose

$$g^{(2)} = \begin{Bmatrix} g_k^{(1)} = \max\limits_{1 \le i \le n} g_i^{(1)} \text{ for } k \ne j \\ g_j(2) \text{ for } k = j \end{Bmatrix} .$$

Find the corresponding $\log P_i(1)$ or $\log P_j(2)$. Compute

$$\log P^{(2)} = \log P^{(1)} - \log P_k(0) + \log P_k(1)$$

or

$$\log P^{(2)} = \log P^{(1)} - \log P_i(1) + \log P_i(2) .$$

Compute

$$C^{(2)} = C^{(1)} + c_k .$$

6. Terminate the process at step N:

$$N = \sum_{1 \le i \le n} x_i ,$$

where, for the minimum cost problem,

$$\log P^{(N-1)} \le \log P_0 < \log P^{(N)} ,$$

or, for the maximum effectiveness problem,

$$C^{(N)} \le C_0 < C^{(N+1)} .$$

20.4.2 Alternative Computational Techniques

For standby redundant units, we can use the tabulated gamma function. For active redundant units, we can use the tabulated beta function. Time required for taking logarithms may be saved. A steepest descent algorithm may be used to obtain a nonstrict solution that is good for practical purposes. The goal function has a multiplicative form:

$$\prod_{1\le i\le n} P_i(x_i)\,.$$

The basic logic is the same as in Section 20.4.1. Some steps yield different results but only some. Values of $g_i(x_i)$ are computed by

$$g_i(x_i)=\frac{P_i(x_i)-P_i(x_i-1)}{c_iP_i(x_i-1)}\,.$$

For $P_i(x_i)$ values close to one, an approximation may be employed:

$$g_i(x_i)\approx\frac{Q_i(x_i-1)-Q_i(x_i)}{c_i}\,. \tag{20.19}$$

The process is terminated at that step N where, for the direct optimal redundancy problem,

$$P^{(N-1)}\le P_0<P^{(N)}\,.$$

The condition for the inverse problem remains the same as before.

EXAMPLE 20.2

A system consists of units with different reliability indexes. To increase reliability, different types of redundancy are used for different subsystems. A diagram of the system is in Figure 20.1.

For the first and third subsystems the computation of $P_i(x_i)$ is made according to the formula

$$P_i(x_i)=1-q_i^{x_i+1}\,,$$

where q_1 is the probability of failure of a single unit during a given time t_0. The second and fourth subsystems have

$$P_i(x_i)=e^{\lambda_i t_0}\sum_{0\le j\le x_i}\frac{(\lambda_i t_0)^j}{j!}\,.$$

The fifth subsystem has

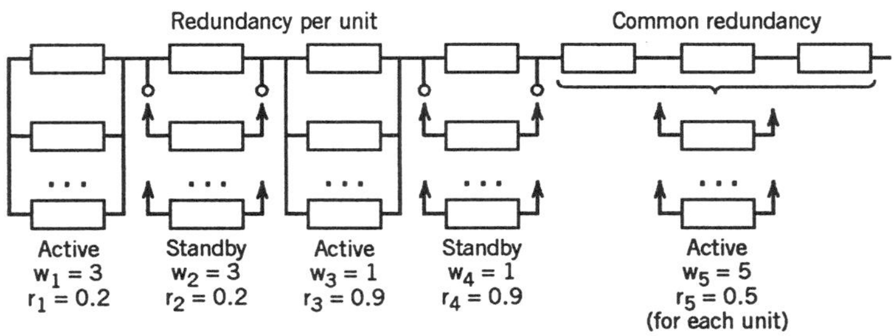

Figure 20.1 System with different types of redundancy, Example 20.2.

$$P_i(x_i) = \sum_{n \le j \le n+x_i} \binom{n+x_i}{j} (1-q)^j q^{n+x_i-j},$$

where $n = 3$.

We want to determine the optimal assignment of redundant units for each subsystem for two cases (1) when the system cost must not exceed 50 and (2) when the probability of failure-free operation must be not less than 0.45.

Solution

We first prepare Table 20.1, which has values of $P_i(x_i)$. On the basis of this table we set up Table 20.2, where we put the values $g_i(x_i)$ computed with (20.19). The final computed results are given in Table 20.3, which allows us to solve both the direct and the inverse problems of optimal redundancy. In this table two facts may be seen.

Table 20.1 Subsystem failure probabilities, Example 20.2

Number of Redundant Units	$P_i(x_i)$				
	$i = 1$	$i = 2$	$i = 3$	$i = 4$	$i = 5$
0	0.200	0.200	0.9000	0.9000	0.125
1	0.360	0.524	0.9900	0.9953	0.312
2	0.488	0.783	0.9990	0.9998	0.500
3	0.590	0.9212	0.9999	1	0.656
4	0.672	0.9763	1	—	0.773
5	0.737	0.9940	—	—	0.8553
6	0.790	0.9987	—	—	0.9103
7	0.831	0.9997	—	—	0.9453
8	0.865	0.9999	—	—	0.9673
9	0.892	1	—	—	0.9807
10	0.9139	—	—	—	0.9888

Table 20.2 $g_i(x_i)$, **Values of Example 20.2**

Number of Redundant Units	$g_i(x_i)$				
	$i=1$	$i=2$	$i=3$	$i=4$	$i=5$
0					
1	0.300(2)	0.812(1)	0.100(8)	0.100(7)	0.300(3)
2	0.118(6)	0.246(4)	0.009	0.004	0.120(5)
3	0.060(11)	0.087(9)	0.001	—	0.062(10)
4	0.046(12)	0.030(15)	—	—	0.035(13)
5	0.033(14)	0.009	—	—	0.021(17)
6	0.024(16)	0.002	—	—	0.013(20)
7	0.017(18)	—	—	—	0.008
8	0.014(19)	—	—	—	0.005
9	0.010(21)	—	—	—	0.003
10	0.008	—	—	—	0.001

Table 20.3 Stepwise selection of redundant units, Example 20.3

Step Number	Number of Redundant Units at the *N*th Step					Probability of Failure-Free Operation of the System, $P^{(N)}$	Cost of the System $C^{(N)}$
	x_1	x_2	x_3	x_4	x_5		
0	0	0	0	0	0	0.004	22
1	0	1	0	0	0	0.011	24
2	1	1	0	0	0	0.019	27
3	1	1	0	0	1	0.048	32
4	1	2	0	0	1	0.071	34
5	1	2	0	0	2	0.114	39
6	2	2	0	0	2	0.155	42
7	2	2	0	1	2	0.171	43
8	2	2	1	1	2	0.188	44
9	2	3	1	1	2	0.221	46
10	2	3	1	1	3	0.290	51
11	3	3	1	1	3	0.351	54
12	4	3	1	1	3	0.400	57
13	4	3	1	1	4	0.471	62
14	5	3	1	1	4	0.517	65
15	5	4	1	1	4	0.548	67
16	6	4	1	1	4	0.587	70
17	6	4	1	1	5	0.649	75
18	7	4	1	1	5	0.683	78
19	8	4	1	1	5	0.711	81

1. The process for the inverse problem stops at the ninth step. Here the probability of failure-free operation of the system is 0.221, and the corresponding values of x_i are $x_1 = 2$, $x_2 = 3$, $x_3 = 1$, $x_4 = 1$, and $x_5 = 2$.
2. The process for the direct problem must be terminated at the thirteenth step, where 62 cost units have been spent. The x_i values are $x_1 = 4$, $x_2 = 3$, $x_3 = 1$, $x_4 = 1$, and $x_5 = 4$.

20.4.3 Distribution of Remaining Resources

In many cases one step is well below the cost limit, but the next step of the steepest descent procedure produces too great a cost. In such a case, the strict steepest descent procedure can be abandoned for the last step. The last step can be the selection of whatever unit or units raise the success probability the most without exceeding the cost limit. This alternative last step may itself be a steepest descent step without the more expensive units. Any comparable situations can be solved as a reliability constraint problem rather than a cost constraint problem.

20.5 DYNAMIC PROGRAMMING: INDEX OF TYPE *R*, ONE CONSTRAINT

20.5.1 Bellman's Algorithm

Dynamic programming is a classical technique for solution of discrete optimization problems. To solve a direct optimal redundancy problem (20.6), we introduce Bellman's function

$$\Psi_k(r) = \min\left\{ \sum_{1 \le i \le k} c_i x_i \,\middle|\, \prod_{1 \le i \le k} R_i(x_i) \ge r \right\}, \tag{20.20}$$

which is developed recursively for $k = 1, \ldots, n$ in correspondence with the equation set

$$\Psi_1(r) = \min\{c_1 x_1 \mid R_1(x_1) \ge r\}, \qquad R_0 \le r < 1,$$
$$\vdots$$
$$\Psi_k(R) = \min\{\Psi_{k-1}(r) + c_k x_k \mid rR_k(x_k) \ge R\}, \qquad R \le r < 1. \tag{20.21}$$

Let $\mathbf{x}_k^*(R)$, $k = 1, \ldots, n$, be the vector of x_k's that deliver minimum in (20.21). It is clear that $\psi_n(R_0)$ represents the optimal value of the goal function. The solution can be found with the equation set

$$x_n^0 = x_n^*(R_0)\,, \quad R^{(n-1)} = \frac{R_0}{R_n(x_n^0)}\;;$$
$$\vdots$$
$$x_k^0 = x_k^*(R_0)\,, \quad R^{(k-1)} = \frac{R^{(k)}}{R_n(x_n^0)}\,. \tag{20.22}$$

To solve the inverse problem, we introduce Bellman's function

$$\tilde{\Psi}_k(r) = \max\left\{ \prod_{1 \le i \le k} R_i(x_i) \,\middle|\, \sum_{1 \le i \le k} c_i x_i \le C \right\}, \tag{20.23}$$

which is developed recursively for $k = 1, \ldots, n$ in correspondence with the equation set

$$\tilde{\Psi}_1(\mathrm{c}) = \max\{R_1(x_1) \,|\, c_1 x_1 \le c\}\,, \qquad 0 \le c \le C_0\,,$$
$$\vdots$$
$$\tilde{\Psi}_k(C) = \max\{\tilde{\Psi}_{k-1}(c) R_k(x_k) \,|\, c + c_k x_k \le C\}\,, \qquad 0 \le c \le C\,. \tag{20.24}$$

The solution of the problem can be found in the way similar to (20.22).

$$x_n^0 = x_n^*(C_0)\,, \quad C^{(n-1)} = C_0 - c_n x_n^0\;;$$
$$\vdots$$
$$x_k^0 = x_k^*(C^{(k)})\,, \quad C^{(k-1)} = C^{(k)} - c_k x_k^0\,,$$
$$k = m-1, , , , 1\,.$$

This algorithm will not be covered in more detail because the next modification of it is more convenient for practical use.

20.5.2 Kettelle's Algorithm

Kettelle's modification of standard dynamic programming usually is faster and easier than a standard algorithm. First, for each ith subsystem one calculates the reliability index $R_i(x_i)$ for each x_i of interest. Next for two subsystems [e.g., the mth and $(m-1)$th] a table such as Table 20.4 is prepared. In the cell for column x_m and row x_{m-1} one enters

$$R_{m-1}^*(x_{m-1}, x_m) = R_{m-1}(x_{m-1}) R_m(x_m)\,,$$
$$C_{m-1}^*(x_{m-1}, x_m) = c_{m-1}(x_{m-1}) + c_m(x_m)\,.$$

Table 20.4 Reliability and cost matrix for Ketelle's algorithm

x_{m-1} \ x_m	0	1	2	...
0	$R^*_{m-1}(0,0)$ $C^*_{m-1}(0,0)$	$R^*_{m-1}(0,1)$ $C^*_{m-1}(0,1)$	$R^*_{m-1}(0,2)$ $C^*_{m-1}(0,2)$	... ...
1	$R^*_{m-1}(1,0)$ $C^*_{m-1}(1,0)$	$R^*_{m-1}(1,1)$ $C^*_{m-1}(1,1)$	$R^*_{m-1}(1,2)$ $C^*_{m-1}(1,2)$	... ...
2	$R^*_{m-1}(2,0)$ $C^*_{m-1}(2,0)$	$R^*_{m-1}(2,1)$ $C^*_{m-1}(2,1)$	$R^*_{m-1}(2,2)$ $C^*_{m-1}(2,2)$	... ...
	...	...	...	...

Finally, values of $R^*_{m-1}(x_{m-1}, x_m)$ and $C^*_{m-1}(x_{m-1}, x_m)$ are placed in a table such as Table 20.5 in increasing order of the value of $C^*_{m-1}(x_{m-1}, x_m)$. In this table the superscripts correspond to the cost order.

Every column containing $R^*_{m-1}(x_{m-1}, x_m)$ and $C^*_{m-1}(x_{m-1}, x_m)$ in this ordered sequence that has a lower reliability index than a column with a lower cost is excluded from future consideration. There is no reason to keep a less reliable object that has a higher cost. All entries that remain go into a table such as Table 20.6. These pairs form an *undominated sequence*.

Table 20.6 may be regarded as the table for a new subsystem, $(m-1)^*$, that replaces the previous two subsystems. The cells are renumbered in the order $y_{m-1} = 1, 2, \ldots$, where the new sequence has only undominated cells. A tilde is placed over each x to show that the entry is in an undominated sequence.

The procedure is repeated for subsystems $(m-1)^*$ and $(m-2)$. A table such as a Table 20.7 is developed. The new table is comparable to the table developed in step 2 (Table 20.4), but y_{m-1} replaces x_m and x_{m-2} replaces x_{m-1}. Steps 2 and 3 are repeated for the new pair, $(m-1)^*$ and $(m-2)$,

Table 20.5 Suites in order of increasing cost, Kettelle's algorithm

C^*_{m-1}	$C^*_{m-1}(0,0)$	$C^*_{m-1}(x^1_{m-1}, x^1_m)$	$C^*_{m-1}(x^2_{m-1}, x^2_m)$
R^*_{m-1}	$R^*_{m-1}(0,0)$	$R^*_{m-1}(x^1_{m-1}, x^1_m)$	$R^*_{m-1}(x^2_{m-1}, x^2_m)$

Table 20.6 Sequence of undominated suites, Kettelle's algorithm

y_{m-1}	1	2
$C^*_{m-1}(y_{m-1})$	$C^*_{m-1}(\tilde{x}^1_{m-1}, \tilde{x}^1_m)$	$C^*_{m-1}(\tilde{x}^2_{m-1}, \tilde{x}^2_m)$
$R^*_{m-1}(y_{m-1})$	$R^*_{m-1}(\tilde{x}^1_{m-1}, \tilde{x}^1_m)$	$R^*_{m-1}(\tilde{x}^2_{m-1}, \tilde{x}^2_m)$

Table 20.7 Second reliability and cost matrix for Kettelle's algorithm

x_{m-2} \ x_{m-1}	1	2	3	...
0	$R^*_{m-2}(0, 1)$ $C^*_{m-2}(0, 1)$	$R^*_{m-2}(0, 2)$ $C^*_{m-2}(0, 2)$	$R^*_{m-2}(0, 3)$ $C^*_{m-2}(0, 3)$	... ...
1	$R^*_{m-2}(1, 1)$, $C^*_{m-2}(1, 1)$	$R^*_{m-2}(1, 2)$, $C^*_{m-2}(1, 2)$	$R^*_{m-2}(1, 3)$ $C^*_{m-2}(1, 3)$	... ...
2	$R^*_{m-2}(2, 1)$, $C^*_{m-2}(2, 1)$	$R^*_{m-2}(2, 2)$ $C^*_{m-2}(2, 2)$	$R^*_{m-2}(2, 3)$ $C^*_{m-2}(2, 3)$	... ...
...	...	...	...	...

and appropriate new tables are compiled. Thus on the basis of values of pairs

$$[R^*_{m-1}(y_{m-1}), C^*_{m-1}(y_{m-1})] ,$$
$$[R_{m-2}(x_{m-2}), C_{m-2}(x_{m-2})] ,$$

a new table is developed. In the cell for column y_{m-1} and row x_{m-2} one enters

$$R^*_{m-2}(x_{m-2}, y_{m-1}) = R_{m-2}(x_{m-2})R^*_{m-1}(y_{m-1}) ,$$
$$C^*_{m-2}(x_{m-2}, y_{m-1}) = C_{m-2}(x_{m-2}) + C^*_{m-1}(y_{m-1}) .$$

After $n - 1$ iterations, the final undominated sequence is obtained. This sequence contains a Pareto set in a "reliability-cost" space: $\{R^*_1(y_1), C^*_1(y_1)\}$.

The solution of the direct optimal redundancy problem can be found from the condition $R^*_1(y_1 - 1) < R_0 \le R^*_1(y_1)$. This condition enables us to find a value $y_1 = (x_1, y_2)$. Thus, the final number of redundant units of the first type, x_1, can be found. In turn, $y_2 = (x_2, y_3)$, so x_2 can be found. This procedure of finding x_1's is terminated after $n - 1$ such steps are performed in inverse order of the development of the final undominated sequence. The solution of the inverse problem of optimal redundancy must satisfy the condition $C^*_1(y_1) \le C_0 < C^*_1(y_1 + 1)$. The procedure for finding the x_1's is the same.

Remark 1 For computational simplicity, it may be more convenient to use a dichotomous procedure to combine the first original pairs (see Figure 20.2*a*), rather than the procedure described above, which corresponds to the

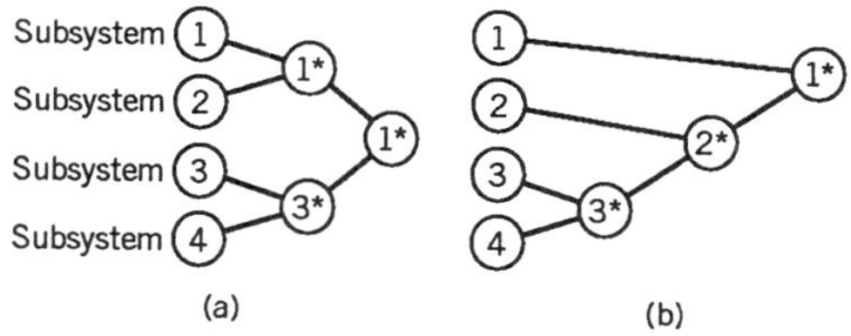

Figure 20.2 Alternative computational procedure, Kettelle's algorithm.

graph in Figure 20.2*b*. The dichotomous procedure is convenient because in a table such as Table 20.4 all pairs in the undominated sequence formed by two initial subsystems consist of the one-link two-dimensional set, as illustrated in Example 20.3, which follows.

Remark 2 Tables 20.4 and 20.7 can be filled step by step as the solution process evolves. If they are filled in advance, one can use the lower and the upper bounds on x_i's described in Section 20.4.1.

Remark 3 The calculation time can be decreased substantially if certain pairs are excluded from each current undominated sequence. One may introduce a value, δ, that distinguishes the reliability index of one pair from another, and a value, ϵ, that distinguishes the cost of one pair from another. Both δ and ϵ could be chosen in accordance with the required accuracy or on the basis of the tolerance of the input data. It may be expedient to eliminate terms of the undominated sequence when there are insignificant differences in reliability or cost. The procedure can be formulated as follows:

If $\{R_j, C_j\}$ and $\{R_{j+1}, C_{j+1}\}$ are neighboring pairs and $R_{j+1} - R_j > \delta$ and at the same time $C_{j+1} - C_j < \epsilon$, the pair $\{R_j, C_j\}$ can be removed from the current undominated sequence.

If $\{R_j, C_j\}$ and $\{R_{j+1}, C_{j+1}\}$ are neighboring pairs and $R_{j+1} - R_j < \delta$ and at the same time $C_{j+1} - C_j > \epsilon$, the pair $\{R_{j+1}, C_{j+1}\}$ can be removed from the current undominated sequence.

If $\{R_j, C_j\}$ and $\{R_{j+1}, C_{j+1}\}$ are neighboring pairs and $R_{j+1} - R_j < \delta$ and at the same time $C_{j+1} - C_j < \epsilon$, the pairs can be considered as not significantly different, so either one can be removed from the current undominated sequence.

EXAMPLE 20.3

Consider the Example 20.2 system. We want to solve the direct optimal redundancy problem for $P_0 \geq 0.45$ and the inverse problem for $C_0 < 50$.

Solution

We will use the $P_i(x_i)$ values in Table 20.1. In Table 20.8 we give the fifth and fourth subsystem composition. In Table 20.9 we give the second and third subsystem composition. The numbers in the lower corners show the order index of the given number in the undominated sequence. As is readily seen, we have discarded terms that are close. We let $y_1 = (x_4, x_5)$ and $y_2 = (x_2, x_3)$. The y_1 and y_2 subsystems are shown by the corner numbers in Tables 20.8 and 20.9, respectively.

Table 20.10 gives the composition of two newly formed subsystems, and Table 20.11 shows the final undominated sequence. Here $y_3 = (y_1, y_2)$ and $y_4 = (x_1, y_3)$. They correspond to the corner numbers in Tables 20.10 and 20.11, respectively.

Tables 20.8 and 20.9 have typical undominated sequences. The first suite is always in the upper left-hand corner of the table. For the second suite one should check all three neighboring cells. In Tables 20.8 and 20.9, suites 2, 3, and 4 happen to be in these cells. The suite below suite 2 in Table 20.8 does not belong to the undominated sequence. This means that no suites below it in the column belong to the sequence. A similar situation is present in Table 20.9. Here the cell to the right of cell 2 does not belong to the undominated

Table 20.8 Kettelle procedure, subsystems 4 and 5

x_4 \ x_5	0	1	2	3	4	5	. . .
0	0.1125 16 **1**	0.2812 21 **3**	0.4500 26 **5**	0.5907 31 **7**	0.6961 36 **9**	0.7698 41	. . .
1	0.1244 17 **2**	0.3109 22 **4**	0.4976 27 **6**	0.6531 32 **8**	0.7695 37 **10**	0.8510 42 **11**	. . .
2	0.1250 18	0.3125 23	0.4999 28	0.6562 33	0.7733 38	0.8551 43	. . .
3	0.1250 19	0.3125 24	0.5000 29	0.6563 34	0.7734 39	0.8553 44	. . .
4	0.1250 20	0.3125 25	0.5000 30	0.6563 35	0.7734 40	0.8553 45	. . .
5	0.1250 21 . . .	0.3125 26 . . .	0.5000 31 . . .	0.6563 36 . . .	0.7734 41 . . .	0.8553 46 . . .	. . . . . .

Table 20.9 Kettelle procedure, subsystems 2 and 3

x_2 \ x_3	0	1	2	3	...
0	0.1800 3 **1**	0.1980 4 **2**	0.1998 5	0.2000 6	...
1	0.4724 5 **3**	0.5197 6 **4**	0.5244 7	0.5249 8	...
2	0.7051 7 **5**	0.7756 8 **6**	0.7826 9	0.7833 10	...
3	0.8291 9 **7**	0.9120 10 **8**	0.9203 11	0.9211 12	...
4	0.8887 11	0.9666 12 **9**	0.9753 13	0.9762 14	...
5	0.8946 13	0.9841 14 **10**	0.9930 15 **11**	0.9939 16	...
6	0.8988 15	0.9888 16	0.9977 17 **12**	0.9986 18	...
7	0.8997 17	0.9898 18	0.9987 19 **13**	0.9996 20	...
	...	...	...	...	...

sequence. This means that no cells to the right of this cell belong to the sequence. Clearly the sequential development can proceed very quickly.

Table 20.10 shows that this simple procedure does not always work. In fact entire rows can have no suites in the undominated sequence (e.g., the fourth column and seventh row).

Now let us find the solution for the system that costs less than 50 units. From Table 20.11 we find that $y_4 = 8$; that is, $x_1 = 3$ and $y_3 = 19$. From Table 20.10 we find that $y_3 = 19$ and that $y_1 = 6$ and $y_2 = 8$. Taking $y_1 = 6$ in Table 20.8, we determine that $x_4 = 1$ and $x_5 = 2$. Taking $y_2 = 8$ from Table 20.9, we find that $x_2 = 3$ and $x_3 = 1$. The probability of system failure-free operation can be found in Table 20.11, and equals 0.2679.

Table 20.10 Kettelle procedure, combining the results of the steps in Tables 20.8 and 20.9

y_1 \ y_2	1	2	3	4	5	6	7	8	9	10	11	12
1	0.0203 19 **1**	0.0223 20	0.0532 21	0.0585 22 **3**	0.0793 23	0.0873 24 **5**	0.0933 25	0.1024 26	0.1087 28	0.1107 30	0.1117 31	0.1122 33
2	0.0224 20 **2**	0.0246 21	0.02588 22 **4**	0.0617 23	0.0877 24 **6**	0.0965 25 **7**	0.1031 26	0.1135 27	0.1202 29	0.1224 31	0.1235 32	0.1241 34
3	0.0506 24	0.0558 25	0.1328 26 **8**	0.1461 27	0.1983 28 **10**	0.2181 29	0.2331 30	0.2565 31	0.2716 33	0.2767 35	0.2792 36	0.2806 38
4	0.0560 25	0.0616 26	0.1469 27 **9**	0.1616 28	0.2192 29 **11**	0.2411 30 **12**	0.2578 31 **13**	0.2835 32 **14**	0.3005 34	0.3060 36	0.3087 37	0.3102 39
5	0.0810 29	0.0891 30	0.2126 31	0.2339 32	0.3173 33 **15**	0.3490 34	0.3731 35	0.4104 36	0.4350 38	0.4428 40	0.4469 41	0.4490 43

6	0.0896 30	0.0985 31	0.2351 32	0.2586 33	0.3509 34 **16**	0.3859 35 **17**	0.1126 36 **18**	0.4538 37 **19**	0.4810 39 **20**	0.4897 41	0.4941 42	0.4965 44
7	0.1063 31	0.1170 35	0.2790 36	0.3070 37	0.4165 38	0.4581 39	0.4897 40	0.5387 41	0.5710 43	0.5813 45	0.5866 46	0.5893 48
8	0.1176 35	0.1293 36	0.3085 37	0.3394 38	0.4605 39	0.5065 40 **21**	0.5115 41 **22**	0.5956 42 **23**	0.6313 44 **24**	0.6427 46 **25**	0.6485 47	0.6516 49
9	0.1253 39	0.1378 40	0.3288 41	0.3618 42	0.4908 43	0.5399 44	0.5771 45	0.6318 46	0.6729 48	0.6850 50	0.6912 51	0.6945 53
10	0.1385 40	0.1524 41	0.3635 42	0.3999 43	0.5426 44	0.5968 45	0.6380 46	0.7018 47 **26**	0.7438 49 **27**	0.7573 51 **28**	0.7641 52	0.7677 54
11	0.1532 45	0.1685 46	0.4020 47	0.4423 48	0.6000 49	0.6600 50	0.7056 51	0.7761 52 **29**	0.6226 54 **30**	0.8375 56	0.8450 57	0.8490 59

Table 20.11 Kettelle procedure, final undominated sequence

x_1 \ y_3	18	19	20	21	22	23	24	25	26
0	0.0825 39 **1**	0.0908 40 **2**	0.0962 42	0.1013 43	0.1083 44	0.1191 45	0.1263 47	0.1285 49	0.1404 50
1	0.1485 42 **3**	0.1634 43 **4**	0.1732 45	0.1823 46	0.1949 47	0.2144 48	0.2273 50	0.2314 52	0.2526 53
2	0.2013 45 **5**	0.2215 46 **6**	0.2347 48	0.2472 49	0.2643 50	0.2907 51	0.3081 53	0.3136 55	0.3425 56
3	0.2436 48 **7**	0.2679 49 **8**	0.2840 51 **9**	0.2900 52	0.3197 53 **11**	0.3516 54 **12**	0.3727 56 **13**	0.3795 58	0.4143 59
4	0.2774 51	0.3051 52 **10**	0.3234 54	0.3406 55	0.3641 56	0.4000 57 **14**	0.4245 59 **15**	0.4322 61	0.4719 62
5	0.3044 54	0.3349 55	0.3549 57	0.3737 58	0.3996 59	0.4395 60 **16**	0.4658 62	0.4742 64	0.5179 65
6	0.3260 57	0.3586 58	0.3801 60	0.4000 61	0.4279 62	0.4706 63	0.4989 65	0.5079 67	0.5546 68
7	0.3432 60	0.3775 61	0.4000 63	0.4214 64	0.4505 65	0.4955 66	0.5252 68	0.5347 70	0.5888 71

With a comparable procedure, we find from Table 20.11 that the solution for the second problem, for which the reliability index must be not less than 0.45, is $y_4 = 17$, where the index is 0.4658. Also, successively, we find that $y_4 = 17$ has $x_1 = 4$, $x_2 = 3$, $x_3 = 1$, $x_4 = 1$, and $x_5 = 4$. The cost of the system is 62.

20.6 COMPUTER PROGRAMS

The methods in Sections 20.5 and 20.6 have software written specifically for them. One such software package that is user friendly is Diana (see Appendix F). Diana includes both the approximate solution by a steepest descent algorithm and the strict solution by a modified Kettelle's algorithm. In the computer screen figures in this section, the first is "the first-level result" and the second is "the second-level result." Both solutions are presented on the screen as final undominated sequences. The user can select any suite in the undominated sequence and call forth the subsolution for that suite. Diana also permits special restrictions on the redundant system, such as restricted numbers of redundant units of certain types.

EXAMPLE 20.4

A series system consists of five units. The reliability and cost of each is presented on the monitor screen as shown in Table 20.12. The unusual cost and reliability values were chosen to make analysis of the numerical results easier. For simplicity, let us assume that all redundant units are active.

Diana lets the user pick a display of either the complete undominated sequence (see the monitor screen in Figure 20.3) or the principal suites in the sequence (see Figure 20.4). Each small square represents a suite in the undominated sequences. The cursor can be moved to any square to display its failure probability, Q, and cost (price). The complete sequence, which can be retrieved if desired, is in Table 20.13.

Table 20.12 Basic system description, Diana computer program

Name	Structure	Failure Probability	Redundancy Cost
Model	Series		
. unit_1	. -----	0.1	5.25521
. unit_2	. -----	0.12	3.751
. unit_3	. -----	0.14	2.11
. unit_4	. -----	0.16	1.1
. unit_5	. -----	0.18	1

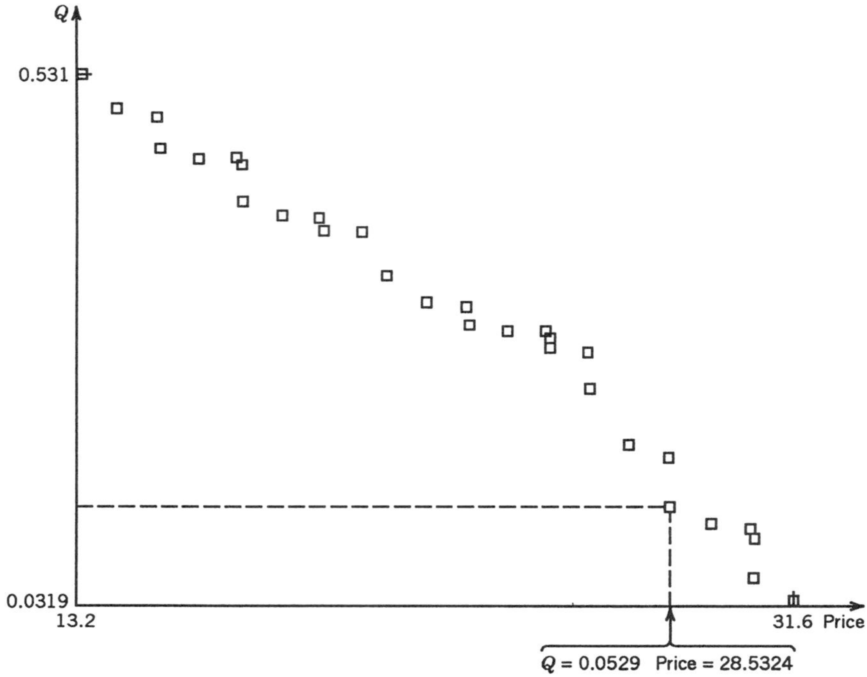

Figure 20.3 Diana computer program: Monitor display of complete undominated sequence from cost = 13.2, $Q = 0.531$ to cost = 28.5324, $Q = 0.0529$.

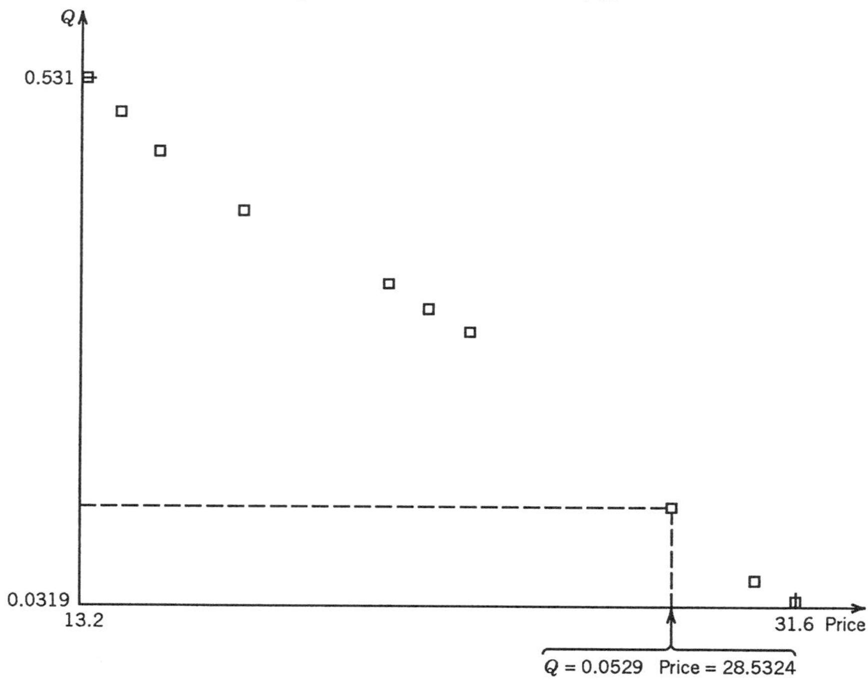

Figure 20.4 Sequence with display of only relatively best suite in each price-reliability range.

Table 20.13 Diana computer program sequence of undominated suites

Number	Unreliability	Cost	Number of Units				
			$x_1 + 1$	$x_2 + 1$	$x_3 + 1$	$x_4 + 1$	$x_5 + 1$
1	0.531	13.2162	1	1	1	1	1[a]
2	0.446	14.2162	1	1	1	1	2[a]
3	0.431	15.2162	1	1	1	1	3
4	0.358	15.3162	1	1	1	2	2[a]
5	0.340	16.3162	1	1	1	2	3
6	0.337	17.3162	1	1	1	2	4
7	0.268	17.4262	1	1	2	2	2[a]
8	0.248	18.4262	1	1	2	2	3
9	0.244	13.4262	1	1	2	2	4
10	0.231	19.5262	1	1	2	3	3
11	0.228	20.5262	1	1	2	3	4[a]
12	0.180	21.1772	1	2	2	2	2[a]
13	0.158	22.1772	1	2	2	2	3[a]
14	0.153	23.1772	1	2	2	2	4
15	0.139	23.2772	1	2	2	3	3[a]
16	0.135	24.2772	1	2	2	3	4
17	0.134	25.2772	1	2	2	3	5
18	0.124	25.3872	1	2	3	3	3
19	0.120	26.3872	1	2	3	3	4
20	0.0981	26.4324	2	2	2	2	2
21	0.733	27.4324	2	2	2	2	3
22	0.688	28.4324	2	2	2	2	4
23	0.529	28.5324	2	2	2	3	3[a]
24	0.783	29.5324	2	2	2	3	4
25	0.475	30.5324	2	2	2	3	5

[a] Relatively best suite in cost-reliability range.

EXAMPLE 20.5

A system consists of two subsystems as shown in Figure 20.5. The first subsystem has two units in parallel with failure probabilities 0.5 and 0.4 and costs 3 and 4, respectively. The second subsystem consists of three units in series with failure probabilities 0.015, 0.02, and 0.03 and costs 0.5, 0.3, and 0.1, respectively. Active redundancy may be used if desired.

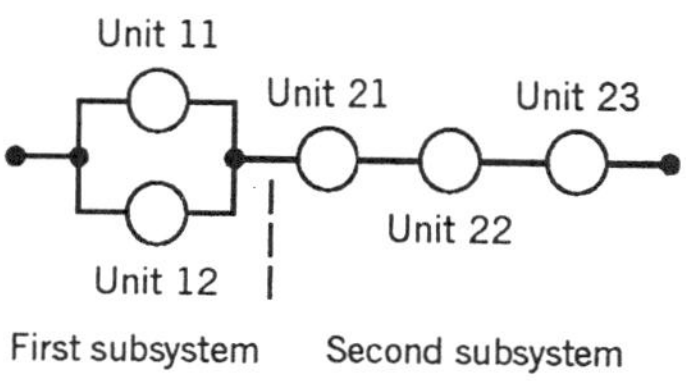

Figure 20.5 System of Example 20.5.

Table 20.14 System with restrictions in Diana

Name	Structure	Failure Probability	Opt	Redundancy Max	Redundancy Cost
Model	Series				
. subsyst_1	. Parallel	0.00001	yes	3	0.00001
. . unit_11	. . -----	0.5	yes	3	3
. . unit_12	. . -----	0.4	yes	2	5
. subsyst_2	. Series	0.00001	no	1	0.00001
. . unit_21	. . -----	0.015	yes	10	0.5
. . unit_22	. . -----	0.02	yes	10	0.3
. . unit_23	. . -----	0.03	yes	4	0.1

There are special restrictions. Subsystem 1 can include no more than three type 11 units or more than two type 12 units. If this redundancy is insufficient, however, either one or two parallel subsystems may be added. For subsystem 2 there is only one restriction: not more than four type 23 units are permitted. The input data are presented in the Diana screen display in Table 20.14.

The problem is selection of the best system for its cost among all the undominated systems in the system failure probability interval [0.01, 0.02]. The first level (steepest descent algorithm) and the second level (Kettelle's algorithm) results are shown in Figures 20.6 and 20.7. Table 20.15 presents the final results.

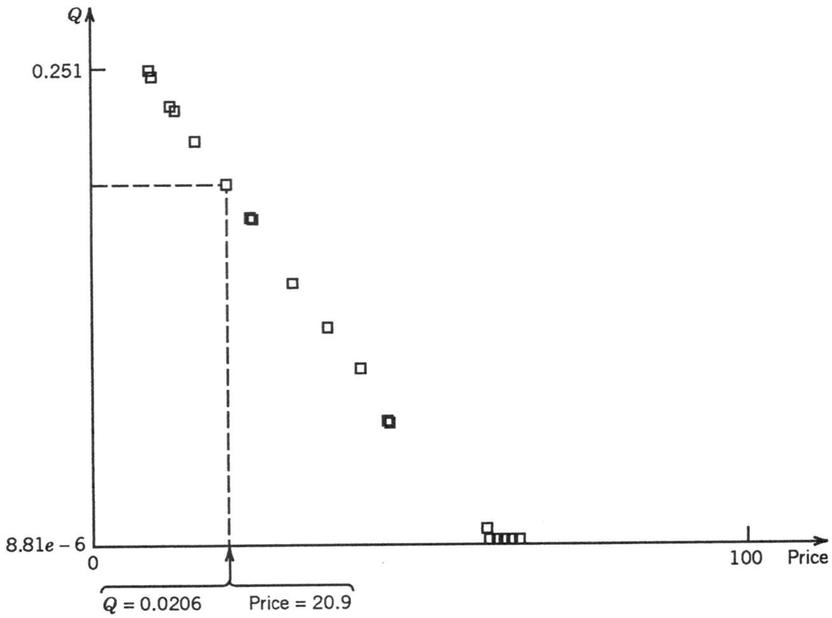

Figure 20.6 Diana computer program first-level results, Example 20.5.

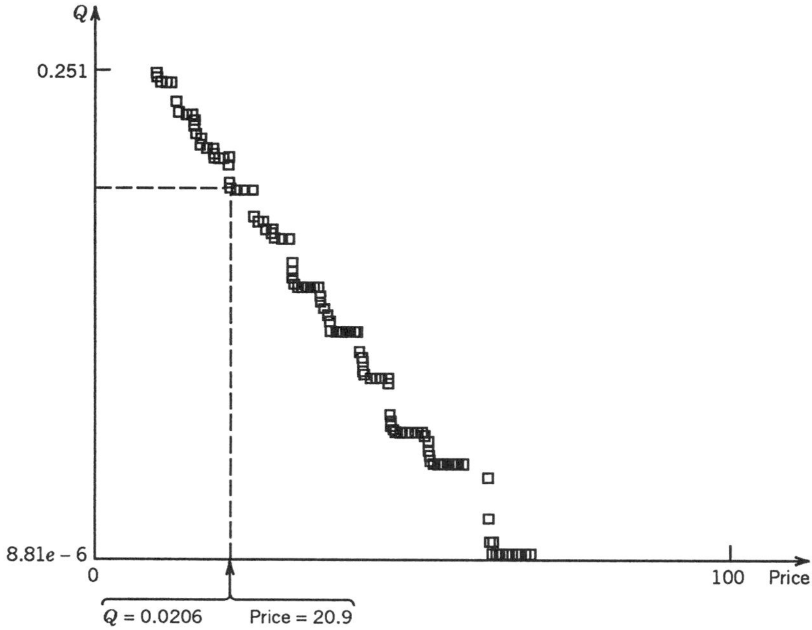

Figure 20.7 Diana second-level results, Example 20.5.

Table 20.15 Final results, Example 20.5

Cost	Probability of Failure	First Subsystem	Number of Units 11	12	21	22	23
23.7	0.02	1	3	2	5	6	4
23.8	0.0115	2	2	1	2	2	2
23.9	0.0106	2	2	1	2	2	3
24.0	0.0106	2	2	1	2	2	4
24.2	0.0103	2	2	1	2	3	3
24.3	0.0102	2	2	1	2	3	4
24.6	0.0102	2	2	1	2	4	4
24.7	0.0100	2	2	1	3	3	3
24.8	0.0100	2	2	1	3	3	4
25.1	0.0100	2	2	1	3	4	4
25.4	0.0100	2	2	1	3	5	4
25.6	0.0100	2	2	1	4	4	4
25.8	0.0095	3	1	1	2	2	2
25.9	0.0087	3	1	1	2	2	3
26.0	0.0086	3	1	1	2	2	4

20.7 THE KNAPSACK PROBLEM—OPTIMIZATION OF BOTH TYPE AND NUMBER

A question that may arise at the very earliest stage of an engineering design is the selection of one out of several possible designs for each unit. We might want to assemble a system from several units $u_1, u_2, \ldots, u_n$. Each unit u_j may have variants

$$u(j_1), u(j_2), \ldots, u(j_{N_j}),$$

where N_j is the number of variants. Each variant has its own reliability index and cost, R_{j_k} and C_{j_k}, respectively. The problem is to find units $u_1(k_1), \ldots, u_n(k_n)$ so that the system will either (1) have a given level of a reliability at the lowest possible cost or (2) be as reliable as possible under a given maximum cost. In other words, we have to solve one of the two following discrete optimization problems:

$$\min_{\mathbf{d}} \left\{ \sum_{1\le j\le n} \sum_{1\le k_j\le N_j} C_j(k_j)d(k_j) \,\middle|\, \sum_{1\le k_j\le N_j} d(k_j) = 1\,; \right.$$

$$\left. \times \prod_{1\le j\le n} \sum_{1\le k_j\le N_j} R_j(k_j)d(k_j) \ge R^0 \right\} \tag{20.25}$$

Table 20.16 Cost and reliability of each variant of each unit in knapsack problem, Example 20.6

Name	Structure	Failure Probability	Redundance Cost
Model	Series		
. unit_1		-------	.000001
. . variant_11	. . -----	0.2	1
. . variant_12	. . -----	0.15	4
. . variant_13	. . -----	0.13	5.5
. . variant_14	. . -----	0.12	6
. unit_2		-------	.000001
. . variant_21	. . -----	0.3	1
. . variant_22	. . -----	0.2	7
. . variant_23	. . -----	0.1	18
. unit_3		-------	.000001
. . variant_31	. . -----	0.25	1
. . variant_32	. . -----	0.21	3
. . variant_33	. . -----	0.19	7
. . variant_34	. . -----	0.18	8
. . variant_35	. . -----	0.11	15

Table 20.17 Best design for 20 cost units, Example 20.6

		Failure	Redundance			(R)Failure	(R)Redundance	
Name	Structure	Probability	No.	Cost	R	Probability	No.	Cost
Model	Series				+	0.43		20
. unit_1		-------	1	.000001	+	0.12	1	6
. . variant_11	. . -----	0.2	--	1	+			
. . variant_12	. . -----	0.15	--	4	+			
. . variant_13	. . -----	0.13	--	5.5	+			
. . variant_14	. . -----	0.12	--	6	+	0.12	1	6
. unit_2		-------	1	.000001	+	0.2	1	7
. . variant_21	. . -----	0.3	--	1	+			
. . variant_22	. . -----	0.2	--	7	+	0.2	1	7
. . variant_23	. . -----	0.1	--	18	+			
. unit_3		------	1	.000001	+	0.19	1	7
. . variant_31	. . -----	0.25	--	1	+			
. . variant_32	. . -----	0.21	--	3	+			
. . variant_33	. . -----	0.19	--	7	+	0.19	1	7
. . variant_34	. . -----	0.18	--	8	+			
. . variant_35	. . -----	0.11	--	15	+			

Table 20.18 Best design for 30 cost units, Example 20.6

		Failure	Redundance			(R)Failure	(R)Redundance	
Name	Structure	Probability	No.	Cost	*R*	Probability	No.	Cost
Model	Series				+	0.373		30
. unit_1		-------	1	.000001	+	0.15	1	4
. . variant_11	. . -----	0.2	--	1	+			
. . variant_12	. . -----	0.15	--	4	+	0.15	1	4
. . variant_13	. . -----	0.13	--	5.5	+			
. . variant_14	. . -----	0.12	--	6	+			
. unit_2		-------	1	.000001	+	0.1	1	18
. . variant_21	. . -----	0.3	--	1	+			
. . variant_22	. . -----	0.2	--	7	+			
. . variant_23	. . -----	0.1	--	18	+	0.1	1	18
. unit_3		------	1	.000001	+	0.18	1	8
. . variant_31	. . -----	0.25	--	1	+			
. . variant_32	. . -----	0.21	--	3	+			
. . variant_33	. . -----	0.19	--	7	+			
. . variant_34	. . -----	0.18	--	8	+	0.18	1	8
. . variant_35	. . -----	0.11	--	15	+			

or

$$\min_{\mathbf{d}}\left\{\prod_{1\le j\le n}\sum_{1\le k_j\le N_j} R_j(k_j)d(k_j)\,\middle|\,\sum_{1\le k_j\le N_j} d(k_j)=1\,;\right.$$

$$\left.\times\sum_{1\le j\le n}\sum_{1\le k_j\le N_j} C_j(k_j)d(k_j)\ge C^0\right\} \tag{20.26}$$

The vector $\mathbf{d}=(d(k_1),\ldots,d(k_n))$ where $d(k_j)=1$ if variant $u_j(k_j)$ of the unit u_j is chosen and $d(k_j)=0$ otherwise. Only one variant of each unit may be used. A sample problem, which was solved with the Diana software package, is in Example 20.6.

EXAMPLE 20.7

A series system consists of three subsystems, and each subsystem has the mutually compatible variants shown in Table 20.16. Two optimal designs are desired, one for a cost limit of 20 and one for a cost limit of 30. Equation (20.26) yields the results shown in Tables 20.17 and 20.18. Both designs are found with the Diana computer program operating at the Kettelle algorithm level. The number 1 in the next to the last column shows the variant of each unit selected. As may be seen, all three variants changed when the limit went from 20 to 30. This is not unusual for knapsack problems.

20.8 SOME SPECIAL CASES

20.8.1 System Performance Effectiveness

A complex system consists of n subsystems. Each subsystem must be in one of two states, operational or failed. Let the ith subsystem state be denoted s_i:

$$s_i=\begin{cases}1 & \text{if the } i\text{th subsystem is operational}\,,\\ 0 & \text{otherwise}\,.\end{cases}$$

Let the system state be denoted $\mathbf{S}=(s_1,\ldots,s_n)$. The system performance effectiveness index is

$$E=\sum_{1\le S\le 2^n} H_S\Phi_S\,, \tag{20.27}$$

where $\mathbf{S}$ is one of 2^n possible system states, Φ_S is the conditional effectiveness index for the state S, and H_S for independent subsystems is

$$H_S=\prod_{1\le n} r_i^{s_i}(1-r_i)^{1-s_i}\,,$$

where r_i is the probability the ith subsystem is failure free, that is, $r_i = Pr\{s_i = 1\}$. For a repairable system, an availability coefficient can be used for each subsystem. For any fixed i we can use the theorem of Boolean Algebra that any logic function $f(S) = f(s_1, \ldots, s_n)$ can be presented in the form

$$f(S) = s_1 f(s_1, \ldots, s_{i-1}, 1, s_{i+1}, \ldots, s_n)$$
$$\vee\ \bar{s}_1 f(s_1, \ldots, s_{i-1}\ 0, s_{i+1}, \ldots, s_n)\ . \qquad (20.28)$$

Using (20.28), (20.27) can be rewritten in the form

$$E = r_i\Big(\sum_{S^*} H_{S_i^*}(\Phi_{(S_i^*,1)} - \Phi_{(S_i^*,0)})$$
$$+ \sum_{S_i^*} H_{S_i^*}\Phi_{(S_i^*,0)}\ ,$$

where S_1^* is state S with the ith subsystem excluded from all consideration; $(S_1^*, 1)$ is the state when the ith subsystem is in an operational state, and $(S_1^*, 0)$ is the state when the ith subsystem is in a failed state. Thus the E is expressed as a linear function of the reliability indexes of certain single subsystems.

For a complex system, pure structural redundancy may not be the only method of performance effectiveness improvement. There also may be a choice of subsystem variants. The problem then can be considered as a modification of the problem in Section 20.7. Let us assume that the variants of each subsystem have different reliability indexes but the same effect on the system performance effectiveness index. Let the ith subsystem have k_1 variants, each of them with reliability index $r_{i(j)}$ and cost $c_{i(j)}$. For each subsystem we arrange these variants in the undominated sequence:

$$\{c_{i(1)}, r_{i(1)}\}, \{c_{i(2)}, r_{i(2)}\}, \ldots, \{c_{i(k_i)}, r_{i(k_i)}\}\ .$$

The goal function E in this problem makes use of a strict method impossible. Also the sequence above does not necessarily have the logarithmic convexity that permits use of the steepest descent algorithm. However, an approximate solution can be found with the steepest descent method. We will describe it briefly.

Compute for all i, $i = 1, \ldots, n$, values

$$g_i = \frac{r_{i(2)} - r_{i(1)}}{c_{i(2)} - c_{i(1)}} \sum_{S_i^*} (\Phi_{(S_i^*,\,1)} - \Phi_{(S_i^*,\,0)}) H_{S_i^*}\ .$$

If all subsystems are highly reliable, the following approximation may be used:

$$\sum_{S_i^*} (\Phi_{(S_i^*,1)} - \Phi_{(S_i^*,0)}) H_{S_i^*} \approx \sum_{\substack{k=1 \\ k \neq j}} (\Phi_I - \Phi_{(I_k,0)})(1 - r_k) ,$$

where I is the totally operational system state, that is, $I = (1, \ldots, 1)$, and $(I_k^*, 0)$ is the state when all subsystems except the kth one are in an operational state.

Thus for highly reliable systems we can write

$$g_i = \frac{r_{i(2)} - r_{i(1)}}{C_{i(2)} - C_{i(1)}} \sum_{\substack{k=1 \\ k \neq i}} (\Phi_I - \Phi_{(I_k^*,0)})(1 - r_k) .$$

The rest of the procedure coincides with that described in Section 20.4. The stopping rules are

$E^{(N-1)} < E_0 \leq E^{(N)}$ for the direct optimization problem,

$C^{(N)} \leq C_0 < C^{(N+1)}$ for the inverse problem.

20.8.2 Optimization of MTTF

Let us consider a series system of n independent units. Let the known variables for each unit include the MTTF, T_i, the standard deviation, σ_i, and the cost, c_i. The approximate solution of the optimization problem will be suitable when $x_i \gg 1$, $i = 1, \ldots, n$. The random time of exhaustion of x_i spare parts of the ith type has an approximately normal distribution.

Let $N'(T, \sigma)$ be the complementary function for the normal distribution function $N(T, \sigma)$, $N'(T, \sigma) = 1 - N(T, \sigma)$. For a series of n independent redundant groups

$$P_s(t) = \prod_{1 \leq i \leq n} \bar{N}_i(x_i T_i, \sigma_i \sqrt{x_i}) . \tag{20.29}$$

It is clear that the system MTTF, T_S, satisfies the inequality

$$T_s \leq \min_{1 \leq i \leq n} x_i T_i . \tag{20.30}$$

An approximate solution of the objective function

$$\min_x \{C(X) \mid T(X) \geq T^0\} , \tag{20.31}$$

can be found with

$$x_i = \min\{x_i^* : \; x_i T_i - b_n \sigma_i \sqrt{x_i} \geq t\} , \tag{20.32}$$

where the coefficient b_n depends only on the number of redundant groups and can be found from

$$N(b_n) = 1 - \sqrt[n]{0.5}\,. \qquad (20.33)$$

A practical procedure for solution of the optimal redundancy problem (20.31) has the following steps.

1. Find x^0 for all i, $1 \le i \le n$, by the rule

$$x_i^0 = \frac{T^0}{T_i}\,. \qquad (20.34)$$

2. With the help of a table for the standard normal distribution, determine b_n using (20.31).
3. Find for all i, $1 \le i \le n$, the first approximation of possible optimal x_1,

$$x_i^1 = x_i^0 + \frac{1}{T_i}\, b_n \sigma_i \sqrt{x_i^0}\,. \qquad (20.35)$$

The vector of solutions given by (20.35) may be denoted with

$$X^1 = (x_1^1, \ldots, x_n^1)\,. \qquad (20.36)$$

Remark 1 The second term in (20.35) gives us the necessary shift of the mean to the left, so at point T^0 a quantile of the distribution of the ith TTF should approximately equal b_n.

Remark 2 Solution (20.36) usually suits practical purposes. If a more accurate result is needed, the following variant of the "greedy" algorithm may be used.

4. Calculate $T(X^1)$:

$$T(X^1) = \int_0^\infty \prod_{1 \le i \le n} N_i(t) \mid T_i x_i^1, \sigma_i \sqrt{x_i^1})\, dt\,. \qquad (20.37)$$

5. If $T(X^1) \ge T^0$, go to (7a).

6a. If $T(X^1) < T^0$, calculate $T(X_i^1)$ using a modified version of (20.37). Let $\hat{X}_i^1 = X_i^1 + e_i$, where e_i is a vector of type $e_i = (0, 0, \ldots, 1_i, \ldots, 0)$,

$$x_i^1 = (x_1^1, \ldots, x_i^1 + 1, \ldots, x_n^1)\,.$$

6b. Calculate

$$d_i = \frac{1}{c_i}\,[T(X_i^1) - T(X^1)]\,.$$

6c. For redundant group i_1 for which

$$i_1 = \{i: \max_{1 \le i \le n} d_i\} ,$$

change $x_i^1 := x_i^1 + 1$.

Repeat the procedure until the first occurrence of $T(\mathbf{X}^k) \geq T^0$. The vector $T(\mathbf{X}^k)$ at the stopping point is the approximate solution.

7a. If $(X^1) \geq T^0$, calculate $T(\hat{X}_i^1)$ with a modification of (20.37). Let $\hat{X}_i^1 = X^1 - e_i$,

$$\hat{X}_i^1 = (x_i^1, \ldots, x_i^1 - 1, \ldots, x_n^1) .$$

7b. Calculate

$$\gamma_i = \frac{1}{c_i} [T(X^1) - T(\hat{X}_i^1)] .$$

7c. For redundant group i_1 for which

$$i_1 = \{i: \max_{1 \le i \le n} \gamma_i\} ,$$

change $x_i^1 := x_i^1 - 1$. Repeat the procedure until $T(\mathbf{X}^{k+1}) \leq T^0$ for the first time. The vector $T(\mathbf{X}^k)$ at the iteration before this stopping point is the approximate solution.

If redundant groups have more than one main unit, one can develop a similar procedure using approximate MTTF (see Chapter 3).

20.8.3 Multifunctional System

Consider a system with k different operations (Section 20.2.5). Each operation requires certain units. Let an operation i require the subset G_i of the system units to be operational. In general, these subsets intersect, $G_i \cap G_j \neq \emptyset$, for at least some system operations. Because the system operations can differ in importance, the reliability requirements for different operations can differ. The minimum cost allocation of redundant units in such a system is difficult to solve precisely. The following procedure enables an approximate optimal solution.

1. Solve k isolated direct optimal redundancy problems (one for each subset of units providing a certain system operation:

$$\min\left\{ \sum_{i \in G_j} c_i x_i \,\middle|\, \prod_{i \in G_j} R_i(x_i) \geq R_j^0 \right\} .$$

Find corresponding optimal values $x_1^j, \ldots, x_m^j$ with the help of any of the preceding methods. We denote the solution for system operations as

$$\mathbf{X}^j = (x_1^j, \ldots, x_{m_j}^j) .$$

2. For each set of units that belongs to an intersection, find the decreasing order of all x_1^j's. We say that subsystem j dominates subsystem k if, for each unit i belonging to both subsystems, $x_i^j \geq x_i^k$ and at least one is a strong inequality.
3. On the base of the order from step 2, construct a tree of partial ordering of all functional subsystems. By the nature of this problem the tree may have several roots. There may be subsystems with strong reliability requirements that have no intersections.
4. Denote as $G^{(1)}$ the set of units of any root subsystem of the tree of partial ordering. Denote as $X^{(1)}$ the optimal solution for this subsystem. Accordingly the tree's construction will show that the $x_1^{(1)}$'s have the largest values of all the x_i's in $G^{(1)}$.
5. Fix all $x_i^{(1)}$, and reformulate the original problem:

$$\min_x \left\{ \sum_{i \in G_j \setminus G^{(1)}} c_i x_i \,\middle|\, \prod_{i \in G_j \setminus G^{(1)}} R_i(x_i) \geq \frac{R_j^0}{R_j^{(1)}} \right\} ,$$

where $R_j^{(1)}$ has fixed value after the first step,

$$R_j^{(1)} = \prod_{i \in (G_j \cap G^{(1)})} R_i(x_i^{(1)}) .$$

6. The new problem is equivalent to the original one but has a reduced dimension. Start from step 1 again.

Remark If there is more than one root in the tree, each is optimized independently.

20.9 RELIABILITY INDEX OF TYPE *R*, SEVERAL RESTRICTIONS

There may be more than one restriction, as discussed in Section 20.2.4. For example, for equipment in an aircraft, missile or submarine, in addition to cost restrictions, there are restrictions on weight, volume, power requirements, and dimensions. This problem sometimes can be solved by reduction of the original problem to one dimension by introducing an "aggregated cost." In some circumstances the precise procedure in Section 20.9.1 will be better. Other circumstances might call for the heuristic procedures in 20.9.2.

20.9.1 Accurate Solution

For simplicity consider the two-dimensional case ($M = 2$). To formulate an aggregate cost, let us introduce nonnegative weighted coefficients a_1 and a_2, where $a_1 + a_2 = 1$. Next, choose an increment, $\Delta = 1/k$, where k is an integer and $k > 1$. Then determine all pairs of coefficients.

Let c_{ij} be the cost of type j (such as dollars or weight) of the ith unit of the system. Now compose for each ith unit of the system $k + 1$ different types of aggregated cost identified with superscripts of a:

$$
\begin{aligned}
&a_1^1 = 0\,, \quad a_2^1 = 1\,,\\
&a_1^2 = \frac{1}{k}\,, \quad a_2^2 = 1 - \frac{1}{k}\,,\\
&a_1^3 = \frac{2}{k}\,, \quad a_2^3 = 1 - \frac{2}{k}\,,\\
&\quad\vdots \qquad\qquad \vdots\\
&a_1^{k+1} = 1\,, a_2^{k+1} = 0\,.
\end{aligned}
$$

$$
\begin{aligned}
c_i^1 &= c_{1i}a_1^1 + c_{2i}a_2^1\,,\\
c_i^2 &= c_{1i}a_1^2 + c_{2i}a_2^2\,,\\
&\;\;\vdots\\
c_i^{k+1} &= c_{1i}a_1^{k+1} + c_{2i}a_2^{k+1}\,,
\end{aligned}
$$

for all i, $1 \le i \le n$. In general, there will be enough different possible variants of aggregated cost because for $\Delta = 1/k$ one has to find k^M values of c_i^l:

$$
c_i^l = \sum_{1 \le j \le M} c_{ij}a_j^l\,, \tag{20.38}
$$

where $\mathbf{a}^l$ is a vector, $\mathbf{a}^l = (a_1^l, a_2^l, \ldots, a_M^l)$, M is a dimension of the problem, all a_j^1 are nonnegative, and for all l,

$$
\sum_{1 \le i \le n} a_i^l = 1\,. \tag{20.39}
$$

Now we can solve each jth one-dimensional problem, $1 \le j \le k + 1$, using any of the methods described above. Each stepping process must be stopped as soon as one of the restrictions $C_1^0, C_2^0, \ldots, C_M^0$ has been violated. This will be the point of the termination of the solution for the given aggregated cost. Let us denote the solution for the lth aggregated cost as $\mathbf{X}^1 =$

$(x_1^1, \ldots, x_n^1)$. For each such solution we have to calculate the value of the reliability index of type $R: R(\mathbf{X}^1)$. Each of the obtained solutions satisfies the restrictions. The final problem is to find the solution with the highest reliability index,

$$\max_{1 \le l \le k^M} R(\mathbf{X}^1) . \tag{20.40}$$

EXAMPLE 20.7

Consider the system in Example 20.2 (Figure 20.1). The system consists of five subsystems. Costs of redundant units of the system are $c_{11} = 3$, $c_{12} = 2$, $c_{13} = 1$, $c_{14} = 5$, and $c_{15} = 5$ in some given measure of monetary cost. Additionally there are costs in some weight measure: $c_{21} = 1$, $c_{22} = 3$, $c_{23} = 5$, $c_{24} = 5$, and $c_{25} = 1$. Find the optimal allocation of redundant units for the following three cases:

1. The total monetary cost of the system must not exceed 55 units, and the total weight cost must not exceed 45 units.
2. The total monetary cost must not exceed 55 units, and the total weight cost must not exceed 40 units.
3. The total monetary cost must not exceed 55 units, and the total weight cost must not exceed 30 units.

Solution

For simplicity we use the steepest descent algorithm. We choose for the coefficients a_1 and a_2 the value $\Delta = 0.25$. The results of the calculation of the aggregated costs are in Table 20.19.

For $a_2^1 = 0$ the problem is equivalent to the one-dimensional problem solved in Example 20.2. We could use that table with $\gamma_i(x_i)$ values and the solution itself. For all values of a^j special tables have been compiled (see Tables 20.20, through 20.24). In these tables the step numbers of the

Table 20.19 Results, a monetary constraint and a weight constraint, Example 20.7

	Number of Units				
Aggregate Cost	1	2	3	4	5
1	3	2	1	1	5
2	2.5	2.25	2	2	4
3	2	2.5	3	3	3
4	2	2.75	4	4	2
5	1	3	5	5	1

Table 20.20 Values of $g^{(1)}$, aggregate cost, Example 20.7

x_1	$g_1(x_1)$	$g_2(x_2)$	$g_3(x_3)$	$g_4(x_4)$	$g_5(x_5)$
0	—	—	—	—	—
1	0.1959(3)	0.4796(1)	0.0953(7)	0.1000(6)	0.1833(4)
2	0.1014(5)	0.2015(2)	0.0090(23)	0.0050	0.0940(8)
3	0.0635(10)	0.0819(9)	0.0009	0.0002	0.0544(11)
4	0.0433(12)	0.0295(15)	~0	~0	0.0329(13)
5	0.0310(14)	0.0091(22)	—	—	0.0202(17)
6	0.0229(16)	0.0024	—	—	0.0124(20)
7	0.0172(18)	0.0006	—	—	0.0076
8	0.0132(19)	0.0001	—	—	0.0046
9	0.0102(21)	~0	—	—	0.0028
10	0.0079(24)	—	—	—	0.0016

Table 20.21 Values of $g^{(2)}$, aggregate cost, Example 20.7

$x_i^{(2)}$	$g_1^{(2)}(x_1)$	$g_2^{(2)}(x_2)$	$g_3^{(2)}(x_3)$	$g_4^{(2)}(x_4)$	$g_5^{(2)}(x_5)$
0	—	—	—	—	—
1	0.2351(2)	0.4230(1)	0.0474(12)	0.0500(11)	0.2290(3)
2	0.1288(5)	0.1790(4)	0.0045	0.0025	0.1175(6)
3	0.0762(7)	0.0727(8)	0.0005	0.0001	0.0679(9)
4	0.0527(10)	0.0262(16)	0	0	0.0412(13)
5	0.0372(14)	0.0081	—	—	0.0252(17)
6	0.0275(15)	0.0021	—	—	0.0155(20)
7	0.0206(18)	0.0005	—	—	0.0095(22)
8	0.0158(19)	0.0001	—	—	0.0058
9	0.0123(21)	0	—	—	0.0035
10	0.0095(23)	—	—	—	0.0020

Table 20.22 Values of $g^{(3)}$, aggregate cost, Example 20.7

$g_i^{(3)}$	$g_1^{(3)}(x_1)$	$g_2^{(3)}(x_2)$	$g_3^{(3)}(x_3)$	$g_4^{(3)}(x_4)$	$g_5^{(3)}(x_5)$
0	—	—	—	—	—
1	0.2940(4)	0.3818(1)	0.0318(16)	0.3333(2)	0.3060(3)
2	0.1520(7)	0.1622(5)	0.0032	0.0017	0.1568(6)
3	0.0953(8)	0.0655(10)	0.0003	0.0001	0.0910(9)
4	0.0650(11)	0.0236(18)	0	0	0.0548(12)
5	0.0465(13)	0.0073	—	—	0.0357(14)
6	0.0344(15)	0.0019	—	—	0.0207(19)
7	0.0258(17)	0.0005	—	—	0.0127(22)
8	0.0198(20)	0.0001	—	—	0.0077
9	0.0153(21)	0	—	—	0.0047
10	0.0118(23)	—	—	—	0.0027

Table 20.23 Values of $g^{(4)}$, aggregate cost, Example 20.7

$x_i^{(4)}$	$g_1^{(4)}(x_1)$	$g_2^{(4)}(x_2)$	$g_3^{(4)}(x_3)$	$g_4^{(4)}(x_4)$	$g_5^{(4)}(x_5)$
0	—	—	—	—	—
1	0.3918(2)	0.3490(3)	0.0240(19)	0.0250(18)	0.4580(1)
2	0.2028(5)	0.1475(6)	0.0022	0.0013	0.2350(4)
3	0.1270(8)	0.0595(12)	0.0002	0.0001	0.1358(7)
4	0.0866(9)	0.0214(20)	0	0.	0.0820(10)
5	0.0620(11)	0.0066	—	—	0.0508(13)
6	0.0458(14)	0.0017	—	—	0.0310(16)
7	0.0344(15)	0.0004	—	—	0.0190(22)
8	0.0264(17)	0.0001	—	—	0.0115
9	0.0204(21)	0	—	—	0.0070
10	0.0158(23)	—	—	—	0.0040

Table 20.24 Values of $g^{(5)}$, aggregate cost, Example 20.7

$x_i^{(5)}$	$g_1^{(5)}(x_1)$	$g_2^{(5)}(x_2)$	$g_3^{(5)}(x_3)$	$g_4^{(5)}(x_4)$	$g_5^{(5)}(x_5)$
0	—	—	—	—	—
1	0.5870(2)	0.3200(4)	0.0191	0.0200	0.9150(1)
2	0.3042(5)	0.1350(9)	0.0018	0.0010	0.4700(3)
3	0.1900(7)	0.0547(15)	0.0002	0	0.2720(6)
4	0.1300(10)	0.0197	0	—	0.1645(8)
5	0.0930(12)	0.0061	—	—	0.1010(11)
6	0.0687(13)	0.0016	—	—	0.0620(14)
7	0.0516(16)	0.0004	—	—	0.0380(18)
8	0.0396(17)	0	—	—	0.0230(21)
9	0.0306(19)	—	—	—	0.0140
10	0.0237(20)	—	—	—	0.0080

optimization process are in parentheses. Table 20.25 gives solutions for each value of a_2^j at different steps of the optimization process. With this table it is possible for any given a_2^j to find the stopping step in the optimization process—the step when, for the first time, at least one restriction has been violated. Knowing the number of this step, one can use the corresponding table ((20.20–20.24)) to find the optimal solution, namely the optimal number of redundant units for each case.

For problem 1, for example, we see that the system weight restriction is not critical because even for $a_2^j = 0$ the first violated restriction is C_1^0, the restriction for the resource that has been spent optimally. This violation occurs at the twelfth step for $a_2^1 = 0$. From Table 20.20 we find that by the eleventh step the allocation was $x_1 = 3$, $x_2 = 3$, $x_3 = 1$, $x_4 = 1$, and $x_5 = 3$. (See the step numbers in parentheses in each of the five columns.)

For problem 3 we see from Table 20.25 that the system monetary cost restriction is not critical because even for $a_2^5 = 1$ the first violated restriction is C_2^0 for the optimally spent type of resource. This violation occurs at the

Table 20.25 Monetary and weight costs, Example 20.7

Number of Steps	$a_2^{(1)}=0$ C_1	C_2	$a_2^{(2)}=0.25$ C_1	C_2	$a_2^{(3)}=0.5$ C_1	C_2	$a_2^{(4)}=0.75$ C_1	C_2	$a_2^{(5)}=1$ C_1	C_2
0	22	17	22	17	22	17	22	17	22	17
1	24	20	24	20	24	20	27	18	27	18
2	26	23	27	21	25	25	30	19	30	19
3	29	24	32	22	30	26	32	22	35	20
4	34	25	34	25	33	27	37	23	37	23
5	37	26	37	26	35	30	40	24	40	24
6	38	31	42	27	40	31	42	27	45	25
7	39	36	45	28	43	32	47	28	48	26
8	44	37	47	31	46	33	50	29	53	27
9	46	40	52	32	51	34	53	30	55	30
10	49	41	55	33	53	37	58	31	58	31
11	54	42	56	38	56	38	61	32	63	32
12	57	43	57	43	61	39	63	35	66	33
13	62	44	62	44	64	40	68	36	69	34
14	65	45	65	45	69	41	71	37	74	35
15	67	48	68	46	72	42	74	38	76	38
16	70	49	70	49	73	47	79	39	79	39
17	75	50	75	50	76	48	82	40	82	40
18	78	51	78	51	78	51	83	45	87	41
19	81	52	81	52	83	52	84	50	90	42
20	86	53	86	53	86	53	86	53	93	43

tenth step for $a_2^4 = 0.75$ and $a_2^5 = 1$. A simple analysis shows that the solution for $a_2^5 = 1$ is preferable. From Table 20.24 we find that up to the ninth step the allocation was $x_1 = 3$, $x_2 = 2$, $x_3 = 0$, $x_4 = 0$ and $x_5 = 4$. (See the step numbers in parentheses.)

Last, we consider the most difficult case, Problem 2. Now we have to find optimal solutions for all a_2^j and compare them. For this purpose we have compiled Table 20.26, where there are all the needed data: the value of a_2^j, the terminal step number, values of x_i's, and, finally, values of $L(\mathbf{X})$ for corresponding $\mathbf{X}$'s. (Note that $L(\mathbf{X})$ has been calculated with the use of

Table 20.26 Solution table for the most difficult case, Example 20.7

$a_2^{(j)}$	Stopping Step	x_1	x_2	x_3	x_4	x_5	L
0	9	2	3	1	1	2	1.509
0.25	10	4	3	0	0	3	2.112
0.50	10	3	3	0	1	3	1.142
0.75	9	4	2	0	0	3	1.275
1.00	9	3	2	0	0	4	1.241

Table 20.1.) We see that the best solution is for $a_2^2 = 0.25$, and it is $x_1 = 4$, $x_2 = 3$, $x_3 = 0$, $x_4 = 0$, and $x_5 = 3$.

The solution is approximate because of the roughness of the lattice of the values of a_1 and a_2. Of course one could try to find a more accurate solution using $a_2 = 0.2$ and $a_2 = 0.3$. However, the discrete nature of the problem can result in no change in the solution even though there is a change in some optimization parameters.

Remark For multidimensional restrictions optimal a_i's can be found by a special procedure. For example, one could choose the vector $a_j = (a_1^j, a_2^j, \ldots, a_M^j)$ using the steepest descent method.

20.9.2 Heuristic Solutions

For multiple restrictions heuristic procedures often give good solutions. These methods are discussed here.

1. *One-dimensional optimization with concurrent control of all restrictions.* First the optimal one-dimensional solution, $R(X_j^*)$, is found for each separate type of resource. But a new stopping rule is used. The process stops when any restriction is violated. Thus, the solution (X_j^*) differs from that obtained with only one restriction on the resource j. The approximate solution is

$$R = \max_{1 \le j \le M} R(\mathbf{X}_j^*)$$

Remark If the first violation with the jth resource occurs with its own constraint, the other constraints are inconsequential and the result is absolutely optimal.

2. "*Reflecting screen*" *method.* This method also starts with a one-dimensional optimization for each resource. Other resources may be spent nonoptimally. If there is a nontrivial restriction, some resource, k, will be spent first. We then exclude the solution steps that are worst in expenditure of resource k. This can be done with a steepest descent method in an inverse direction. If necessary, the process can be continued for additional resources.

3. *Choice of the most sensitive restriction.* We approximately evaluate the number of steps to the restriction for each type of resource. For this purpose we calculate in advance for each resource one of the following values:

$$D_j = \frac{1}{M} \sum_{1 \le i \le n} c_{ij} \tag{20.41}$$

or

$$D_j^* = (0.5)[\max_{1 \le i \le n} c_{ij} - \min_{1 \le i \le n} c_{ij}] \tag{20.42}$$

Then, using D_j or D_j^*, we can approximate the number of steps until the stop for resource j. For step s and resource j we can use:

$$K_j^s = \frac{1}{D}(C_j^0 - C_j^s) \tag{20.43}$$

where D is defined by (20.41) or (20.42).

Before making any step, we can compare all K_j^2 and choose the smallest one. The subscript will show the most critical type of resource, which is chosen for the optimization at the next step.

CHAPTER 21

OPTIMAL SUPPLY OF SPARE PARTS

21.1 SPARE PARTS FOR ONE PERIOD

A system often is supplied with spare parts from one supply facility that is close enough for the spare parts to be used for immediate repair if the system fails. The spare units themselves may be unrepairable. The stock becomes exhausted and has to be refilled from time to time. We will consider the optimal allocation of different types of spare units for a given period of time, t. The problem is very close to the optimal redundancy problem for standby redundancy without renewal, but is treated separately in this chapter because of its importance and its special characteristics.

21.1.1 A General Purpose Steepest Descent Procedure

For many practical purposes an exponential distribution of time to failure is an acceptable approximation even if some distributions are not purely exponential. In such a case, one may use a steepest decent method and standard Poisson distribution tables for optimization of stocks of spare parts. There are two classical optimization problems: (1) minimization of the total stock cost under the condition that the probability of the absence of a needed spare does not exceed a given value, Q_0, and (2) minimization of the probability of the absence of a needed spare part under the condition that the total stock cost does not exceed a given value, C_0. The following is a general four-step procedure for the optimization:

1. Determine the total failure rate of type i units:

$$\Lambda_i = m_i \lambda_i ,$$

where m_i is a number of units and λ_i is the failure rate of the unit of type i.

2. Determine the mean number of failures of type i units for a time t:

$$a_i = \Lambda_i t = m_i \lambda_i t .$$

3. For each type i unit, set up tables of values of $g_i^{(k)}$ for various $k = 1, 2, \ldots$, using

$$g_i^{(k)} = \frac{p_k(a_i)}{c_i[1 - P_k(a_i)]} ,$$

where

$$p_k(a_i) = \frac{a^k}{k!} e^{-a} ,$$

$$P_k(a_i) = \sum_{k \le j < \infty} \frac{a^j}{j!} e^{-a} .$$

Numerical values can be found in Poisson distribution tables (see Appendix E). For small a, one may use an approximation

$$p_k(a) \approx \frac{a^k}{k!} .$$

For $k > 10$ the values of $k!$ can be computed approximately with Stirling's formula (see Appendix D).

For $a > 20$, a Poisson d.f. can be approximated with a normal d.f. with mean a and standard deviation $\sigma = \sqrt{a}$. Then

$$P_k(a) \sim 1 - \Phi\left(\frac{k - a}{\sqrt{a}}\right) ,$$

where $\Phi(u)$ is a standard normal distribution (see Appendix E).

4. Follow a steepest descent search using calculated values of $g_i^{(k)}(a_i)$. For the first optimization problem, the search stops at a step k when the probability of a shortage is smaller than Q_0 for the first time. For the second problem, the stopping step k is reached when the total system cost is below C_0 for the first time.

A reader who wants a numerical example may refer back to Chapter 20.

21.1.2 Sparekit Computer Program

A technique often useful for solution of both optimization problems is available as the Sparekit Computer Program (see Appendix F). Each additional spare of each component is given a cost-effectiveness ranking determined by the reliability multiplication index of each cost unit spent on the spare:

$$E_{in} = \sqrt[c_i]{\frac{R_i(x_i + 1)}{R_i(x_i)}} ,$$

where E_{in} is the cost-effectiveness index of the nth spare of type i, and $R_i(x)$ is the probability of no shortage of spare units of type i. All E_{in} are placed in numerical order. The ranked list can then be used to find either the least expensive suite of spares that yields a desired reliability or the most effective suite of spares that costs less than the budgetary limit.

21.2 MULTISTAGE AND MULTILEVEL SUPPLY OF SPARE UNITS

The method described in this section contains a heuristic model of a multistage and multilevel hierarchical supply system (HSS) with several different spare units. This model is very useful for the analysis of territorially dispersed technical systems. Let us consider several different technical systems. Each of them has an individual warehouse of spare units for current repairs. This warehouse can be refilled under a schedule or by request. The request is based on the current state of the warehouse. This is an inventory control problem considered in Chapter 22. We will investigate the hierarchical system of warehouses with regular periodic refilling in this chapter.

Local warehouses are refilled from regional warehouses. Regional warehouses in turn are refilled from a central warehouse with its own refilling schedule. We investigate a hierarchical supply system (HSS) of three levels, which is typical in practice. All results can be applied with slight modification to a HSS with a different number of levels.

We will analyze the whole process of supply. If a system or warehouse at a lower level requests a spare unit and is rejected, we will consider this event as a failure of the supply system. It is clear that rejection of a spare request does not always lead to failure of a system, but this decomposition is a conceptual and computational convenience. For the HSS designer, the goal of the supply system operation is to keep all systems operable. The HSS designer usually does not know the technical object reliability indexes and requests. His goal is to create an HSS that can support the technical object in its given state so he needs to find his own goals and his own indexes of reliability of the HSS itself.

The goal of each warehouse is to supply the lower level. If a request to refill a lower-level warehouse meets a rejection, it will be considered a failure of the lower-level warehouse. The moment of failure is the beginning of an idle period for the warehouse.

We will not consider any procedure for optimal allocation of spare units. We will only construct the goal function. We will use the steepest descent method for finding the optimal solution. This method is convenient if one has analytical expressions of the goal function in closed form. We provide such expressions below.

We will consider an exponential distribution of TTF for all units. If there are nonexponential distributions, but the investigated HSS is a large-scale system with a large number of units of each type, the main results will be valid. The theoretical explanation of this fact lies in Grigelionis' theorem of superposition of random point processes. For better understanding of the suggested method, we first consider an artificial case when all stocks of the warehouses consist of only one type of spare unit.

22.2.1 One Type of Spare Unit

Let us consider the kth system in the jth zone of service. Later we will denote such a system with the subscript kj. This system consists of n_{ijk} identical units of the only type, i. The subscript i is included for convenience of presentation of a general case in the next section.

For each zone, j, a system subscript, k, changes from 1 to N_j; in other words, the jth regional warehouse serves N_j systems. There are N regions served by the central warehouse, $1 \le j \le N$. We introduce the following notation:

t_1 = refilling period for local warehouses.
t_2 = refilling period for regional warehouses.
t_3 = refilling period for central warehouses.
$b_1 = t_2/t_1$, $b_2 = t_3/t_1$.
λ_i = failure rate of a unit of the ith type.
Λ_i = a parameter $\Lambda_i = \lambda_i t_1$.
n_{ijk} = number of units of the ith type in the kth system of the jth serving zone, $1 \le k \le N_j$, $1 \le j \le N$.
n_{ij} = number of units of the ith type in all systems of the jth zone, $1 \le j \le N$,

$$n_{ij} = \sum_{1 \le k \le N_j} n_{ijk} .$$

λ_{ikj} = failure rate of units of the ith type of the kjth system (kth system in zone j), $1 \le k \le N_j$, $1 \le j \le N$,

$$\lambda_{ikj} = \lambda_i n_{ikj} .$$

Λ_{ikj} = a parameter, $1 \le k \le N_j$, $1 \le j \le N$,

$$\Lambda_{ikj} = \lambda_{ikj} t_1 .$$

λ_{ij} = total failure rate of units of the ith type of all systems served by the jth warehouse,

$$\lambda_{ij} = \sum_{1 \le k \le N_j} \lambda_i n_{ikj} = \sum_{1 \le k \le N_j} \lambda_{ikj} = \lambda_i n_{ij} .$$

Λ_{ij} = a parameter, $1 \le j \le N$,

$$\Lambda_{ij} = \lambda_{ij} t_2 .$$

λ_i^* = failure rate of all units of the ith type of all systems served by the HSS,

$$\lambda_i^* = \sum_{1 \le j \le N} \lambda_{ij} .$$

Λ_i^* is a parameter, $\Lambda_i^* = \lambda_i^*(t_3 - t_2)$.

x_{ikj} = number of spare units of the ith type in the individual warehouse of the kth system of the jth serving zone, $1 \le k \le N_j$, $1 \le j \le N$.

x_{ij} = number of spare units of the ith type in the jth regional warehouse, $1 \le j \le N$,

$$x_{ij} = \sum_{1 \le k \le N_j} x_{ikj} .$$

x_i = number of spare units of the ith type in the central warehouse,

$$x_i = \sum_{1 \le j \le N} x_{ij} .$$

n_i = number of units of the ith type in all systems served by the HSS,

$$n_i = \sum_{1 \le j \le N} \sum_{1 \le k \le N_j} n_{ikj} = \sum_{1 \le j \le N} n_{ij} .$$

$\pi_{s(\Lambda)}$ = a Poisson probability,

$$\pi_s(\Lambda) = \frac{\Lambda^s}{s!} \exp^{-\Lambda} .$$

$\Pi_s(\Lambda)$ = a cumulative Poisson probability,

$$\Pi_s(\Lambda) = \sum_{0 \le i \le s} \pi_q(\Lambda) .$$

N_j = number of different systems served by the jth regional warehouse;
N = number of regional warehouses in the HSS.

If there are x_{ikj} spare units in an individual warehouse, kj, the probability that the system has not failed during time t_1 equals

$$p_{ikj}(t_1) = \sum_{0 \le z \le x_{ikj}} \pi_z(\Lambda_{ikj}) = \Pi_{x_{ikj}}(\Lambda_{ikj}) . \tag{21.1}$$

If there is no failure for the current period, t_1, then a random number of failures, ν_{ikj}, has occurred, and the warehouse of the kjth system has to be refilled on this value. This random value has a Poisson distribution truncated from the right. But if $p_{ikj}(t_1)$ is close to 1, this truncation can be neglected. For an approximate solution of the problem one may disregard truncation and use a common Poisson distribution.

Let the jth regional warehouse serve N_j different independent systems. Consider failure-free operation of this warehouse with N_j served systems during period t_2. Failure of the warehouse means a shortage in spare units for at least one system in at least one interval, t_1.

The probability of successful operation of all systems during time t_1 can be written as

$$p_{ij}(t_1) = \prod_{1 \le k \le N_j} p_{ikj}(t_1) , \tag{21.2}$$

where $p_{ikj}(t_1)$ is defined in (21.1). The probability of successful operation of all systems during time t_2 can be written as

$$p_{ij}(t_2) = [p_{ij}(t_1)]^{b_1} . \tag{21.3}$$

The jth regional warehouse itself should refill all its own warehouses. Let there be x_{ij} spare units at this warehouse for the period t_2. Under the assumptions we can consider the process to be a Poisson failure process for all kjth systems. The failure rate of all of these systems equals λ_{ij}. The average number of requested spare units from the jth regional warehouse equals Λ_{ij}. Successful operation of the regional warehouse means that it can refill all individual warehouses b_1 times.

Remark Strictly speaking, if in an isolated interval of time, t_3, of an HSS function as a whole, there are $b_1 b_2$ intervals of t_1, but there is no need to supply each warehouse with spare units for the last period t_1. We assume that $b_1 b_2$ is large enough to neglect the fact that the last interval, t_1, is different from the rest. If this fact is not taken into account, the final equations will be a little less cumbersome.

Let us additionally assume that a successful refilling of the kjth individual warehouse occurs if it is up to the initial size x_{ikj}; otherwise, it will be considered a failure. This assumption is not strictly true, but any error is negligible if $b_1 \gg 1$. Thus the probability of successful operation during a period, t_2, for the kjth regional warehouse with only type i spare units can be written

$$r_{ij}(t_2) = \Pi_{x_{ij}}(\Lambda_{ij}) , \tag{21.4}$$

where x_{ij} is the number of spare units of the ith type in the jth regional warehouse. The complete probability that the jth regional system as a whole will work successfully during time t_2 is

$$R_{ij}(t_2) = r_{ij}(t_2) p_{ij}(t_2) . \tag{21.5}$$

Finally, during a period, t_3, the central warehouse should be able to refill all N regional warehouses $b_2 - 1$ times with its stock of x_{ij} spare units of type i. This occurs with probability

$$r_i(t_3) = \Pi_{x_i}(\hat{\Lambda}_i) . \tag{21.6}$$

Definition of Performance Effectiveness Indexes

Let us consider four indexes of successful operation of the HSS:

1. R_1 is the probability of successful operation of the HSS as a whole during time t_3.
2. R_2 is an analogue of the availability coefficient of the HSS as a whole.
3. R_3 is the probability of successful operation of the ith system served by the HSS during time t_3.
4. R_4 is an analogue of an availability coefficient for the ith system served by the HSS.

R_1 Derivation

The highest level of the HSS—the central warehouse—has x_i spare units and serves N regional warehouses. Let the central warehouse have to work without refilling for period t_3. It should refill all regional warehouses $b_2 - 1$ times. No regional warehouse should have a shortage during b_2 of its own periods, and no individual warehouse should fall short during $b_1 b_2$ periods of performance. The final expression will be

$$P(t_3) = r_i(t_3) \prod_{1 \le j \le N} [R_{ij}(t_2)]^{b_2} . \tag{21.7}$$

The goal function gives a lower limit of the probability of system successful operation because we regarded some potential failure states of the HSS to be equivalent to system failures.

R_2 Derivation

We consider warehouse exhaustion and the start of an idle period for a lower-level warehouse or the system itself. Assume a Poisson process of unit failures, and consider the fact that a conditional joint distribution of n arrival moments during a fixed interval is an n-dimensional uniform distribution. If there have been z failures, the failure moments divide the interval of observation t into $z+1$ subintervals. All of these subintervals have the same distribution, so the conditional mean of each is $t/(z+1)$.

Now we can write an expression for the mean idle time, $a_{ikj}(t_1 t, x_{ikj})$, for the kjth served system:

$$a_{ikj}(t_1, x_{ikj}) = \sum_{z>x_i} P(\nu_{ikj} = z \mid z > x_{ikj}) t_1 \frac{z - x_{ikj}}{z+1}, \tag{21.8}$$

where $P(\nu_{ikj} = z \mid z > x_{ikj})$ is the conditional probability of z events of a Poisson process in the interval t_1 under the condition that $z > x_{ikj}$; in other words, the distribution is truncated from the right.

Again for a highly reliable supply system, we assume that we can neglect the truncation. As a result we have the simple expression

$$a_{ikj}(t_1, x_{ikj}) \approx \frac{t_1}{x_{ikj}+2} \pi_{x_{ikj}+1}(\Lambda_{ikj}). \tag{21.9}$$

Expression (21.9) gives the lower bound for (21.8) because the portion of idle time will be larger for a larger number of failures. Of course the simplicity of (21.9) in comparison with (21.8) is not significant if one uses a computer. Similar expressions can be written for a regional warehouse and for the central warehouse of the HSS. For the jth regional warehouse we have for each period t_2 an approximate expression

$$a_{ij}(t_2, x_{ij}) \approx \frac{t_2}{x_{ij}+2} \pi_{x_{ij}+1}(\Lambda_{ij}). \tag{21.10}$$

Remark The derivation above is for a highly reliable HSS. This means that if there is a minor shortage of spare units in the jth regional warehouse, only one individual warehouse will not be refilled. The possibility of shortages at more than one warehouse is considered negligible.

For the central warehouse the expression has a similar form:

$$a_i(t_3, x_i) = \frac{t_3}{x_i+2} \pi_{x_i+1}(\Lambda_i) \tag{21.11}$$

The preceding remark is applicable here as well.

Now let us consider an availability coefficient. Strictly speaking, the considered index is not a true availability coefficient because failure of a warehouse happens only at the end of a refilling period. The index is an availability coefficient in that it is the average portion of the operating time without a shortage of spares.

A failure of a given kjth system can occur if one of the following three events occurs:

1. A failure can occur if the number of failed units of the kjth system is more than x_{ikj} in an interval, t_1. In this case this system has the average idle time $a_{ikj}(t_1, x_{ikj})$ in accordance with (21.9). If a failure occurs, it has to occur before a moment kt_1, $k = 1, 2, \ldots$. The failed state persists until the warehouse is refilled. The expected value of the idle time, $a_{ikj}(t_1, x_{ikj})$, is defined by (21.9). The probability of such a failure is $q_{ikj}(t_1) = 1 - p_{ikj}(t_1)$, where $p_{ikj}(t_1)$ is defined by (21.1).
2. A failure can occur if the total number of spare units requested for all N_j systems from the jth regional warehouse during the interval t_2 exceeds x_{ij}. In this case the kjth system has an expected idle time of $a_{ij}(t_2, x_{ij})$ in correspondence with (21.10). A failure can occur just before a moment kt_2, $k = 1, 2, \ldots$. The probability of such an event equals $q_{ij}(t_2) = 1 - r_{ij}(t_2)$, where $r_{ij}(t_2)$ can be calculated with (21.4).
3. A failure can occur if the total number of spare units requested by all N regional warehouses during the interval t_3 exceeds x_i. In this case the central warehouse has an expected idle time of $a_i(t_3, x_i)$ in accordance with (21.11). A failure can occur at the end of interval t_3. The expected duration of the downtime is $a_i(t_3, x_i)$, as determined by (21.11). The probability is $q_i(t_3) = 1 - r_i(t_3)$, where $r_i(t_3)$ can be calculated with (21.6).

A typical HSS may consist of three discrete geometrical random sequences of events with associated parameters. For example, the top sequence may be factory-to-regional depot/warehouses, the second level of sequences may be from warehouses to local supply facilities, and the third level may be from facilities to individual systems. If the HSS is very reliable, each process can be presented as an alternating stochastic process consisting of long intervals of successful operation and short idle intervals.

Disregarding the lattice structure of these processes, one can use Grigelionis' Theorem to approximate the supply process in the HSS as a whole. Let us denote the HSS failure rate as $\tilde{\Lambda}$. $\tilde{\Lambda}$ can be presented as a sum of corresponding rates:

$$\tilde{\Lambda} = \tilde{\Lambda}_1 + \tilde{\Lambda}_2 + \tilde{\Lambda}_3 , \tag{21.12}$$

where

$$\tilde{\Lambda}_1 = \frac{1}{t_1} \sum_{1 \le j \le N} \sum_{1 \le k \le N_j} q_{ikj}(t_1) .$$

The average idle time of the HSS can be calculated as a weighted value:

$$\tilde{\Lambda}_2 = \frac{1}{t_2} \sum_{1 \le j \le N} q_{ij}(t_2) ,$$

$$\tilde{\Lambda}_3 = \frac{1}{t_3} q_i(t_3) ,$$

$$\tilde{A} = \tilde{\Lambda}_1 \tilde{A}_1 + \tilde{\Lambda}_2 \tilde{A}_2 + \tilde{\Lambda}_3 a_i(t_3, x_i) ,$$

where $\tilde{A}^1$ and $\tilde{A}^2$ also are weighted values,

$$\tilde{A}_1 = \tilde{Q}_1^{-1} \sum_{1 \le j \le N} \sum_{1 \le j \le N_j} q_{ikj}(t_1) a_{ikj}(t_1, x_{ikj}) ,$$

$$\tilde{A}_2 = \tilde{Q}_2^{-1} \sum_{1 \le j \le N} q_{ij}(t_2) a_{ij}(t_2, x_{ij}) ,$$

and a_i is defined by (21.11). $\tilde{Q}_1$ and $\tilde{Q}_2$ are normalization factors:

$$\tilde{Q}_1 = \sum_{1 \le j \le N} \sum_{1 \le k \le N_j} q_{ikj}(t_1) , \tag{21.13}$$

$$\tilde{Q}_2 = \sum_{1 \le j \le N} q_{ij}(t_2) . \tag{21.14}$$

Now the final result can be easily obtained for the chosen reliability index R for the HSS as a whole:

$$R = \frac{1}{1 + \tilde{A}\tilde{\Lambda}} . \tag{21.15}$$

R_3 Derivation

Let us consider a case when no system has priority. All shortages of spare units influence the system with equal probability. From obvious arguments we obtain for the kjth system:

$$R_{3(i)} = [p_{ikj}(t_1, x_{ikj})]^{b_1 b_2} B^{b_2} B_{j(i)}^{b_2 - 1} B_{(i)} , \tag{21.16}$$

where $B_{j(i)}$ is the probability that the fixed kjth system has not failed during period t_2 because of a shortage of the ith spare unit in the jth regional warehouse, and $B_{(i)}$ is the probability that it has not failed during period t_3 because of a similar shortage at the central warehouse.

By assumption the kjth system has no priority over others. So if a failure of the jth warehouse has occurred, the probability is $1/N_j$ that it is a failure of the kjth system. A failure of the jth warehouse does not necessarily mean a failure of this system. For the kjth system, the influence of the jth warehouse can be evaluated with a probability:

$$B_{j(i)} = r_{ij}(t_2, x_{ij}) + \left(1 - \frac{1}{N_j}\right) q_{ij}(t_2, x_{ij})$$

$$= 1 - \frac{1}{N_j} q_{ij}(t_2, x_{ij}) . \tag{21.17}$$

Similar arguments permit us to write the same "coefficient of influence" of the central warehouse on the kjth system:

$$B_{(i)} = 1 - \frac{1}{N_j N} q_i(t_3, x_i) . \tag{21.18}$$

***R_4* Derivation**

With a highly reliable HSS, idle time at any warehouse will usually influence only one system at the lower level. In this situation, for a type i unit in a fixed system kj we can write

$$R_{4kj(i)} = [t_1 - q_{ikj}(t_1, x_{ikj}) a_{ikj}(t_1, x_{ikj})] t_1^{-1} \tilde{B}_{j(i)} \tilde{B}_{(i)} , \tag{21.19}$$

where $\tilde{B}_{j(i)}$ is the component of the availability coefficient of the fixed kth system determined by a shortage at the jth regional warehouse, and $\tilde{B}_{(i)}$ is a similar component determined by a shortage at the central warehouse,

$$\tilde{B}_{j(i)} = [t_2 - q_{ij}(t_2) a_{ij}(t_2, x_{ij})] t_2^{-1} \tag{21.20}$$

and

$$\tilde{B}_{(i)} = [t_3 - q_i(t_3) a_i(t_3, x_i)] t_3^{-1} . \tag{21.21}$$

To calculate the index for an "average" availability coefficient rather than that of a particular system, we write

$$R_{4(i)} = \frac{\sum_{1 \le j \le N} \sum_{1 \le j \le N_j} R_{kj(i)}}{\sum_{1 \le j \le N} \sum_{1 \le k \le N_j} n_{kj}} . \tag{21.22}$$

21.2.2 Several Different Types of Units

In the preceding discussion we considered the simple case of a ones-unit system. Now we turn to the case where each technical system includes several different types of units. The results of this section are based on those of the previous section. The following notation supplements that of Section 21.2.1:

m = number of different units.

$\boldsymbol{\lambda}_{kj}$ = a vector of the failure flow parameters of units of different kinds in the kjth system (for period t_1),

$$\mathbf{\Lambda}_{kj} = (\Lambda_{ikj}; i = 1, \ldots, m) = (\Lambda_{1kj}, \ldots, \Lambda_{mkj}) .$$

$\mathbf{\Lambda}_j$ = a vector of the failure flow parameters of units of different kinds in the jth region (for period t_2),

$$\mathbf{\Lambda}_j = (\Lambda_{ij}; i = 1, \ldots, m) = (\Lambda_{1j}, \ldots, \Lambda_{mj}) .$$

$\mathbf{\Lambda}$ = a vector of the failure flow parameters of units of different kinds in the whole HSS (for period t_3),

$$\mathbf{\Lambda} = (\Lambda_i; i = 1, \ldots, m) = (\Lambda_1, \ldots, \Lambda_m) .$$

X_{kj} = a vector of the numbers of spare units of different kinds at the kjth system warehouse,

$$\mathbf{X}_{kj} = (x_{ikj}; i = 1, \ldots, m) = (x_{1kj}, \ldots, x_{mkj}) .$$

X_j = a vector of the numbers of spare units of different kinds at the jth regional warehouse,

$$\mathbf{X}_j = \sum_{1 \le k \le N_j} \mathbf{X}_{jk} = \Big(\sum_{1 \le k \le N_j} x_{1kj}, \ldots, \sum_{1 \le k \le N_j} x_{mkj} \Big) = (x_{1j}, \ldots, x_{mj}) .$$

X = a vector of the numbers of spare units of different kinds at the central warehouse,

$$\mathbf{X} = \sum_{1 \le j \le N} \mathbf{X}_j = \Big(\sum_{1 \le j \le N} \sum_{1 \le k \le N_j} x_{1kj}, \ldots, \sum_{1 \le j \le N} \sum_{1 \le k \le N_j} x_{mkj} \Big)$$
$$= (x_i; 1 \le i \le m) = (x_1, \ldots, x_m) .$$

Now we will write the expressions for m different types of units following the same order as before. We will omit all explanations similar to those given above.

The probability of successful operation for the kjth system during period t_2 can be written

$$p_{kj}(t_1) = \prod_{1 \le i \le m} p_{ikj}(t_1) = \prod_{1 \le i \le m} \Pi_{x_{ikj}}(\Lambda_{ikj}) . \tag{21.23}$$

The probability of successful operation for all systems served by the jth regional warehouse during interval t_2 can be written

$$p_j(t_2) = \prod_{1 \le k \le N_j} [p_{kj}(t_1)]^{b_1} = \prod_{1 \le k \le N_j} \prod_{1 \le i \le m} [\Pi_{x_{ikj}}(\Lambda_{ikj})]^{b_1} . \tag{21.24}$$

The probability of successful operation for the jth regional warehouse during a period t_2 can be written

$$\tilde{r}_j(t_2) = \prod_{1 \le i \le m} \Pi_{x_{ij}}(\Lambda_{ij}) .$$

The complete probability that the regional system as a whole will work successfully is

$$R_j(t_2) = \tilde{r}_j(t_2) p_j(t_2) . \tag{21.25}$$

Finally, the central warehouse should be able to refill all N regional warehouses $b_2 - 1$ times with its stock of X_j spare units of different types. This occurs with probability

$$r_i(t_3) = \prod_{1 \le i \le m} \Pi_{x_i}(\Lambda_i^*) . \tag{21.26}$$

Now let us consider the same four indexes of successful operation of the HSS that were considered in Section 21.2.1: R_1, R_2, R_3, and R_4.

R_1 Derivation

The probability that there is no shortage at the central warehouse is

$$r(t_3) = \prod_{1 \le i \le m} r_i(t_3) = \prod_{1 \le i \le m} \Pi_{x_i}(\Lambda_i^*) \tag{21.27}$$

For N regions (all warehouses including all systems) during the same period t_3 the probability of success is

$$P(t_3) = \prod_{1 \le j \le N} [R_j(t_2)]^{b_2} , \tag{21.28}$$

where $R_j(t_2)$ is determined by (21.25). The probability of success of the three-level HSS as a whole is

$$R(t_3) = r(t_3) P(t_3) . \tag{21.29}$$

The goal function, (21.29), gives a low estimate of the probability of successful system operation because some of the beneficial states of the HSS were considered as harmful.

R_2 Derivation

An approximate expression for the mean idle time for the jth system is $a_{kj}(t_1, X_{kj})$. Difficulties arise with several types of units.

We consider the case in detail. First of all, we assume that only one type of unit is apt to be exhausted at the end of interval t_1 if there is a very reliable supply. The average idle time for a system then can be found as a "weighted" value of the kind

$$a_{kj}(t_1, X_{kj}) \approx Q_{kj}^{-1} \sum_{1 \le i \le = m} q_{ikj}(t_1, x_{ikj}) \frac{t_1}{X_{ikj} + 2}, \qquad (21.30)$$

where

$$Q_{kj} = 1 - \prod_{1 \le i \le = m} [1 - q_{ikj}(t_1, x_{ikj})] \approx \sum_{1 \le i \le m} q_{ikj}(t_1, x_{ikj}), \qquad (21.31)$$

$$q_{ikj}(t_1, x_{ikj}) = 1 - \Pi_{x_{ikj}}(\Lambda_{ikj}) \approx \pi_{x_{ikj}+1}(\Lambda_{ikj}).$$

Similar expressions can be written for a regional warehouse and for the central warehouse of the HSS. For the jth regional warehouse, we have for each period t_2 an approximate expression

$$a_j(t_2, \mathbf{X}_j) \approx Q_j^{-1} \sum_{1 \le i \le M} q_{ij}(t_2, x_{ij}) \frac{t_2}{x_{ij} + 2}, \qquad (21.32)$$

where

$$Q_j = 1 - \prod_{1 \le i \le m} [1 - q_{ij}(t_2, x_{ij})] \approx \sum_{1 \le i \le m} q_{ij}(t_2, x_{ij}), \qquad (21.33)$$

and

$$q_{ij}(t_2, x_{ij}) = 1 - \Pi_{x_{ij}}(\Lambda_{ij}) \approx \pi_{x_{ij}+1}(\Lambda_{ij}).$$

For the central warehouse the comparable expression is

$$a(t_3, \mathbf{X}) = Q^{-1} \sum_{1 \le i \le m} q_i(t_3, x_i) \frac{t_3}{x_i + 2}, \qquad (21.34)$$

where

$$Q = 1 - \prod_{1 \le i \le m} [1 - q_i(t_3, x_i)] \approx \sum_{1 \le i \le m} q_i(t_3, x_i), \qquad (21.35)$$

and

$$q_j(t_3, x_i) = \pi_{x_i+1}(\Lambda_i^*).$$

Remark An insignificant shortage of spare units in the central warehouse in a highly reliable HSS is very possible. There may be only one regional warehouse that will not be refilled. The possibility of a shortage with $x > \min x_{ji}$ and more than one regional warehouse not refilled is again considered practically negligible.

Consider the availability coefficient for the kjth system served by the jth regional warehouse. The system can become unavailable if one of the three following events has occurred:

1. A kjth system will become unavailable if failures of an ith type of unit exceed x_{ikj} during an interval t_1. In this case the system has an expected idle time of $a_{kj}(t_1, X_{kj})$ in accordance with (21.30). If a failure occurs, it has to appear before the discrete moments kt_1, $k = 1, 2, \ldots$, and will endure until the warehouse is refilled. A failure of any unit occurs with probability Q_{kj} calculated by (21.31).
2. A system will become unavailable if the total number of spare units of at least one type requested to refill all N_j systems from the jth regional warehouse during the interval t_2 exceeds x_{ij}. In this case the jth warehouse has an expected idle time of $a_j(t_2, X_j)$ in accordance with (21.32). A failure can occur just before the moments kt_2, $k = 1, 2, \ldots$, with probability Q_j calculated using (21.33).
3. A system can become unavailable if the total number of spare units of at least one type requested by all N regional warehouses during the interval t_3 exceeds x_i. In this case the central warehouse has an expected idle time of $a(t_3, X)$. A failure can only occur at the end of interval t_3, and its probability, Q, can be calculated with (21.35).

Thus we have three discrete geometrical random sequences of events with the parameters given above. Disregarding the lattice nature of these processes, we can use Grigelionis' Theorem to compile the resulting process of supply failures for the whole HSS. Let us denote the HSS failure rate with $\tilde{\Lambda}$. $\tilde{\Lambda}$ can be presented as a sum of corresponding rates:

$$\tilde{\Lambda} = \tilde{\Lambda}_1 + \tilde{\Lambda}_2 + \tilde{\Lambda}_3 , \tag{21.36}$$

where

$$\tilde{\Lambda}_1 = t_1^{-1} \sum_{1 \le j \le N} \sum_{1 \le k \le N_j} Q_{kj} , \tag{21.37}$$

$$\tilde{\Lambda}_2 = t_2^{-1} \sum_{1 \le j \le N} Q_j , \tag{21.38}$$

$$\tilde{\Lambda}_3 = t_3^{-1} Q . \tag{21.39}$$

The expected idle time of the HSS can be calculated as a "weighted" value:

$$\tilde{A} = \tilde{\Lambda}_1 a_{kj}(t_1, X_{kj}) + \tilde{\Lambda}_2 a_j(t_2, X_j) + \tilde{\Lambda}_3 a(t_3, X) . \tag{21.40}$$

Now we can easily obtain the final result for the chosen reliability index, R, for the HSS as a whole:

$$R = \frac{1}{1 + \tilde{A}\tilde{\Lambda}} . \tag{21.41}$$

R_3 Derivation

We will consider the case where no system has service priority, so all systems have an equal probability of being affected by a shortage of spare units. From obvious arguments we obtain for the kjth system,

$$R_3 = [p_{kj}(t_1, \mathbf{X}_{kj})]^{b_1 b_2} B_j^{b_2} B \,, \tag{21.42}$$

where B_j is the probability that the kjth system has not failed during period t_2 because of a shortage in the jth regional warehouse, and where B is the probability that it has not failed during period t_3 because of a shortage in the central warehouse.

Using arguments similar to those used for (21.17) and (21.18), we write immediately

$$B_j = 1 - \frac{1}{N_j} q_j(t_2, \mathbf{X}_j) \tag{21.43}$$

and

$$B = 1 - \frac{1}{N_j N} q(t_3, \mathbf{X}) \,. \tag{21.44}$$

R_4 Derivation

In a very reliable HSS, idle time at any warehouse is apt to influence only one system at the lower level. Under this assumption for any kjth system chosen at random we write

$$R_4 = \frac{t_1}{t_1 - Q_{kj} a_{kj}(t_1, X_{kj})} \tilde{B}_1 \tilde{B} = 1 - Q_{kj} a_{kj}(t_1, X_{kj}) t_1^{-1} \tilde{B}_1 \tilde{B} \,, \tag{21.45}$$

where $a_{kj}(t_1, X_{kj})$ is determined with (21.30), $\tilde{B}_1$ is the component of the availability coefficient of the kjth system determined by a shortage in the jth regional warehouse, and $\tilde{B}$ is a similar component determined by the shortage in the central warehouse.

Considering the absence of priorities, and taking into account previously obtained results (21.43) and (21.44), we write

$$\tilde{B}_1 \approx 1 - \frac{Q_j a_j(t_2, X_j)}{N_j t_2} \tag{21.46}$$

and

$$\tilde{B} = 1 - \frac{Q a(t_3, X)}{N_j N t_3} \,. \tag{21.47}$$

Remark For this complex engineering problem we have developed a simple mathematical model that yields reliability indexes for an HSS. This permits the use of a steepest descent method for optimization of the HSS stocks with a given structure and rule of warehouses refilling. The expressions are accurate, however, only for very reliable supply systems.

Historically even attempted descriptions of this problem with mathematical rigor have been unsuccessful. Monte Carlo simulation can help in this case as well as in any other case where there is only one suitable algorithmic description of the system. But stochastic simulation does not by itself yield reasonable procedures for optimization.

21.3 SUPPLY OF RENEWABLE MODULES AND UNREPAIRABLE UNITS

21.3.1 Description of a Maintenance Support System

A technical system is supplied with standby modules of different types that are ready to replace failed modules. The failed modules are sent to a repair workshop. The repair workshop needs repair spare units of different types. The units themselves are unrepairable and form the workshop's stock. Module repair at the workshop can continue until the stock of spare units is exhausted. The stock is intended for a period of time t.

We assume complete repair; a repaired module is identical to a new one. We also assume that all types of modules can be repaired independently. Thus the maintenance support system (MSS) consists of standby modules, the warehouse, and a stock of spare units needed for repair of failed modules.

Replacement of a failed module with a standby module takes a finite time. Usually replacement times are substantially shorter than repair times and thus decrease idle time. A standby redundancy model therefore can be used for a rough mathematical model of module substitution. A served system becomes unavailable if it fails when there are no standby modules of the type needed because all have been sent to the workshop and have not yet been returned.

Remark We do not take into account possible internal system redundancy. If module repair requires a type of spare unit that has been exhausted by previous repairs, a failure will occur because of workshop stock exhaustion. Let us assume that all time-to-failure distributions are exponential. The means of the repair time distributions are very small in comparison to the MTTF. Also the numbers of standby modules and spare units are such that the probability of unavailability of a system is very small. The results presented here are based on a heuristic model developed from the Renyi theorem.

21.3.2 Notation

m = number of different types of modules.

M_j = number of operating modules of the jth type.

$\mathbf{M}$ = a vector of a total set of different modules, $\mathbf{M} = (M_1, \ldots, M_m)$.

y_i = number of standby modules of the jth type.

$\mathbf{Y}$ = a vector of a set of different standby modules, $\mathbf{Y} = (y_1, \ldots, y_m)$.

Λ_j = failure rate of a module of the jth type.

a_j = rate of replacement of the jth module (a_j equals the inverse value of the mean time of replacement of the module, A_j).

b_j = repair rate of the jth module at the workshop (b_j equals the inverse value of the mean time of repair of the module, B_j). Repair time includes the time for delivery to the workshop, waiting in line if there is a line, repair itself, and return to the system warehouse.

n = number of different types of units.

b_j = repair rate of module j at the workshop (b_j equals the inverse of the mean time of an actual repair of the module, B_j).

M_i = number of operating units of the ith type in the system.

$\mathbf{N}$ = a vector,

$$\mathbf{N} = (N_1, \ldots, N_n) ,$$

that represents the complete set of units in all modules of the system.

x_i = number of spare units of type i in the warehouse for a total period t.

$\mathbf{X}$ = a vector,

$$\mathbf{X} = (x_1, \ldots, x_n) ,$$

that represents the set of spare units of all types in the warehouse.

λ_i = failure rate of a type i unit.

$\Pi_y(\lambda)$ = a Poisson d.f.,

$$\Pi_y(\Lambda) = \sum_{1 \le i \le y} \pi_i(\Lambda) ,$$

where

$$\pi_i(\Lambda) = \frac{\Lambda^i}{i!} e^{-\Lambda} .$$

21.3.3 Probability of Successful Service

We consider the probability of successful supply but not of successful operation of the system. For simplicity we will assume that different types of failed modules are repaired independently. For a highly reliable system then, the errors are negligible.

The investigated probability, R, can be approximated with a product of probabilities:

$$R \approx R_1(\mathbf{M}, \mathbf{Y}; t) R_2(\mathbf{N}, \mathbf{X}; t) , \tag{21.48}$$

where $R_1(\mathbf{M}, \mathbf{Y}; t)$ is the probability of successful operation of the system during time t if it consists of $\mathbf{M}$ operating modules with $\mathbf{Y}$ standby modules and $R_2(\mathbf{N}, \mathbf{X}; t)$ is the probability that there will be no deficit of spare units of any type at the workshop that might lead to an idle system during time t if there are $\mathbf{N}$ operating units and $\mathbf{X}$ spare ones.

Remark Equation (21.48) is not strictly correct because all multipliers are dependent. The more modules fail, the more spare units are spent. For very reliable systems, however, this dependence can be neglected.

Consider the first term of (21.48). The assumption of independent repair of each type of module permits us to write

$$R_1(\mathbf{M}, \mathbf{Y}; t) = \prod_{1 \le j \le m} R_1(M_j, y_j; t) , \tag{21.49}$$

where $R_1(M_j, y_j; t)$ is the corresponding probability for the jth type of module. Each of these probabilities for $j = 1, \ldots, m$ can be found by standard methodology, but we will use an approximation based on the Renyi Theorem of thinning of a renewal point process.

Each failure of a module of the jth type can cause system failure if during the repair time y_j modules of the same type also fail. The probability of this is

$$Q_1(M_j, y_j; B_j) = 1 - \Pi_{y_j}(B_j M_j \Lambda_j) \approx \pi_{y_j+1}(B_j M_j \Lambda_j) . \tag{21.50}$$

Here $M_j \Lambda_j$ is the failure rate for modules of the jth type, $B_j M_j \Lambda$ is the parameter of the corresponding Poisson distribution, and $\pi(B_j M_j \Lambda_j)$ is the probability that y_j failures of type j modules occur during the type j module repair time.

Remark The repair time is a random variable. The precise expression for (21.50) is

$$Q_1(M_j, y_j; t) = \int_0^\infty p_{y_j+1}(\pi M_j \Lambda_j)\, dF_{B_j}(\tau) , \tag{21.51}$$

where $F_B j(t)$ is the distribution of the repair time. However, if a system is highly reliable, for a rough calculation we may regard B_j as the delivery time for a spare module from the workshop.

Using the Renyi Theorem, we write the parameter of the system failure flow caused by the jth module as

$$\tilde{\Lambda}_j = \pi_{y_j}(B_j M_j \Lambda_j) M_j \Lambda_j . \tag{21.52}$$

Thus, the approximate expression for probability $R_1(\mathbf{M}_j, \mathbf{y}_j; t)$ is

$$R_1(M_j, y_j; t) = \exp^{-\tilde{\Lambda}_j t} . \tag{21.53}$$

For the whole system, (21.49) now can be written

$$R_1(M, y; t) = \prod_{1 \le j \le m} \exp(-\tilde{\Lambda}_j t) = \exp(t \sum_{1 \le j \le m} \tilde{\Lambda}_j) = \exp(-t\tilde{\Lambda}) , \tag{21.54}$$

where

$$\tilde{\Lambda} = \sum_{1 \le j \le m} \tilde{\Lambda}_j .$$

The second term of (21.48) expresses the probability that there will be no shortage of required modules at the workshop during period t. If spare units of some kind are exhausted, each failure of a unit of this type leads to exclusion of one of the spare modules because of the impossibility of repair. Strict solution of the problem is practically hopeless. We prefer an approximate method that gives a lower boundary for the probability.

Let us introduce a value, z_i, that equals the smallest number of modules of one type that includes the ith type of unit. Let us also assume that, when there are no spare units of the ith type for repair, all forthcoming failures of the system will occur because of type i units. Moreover, these failures will occur in the "weakest" module type, which is characterized by a value, z_i. Each forthcoming failure leads to exclusion of one of these modules. Thus, z_i will decrease.

Remark This assumption gives the lower bound of the probability.
If all operating units are independent, we can write

$$R_2(\mathbf{N}, \mathbf{x}; t) = \prod_{1 \le i \le n} [1 - \Pi_{x_i + z_i}(tN_i\lambda_i) \approx \prod_{1 \le i \le n} \pi_{x_i + z_i + 1}(tN_i\lambda_i)] . \tag{21.55}$$

21.3.4 System Availability Coefficient

Again we exclude the possibility of internal redundancy. In this sense we consider a system that has a series structure. The reliability index for the kth system is

$$K = K_1 K_2 K_3 \tag{21.56}$$

where K_1 is determined by the substitution time for failed modules (ignoring the short time needed to replace a failed module with a standby module $K_1 = 1$), K_2 is determined by idle time due to the absence of available standby modules of any type because all modules of this type are in repair, and K_3 is determined by the idle time at the end of a period, t, because of exhaustion of spare units of any type at the repair workshop.

The total failure rate of the jth type of module for the system can be written :

$$\tilde{\Lambda}_j = \Lambda_j M_j \ . \tag{21.57}$$

After each such failure, the system is idle during the replacement time. The availability coefficient can be written as

$$K_j = \frac{1}{1 + B_j \tilde{\Lambda}_j} \ .$$

We then write

$$K_1 \approx \prod_{1 \le j \le m} K_j \ . \tag{21.58}$$

Remark In practice the time for module replacement is ignored in engineering calculations. But occasionally it is reasonable to consider ideal standby redundancy with repair (see Chapter 4).

The term K_2 depends on the repair regime at the individual workshop and the type of repair time distribution.

Remark Time is required to deliver a failed module to the repair workshop and to return it. We assume that the workshop is located within the system, so we ignore these two additional units of idle time.

Let us consider the system's idle time in two important cases, constant repair time and exponential repair time. We assume that the repair time is constant, and that the failure flow is a Poisson process. We observe y failures during the period A, and we divide this interval into $y + 1$ random parts. All of these new random values are identically distributed with a mean value of $A/(y + 1)$. Since only the $(y_j + 1)$th failure of type j leads to a system failure, the system idle time is approximately $A_j/(y_j + 1)$.

Remark In an approximation we ignore the possibility of more than y failures in the time interval A. This is not necessary in approximate solutions.

If the repair time is distributed exponentially and there is only one repair facility, then, by the memoryless property, the residual time has the same distribution and the same mean, A_j. If there are enough repair facilities for all failed modules of all types, all can be repaired simultaneously. Then one will observe the minimum value of $y_j + 1$ exponentially distributed random variables, $A_j/(y_j + 1)$.

The rate of occurrence of the system's idle periods Λ_j is determined with (21.52). We will assume that repairs of modules of different types are independent. We may then instantly write the expression for the second term of (21.56):

$$K_2 = \prod_{j=1}^{m} \frac{1}{1 + \tilde{\Lambda}_j \tilde{A}_j} . \tag{21.59}$$

The third term of (21.56) is determined by the shortage of spare units during the interval t. We note that in a Poisson process all intervals between arrivals have the same distribution. For the ith type of unit we write

$$K_{3i} = P_{x_i}(\tilde{\lambda}_i t) + \sum_{z > x_i} p_z(\tilde{\lambda}_i t) \frac{x_i + 1}{z + 1} , \tag{21.60}$$

where $\tilde{\lambda}_i$ is the parameter of the total failure flow for the ith type of unit, $\tilde{\lambda}_i = \lambda_i N_i$. Again, if

$$p_{x_i+1}(\tilde{\lambda}_i t) \gg \sum_{z > x_i} p_z(\tilde{\lambda}_i t) ,$$

expression (21.60) can be rewritten as

$$K_{3i} \approx P_{x_i}(\tilde{\lambda}_i t) + \frac{x_i + 1}{x_i + 2} p_{x_i+1}(\tilde{\lambda}_i t) \approx 1 - p_{x_i+1}(\tilde{\lambda}_i t) \frac{1}{x_i + 2} . \tag{21.61}$$

For a highly reliable system we can assume that the probability of a deficit of any spare unit during interval t is low. We can ignore the possibility of several spare units failing at the same time. This assumption is very important because all idle times appear at the very end of interval t, and hence are not independent. This means that K_3 is greater than the product of the K_{3i} from $i = 1 - n$.

Remark If spare units are not delivered periodically, we have a more complex situation. There is no precise practical solution for a large multi-dimensional case. Individual cases can be handled by development of approximate solutions of the type given here.

CHAPTER 22

OPTIMAL CONTROL OF INVENTORIES OF SPARE PARTS

22.1 PURPOSES AND VARIABLES

The main subject of investigation in inventory control is $Z(t)$, the number of available spare units at the moment t. The value of $Z(t)$ is called the "level" of the stock. This stochastic process, $Z(t)$, decreases when spare units are used and increases when they are acquired. The expenditure of spare units is determined by the system's failure flow.

Certain parameters of the process $Z(t)$ are of particular interest. These include the distribution of the time of stock exhaustion (or the probability of no interruption of the supply of spare units), the stationary distribution of the stock level, the mean level of the stock, and the rate of stock resupply.

Optimal inventory control policies are determined by some selected objective. Usually this objective is the attainment of the desired level of a performance effectiveness index at minimum cost. Alternatively, the objective may be attainment of the highest index level possible under fixed cost levels. These factors form controversial problems that are difficult to solve and often not even clearly formulated.

Mathematical inventory models have what are called "penalty functions." A penalty function f_i, where i is an integer, shows the total cost (for $i > 0$) or the total loss (for $i < 0$) of the maintenance of state $Z(t)$. A variable $c(n)$, where n is an integer >0, is the total expense of provision and delivery of n spare units. Integration of all expenses over the track of $Z(t)$ and computation of the average can be the principal steps in the determination of an optimal inventory policy.

22.2 DEFINITIONS AND NOTATION

22.2.1 Main Stochastic Processes

$Z(t)$ is a stock of identical units at the moment t, a controlled stochastic process with integer values. Negative values of $Z(t)$ correspond to numbers of demands not filled or not filled on time.

$\zeta(t_1, t_2)$ is either the expenditure of spare units, the number of requests, or both, in the time interval $[t_1, t_2]$. This process is assumed to have known parameters.

$\{\xi_i\}$, $1 \le i < \infty$, is a sequence of independent and identically distributed intervals between requests or arrivals for a recurrent point process.

$\{\nu_i\}$, $1 \le i < \infty$, is a sequence of independent and identically distributed random integers that gives the number of each supply request. If failures are independent, each request can be for exactly one unit (i.e., $\nu_i = 1$ with probability 1).

$Y(t)$ is the nominal stock, which includes all available units plus those in the process of delivery that can not be used at the moment t:

$$F(x) = Pr\{\xi_i \le x\} ,$$

$$\bar{F}(x) = 1 - F(x) ,$$

$$a = \int_0^\infty \bar{F}(x)\, dx .$$

$F_m(x)$ is the m-order convolution of $F(x)$:

$$F_{m+1}(x) = \int_0^x F_m(x - y)\, dF(y) ,$$

$$r_i = Pr\{\nu_j = i\} ,$$

$$\pi(z) = \sum_{1 \le i < \infty} r_i z^i .$$

22.2.2 Stock Parameters

All stock supply is available but there can be delay. τ is the supply delivery delay; it can be a random or nonrandom variable:

$$G(x) = Pr\{\tau \le x\} ,$$

$$\bar{G}(x) = 1 - G(x) ,$$

$$h = E\tau = \int_0^\infty \bar{G}(x)\, dx .$$

The supply delivery rule is determined by the following parameters:

m = supply level at which resupply is ordered.
n = supply delivery size.
$k = m + n$ = maximum stock level.

22.2.3 Cost Parameters

$c(n)$, $n \geq 1$ is the cost of the supply ordered. f_i is a penalty function that is based on the cost if the stock is $Z(t) = i$; $-\infty < i < \infty$. The costs include storage expenses, penalties for rejection, and other variables.

V is the total cost rate, or "intensity" after stationarity of the process.

22.2.4 Inventory Control System Parameters

μ is the frequency of supply deliveries, orders, or both, for a stationary process.

Θ is a period of time during which $Z(t) > 0$.

K is the stationary probability that the stock is empty.

EZ is the expected value of Z, the average stock level.

$p_i = Pr\{Z = i\}$ is the probability of stock level i during stationarity.

$s_i = Pr\{Y = k - i\}$ is the probability of a nominal stock of $h - i$ during stationarity.

22.3 MODEL DESCRIPTIONS

22.3.1 An Interrupted Flow of Requests (Model 1)

Ordinarily, single requests appear in independent and identically distributed time intervals ξ_i. When the level of stock drops below m, an order is generated for supply delivery, and in a random time τ a new supply of units is delivered. If the stock becomes exhausted, new requests do not appear until the stock is resupplied. In other words, the flow of requests is interrupted. In practice this can happen if the lack of spare units leads to system stoppage, and consequently no new failures or requests for spare units. Usually $n \geq m$.

22.3.2 Single Requests and Constant Delivery (Model 2)

A queue of unsatisfied requests of form $Z(t)$ can be negative. An order for a delivery of n spare units is generated after n spare units from the stock have been used. The delivery delay equals τ. The maximum stock level equals k. In contrast to model 1, the $Z(t)$ level at the moment of an order generation can vary as can the number of units ordered. A value, $Y(t)$, always equals $m = k - n$ at the time or an order.

22.2.3 Random Request Sizes (Model 3)

The difference between this model case and Model 2 is that requests are independent and identically distributed values. An order is generated when $Y(t)$ intersects the order level from above; the order size makes $Y(t)$ increase to the level k. The parameter $n = k - m$ is both the order size and the width of the variation of the process $Y(t)$.

22.2.4 Requests from Several Sources (Model 4)

All conditions except the number of sources are the same as for Model 2. If the stock is used to provide several systems with spare units, one can consider a superposition of several recurrent flows. If the number of such flows is large, one can use with success a Poisson approximation for the resulting flow. Here we consider a general case when such an approximation is not correct.

22.4 MAIN RESULTS

Model 1

$$\mu = \frac{1}{na + b_m},$$

where

$$b_m = \int_0^\infty \bar{G}(x) F_m(x)\, dx,$$

$$E\Theta = \frac{na}{d_m},$$

where

$$d_m = \int_0^\infty \bar{G}(x)\, dF_m(x),$$

$$R(t) = Pr\{\Theta > t\} \approx e^{-d_m(t/na)} \qquad \text{for } d_m \ll 1,$$

$$K = \frac{na}{na + b_m},$$

$$EZ = \frac{[na(n + 2m + 1)/2] - n(h - b_m)}{na + b_m},$$

$$V = f_0(1 - K) + f_1 EZ + \mu c(n) \qquad \text{for } f_i = f_1 i,\ i \geq 1,$$

where $\bar{G} = 1 - G$.

Model 2

$$\mu = \frac{1}{na} ,$$

$$s_i = \frac{1}{n} , \qquad i = 0, 1, \ldots , n-1 ,$$

$$p_i = \frac{1}{n} \sum_{m+1 \le j \le k} q_{j-i} , \qquad i \le k ,$$

where

$$q_0 = 1 - \frac{1}{a} \int_0^{\tau} \bar{F}(x)\, dx .$$

$$q_i = \frac{1}{a} \int_0^{\tau} [F_{i-1}(\tau - x) - F_i(\tau - x)]F(x)\, dx , \qquad i \ge 1 ,$$

$$EZ = \frac{n + 2m + 1}{2} - \frac{\tau}{a} ,$$

$$V(k, n) = \frac{1}{n} \sum_{0 \le j \le n-1} g_{k-j} + \frac{c(n)}{na} ,$$

where

$$g_j = \sum_{-\infty < i \le j} q_{j-i} f_i ,$$

and

$$\bar{F} = 1 - F .$$

Model 3

$$s_i = l_i \Bigl(\sum_{0 \le j \le n-1} l_j \Bigr)^{-1} , \qquad i = 0, 1, \ldots , n-1 ,$$

where the sequence $\{l_i\}$, $1 \le i < \infty$ has the following generating function:

$$\varphi(z) = \sum_{0 \le i \le n} l_i z^i = \frac{1}{1 - \pi(z)} ,$$

$$\mu = \frac{1}{a \sum_{0 \le j \le n-1} l_j} ,$$

$$p_i = \sum_{0 \le j \le n-1} s_j w_{k-j} ,\ i \le k ,$$

where

$$w_j = \lim_{t\to\infty} Pr\{\zeta(t-\tau, t) = j\}\ ,$$

$$\sum_{0\le j<\infty} w_j z^j = \sum_{0\le i<\infty} q_i \pi^i(z)\ .$$

For $c(n) = c_0 + c_i i,\ i \ge 1$,

$$V(k, n) = \frac{\Sigma_{0\le i\le n-1}\, l_i g_{k-i} + c_0/a}{\Sigma_{0\le i\le m-1}\, l_i} + c_1\,\frac{E\nu}{a}\ ,$$

where

$$g_j = \sum_{-\infty<i\le j} f_i w_{j-l}\ .$$

Model 4

$$s_i = 1/n,\ i = 0, \ldots, n-1\ .$$

(as in Model 2).

22.5 OPTIMIZATION OF THE CONTROL PARAMETER

Consider Model 1 and let:

$$V(m, n) = \frac{f_0 b_m + f_1 n[(n + 2m + 1)/2 - h + b_m] + c(n)}{na + b_m}\ . \tag{22.1}$$

It is necessary to minimize this function on the integer grid $n \ge 0$, $m \ge 0$, and $n \ge m$. For any $F(x)$ and $G(x)$, a sequence, b_m, is nonincreasing and convex. The first derivative of b_m is negative, and the second derivative is positive. Using this fact, we show that for a constant n, the function $V(m, n)$ is unimodal; that is, there is one minimum for all m. For $c(n) = c_0 + c_1 n$, $n \ge 1$, the function $V(m, n)$ also is unimodal for all n if m is constant. The minimum can be found in closed form:

$$n^*(m) = \max\left\{m, \left\lceil \sqrt{\frac{1}{4} + \frac{2c_0}{f_1 a} + \frac{2b_m}{a}\left(-m - \frac{b_m}{2a} + \frac{f_0}{f_1} - \frac{1}{2} + \frac{h}{a} - \frac{c_1}{f_1 a}\right)} - \frac{b_m}{a} - \frac{1}{2}\right\rceil\right\}, \tag{22.2}$$

where $\lceil x \rceil$ means that x is approximated with the closest larger integer.

The main calculation difficulties come in finding b_m with convolutions of $F(x)$, but in some cases there is a closed form. For example, if

$$G(x) = 1 - e^{-\alpha x} ,$$

then

$$b_m = \frac{1}{\alpha} [\tilde{F}(\alpha)]^m ,$$

where

$$\tilde{F}(s) = \int_0^\infty e^{-sx} \, dF(x) .$$

For $\tau = h =$ a constant,

$$F(x) = 1 - e^{-\lambda^x} ,$$

$$b_m = hQ(m-1, \lambda h) - \left(\frac{1}{\lambda}\right) mQ(m, \lambda h) , \tag{22.3}$$

where $Q(d, a)$ may be found in tables similar to Table 22.3 for a not exceeding 100–150. For larger values, one can use a normal approximation for a Poisson distribution.

If $G(x)$ is an l-order Erlang distribution with parameter λ, and $\tau = h =$ a constant, then

$$b_m = hQ(lm-1, \lambda h) - \left(\frac{1}{\lambda}\right) mlQ(ml, \lambda h) .$$

If closed expressions for b_m are not available, one can make a Monte Carlo simulation based on the following:

$$b_m = E\omega ,$$

where

$$\omega = \left[\tau - \sum_{1 \le i \le m} \xi_i\right]^+ ,$$

$$[x] = \begin{cases} x & \text{if } x \ge 0 , \\ 0 & \text{otherwise} . \end{cases}$$

In correspondence with distributions $F(x)$ and $G(x)$ random values τ and $\{\zeta_i\}_{i=1}^m$ are generated, and the related value of ω is calculated. This procedure is repeated the needed number of times, and an estimate of b_m, $\hat{b}_m$, is calculated:

$$\hat{b}_m = \frac{1}{N} \sum_{1 \le i \le N} \omega_i ,$$

where N is the number of Monte Carlo experiments.

The accuracy of the estimates depends on N. One has to be able to evaluate the variance Var_m to choose N. It is easy to show that

$$\mathrm{Var}_\omega \leq (h + ma)^2 d_m + \mathrm{Var}_\tau + m^2 \,\mathrm{Var}\,\xi \,.$$

Table 22.1 gives the results of a Monte Carlo simulation for the case where $\tau = 5$, $G(x) = 1 - e^{-k}$, and related accurate values of b_m are calculated with (22.3). For $N \leq 100$ the estimate $\hat{b}_{10}$ is equal to 0. For $N = 1000$, $\hat{b}_{10} = 0.0109$, which is below the actual b_{10} value of 0.0227. From the table we see that for a fixed N the error decreases as b_m increases. N should be increased enough to yield satisfactory results.

The simulation method of estimating b_m can be extended to situations where distributions $F(x)$ and $G(x)$ are unknown, but some values of ξ and τ are available. In this case we would estimate a and h and then evaluate b_m. To this end we would generate random values with empirical distributions and evaluate the data base for possible sources of error, such as outliers or nonrandomness. Using this procedure, we can determine optimal values of m and n with no knowledge of distributions $F(x)$ and $G(x)$.

EXAMPLE 22.1

Assume that the flow of requests is Poisson and that the delay time is constant. For this case b_m can be found with (22.3). Let $a = 1$ ($\lambda = 1/a$), $\tau = 5$, $f_0 = 50$, $f_1 = 0.1$, and $c(n) = c_0 + c_1 n$ ($n > 0$), $c_0 = 20$, $c_1 = 10$. All values of course have the same dimension.

Solution

Equation (22.1) yields

$$V(m, n = \text{const})$$
$$= 50 + 0.1n + \frac{0.1mn + 0.1n + [(n+1)/2 - 5] + 20 + 10n - (50 + 0.1n)n}{b_m + n}, \qquad (22.4)$$

and (22.2) gives

Table 22.1 Results of the Monte Carlo simulation

	N					
$\hat{b}$	20	40	60	80	100	Accurate Value
$\hat{b}_1$	3.8999	3.8956	3.8738	3.8889	3.9430	4.0067
$\hat{b}_5$	0.8445	0.7892	0.8318	0.7963	0.7422	0.8773
$\hat{b}_8$	0.0124	0.0709	0.0473	0.0355	0.0514	0.1221

$$n^*(m) = \max\{m, \lceil\sqrt{400.25 + 2b_m(-m - 0.5b_m + 404.5)} - b_m - 0.5\rceil\} . \tag{22.5}$$

Values of b_m calculated with (22.3) are given in Table 22.2.

For $m > 15$, b_m values can be estimated with satisfactory accuracy to be zero. A search for the minimum can be made by the steepest descent method, starting at the point $(0, 0)$, where $V = 50$. By (22.5),

$$n^*(0) = \lceil\sqrt{4420.25} - 5.5\rceil = 61 .$$

Letting $n = 61$ in (22.4),

$$V(m, n = 61) = 56.1 + \frac{6.1m - 2633.5}{b_m + 61} . \tag{22.6}$$

From the point $(0, 61)$, where $V = 16.2$, descent is by the axis. The unimodal function minimum can be found by the Fibonacci Method. The ith term of the Fibonacci sequence can be found with the formula $F_i = F_{i-1} + F_{i-2} + \cdots$, $i \geq 3$.

Assume that the minimum is located at one of the integer points of the interval whose length equals one of terms of the Fibonacci sequence. In this case we take $(-1, 20)$ as such an interval. The length of this interval is denoted F_8. For $m > 20$ function $V(m)$ increases linearly with m since $b_m \approx 0$. Hence the minimum of the function is in the interval $(-1, 20)$. The Fiboncci Method leads to a systematic decrease of the interval of uncertainty, F_8.

At the first step of the calculation procedure, one calculates (22.6) for two points located at the distance $F_7 = 13$ from the ends of F_8, the points $m = 7$ and $m = 12$:

$$V(7) = 13.805 ,$$

$$V(12) = 14.123 .$$

As $V(12) > V(7)$, the minimum cannot belong to the interval $(12, 20)$. We therefore consider the interval $(-1, 12)$ with length F_7.

Table 22.2 Values of b_m

m	b_m	m	b_m	m	b_m	m	b_m
0	5.0000	4	1.4368	8	0.1221	12	0.0030
1	4.0067	5	0.8773	9	0.0531	13	0.0010
2	3.0472	6	0.4933	10	0.0227	14	0.0003
3	2.1718	7	0.2555	11	0.0085	15	0.0001

The two points located at the distance $F_6 = 8$ from the ends of F_7 are $m = 4$ and $m = 7$. The value $V(7)$ already has been calculated. $V(4) = 14.312$. Because $V(4) > V(7)$, we choose the interval $(4, 12)$ with length F_6. The new F_5 value is 5, and the new points are $m = 7$ and $m = 9$. $V(6) = 13.865$.

Since $V(7) < V(9)$, we choose the interval $(4, 9)$ with length F_5, $F_4 = 3$, $m = 6$, and $m = 7$. We find that $V(6)–13.869$, so $V(7) < V(6)$. The last interval is $(6, 9)$ which contains 8, the only point with an unknown value of the function (22.6). We calculate $V(8) = 13.813$ and find that the minimum of (22.6) is at the point $m = 7$. The two-dimensional point $(7, 61)$ is found to have the value $V(7, 61) = 13.805$.

Further search is performed via the s axis. From (22.5) we have

$$n_*(7) = \lceil \sqrt{603.3072} - 0.7555 \rceil = 24 .$$

From the point $(7, 24)$ the descent is performed via the m axis. Equation (22.4) yields

$$V(m, n = 24) = 52.4 + \frac{2.4m - 979.6}{b_m + 24} . \tag{22.7}$$

Further searching for V in the interval $(-1, 20)$ with the Fibonacci Method gives the following sequence:

$$V(7) = 12.706 , \quad V(12) = 12.788 , \quad V(4) = 14.266 ,$$
$$V(9) = 12.5715 , \quad V(10) = 12.621 , \quad V(8) = 12.586 .$$

The minimum is at $m = 9$.

From the point $(9, 24)$ the search axis is n. Equation (22.5) yields

$$n_*(9) = \lceil \sqrt{442.25} - 0.5531 \rceil = 21 ,$$

and the process reaches the point $(9, 21)$. Equation (22.4) gives

$$V(m, n = 21) = 52.1 + \frac{2.1m - 851.1}{b_m + 21} .$$

By the Fibonacci Method we find that the minimum of the function is at $m^* = 9$. The point $(9, 21)$ has been obtained twice, so it is the solution. $V(9, 21) = 12.552$.

Now consider Model 2. Let

$$V(m, n) = V_1(m, n) + V_2(n) ,$$

where

$$V_1(m) = \frac{1}{n} \sum_{m+1 \le j \le m+n} g_j ,$$

$$V_2(n) = \frac{c(n)}{na} .$$

Let the sequence g_j have its only minimum at the point j_0. When we fix the sequential values of n, the conditionally optimal zones Γ_n are constructed by the rule

$$\Gamma_1 = j_0 ,$$

$$\Gamma_2 = (j_0, j_1) ,$$

where

$$j_1 = \begin{cases} j_0 + 1 & \text{if } g_{j_0+1} < g_{j_0-1} , \\ j_0 - 1 & \text{if } g_{j_0+1} \ge g_{j_0-1} . \end{cases}$$

Thus Γ_n is found from the solution of the new point that is either the one from the left or the one from the right of zone Γ_{n-1}, depending on where g_j is smaller.

Denote the terms of the sequence g_j in increasing order as $a_1, a_2, \ldots$. Let

$$A_n = \sum_{1 \le i \le n} a_i .$$

The optimal value of n then is found by minimization of

$$\frac{A_n}{n} + \frac{c(n)}{na} . \tag{22.8}$$

For $c(n) = c_0 + c_1 n$ this function is unimodal, and its minimum is reached at an n where $na_{n+1} - A_n \ge c_0/a$. If the flow of requests is Poisson with a parameter λ and:

$$f_1 = \begin{cases} k_1 i , & i \le 0 , \\ k_2 i , & i > 0 , \end{cases}$$

then

$$g_j = (k_1 + k_2)[\lambda h Q(j, \lambda h) - jQ(j+1, \lambda h)] + k_2(j - \lambda h) ,$$

and

$$\Delta g_j = k_2 - (k_1 + k_2) Q(j+1, \lambda h) . \tag{22.9}$$

Table 22.3 Values of $Q(j, 10)$

j	9	10	11	12	13	14	15
$Q(j, 10)$	0.542	0.417	0.303	0.208	0.136	0.083	0.049
g_j	—	12.5	10.4	9.3	9.1	10	11.1

Table 22.4 Results for Example 22.2

n	1	2	3	4	5
a_n	9.1	9.3	10	10.4	11.1
A_n	9.2	18.4	28.4	38.4	49.9
$na_{n+1} - A_n$	0.2	1.6	2.8	5.6	

EXAMPLE 22.2

Let $c_0 = 1$, $\lambda = 5$, $h = 2$, $k_1 = 8$, and $k_2 = 2$. The value of c_1 does not play any role because it connects only with a constant λc_1 in the expression for V.

Solution

To find the minimum g_j, we obtain from values of $Q(j, 10)$ in the table a minimal j_0 such that

$$Q(j, 10) \leq = \frac{k_2}{k_1 + k_2} = 0.2 .$$

We find that $j_0 = 13$. Table 22.3 has $Q(j, 10)$ values surrounding this point. In the third line we find values g_j calculated with (22.9). It is clear that $\Gamma_1 = 13$, $\Gamma_2 = (12, 13)$, $\Gamma_3 = (12, 13, 14)$, $\Gamma_4 = (11, 12, 13, 14)$, and $\Gamma_5 = (11, 12, 13, 14, 15)$.

Now we can minimize (22.8). The results of the calculations are put in Table 22.4. In the lowest line we find that $na_{n+1} - A_n > \lambda c_0$. This gives $n^* = 4$. Hence the $Y(t)$ search zone is $\Gamma_4 = (11, 12, 13, 14)$. In other words, the optimal parameters are $m^* = 10$ and $n^* = 4$. For this case $V = A_4/4 + \lambda c_0/4 = 10.85$.

22.6 APPROXIMATE SOLUTION FOR CENTRALIZED SYSTEM

Two cases are considered for a central facility that supplies facilities at a lower level.

22.6.1 Case 1

Central stock has N units for the supply of k stocks at the lower level during a fixed period of time, t_0. Each unit has a random time to failure, ξ, with distribution function $F(x)$. The delivery time is constant and equals τ. Each lower level uses the (M, n) rule for orders. Let $P(t_0)$ be the probability of uninterrupted supply of all the lower-level stocks. Then

$$P(t_0) \approx e^{-kt_0 F^{*m}(\tau)/na} \Phi\left(\frac{N - (kt_0/a) - M}{\sqrt{(\sigma^2 kt_0/a^3) + D}} \right), \tag{22.10}$$

where $M = (1/2)(n+1)(k-1)$, $D = (1/12)(k-1)(n_2 - 1)$, and $\Phi(x)$ is the normal distribution function. This formula gives a good approximation when

1. $F_m(\tau) \ll 1$, namely the probability of more than m requests during a delivery period is small,
2. The total number of requests from all lower-level stocks is large enough to use a normal approximation.

22.6.2 Case 2

Central stock sends orders for its own resupply that have a delivery time of T. Central stock uses the (M, N) rule to order, and lower levels use the (M, n) rule to order. As before, $n \geq m$ and $N \geq M$. Central stock has only requests for deliveries of n spare units, so it is more convenient to use new integer parameters $M_* = M/n$ and $N_* = N/n$.

When $k \gg 1$, a Poisson approximation can be used to find $P(t)$. The parameter of this approximate Poisson process Λ^* can be calculated with $\Lambda^* = k/na^*$, where $a^* \approx a[1 + (m/n)F^{*m}(T)]$. It is assumed that $F^{*m}(T) \leq 1$. If we additionally assume that

$$A_{M^*-1}(\Lambda^* T) = 1 - e^{-\Lambda^* T} \sum_{0 \leq j \leq M^*-1} \frac{(\Lambda^* T)^j}{j!} \ll 1 ,$$

then the function $F(t)$ can be successfully approximated with an exponential function

$$P(t) \approx e^{-\Lambda_\Sigma t} ,$$

where

$$\lambda_\Sigma = \frac{k}{na} F^{*m}(T) + \frac{\Lambda^*}{N^*} A_{M^*-1}(\Lambda^* T) ,$$

and

$$A_{M^*-1}(\Lambda^* T) = 1 - e^{-\Lambda^* T} \sum_{j=0}^{M^*-1} \frac{(\Lambda^* T)^j}{j!} < 1 .$$

CHAPTER 23

OPTIMAL MAINTENANCE

During operation, all equipment declines. Failures, degradation, and aging occur. Monitoring and maintenance are required, and there are expenses for inspection, preventive maintenance, and repair.

The expenses lead to the problem of optimal maintenance. An optimal maintenance plan includes both the type of maintenance and the period that minimizes total expenses. The total expenses may include not only maintenance cost but system losses because of idle time.

Usually there are two broad categories of maintenance—planned preventive maintenance and repair after discovery of system failure. The failure may be discovered instantly by continuous monitoring or after a delay by periodic inspection. In most mathematical models, maintenance and repair produce an item that is "as good as new."

The main indexes in maintenance analysis are the availability coefficient, the operational availability coefficient, and two economic indexes—the average profit from the system per unit of time, and the average total cost per unit of time. The first economic index is very convenient when both system outcome and maintenance expenses are measured in the same cost units.

Let us introduce the following notation:

C = average system profit per unit of calendar time.

C^* = average total system loss plus expense per unit of useful time.

c_a = loss in a unit of time during a repair after a failure.

c_0 = profit from the system per unit of time during a failure-free period.

c_π = loss in a unit of time during a planned maintenance.

c_c = loss in a unit of time during the existence of "hidden" (unrecognized) system failure.
$F(t)$ = distribution function (d.f.) of system time to failure.
$H_i(t)$ = d.f. of the repair time of the ith unit.
t_i = average duration of the ith unit repair time after failure.
t_0 = operational time a system needs to perform its operation.
t_c = average time before discovery of a hidden failure.
T = system mean time to failure (MTTF).
T_a = mean repair time after a failure.
$T_{a\pi}$ = mean time of unplanned maintenance of the system after a failure.
T_π = mean time of planned maintenance.
$\lambda(t)$ = system failure rate.
τ_0 = optimal period of planned maintenance.

23.1 LIST OF OPTIMAL MAINTENANCE MODELS

Five models are presented in this chapter:

Model 1 (*Section 23.2*). There are repairs only after system failures; the failures are recognized immediately.

Model 2 (*Section 23.3*). Planned maintenance and repairs after system failures restore the system completely; the failures are recognized immediately.

Model 3 (*Section 23.4*). Only planned maintenance is performed. If a failure is found during this procedure, the system is repaired.

Model 4 (*Section 23.5*). Repair of a unit in a series system is performed after each failure.

Model 5 (*Section 23.6*). Repairs of a unit in a series system is performed after each failure, and the entire system has periodic planned maintenance.

23.2 MODEL 1, REPAIRS AFTER FAILURES

Let us consider a system in which repairs are made only after system failures. The average delay before discovery of a failure is t_c. Repair starts after the failure is recognized and has an average duration of T_a. After repair, the system is as good as new, and system operation and maintenance continue.

The data needed are $F(t)$, T, t_π, T_a, c_c, c_a, c_0, and t_0. The four main system performance indexes are

$$K = \frac{T}{T + t_c + T_a},$$

$$R(t_0) = \frac{T - \int_0^{t_0} \bar{F}(x)\, dx}{T + t_c + T_a},$$

$$C^* = \frac{c_c t_c + c_a T_a}{T},$$

$$C = K(c_0 - C^*) = \frac{c_0 T - c_c t_c - c_a T_a}{T + t_c + t_a}.$$

For this model there is no optimization problem. If a system failure is discovered instantly, $t_0 = 0$. If a system failure is undiscovered, ($t_0 = \infty$), and the model is not applicable.

EXAMPLE 23.1

The system input data are $F(t) = e^{-0.1\lambda t}$, $T = 10$ hours, $T_a = 1$ hour, $t_c = 1$ hour, $c_a = 1$ cost unit per hour; $c_c = 2$ cost units per hour; $c_0 = 2$ cost units per hour; and $t_0 = 2$ hours.

The availability coefficient is

$$K = \frac{10}{10 + 1 + 1} = 0.8333.$$

The operational availability coefficient is

$$R(2) = \frac{10 - \int_0^2 e^{-0.1x}\, dx}{10 + 1 + 1} = 0.8333e^{-0.2} = 0.6822.$$

The average expense per unit of time is

$$C^* = \frac{2 \cdot 1 + 1 \cdot 1}{10} = 0.3.$$

The average profit per unit of time is

$$C = \frac{10}{12}(2 - 0.3) = 1.415.$$

23.3 MODEL 2, PLANNED MAINTENANCE AND REPAIRS AFTER FAILURES

Let us consider a system that has planned maintenance and repairs after system failures. Every failure can be recognized immediately. All renewal procedures are completed in the same order:

At the very beginning the first maintenance is planned. If the system does not fail before this time, the planned maintenance is performed as planned and has a duration of T_π. If a system failure occurs before the time, repair is started and continues for time T_a. After any renewal the system is as good as new. After any such procedure the process of maintenance planning is repeated. We assume that during maintenance as well as repair the system is not in an operational state and cannot perform its functions.

The data needed are $F(t)$, $\lambda(t)$, T_π, T_a, c_π, c_a, c_0, and t_0.

Table 23.1 gives the equations for the system's indexes. Preventive maintenance, of course, is performed only if a system has an increasing failure rate. The equations in Table 23.1 can be solved only with numerical methods.

EXAMPLE 23.2

The system input data are $F(t) = \exp(-0.00025t^2)$, $\lambda(t) = 0.0005$ per hour, $T = 56.05$ hours, $T_\pi = 5$ hours, $T_a = 30$ hours, $c_\pi = 1$ cost unit per hour, $c_a = 2$ cost units per hour, $c_0 = 2$ cost units per hour, $t_0 = 1$ hour. We want to find the optimal system maintenance parameters.

1. *Availability coefficient.* By substituting input data into the equation for finding the τ_0 that optimizes K, we have

$$\phi(\tau) = -1 + e^{-0.00025\tau^2} + \int_0^\tau e^{-0.00025x^2}\,dx$$
$$= \frac{5}{30-5} = 0.2\,.$$

The equation has only one root. This root can be found with the graph of the function $\varphi(\Theta)$. From this graph we find that $\tau_0 \approx 29$ hours.

We then can find the optimal value of the system availability coefficient:

$$K(\tau_0) = \frac{1}{1 + (T_a - T_\pi)\lambda(\tau_0)} = \frac{1}{1 + (30-5)\cdot 0.0005 \cdot 29} = 0.734\,.$$

For comparison we look at the value of the same index in the absence of planned maintenance:

$$K(\infty) = K = \frac{T}{T + T_a} = \frac{56.05}{56.05 + 30} = 0.65\,.$$

Notice that K increases as mean maintenance time decreases. For example, for $T_\pi = 2$ h, we have $\tau_0 = 17$ h and $K(17) = 0.808$.

Table 23.1 Main performance effectiveness indexes of model 2: Planned and unplanned maintenance

Index	Formula for Root τ_0 Equal to the Optimal Period of Maintenance	Formula for Calculation of Reliability Index if there is a Single Root τ_0	Formula for Calculation of Reliability Index if $\tau_0 = \infty$
K	$\frac{T_\pi}{T_a - T_\pi} = -F(\tau) + \lambda(\tau) \int_0^\tau P(x)\,dx$	$\frac{1}{1 + (T_a - T_\pi)\lambda(\tau_0)}$	$\frac{T}{T + T_a}$
$R(t_0)^a$	$\frac{T_\pi}{T_a - T_\pi + t_0} = -F(\tau) + \lambda(\tau) \int_0^\tau P(x)\,dx - \frac{T_\pi t_0 \lambda(\tau)}{T_a - T_\pi + t_0}$	$\frac{\int_0^{\tau_0} P(x + t_0)\,dx}{\int_0^{\tau_0} P(x)\,dx + T_\pi + (T_a - T_\pi)F(\tau_0)}$	$\frac{T - \int_0^{t_0} P(x)\,dx}{T + T_a}$
C^*	$\frac{c_\pi T_\pi}{c_a T_a - c_\pi T_\pi} = -F(\tau) + \lambda(\tau) \int_0^\tau P(x)\,dx$	$(c_a T_a - c_\pi T_\pi)\lambda(\tau_0)$	$\frac{c_a T_a}{T}$
C	$\frac{(c_0 + c_\pi)T_\pi}{(c_0 + c_a)T_a - (c_0 + c_\pi)T_\pi} = -F(\tau) + \lambda(\tau) \int_0^\tau P(x)\,dx + \lambda(\tau) \frac{(c_a - c_\pi)T_\pi T_a}{(c_0 + c_a)T_a - (c_0 + c_\pi)T_\pi}$	$\frac{c_0 - (c_a T_a - c_\pi T_\pi)\lambda(\tau_0)}{1 + (T_a - T_\pi)\lambda(\tau_0)}$	$\frac{c_0 T - c_a T_a}{T + T_a}$

Note: If there are several roots $\tau_{01}, \ldots, \tau_{0s}$, the optimal value τ_{0i} must be found by directly substituting each of them into the formula and choosing the root delivering the maximum reliability index or the minimum cost.

[a]Equation is aproximate for the condition $t_0 \ll T$.

2. *Operational availability coefficient.* From the table of input data, we see that $t_0 \ll T$, so we use the approximate equation for $R(t_0)$ to find the appropriate maximum. After substitution of the input data, we have

$$\frac{5}{30-5+1} = -1 + e^{-0.00025\tau^2} + 0.0005\tau \int_0^\tau e^{-0.00025x^2}\,dx - \frac{5\cdot 1}{30-5+1}\,0.0005\tau\,,$$

or

$$0.192 = \varphi(\tau) - 0.0001\tau\,.$$

The plot of the function $\varphi(\tau) - 0.0001\tau$ in the range of interest practically coincides with $\varphi(\Theta)$. The optimal value is $\Theta_0 = 28.5$ hours, and the corresponding value of the operational availability coefficient is

$$R_{\tau_0}(1) = \frac{\int_0^{28.5} e^{0.00025(x+1)^2}\,dx}{\int_0^{28.5} e^{-0.00025x^2}\,dx + 5 + 25\cdot(1 - e^{0.00025\cdot(28.5)^2})} = 0.73\,.$$

For comparison we find the probability for no planned maintenance:

$$R_\infty(t_0) = \frac{56.05 - \int_0^1 e^{-0.00025x^2}\,dx}{56.05 + 30.0} = \frac{55.048}{86.05} = 0.64\,.$$

Notice that the small t_0 brings the optimal value of the operational availability coefficient very close to the optimal value of the availability coefficient. The optimal periods are also close.

3. *Average cost per unit of time.* After substituting input data in the equation for C^* in Table 23.1 to find τ_0, we write

$$\frac{1\cdot 5}{2\cdot 30 - 1\cdot 5} = -1 + e^{-0.00025\tau^2} + \int_0^\tau e^{-0.00025x^2}\,dx\,,$$

or $\varphi(\tau) = 0.091$. Because the function $\lambda(t)$ is monotonic, the equation has only one root that can be found. The optimal period is 19 h. The minimum cost is given by

$$\min C^*(\tau) = C^*(\tau_0) = (2\cdot 30 - 1\cdot 5)0.0005\cdot 19 = 0.5225\,.$$

For comparison we find the value for no planned maintenance.

$$C^* \frac{2 \cdot 30}{56 \cdot 05} = 1.07 .$$

4. *Average profit per unit of time*. Using the equation for C from Table 23.1 to find τ_0, we write

$$0.143 = -1 + e^{-0.00025\tau^2} + 0.0005\tau \int_0^{\tau} e^{-0.00025x^2}\, dx + 0.000715\tau ,$$

or $0.143 = \varphi(\tau) + 0.000715\tau$. In this case we have a situation similar to the previous one. We can neglect the second term because it is so small. The root of the equation can be found directly. The optimal period is $\tau_0 = 23.5$ hours, and the average profit equals

$$\max C(\tau) = C(\tau_0) = \frac{2 - (2 \cdot 30 - 1 \cdot 5)0.0005 \cdot 24.5}{1 + (30 - 5)0.0005 \cdot 24.5} = 1.020 .$$

For comparison, we find the average profit with no planned maintenance:

$$C(\infty) = \frac{2 \cdot 56.05 - 2 \cdot 30}{56.05 + 30} = \frac{52.10}{86.05} = 0.605 .$$

23.4 MODEL 3, PLANNED MAINTENANCE

For some systems, failures are discovered only with special testing. Such systems require planned maintenance. An ordinary planned maintenance takes time T_π, and maintenance with additional repair takes time T_a.

If a failure occurs, the system remains in a failure state until the planned maintenance. After each maintenance, whether regular or with repair, the next maintenance is scheduled.

The data needed are $F(t)$, $f(t)$, $\lambda(t)$, T_π, T_a, c_π, $c_a c_c$, c_0, and t_0. Equations for the maintenance parameters are given in Table 23.2.

EXAMPLE 23.3

The system input data are $F(t) = \exp(-0.01t^2)$, $\lambda(t) = 0.01$ per hour, $T_\pi =$ 20 hours, $T_a = 30$ hour, $c_\pi = 1$ cost unit per hour, $c_a = 2$ cost units per hour, $c_c = 2$ cost units per hour, $c_0 = 3$ cost units per hour, and $t_0 = 10$ hours. Let us find the parameters for optimal system maintenance:

Table 23.2 Main performance effectiveness indexes of Model 3: Planned maintenance

Index	Formula for Root τ_0 Equal to the Value of Period of Optimal Maintenance	Formula for Index Calculation when there is a Single Root τ
K	$\frac{T_\pi}{T_a - T_\pi} = -F(\tau) + \lambda(\tau)\int_0^\tau P(x)\,dx + \frac{\int_0^\tau x\,dF(x)}{(T_a - T_\pi)P(\tau)}$	$\frac{P(\tau_0)}{1+(T_a - T_\pi)f(\tau_0)}$
$R(t_0)$	$\frac{T_\pi}{T_a - T_\pi} = -F(\tau) + \lambda(\tau)\frac{P(\tau)}{P(\tau + t_0)} \cdot \int_0^\tau P(x + t_0)\,dx + \frac{\int_0^\tau x\,dF(x + t_0)}{(T_a - T_\pi)P(\tau + t_0)}$	$\frac{P(\tau_0 + t_0)}{1+(T_a - T_\pi)f(\tau_0)}$
C^*	$\frac{c_\pi T_\pi}{(c_a T_a - c_\pi T_\pi}) = -F(\tau) + \lambda(\tau)\int_0^\tau P(x)\,dx + \frac{c_0\int_0^\tau x\,dF(x)}{(c_a T_a - c_\pi T_\pi)P(\tau)}$	$c_0\frac{F(\tau_0)}{P(\tau_0)} + (c_0 T_a - c_\pi T_\pi)\lambda(\tau_0)$
C	$\frac{(c_0 + c_\pi T_\pi)}{(c_0 + c_0)(T_a - T_\pi)} = -F(\tau) + \lambda(\tau)\int_0^\tau P(x)\,dx + \frac{\int_0^\tau x\,dF(x)}{(T_a - T_\pi)P(\tau)} + \frac{(c_0 - c_\pi)T_a F(\tau)}{(c_0 + c_c)(T_a - T_\pi)P(\tau)} + \lambda(\tau)\frac{T_a T_\pi(c_a - c_\pi) + \tau[T_a(c_a - c_c) + T_\pi(c_c - c_\pi)]}{(c_0 + c_c)(T_a - T_\pi)(c_0 + c_c)(T_a - T_\pi)}$	$\frac{c_0 P(\tau_0) - c_c F(\tau_0) - (c_a T_a - c_\pi T_\pi)f(\tau_0)}{1+(T_a - T_\pi)f(\tau_0)}$

Note: In this case roots are always finite. If there are several roots $\tau_{01}, \ldots, \tau_{0s}$, the optimal one, τ_{0i}, must be found by directly substituting each of them and choosing a single one delivering the maximum reliability index or the minimum cost.

1. *Availability coefficient.* By substituting an exponential distribution, $F(t) = 1 - e^{-\lambda t}$, into the equation for the τ_0 to optimize $R(t_0)$ and putting $t_0 = 0$, we find that from Table 23.3,

$$\frac{T_\pi}{T_a - T_\pi} = \frac{\int_0^\tau x\, dF(x)}{(T_a - T_\pi)P(\tau)} ,$$

or

$$T_\pi = \frac{1}{P(\tau)} \int_0^\tau x\, dF(x) = e^{\lambda t} \int_0^\tau xd(1 - e^{-\lambda x}) = \frac{1}{\lambda} (e^{\lambda\tau} - 1) - \tau .$$

Finally,

$$\lambda T_\pi = (e^{\lambda\tau} - 1) - \lambda\tau = \phi(\lambda\tau) .$$

The optimal period of maintenance is given by

$$\tau_0 = \frac{0.575}{0.010} = 57.5 \text{ h} .$$

Now we can find the maximum value of the system availability coefficient that corresponds to the optimal maintenance period:

$$K(\tau_0) = \frac{e^{-0.0575}}{1 + (30 - 20) \cdot 10^{-2} e^{-0.575}} = 0.530 .$$

2. *Operational availability coefficient.* For an exponential failure time distribution, the optimal value of the maintenance period, τ_0, for the operational availability coefficient coincides with the one obtained for the availability coefficient, 57.5 hours. After substituting this value into the expression for $R(t_0)$ in Table 23.3 we find the maximum probability:

Table 23.3 Main performance effectiveness indexes of model 4: Unplanned maintenance for units

Index	Formula
K	$\dfrac{1}{1 + \Sigma_{i=1}^m (1/T_i)(t_i + t_c)}$
C^*	$\sum_{i=1}^m \dfrac{c_i t_i + c_0 t_0}{T_i}$
C	$(c_0 - C^*)K = \dfrac{c_0 - \Sigma_{i=i}^m (1/T_i)(c_i t_i + c_c t_c)}{1 + \Sigma_{i=1}^m (1/T_i)(t_i + t_c)}$

$$R_{\tau_0} = \frac{e^{-0.575}e^{-0.10}}{1 + (30 - 20)\cdot 10^{-2}e^{-0.575}} = 0.530 \cdot 0.905 = 0.480\ .$$

3. *Average cost per unit of time*. Substituting the d.f., $F(t) = 1 - e^{-\lambda t}$, into the equation for C^* in Table 23.3 to find τ_0, we write

$$\frac{c_\pi}{c_0} t_\pi = \frac{\int_0^\tau x\,dF(x)}{P(\tau)} = \frac{\int_0^x xd[1 - e^{-\lambda x}]}{e^{-\lambda\tau}} = \frac{1}{\lambda}(e^{\lambda\tau} - 1) - \tau\ ,$$

or

$$\frac{c_\pi}{c_0}\lambda T_\pi = (e^{\lambda\tau} - 1) - \lambda\tau = \phi(\lambda\tau)\ .$$

For $T_\pi = 20$ hours, $c_\pi = 1$ cost unit per hour, $c_c = 2$ cost units per hour, and $\lambda = 0.01$ per hour, we have

$$\frac{c_\pi}{c_0}\lambda T_\pi = 0.1\ .$$

From the graph of $\phi(\lambda\tau)$ we find that $\lambda\tau_0 = 0.420$. The optimal maintenance period minimizing the average loss per time unit is

$$\tau_0 = \frac{0.420}{0.10} = 42\ \text{h}\ .$$

In accordance with the formula for C^* from Table 23.3, the minimum cost is

$$C^*(42) = 2\,\frac{1 - e^{-0.420}}{e^{-0.420}} + (2\cdot 30 - 1\cdot 20)\cdot 0.01 = 1.444\ .$$

4. *Average profit per time unit*. Substituting the d.f. $F(t) = 1 - e^{-\lambda t}$, and using the equation for C from Table 23.3 to find τ_0, we write

$$\frac{(c_0 + c_\pi)T_\pi - \lambda T_a T_\pi(c_a - c_\pi)}{[(c_0 + c_c)/\lambda] + (c_c + c_\pi)T_a} = (e^{\lambda\tau} - 1) - \lambda\tau + \lambda\tau\,\frac{T_a(c_a - c_\pi) + T_\pi(c_c - c_\pi)}{[(c_0 + c_c)/\lambda] + T_a(c_c - c_\pi)}\ . \quad (23.1)$$

If the coefficient

$$\frac{T_a(c_a - c_\pi) + T_\pi(c_c - c_\pi)}{[(c_0 + c_c)/\lambda] + T_a(c_c - c_\pi)}$$

is much less than 1, the last term in the right-hand side of (23.1) may be neglected. Then we can write

$$\frac{(c_0 + c_\pi)T_\pi - \lambda T_a T_\pi (c_a - c_\pi)}{[(c_0 - c_c)/\lambda] + (c_c - c_\pi)T_a} = (e^{\lambda\tau} - 1) - \lambda\tau = \phi(\lambda\tau)\,. \tag{23.2}$$

In this example

$$\frac{(c_0 + c_\pi)T_\pi - \lambda T_a T_\pi (c_a - c_\pi)}{[(c_0 - c_c)/\lambda] + (c_c - c_\pi)T_a} = \frac{30(2-1) + 20(2-1)}{(3 + 2/0.01) + 30(2-1)} = \frac{50}{530}$$
$$= 0.0943 \ll 1\,.$$

We find an approximate solution for (23.2):

$$\phi(\lambda\tau) = \frac{4 \cdot 20 - 0.01 \cdot 20 \cdot 30 \cdot 1}{100 \cdot 5 + 30 \cdot 1} = \frac{74}{530} = 0.140\,.$$

Using a graph of $\phi(\lambda\tau)$, we find the root of the equation: $\lambda\tau = 0.490$. The optimal maintenance period is

$$\tau_0 = \frac{0.490}{0.01} = 49\text{ h}\,.$$

The maximum value of the average profit per unit of time is

$$C(49) = \frac{3e^{-0.49} - 2(1 - e^{0.49}) - (2 \cdot 30 - 1 \cdot 20) \cdot 0.01e^{-0.49}}{1 + (30 - 20) \cdot 0.01e^{-0.49}} = 0.771\,.$$

23.5 MODEL 4, REPAIRS OF FAILED UNITS OF SERIES SYSTEM

A series system consists of m units. Failure of any unit results in system failure. A failure is discovered after a time, τ_h, that is independent of the type of failed unit. The system is then repaired and the duration of the repair process of the ith unit failure equals t_i, $i = 1, 2, \ldots, m$. After repair the system continues its operation. During the repair all units of the system are switched off and their reliability parameters do not worsen. The main feature of the system is no total renewal after failure.

The data needed are T_i, t_i, t_c, c_i, and c_0. Expressions for calculation of optimal maintenance parameters are in Table 23.3. In this model no preventive maintenance is planned. Thus the model is used only for calculation of numerical values of the system indexes.

EXAMPLE 23.4

The following are the input data for the four units:

i	T_i hours	t_i hours	t_c hours	c_i cost/hour	c_c cost/hour	c_0 cost/hour
1	15	2	1	0.5	1.5	2
2	10	1	1	1.0	1.5	2
3	8	1	1	0.8	1.5	2
4	5	1	1	0.3	1.5	2

Let us calculate the system parameters:

1. *Availability coefficient.*

$$K = \frac{1}{1 + (1+2)/15 + (1+1)/10 + (1+1)/8 + (1+1)/5} = \frac{20}{37}$$
$$= 0.541 .$$

2. *Average expense per unit of time.*

$$C^* = \frac{0 \cdot 2 + 1.5 \cdot 1}{15} + \frac{1 \cdot 1 + 1.5 \cdot 1}{10} + \frac{0.8 \cdot 1 + 1.5 \cdot 1}{8}$$
$$+ \frac{0.3 \cdot 1 + 1.5 \cdot 1}{5} = 1.06 \text{ cost units per hour} .$$

3. *Average profit per unit of time.*

$$C = (2 - 1.06) \cdot 0.541 = 0.509 \text{ cost unit per hour} .$$

23.6 MODEL 5, REPAIRS OF FAILED UNITS AND PLANNED MAINTENANCE OF SERIES SYSTEM

A system consists of m units or subsystems in series. Repair starts immediately after failure. Repair or planned maintenance renews the system completely. The first maintenance is planned after some given period of operation.

During unit repair the system is off. Thus the calendar time of a current maintenance period is random and depends on the number of failures. After maintenance the system is completely new, and the process starts from the beginning.

The data needed are $F_i(t)$, T_i, T_π, t_i, c_π, c_i, c_0, and t_0. Some additional mathematical operations are necessary. For each ith unit, there is a renewal function, $H_i(t)$, given by the following integral equation:

$$H_i(t) = F_i(t) + \int_0^t H_i(t-x)\, dF_i(x)\,.$$

We can find the solution with the help of a Laplace-Stieltjes transform (LST) equation:

$$H_i^*(s) = \frac{F_i^*(s)}{1 - F_i^*(s)}\,,$$

where $H_i^*(s)$ is the LST of $H_i(t)$, and $F_i^*(s)$ is the LST of $F_i(t)$. Equations for calculating the optimal maintenance parameters are in Table 23.4.

EXAMPLE 23.5

The system input data are $F(t) = (1 - e^{-\lambda t})^2$, $\lambda = 0.01$ per hour, $T_\pi = 1$ hour, $T_a = 10$ hours, $c_\pi = 1$ cost unit per hour, $c_a = 2$ cost units per hour, $c_0 = 3$ cost units per hour, and $t_0 = 2$ hours. We want to find the optimal planned maintenance period.

Solution

We first determine some additional expressions. We define

$$F^*(s) = s\int_0^\infty e^{-st}F(t)\, dt = \frac{2\lambda^2}{s^2 + 3\lambda s + 2\lambda^2}\,.$$

Then

$$H^*(s) = \frac{F^*(s)}{1 - F^*(s)} = \frac{2\lambda^2}{s(s+3\lambda)} = \frac{2\lambda}{3s} - \frac{2\lambda}{3(s+3\lambda)}\,,$$

and

$$H(t) = \frac{2}{3}\lambda t - \frac{2\lambda}{3}\,\frac{1 - e^{-3\lambda t}}{3\lambda} = \frac{2}{3}\lambda t - \frac{2}{9} + \frac{2}{9}e^{-3\lambda t}\,,$$

$$H'(t) = \frac{2}{3}\lambda - \frac{2}{3}\lambda e^{-3\lambda t} = \frac{2}{3}\lambda(1 - e^{-3\lambda t})\,.$$

1. *Availability coefficient*. By substituting input data into the equation for finding the τ_0 that optimizes K in Table 23.4, we have

$$\tau T_a \frac{2}{3}\lambda(1 - e^{-3\lambda t}) - T_a\left(\frac{2}{3}\lambda\tau - \frac{2}{9} + \frac{2}{9}e^{-3\lambda t}\right) = T_\pi\,,$$

or

Table 23.4 Main performance effectiveness indexes of model 5: Unplanned maintenance for units, planned maintenance for the system

Index	Formula for Root τ_0 Equal to the Value of the Period of the Optimal Maintenance	Formula for Index Calculation for a Single Root τ_0	Formula for $\tau_0 = \infty$
K	$\tau \sum_{i=1}^{m} t_i H_i'(\tau) - \sum_{i=1}^{m} t_i H_i(\tau) = T_\pi$	$\frac{1}{1 + \Sigma_{i=1}^{m} t_i H_i'(\tau_0)}$	$\frac{1}{1 + \Sigma_{i=1}^{m} t_i / T_i}$
$R(t_0)$	$\frac{\int_0^\tau \Sigma_{i=1}^{m} [P_i(x + t_0) + \int_0^x P_i(x + t_0 - y)\, dH_i(y)]\, dx}{\tau + T_\pi + \Sigma_{i=1}^{m} t_i H_i(\tau)}$	$\frac{\Pi_{i=1}^{m} [P(\tau_0 + t_0) + \int_0^{\tau_0} P_i(\tau_0 + t_0 - y)\, dH_i(y)]}{1 + \Sigma_{i=1}^{m} t_i H_i'(\tau_0)}$	$\frac{\Pi_{i=1}^{m} [\int_0^\infty P_i(x + t_0)\, dx] / T_i}{1 + \Sigma_{i=1}^{m} t_i / T_i}$
	$= \frac{\Sigma_{i=1}^{m} [P_i(x + t_0) + \int_0^x P_i(\tau + t_0 - y)\, dH_i(y)]}{1 + \Sigma_{i=1}^{m} t_i H_i'(\tau)}$		
C^*	$\tau \sum_{i=1}^{m} c_i t_1 H_i'(\tau) - \sum_{i=1}^{m} c_i t_i H_i(\tau) - c_\pi T_\pi$	$\sum_{i=1}^{m} c_i t_i H_i'(\tau_0)$	$\sum_{i=1}^{m} \frac{c_\pi i_i}{T_i}$
C	$\sum_{i=1}^{m} \sum_{j=1}^{m} t_i t_j H_i(\tau) H_j'(\tau)(c_j - c_i)$	$\frac{c_0 - \Sigma_{i=1}^{m} T_i c_i H_i'(\tau_0)}{1 + \Sigma_{i=1}^{m} t_i H_i'(\tau_0)}$	$\frac{c_0 - \Sigma_{i=1}^{m} c_\pi (t_i / T_i)}{1 + \Sigma_{i=1}^{m} (t_c / T_i)}$
	$+ T_{\pi\pi} \sum_{i=1}^{m} t_i H_i'(\tau)(c_i - c_\pi) + \tau \sum_{i=1}^{m} t_i H_i'(\tau)(c_0 + c_1)$		
	$- \sum_{i=1}^{m} t_i H_i(\tau)(c_0 + c_i) = T_\pi (c_0 + c_\pi)$		

Note: If there are several roots $\tau_{01}, \ldots, \tau_{0s}$, the optimal one must be found by directly substituting each root into the formula and choosing the maximum reliability index or the minimum cost.

$$\frac{9T_\pi}{2T_a} = 1 - e^{-3\lambda t}(1 + 3\lambda t) .$$

We have the equation

$$\phi(x) = [1 - (1 + x)e^{-x}] .$$

We also have

$$\frac{9T_\pi}{2T_a} = 0.45 ,$$

$$x_0 = 1.524 ,$$

$$\tau_0 = \frac{x_0}{3\lambda} \frac{1.524}{0.03} = 50.8 \text{ h} .$$

The maximum value of the system availability coefficient is attained at $\tau_0 = 50.8$ hours:

$$K_{\tau_0} = \frac{1}{1 + T_a H'(\tau_0) = 0.9505} .$$

For comparison, let us find the system availability coefficient for $\tau = \infty$, that is, when there is no planned maintenance:

$$K = \frac{T}{T + T_a} = \frac{15}{16} = 0.9375 .$$

2. *Operational availability coefficient.* After substitution of functions $H(t)$ and $H'(t)$ into the expression for $R(t_0)$ in Table 23.4, we write

$$\frac{6\lambda T_\pi(1 - 0.25e^{-\lambda t_0}) - (4/3)\lambda T_a(1 - e^{-\lambda t_0})}{(1 + 2\lambda T_a) - (1 + \lambda T_a)e^{-\lambda t_0}}$$

$$= 1 - (1 + 3\lambda\tau)e^{-3\lambda\tau} - \frac{3\lambda T_\pi(1 - e^{-\lambda t_0})}{(1 + 2\lambda T_a) - (1 + \lambda T_a)e^{-\lambda t_0}} .$$

We can use an approximate value of the root $x_0 = 3\lambda\tau_0$. The left-hand side of the preceding equation equals

$$\frac{6\lambda T_a(1 - 0.25e^{-\lambda t_0}) - (4/3)\lambda T_a(1 - e^{-\lambda t_0})}{(1 + 2\lambda T_a) - (1 + \lambda T_a)e^{-\lambda t_0}} = 0.3502 .$$

We have $x_0 = 3\lambda t_0 = 1.236$. Hence $t_0 = 41.2$ hours. The maximum value of $R(t_0)$ can be calculated with the equation in Table 23.3 and equals 0.973.

For comparison, let us calculate the probability of failure-free operation without planned maintenance, using the expression for $R(t_0)$ from Table 23.4: $R(t_0) = 0.925$. Notice that if the operation is performed by a completely new system, $R(t_0) = \exp(-\lambda t_0) = 0.9802$.

3. *Average expense per unit of time*. Substituting input data in the equation for C^* in Table 23.3 to find τ_0, we write

$$\frac{9}{2}\frac{c_\pi T_\pi}{c_a T_a} = 1 - e^{-3\lambda\tau}(1 + \lambda\tau) .$$

With

$$\frac{9}{2}\frac{c_\pi T_\pi}{c_a T_a} = 0.225 ,$$

we find that $x_0 = 3\lambda\tau_0 = 0.888$, or

$$\tau_0 = \frac{0.888}{0.03} = 29.6 \text{ h} .$$

Hence the optimal maintenance period equals 29.6 hours and $C^*(29.6) = 0.0786$ cost units per hour. Without planned maintenance (i.e., for $\tau = \infty$) the average cost per unit of time is

$$C^*(\infty) = \frac{c_a T_\pi}{T} = \frac{2 \cdot 10}{150} = 0.1333 \text{ cost unit per hour} .$$

4. *Average profit per unit of time*. Substituting input data into the equation for C in Table 23.4, we find the optimal value of the root: $\lambda\tau = x = 1.248$. Hence

$$\tau_0 = \frac{1.248}{3 \cdot 10^{-2}} = 41.6 \text{ h} .$$

For this period of optimal maintenance:

$$\max_\tau C(\tau) = C(41.6) = \frac{c_0 - c_a T_a H'(\tau_0)}{1 + T_a H'(\tau_0)} = 2.78 \text{ cost units per hour} .$$

For comparison, let us present the corresponding value with no planned maintenance ($\tau = \infty$):

$$C(\infty) = \frac{3 - 2(10/150)}{1 + (10/150)} = 2.68 \text{ cost units per hour} .$$

CHAPTER 24

OPTIMAL DIAGNOSIS

Technical diagnostic problems, such as finding system failure states or searching for failed units, can be difficult. An electronic system usually is regarded as an object of diagnosis represented by n units (set Ω) connected with each other by functional ties. Each unit has two possible states: operability or failure. The notation p_i is the probability of an operational state of the ith unit, and q_i is the probability of a failure state of the ith unit ($q_1 = 1 - p_i$). All units are considered independent. System diagnosis is based on tests, with each test determining the state of a subset of units. The diagnosis has two main goals: (1) to find if the system is up or down, and (2) to identify all failed units if the system is down.

Overall failure or operability usually is found with a general test. However, if the general test requires too much time or other resources, one can use a sequence of several ordinary tests.

For both tasks, one can use a set of tests T_i, $i = 1, \ldots, m$, each of which checks a subset of units, Ω_i. The results of each test are classified as "successful" if all units belonging to the subset Ω_i are operational, or "unsuccessful" if at least one unit of the set has failed. Each test has a cost, c_i, that can be in time, money, test equipment, or a combination of resources. A set of tests can be given in the form of the matrix $\mathbf{T} = |T_{ij}|$, $i = 1, \ldots, m, j + 1, \ldots, n$, with columns corresponding to types of tests and rows corresponding to units of a set Ω. A T_{ij} is 0 if the row j unit is not tested in the column i test; the T_{ij} is one if the unit is tested in the test. In symbols,

$$T_{ij} = \begin{cases} 1 & \text{if } j \in \Omega_i , \\ 0 & \text{if } j \in \bar{\Omega}_i , \end{cases} \qquad i = 1, \ldots, m, \quad j = 1, \ldots, n .$$

A row-vector $c = \{c_1, \ldots, c_m$ consists of the test costs. The complete set of tests must be able to isolate any possible failure.

A diagnosis can be classified by several criteria, as discussed in the following paragraphs.

1. The diagnosis procedures can be characterized by depth. The diagnosis can be intended to determine only whether the whole system has failed. A deep diagnosis is intended to identify the state of each unit.
2. The diagnosis procedures can be either sequential or combinatorial. In the first case, the order of the future tests is determined by the results of the previous tests. In the second case, the entire sequence of tests is planned in advance and final conclusions come only after the entire sequence.
3. The diagnosis procedures can have either of two types of optimization: (1) minimization of the expected cost of the whole diagnosis, or (2) minimization of the highest possible cost of the tests.
4. The diagnosis procedure depends on whether only unit failure or any combination of failed units is possible in the system.
5. In the latter case the diagnosis can be combined with repair of the failed units. The diagnosis procedure continues until the system is completely repaired. If there is no repair during the diagnosis, the procedure stops at the moment of complete identification of all of the system unit states.
6. The diagnosis procedure can cover some or all system units.
7. The diagnosis procedure can be performed with full confidence tests or with partial confidence tests.
8. Finally, input data for decision making (information available during the diagnosis or after it) can be complete or incomplete.

Optimization of the diagnosis procedure is a specialized engineering problem that cannot be covered in detail in one chapter. This chapter gives only the simplest mathematical models and algorithms used to diagnose or to design test equipment.

24.1 ONE FAILED UNIT

24.1.1 Notation

One can know with certainty that a single failed unit is in the system under diagnosis. Such a situation exists, for example, if system operation stops after any failure, as with a series system. Let the test matrix, **T**, and probabilities, q_i, be given. The optimization problem then is selection of a set of tests and selection of an order of performance that will find the failed unit at minimum expense.

Performance of a test T_i can be considered as a division of the set Ω into two subsets, Ω_i and $\tilde{\Omega}_i$. If the test is "unsuccessful," the failed unit is in the subset Ω_i, and if the test is "successful," the failed unit is in the subset $\tilde{\Omega}_i$. As a follow-up one can use tests $T_* \in T$ (below we refer to these tests as "essential") that divide the subset Ω_i or $\tilde{\Omega}_i$ to isolate the failed unit. A test T_* is essential for a subset Ω_i if and only if

$$\Omega_* \wedge \Omega_i \neq \Omega_i$$

and

$$\Omega_* \wedge \Omega_i \neq \emptyset ,$$

where $\wedge$ is a symbol for the intersection of two sets and Ω_* is the subset covered by the test T_*. A list of essential tests for the subset Ω_i is denoted T_i. If, for two tests, T_u and T_v from T_i,

$$\Omega_* \wedge \Omega_i = \Omega_v \wedge \Omega_i ,$$

then the list T_i should be left only with the test with the lowest cost. List T_i can be represented by a matrix whose rows correspond to units of the set Ω_i. Each kth line of this matrix corresponds to a test, $T^k_{u(i)}$, $k = 1, \ldots, m_i$ that checks the operational state of the subset

$$\Omega^k_{u(i)} = \Omega_u \wedge \Omega_i$$

with costs $c^k_{u(i)} = c_u$. A subscript without parentheses, which may be omitted for simplicity, identifies each test position in the original matrix T. A test matrix for subset $\tilde{\Omega}_i$ can be formed analogously if a failed unit occurs in this subset.

If there is a single failed unit in the system, the matrix **T** is sufficient and all its pairs of rows are tested.

24.1.2 Tests with Equal Costs

In some situations one can arrange tests to check any needed subset of units and all the tests are similar in cost. For example, an automation test procedure can have practically the same checking time for any test. The units in the system have different probabilities of failure. The problem is to combine subsets of units to minimize expected expenses.

For this purpose we can perform the following steps:

1. Order units by nonincreasing values of $\tilde{q}_i$.
2. Combine the last two units into a new conditional one with

$$\tilde{q} = \tilde{q}_n + \tilde{q}_{n-1} \cdot$$

3. Put the new conditional unit at the appropriate place in the sequence of the other $n-2$ units with the same rule.
4. Continue the process of forming equivalent units until all system units have been combined into a single conditional unit.
5. The optimal diagnostic program follows in reverse order.

EXAMPLE 24.1

Let a system consist of seven units with the following parameters:

$$q_1^* = 0.3\,, \quad q_2^* = 0.3\,, \quad q_3^* = 0.1\,, \quad q_4^* = 0.1\,,$$
$$q_5^* = 0.5\,, \quad q_6^* = 0.05\,, \quad q_7^* = 0.03\,.$$

Construct a diagnosis program.

Solution

Figure 24.1 shows a diagnosis program made in accordance with the preceding procedure. As shown in the diagnostic tree, the first test, T_1, divides the set into two subsets: $\{1, 2\}$ and $\{3, 4, 5, 6, 7\}$. If a failed unit is from the first subset, we use test T_2. If a failed unit is from the second subset, we use test T_3, which divides the subset in two: $\{3, 4\}$ and $\{5, 6, 7\}$. The procedure continues under the same logic.

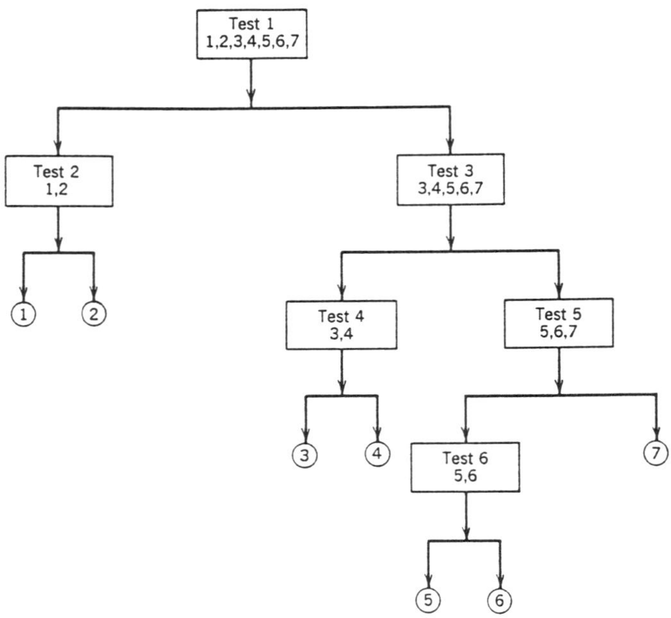

Figure 24.1 Complete diagnostic procedure for Example 24.1.

24.1.3 Approximate Solution with the Use of Information Quantity

The approximate solution method is used when test costs differ and it is not possible to use flexible tests with any desired subsets. The algorithm is the following:

1. Consider all possible tests T_i with a subset of units Ω_i. For each test calculate

$$g_i = \frac{\Delta H_i}{c_i} ,$$

$$\Delta H_i = P_i \ln P_i + Q_i \ln Q_i ,$$

where Q_i is the probability that a failed unit belongs to the subset Ω_i and $P_i = 1 - Q_i$.
2. Choose the test with the maximum g_i.
3. At the next step of the procedure, apply the same rule to the remaining subset of units, which includes a failed unit. The choice of the subset depends on the result of the previous test.

24.1.4 Dynamic Programming Method

First, let us introduce some notation.

The probability that the test of subset Ω_i (or $\bar{\Omega}_i$) will be unsuccessful is

$$Q_i = \sum_{j \in \Omega_i} \tilde{q}_j .$$

For $\bar{\Omega}_i$ we have

$$Q_{\bar{i}} = \sum_{j \in \Omega_i} \tilde{q}_j .$$

If previous steps have confirmed that the failed unit is in subset Ω_i, the probability that the unit is in subset Ω_u, $\Omega_u \subset \Omega_i$, is

$$\Omega_{u(i)} = \frac{Q_u}{Q_i} .$$

Let a procedure for failed unit localization be determined for a subset Ω_i containing a failed unit. Denote this program as $\sigma(\Omega_i)$. If the first step includes the test $T^k_{(i)} \in T_i$, then the program $\sigma(\Omega_i)$ can be presented as

$$\sigma(\Omega_i) = [T^k_{(i)}, \sigma(\Omega^k_{(i)}), \sigma(\bar{\Omega}^k_{(i)})] ,$$

where $\Omega^k_{(i)}$ and $\bar{\Omega}^k_{(i)}$ are the subsets into which the set Ω_i is divided by the test

$T^k_{(i)}$, and $\sigma(x)$ is the sequence of tests for subset x. The expected value of the cost of program $\sigma(\Omega_i)$ can be found from a recurrent expression

$$C[\sigma(\Omega_i)] = c^k_{(i)} - Q^k_{(i)} C[\sigma(\Omega^k_{(i)})] + (1 - Q^k_{(i)}) C[\sigma(\bar{\Omega}^k_{(i)})] , \qquad (25.1)$$

where all calculations start from two-unit sets for which the cost of the failed unit localization equals the minimal cost of dividing the subset into two one-unit subsets.

The dynamic programming method is based on systematic consideration of subsets $\Omega_i \in \Omega$ and determination of the test $T^k_{(i)} \in T_i$ that divides this subset optimally.

If $\sigma^*(\Omega_i)$ is the optimal program for the subset Ω_i containing the failed unit, and $C[\sigma^*(\Omega_i)]$ is the cost of this program, the procedure for finding $\sigma^*(\Omega)$ can be represented as a recurrent procedure for finding $\sigma^*(\Omega_i)$ for all subsets of x units, where $x = 2$ for the first iteration and increases by one with each iteration. At each step of the procedure, one finds a test $T^k_{(i)} \in T_i$ such that

$$C[\sigma^*(\Omega_i)] = \min_k \{C^k_{(i)} + Q^k_{(i)} C[\sigma^*(\Omega^k_{(i)})] + (1 - Q^k_{(i)}) C[\sigma^*(\bar{\Omega}^k_{(i)})]\} .$$

To construct the optimal diagnostic program by the dynamic programming method, one has to find and remember all tests that divide all current subsets of the set Ω optimally. The dynamic programming method provides the optimal solution to the diagnostic problem, but its direct use for problems of practical dimensions can be unrealistic. These circumstances may lead to algorithms for approximate solutions or to some other strict procedure.

24.1.5 The Perturbation Method for One-Unit Tests

If one-unit tests are available, a simple form of dynamic programming can be used. The perturbation method is based on the fact that pair transpositions of tests can be used to obtain any possible sequence of tests, including the optimal sequence. The problem is to find a monotonic procedure of test transposition that produces the optimal procedure for a finite number of steps.

In a form suitable for any set of tests, the objective function for this procedure may be expressed with

$$C[\sigma(\Omega)] = c_1 + \tilde{q}_1 C[\sigma(e_1)] + \tilde{p}_1 C[\sigma(\Omega \backslash e_1)] ,$$

where

$$C[\sigma(e_1)] = 0 ,$$

e_1 is a one-unit subset that does not affect costs of future checking, and

$$C[\sigma(\Omega\backslash e_1)] = c_2 + C[\sigma(\Omega\backslash(e_1 \vee e_2))] .$$

Therefore one can write

$$C = c_1 + \tilde{p}_1(c_2 + \tilde{p}_2(c_3 + \tilde{p}_3(c_4 + \cdots))) .$$

An analogous procedure can be performed for the fixed test sequences for units with subscripts k and $k+1$. The total expected cost is calculated. The order of only these two units then is changed for these sequences, and the new total expected cost is calculated. The cost difference for the two diagnostic procedures will depend only on two sets of values corresponding to units k and $k+1$. The conclusion is that one-unit tests can be performed in the order that corresponds to the multiple inequality

$$\frac{c_1}{q_1} \leq \frac{c_2}{q_2} \leq \cdots \leq \frac{c_a}{q_n} .$$

When two ratios are equal, the order is immaterial.

24.1.6 Recursive Method

For a strict solution one can use the recursive method, which in practice can be more effective than the dynamic programming method. The search for a failed unit in the set Ω can be regarded as a multistep process of sequential division of its subsets until one-unit subsets are obtained. The core of the method is construction of a "dividing tree" by sequential division of Ω. Each branch of this tree is divided until the best are on the last layers of one-unit subsets. A return is then made to the previous layer of the dividing tree.

For division of the set Ω at the first step (layer), one can use any test from matrix T. Consider the first test in any number of tests. This test divides the set Ω into Ω_1 and $\tilde{\Omega}_1$. Suppose that the optimal programs for further searching for the failed unit:

$$\sigma_1^*(\Omega_1) \qquad \text{if the test } T_1 \text{ is unsuccessful ,}$$

$$\sigma_1^*(\tilde{\Omega}_1) \qquad \text{if the test } T_1 \text{ is successful ,}$$

are known. Then one can construct the conditionally optimal program

$$\sigma_1^*(\Omega) = [T_1, \sigma^*(\Omega_1), \sigma^*(\bar{\Omega}_1)]$$

and calculate the expected cost of this procedure with (24.1). One can similarly find the conditionally optimal program $\sigma_2^*(\Omega)$. The program with the lowest cost can be discovered.

Assume that for some subset Ω_1 the program $\sigma^*(\Omega_i)$ needed for the construction of $\sigma_i^*(\Omega_i)$ is unknown. To solve this, first select the best of the previously discovered $i-1$ conditionally optimal programs, and consider all possible divisions of the subset Ω_i. Then format the matrix of essential tests, $\mathbf{T}_i$, and for each $t_{(i)}^k \in T_i$ find the conditionally optimal program:

$$\sigma^{k*}(\Omega_i) = \{T_{(i)}^{(k)}, \sigma^*(\Omega_{(i)}^k), \sigma^*(\bar{\Omega}_{(i)}^k)\}, \qquad k = 1, \ldots, m.$$

Then find the optimal $\sigma^*(\Omega)$:

$$C[\sigma^*(\Omega_i)] = \min_{1 \le k \le m_i} C[\sigma^{k*}(\Omega_i)].$$

As the previous calculations were interrupted only because of the unknown $\sigma^*(\Omega_i)$, they can now be resumed.

A systematic application of the procedure for each subset obtained by division of the set Ω leads to the construction of the optimal program for the entire system. The efficiency of this method can be increased by the elimination of suboptimal programs and by elimination of branches that are not candidates. This can be done by a comparison of the best candidate with the best program obtained up to the current step.

As follows from the foregoing, the conditionally optimal programs $\sigma^{k*}(\Omega_i)$, $k = 1, \ldots, m_i$, for the determination of the optimal program $\sigma^*(\Omega_i)$ may be developed conveniently with development of the sequence of tests $t_{(i)}^k$ of the matrix $\mathbf{T}_1$. To improve the procedure for elimination of nonprospective branches, we should try to find order tests in approximate order of cost. Then any simple method for the estimation of the lower bound of the unknown continuations of any branch can be used to eliminate nonprospective branches from the very beginning. At some step we may have found $k-1$ conditionally optimal programs and have chosen the best one, $\sigma^*(\Omega_i)$, with the cost $C[\sigma^*(\Omega_i)]$. Let $\Gamma_{(i)}^k$ be the lower bound of the cost of the conditionally optimal program $\sigma^{k*}(\Omega_i)$. It is obvious that the program $\sigma^{k*}(\Omega_i)$ cannot be the best one if

$$C[\sigma^{v*}(\Omega_i)] \le \Gamma_{(i)}^k.$$

The program $\sigma^{k*}(\Omega_i)$ is not potentially optimal and can be eliminated without further analysis.

The best procedures for selecting the test sequence depend on the equipment in question. But some simple general recommendations can be made. The tests should be ranked in order of nondecreasing cost. To determine $\Gamma_{(i)}^k$, we can use a simple corollary of (24.1). The cost of a conditionally optimal program, $\sigma^{k*}(\Omega_i)$, cannot be less than the cost of the test $T_{(i)}^k$ used at the first step of division of the set Ω_i. In other words, one can take $\Gamma_{(i)}^k = c_{(i)}^k$.

The logic behind the diagnostic program construction is the same for all layers, as is clear from the recurrent structure of the problem. To use this property, we rank tests of the original matrix $\mathbf{T}$ in order of the number of units in each line of the test table.

24.1.7 Minimax Criterion

In practice diagnostic procedures often are planned with no reliability data or with inadequate data. In this case a different optimality criterion will be different. Under the minimax criterion, we find the highest possible cost for each program, which might be called "maxposs," and select the program with the lowest maxposs.

Consider a program, $\sigma(\Omega_i)$. Let the first test used be $t^k_{(i)} \in T_1$. The cost of this program is obtained using

$$R[\sigma(\Omega_i)] = c^k_{(i)} + \max\{R[\sigma(\Omega^k_{(i)})], R[\sigma(\bar{\Omega}^k_{(i)})]\} . \tag{24.2}$$

The procedure can be applied in this case with success if one substitutes the expected costs $C[\sigma(\Omega_i)]$ with the maximal costs found with the help of 24.1.

24.2 SEQUENTIAL SEARCHES FOR MULTIPLE FAILED UNITS

24.2.1 Description of the Procedure

There often is a possibility of more than one failure. In some cases any combination of failed units may be possible. A succession of alternating replacement and diagnosis may be followed. This starts when sequential testing has found the first of several failed units. The failed unit is replaced. Then the minimal subset that includes the new unit, Ω_*, is tested. If the subset still is failed, the search continues for this subset in accordance with the same rules. If the test is successful, however, the searching process is continued for the minimal subset such that $\Omega_* \subset \Omega_*$. When a failure is found, it is corrected. The procedure continues until the system is repaired. The replacement cost does not affect the diagnosis procedure.

In many cases it is possible to perform a general (or global) test to determine whether the system has any failed units. Sometimes, however, a general test may cost too much relative to the subset tests, especially toward the end of the diagnosis.

24.2.2 Method of Sequential One-Unit Tests

A system without continuous monitoring can have several failures by the time the system failure has been discovered. It may be possible to use the general test to check the state of the whole system as well as to perform one-unit tests. Cost of the ith test equals c_i.

One first applies the general test. If the system has failed, the search begins. After the failed unit has been found and replaced, the general test is applied again. If at least one failed unit remains, the procedure continues.

The task is to find the sequence of unit checks to minimize the expected cost. For each unit one calculates

$$g_j = \frac{c_j p_j}{q_j} .$$

The checking procedure follows the order of increasing values of g_j.

EXAMPLE 24.2

Consider a system consisting of five units with the following parameters:

$p_1 = 0.2,\ c_i = 10$ [min],
$p_2 = 0.3,\ c_2 = 20$ [min],
$p_3 = 0.4,\ c_3 = 10$ [min],
$p_4 = 0.5,\ c_4 = 5$ [min],
$p_5 = 0.6,\ c_5 = 10$ [min].

Find the optimal sequence of unit tests.

Solution

Calculate the values g_j:

$$g_1 = \frac{0.2}{0.8}\ 10 = 2.5 ,$$

$$g_2 = \frac{0.3}{0.7}\ 20 = 8.6 ,$$

$$g_3 = \frac{0.4}{0.6}\ 10 = 6.7 ,$$

$$g_4 = \frac{0.5}{0.5} = 5.0 ,$$

$$g_5 = \frac{0.6}{0.4}\ 10 = 15.0 .$$

The best order is T_1, T_4, T_3, T_2, T_5.

24.2.3 Recursive Method

Let test T_1 divide set Ω into Ω_i and $\tilde{\Omega}_i$. If the test is successful, Ω_i can be divided only with essential tests $T_* \in T$ for which $\Omega_* \in \Omega$. The matrix of these tests is denoted T_i. If the first test is unsuccessful, one should choose tests from T that intersect with units of the set $\tilde{\Omega}_i$. For this task pair differences between rows of the initial matrix T are not sufficient.

A subset Ω_i subjected to the test T_i is called "localized" if tests in matrix T permit identification of all failed units when all discovered failed units are replaced. Matrix T is "sufficient" when it is possible to cover the set Ω with subsets Ω_i so that each subset is localized in turn. Localization can include not more than one unit localized by exclusion.

Localization in a two-unit subset requires a test that can be used to check either unit. A three-unit subset requires the capability to localize any of its two-unit subsets. Requirements for larger subsets may be found by inference.

Sufficiency of a matrix **T** can be checked systematically. First, consideration is given to a test, T_i, that can cover more than one unit. A new matrix, $\mathbf{T}_i$, of essential tests is then developed in accordance with the above procedure. If this new matrix has no lines, the subset Ω_i is not localized. One must add new tests or decide not to localize the subset Ω_i. If $\mathbf{T}_i$ contains several lines, choose the subset with the smallest number of lines. If the localization property of this subset is not obvious, construct a matrix of essential tests and check again. Continue this process until finding that the subset Ω_i can be localized, cannot be localized, or can be localized only in part.

Let $\Psi(\Omega_i)$ denote a program for sequential searching for failed units in subset Ω_i. Assume that at the first step of this program one uses the test $T_i^k \in T_i$. One also knows the further programs $\Psi(\Omega_{(i)}^k)$ and $\Psi(\tilde{\Omega}_{(i)}^k)$ to continue the optimal diagnostic procedure in corresponding subsets $\Omega_{(i)}^k$ and $\tilde{\Omega}_{(i)}^k$. Then $\Psi(\Omega_i)$ can be expressed:

$$\Psi(\Omega_i) = [T_{(i)}^k, \Psi(\Omega_{(i)}^k), \Psi(\tilde{\Omega}_{(i)}^k)] . \tag{24.3}$$

The cost of this program is a random variable depending on the current state of the system units. The expected value of the cost is

$$C[\Psi(\Omega_i)] = c_{(i)}^k + \tilde{Q}_{(i)}^k\{C[\Psi(\Omega_{(i)}^k)] + c_{0(i)}\} + (1 - \tilde{Q}_{(i)}^k)C[\Psi(\tilde{\Omega}_{(i)}^k)] ,$$

where $\tilde{Q}_{(i)}^k$ is the posterior probability of failure of at least one unit in the subset $\Omega_{(i)}^k$ when a failure has been detected in subset Ω_i:

$$\tilde{Q}_{(i)}^k = \frac{1 - \Pi_{j\in\Omega_{(i)}^k} p_i}{1 - \Pi_{j\in\Omega_i} p_i} .$$

This process practically coincides with that described in the previous section. The only important difference is in checking the potential of conditionally optimal programs. In this algorithm each conditionally optimal program is checked whether the previous one was prospectively optimal or not.

To test for rejection of conditionally optimal programs $\Psi^{k*}(\Omega_i)$, one uses the lower bound of the program cost calculated as

$$\Gamma_{(i)}^k = c_{(i)}^k + c_{(i)}\tilde{Q}_{(i)}^k .$$

The methodology can be used to develop a minimax program of diagnosis.

The maximal cost of program $\Psi^{k*}(\Omega_i)$ represented in the form of (24.3) is expressed with

$$R[\Psi^k(\Omega_i)] = c^k_{(i)} + c_{0(i)} + R[\Psi(\Omega^k_{(i)})] + R[\Psi(\bar{\Omega}^k_{(i)})] .$$

24.3 FAILURE DETERMINATION WITHOUT A TOTAL SYSTEM TEST

A total system test sometimes is too expensive or even nonexistent. There may be, however, tests for units and/or groups of units. The group tests may or may not overlap. A diagnostic procedure is wanted to determine if the system has failed or not, that is, to find at least one failed unit and to do this with minimal expected cost.

24.3.1 Approximate Algorithm for Arbitrary Intersected Tests

This section covers the procedure when group tests overlap. At least one unit is in more than one test.

Let there be a sequence of $m-1$ previous tests:

$$\sigma^{(M-1)} = \{T_1, \ldots, T_{(m-1)}\} .$$

The problem is to find at least one failed unit in subset $\Omega^{(M-1)}$. An algorithm for this consists of the following steps.

1. Find the probability, $Q_i^{(0)}$, of an unsuccessful outcome for each test T_i in the set $\Omega^{(0)}$, given the condition that a sequence of tests, $\sigma^{(0)}$, has been performed up to the moment:

$$\tilde{Q}_i^{(0)} = 1 - \prod_{j \in \{\Omega_j \wedge \Omega^{(0)}\}} p_j .$$

2. Find the cost, $c_i^{(0)}$, of each test under the condition that the sequence of tests $\sigma^{(0)}$ has been performed up to this moment. Before the checking process $c_i^{(0)} = c_i$, but the cost of the ith test can decrease or increase after previously performed tests.
3. For each test T_i determine

$$g_i^{(0)} = \frac{c_i^{(0)}}{\hat{Q}_i^{(0)}} .$$

4. Apply the test with minimum

$$g_k^{(0)} = \min_{1 \le i \le m} q_i^{(0)} .$$

5. If this test is successful, continue the procedure for the subset $\Omega^{(1)} = \Omega^{(0)} \backslash \Omega_k$. If this test is unsuccessful, stop the diagnostic procedure because the subset Ω_k contains at least one failed unit.

When the procedure is continued for subset $\Omega^{(i)}$, return to step 1 and follow the steps again for $M := M + 1$. The procedure continues until the first failed unit is found.

24.3.2 Nonintersecting Tests

When each unit can be checked with only one test, the tests are given in decreasing value of g_i, where

$$g_i = \frac{c_i}{\hat{Q}_i} \; .$$

EXAMPLE 24.3

The system is the one in Example 24.2. Find the optimal test sequence.

Solution

Calculate the g_1 values:

$$g_1 = \frac{1}{0.8} 10 = 12.5 \; ,$$

$$g_2 = \frac{1}{0.7} 20 = 28.6 \; ,$$

$$g_3 = \frac{1}{0.6} 10 = 16.7 \; ,$$

$$g_4 = \frac{1}{0.5} 5 = 10.0 \; ,$$

$$g_5 = \frac{1}{0.4} 10 = 25.0 \; .$$

The test order is 4, 1, 3, 5, 2.

APPENDIX A

FUNDAMENTAL PROBABILITY THEORY AND MATHEMATICAL STATISTICS

This appendix summarizes the fundamental concepts of probability and statistics that are used frequently in reliability analysis. Proofs and other details may be found in elementary textbooks.

A.1 MAIN THEOREMS

The *conditional probability* of event A_1 given the occurrence of the event A_2 is the probability of event A_1 computed under the assumption that the event A_2 has occurred:

$$Pr\{A_1 \mid A_2\} = \frac{P\{A_1 A_2\}}{Pr\{A_2\}} .$$

The *regression coefficient* of an event A_1 relative to an event A_2, $\rho(A_1, A_2)$, is a measure of their dependence:

$$\rho(A_1, A_2) = Pr\{A_1 \mid A_2\} - Pr\{A_1 \mid A_2'\} = \frac{Pr\{A_1 \wedge A_2\} - Pr\{A_1\}Pr\{A_2\}}{Pr\{A_2\}Pr\{A_2'\}} . \tag{A.1}$$

The *correlation coefficient* of events (A_1, A_2) is a measure of their dependence:

$$r(A_1, A_2) = \sqrt{\rho(A_1, A_2)\rho(A_2, A_1)} = \frac{Pr\{A_1 \wedge A_2\} - Pr\{A_1\}Pr\{A_2\}}{\sqrt{Pr\{A_1\}Pr\{A_1'\}Pr\{A_2\}Pr\{A_2'\}}} . \tag{A.2}$$

The *theorem of the product of probabilities* is the probability of the joint appearance of two events A_1 and A_2 in the same experiment, $Pr(A_1 \wedge A_2)$, which equals the probability of one of them multiplied by the conditional probability (computed under the assumption that the first event has occurred) of the other:

$$Pr\{A_1 \wedge A_2\} = Pr\{A_1 | A_2\} Pr\{A_2\} = Pr\{A_2 | A_1\} Pr\{A_1\} . \quad (A.3)$$

In the general case the theorem can be written

$$\begin{aligned} Pr\Big\{\bigcap_{1 \le i \le n} A_i\Big\} &= Pr\{A_1 \wedge A_2 \wedge \cdots \wedge A_n\} \\ &= Pr\{A_1 | A_2 \wedge \cdots \wedge A_n\} Pr\{A_2 | A_3 \wedge \cdots \wedge A_n\} \cdots \\ &\quad Pr\{A_{n-1} | A_n\} . \end{aligned} \quad (A.4)$$

The *theorem of the product of the probabilities of independent events*. For independent events in a set

$$Pr\Big\{\bigcap_{1 \le i \le n} A_i\Big\} = \prod_{1 \le i \le n} Pr\{A_i\} . \quad (A.5)$$

The *theorem of summation of probabilities of joint random events*. For joint events A_1 and A_2,

$$Pr\{A_1 \vee A_2\} = Pr\{A_1\} + Pr\{A_2\} - Pr\{A_1 \wedge A_2\} . \quad (A.6)$$

For n random events $A_1, A_2, \ldots, A_n$ the final result is

$$\begin{aligned} Pr\Big\{\bigcup_{1 \le i \le n} A_i\Big\} &= \sum_{1 \le I \le n} Pr\{A_i\} - \sum_{1 \le i \le n-1} \sum_{j \le i+1 \le n} Pr\{A_i \wedge A_j\} + \cdots \\ &\quad + \sum_{1 \le i \le n-2} \sum_{i+1 \le j \le n-1} \sum_{j+1 \le k \le n} Pr\{A_i \wedge A_j \wedge A_k\} \\ &\quad - \cdots + (-1)^{n+1} Pr\Big\{\bigcap_{1 \le i \le n} A_i\Big\} . \end{aligned}$$

For independent events in a set:

$$Pr\Big\{\bigcup_{1 \le i \le n} A\Big\} = 1 - Pr\Big\{\bigcap_{1 \le i \le n} A'_i\Big\} .$$

Finally

$$Pr\Big\{\bigcup_{1 \le i \le n} A_i\Big\} = 1 - \prod_{1 \le i \le n} (1 - Pr\{A_i\}) .$$

If $A_1, A_2, \ldots, A_n$ are disjoint events and A is the sum of these events, then:

$$Pr\{A\} = Pr\left\{\bigcup_{1 \le i \le n} A_i\right\} = \sum_{1 \le i \le n} Pr\{A_i\} .$$

The *formula of complete probability*. If event A_0 can occur if and only if one of the disjoint events $A_1, A_2, \ldots, A_n$ has occurred, the probability of the event A_0 equals

$$Pr\{A_0\} = \sum_{1 \le i \le n} Pr\{A_i\} Pr\{A_0 | A_i\} . \tag{A.7}$$

Bayes' formula (the *formula of hypothetical probabilities of hypotheses*) shows that disjoint events $A_1, A_2, \ldots, A_n$ such that an event of interest, A_0, can occur are called hypotheses relative to the event A_0. The probability, $P(A_i)$, that the hypothesis A_1 has occurred with no relation to an occurrence of A_0 is called a *prior probability*. The conditional probability of the hypothetical A_1 given that A_0 has occurred is called a *posterior probability*. The latter probability can be computed with

$$Pr\{A_i | A_0\} = \frac{Pr\{A_i\} Pr\{A_0 | A_i\}}{\sum_{1 \le i \le n} Pr\{A_i\} Pr\{A_0 | A_i\}} = \frac{Pr\{A_i\} Pr\{A_0 | A_i\}}{Pr\{A_0\}} . \tag{A.8}$$

A.2 RANDOM VARIABLES AND THEIR DISTRIBUTION FUNCTIONS

The *distribution function* (d.f.) of a random variable (r.v.), X, is a function, $F(x)$, that gives the probability that X will be less than or equal to any given value x:

$$F(x) = Pr\{X \le x\} . \tag{A.9}$$

If a discrete r.v. takes a value x_1 with a probability p_1 from a set of possible values $x_1 \le x_2 \le \cdots \le x_n$, then a d.f. can be written as

$$F(x_i) = Pr\{X \le x_i\} = \sum_{1 \le k \le i} p_k . \tag{A.10}$$

The *density function* is the first derivative of the d.f.:

$$f(x) = \lim_{\Delta x \to 0} \frac{F(x + \Delta x) - F(x)}{\Delta x} = \frac{d}{dx} F(x) = F'(x) . \tag{A.11}$$

The *intensity function*, or the *instant conditional density*, of a r.v. at the moment x is the probability density of the considered r.v. computed under the assumption that the r.v. is greater than x:

$$\lambda(x) = \frac{f(x)}{1 - F(x)} . \tag{A.12}$$

From (A.12) one can easily verify that, for any positive continuous r.v.,

$$F(x) = 1 - e^{-\int_0^x \lambda(y)\, dy} .$$

The mean is an initial moment of the first order of a d.f., $F(x)$. An initial moment of the nth order is defined as

$$E\{X^n\} = M_n = \int_{-\infty}^{\infty} x^n \, dF(x) . \tag{A.13}$$

The *variance* of a r.v., which is a measure of its dispersion, is defined as

$$\mathrm{Var}\{X\} = E\{(X - E\{X\})^2\} = E\{X^2\} - (E\{X\})^2 . \tag{A.14}$$

The variance also is given by

$$\mathrm{Var}\{X\} = \int_{-\infty}^{\infty} (x - E\{X\})^2 \, dF(x) . \tag{A.15}$$

The variance is a central moment of the second order of a d.f., $F(x)$. In general, a central moment of the nth order can be written as

$$D_n = E\{(X - E\{X\})^n\} = \int_{-\infty}^{\infty} (x - E\{X\})^n \, dF(x) . \tag{A.16}$$

From the definition of the variance one can find the following properties:

$$\mathrm{Var}(c) = 0 ,$$

$$\mathrm{Var}\{cX\} = c^2 \, \mathrm{var}\{X\} ,$$

$$\mathrm{Var}\{c + X\} = \mathrm{Var}\{X\} ,$$

and for a sum of independent r.v.'s,

$$\mathrm{Var}\left\{ \sum_{1 \le i \le n} X_i \right\} = \sum_{1 \le i \le n} \mathrm{Var}\{X_i\} . \tag{A.17}$$

Initial and *central moments* have the following interrelations (a sample for several first moments):

$$D_2 = M_2 - M_1^2 ,$$
$$D_3 = M_3 - 3M_2M_1 - 2M_1^3 ,$$
$$D_4 = M_4 - 4M_3M_1 + 6M_2M_1^2 - 3M_1^4 ,$$

where M_k is the initial moment of the kth order and D_k is the central moment of the kth order. The *standard deviation* is a value determined with

$$\sigma = |\sqrt{\mathrm{Var}\{X\}}| . \tag{A.18}$$

A *centered r.v.* is an r.v. obtained from the original one by extracting the mean:

$$X^0 = X - E\{X\} . \tag{A.19}$$

A *normalized r.v.* is an r.v. obtained from the original one with:

$$X^* = \frac{X}{\sigma} . \tag{A.20}$$

Remark Centered and normalized r.v.'s are used to make standard tables for different distributions.

A.3 BRIEF REVIEW OF STOCHASTIC PROCESSES

A.3.1 Definitions

A flow of homogeneous events is a stochastic process formed by a set of random moments of arrival $t_1, t_2, \ldots, t_k, t_{k+1}, \ldots$ where $t_{k+1} > t_k$, $k \geq 1$.

In general, a stochastic point process yields a distribution of a random vector $(z_1, z_2, \ldots, z_n)$ for any $n \geq 1$ where

$$z_k = t_k - t_{k-1} , \quad k \geq 1 , \quad t_0 = 0 .$$

If r.v.'s $(z_1, z_2, \ldots, z_n)$ are independent, a corresponding point process is called *a process with restricted memory*. Such a process is described by a set of d.f.'s:

$$F_k(t) = Pr\{z_k \leq t\} , \quad k \geq 1 .$$

A process with a restricted memory for which

$$F_2(t) = F_3(t) = \cdots = F(t) ,$$

is called a *recurrent point process with a delay*. It is determined by d.f.'s

$F_1(t)$ and $F(t)$. When $F_1(t) = F(t)$ the process is called simply a *recurrent point process*. In this case $F(t)$ is the d.f. of the time interval between any two adjacent points (arrivals).

A recurrent point process for which

$$F(t) = 1 - e^{-\alpha t}, \qquad \alpha > 0, \tag{A.21}$$

is called a *Poisson process*. The value α is called the *intensity* of the Poisson process and equals the average number of arrivals during a unit of time. For a Poisson process, the probability $p_k(t_0, t)$, $k \geq b$, is the probability of k events in the time interval between t_0 and $t_0 + t$,

$$p_k(t_0, t) = p_k(t) = \frac{(\alpha t)^k}{k!} e^{-\alpha t}, \tag{A.22}$$

and it does not depend on t_0. Thus, in a Poisson process, a new arrival time does not depend on the time of any previous arrival. This process is called *memoryless*.

For a Poisson process the mean of a random number of events, $\eta(t)$, arriving during an interval of length t is

$$E\{\eta(t)\} = \sum_{0 \leq k < \infty} kp_k(t) = \alpha t. \tag{A.23}$$

An *alternative renewal process* is an alternation of two mutually independent r.v.'s:

$$\theta_1, \xi_1, \theta_2, \xi_2, \ldots, \theta_k, \xi_k, \ldots,$$

where all r.v.'s θ_k have the distribution $F(t)$ and all r.v.'s ξ_k have the distribution $G(t)$. Intervals of operational and idle times are an example from reliability analysis.

A.3.2 Point Stochastic Process Thinning

Let us consider a recurrent point process, and assign to each arrival moment t_k a r.v., $\delta_k = \delta(t_k)$, which equals 0 with probability q_k and 1 with probability $p_k = 1 - q_k$. The value δ_k is called an *event indicator*. Let us apply this indicator to some sample of a point stochastic process. If for some current arrival a r.v. δ_k takes a value 1, this event is included in a new point process. If it is 0, it is excluded. The application of an operator $\delta = (\delta_1, \delta_2, \ldots)$ to the point process forms a thinned, new process in which some arrivals are excluded randomly.

The *Renyi theorem* states that if the thinning operator with $p_k = p$ is applied to a regularly normated renewal point process, the thinned process will go to a Poisson process asymptotically as p goes to 1. The parameter of this new Poisson process will be $1/p$. The Renyi theorem can be generalized to unequal probabilities of thinning if all probabilities tend to 0 uniformly. This property is very important in reliability theory. Different methods lead to different results. For example, if $p_k = 1$ for all even k and $p_k = 0$ for all odd k, the new point process will be a renewal one with an Erlang d.f.

A.3.3 Superposition of Point Stochastic Processes

If a system failure occurs because of several reasons, in particular as a result of failure of different groups of units, one may regard the system failure flow as a superposition of several separate event flows. In this subsection we consider n sources of independent point stochastic processes. Each of them has its own intensity. We will consider a new process, formed by superposition of all these processes. The *Khinchine Theorem* states that under some conditions such a superposition converges to a Poisson process when n tends to infinity. In general, the conditions are not too restrictive.

Grigelionis has generalized Khinchine's theorem and shown that a superposition of nonstationary point processes converges to a nonstationary Poission process. This statement is very important in reliability applications if one considers nonstationary reliability indexes, such as those defined by a random system TTF. The conditions of Grigelionis' theorem are not restrictive; they permit even slight dependence on the original point processes.

APPENDIX B

HOMOGENEOUS MARKOV MODELS

This appendix covers only the Markov techniques that are most relevant to reliability analysis and is intended for the reader with a background in Markov theory. Explanations of the concepts in this appendix may be found in any introductory text.

B.1 NONSTATIONARY COEFFICIENT OF AVAILABILITY

For each state of a Markov process, one can write a differential equation

$$\frac{d}{dt} p_k(t) = -p_k(t) \sum_{i \in e(k)} \Lambda_{ik} + \sum_{i \in E(k)} \Lambda_{ik} p_i(t) , \qquad \text{(B.1)}$$

where Λ_{ij} is the intensity of transition from state i to state j, and $p_i(t)$ is the probability that the system is in state H_1 at moment t. If there are n states, one can construct n differential equations. To find a nonstationary coefficient of availability, one should take any $n-1$ equations and add the norm condition equation, which requires the sum of the probabilities to equal one:

$$\sum_{1 \le i \le n} p_i(t) = 1 . \qquad \text{(B.2)}$$

Then can be added the initial conditions of the type $p_i(0) = p_i$, where $p_i(0)$ is the probability that the system is in state i at the moment $t = 0$. In most problems the initial system state is one of complete operability.

To find a nonstationary availability index, we can use the Laplace-

Stieltjes transform (LST). We then have a system of n linear algebraic equations

$$s\varphi_k(s) - p_k = -\varphi_k(s) \sum_{i\in e(k)} \Lambda_{ki} + \sum_{i\in E(k)} \Lambda_{ik}\varphi_i(s) \tag{B.3}$$

$$\sum_{1\leq i\leq n} s\varphi_i(s) = 1\,, \tag{B.4}$$

where $\varphi_i(s)$ is the LST for $p_i(t)$,

$$\varphi_i(s) = \int_0^\infty p_i(t)\, e^{-st}\, dt\,. \tag{B.5}$$

In (B.3) only any $n-1$ equations are used because the total group of n equations is linearly dependent.

This system of equations should be written in canonical form ordered by subscripts:

$$\begin{aligned} b_{11}\varphi_1(s) + b_{11}\varphi_2(s) + \cdots + b_{1n}\varphi_n(s) &= c_1\,, \\ b_{21}\varphi_1(s) + b_{22}\varphi_2(s) + \cdots + b_{2n}\varphi_n(s) &= c_2\,, \\ \vdots \\ b_{n1}\varphi_1(s) + b_{n2}\varphi_2(s) + \cdots + b_{nn}\varphi_n(s) &= c_n\,, \end{aligned} \tag{B.6}$$

where b_{ij} is the coefficient of the jth term of the ith row of the matrix, and c_i is the sum for the row.

To solve this system of linear algebraic equations, one can apply Cramer's rule:

$$\varphi_i(s) = \frac{D_i(s)}{D(s)}\,, \tag{B.7}$$

where $D(s)$ is the determinant of the system and $D_i(s)$ is the determinant with a column of c_i constants substituted for the ith column.

We then find the LST of the avialability coefficient,

$$\varphi(s) = \sum_{i\in E} \varphi_i(s) = \frac{1}{D(s)} \sum_{i\in E} D_i(s)\,. \tag{B.8}$$

We can use the following simple procedure to invert this LST.

1. Write $\varphi(s)$ in the form

$$\varphi(s) = \frac{A_0 + A_1 s + A_2 s^2 + \cdots + A_m s^m}{B_0 + B_1 s + B_2 s^2 + \cdots + B_{m+1} s^{m+1}}\,, \tag{B.9}$$

where A_j and B_i are known coefficients.

2. Find the polynomial roots

$$B_0 + B_1 s + B_2 s^2 + \cdots + B_{m+1} s^{m+1} = 0 .$$

Let these roots equal $b_1, b_2, \ldots, b_{m+1}$. This means that

$$B_0 + B_1 s + \cdots + B_{m+1} s^{m+1} = \prod_{1 \le j \le m+1} (s - b_j) .$$

3. Write $\varphi(s)$ in the form of a sum of simple fractions:

$$\varphi(s) = \frac{\beta_1}{s - b_1} + \frac{\beta_2}{s - b_2} + \cdots + \frac{\beta_{m+1}}{s - b_{m+1}} , \qquad \text{(B.10)}$$

where β_i are coefficients to be found.

4. Rewrite $\varphi(s)$ in the form

$$\varphi(s) = \frac{\Sigma_{1 \le j \le n} \beta_j \Pi_{i \ne j} (s - b_i)}{(x - b_1)(s - b_2) \cdots (s - b_{m+1})} .$$

Elementary transformations yield

$$\varphi(s) = \frac{\alpha_0 + \alpha_1 s + \alpha_2 s^2 + \cdots + \alpha_m s^m}{(s - b_1)(s - b_2) \cdots (s - b_{m+1})} ,$$

where α_i are expressed through different β_i's and b_j's.

5. Polynomials of the form $\varphi(s)$ and of the form (B.9) are equal if and only if

$$A_0 = \alpha_0 , \quad A_1 = \alpha_1 , \quad A_2 = \alpha_2, \ldots , A_m = \alpha_m .$$

The α_1 coefficients of interest are defined from these equations.

6. After the α_i's have been found, the inverse LST is applied to the $\varphi(s)$ using (B.10):

$$\varphi(s) = \sum_{1 \le j \le m+1} \frac{\beta_j}{s - b_j} \leftrightarrow K(t) = \sum_{1 \le j \le m+1} \beta_j e^{-b_j t} .$$

Remark If $\alpha(s)$ has multiple roots for the denominator (i.e., if several b_j's are equal), then (B.10) should be rewritten:

$$\varphi(s) = \sum_{1 \le i \le n'} \frac{\beta_j}{(s - b_j)^k} ,$$

where k is the number of roots equal to b_j, and n' is the number of different roots. To all terms of the form

$$\frac{\beta_j}{(s-b_j)^k},$$

the corresponding inverse LST is applied:

$$\frac{\beta_j}{(s-b_j)^k} \leftrightarrow \beta_j \frac{t^{k-1}}{(k-1)!} e^{-b_j t}.$$

B.2 PROBABILITY OF FAILURE-FREE OPERATION

To determine the probability of failure-free operation, absorbing system failure states are introduced into the transition graph. All transition intensities out of an absorbing system state are 0. In the previous equations one should change the domain of summation in a way that is equivalent to eliminating the zero transition rates. One can write for an operational state k:

$$\frac{d}{dt} p_k(t) = -p_k(t) \sum_{i \in e(k)} \Lambda_{ki} + \sum_{i \in E(k) \cap E_+} \Lambda_{ik} p_i(t).$$

If the transition graph has m operational states, one can construct m differential equations. (In this case the equations are not linearly dependent.) These equations and initial conditions are used to find the probability of failure-free operation of the system.

One again uses a LST to find the following system of linear differential equations:

$$s\varphi_k(s) - p_k = -\varphi_k(s) \sum_{i \in e(k)} \Lambda_{ki} + \sum_{i \in E(k) \cap E_+} \Lambda_{ik} \varphi_i(s) \qquad \text{(B.11)}$$

for all k operational states.

The solution of this system of equations can be found as before.

B.3 DETERMINATION OF THE MTTF AND MTBF

Recall that

$$T = \int_0^\infty P(t)\,dt.$$

If $\varphi(s)$ is the LST for the probability of failure-free operation of the system, then

$$T = \left[\int_0^\infty e^{-st} P(t)\,dt\right]_{s=0} = \varphi(s)|_{s=0}.$$

Thus, the MTTF (or MTBF) is found by solving the following system:

$$-p_k = \varphi_k(0) \sum_{i \in e(k)} \Lambda_{ki} + \sum_{i \in E(k) \cap E_+} \Lambda_{ik} \varphi_i(0)$$

for all $k \in E_+$. Note once more that this system was derived from (B.11) by substituting into it $s = 0$. To find the MTTF, we set the initial conditions as $p_i(0) = 1$, for the totally operably system state, and $p_j(0) = 0$ for all the other states. To find the MTBF, we set the initial conditions in the form $p_i^*(0) = p_i^*$, where the p_i^*'s, in this case, are the conditional stationary probabilities of the states i that belong to the subset of states, E^*, first visited after system renewal:

$$E^* = \bigcup_{k \in e_-} e(k) \, .$$

The conditional stationary probabilities, the p_i^*'s, can be obtained from the unconditional probabilities with

$$p_i^* = \frac{p_i(0)}{\sum_{j \in E^*} p_j(0)} \, .$$

EXAMPLE B.1

Consider the repairable, dual system of two different units shown in Figure B.1, with parameters λ_1, λ_2, μ_1, and μ_2. A transition graph is presented in Figure B.2. Here H_0 is the state when both units are operational, $H_1(H_2)$ is the state when the first (second) unit has failed, and H_{12} is the state when both units have failed. Let $p_k(t)$ equal the probability of the kth state. There are then two systems of equations:

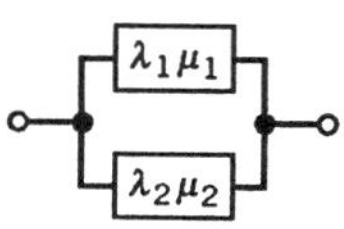

Figure B.1 Repairable two-system unit with main and redundant units that are not identical.

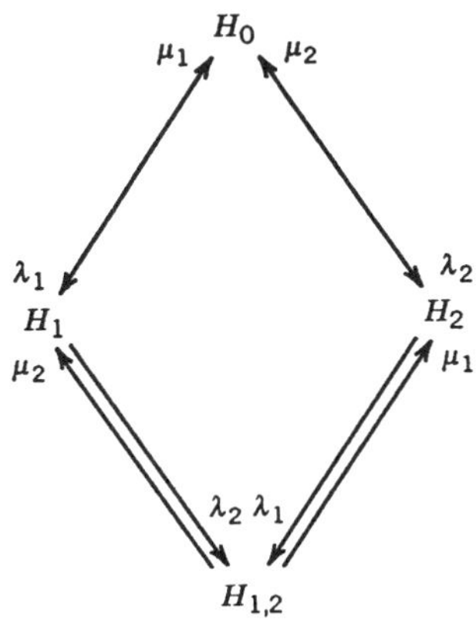

Figure B.2 Transition diagram for the system in Figure B.1.

If H_{12} is reflecting,

$$\frac{d}{dt} p_0(t) = -(\lambda_1 + \lambda_2)p_0(t) + \mu_1 p_1(t) + \mu_2 p_2(t)\,,$$

$$\frac{d}{dt} p_1(t) = \lambda_1 p_0(t) - (\lambda_2 + \mu_1)p_1(t) + \mu_2 p_{12}(t),$$

$$\frac{d}{dt} p_2(t) = \lambda_2 p_0(t) - (\lambda_1 + \mu_2)p_2(t) + \mu_1 p_{12}(t),$$

$$p_0(t) + p_1(t) + p_3(t) + p_{12}(t) = 1.$$

If H_{12} is absorbing,

$$\frac{d}{dt} p_0(t) = -(\lambda_1 + \lambda_2)p_0(t) + \mu_1 p_1(t) + \mu_2 p_2(t),$$

$$\frac{d}{dt} p_1(t) = \lambda_1 p_0(t) - (\lambda_2 + \mu_1)p_1(t),$$

$$\frac{d}{dt} p_2(t) = \lambda_2 p_0(t) - (\lambda_1 + \mu_2)p_2(t).$$

The corresponding solutions in matrix form can be found easily.

B.4 INDEXES FOR A BIRTH-AND-DEATH PROCESS

B.4.1 Model Description

The behavior of a number of real systems can be described with the birth-and-death model. A system has k operating and n redundant units. The system has one state for each possible number of failed units: $H_0, H_1, \ldots, H_{n+1}$, where H_j is a state with j failed units. In state H_{n+1} all redundant units and one operating unit have failed.

If the system is in state H_j at the moment t, there are three probabilities for the next interval Δt: probability $\Lambda_j\,\Delta(t) + o(\Delta t)$ that it will pass to state H_{j+1}, probability $M_j\,\Delta(t) + o(\Delta t)$ that it will pass to state H_{j-1}, and probability $1 - (\Lambda_j + M_j)\,\Delta t + o(\Delta t)$ that it will remain in state H_j.

State H_j may be modeled as being either absorbing or reflecting. The first condition is used to find the probability of failure-free operation, the MTTF, and the MTBF. The second condition is used to find availability coefficients, both stationary and nonstationary. The system transition graph is presented in Figure B.3, where the notation follows the preceding text.

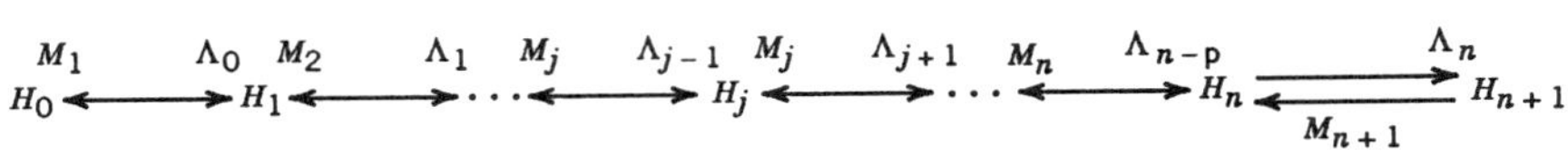

Figure B.3 Transition diagram for a repairable system consisting of k main and n redundant identical and independent units with one down state.

B.4.2 Absorbing State H_{n+1}

The birth-and-death process with an absorbing state can be described with the help of the following system of differential equations:

$$\frac{d}{dt} p_j(t) = \Lambda_{j-1} p_{j-1}(t) - (\Lambda_j + M_j) p_j(t) + M_{j+1} p_{j+1}(t)\,, \qquad 0 \le j \le n+1\,,$$

$$\sum_{0 \le j \le n+1} p_j(t) = 1\,,$$

$$\Lambda_{-1} = \Lambda_{n+1} = M_0 = M_{n+1} = M_{n+2} = 0\,, \tag{B.12}$$

where $p_j(t)$ is the probability that the system is in the state H_j at the moment t.

Let the $p_j(t)$ probabilities satisfy the initial conditions:

$$p_j(t=0) = p_j(0)\,, \qquad 0 \le j \le n+1\,. \tag{B.13}$$

To find the random variable distribution function $Pr\{\theta_{n+1} \le t\} = p_{n+1}(t)$, the probability that the system has not reached the absorbing state H_{n+1} at time t, one applies a LST to (B.12). Notice that a r.v. θ_{n+1} is the duration of time before the system reaches the absorbing state H_{n+1} the first time. Then one finds a system of linear algebraic equations:

$$\Lambda_{j-1}\varphi_{j-1}(s) - (\Lambda_j + M_j + s)\varphi_j(s) + M_{j+1}\varphi_{j+1}(s) = -p_j(0)\,,$$
$$0 \le j \le n+1\,, \qquad \Lambda_1 = \Lambda_{n+1} = M_0 = M_{n+1} = M_{n+2} = 0\,. \tag{B.14}$$

From Cramer's rule,

$$\varphi_{n+1}(s) = \frac{D_{n+1}(s)}{s\Delta_{n+1}(s)}\,, \tag{B.15}$$

where

$$\Delta_{n+1}(s) =$$
$$\begin{vmatrix} -(\Lambda_0+s) & M_1 & 0 & \vdots & 0 & 0 \\ \Lambda_0 & -(\Lambda_1+M_1+s) & M_2 & \vdots & 0 & 0 \\ 0 & \Lambda_1 & -(\Lambda_2+M_2+s) & \vdots & 0 & 0 \\ \vdots & \vdots & \vdots & \vdots & \vdots & \vdots \\ 0 & 0 & 0 & \vdots & -(\Lambda_{n-1}+M_{n-1}+s) & M_n \\ 0 & 0 & 0 & \vdots & \Lambda_{n-1} & -(\Lambda_n+M_n+s) \end{vmatrix}$$
$$= (-1)^{n+1} \prod_{1 \le k \le n+1} (s + s_k^{n+1})\,, \tag{B.16}$$

$$D_{n+1}(s) =$$
$$\begin{vmatrix} -(\Lambda_0+s) & M_1 & 0 & \cdots & 0 & 0 & -p_0(0) \\ \Lambda_0 & -(\Lambda_1+M_1+s) & M_2 & \cdots & 0 & 0 & -p_1(0) \\ 0 & \Lambda_1 & -(\Lambda_2+M_2+s) & \cdots & 0 & 0 & -p_2(0) \\ \vdots & \vdots & \vdots & & \vdots & \vdots & \vdots \\ 0 & 0 & 0 & \cdots & -(\Lambda_{n-1}+M_{n-1}+s) & M_n & -p_{n-1}(0) \\ 0 & 0 & 0 & \cdots & \Lambda_{n-1} & -(\Lambda_n+M_n+s) & -p_n(0) \\ 0 & 0 & 0 & \cdots & 0 & \Lambda_n & -p_{n+1}(0) \end{vmatrix}$$

(B.17)

It can be shown that $\Delta_{n+1}(s)$ is a polynomial of power n and that all its roots

$$x_k^{(n+1)} = -s_k^{(n+1)}, \qquad s_k^{(n+1)} > 0, \quad 1 \le k \le n+1, \tag{B.18}$$

are different and negative; all roots of the polynomials of the neighbors of n are alternating.

Decomposition of the determinant $D_{n+1}(s)$ by the last row yields a recurrent equation

$$D_{n+1}(s) = -p_{n+1}(0)\Delta_{n+1}(s) + \sum_{0 \le i \le n} (-1)^{n+1-i} p_i(0) \times \Delta_i(s) \prod_{i \le k \le n} \Lambda_k . \tag{B.19}$$

The probability of interest, $p_{n+1}(t)$, is found with the help of an inverse LST:

$$p_{n+1}(t) = \frac{1}{2\pi i} \int_L \varphi_{n+1}(s) e^{st}\, ds = \frac{1}{2\pi i} \int_L \frac{D_{n+1}(s) e^{st}\, ds}{-s\Delta_{n+1}(s)}, \quad i = \sqrt{-1}, \tag{B.20}$$

where a counter L includes all zero roots of the denominator of the function under the integral.

Using (B.11), we write

$$p_{n+1}(t) = -\frac{D_{n+1}(0)}{\Delta_{n+1}(0)} + (-1)^{n+1} \sum_{1 \le i \le n+1} \frac{D_{n+1}(-s_i^{(n+1)}) e^{-s_i^{(n+1)} t}}{s_i^{(n+1)} \prod_{\substack{1 \le k \le n+1, \\ k \ne i}} (s_k^{(n+1)} - s_i^{(n+1)})} . \tag{B.21}$$

Simple transformation yields

$$\Delta_{n+1}(0) = (-1)^{n+1} \prod_{0 \le i \le n} \Lambda_i . \tag{B.22}$$

Equation (B.19) then gives

$$D_{n+1}(0) = -\Delta_{n+1}(0), \tag{B.23}$$

and (B.21) now has the form

$$p_{n+1}(t) = 1 - (-1)^n \sum_{1 \le i \le n+1} \frac{D_{n+1}(-s_j^{(n+1)}) e^{-s_i^{(n+1)} t}}{s_i^{(n+1)} \prod_{\substack{1 \le k \le n+1 \\ k \ne i}} (s_k^{(n+1)} - s_i^{(n+1)})} . \tag{B.24}$$

Using (B.19), the probability of failure-free operation of the system, $P(t) = 1 - p_{n+1}(t)$, can be written

$$P(t) = \sum_{0 \le i \le n} p_i(0) \prod_{i \le j \le n} \Lambda_j \sum_{i \le k \le n+1} \frac{e^{-s_k^{(n+1)}t}}{s_k^{(n+1)}} \times \frac{\prod_{1 \le m \le i} (s_m^{(i)} - s_k^{(n+1)})}{\prod_{\substack{1 \le i \le n \\ i \ne k}} (s_i^{(n+1)} - s_k^{(n+1)})} . \tag{B.25}$$

Two cases are of special interest: (1) when all units are in up states at the moment $t = 0$, and (2) when the system has just come out of system failure at $t = 0$.

If the system starts with all units completely operational:

$$p_0(0) = 1 ,$$
$$p_i(0) = 0 , \qquad 1 \le i \le n+1 ,$$

the application of (B.17) and (B.25) yields

$$p^{(0)}(t) = \Lambda_0 \Lambda_1 \cdots \Lambda_n \sum_{1 \le i \le n+1} \frac{e^{-s_i^{(n+1)}t}}{s_i^{(n+1)} \prod_{\substack{1 \le k \le n+1 \\ k \ne i}} (s_k^{(n+1)} - s_i^{(n+1)})} . \tag{B.26}$$

If the system starts operating just after coming out of system failure:

$$p_n(0) = 1 ,$$
$$p_i(0) = 0 , \qquad i = 0, 1, \ldots, n-1, n+1 ,$$

then

$$P^{(n)}(t) = (-1)^n \Lambda_n \sum_{1 \le i \le n+1} \Delta_n(-s_i^{(n+1)}) \frac{e^{-s_i^{(n+1)}t}}{s_i^{(n+1)} \prod_{\substack{1 \le i \le n+1, \\ k \ne i}} (s_k^{(n+1)} - s_i^{(n+1)})} . \tag{B.27}$$

Now let us determine for any initial conditions (B.13) the mean time T until the first system transition into the absorbing state. In accordance with (B.15) and (B.27),

$$T = (-1)^{n+1} \frac{\begin{vmatrix} 1 & \cdots & 1 & 0 \\ & & & p(0) \\ & \Delta & & \vdots \\ & & & p(0) \end{vmatrix}}{\Delta} , \tag{B.28}$$

where the determinant Δ, in correspondence with (B.9), is defined as

$$\Delta = \begin{vmatrix} -\Lambda_0 & M_1 & 0 & \cdots & 0 & 0 \\ \Lambda_0 & -(\Lambda_1 + M_1) & M_2 & \cdots & 0 & 0 \\ 0 & \Lambda_1 & -(\Lambda_2 + M_2) & \cdots & 0 & 0 \\ \vdots & \vdots & \vdots & & \vdots & \vdots \\ 0 & 0 & 0 & \cdots & -(\Lambda_{n-1} + M_{n-1}) & M_n \\ 0 & 0 & 0 & \cdots & \Lambda_{n-1} & -(\Lambda_n + M_n) \end{vmatrix} . \tag{B.29}$$

After transformations, (B.28) and (B.29) give

$$T = \sum_{0 \le j \le n} \frac{\pi_j(0) \, \Sigma_{0 \le k \le j} \, \theta_k}{\Lambda_j \theta_j} , \tag{B.30}$$

where

$$\pi_j(0) = \sum_{0 \le i \le j} p_i(0) ,$$

$$\theta_j = \frac{\Lambda_0 \Lambda_1 \cdots \Lambda_{j-1}}{M_1 M_2 \cdots M_j} , \qquad \theta_0 = 1 . \tag{B.31}$$

This general formula may be used to find the system MTTF, $T^{(0)}$, and the system MTBF, $T^{(n)}$, in two particular cases. In the first case, for $p_0(0) = 1$, (B.30) yields

$$T^{(0)} = \sum_{0 \le j \le n} \frac{\Sigma_{0 \le k \le j} \, \theta_k}{\Lambda_j \theta_j} .$$

In the second case, for $p_0(0) = 1$, (B.30) yields

$$T^{(n)} = \frac{\Sigma_{0 \le k \le j} \, \theta_k}{\Lambda_n \theta_n} .$$

Note that for this model the mean time between the ith and $(i+1)$th failures does not depend on i if $i \ge 1$.

B.5.3 Reflecting System Failure State H_{n+1}

Consider a system with a reflecting H_{n+1} state. Behavior of this system can be described with (B.12), where the last equations have the form

$$\frac{d}{dt} p_n(t) = \Lambda_{n-1} p_{n-1}(t) - (\Lambda_n + M_n) p_n(t) + M_{n+1} p_{n+1}(t) ,$$

$$\frac{d}{dt} p_{n+1}(t) = \Lambda_n p_n(t) - M_{n+1} p_{n+1}(t) .$$

The stationary availability coefficient by definition equals

$$K = 1 - \lim_{t\to\infty} p_{n+1}(t) .$$

With simple transformations

$$K = \left(1 + \frac{\theta_{n+1}}{\sum_{0\le i\le n} \theta_i}\right)^{-1} .$$

In general, stationary probabilities can be easy to find. As t tends to infinity, the system of differential equations can be replaced with

$$(\Lambda_j + M_j)p_j = \Lambda_{j-1}p_{j-1} + M_{j+1}p_{j+1} , \qquad 0\le j\le n+1 ,$$
$$\Lambda_{-1} = M_0 = \Lambda_{n+1}, = M_{n+2} = 0 .$$

Hence

$$p_j = \theta_j p_0 ,$$

where the θ_j, as before, is defined by (B.31).

From the norm condition, it follows that

$$p_0 = \frac{1}{\sum_{0\le i\le n+1} \theta_j} .$$

Thus,

$$p_j = \frac{\theta_j}{\sum_{0\le i\le n+1} \theta_i} , \qquad 0\le j\le n+1 . \tag{B.32}$$

The system MTBF can be found when the system starts to operate with the initial conditions determined by

$$p_j(0) = \lim_{t\to\infty} p_j(t) = p_j , \qquad 0\le j\le n+1 ,$$

where p_j's are defined by (B.32). Finally,

$$T^{(R)} = \frac{1}{\sum_{0\le k\le n+1} \theta_k} \sum_{0\le j\le n} \frac{(\sum_{0\le i\le j})^2}{\Lambda_j \theta_j} .$$

B.4.4 A System with Several Failure States

If a system has k operating and n redundant units, it has several system failure states: $H_{n+1}, H_{n+2}, \ldots, H_{k+n}$. The notation $N = k + n$ may be used for compactness. The reswitching circuits can be operational even though the system is in one of the system failure states. Calculations for the MTTF,

$T^{(0)}$, the MTBF, $T^{(n)}$, and the probability of system failure-free operation, $P(t)$, are the same as given previously. In addition

$$p_j = \lim_{t \to \infty} p_j(t) = \frac{\theta_j}{\sum_{0 \le k \le N} \theta_k} , \qquad 0 \le j \le N ,$$

where θ_j is defined by (B.31). The availability coefficient is given by

$$K = \sum_{0 \le j \le n} p_j = \frac{\sum_{0 \le j \le n} \theta_j}{\sum_{0 \le k \le N} \theta_k} .$$

APPENDIX C

DISTRIBUTIONS WITH MONOTONIC INTENSITY FUNCTIONS

Failure distributions usually are less predictable than repair time distributions. Nevertheless, practical experience permits prediction of some of the general characteristics of any failure distribution. The "aging" processes of fatigue, wear, and accumulated damage are especially predictable, particularly for mechanical equipment.

These physical phenomena result in a preponderance of failure rates that increase. There are, however, a few items with decreasing failure rates—on a physical level, reliability properties improve with time. Some metals increase in strength under a constant load. Weak units and manufacturing defects often are eliminated during a high-failure "infant mortality" period. Of course, there are no "immortal" items; an increase in the failure rate at some point is almost inevitable. Many items have a failure rate that decreases initially and then increases; these "bathtub" and "U-shaped" functions are not the subject of this appendix. This appendix covers only the main characteristics of the most important class of failure rates, monotonic failure rates. More detailed coverage may be found in Barlow and Proschan (Appendix G).

The exponential failure rate is a boundary one between increasing and decreasing rates. The rate could be regarded as hypothetical for many purposes because there are virtually no manufactured items that time does not change. A basic parameter of this branch of reliability analysis is the "conditional failure rate," which sometimes is called the "intensity function".

$$\lambda(t) = \frac{f(t)}{1 - F(t)} . \tag{C.1}$$

Here $\lambda(t)$ is the instantaneous density of the time-to-failure distribution and $\lambda(t)\Delta$ is the probability of failure in the interval of time $[t, t+\Delta]$ of an unrepairable unit that has not failed by moment t. This density changes continually through time.

Sometimes it is useful to consider the *hazard function*:

$$\Lambda(t) = \int_0^t \lambda(x)\, dx\,. \tag{C.2}$$

Integration yields

$$\Lambda(t) = -\log \bar{F}(t)\,, \tag{C.3}$$

where

$$\bar{F}(t) = 1 - F(t)\,.$$

This appendix has several descriptions of distributions with monotonic intensity functions.

C.1 CLASSES OF MONOTONIC INTENSITY FUNCTION DISTRIBUTIONS

C.1.1 IFR (DFR) Distributions

A distribution, F, is called a distribution with an increasing (decreasing) failure rate (corresponding abbreviations are IFR and DFR) if it is characterized by an increasing (decreasing) $\lambda(t)$.

C.1.2 IFRA (DFRA) Distributions

A distribution, F, is called a distribution with an increasing (decreasing) failure rate average (corresponding abbreviations are IFRA and DFRA) if it is characterized by an $\lambda(t)$ whose cumulative average value increases (decreases) monotonically. Thus

$$\frac{1}{t}\log \bar{F}(t) \uparrow (\downarrow) \qquad \text{with } t\uparrow,\ t\geq 0\,.$$

C.1.3 NBU (NWU) Distributions

A d.f. is called "new is better (worse) than used" (NBU (NWU)) if

$$\bar{F}(x+y) < (>) \bar{F}(x)\bar{F}(y) \qquad \text{for } x\geq 0,\ y\geq 0\,. \tag{C.4}$$

This expression is equivalent to the statement that the probability of failure-free operation $\bar{F}(x+y)/\bar{F}(x)$ of a unit with age x is smaller (larger) than the probability for a new unit.

C.1.4 NBUE (NWUE) Distributions

A distribution F is called "new is better (worse) than used in expectation" (NBUE (NWUE)) if

$$\int_0^\infty \bar{F}(x)\, dx \leq (\geq) \mu \bar{F}(t) \qquad \text{for } t \geq 0 . \tag{C.5}$$

Thus, a NBUE F must have a finite MTTF, but the MTTF of a NWUE F theoretically can be either finite or infinite.

Notice that an integral

$$\int_t^\infty \frac{\bar{F}(x)}{\bar{F}(t)}\, dx \tag{C.6}$$

represents the expected residual life (the conditional mean time to failure) of a unit that has been operating without failure until moment t. Thus, for a NWUE distribution the residual time is less than the MTTF of a new unit.

Now we can write the following relations between these classes of distributions:

$$\begin{aligned} &\text{IFR} \subseteq \text{IFRA} \subseteq \text{NBU} \subseteq \text{NBUE} ,\\ &\text{DFR} \subseteq \text{DFRA} \subseteq \text{NWU} \subseteq \text{NWUE} . \end{aligned} \tag{C.7}$$

Notice that these distributions do not include all classes of distributions with monotonic intensity functions. NBU, NBUE, NWU, and NWUE distributions can be useful in models of accumulations of wear, stress, damage, or losses.

C.2 PRESERVATION OF MONOTONICITY UNDER TRANSFORMATIONS

C.2.1 Convolutions of d.f.'s

A convolution of IFR d.f.'s produces a d.f. of the same type. If all standby units are IFR, for example, the system also will be IFR. A convolution of DFR d.f.'s does not produce a d.f. of the same type. For example, if a unit has a gamma d.f. of order α with $0.5 < \alpha < 1$, a two-unit system will have a gamma d.f. with parameter 2α, $2\alpha > 1$, and hence will be strictly IFR.

C.2.2 Mixtures of d.f.'s

A mixture of distributions can appear in practice when we take several different groups of homogeneous units, where all units within a group have the same d.f., and then mix them. A unit chosen at random then will have a "mixed" distribution. In other words, if there are distributions $F_i(t)$, $i = 1, \ldots, n$, then:

$$F_{\text{mix}}(t) = \sum_{1 \le i \le n} p_i F_i(t)$$

represents the distribution of the mixture. The p_i's are the probabilistic "weight" coefficients, $p_i > 0$, and

$$\sum_{1 \le i \le n} p_i = 1 .$$

A mixture of DFR d.f.'s has a DFR distribution. A mixture of IFR d.f.'s, however, is not necessarily an IFR d.f.

C.2.3 Formation of Monotonic Structures

A structure is monotonic if a structural function, ϕ, of the system has the following property:

$$\varphi(z) \ge \varphi(x^*) , \qquad \text{when } x \ge x^* , \tag{C.8}$$

where $\mathbf{X} = (x_1, \ldots, x_n)$ and $x_i = 1$ if the ith unit is operational and 0 otherwise. $\mathbf{0} = (0, \ldots, 0)$, $\mathbf{1} = (1, \ldots, 1)$, $\varphi(0) = 0$, and $\varphi(1) = 1$. In a monotonic structure a change in the reliability of a unit must result in either no change or a change in the same direction in structural reliability.

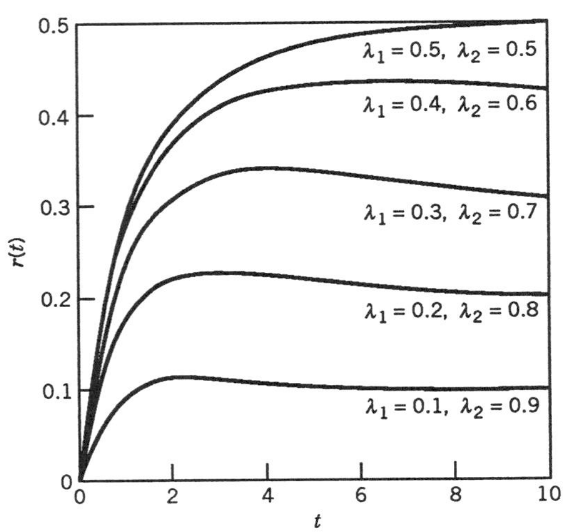

Figure C.1 Nonmonotonicity produced by two exponential components in parallel.

Table C.1 Preservation of the type of distribution after transformation of the system structure

	Transformation		
	Formation of Monotonic Structure	Convolution of d.f.'s	Mixture of d.f.'s
IFR	No	Yes	No
IFRA	Yes	Possibly	No
DFR	No	No	Yes
DFRA	No	No	Yes
NBU	Yes	Yes	No
NBUE	No	Yes	No
NWU	No	No	No
NWUE	No	No	Possibly

A system with a monotonic structure and IFR units usually has an IFR d.f. itself. An example of nonmonotonicity, however, is a system of two exponential units with parameters λ_1 and λ_2, $\lambda_1 \neq \lambda_2$. The intensity of the system equals

$$\lambda(t) = \frac{\lambda_1 e^{-\lambda_1 t} + \lambda_2 e^{-\lambda_2 t} - (\lambda_1 + \lambda_2) e^{-(\lambda_1+\lambda_2)t}}{e^{-\lambda_1 t} + e^{-\lambda_2 t} - e^{-(\lambda_1-\lambda_2)t}} . \tag{C.9}$$

Function $\lambda(t)$ is not monotonically increasing for all λ_i, as may be seen in Figure C.1.

Results of this section are summarized in Table C.1.

C.3. BOUNDARY ESTIMATES FOR RELIABILITY INDEXES

Upper and lower bounds can be very useful in different applications when there is reason to believe that an intensity function is monotonic. An example might be an object influenced by aging or stress.

C.3.1 Bounds for Probability When Only a Quantile Is Known

If F is an IFRA (DFRA) distribution with a known quantile ϵ_p of level p [i.e., if $F(\epsilon_p) = p$],

$$\begin{aligned} \bar{F}(t) &\geq (\leq) e^{-\alpha t} \qquad \text{for } 0 \leq t \leq \epsilon_p , \\ \bar{F}(t) &\leq (\geq) e^{-\alpha t} \qquad \text{for } t \leq \epsilon_p , \end{aligned} \tag{C.10}$$

where

$$\alpha = -\epsilon_p^{-1} \log(1-p) . \tag{C.11}$$

C.3.2 Probability Bounds When Only a Distribution Moment Is Known

A star-shaped function is a function such that an arrow from the origin to any point in the function lies above the function. In other words, $(1/x)F(x)$ increases for $x \geq 0$. All concave functions belong to a subclass of star-shaped functions. If, for two functions, F and G, an inverse function $G^{-1}(F(x))$ is star-shaped, we write $F <_{*} G$. We then define bounds by stating that if $F <_{*} G$, $F(0) = G(0) = 0$, and F, G are continuous d.f.'s with equal moments of the rth order,

$$m_r = \int_0^\infty t^r \, dF(t) = \int_0^\infty t^r \, dG(t) , \qquad r \geq 1 , \tag{C.12}$$

then

$$\bar{F}(t) \geq \begin{cases} \bar{G}(t) & \text{for } t < \sqrt[r]{m_r} , \\ 0 & \text{for } t \geq \sqrt[r]{m_r} . \end{cases} \tag{C.13}$$

If F is an IFRA (DFRA) distribution with a known mean T, the bounds for the rth initial moment of the distribution:

$$m_r = \int_0^\infty t^r \, dF(t)$$

are defined as

$$\begin{aligned} m_r &\geq (\leq) \Gamma(r+1) T^r \qquad \text{for } 0 \leq r < 1 , \\ m_r &\leq (\geq) \Gamma(r+1) T^r \qquad \text{for } 1 \leq r < \infty , \end{aligned} \tag{C.14}$$

In particular, for $r = 2$,

$$M_2 \leq (\geq) 2T^2 \tag{C.15}$$

for IFRA (DFRA) distributions, respectively.

If F is a continuous IFR distribution with a known rth initial moment m_r and

$$\mu_r = \frac{m_r}{\Gamma(r+1)} , \qquad r \geq 1 ,$$

then

$$\tilde{F}(t) \geq \begin{cases} e^{-t/\sqrt[r]{\mu_r}} & \text{for } t < \sqrt[r]{m_r} , \\ 0 & \text{for } t \geq \sqrt[r]{m_r} . \end{cases} \tag{C.16}$$

Let the first initial moment be known. It is clear that

$$\bar{F}(t) \geq \begin{cases} e^{-t/T} & \text{for } t < T, \\ 0 & \text{for } t \geq T. \end{cases} \tag{C.17}$$

If F is an IFR distribution with mean T, then for any fixed $t > 0$,

$$\bar{F}(t) \geq \begin{cases} 1 & \text{for } t < T, \\ e^{-\omega_t t} & \text{for } t \geq T, \end{cases} \tag{C.18}$$

where $\omega_t > 0$ is a function of t that satisfies

$$1 - \omega_t T = e^{-\omega_t t}. \tag{C.19}$$

Values of $F(t)$ for $t > T$ are given in the tables of Chapter 2. The upper and lower bounds for an IFR distribution are shown in Figure 2.1.

APPENDIX D

MATHEMATICAL CONSTANTS AND FUNCTIONS

D.1 CONSTANTS

$$e = 2.718\,, \quad \frac{1}{e} = 0.368 \tag{D.1}$$

$$\pi = 3.142\,, \quad \frac{1}{\pi} = 0.318\,. \tag{D.2}$$

Euler's constant

$$C = \lim_{n\to\infty}\left(\sum_{k=1}^{n}\frac{1}{k} - \ln n\right) \approx 0.577\,. \tag{D.3}$$

D.2 ELEMENTARY COMBINATORIAL EXPRESSIONS

Stirling's formula:

$$n! \approx \left(\frac{n}{e}\right)^{n}\sqrt{2\pi n}\left(1 + \frac{1}{12}\,n\right),$$

$$\ln(n!) \approx \left(n + \frac{1}{2}\right)\ln n - n + 0.92\,, \tag{D.4}$$

$$C_n^m = \frac{n!}{m!}\,(n-m)!\,, \tag{D.5}$$

$$\sum_{k=0}^{s} C_n^k C_m^{s-k} = C_{n+m}^{s}\,, \tag{D.6}$$

$$\sum_{k=0}^{m} C_{n+k}^{n} = C_{n+m+1}^{n+1} . \tag{D.7}$$

D.3 FINITE SUMS

Arithmetic set:

$$\sum_{k=0}^{n-1} (a + kr) = \frac{n}{2} [2a + (n-1)r] = \frac{n}{2} (a + l) , \tag{D.8}$$

where L is the last term of the set.
Geometric set:

$$\sum_{k=1}^{n} aq^{k-1} = \frac{a(q^n - 1)}{q - 1} , \tag{D.9}$$

$$\sum_{k=1}^{n} k = \frac{n(n+1)}{2} , \tag{D.10}$$

$$\sum_{k=1}^{n} \frac{1}{k} \approx 0.577 + \ln n + \frac{1}{2n} - \frac{1}{12n(n+1)} . \tag{D.11}$$

D.4 SETS

Taylor's set for a function of a single variable:

$$f(x) = f(a) + \frac{x-a}{1!} f'(a) + \frac{(x-a)^2}{2!} f''(a) + \cdots + \frac{(x-a)^n}{n!} f^{(n)}(a) + \cdots ,$$

or

$$f(a+h) = f(a) + \frac{h}{1!} f'(a) + \frac{h^2}{2!} f''(a) + \cdots + \frac{h^n}{n!} f^{(n)}(a) + \cdots , \tag{D.12}$$

$$(1+x)^q = 1 + qx + \frac{q(q-1)}{2!} x^2 + \cdots + \frac{q(q-1)\ldots(q-k+1)}{k!} x^k + \cdots , \qquad \text{where } |x| \le 1 , \tag{D.13}$$

$$e^{-x} = 1 - \frac{x}{1!} + \frac{x^2}{2!} - \frac{x^3}{3!} + \cdots = \sum_{k=0}^{\infty} (-1)^k \frac{x^k}{k!} , \tag{D.14}$$

$$a^x = \sum_{k=0}^{\infty} \frac{(x \ln a)^k}{k!} . \tag{D.15}$$

D.5 INTEGRALS

$$\int e^{ax}\,dx = \frac{1}{a}\,e^{ax}\,, \tag{D.16}$$

$$\int xe^{ax}\,dx = e^{ax}\left(\frac{x}{a} - \frac{1}{a^2}\right), \tag{D.17}$$

$$\int x^n e^{ax}\,dx = e^{ax}\left[\frac{x^n}{a} + \sum_{k=1}^{n} (-1)^k \frac{1}{a^{k+1}} \frac{n!}{(n-k)!}\,x^{n-k}\right], \tag{D.18}$$

$$\int_{t_1}^{t_2} e^{-ax}\,dx = \frac{1}{a}\,e^{-at_1}(1 - e^{-a(t_2-t_1)})\,, \tag{D.19}$$

$$\int_{t_1}^{t_2} xe^{-ax}\,dx = \frac{e^{-at_1}}{a^2}\,[1 + at_1 - e^{-a(t_2-t_1)}(1 + at_2)]\,, \tag{D.20}$$

$$\int_0^t x^n e^{-ax}\,dx = \frac{n!}{a^{n+1}} - e^{-at} \sum_{k=0}^{n} \frac{n!}{k!} \frac{t^k}{a^{n-k+1}}\,, \tag{D.21}$$

$$\int_0^\infty x^n e^{-ax}\,dx = n!a^{-(n+1)}\,. \tag{D.22}$$

D.6 LAPLACE TRANSFORMATIONS

The Laplace transformation changes an original function $p(t)$ of a real variable into a transform $\varphi(s)$ of a complex variable,

$$\varphi(s) = \int_0^\infty p(t)e^{-st}\,dt\,, \qquad t > 0\,.$$

For the original function and its transform,

$$\lim_{s\to 0} s\varphi(s) = \lim_{t\to 0} p(t)\,,$$

$$\lim_{s\to\infty} s\varphi(s) = p(0)\,.$$

Table D.1 Values of $N!$ and $1/N!$

n	$n!$	$1/n!$	n	$n!$	$1/n!$
1	1	1	7	5040	0.0^32
2	2	0.5	8	40320	0.0^42
3	6	0.167	9	362880	0.0^53
4	24	0.042	10	3628800	0.0^63
5	120	0.0^28	11	39916800	0.0^73
6	720	0.0^21	12	479001600	0.0^82

Table D.2

Transform	Original	Transform	Original
$\alpha a_1(s) + \beta a_2(s)$	$\alpha p_1(t) + \beta p_2(t)$	$a(s-a)$	$e^{\alpha t}p(t)$
$-p(0) + sa(s)$	$p'(t)$	$\dfrac{1}{s^n}$, $n = 1, 2, \ldots$	$\dfrac{t^{n-1}}{(n-1)!}$
$\dfrac{a(s)}{s} + \dfrac{C}{s}$	$\int_0^t p(x)\,dx + C$	$\dfrac{1}{(s-b)^n}$, $n = 1, 2, \ldots$	$\dfrac{t^{n-1}}{(n-1)!}e^{bt}$
$\dfrac{1}{b}a\left(\dfrac{s}{b}\right)$, $b > 0$	$p(bt)$	$\dfrac{1}{s^2 + b^2}$	$\dfrac{1}{b}\sin bt$
$e^{-bs}a(s)$, $b \geq 0$	$p(t-b)$	$\dfrac{s}{s^2 + b^2}$	$\cos bt$
$a_1(s)a_2(s)$	$\int_0^t p_1(x)p_2(t-x)\,dx$	$\dfrac{1}{s^2 - b^2}$	$\dfrac{1}{b}\operatorname{sh} bt$
$a^{(k)}(s)$	$(-1)^k t^k p(t)$	$\dfrac{s}{s^2 - b^2}$	$\operatorname{ch} bt$

Table D.3 e^{-x} **functions, 0 to 1**

	0	1	2	3	4	5	6	7	8	9
0	1.000	0.999	0.998	0.997	0.996	0.995	0.994	0.993	0.992	0.991
0.0	1.000	0.990	0.980	0.970	0.961	0.951	0.942	0.932	0.923	0.914
0.1	0.905	0.896	0.887	0.878	0.869	0.861	0.852	0.844	0.835	0.827
0.2	0.819	0.811	0.802	0.795	0.787	0.779	0.771	0.763	0.756	0.748
0.3	0.741	0.733	0.726	0.719	0.712	0.705	0.698	0.691	0.684	0.677
0.4	0.670	0.664	0.657	0.651	0.644	0.638	0.631	0.625	0.619	0.613
0.5	0.606	0.600	0.584	0.589	0.583	0.577	0.571	0.566	0.560	0.554
0.6	0.540	0.543	0.538	0.533	0.527	0.522	0.517	0.512	0.507	0.502
0.7	0.497	0.492	0.487	0.482	0.477	0.472	0.468	0.463	0.458	0.454
0.8	0.449	0.445	0.440	0.436	0.432	0.427	0.423	0.419	0.415	0.411
0.9	0.407	0.403	0.398	0.395	0.391	0.387	0.383	0.379	0.375	0.372
1.0	0.368	0.333	0.301	0.273	0.247	0.223	0.202	0.183	0.165	0.150
2.0	0.135	0.122	0.111	0.100	0.091	0.082	0.074	0.067	0.061	0.055
3.0	0.050	0.045	0.041	0.037	0.033	0.030	0.027	0.025	0.022	0.020
4.0	0.018	0.017	0.015	0.014	0.012	0.011	0.010	0.009	0.008	0.007
5.0	0.0067	0.0061	0.0055	0.0058	0.0045	0.0041	0.0037	0.0033	0.0030	0.0027
6.0	0.0025	0.0022	0.0020	0.0018	0.0017	0.0015	0.0014	0.0012	0.0011	0.0010

APPENDIX E

STATISTICAL TABLES

Table E.1 Values of e^{-x}

x	0	1	2	3	4	5	6	7	8	9
0.00	1.000	0.999	0.998	0.997	0.996	0.995	0.994	0.993	0.992	0.991
0.0	1,000	0.990	0.980	0.970	0.961	0.951	0.942	0.932	0.923	0.914
0.1	0.905	0.896	0.887	0.878	0.869	0.861	0.852	0.844	0.835	0.827
0.2	0.819	0.811	0.802	0.795	0.787	0.779	0.771	0.763	0.756	0.748
0.3	0.741	0.733	0.726	0.719	0.712	0.705	0.698	0.691	0.684	0.677
0.4	0.670	0.664	0.657	0.651	0.644	0.638	0.631	0.625	0.619	0.613
0.5	0.606	0.600	0.594	0.589	0.583	0.577	0.571	0.566	0.560	0.554
0.6	0.540	0.543	0.538	0.533	0.527	0.522	0.517	0.512	0.507	0.502
0.7	0.497	0.492	0.487	0.482	0.477	0.472	0.468	0.463	0.458	0.454
0.8	0.449	0.445	0.440	0.436	0.432	0.427	0.423	0.419	0.415	0.411
0.9	0.407	0.403	0.398	0.395	0.391	0.387	0.383	0.379	0.375	0.372
1.0	0.368	0.333	0.301	0.273	0.247	0.223	0.202	0.183	0.165	0.150
2.0	0.135	0.122	0.111	0.100	0.091	0.082	0.074	0.067	0.061	0.055
3.0	0.050	0.045	0.041	0.037	0.033	0.030	0.027	0.025	0.022	0.020
4.0	0.018	0.017	0.015	0.014	0.012	0.011	0.010	0.009	0.008	0.007
5.0	0.0067	0.0061	0.0055	0.0050	0.0045	0.0041	0.0037	0.0033	0.0030	0.0027
6.0	0.0025	0.0022	0.0020	0.0018	0.0017	0.0015	0.0014	0.0012	0.0011	0.0010

Table E.2 Absolute values of $\ln(x)$, $0 < x < 1$

x	0	1	2	3	4	5	6	7	8	9
0.999	0.0010	0.0009	0.0008	0.0007	0.0006	0.0005	0.0004	0.0003	0.0002	0.0001
0.99	0.010	0.009	0.008	0.007	0.006	0.005	0.004	0.003	0.002	0.001
0.9	0.105	0.094	0.083	0.073	0.062	0.051	0.041	0.030	0.020	0.010
0.8	0.223	0.211	0.198	0.186	0.174	0.162	0.151	0.139	0.128	0.117
0.7	0.357	0.342	0.329	0.314	0.301	0.288	0.274	0.261	0.248	0.236
0.6	0.512	0.494	0.478	0.462	0.446	0.431	0.416	0.400	0.386	0.371
0.5	0.693	0.673	0.654	0.635	0.616	0.598	0.580	0.562	0.545	0.528
0.4	0.916	0.892	0.867	0.844	0.821	0.798	0.777	0.755	0.734	0.713
0.3	1.204	1.171	1.139	1.109	1.079	1.050	1.022	0.994	0.968	0.942
0.2	1.609	1.561	1.514	1.470	1.427	1.386	1.347	1.309	1.273	1.238
0.1	2.303	2.207	2.120	2.040	1.966	1.897	1.833	1.772	1.715	1.661
0.0	∞	4.605	3.912	3.506	3.219	2.996	2.813	2.659	2.526	2.408
0.00	∞	6.908	6.215	5.809	5.521	5.298	5.116	4.962	4.828	4.711
0.000	∞	9.210	8.517	8.111	7.824	7.601	7.419	7.264	7.131	7.013

Table E.3 Cumulative percentages of normal distribution, $0 \leq \sigma < 4$

$$\Phi(x) = \frac{1}{\sqrt{2\pi}} \int_{-\infty}^{x} e^{-u^2/2}\, du$$

x	0	1	2	3
0.00	0.500	0.841	0.9773	0.9^2865
0.05	0.520	0.853	0.798	0.886
0.10	0.540	0.864	0.821	0.9^3032
0.15	0.560	0.875	0.842	0.184
0.20	0.573	0.885	0.861	0.313
0.25	0.599	0.894	0.878	0.423
0.30	0.618	0.9032	0.893	0.517
0.35	0.637	0.115	0.906	0.596
0.40	0.655	0.192	0.9^2180	0.663
0.45	0.674	0.265	0.286	0.720
0.50	0.692	0.332	0.379	0.767
0.55	0.709	0.394	0.461	0.807
0.60	0.726	0.452	0.534	0.841
0.65	0.742	0.505	0.598	0.869
0.70	0.758	0.554	0.653	0.892
0.75	0.773	0.599	0.702	0.912
0.80	0.788	0.641	0.745	0.9^4277
0.85	0.802	0.678	0.781	0.409
0.90	0.816	0.713	0.813	0.519
0.95	0.829	0.744	0.841	0.609

Table E.4 Probability densities of normal distribution

$$\varphi(x) = \frac{1}{\sqrt{2\pi}} e^{-x^2/2}$$

x	0	1	2
0.00	0.399	0.242	0.0540
0.05	0.398	0.230	0.488
0.10	0.397	0.218	0.440
0.15	0.395	0.206	0.396
0.20	0.391	0.194	0.355
0.25	0.387	0.183	0.317
0.30	0.381	0.171	0.283
0.35	0.375	0.160	0.252
0.40	0.368	0.150	0.224
0.45	0.361	0.140	0.198
0.50	0.352	0.130	0.175
0.55	0.343	0.120	0.155
0.60	0.333	0.111	0.136
0.65	0.323	0.102	0.119
0.70	0.312	0.0941	0.104
0.75	0.301	0.863	0.0^2910
0.80	0.290	0.790	0.792
0.85	0.278	0.721	0.687
0.90	0.266	0.656	0.595
0.95	0.254	0.596	0.514

Table E.5 Quantiles of normal distribution

$$P = \frac{1}{\sqrt{2\pi}} \int_{-\infty}^{U_p} e^{-x^2/2}\,dx = \Phi(U_p)$$

P	U_p	P	U_p
0.50	0	0.98	2.054
0.55	0.126	0.99	2.326
0.60	0.253	0.9^25	2.58
0.65	0.385	0.9^3	3.09
0.70	0.524	0.9^35	3.29
0.75	0.674	0.9^4	3.72
0.80	0.842	0.9^45	3.89
0.85	1.036	0.9^5	4.27
0.90	1.282	0.9^55	4.42
0.91	1.341	0.9^6	4.75
0.92	1.405	0.9^65	4.89
0.93	1.476	0.9^7	5.20
0.94	1.555	0.9^75	5.33
0.95	1.645	0.9^8	5.61
0.96	1.751	0.9^85	5.73
0.97	1.881	0.9^9	6.00

Table E.6 Weibull-Gnedenko distribution

$$P(t) = \exp\left[-\left(\frac{t}{a}\right)^b\right]$$

$\frac{t}{a}$	b = 0.1	0.3	0.5	0.7	0.9	1.1
0.1	0.452	0.606	0.729	0.819	0.882	0.924
0.2	0.427	0.540	0.639	0.723	0.791	0.843
0.3	0.412	0.498	0.578	0.650	0.713	0.766
0.4	0.402	0.468	0.531	0.591	0.645	0.694
0.5	0.393	0.444	0.493	0.540	0.585	0.627
0.6	0.387	0.424	0.461	0.497	0.532	0.566
0.7	0.381	0.407	0.433	0.459	0.484	0.509
0.8	0.376	0.392	0.409	0.425	0.441	0.457
0.9	0.372	0.380	0.387	0.395	0.403	0.410
1.0	0.368	0.368	0.368	0.368	0.368	0.368
1.5	0.353	0.323	0.294	0.265	0.237	0.210
2.0	0.342	0.292	0.243	0.197	0.155	0.117
2.5	0.334	0.268	0.206	0.150	0.102	0.065

$\frac{t}{a}$	b = 1.3	1.5	1.7	2.0	3.0	4.0
0.1	0.951	0.969	0.980	0.990	0.999	1.000
0.2	0.884	0.914	0.937	0.961	0.992	0.998
0.3	0.811	0.848	0.879	0.914	0.973	0.992
0.4	0.738	0.776	0.810	0.852	0.938	0.975
0.5	0.666	0.702	0.735	0.779	0.882	0.939
0.6	0.598	0.628	0.657	0.698	0.806	0.878
0.7	0.533	0.557	0.580	0.613	0.710	0.786
0.8	0.473	0.489	0.504	0.527	0.599	0.664
0.9	0.418	0.426	0.433	0.445	0.482	0.519
1.0	0.368	0.368	0.368	0.368	0.368	0.368
1.5	0.184	0.159	0.136	0.105	0.034	0.006
2.0	0.085	0.059	0.039	0.018	0.000	—
2.5	0.037	0.019	0.009	0.002	—	—

Table E.7 Rayleigh distribution

$$F(x) = 1 - \exp\left(-\frac{x^2}{2\sigma^2}\right)$$

$\frac{x}{\sigma}$	$F(x)$	$\frac{x}{\sigma}$	$F(x)$	$\frac{x}{\delta}$	$F(x)$
0.0	0.000	1.5	0.675	3.0	0.9889
0.1	0.005	1.6	0.722	3.1	0.9918
0.2	0.020	1.7	0.764	3.2	0.9940
0.3	0.044	1.8	0.802	3.3	0.9957
0.4	0.077	1.9	0.836	3.4	0.9969
0.5	0.117	2.0	0.865	3.5	0.9978
0.6	0.165	2.1	0.890	3.6	0.9985
0.7	0.217	2.2	0.911	3.7	0.9990
0.8	0.274	2.3	0.929	3.8	0.9993
0.9	0.333	2.4	0.944	3.9	0.9995
1.0	0.393	2.5	0.956	4.0	0.9997
1.1	0.454	2.6	0.966	4.01	0.99978
1.2	0.513	2.7	0.974	4.02	0.99985
1.3	0.570	2.8	0.980	4.04	0.99994
1.4	0.625	2.9	0.985	4.06	0.99998

Table E.8 Binomial distribution

$$Q_d = \sum_{k=d}^{n} C_n^k p^k q^{n-k}$$

		p									
n	d	0.01	0.02	0.03	0.4	0.05	0.06	0.07	0.08	0.09	0.10
	1	0.0490	0.0961	0.1413	0.1846	0.2262	0.2661	0.3043	0.3409	0.3760	0.4095
5	2	0.0010	0.0038	0.0085	0.0148	0.0226	0.0319	0.0425	0.0544	0.0674	0.0815
	3	0.0000	0.0001	0.0003	0.0006	0.0012	0.0020	0.0031	0.0045	0.0063	0.0086
	4	0.0000	0.0000	0.0000	0.0000	0.0000	0.0001	0.0001	0.0002	0.0003	0.0005
	1	0.0585	0.1142	0.1670	0.2172	0.2649	0.3101	0.3530	0.3936	0.4321	0.4686
	2	0.0015	0.0057	0.0125	0.0216	0.0328	0.0459	0.0608	0.0773	0.0952	0.1143
6	3	0.0000	0.0002	0.0005	0.0012	0.0022	0.0038	0.0058	0.0085	0.0118	0.0159
	4	0.0000	0.0000	0.0000	0.0000	0.0001	0.0002	0.0003	0.0005	0.0009	0.0013
	5	0.0000	0.0000	0.0000	0.0000	0.0000	0.0000	0.0000	0.0000	0.0000	0.0001
	1	0.0679	0.1319	0.1920	0.2486	0.3017	0.3515	0.3983	0.4422	0.4832	0.5217
	2	0.0020	0.0079	0.0171	0.0294	0.0444	0.0618	0.0813	0.1026	0.1255	0.1497
7	3	0.0000	0.0003	0.0009	0.0020	0.0038	0.0063	0.0097	0.0140	0.0193	0.0257
	4	0.0000	0.0000	0.0000	0.0001	0.0002	0.0004	0.0007	0.0012	0.0018	0.0027
	5	0.0000	0.0000	0.0000	0.0000	0.0000	0.0000	0.0000	0.0001	0.0001	0.0002
	1	0.0773	0.1492	0.2163	0.2786	0.3366	0.3904	0.4404	0.4868	0.5298	0.5695
	2	0.0027	0.0103	0.0223	0.0382	0.0572	0.0792	0.1035	0.1298	0.1577	0.1869
8	3	0.0001	0.0004	0.0014	0.0031	0.0058	0.0096	0.0147	0.0211	0.0289	0.0381
	4	0.0000	0.0000	0.0001	0.0002	0.0004	0.0008	0.0013	0.0022	0.0034	0.0050
	5	0.0000	0.0000	0.0000	0.0000	0.0000	0.0000	0.0001	0.0002	0.0003	0.0004
	1	0.0865	0.1663	0.2398	0.3075	0.3698	0.4270	0.4796	0.5278	0.5721	0.6126
	2	0.0034	0.0131	0.0282	0.0478	0.0712	0.0978	0.1271	0.1583	0.1912	0.2252
9	3	0.0001	0.0006	0.0020	0.0045	0.0084	0.0138	0.0209	0.0298	0.0405	0.0530
	4	0.0000	0.0000	0.0001	0.0003	0.0006	0.0013	0.0023	0.0037	0.0057	0.0083
	5	0.0000	0.0000	0.0000	0.0000	0.0000	0.0001	0.0002	0.0003	0.0006	0.0009

	1	0.0956	0.1829	0.2626	0.3352	0.4013	0.4614	0.5160	0.5656	0.6106	0.6513
	2	0.0043	0.0162	0.0345	0.0582	0.0861	0.1176	0.1517	0.1879	0.2255	0.2639
10	3	0.0001	0.0009	0.0028	0.0062	0.0115	0.0188	0.0283	0.0401	0.0540	0.0702
	4	0.0000	0.0000	0.0002	0.0004	0.0010	0.0020	0.0036	0.0058	0.0088	0.0128
	5	0.0000	0.0000	0.0000	0.0000	0.0001	0.0002	0.0003	0.0006	0/0010	0.0016
	1	0.1047	0.1993	0.2847	0.3618	0.4312	0.4937	0.5499	0.6004	0.6456	0.6862
	2	0.0052	0.0195	0.0414	0.0692	0.1019	0.1382	0.1772	0.2181	0.2601	0.3026
11	3	0.0002	0.0012	0.0037	0.0083	0.0152	0.0248	0.0370	0.0519	0.0695	0.0896
	4	0.0000	0.0000	0.0002	0.0007	0.0016	0.0030	0.0053	0.0085	0.0129	0.0185
	5	0.0000	0.0000	0.0000	0.0000	0.0001	0.0003	0.0005	0.0010	0.0017	0.0028
	1	0.1136	0.2153	0.3062	0.3873	0.4596	0.5241	0.5814	0.6323	0.6775	0.7176
	2	0.0062	0.0231	0.0487	0.0809	0.1184	0.1595	0.2033	0.2487	0.2948	0.3410
12	3	0.0002	0.0015	0.0049	0.0107	0.0196	0.0316	0.0468	0.0652	0.0866	0.1109
	4	0.0000	0.0001	0.0003	0.0010	0.0022	0.0043	0.0075	0.0120	0.0180	0.0256
	5	0.0000	0.0000	0.0000	0.0001	0.0002	0.0004	0.0009	0.0016	0.0027	0.0043
	1	0.1225	0.2310	0.3270	0.4118	0.4867	0.5526	0.6107	0.6618	0.7066	0.7458
	2	0.0073	0.0270	0.0564	0.0932	0.1354	0.1814	0.2298	0.2794	0.3293	0.3787
13	3	0.0003	0.0020	0.0062	0.0135	0.0245	0.0393	0.0577	0.0799	0.1054	0.1339
	4	0.0000	0.0001	0.0005	0.0014	0.0031	0.0060	0.0103	0.0163	0.0242	0.0342
	5	0.0000	0.0000	0.0000	0.0001	0.0003	0.0007	0.0013	0.0024	0.0041	0.0065
	1	0.1313	0.2464	0.3472	0.4353	0.5123	0.5795	0.6380	0.6888	0.7330	0.7712
	2	0.0084	0.0310	0.0645	0.1059	0.1530	0.2037	0.2565	0.3100	0.3632	0.4154
14	3	0.0003	0.0025	0.0077	0.0167	0.0301	0.0478	0.0698	0.0958	0.1255	0.1584
	4	0.0000	0.0001	0.0006	0.0019	0.0042	0.0080	0.0136	0.0214	0.0315	0.0441
	5	0.0000	0.0000	0.0000	0.0002	0.0004	0.0010	0.0020	0.0035	0.0059	0.0092
	1	0.1399	0.2614	0.3668	0.4579	0.5367	0.6047	0.6633	0.7137	0.7570	0.7941
	2	0.0096	0.0353	0.0730	0.1191	0.1710	0.2262	0.2832	0.3403	0.3965	0.4510
15	3	0.0004	0.0030	0.0094	0.0203	0.0362	0.0571	0.0829	0.1130	0.1469	0.1841
	4	0.0000	0.0002	0.0009	0.0025	0.0055	0.0194	0.0175	0.0273	0.0399	0.0556
	5	0.0000	0.0000	0.0001	0.0002	0.0006	0.0014	0.0028	0.0050	0.0082	0.0127

Table E.9 Poisson probabilities

$$p_k = \frac{a^k}{k!} e^{-a} = 0$$

	a						
k	0.001	0.002	0.003	0.004	0.005	0.006	0.007
0	9990	9980	9970	9960	9950	9940	9930
1	$0^{3}999$	$0^{2}200$	$0^{2}299$	$0^{2}398$	$0^{2}498$	$0^{2}596$	$0^{2}695$
2	$0^{5}1$	$0^{5}200$	$0^{5}449$	$0^{5}797$	$0^{4}124$	$0^{4}179$	$0^{4}243$

	a						
k	0.008	0.009	0.010	0.020	0.030	0.040	0.050
0	9920	9910	9900	9802	9704	9608	9512
1	$0^{2}794$	$0^{2}892$	$0^{2}990$	0196	0291	0384	0476
2	$0^{4}317$	$0^{4}401$	$0^{4}495$	$0^{3}196$	$0^{3}437$	$0^{3}769$	$0^{2}119$
3	—	—	—	$0^{5}131$	$0^{5}437$	$0^{4}102$	$0^{4}198$

	a						
k	0.06	0.07	0.08	0.09	0.10	0.20	0.30
0	9418	9324	9231	9139	9048	8187	7408
1	0565	0653	0738	0823	0905	1637	2222
2	$0^{2}170$	$0^{2}228$	$0^{2}295$	$0^{2}370$	$0^{2}452$	0164	0333
3	$0^{4}339$	$0^{4}533$	$0^{4}788$	$0^{3}111$	$0^{3}151$	$0^{2}109$	$0^{2}333$
4	$0^{5}1$	$0^{6}93$	$0^{5}158$	$0^{5}250$	$0^{5}377$	$0^{4}546$	$0^{3}250$
5	—	—	—	—	—	$0^{5}218$	$0^{4}150$

	a						
k	0.4	0.5	0.6	0.7	0.8	0.9	1.0
0	6703	6065	5488	4966	4493	4066	3679
1	2681	3033	3293	3476	3595	3659	3679
2	0536	0758	0988	1217	1438	1647	1839
3	$0^{2}715$	0126	0198	0284	0383	0494	0613
4	$0^{3}715$	$0^{2}158$	$0^{2}296$	$0^{2}497$	$0^{2}767$	0111	0153
5	$0^{4}572$	$0^{3}158$	$0^{3}356$	$0^{3}696$	$0^{2}123$	$0^{2}200$	$0^{2}307$
6	$0^{5}381$	$0^{4}132$	$0^{4}356$	$0^{4}811$	$0^{3}164$	$0^{3}300$	$0^{3}511$
7	—	$0^{5}1$	$0^{5}305$	$0^{5}811$	$0^{4}187$	$0^{4}386$	$0^{4}730$
8	—	—	—	$0^{5}1$	$0^{5}187$	$0^{5}434$	$0^{5}912$
9	—	—	—	—	—	—	$0^{5}101$

Table E.9 (*Continued*)

$$p_k = \frac{a^k}{k!} e^{-a} = 0$$

	a								
k	2	3	4	5	6	7	8	9	10
0	1353	0498	0183	$0^{2}674$	$0^{2}248$	$0^{3}912$	$0^{3}335$	$0^{3}123$	$0^{4}454$
1	2707	1494	0733	0337	0149	$0^{2}638$	$0^{2}268$	$0^{2}111$	$0^{3}454$
2	2707	2240	1465	0842	0446	0223	0107	$0^{2}500$	$0^{2}227$
3	1804	2240	1954	1404	0892	0521	0286	0150	$0^{2}757$
4	0902	1680	1954	1755	1339	0912	0573	0337	0189
5	0361	1008	1563	1755	1606	1277	0916	0607	0378
6	0120	0504	1042	1462	1606	1490	1221	0911	0631
7	$0^{2}344$	0216	0595	1044	1377	1490	1396	1171	0901
8	$0^{3}859$	$0^{2}810$	0298	0653	1033	1304	1396	1318	1126
9	$0^{3}191$	$0^{2}270$	0132	0363	0688	1014	1241	1318	1251
10	$0^{4}382$	$0^{3}810$	$0^{2}529$	0181	0413	0710	0993	1186	1251
11	$0^{5}694$	$0^{3}221$	$0^{2}192$	$0^{2}824$	0225	0452	0722	0970	1137
12	$0^{5}116$	$0^{4}524$	$0^{3}642$	$0^{2}343$	0113	0263	0481	0728	0948
13	—	$0^{4}127$	$0^{3}197$	$0^{2}132$	$0^{2}520$	0142	0296	0504	0729
14	—	$0^{5}273$	$0^{4}564$	$0^{3}472$	$0^{2}223$	$0^{2}709$	0169	0324	0521
15	—	$0^{5}1$	$0^{4}150$	$0^{3}157$	$0^{3}891$	$0^{2}331$	$0^{2}903$	0194	0347
16	—	—	$0^{5}376$	$0^{4}491$	$0^{3}334$	$0^{2}145$	$0^{2}451$	0109	0217
17	—	—	$0^{5}1$	$0^{4}144$	$0^{3}118$	$0^{3}596$	$0^{3}212$	$0^{2}579$	0128
18	—	—	—	$0^{5}401$	$0^{4}393$	$0^{3}232$	$0^{3}944$	$0^{2}289$	$0^{2}709$
19	—	—	—	$0^{5}106$	$0^{4}124$	$0^{4}854$	$0^{3}397$	$0^{2}137$	$0^{2}373$
20	—	—	—	—	$0^{5}373$	$0^{4}299$	$0^{3}159$	$0^{3}617$	$0^{2}187$
21	—	—	—	—	$0^{5}106$	$0^{5}997$	$0^{4}606$	$0^{3}264$	$0^{3}889$
22	—	—	—	—	—	$0^{5}317$	$0^{4}220$	$0^{3}108$	$0^{3}404$
23	—	—	—	—	—	$0^{5}1$	$0^{5}766$	$0^{4}423$	$0^{3}176$
24	—	—	—	—	—	—	$0^{5}255$	$0^{4}159$	$0^{4}732$
25	—	—	—	—	—	—	$0^{5}1$	$0^{5}571$	$0^{4}293$
26	—	—	—	—	—	—	—	$0^{5}198$	$0^{4}113$
27	—	—	—	—	—	—	—	$0^{5}1$	$0^{5}417$
28	—	—	—	—	—	—	—	—	$0^{5}149$
29	—	—	—	—	—	—	—	—	$0^{5}1$

Table E.9 (*Continued*)

$$p_k = \frac{a^k}{k!} e^{-a} = 0$$

	a			*a*	
k	15.0	20.0	*k*	15.0	20.0
0	—	—	23	0.013	0.067
1	$0.0^{5}500$	—	24	$0.0^{2}830$	0.056
2	$0.0^{4}340$	—	25	$0.0^{2}498$	0.044
3	$0.0^{3}172$	$0.0^{5}300$	26	$0.0^{2}287$	0.034
4	$0.0^{3}645$	$0.0^{4}140$	27	$0.0^{2}160$	0.025
5	$0.0^{2}194$	$0.0^{4}550$	28	$0.0^{3}855$	0.018
6	$0.0^{2}484$	$0.0^{3}183$	29	$0.0^{3}442$	0.012
7	0.010	$0.0^{3}523$	30	$0.0^{3}221$	$0.0^{2}834$
8	0.019	$0.0^{2}131$	31	$0.0^{3}107$	$0.0^{2}538$
	0.032	$0.0^{2}291$	32	$0.0^{4}500$	$0.0^{2}336$
10	0.047	$0.0^{2}582$	33	$0.0^{4}230$	$0.0^{2}204$
11	0.066	0.011	34	$0.0^{4}100$	$0.0^{2}120$
12	0.083	0.018	35	$0.0^{5}400$	$0.0^{3}685$
13	0.096	0.027	36	$0.0^{5}200$	$0.0^{3}381$
14	0.102	0.039	37	$0.0^{5}100$	$0.0^{3}206$
15	0.102	0.052	38	—	$0.0^{3}108$
16	0.096	0.064	39	—	$0.0^{4}560$
17	0.085	0.076	40	—	$0.0^{4}280$
18	0.071	0.084	41	—	$0.0^{4}140$
19	0.056	0.089	42	—	$0.0^{5}600$
20	0.042	0.089	43	—	$0.0^{5}300$
21	0.030	0.085	44	—	$0.0^{5}100$
22	0.020	0.077	45	—	$0.0^{5}100$

Table E.10 Cumulative Poisson probabilities

$$Q_d = \sum_{k=d}^{\infty} \frac{a^k}{k!} e^{-a} = 0$$

	a						
d	0.001	0.002	0.003	0.004	0.005	0.006	0.007
1	$0^{3}999$	$0^{2}200$	$0^{2}300$	$0^{2}399$	$0^{2}499$	$0^{2}598$	$0^{2}698$
2	$0^{5}10$	$0^{5}200$	$0^{5}449$	$0^{5}798$	$0^{4}125$	$0^{4}179$	$0^{4}244$

	a						
d	0.008	0.009	0.010	0.020	0.030	0.040	0.050
1	$0^{2}797$	$0^{2}896$	$0^{2}995$	0198	0296	0392	0488
2	$0^{4}318$	$0^{4}403$	$0^{4}497$	$0^{3}197$	$0^{3}441$	$0^{3}779$	$0^{2}121$
3	—	—	—	$0^{5}131$	$0^{5}440$	$0^{4}104$	$0^{4}201$

	a						
d	0.06	0.07	0.08	0.09	0.10	0.20	0.30
1	0582	0676	0769	0861	0952	1813	2592
2	$0^{2}173$	$0^{2}234$	$0^{2}303$	$0^{2}382$	$0^{2}468$	0175	0369
3	$0^{4}344$	$0^{4}542$	$0^{4}804$	$0^{3}114$	$0^{3}155$	$0^{2}115$	$0^{2}360$
4	$0^{5}1$	$0^{5}1$	$0^{5}160$	$0^{5}254$	$0^{5}385$	$0^{4}568$	$0^{3}266$
5	—	—	—	—	—	$0^{5}226$	$0^{4}158$
6	—	—	—	—	—	—	$0^{5}1$

	a						
d	0.4	0.5	0.6	0.7	0.8	0.9	1.0
1	3297	3935	4512	5034	5507	5934	6321
2	0616	0902	1219	1558	1912	2275	2642
3	$0^{2}793$	0144	0231	0341	0474	0629	0803
4	$0^{3}776$	$0^{2}175$	$0^{2}336$	$0^{2}575$	$0^{2}908$	0135	0190
5	$0^{4}612$	$0^{3}172$	$0^{3}394$	$0^{3}786$	$0^{2}141$	$0^{2}234$	$0^{2}366$
6	$0^{5}404$	$0^{4}142$	$0^{4}388$	$0^{4}900$	$0^{3}184$	$0^{3}343$	$0^{3}594$
7	—	$0^{5}100$	$0^{5}329$	$0^{5}888$	$0^{4}207$	$0^{4}434$	$0^{4}832$
8	—	—	—	$0^{5}1$	$0^{5}205$	$0^{5}482$	$0^{4}102$
9	—	—	—	—	—	—	$0^{5}113$

Table E.10 (*Continued*)

$$Q_d = \sum_{k=d}^{\infty} \frac{a^k}{k!} e^{-a} = 0$$

	a								
d	2	3	4	5	6	7	8	9	10
1	8647	9502	9817	9933	9975	9991	9997	9999	—
2	5940	8009	9084	9596	9826	9927	9970	9988	9995
3	3233	5768	7619	8753	9380	9704	9862	9938	9972
4	1429	3528	5665	7350	8488	9182	9576	9788	9897
5	0527	1847	3712	5595	7149	8270	9004	9450	9707
6	0166	0839	2149	3840	5543	6993	8088	8843	9329
7	$0^{2}453$	0335	1107	2378	3937	5503	6866	7932	8699
8	$0^{2}110$	0119	0511	1334	2560	4013	5470	6761	7798
9	$0^{3}237$	$0^{2}380$	0214	0681	1528	2709	4075	5443	6672
10	$0^{4}465$	$0^{2}110$	$0^{2}813$	0318	0839	1695	2834	4126	5421
11	$0^{5}831$	$0^{3}292$	$0^{2}284$	0137	0426	0985	1841	2940	4170
12	$0^{5}136$	$0^{4}714$	$0^{3}915$	$0^{2}545$	0201	0533	1119	1970	3032
13	—	$0^{4}162$	$0^{3}274$	$0^{2}202$	$0^{2}883$	0270	0638	1242	2084
14	—	$0^{5}340$	$0^{4}763$	$0^{3}698$	$0^{2}363$	0128	0342	0739	1355
15	—	$0^{5}1$	$0^{4}199$	$0^{3}226$	$0^{2}140$	$0^{2}572$	0173	0415	0835
16	—	—	$0^{5}489$	$0^{4}690$	$0^{3}509$	$0^{2}241$	$0^{2}823$	0220	0487
17	—	—	$0^{5}113$	$0^{4}199$	$0^{3}175$	$0^{3}958$	$0^{2}372$	0111	0270
18	—	—	—	$0^{5}542$	$0^{4}569$	$0^{3}362$	$0^{2}159$	$0^{2}532$	0143
19	—	—	—	$0^{5}140$	$0^{4}176$	$0^{3}130$	$0^{3}650$	$0^{2}243$	$0^{2}719$
20	—	—	—	—	$0^{5}518$	$0^{4}444$	$0^{3}253$	$0^{2}106$	$0^{2}345$
21	—	—	—	—	$0^{5}146$	$0^{4}145$	$0^{4}940$	$0^{3}439$	$0^{2}159$
22	—	—	—	—	—	$0^{5}453$	$0^{4}334$	$0^{3}175$	$0^{3}700$
23	—	—	—	—	—	$0^{5}135$	$0^{4}114$	$0^{4}668$	$0^{3}296$
24	—	—	—	—	—	—	$0^{5}373$	$0^{4}245$	$0^{3}120$
25	—	—	—	—	—	—	$0^{5}117$	$0^{5}865$	$0^{4}470$
26	—	—	—	—	—	—	—	$0^{5}294$	$0^{4}177$
27	—	—	—	—	—	—	—	$0^{5}1$	$0^{5}642$
28	—	—	—	—	—	—	—	—	$0^{5}225$
29	—	—	—	—	—	—	—	—	$0^{5}1$

Table E.10 (*Continued*)

$$Q_d = \sum_{k=d}^{\infty} \frac{a^k}{k!} e^{-a} = 0$$

	a			*a*	
k	15.0	20.0	*k*	15.0	20.0
0	—	—	23	0.0327	0.279
1	1.0	—	24	0.0195	0.212
2	$0.9^{5}5$	—	25	0.0112	0.157
3	$0.9^{4}61$	1.0	26	$0.0^{2}618$	0.112
4	$0.9^{3}789$	$0.9^{5}7$	27	$0.0^{2}331$	0.0779
5	$0.9^{3}143$	$0.9^{4}83$	28	$0.0^{2}172$	0.0525
6	$0.9^{2}721$	$0.9^{4}28$	29	$0.0^{3}861$	0.0343
7	$0.9^{2}237$	$0.9^{3}745$	30	$0.0^{3}418$	0.0218
8	0.9820	$0.9^{3}221$	31	$0.0^{3}197$	0.0135
9	0.9625	$0.9^{2}791$	32	$0.0^{4}90$	$0.0^{2}809$
10	0.9301	$0.9^{2}500$	33	$0.0^{4}40$	$0.0^{2}472$
11	0.881	0.9892	34	$0.0^{4}17$	$0.0^{2}269$
12	0.815	0.9786	35	$0.0^{5}7$	$0.0^{2}149$
13	0.732	0.9610	36	$0.0^{5}3$	$0.0^{3}804$
14	0.637	0.9339	37	$0.0^{5}1$	$0.0^{3}423$
15	0.534	0.895	38	—	$0.0^{3}217$
16	0.432	0.843	39	—	$0.0^{3}109$
17	0.336	0.779	40	—	$0.0^{4}53$
18	0.251	0.703	41	—	$0.0^{4}25$
19	0.180	0.618	42	—	$0.0^{4}12$
20	0.120	0.530	43	—	$0.0^{5}5$
21	0.0830	0.441	44	—	$0.0^{5}2$
22	0.0531	0.356	45	—	$0.0^{5}1$

Table E.11 Values for 0.10, 0.05, and 0.01 probability levels of deviation of empirical and theoretical distributions (Kolmogorov criterion)

$$P\{\sup|F_n(x) - F(x)| > A(n, q)\} = q$$

	q				q		
n	0.1	0.05	0.01	n	0.1	0.05	0.01
1	0.950	0.975	0.995	26	0.233	0.259	0.311
2	0.776	0.842	0.929	27	0.229	0.254	0.305
3	0.636	0.708	0.829	28	0.225	0.250	0.300
4	0.565	0.624	0.734	29	0.221	0.246	0.295
5	0.509	0.563	0.668	30	0.218	0.242	0.290
6	0.468	0.519	0.617	31	0.214	0.238	0.285
7	0.436	0.483	0.576	32	0.211	0.234	0.281
8	0.410	0.454	0.542	33	0.208	0.231	0.277
9	0.387	0.430	0.513	34	0.205	0.227	0.273
10	0.369	0.409	0.489	35	0.202	0.224	0.269
11	0.352	0.391	0.468	36	0.199	0.221	0.265
12	0.338	0.375	0.449	37	0.196	0.218	0.262
13	0.325	0.361	0.432	38	0.194	0.215	0.258
14	0.314	0.349	0.418	39	0.191	0.213	0.255
15	0.304	0.338	0.404	40	0.189	0.210	0.252
16	0.295	0.327	0.392	41	0.187	0.208	0.249
17	0.286	0.318	0.381	42	0.185	0.205	0.246
18	0.279	0.309	0.371	43	0.183	0.203	0.243
19	0.271	0.301	0.361	44	0.181	0.201	0.241
20	0.265	0.294	0.352	45	0.179	0.198	0.238
21	0.259	0.287	0.344	46	0.177	0.196	0.235
22	0.253	0.281	0.337	47	0.175	0.194	0.233
23	0.247	0.275	0.330	48	0.173	0.192	0.231
24	0.242	0.269	0.323	49	0.171	0.190	0.228
25	0.238	0.264	0.317	50	0.170	0.188	0.226

Table E.12 Values of Kolmogorov function

$(K(y_q) = 1 - q)$

q	0.50	0.40	0.30	0.20	0.10	0.05	0.02	0.01	0.001
y_q	0.828	0.895	0.974	1.073	1.224	1.358	1.520	1.627	1.950

Table E.13 Confidence limits for λ, Poisson distribution

n	0.999		0.99		0.95		0.90	
0	7.60	0.0000	5.30	0.0000	3.69	0.0000	3.00	0.0000
1	10.00	0.0005	7.43	0.0050	5.57	0.0253	4.74	0.0513
2	12.05	0.0320	9.27	0.103	7.22	0.242	6.30	0.355
3	13.93	0.149	10.98	0.338	8.77	0.619	7.75	0.818
4	15.71	0.355	12.59	0.672	10.24	1.09	9.15	1.37
5	17.41	0.63	14.15	1.08	11.67	1.62	10.51	1.97
6	19.05	0.97	15.66	1.54	13.06	2.20	11.84	2.61
7	20.65	1.35	17.13	2.04	14.42	2.81	13.15	3.29
8	22.22	1.77	18.58	2.57	15.76	3.45	14.43	3.98
9	23.75	2.22	20.00	3.13	17.08	4.12	15.71	4.70
10	25.26	2.70	21.40	3.72	18.39	4.80	16.96	5.43
11	26.74	3.20	22.78	4.32	19.68	5.49	18.21	6.17
12	28.20	3.73	24.14	4.94	20.96	6.20	19.44	6.92
13	29.65	4.27	25.50	5.58	22.23	6.92	20.67	7.69
14	31.08	4.83	26.84	6.23	23.49	7.65	21.89	8.46
15	32.50	5.40	28.16	6.89	24.74	8.40	23.10	9.25
16	33.90	5.99	29.48	7.57	25.98	9.15	24.30	10.04
17	35.29	6.59	30.79	8.25	27.22	9.90	25.50	10.83
18	36.68	7.20	32.09	8.94	28.45	10.67	26.69	11.63
19	38.05	7.82	33.38	9.64	29.67	11.44	27.88	12.44
20	39.41	8.45	34.67	10.35	30.89	12.22	29.06	13.25
21	40.76	9.09	35.95	11.07	32.10	13.00	30.24	14.07
22	42.11	9.74	37.22	11.79	33.31	13.79	31.42	14.89
23	43.45	10.40	38.48	12.52	34.51	14.58	32.59	15.72
24	44.78	11.06	39.74	13.25	35.71	15.38	33.75	16.55
25	46.11	11.73	41.00	14.00	36.90	16.18	34.92	17.38
26	47.42	12.41	42.25	14.74	38.10	16.98	36.08	18.22
27	48.74	13.09	43.50	15.49	39.28	17.79	37.23	19.06
28	50.05	13.78	44.74	16.24	40.47	18.61	38.39	19.90
29	51.35	14.47	45.98	17.00	41.65	19.42	39.54	20.75
30	52.64	15.17	47.21	17.77	42.83	20.24	40.69	21.59

Table E.14 0.99 confidence limits for p, binomial distribution

	n						
$N-n$	0	1	2	3	4	5	6
1	0.995	0.997	0.998	0.999	0.999	0.999	0.999
	0.000	0.003	0.041	0.111	0.185	0.254	0.315
2	0.929	0.959	0.971	0.977	0.981	0.984	0.986
	0.000	0.002	0.029	0.083	0.144	0.203	0.258
3	0.829	0.889	0.917	0.934	0.945	0.953	0.958
	0.000	0.001	0.023	0.066	0.118	0.170	0.219
4	0.734	0.815	0.856	0.882	0.900	0.913	0.923
	0.000	0.001	0.019	0.055	0.100	0.146	0.191
5	0.653	0.746	0.797	0.830	0.854	0.872	0.886
	0.000	0.001	0.016	0.047	0.087	0.128	0.169
6	0.586	0.685	0.742	0.781	0.809	0.831	0.848
	0.000	0.001	0.014	0.042	0.077	0.114	0.152
7	0.531	0.632	0.693	0.735	0.767	0.791	0.811
	0.000	0.001	0.012	0.037	0.069	0.103	0.138
8	0.484	0.585	0.648	0.693	0.728	0.755	0.777
	0.000	0.001	0.011	0.033	0.062	0.094	0.127
9	0.445	0.544	0.608	0.655	0.691	0.720	0.744
	0.000	0.001	0.010	0.030	0.057	0.087	0.117
10	0.411	0.509	0.573	0.621	0.658	0.688	0.714
	0.000	0.000	0.009	0.028	0.053	0.080	0.109
11	0.382	0.477	0.541	0.589	0.627	0.658	0.685
	0.000	0.000	0.008	0.026	0.049	0.075	0.101
12	0.357	0.449	0.512	0.561	0.599	0.631	0.658
	0.000	0.000	0.008	0.024	0.045	0.070	0.095
13	0.335	0.424	0.486	0.534	0.573	0.605	0.633
	0.000	0.000	0.007	0.022	0.043	0.065	0.090
14	0.315	0.402	0.463	0.510	0.549	0.582	0.610
	0.000	0.000	0.007	0.021	0.040	0.062	0.085
15	0.298	0.381	0.441	0.488	0.527	0.560	0.588
	0.000	0.000	0.006	0.020	0.038	0.058	0.080
16	0.282	0.363	0.422	0.468	0.507	0.539	0.567
	0.000	0.000	0.006	0.019	0.036	0.055	0.076
17	0.268	0.346	0.404	0.449	0.488	0.520	0.548
	0.000	0.000	0.006	0.018	0.034	0.053	0.073
18	0.255	0.331	0.387	0.432	0.470	0.502	0.531
	0.000	0.000	0.005	0.017	0.032	0.050	0.069
19	0.243	0.317	0.372	0.416	0.453	0.485	0.514
	0.000	0.000	0.005	0.016	0.031	0.048	0.066
20	0.233	0.304	0.358	0.401	0.438	0.470	0.498
	0.000	0.000	0.005	0.015	0.029	0.046	0.064
21	0.223	0.292	0.345	0.387	0.423	0.455	0.483
	0.000	0.000	0.005	0.015	0.028	0.044	0.061
22	0.214	0.281	0.332	0.374	0.410	0.441	0.469
	0.000	0.000	0.004	0.014	0.027	0.042	0.059

Table E.14 (*Continued*)

	n						
$N - n$	0	1	2	3	4	5	6
23	0.206	0.271	0.321	0.362	0.397	0.428	0.455
	0.000	0.000	0.004	0.013	0.026	0.041	0.056
24	0.198	0.262	0.310	0.351	0.385	0.416	0.443
	0.000	0.000	0.004	0.013	0.025	0.039	0.054
25	0.191	0.253	0.300	0.340	0.374	0.404	0.431
	0.000	0.000	0.004	0.012	0.024	0.038	0.053
26	0.184	0.245	0.291	0.330	0.363	0.393	0.420
	0.000	0.000	0.004	0.012	0.023	0.037	0.051
27	0.178	0.237	0.282	0.320	0.353	0.382	0.409
	0.000	0.000	0.004	0.012	0.023	0.035	0.049
28	0.172	0.230	0.274	0.311	0.344	0.373	0.398
	0.000	0.000	0.004	0.011	0.022	0.034	0.048
29	0.167	0.223	0.266	0.303	0.335	0.363	0.389
	0.000	0.000	0.003	0.011	0.021	0.033	0.046
30	0.162	0.216	0.259	0.295	0.326	0.354	0.379
	0.000	0.000	0.003	0.011	0.020	0.032	0.045
35	0.140	0.189	0.227	0.260	0.289	0.315	0.339
	0.000	0.000	0.003	0.009	0.018	0.028	0.039
40	0.124	0.168	0.203	0.233	0.260	0.284	0.306
	0.000	0.000	0.002	0.008	0.016	0.025	0.035
45	0.111	0.151	0.183	0.210	0.235	0.258	0.279
	0.000	0.000	0.002	0.007	0.014	0.022	0.031
50	0.101	0.137	0.166	0.192	0.215	0.237	0.257
	0.000	0.000	0.002	0.006	0.013	0.020	0.028
60	0.085	0.116	0.141	0.164	0.184	0.203	0.229
	0.000	0.000	0.002	0.005	0.011	0.017	0.024
80	0.064	0.088	0.108	0.126	0.143	0.158	0.173
	0.000	0.000	0.001	0.004	0.008	0.013	0.018
100	0.052	0.071	0.088	0.103	0.116	0.129	0.142
	0.000	0.000	0.001	0.003	0.007	0.010	0.015
200	0.026	0.036	0.045	0.053	0.061	0.068	0.075
	0.000	0.000	0.001	0.002	0.003	0.005	0.008
500	0.011	0.015	0.018	0.022	0.025	0.028	0.031
	0.000	0.000	0.001	0.001	0.001	0.002	0.003

Table E.15 0.95 confidence limits for p, binomial distribution

	n						
$N - n$	0	1	2	3	4	5	6
1	0.975	0.987	0.992	0.994	0.995	0.996	0.996
	0.000	0.013	0.094	0.194	0.284	0.359	0.421
2	0.842	0.906	0.932	0.947	0.957	0.963	0.968
	0.000	0.008	0.068	0.147	0.223	0.290	0.349
3	0.708	0.806	0.853	0.882	0.901	0.915	0.925
	0.000	0.006	0.053	0.118	0.184	0.245	0.299
4	0.602	0.716	0.777	0.816	0.843	0.863	0.878
	0.000	0.005	0.043	0.099	0.157	0.212	0.262
5	0.522	0.641	0.710	0.755	0.788	0.813	0.833
	0.000	0.004	0.037	0.085	0.137	0.187	0.234
6	0.459	0.579	0.651	0.701	0.738	0.766	0.789
	0.000	0.004	0.032	0.075	0.122	0.167	0.211
7	0.410	0.527	0.600	0.652	0.692	0.723	0.749
	0.000	0.003	0.028	0.067	0.109	0.151	0.192
8	0.369	0.483	0.556	0.610	0.651	0.684	0.711
	0.000	0.003	0.025	0.060	0.099	0.139	0.177
9	0.336	0.445	0.518	0.572	0.614	0.649	0.677
	0.000	0.003	0.023	0.055	0.091	0.128	0.163
10	0.308	0.413	0.484	0.538	0.581	0.616	0.646
	0.000	0.002	0.021	0.050	0.084	0.118	0.152
11	0.285	0.385	0.454	0.508	0.551	0.587	0.617
	0.000	0.002	0.019	0.047	0.078	0.110	0.142
12	0.265	0.360	0.428	0.481	0.524	0.560	0.590
	0.000	0.002	0.018	0.043	0.073	0.103	0.133
13	0.247	0.339	0.405	0.456	0.499	0.535	0.565
	0.000	0.002	0.017	0.040	0.068	0.097	0.126
14	0.232	0.319	0.383	0.434	0.476	0.512	0.543
	0.000	0.002	0.016	0.038	0.064	0.091	0.119
15	0.218	0.302	0.364	0.414	0.456	0.491	0.522
	0.000	0.002	0.015	0.036	0.061	0.087	0.133
16	0.206	0.287	0.347	0.396	0.437	0.471	0.502
	0.000	0.001	0.014	0.034	0.057	0.082	0.107
17	0.195	0.273	0.331	0.379	0.419	0.453	0.484
	0.000	0.001	0.013	0.032	0.054	0.078	0.102
18	0.185	0.260	0.317	0.363	0.403	0.436	0.467
	0.000	0.001	0.012	0.030	0.052	0.075	0.098
19	0.176	0.249	0.304	0.349	0.388	0.421	0.451
	0.000	0.001	0.012	0.029	0.050	0.071	0.094
20	0.168	0.238	0.292	0.336	0.374	0.407	0.436
	0.000	0.001	0.011	0.028	0.047	0.068	0.090
21	0.161	0.228	0.280	0.324	0.360	0.393	0.423
	0.000	0.001	0.010	0.027	0.045	0.066	0.086
22	0.154	0.219	0.270	0.312	0.349	0.381	0.410
	0.000	0.001	0.010	0.025	0.044	0.063	0.083

Table E.15 (*Continued*)

	n						
$N - n$	0	1	2	3	4	5	6
23	0.148	0.211	0.260	0.301	0.338	0.369	0.397
	0.000	0.001	0.010	0.024	0.042	0.061	0.080
24	0.142	0.203	0.251	0.292	0.327	0.358	0.386
	0.000	0.001	0.009	0.024	0.040	0.058	0.077
25	0.137	0.196	0.243	0.282	0.317	0.347	0.375
	0.000	0.001	0.009	0.023	0.039	0.056	0.075
26	0.132	0.190	0.235	0.274	0.307	0.337	0.364
	0.000	0.001	0.009	0.022	0.030	0.055	0.072
27	0.127	0.184	0.228	0.265	0.298	0.328	0.355
	0.000	0.001	0.008	0.021	0.036	0.053	0.070
28	0.123	0.178	0.221	0.257	0.290	0.319	0.345
	0.000	0.001	0.008	0.020	0.035	0.051	0.068
29	0.119	0.172	0.215	0.250	0.282	0.311	0.336
	0.000	0.001	0.008	0.020	0.034	0.050	0.066
30	0.116	0.167	0.208	0.243	0.275	0.303	0.328
	0.000	0.001	0.008	0.019	0.033	0.048	0.064
35	0.100	0.145	0.182	0.214	0.242	0.268	0.292
	0.000	0.001	0.007	0.017	0.029	0.042	0.056
40	0.088	0.129	0.162	0.191	0.217	0.241	0.263
	0.000	0.001	0.006	0.015	0.025	0.037	0.049
45	0.079	0.115	0.145	0.172	0.196	0.218	0.239
	0.000	0.001	0.005	0.013	0.023	0.033	0.044
50	0.071	0.104	0.132	0.157	0.179	0.200	0.219
	0.000	0.001	0.005	0.012	0.021	0.030	0.040
60	0.060	0.088	0.112	0.133	0.152	0.170	0.187
	0.000	0.000	0.004	0.010	0.017	0.025	0.034
80	0.045	0.067	0.085	0.102	0.118	0.132	0.145
	0.000	0.000	0.008	0.008	0.013	0.019	0.026
100	0.036	0.054	0.069	0.083	0.096	0.108	0.119
	0.000	0.000	0.002	0.006	0.011	0.016	0.021
200	0.018	0.027	0.035	0.043	0.049	0.056	0.062
	0.000	0.000	0.001	0.003	0.005	0.008	0.011

Table E.16 Critical values $\chi^2(r, q)$ for a chi-square distribution

$$q = \frac{1}{2^{r/2}\Gamma(r/2)} \int_{\chi^2(r,\,q)}^{\infty} x^{r/2-1} e^{-x/2}\, dx$$

	p													
r	0.99	0.98	0.95	0.90	0.80	0.70	0.50	0.30	0.20	0.10	0.05	0.02	0.01	0.001
1	0.000	0.001	0.004	0.016	0.064	0.148	0.455	1.074	1.642	2.71	3.84	5.41	6.63	10.83
2	0.020	0.040	0.103	0.211	0.446	0.713	1.386	2.41	3.22	4.60	5.99	7.82	9.21	13.82
3	0.115	0.185	0.352	0.584	1.005	1.424	2.37	3.66	4.64	6.25	7.81	9.84	11.34	16.27
4	0.297	0.429	0.711	1.064	1.649	2.20	3.36	4.88	5.99	7.78	9.49	11.67	13.28	18.46
5	0.554	0.752	1.145	1.610	2.34	3.00	4.35	6.06	7.29	9.24	11.07	13.39	15.09	20.5
6	0.872	1.134	1.635	2.20	3.07	3.83	5.35	7.23	8.56	10.64	12.59	15.03	16.81	22.5
7	1.239	1.564	2.17	2.83	3.82	4.67	6.35	8.38	9.80	12.02	14.07	16.62	18.48	24.3
8	1.646	2.03	2.73	3.49	4.59	5.53	7.34	9.52	11.03	13.36	15.51	18.17	20.1	26.1
9	2.09	2.53	3.33	4.17	5.38	6.39	8.34	10.66	12.24	14.68	16.92	19.68	21.7	27.9
10	2.56	3.06	3.94	4.86	6.18	7.27	9.34	11.78	13.44	15.99	18.31	21.2	23.2	29.6
11	3.05	3.61	4.58	5.58	6.99	8.15	10.34	12.90	14.63	17.28	19.68	22.6	24.7	31.3
12	3.57	4.18	5.23	6.30	7.81	9.03	11.34	14.01	15.81	18.55	21.0	24.1	26.2	32.9

13	4.11	4.76	5.89	7.04	8.63	9.93	12.34	15.12	16.98	19.81	22.4	25.5	27.7	34.6
14	4.66	5.37	6.57	7.79	9.47	10.82	13.34	16.22	18.15	21.1	23.7	26.9	29.1	36.1
15	5.23	5.98	7.26	8.55	10.31	11.72	14.34	17.32	19.31	22.3	25.0	28.3	30.6	37.7
16	5.81	6.61	7.96	9.31	11.15	12.62	15.34	18.42	20.5	23.5	26.3	29.6	32.0	39.3
17	6.41	7.26	8.67	10.08	12.00	13.53	16.34	19.51	21.6	24.8	27.6	31.0	33.4	40.8
18	7.01	7.91	9.39	10.86	12.86	14.44	17.34	20.6	22.8	26.0	28.9	32.3	34.8	42.3
19	7.63	8.57	10.11	11.65	13.72	15.35	18.34	21.7	23.9	27.2	30.1	33.7	36.2	43.8
20	8.26	9.24	10.85	12.44	14.58	16.27	19.34	22.8	25.0	28.4	31.4	35.0	37.6	45.3
21	8.90	9.92	11.59	13.24	15.44	17.18	20.3	23.9	26.2	29.6	32.7	36.3	38.9	46.8
22	9.54	10.60	12.34	14.04	16.31	18.10	21.3	24.9	27.3	30.8	33.9	37.7	40.3	48.3
23	10.20	11.29	13.09	14.85	17.19	19.02	22.3	26.0	28.4	32.0	35.2	39.0	41.6	49.7
24	10.86	11.99	13.85	15.66	18.06	19.94	23.3	27.1	29.6	33.2	36.4	40.3	43.0	51.2
25	11.52	12.70	14.61	16.47	18.94	20.9	24.3	28.2	30.7	34.4	37.7	41.7	44.3	52.6
26	12.20	13.41	15.38	17.29	19.82	21.8	25.3	29.2	31.8	35.6	38.9	42.9	45.6	54.1
27	12.88	14.12	16.15	18.11	20.7	22.7	26.3	30.3	32.9	36.7	40.1	44.1	47.0	55.5
28	13.56	14.85	16.93	18.94	21.6	23.6	27.3	31.4	34.0	37.9	41.3	45.4	48.3	56.9
29	14.26	15.57	17.71	19.77	22.5	24.6	28.3	32.5	35.1	39.1	42.6	46.6	49.6	58.3
30	14.95	16.31	18.49	20.6	23.4	25.5	29.3	33.5	36.2	40.3	43.8	48.0	50.9	59.7

Table E.17 Student's *t* values distribution

$$q = 2\,\frac{\Gamma((n+1)/2)}{\Gamma(n/2)\sqrt{\pi n}} \int_{t(n,q)}^{\infty} \left(1 + \frac{x^2}{n}\right)^{-(n+1)/2} dx$$

	q							
n	0.80	0.40	0.20	0.10	0.05	0.02	0.01	0.001
1	0.325	1.376	3.078	6.314	12.706	31.821	63.657	636.619
2	0.289	1.061	1.886	2.920	4.303	6.965	9.925	31.598
3	0.277	0.978	1.638	2.353	3.182	4.541	5.841	12.941
4	0.271	0.941	1.533	2.132	2.776	3.747	4.604	8.610
5	0.267	0.920	1.476	2.015	2.571	3.365	4.032	6.859
6	0.265	0.906	1.440	1.943	2.447	3.143	3.707	5.959
7	0.263	0.896	1.415	1.895	2.365	2.998	3.499	5.405
8	0.262	0.889	1.397	1.860	2.306	2.896	3.355	5.041
9	0.261	0.883	1.383	1.833	2.262	2.821	3.250	4.781
10	0.260	0.879	1.372	1.812	2.228	2.764	3.169	5.587
11	0.260	0.876	1.363	1.796	2.201	2.718	3.106	4.437
12	0.259	0.873	1.356	1.782	2.179	2.681	3.054	4.318
13	0.259	0.870	1.350	1.771	2.160	2.650	3.012	4.221
14	0.258	0.868	1.345	1.761	2.145	2.624	2.977	4.140
15	0.258	0.866	1.341	1.753	2.131	2.602	2.947	4.073
16	0.258	0.865	1.337	1.746	2.120	2.583	2.921	4.015
17	0.257	0.863	1.333	1.740	2.110	2.567	2.898	3.965
18	0.257	0.862	1.330	1.734	2.101	2.552	2.878	3.922
19	0.257	0.861	1.328	1.729	2.093	2.539	2.861	3.883
20	0.257	0.860	1.325	1.725	2.086	2.528	2.845	3.850
21	0.257	0.859	1.323	1.721	2.080	2.518	2.831	3.819
22	0.256	0.858	1.321	1.717	2.074	2.508	2.819	3.792
23	0.256	0.858	1.319	1.714	2.069	2.500	2.807	3.767
24	0.256	0.857	1.318	1.711	2.064	2.492	2.797	3.745
25	0.256	0.856	1.316	1.708	2.060	2.485	2.787	3.725
26	0.256	0.856	1.315	1.706	2.056	2.479	2.779	3.707
27	0.256	0.855	1.314	1.703	2.052	2.473	2.771	3.690
28	0.256	0.855	1.313	1.701	2.048	2.467	2.763	3.674
29	0.256	0.854	1.311	1.699	2.045	2.462	2.756	3.659
30	0.256	0.854	1.310	1.697	2.042	2.457	2.750	3.646
40	0.255	0.851	1.303	1.684	2.021	2.423	2.704	3.551
60	0.254	0.848	1.296	1.671	2.000	2.390	2.660	3.460
120	0.254	0.845	1.289	1.658	1.980	2.358	2.617	3.373
	0.253	0.842	1.282	1.645	1.960	2.326	2.576	3.291

Table E.18 Random numbers

218	167	737	498	237
155	844	335	308	003
512	480	704	201	345
635	342	658	752	481
331	485	402	964	398
205	360	181	402	657
682	674	459	904	295
831	304	043	132	766
705	767	185	517	215
564	532	937	594	319
825	354	514	505	152
613	739	043	284	528
131	485	232	356	453
515	248	325	786	176
424	429	930	479	394
340	716	108	056	839
654	958	790	703	499
805	154	202	368	925
337	376	471	473	701
931	069	322	301	595
217	258	688	308	632
684	352	164	744	611
148	723	383	913	102
634	633	136	102	389
048	758	603	204	857
360	914	378	194	260
674	579	436	939	724
304	936	506	651	677
767	672	239	422	936
532	797	366	819	204
354	813	648	347	392
739	345	692	352	142
207	334	851	934	837
512	290	316	645	283
406	054	726	965	031
008	877	841	295	475
561	917	428	279	133
707	354	381	848	370
475	090	376	637	997
109	831	031	151	018
982	472	819	659	881
569	793	888	980	052
094	880	615	538	288
217	508	912	020	713
670	290	636	701	612
321	425	136	884	677
989	102	242	634	606
121	192	026	719	760
323	707	189	490	653
597	307	656	992	197

APPENDIX F

RELIABILITY SOFTWARE

F.1 PROGRAMS FOR THIS HANDBOOK

Computer software developed by the authors and editors for all the major reliability programs in this book is available from Harrison and Company, Post Office Box 2732, Merrifield, Virginia 22116, (703) 280-2202.

F.1.1 Programs for Individual Chapters and Appendixes

Standard programs are available for IBM-compatible personal computers for specific portions of this handbook, and special programs can be ordered for main frame and other computers. A reader may order either individual programs or a set of programs for the entire handbook. Work on the special programs often can be performed in Russia at a lower cost than the cost of comparable work performed in the United States. Enquiries on both standard and special programs may be sent by mail, phone, or FAX to the address above.

Students and professors may be interested in the educational versions of the Harrison personal computer programs. The versions are menu driven, and require no knowledge of engineering or mathematics except elementary calculus, and, in some cases, elementary statistics. The programs are written to be suitable for self-teaching, but can be used in the classroom. Special rates are available to educational institutions.

F.1.2 DIANA, A Design Engineering and Cost Analysis Program

Harrison and Company also offers DIANA, a set of reliability programs intended for users who want to work interactively with the calculation programs, the simulation programs, and their own records on such subjects as costs, maintenance, design characteristics, and equipment history. DIANA is intended to help engineers, administrators, and operations research analysts during the conception, feasibility analysis, design, development, and initial use of complex new items. Subprograms include the following.

- Reliability indexes for existing equipment and for combinations of existing equipment or components.
- Optimization of the structure and the redundancy of a proposed design. The user may find either the least expensive design meeting an input reliability requirement or the most reliable design possible within an input cost limit. Constraints other than cost, such as weight or volume, also may be programmed.
- Reliability forecasts for three types of operating situations—no repair, unlimited repair capability, and limited repair capability. In addition to being used for equipment evaluation, these calculations may aid in the determination of maintenance plans and policies.
- Analysis of existing inventories of spare parts and optimization of proposed inventories of spare parts.
- Rule-of-thumb cost-reliability indexes for the conceptual phase of a design project or for "what if" questions.

DIANA includes the program for Monte Carlo Simulations.

F.1.3 RELT Product Reliability and Accelerated Test Programs

Four RELT programs sold by Harrison and Company offer improved quality assurance to developers of new products and manufacturers of existing products. They may be used either independently or with other computer programs based on documentation such as MIL HDBK 217, MIL STD 1629, MIL HDBK 472, or Bellcore. When used with other programs for new designs, they provide empirical tests for defective components or production processes. Requirements are PC DOS or MS DOS 3.0 or higher, 512,000 RAM, and an EGA or VGA monitor.

- RELT 1 finds reliability characteristics of items from tests of the entire item. Mean and confidence limits for each tested parameter are found. The type of failure time distribution does not have to be known. Samples may be small and may be censored. There are three test options: Bernoulli tests, time-to-failure tests, and load-strength tests.
- RELT 2 finds reliability characteristics of items from tests of item elements. Element test data or specifications are integrated with the system structural reliability diagram. Results can be displayed graphically on the screen.

- RELT 3 is designed for automatic solution of reliability problems with data from either tests or field use. Both whole system tests and component tests can be handled, as can both normal and forced tests.
- RELT 4 analyzes accelerated test results and predicts results of future normal operation. Relationships are established between stress parameters and lifetime distributions.

F.1.4 RELMAIN Program for Experimental and Aging Equipment

The Harrison RELMAIN programs may be used with either developmental equipment or aging equipment whose failure time distributions must be predicted when there are no data available except previous failure times. Probabilistic and statistical analyses may be applied to such questions as expected lifetime, the possibility of manufacturing defects, and the optimization of replacement schedules. As a failure time prediction program, RELMAIN is particularly valuable for equipment that cannot be monitored, inspected visually, or tested, but should be replaced or repaired before it fails.

F.1.5 SPAREKIT Program for Optimization of Stocks of Spare Parts

The Harrison SPAREKIT program, which is described in Section 21.1, is a very small and very inexpensive program for IBM and IBM-compatible personal computers. It is a menu-driven program that requires neither technical expertise nor computer expertise. SPAREKIT is intended primarily for small organizations that have neither the funds nor the requirement for a standard, full-scale inventory optimization program. Uses include:

- Selection of a kit of spare parts to be kept in the field with a manufactured item.
- Selection of a kit of spare parts to be kept on hand at each of several service facilities rather than at a factory or at distribution centers.
- Selection of inventories of spares to be ordered in advance to be kept for emergency repairs.

F.2 MAINTENANCE SOFTWARE

Maintenance software generally falls into three broad categories: (1) programs with data that are input automatically or semiautomatically from instruments that monitor indicators of equipment condition, such as ultrasound or infrared readings, lubricants, wear particles, or vibration, (2) programs with manual inputs, and (3) integration programs with multiple inputs that may be either automatic or manual. The principal functions of the programs are to facilitate record-keeping, facilitate maintenance scheduling, predict costs, diagnose causes of failures, and optimize preventive maintenance.

Maintenance software costs can range from a few hundred to several hundred thousand dollars. Properly selected programs, however, can pay for themselves several times over through reduction of equipment downtime, prevention of failures, and increased maintenance efficiency. Prospective purchasers should contact software firms that advertise in the industrial and technical magazines in their fields. *Maintenance Technology*, Section F.4, updates several comprehensive surveys throughout the year.

F.3 SOURCES OF CURRENT INFORMATION

Data Sources is a semiannual publication with a Software Volume that covers the spectrum of reliability and related fields. The listings are extensively cross-referenced, ordered by hardware requirements, and usually include capabilities, cost, previous history, and producer information. Most of the reliability programs are listed either under Engineering/Scientific or under Statistics & Mathematics. The periodical, which is carried by many public libraries, is published by Data Sources, One Park Avenue, New York, New York 10016, (212) 503-5388.

Maintenance Technology is a monthly magazine with data bases on maintenance software that are updated frequently. The periodical is published by Applied Technology Publications, 1300 S. Grove Avenue, Suite 205, Barrington, IL 60010, (708) 382-8100.

Some software can be acquired for no charge or a nominal charge. Sector Systems Co. has a variety of engineering and computer-assisted design programs that are in the public domain and are offered at prices below $10.00 per disk. Sector is at 416 Ocean Avenue, Marblehead, MA 01945, (617) 639-2625. "Shareware" programs also are available without payment of royalties. There are several mail-order shareware distributors, including Public Software Library (1-800-242-4775), and Public Brand Software (1-800-426-3475). Both send free catalogs.

BIBLIOGRAPHY

Abdel-Hameed, M.S., et al. (1984). *Reliability Theories and Models*. Academic Press, San Diego, CA.

Anderson, R. T. (1976). *Reliability Design Handbook*. Reliability Analysis Center, NO RDH 376.

Anderson, R. T. (1990). *Reliability Centered Maintenance*. Elsevier, New York.

Asher, H., and Feingold, H. (1984). *Repairable Systems Reliability*. Marcel Dekker, New York.

Bakay, A. J. (1982). *A Guide for Auditing of Maintainability and Maintenance*. Institute of Internal Auditors, Altamonte Springs, Florida.

Barlow, R. E., and Proschan, F. (1965). *Mathematical Theory of Reliability*. Wiley, New York.

Barlow, R.E., and Proschan, F. (1981). *Statistical Theory of Reliability and Life Testing: Probability Models*, 2nd ed. To Begin With, Silver Spring, MD.

Bazovsky, I. (1961). *Reliability Theory and Practice*. Prentice Hall, Englewood Cliffs, NJ.

Becker, P. W., and Jensen, F. (1977). *Design of Systems and Circuits for Maximum Reliability or Maximum Production Yield*. McGraw-Hill, New York.

Billinton, R., and Allan, R. N. (1983). *Reliability Evaluation of Engineering Systems: Concepts and Techniques*. Pitman Advanced Publishing Program, London.

Billinton, R., and Allan, R. N. (1984). *Reliability Evaluation of Power Systems*. Plenum Press, New York.

Blockley, D. I. (1992). *Engineering Safety*. McGraw-Hill, New York.

Breyfogle, F. W. (1992). *Statistical Methods for Testing, Development and Manufacturing*. Wiley, New York.

Bullock, J. H. (1979). *Maintenance Planning and Control*. National Association of Accountants.

Casciata, F. (1991). *Fragility Analysis of Complex Structural Systems*. Wiley, New York.

Cox, D. R. (1962). *Renewal Theory*. Methuen, London.

Cox, D. R., and Isham, V. (1980). *Point Processes*. Chapman and Hall, London.

Cox, D. R., and Oakes, D. (1984). *Analysis of Survival Data*. Chapman and Hall, London.

Crowder, M. J., Kimber, A. C., Smith, R. L., and Sweeting, T. J. (1991) *Statistical Analysis of Reliability Data*. Chapman and Hall, London.

Dhillon, B. S. (1983). *Systems Reliability, Maintainability, and Management*. Princeton, NJ.

Dhillon, B. S. (1983). *Reliability Engineering in System Design and Operation*. Van Nostrand Reinhold, New York.

Dhillon, B. S. (1988). *Mechanical Reliability: Theory, Models and Applications*, American Institute of Aeronautics and Astronautics, Washington.

Dhillon, B. S., and Reiche, H. (1985). *Reliability and Maintainability Management*. Van Nostrand Reinhold, New York.

Dhillon, B. S., and Singh, C. (1981). *Engineering Reliability: New Techniques and Applications*. Wiley, New York.

Elandt-Johnson, R. C., and Johnson, N. L. (1980). *Survival Models and Data Analysis*. Wiley, New York.

Enrick, N. L. (1977). *Quality Control and Reliability*. 7th ed. Industrial Press, New York.

Flamm, J., and Luisi, T. (1992). *Reliability Data Collection and Analysis*. Kluwer Academic Publishers, Norwell, MA.

Gertsbakh, I. B. (1977). *Models of Preventive Maintenance*. North-Holland, Amsterdam.

Gheorghe, A. (1982). *Applied System Engineering*. Wiley, New York.

Gilmore, H. L., and Schwartz, H. C. (1986). *Integrated Product Testing and Evaluation*. Marcel Dekker/ASQC Quality Press, New York.

Gnedenko, B. V., Belyayev, Uy. K., and Solovyev, A. D. (1969). *Mathematical Methods of Reliability Theory*. Academic Press, San Diego, CA.

Goldberg, H. (1981). *Extending the Limits of Reliability Theory*. Wiley, New York.

Gordon, G. (1978). *System Simulation*, 2d ed. Prentice Hall, Englewood Cliffs, NJ.

Green, A. E., and Bourne, A. J. (1972). *Reliability Technology*. Wiley Interscience, New York.

Grosh, D. L. (1989). *A Primer of Reliability Theory*. Wiley, New York.

Guide on Reliability of Systems, Equipments and Components, BS 5760.

Halpern, S. (1978). The Assurance Sciences: An Introduction to Quality Control and Reliability. Prentice Hall, Englewood Cliffs, NJ.

Harr, M. E. (1987). *Reliability-Based Design in Civil Engineering*. McGraw-Hill, New York.

Henley, E. J., and Kumamoto, H. (1981). *Reliability Engineering and Risk Assessment*. Prentice Hall, Englewood Cliffs, NJ.

Henley, E. J., and Kumamoto, H. (1985). *Designing for Reliability and Safety Control*. Prentice Hall, Englewood Cliffs, NJ.

International Electrotechnical Commission. (1974). List of Basic Terms, Definitions and Related Mathematics for Reliability. IEC 271. *International Vocabulary*, Chapter 191: *Dependability and Quality Service*, IEC 50 (191).

International Electrotechnical Commission. (1991). *International Vocabulary*. Chapter 191: *Dependability and Quality Service*. IEC 50 (191).

Jones, J. V. (1988). Engineering Design: Reliability, Maintainability, and Testability. TAB Professional and Reference Books, Blue Ridge Summit, PA.

Kapur, K.C., and Lamberson, L. R. (1977). *Reliability in Engineering Design*. Wiley, New York.

Kececioglu, D. (1991). Reliability Engineering Handbook. Prentice Hall, Englewood Cliffs, NJ.

Kozlov, B. A., and Ushakov, I. A. (1970). *Reliability Handbook*. Holt, Rienhart and Winston, New York.

Lawless, J. F. (1983). *Statistical Models and Methods for Lifetime Data*. Wiley, New York.

Lloyd, D. K., and Lipov, M. (1962). *Reliability Management, Methods and Mathematics*. Prentice Hall, Englewood Cliffs, NJ.

Longbottom, R. (1980). *Computer System Reliability*, Wiley, New York.

Mann, N. R., Schafer, R. E., and Singpurwalla, N. D. (1974). *Methods for Statistical Analysis of Reliability and Lifetime Data*. Wiley, New York.

Moss, M. A. (1985). Designing for Minimum Maintenance Expense. Marcel Dekker, New York.

Myers, R. H., Wong, K. L., and Gordy, H. M. (1964). *Reliability Engineering for Electronic Systems*. Wiley, New York.

O'Connor, P. (1990). *Practical Reliability Engineering*. 3rd ed. Wiley Interscience, New York.

McCormick, N. J. (1981). *Reliability and Risk Analysis*. Academic Press, San Diego, CA.

Moubry, J. (1991). Reliability-Centered Maintenance. Butterworth-Heinemann, Boston.

Nelson, W. (1982). *Applied Life Data Analysis*. Wiley, New York.

Nelson, W. (1990). *Accelerated Testing: Statistical Models, Test Plans, and Data Analyses*. Wiley, New York.

O'Connor, P. (1991). *Practical Reliability Engineering*. 3rd ed. Wiley, New York.

Osaki, S. (1984). *Stochastic System Reliability Modeling*. World Scientific Publishers, Singapore.

Patton, J. D. (1983). *Preventive Maintenance*. Instrument Society of America. Research Triangle Park, NC.

Pau, L. F. (1981). *Failure Diagnosis and Performance Monitoring*. Marcel Dekker, New York.

Polovko, A. M. (1968). *Fundamentals in Reliability Theory*. Academic Press, San Diego, CA.

Proceedings, Reliability and Maintainability Symposium (Annual). Institute of Electrical and Electronics Engineers, New York.

Raheja, D. (1991). *Assurance Technologies: Principles and Practices*. McGraw-Hill, New York.

Ramamurthy, K. G. (1990). *Coherent Structures and Simple Games*. Kluwer Academic Publishers, Norwell, MA.

Ravichandran, N. (1990). Stochastic Methods in Reliability Theory. Wiley, New York.

Roberts, N. H. (1964). *Mathematical Models in Reliability Engineering*. McGraw-Hill, New York.

Sander, P., and Badoux, R. (1991). *Bayesian Methods in Reliability*. Kluwer Academic Publishers, Norwell, MA.

Sandler, G. H. (1964). *System Reliability Engineering*. McGraw-Hill, New York.

Shooman, M. L. (1968). *Probabilistic Reliability: An Engineering Approach*. McGraw-Hill, New York.

Singh, C., and Billinton, R. (1977). *System Reliability Modelling and Evaluation*. Hutchinson, London.

Sinhaa, S. K., and Kale, B. K. (1980). *Life Testing and Reliability Estimation*. Wiley-Eastern, New Delhi.

Spencer, B. F. (1986). *Reliability of Randomly Excited Hysteretic Structures*. Springer-Verlag, New York.

Sundararajan, C. (1991). *Guide to Reliability Engineering*. Van Nostrand Reinhold, New York.

Thoft-Christensen, P., and Yoshisada, M. (1986). *Application of Structural Systems Reliability Theory*. Springer-Verlag, New York.

Thompson, W. A., Jr. (1988). *Point Process Models with Applications to Safety and Reliability*. Chapman and Hall, London.

Tillman, Frank A., et al. (1980). *Optimization of Systems Reliability*. Marcel Dekker, New York.

Tobias, P. A., and Trindade, D. C. (1986). *Applied Reliability*. Van Nostrand Reinhold, New York.

U.S. Military Standard: Reliability Demonstration, Exponential Distribution. MIL STD, 781.

U.S. Military Standard: Reliability Programs for System and Equipment. MIL STD, 785.

U.S. Military Standard: System Safety Programs for Systems and Associated Subsystems and Equipment: Requirements for. MIL STD, 882.

Villemeur, A. (1992). *Reliability, Availability, Maintainability and Safety Assessment*. Vol. 1: *Methods and Techniques*. Wiley, New York.

Villemeur, A. (1992). *Reliability, Availability, Maintainability and Safety Assessment*. Vol. 2: *Assessment, Hardware, Software and Human Factors*. Wiley, New York.

Vinogradov, O. G. (1991). *Introduction to Mechanical Reliability: A Designer's Approach*. Hemisphere, New York.

Wen, Y.-K. (1990). *Structural Load Modeling and Combination for Performance and Safety Evaluation*. Elsevier, Amsterdam.

Zacks, S. (1992). *Introduction to Reliability Analysis: Probability Models and Statistical Models*. Springer-Verlag, New York.

PRINCIPAL BOOKS IN RUSSIAN

Barzlovich, Ye. Yu., and Kashtanov, V. A. (1975). *Maintenance Management under Incomplete Information* (in Russian). Soviet Radio, Moscow.

Belyaev, Yu. K. (1975). *Probabilistic Methods of Sampling Control* (in Russian). Nauka, Moscow.

Bolotin, V. V. (1971). *Application of Probabilistic Methods and Reliability Theory in Construction Design* (in Russian). Stroiizdat, Moscow.

Buslenko, N. P. (1978). *Monte Carlo Simulation of Complex System* (in Russian). Nauka, Moscow.

Buslenko, N. P., Kalashnikov, V. V., and Kovalenko, I. N. (1973). *Lectures on Theory of Complex systems* (in Russian). Sovietskoe Radio, Moscow.

Cherkesov, G. N. (1974). *Reliability of Technical Systems with Time Redundancy* (in Russian). Sovietskoe Radio, Moscow.

Dzirkal, E. V. (1981). *Setting and Estimation of Reliability Requirements for Complex Systems*. Sovietskoe Radio, Moscow.

Gadasin, V. A., and Ushakov, I. A. (1975). *Reliability of Complex Control Information Systems* (in Russian). Sovietskoe Radio, Moscow.

Gnedenko, B. V., ed. (1983). *Aspects of Mathematical Theory of Reliability* (in Russian). Authors: Barzilovich, Ye. Yu., Belyaev, Yu. K., Kashtanov, V. A., Kovalenko, I. N., Solovyev, A. D., and I. A. Ushakov. Radio i Svyaz, Moscow.

Konyonkov, Yu. K., and Ushakov, I. A. (1975). *Aspects of Reliability of Electronic Equipment under Mechanical Stress*. Sovietskoe Radio, Moscow.

Korolyuk, V. S., and Turbin, A. F. (1982). *Markov Renewal Processes in Reliability*. Naukova Dumka, Kiev.

Kovalenko, I. N. (1980). *Analysis of Rare Events for Reliability and Performance Effectiveness of Complex Systems Investigation*. Sovietskoe Radio, Moscow.

Kozlov, B. A. (1969) *Redundancy with Renewal*. Sovietskoe Radio, Moscow.

Kozlov, B. A., and Ushakov, I. A. (1966). *Brief Handbook on Electronic Equipment Reliability Estimations*. Sovietskoe Radio, Moscow.

Kozlov, B. A., and Ushakov, I. A. (1975). *Handbook on Reliability of Automation and Electronic Equipment*. Sovietskoe Radio, Moscow.

Pavlov, I. V. (1982). *Statistical Methods of Reliability and Performance Estimation of Complex Systems on Basis of Experimental Data* (ed. by I. A. Ushakov). Sovietskoe Radio, Moscow.

Pashkovsky, G. S. (1981). *Aspects of Failure Searching and Finding for Electronic Equipment* (ed. by I. A. Ushakov). Sovietskoe Radio, Moscow.

Pollyak, Yu. G. (1971). *Monte Carlo Simulation with Computers*. Sovietskoe Radio, Moscow.

Raykin, A. L. (1978). *Elements of Reliability Theory* (ed. by I. A. Ushakov). Sovietskoe Radio, Moscow.

Reinschke, K., and Ushakov, I. A. (1988). *Application of Graph Theory in Reliability Analysis*. Radio i Svyaz, Moscow.

Rubalsky, G. B. (1977). *Inventory Control under Stochastic Requests* (ed. by I. A. Ushakov). Sovietskoe Radio, Moscow.

Rudenko, Yu. N., and Cheltsov, M. B. (1974). *Reliability of and Redundancy in Electric Power Systems*. Nauka, Novosibirsk.

Rudenko, Yu. N., and Ushakov, I. A. (1989). *Reliability of Power Systems* (ed. by B. V. Gnedenko). Nauka, Novosibirsk.

Ushakov, I. A. (1969). *Methods of Solving Optimal Redundancy Problems under Restrictions*. Sovietskoe Radio, Moscow.

Ushakov, I. A., ed. (1985). Reliability of Technical Systems: Handbook. Radio i Svyaz, Moscow.

REFERENCES

Chapter 1

Dzirkal, E. V. (1981). *Setting and Estimation of Reliability Requirements for Complex Systems* (in Russian). Sovietskoe Radio, Moscow.

Kozlov, B. A. (1969). *Redundancy with Renewal* (in Russian). Sovietskoe Radio, Moscow.

Kozlov, B. A., and Ushakov, I. A. (1970). *Reliability Handbook*. Holt, Rinehart and Winston, New York.

Ushakov, I. A., ed. (1985). *Reliability Handbook* (in Russian). Radio i Svyaz, Moscow.

Chapter 4

Kozlov, B. A. (1969). *Redundancy with Renewal* (in Russian). Sovietskoe Radio, Moscow.

Kozlov, B. A., and Ushakov, I. A. (1970). *Reliability Handbook*. Holt, Rinehart and Winston, New York.

Ushakov, I. A. (1980). An approximate method of calculating complex systems with renewal. *Eng. Cybernetics* (USA) **18**(6).

Ushakov, I. A., ed. (1985). *Reliability Handbook* (in Russian). Radio i Svyaz, Moscow.

Chapter 5

Gnedenko, B. V. (1964). Idle duplication. *Eng. Cybernetics* (USA) **2**(4).

Gnedenko, B. V. (1964). Duplication with repair. *Eng. Cybernetics* (USA) **2**(5).

Ushakov, I. A. (1980). An approximate method of calculating complex systems with renewal. *Eng. Cybernetics* (USA) **18**(6).

Chapter 6

Genis, Ya. G., and Ushakov, I. A. (1988). Estimation of reliability of redundant restorable with known test and switching characteristics. *Autom. Remote Control* (USA) **49**(6).

Gnedenko, B. V., ed. (1983). *Aspects of Mathematical Theory of Reliability* (in Russian). Authors: Barzilovich, Ye. Yu., Belyaev, Yu. K., Kashtanov, V. A., Kovalenko, I. N., Solovyev, A. D., and I. A. Ushakov. Radio i Svyaz, Moscow.

Gnedenko, B. V., Belyayev, Yu. K., and Solovyev, A. D. (1969). *Mathematical Methods of Reliability Theory*. Academic Press, San Diego, CA.

Solovyev, A. D. (1970). Asymptotic behavior of the time of the first occurrence of rare event in a regenerating process. *Eng. Cybernetics* (USA) **9**(6).

Solovyev, A. D. (1972). Asymptotic distribution of the moment of the first crossing of a high level by birth and death process. *Proc. Sixth Berkeley Symp. Math. Statist. Probab.* **3**. Univ. of California.

Solovyev, A. D. (1972). Press occurrence of rare event in a regenerating process. *Eng. Cybernetics* (USA) **9**(6).

Chapter 7

Ball, M., and Provan, J. (1982). Bounds on the reliability polynomial for shellable independence system. *SIAM J. Alg. Disc. Meth.* **3**.

Barlow, R. E., and Proschan, F. (1975). *Mathematical Theory of Reliability*. Holt, Rihehart and Winston, New York.

Barlow, R. E., and Proschan, F. (1981). *Statistical Theory of Reliability and Life Testing: Probability Models*. To Begin With, Silver Spring.

Bodin, L. (1970). Approximation of system reliability using a modular decomposition. *Technometrics* **12**.

Colburn, C. (1987). *The Combinatorics of Network Reliability*. Oxford University Press.

Gadasin, V. A., and Ushakov, I. A. (1975). *Reliability of Complex Control Information Systems* (in Russian). Sovietskoe Radio, Moscow.

Gadasin, V. A., and Lakaev, A. S. (1978). Estimation of the reliability of retranslation grids by the method of decomposition. *Eng. Cybernetics* (USA) **16** (4).

Kaustov, V. E., Litvak, E. I., and Ushakov, I. A. (1986). The computational effectiveness of reliability estimates by the method of nonedge-intersecting chains and cuts. *Soviet J. on Comp. and Syst. Science* (USA) **24**(4).

Litvak, E. I. (1981). A generalized triangle-star transformation in the investigation of properties of complex networks. *Eng. Cybernetics* (USA) **19**(1).

Litvak, E. I. (1981). Two-sided estimates of the minimum cost of a flow in a two-terminal network. *Engr. Cybernetics* (USA) **19**(3).

Litvak, E. I., and Ushakov, I. A. (1984). Estimation of parameters of structurally complex systems. *Eng. Cybernetics* (USA) **22**(4).

Reinschke, K., and Ushakov, I. A. (1988). *Application of Graph Theory in Reliability Analysis* (in Russian). Radio i Svyaz, Moscow.

Reinschke, K., and Uschakov, I. A., (1988). *Zuverlassigkeitsstructuren*. Oldenberg, Munich.

Ushakov, I. A., and Litvak, E. I. (1977). An upper and lower estimate of the parameters of two-terminal networks. *Eng. Cybernetics* (USA) **15**(1).

Chapter 8

Dzirkal, E. V. (1981). *Setting and Estimation of Reliability Requirements for Complex Systems* (in Russian). Sovietskoe Radio, Moscow.

Kozlov, B. A., and Ushakov, I. A. (1970). *Reliability Handbook*. Holt, Rinehart and Winston, New York.

Rudenko, Yu. N., and Ushakov, I. A. (1985). Safety as one property of power-system reliability. *Power Eng.* (USA) **23**(5).

Rudenko, Yu. N., and Ushakov, I. A. (1989). *Reliability of Power Systems* (in Russian), ed. by B. V. Gnedenko. Nauka, Novosibirsk.

Ushakov, I. A., and Konenkov, Yu. K. (1964). Evaluation of the operational effectiveness of branching systems with symmetrical structure. *CORS J.* **2**(2).

Chapter 9

Cherkesov, G. N. (1974). *Reliability of Technical Systems with Time Redundancy* (in Russian). Sovietskoe Radio, Moscow.

Kredenttser, B. P. (1978). *Forecasting Reliability with Time Redundancy* (in Russian). Soviet Radio, Moscow.

Rudenko, Yu. N., and Ushakov, I. A. (1984). On evaluating the operational viability of complex power systems. *Power Eng.* (USA) **17**(1).

Rudenko, Yu. N., and Ushakov, I. A. (1989). Reliability of Power Systems (in Russian), ed. by B. V. Gnedenko. Nauka, Novosibirsk.

Chapter 11

Bolotin, V. V. (1971). *Application of Probabilistic Methods and Reliability Theory in Construction Design* (in Russian). Stroiizdat, Moscow.

Kapur, K. C. and Lamberson, L. R. (1977). *Reliability in Engineering Design.* Wiley, New York.

Konenkov, Yu. K., and Ushakov, I. A. (1975). *Aspects of Reliability of Electronic Equipment under Mechanical Stress* (in Russian). Sovietskoe Radio, Moscow.

Chapter 12

Kaplan, E. L., and Meier, P. (1958). Nonparametric estimation from incomplete observations. *J. Amer. Statist. Assoc.* **53**.

Pavlov, I. V., and Ushakov, I. A. (1985). Unbiased estimation of a distribution function from a multiple truncated sample. *Theory Probab. Appl.* (USA) **29**(3).

Ushakov, I. A. (1980). Estimation of reliability from the results of truncated observations. *Eng, Cybernetics* (USA) **18**(5).

Chapter 13

Dzirkal, E. V. (1981). Setting and Estimation of Reliability Requirements for Complex Systems (in Russian). Sovietskoe Radio, Moscow.

Chapter 14

Dzirkal, E. V. (1981). *Setting and Estimation of Reliability Requirements for Complex Systems* (in Russian). Sovietskoe Radio, Moscow.

Pavlov, I. V. (1980). Confidence bounds for convex functions of many unknown parameters. *Theory Probab. Appl.* (USA) **25**.

Pavlov, I. V. (1982). *Statistical Methods of Estimation of Reliability of Complex System Based on Experimental Results* (in Russian), ed. by I. A. Ushakov. Radio i svyaz, Moscow.

Sonkina, T. P., and Tyoskin, O. I. (1984). Confidence limits for the reliability of system connected series. *Eng. Cybernetics* (USA) **22**(4).

Sudakov, R. S. (1974). A problem of interval estimation of the reliability index of a series system. *Eng. Cybernetics* (USA) **12**(3).

Sudakov, R. S. (1980). On interval estimation. *Eng. Cybernetics* (USA) **18**(1).

Tyoskin, I. I. (1979). Precise confidence limits for the reliability of redundant systems in the case of failure-free tests of their elements. *Eng. Cybernetics* (USA) **17**(4).

Chapter 16

Savchuk, V. P. (1989). *Bayesian Methods of Statistical Estimating. Reliability of Technical Systems* (in Russian). Nauka, Moscow.

Chapter 18

Barlow, R. E., and Scheuer, E. M. (1971). *Estimation from accelerated life tests. Technometrics*, 13-1.

Clarke, J. M. (1979). No-growth growth curves. *IEEE Proc. Rel. & Maint. Symp.*

Collas, Gerard (1991). Dynamic reliability prediction: How to adjust modeling and reliability growth. *IEEE Proc. Ann. Rel. & Maint. Symp.*

Crow, L. H. (1983). Reliability growth projection from delayed fixes. *IEEE Proc. Ann. Rel. & Maint. Symp.*

Crow, L. H. (1984). Methods for assessing reliability growth potential. *IEEE Proc. Annual Rel. & Maint. Symp.*

Crow, L. H. (1986) On the initial system reliability. *Proc. Ann. Rel. & Maint. Symp.*

Demko, E. (1986). True reliability growth measurement. *IEEE Proc. Ann. Rel. & Maint. Symp.*

Dhillon, B. S. (1983). *Reliability Systems Design and Operation.* Van Nostrand & Reinhold, New York.

Fard, N. S. and Dietrich, D. L. (1983). Comparison of attribute reliability growth models. *IEEE Proc. Ann. Rel. & Maint. Symp.*

Gray, C. T. (1986). A modeling framework for reliability growth. *IEEE Proc. Ann. Rel. & Maint. Symp.*

Jewell, W. S. (1984). A general framework for learning curve reliability growth models. *Operations Research*, **32**.

Kaplan, S., Cunha, G. D. M., Dykes, A. A. and Shaver, D. (1990). A Bayesian methodology for assessing reliability during product development. *IEEE Proc. Ann. Rel. & Maint. Symp.*

Lawless, J. F. (1982). *Statistical Models and Methods for Lifetime Data.* Wiley, New York.

Mann, N. R. et al. (1973). *Methods for Statistical Analysis of Reliability and Life Data.* Wiley, New York.

Mazzuchi, T. A. and Soyer, R. (1991). A Bayesian attribute reliability growth model. *Proc. Ann. Rel. and Maint. Symp.*

Wronka, J. C. (1988). Tracking of reliability growth in early development. *IEEE Proc. Ann. Rel. & Maint. Symp.*

Chapter 19

Buslenko, N. P. (1978). *Monte Carlo Simulation of Complex System* (in Russian). Nauka, Moscow.

Buslenko, N. P., Kalashnikov, V. V., and Kovalenko, I. N. (1973). *Lectures on Theory of Complex Systems* (in Russian). Sovietskoe Radio, Moscow.

Kovalenko, I. N. (1980). *Analysis of Rare Events for Reliability and Performance Effectiveness of Complex Systems Investigation* (in Russian). Sovietskoe Radio, Moscow.

Pollyak, Yu. G. (1971). *Monte Carlo Simulation with Computers* (in Russian). Sovietskoe Radio, Moscow.

Chapter 20

Barlow, R. E., and Proschan, F. (1981). *Statistical Theory of Reliability and Life Testing: Probability Models*. To Begin With, Silver Spring, MD.

Genis, Ya. G., and Ushakov, I. A. (1983). Optimization of the reliability of the multipurpose system. *Eng. Cybernetics* (USA) **21**(3).

Gnedenko, B. V., ed. (1983). *Aspects of Mathematical Theory of Reliability* (in Russian). Authors: Barzilovich, Ye. Yu., Belyaev, Yu. K., Kashtanov, V. A., Kovalenko, I. N., Solovyev, A. D., and Ushakov, I. A. Radio i Svyaz, Moscow.

Kettelle, J. D., Jr. (1962). Least-cost allocation of reliability investment. *Opns Res.* **10**.

Raykin, A. L. (1978). *Elements of Reliability Theory*, ed. by I. A. Ushakov (in Russian). Sovietskoe Radio, Moscow.

Ushakov, I. A. (1969). *Methods of Solving Optimal Redundancy Problems under Restrictions* (in Russian). Sovietskoe Radio, Moscow.

Ushakov, I. A. (1986). A universal generating function. *Sov. J. Comput. Syst. Science* (USA) **24**(5).

Ushakov, I. A. (1987). Solution of multicriterial discrete optimization problems using a universal generating function. *Sov. J. Comput. Syst. Science* (USA) **25**(5).

Ushakov, I. A. (1987). Optimal standby problems and a universal generating function. *Sov. J. Comput. Syst. Science* (USA) **25**(4).

Ushakov, I. A. (1988). Solving an optimal redundancy problem by the mean of generalized generating functions. *J. Inf. Process. Cybern.* (East Germany) **24**(3).

Chapter 22

Rubalsky, G. B. (1977). *Inventory Control under Stochastic Requests* (in Russian), ed. by I. A. Ushakov. Sovietskoe Radio, Moscow.

Rubalsky, G. B. (1984). Mixed renewal and restocking models for product of a single type. *Eng. cybernetics* (USA) **22**(2).

Rubalsky, G. B., and Ushakov, I. A. (1976). Multiechelon inventory control system with dense branches. *Kybernetes* **5**(2).

Barziovich, Ye. Yu., and Kashtanov, V. A. (1975). *Maintenance Management under Limited Information about System Reliability* (in Russian). Soviet Radio, Moscow.

Barziovich, Ye. Yu., and Kashtanov, V. A. (1975). *Some Aspects of Mathematical Theory of Maintenance of Technical Systems* (in Russian). Soviet Radio, Moscow.

Gertsbakh, I. B. (1977). *Models of Preventive Maintenance*, North-Holland, Amsterdam.

Gnedenko, B. V., ed (1983). *Aspects of Mathematical Theory of Reliability* (in Russian). Authors: Barzilovich, Ye. Yu., Belyaev, Yu. K., Kashtanov, V. A., Kovalenko, I. N., Solovyev, A. D., and I. A. Ushakov. Radio i Svyaz, Moscow.

Chapter 24

Kozlov, B. A., and Ushakov, I. A. (1975). *Handbook on Reliability of Automation and Electronic Equipment* (in Russian). Sovietskoe Radio, Moscow.

Pashkovsky, G. S. (1981). *Aspects of Failure Searching and Finding for Electronic Equipment* (in Russian), ed. by I. A. Ushakov. Sovietskoe Radio, Moscow.

INDEX